Understanding Electromagnetic Transients in Power Systems

IEEE Press
445 Hoes Lane
Piscataway, NJ 08854

Understanding Electromagnetic Transients in Power Systems

Luiz Cera Zanetta Jr.
University of São Paulo
Cidade Universitária, São Paulo, Brazil

IEEE Press Series on Power and Energy Systems
Ganesh Kumar Venayagamoorthy, Series Editor

IEEE*Press*

WILEY

Published by John Wiley & Sons, Inc., Hoboken, New Jersey.
Published simultaneously in Canada.

Library of Congress Cataloging-in-Publication Data Applied for

Hardback ISBN: 9781394240555

Cover Design: Wiley
Cover Images: © STK_08/Shutterstock, Courtesy of Luiz Cera Zanetta Jr.

Set in 9.5/12.5pt STIXTwoText by Straive, Pondicherry, India

To my family.

Contents

About the Author

Luiz Cera Zanetta, Jr., PhD

Luiz Cera Zanetta, Jr., PhD, a senior member of IEEE and a full professor at the University of São Paulo, has significantly contributed to advancing and strengthening the field of electromagnetic transient analysis and equipment specifications for major power plant projects and long-distance interconnections in Brazil, both in consulting engineering and in academia.

His interests range from electromagnetic transients to Flexible AC Transmission Systems, including dynamic stability analysis, focusing on research domains with distinct peculiarities. Currently his primary interest is to contribute to the enhancement of engineering education.

Preface

First and foremost, we clarify that the use of the term *understanding* in the book's title is meant to offer a gentle introduction to a complex subject, rather than an arrogant claim.

This is mainly addressed to the student who intends to begin his training in electromagnetic transients, but it can also be useful for the professional who seeks the maturation of his evolving and as yet unresolved ideas.

The book retains the essence of its first edition, originally published in Portuguese in 2003, but only recently we have been able to release an English version, incorporating adaptations to the chapters along with some timely updates.

This book arose from the author's desire to share his experiences with transmission line and high-voltage equipment projects, drawing from his work in the electrical power industry and later in academia, with the hope that these insights can be valuable to both students and professionals in power systems. Some of the interpretations gathered here can also be found in the literature, and the contribution of this collection lies in bringing together and organizing this information. We apologize in advance if we did not adequately credit any sources for the results we used.

The aim of this book is to present the fundamentals of transients step-by-step, providing the necessary background for understanding the topic while minimizing the need for external references. Though it does not seek to exhaustively cover the subject, it is designed to make the material more accessible, helping to clarify the relationship between fundamental transient concepts, their calculations, and their impact on equipment.

Our intention is not to explore the general aspects of electromagnetic transients theory, but rather to focus on circuits and on the analysis of wave propagation in transmission lines. This book also does not cover the evolving technological features of substation equipment in detail, as we believe specialized publications are more appropriate for in-depth information on these topics. However, when discussing transients, it is inevitable to consider their effects on the equipment components that make up a power system. We do this without straying from the core principles, always aiming to deepen the understanding of the key concepts behind transients.

Although the topic seems at first sight a bit dry, we try to present it in simple language, so as to expeditiously cover phenomena that change rapidly in time.

We aim, whenever possible, to supplement the text with detailed examples to clarify the theoretical explanations. It's clear that complex networks involving many differential equations can only be analyzed with the help of electromagnetic transient programs. However, our focus on the analytical treatment of these transients comes from the need to understand the underlying principles, at least on a qualitative level.

In the opening Chapters 1 and 2, we present basic concepts of electrical circuits and wave propagation, which may seem very basic at first glance, but we believe they are relevant not only for students but also for professionals who do not use them frequently.

In Chapter 1, the essential information for addressing transients in electrical circuits has been condensed, and the Laplace transform is introduced as a powerful method for converting differential equations into algebraic equations, enhancing the understanding of transients through pole-residue analysis. In Chapter 2, we present the fundamental concepts of wave propagation in single-phase transmission lines, which serve as a smooth introduction to understanding multiphase lines. The relationships between voltages and currents are explained, along with key transmission line characteristics, and an analysis of discontinuities is also presented. The Thévenin equivalent, an important concept, is also approached and recurs throughout the book.

The relationship between multiphase voltage and current waves in transmission lines is analyzed in Chapter 3, in both the phase and modal domain. This approach is useful to understand basic transmission line models in EMT programs, which, although simplified, perform effectivelly in standard studies.

In Chapter 4, we introduce the most efficient numerical method applied in electromagnetic transient calculations. The elementary formulation of an EMTP using nodal analysis, while much simpler than existing commercial or free softwares, aims to introduce students to the core of developing an electromagnetic transient calculation program. Numerical examples with simple networks were prepared, detailing some integration steps that clarify nodal solutions and the updating of history terms.

Chapter 5 addresses the representation of frequency-dependent parameters of transmission lines, a subject that, to this day, can still benefit from further contributions despite much effort and sophisticated models. We show how rational functions are fitted in the frequency domain to enable efficient recursive convolutions in the time domain, aiming to make more advanced transmission line models in electromagnetic transient programs more accessible. To complement the modeling of transmission lines, the use of the nodal admittance matrix is approached, as well as the frequency dependent network equivalents (FDNEs).

The role of power electronic converters, specifically voltage source converters (VSCs), connected to the power grid is addressed in Chapter 6. The elements of power electronics complement the information on three-phase voltage sources interacting with the power system. The chapter reviews the basic equations, starting from thyristor-based converters to the latest developments in VSCs and modular multilevel converters (MMCs). We focus on the basic elements of space-vector control and the synchronization device phase-locked loop (PLL), which are still not straightforward for power system students, in order to clarify the control structure of VSCs.

Chapters 7–9 deal with the basic overvoltages in power systems. Chapter 7 approaches the temporary overvoltages and steady-state conditions on an electrical grid in situations in which there are topological changes in the network caused by maneuvers or short circuits. Mastery of these conditions is essential for analyzing transient phenomena because they are part of the overall transient solution right from the start. In this chapter, we also deal with saturation in transformers and introduce the complex ferroresonance phenomenon in power systems.

The sizing of transmission lines and substations is largely determined by the influence of switching and lightning surges, which significantly impact the dimensions and costs of these installations. Consequently, it is essential to thoroughly analyze these transients, with the core analysis presented in Chapters 8 and 9.

In Chapter 8, we present a general description of the main transient phenomena in transmission lines, related to slow wavefronts (switching surges). The chapter analyzes the primary cases of overvoltages and also discusses the mitigation measures for controlling them as surge arresters, pre-insertion resistors, and controlled switching in circuit breakers. Chapter 9 addresses fast wavefronts (lightning surges) that affect the lightning performance of transmission lines and the insulation levels of substation equipment. The chapter provides information on modeling lightning currents and their probabilistic behavior, along with models for transmission lines, towers, and grounding systems. It presents the methodology for calculating lightning surges caused by direct strokes and backflashovers, analyzes the effects of line arresters, and covers the calculation of induced overvoltages. Additionally, the procedure for calculating lightning surges on substation equipment is also discussed.

In Chapters 10 and 11, we address basic transients involving shunt and series capacitors. In Chapter 10, the analysis of shunt capacitor switching focuses primarily on high-frequency transients, such as energization, restrike, back-to-back switching, and inrush currents. The chapter examines high-frequency stresses on equipment, emphasizing key considerations for equipment specification. In Chapter 11, electromagnetic transient studies are presented to define protection level settings for both the capacitor bank and the MOV protection scheme. These settings are specified to respond effectively to both internal and external faults related to the line where the bank is installed.

Chapter 12 examines topics related to the opening of circuit breakers due to faults, focusing on transient recovery voltages (TRVs). The main parameters influencing TRVs are explained using basic models that clarify the principles governing these voltages. After calculating the TRV, the chapter discusses the appropriate selection of circuit breakers in line with IEEE or IEC standards.

Surge arresters are studied in Chapter 13 due to their importance in overvoltage control and their frequent mention throughout the book. The chapter clarifies the principles of interaction between surge arresters and transmission lines for both switching and lightning surges, along with the appropriate rating selection based on standards' recommendations.

In Chapter 14, we supplement the previous chapters addressing overvoltages by detailing the engineering procedure for establishing insulation levels for equipment, referred to as insulation coordination. Both deterministic and statistical methods are discussed, applicable to insulations classified as non-self-restoring and self-restoring, respectively, to achieve an optimal balance between the lowest withstand levels of insulation and the most representative overvoltages. Basic probabilistic concepts are reviewed to introduce the statistical method. To conclude the chapter, we also mention some recommendations from IEEE and IEC standards for insulation coordination.

Given the expansive nature of the subject matter covered in this book, it is recognized that experts may draw on more specialized technical publications for a deeper examination of specific items. Some topics were not explored due to limitations on the length of the text and others because they require us further investigation to be addressed consistently.

We hope this text will serve as a valuable resource for readers, providing meaningful insights and practical guidance in the field. Feedback and suggestions are welcomed, as they can further enhance the quality and clarity of the work.

Luiz Cera Zanetta Jr.
University of São Paulo
Cidade Universitária, São Paulo, Brazil

1

Transients in Elementary Circuits and the Laplace Transform

1.1 Introduction

Electromagnetic transient analysis is essential for specifying equipment in substations and transmission lines because the voltage and current amplitudes are much greater than those found in a sinusoidal steady state. Therefore, it defines the dimensions and materials of the equipment, significantly impacting the costs of an electrical power system.

This chapter deals with the solution of transients in simple circuits, which correspond to several phenomena related to power grids. A summary of some of the properties of the Laplace transform will be presented with applications to basic circuits typical of electrical networks. The discussion explores two fundamental circuit concepts: the RL (resistor-inductor) circuit, illustrating transients characterized by high currents, and the RLC (resistor-inductor-capacitor) circuit, which demonstrates oscillations in currents and voltages, possibly leading to overvoltages.

The solution of differential equations will preferably be based on the Laplace transform, but a detailed description of this transform is beyond the scope of this chapter. This material on complex variables is present in a broad literature concerning integral transformations [1–3], as well as the relationship between the Laplace transform and the Fourier transform. The text is also intended to review fundamental concepts of circuit theory, which will be useful throughout this book.

The Laplace transform serves as a potent method for converting differential equations from the time domain to the frequency domain, in which they are solved algebraically, with their subsequent inversion to the time domain.

The analysis of complex networks with nonliarities requires tools such as electromagnetic transient programs, and only simple circuits with a limited number of differential equations can be solved manually. However, even with complex networks, analysis using elementary circuits provides crucial support in the physical interpretation of the concerned phenomenon, at least in qualitative terms.

Understanding Electromagnetic Transients in Power Systems, First Edition. Luiz Cera Zanetta Jr.

Published 2025 by John Wiley & Sons, Inc.

1.2 Laplace Transform

1.2.1 Definition

The Laplace transform of a function $f(t)$, denoted by the function $F(s)$, is defined by the integral:

$$\mathscr{L}[f(t)] = F(s) = \int_{0^-}^{\infty} f(t)e^{-st}dt. \tag{1.1}$$

In this integral, t is a variable on the real axis, representing time, and s is a variable in the complex frequency domain. Although the notation at the lower limit of the integral in 0^- introduces greater rigor in cases of impulse excitation, this notation will not always be adopted in the text.

Some elementary transforms are presented below, assuming that the conditions of convergence are satisfied in all the analyzed cases.

1.2.2 Some Transforms and Their Elementary Properties

Since the Laplace transform is a linear operation, the functions are treated as unitary in amplitude, and the transforms for other amplitudes are proportionally obtained.

- Step Function Transform

$$f(t) = H(t) \leftrightarrow F(s) = \frac{1}{s}. \tag{1.2}$$

- Exponential Function Transform

$$f(t) = e^{-at}H(t) \leftrightarrow F(s) = \frac{1}{s+a}. \tag{1.3}$$

This is a fundamental transform for analyzing transients and will be frequently used because it is intrinsically connected to the basics of transient behavior in electrical systems.

- Sine Function Transform
  Since sinusoidal functions can be broken down into sums of exponentials using Euler's formula, calculating their transforms is straightforward.

$$f(t) = \sin(\omega t) \leftrightarrow F(s) = \frac{\omega}{s^2+\omega^2}. \tag{1.4}$$

- Cosine Function Transform

$$f(t) = \cos(\omega t) \leftrightarrow F(s) = \frac{s}{s^2 + \omega^2}. \tag{1.5}$$

- Impulse Function Transform

$$f(t) = \delta(t) \leftrightarrow F(s) = 1. \tag{1.6}$$

The properties described below are important because they allow the transformation of differential and integral equations in the time domain into algebraic equations in the frequency domain. Some fundamental formulas for the transformation of integrals and derivatives of functions are presented below.

- Transform of the Derivative of a Function
  Let $f(t)$ be a function in the time domain, with Laplace Transform $F(s)$ and $f'(t)$ its derivative. The Laplace transform of $f'(t)$, represented by, $\mathscr{L}\,[f'(t)]$, is obtained by applying the definition:

$$\mathscr{L}\,[f'(t)] = \int_{0^-}^{\infty} f'(t)e^{-st}dt \leftrightarrow \mathscr{L}[f'(t)] = sF(s) - f(0^-). \tag{1.7}$$

The result in (1.7) can be easily extended to the second derivative of a function:

$$\mathscr{L}[f''(t)] = s^2F(s) - sf(0^-) - f'(0^-). \tag{1.8}$$

- Transform of the Integral of a Function
  The transform of the integral of a function is obtained as follows:

$$\mathscr{L}\left[\int_{-\infty}^{t} f(t)dt\right]. \tag{1.9}$$

defining:

$$g(t) = \int_{-\infty}^{t} f(t)dt, \tag{1.10}$$

$$g'(t) = f(t). \tag{1.11}$$

Applying the Laplace transform to (1.11):

$$sG(s) - g(0^-) = F(s) \tag{1.12}$$

$$G(s) = \frac{F(s)}{s} + \frac{g(0^-)}{s}. \tag{1.13}$$

- First Shifting Theorem

$$\mathscr{L}[e^{-at}f(t)] = F(s+a). \tag{1.14}$$

Note that (1.14) generalizes the results from (1.2) and (1.3).

- Second Shifting Theorem

$$\mathscr{L}[f(t-a)] = e^{-as}F(s). \tag{1.15}$$

This property will be useful in interpreting delays in the propagation of electromagnetic waves, to be analyzed in Chapter 2.

- Multiplication of the Argument of $f(t)$ by a Constant
  Let b be a real constant and $F(s) = \mathscr{L}[f(t)]$; then, the transform of $f(bt)$,

$$\mathscr{L}[f(bt)] = \int_{0^-}^{\infty} e^{-st}f(bt)dt, \tag{1.16}$$

$$\mathscr{L}[f(bt)] = \frac{1}{b}F\left(\frac{s}{b}\right). \tag{1.17}$$

- Derivative of $F(s)$
  By differentiating (1.1) with respect to the complex variable s, it is immediate that:

$$-\frac{d}{ds}F(s) = \mathscr{L}[tf(t)]. \tag{1.18}$$

Example 1.1 Calculate the Laplace transform of the function $t \sin \omega t$.

Solution

Considering the Laplace transform of the function $f(t) = \sin(\omega t)$:

$$F(s) = \frac{\omega}{s^2 + \omega^2},$$

as in (1.18), calculating the derivative of $F(s)$ and reversing the sign:

$$-\frac{d}{ds}F(s) = \mathscr{L}[tf(t)],$$

thus,

$$-\frac{d}{ds}F(s) = \frac{2\omega s}{(s^2 + \omega^2)^2},$$

therefore:

$$\mathscr{L}[t\sin \omega t] = \frac{2\omega s}{(s^2 + \omega^2)^2}.$$

This transform is useful for understanding resonant conditions. The double pole in the denominator explains how sinusoidal signals can grow indefinitely with time.

1.2.3 Inversion of the Laplace Transform

It was shown that differential equations in the time domain can be transformed into algebraic expressions in the frequency domain. It is now of interest to obtain the inverse transform of $F(s)$ to recover the corresponding $f(t)$.

The inverse transform of $F(s)$, designated $f(t) = \mathscr{L}^{-1}[F(s)]$, is defined by the integral:

$$f(t) = \int_0^{\infty} F(s)e^{st}ds. \tag{1.19}$$

Using the uniqueness and linearity properties of the Laplace transform, the inversion will be implemented in its simplest form, that is, by splitting a more complex function into elementary terms whose inverse transforms can be obtained from tables.

- Inverse Transform of Rational Functions
  In the transient treatment of electrical circuits with the application of the Laplace transform, rational functions in the complex variable s are usually found.

  Let $F(s)$ be a rational function given by the quotient of two polynomials in frequency domain s:

$$F(s) = \frac{N(s)}{D(s)} = \frac{b_0 s^m + b_1 s^{m-1} + \ldots + b_{m-1}s + b_m}{s^n + a_1 s^{n-1} + \ldots + a_{n-1}s + a_n}, \tag{1.20}$$

which is strictly proper, assuming that $m < n$.

$N(s)$ and $D(s)$ are polynomials with real coefficients that can be factored.

In this case, $D(s)$ can be written as:

$$D(s) = (s - \lambda_1)\ldots(s - \lambda_n), \tag{1.21}$$

where λ_i are the roots of the monic polynomial $D(s)$ in (1.21) considered without multiplicity ($\lambda_i \neq \lambda_j$, $i \neq j$), which is the most frequent and important case. Cases with duplicity, under resonant conditions, will be addressed specifically in Section 1.6:

$$D(s) = \prod_{i=1}^{n} (s - \lambda_i), \tag{1.22}$$

and similarly, for the numerator:

$$N(s) = b_0 \prod_{j=1}^{m} (s - z_j). \tag{1.23}$$

The poles of the rational function are represented by λ_i and the zeros by z_j. Rational functions can be expanded into partial fractions:

$$F(s) = \frac{N(s)}{D(s)},$$

$$\frac{N(s)}{D(s)} = \frac{k_1}{s - \lambda_1} + \frac{k_2}{s - \lambda_2} + \cdots \frac{k_n}{s - \lambda_n}. \tag{1.24}$$

The residues k_i are real or complex numbers, depending on whether the poles λ_i are real or complex. Complex poles always appear in conjugate pairs since the polynomial $D(s)$ has real coefficients.

- Inverse Transform of Rational Functions with Single Poles

  Considering the same case above, with the roots of $D(s)$ without multiplicity, the expansion of rational functions into partial fractions needs the calculation of the residues.

  The calculation of the residues can be understood by multiplying both sides of Eq. (1.24) by $(s - \lambda_i)$ and substituting the polynomial $D(s)$ from (1.21) on the left-hand side of this equation. It is observed that this polynomial is canceled out in the denominator of the equation. On the right-hand side, only the denominator related to the residue k_i is canceled, and all other terms are multiplied by $(s - \lambda_i)$.

$$\left.\frac{N(s)}{(s-\lambda_1) \times \cdots (s-\lambda_{i-1}) \times (s-\lambda_{i+1}) \times \cdots (s-\lambda_n)}\right|_{s=\lambda_i} =$$
$$= \left.\left[\frac{k_1 \times (s-\lambda_i)}{s-\lambda_1} + \cdots k_i + \cdots \frac{k_n \times (s-\lambda_i)}{s-\lambda_n}\right]\right|_{s=\lambda_i}.$$

Calculating both sides of the equation at the point $s = \lambda_i$, it is concluded that the residue k_i can be obtained using the expression:

$$k_i = \left.\frac{N(s)}{(s-\lambda_1) \times \cdots (s-\lambda_{i-1}) \times (s-\lambda_{i+1}) \times \cdots (s-\lambda_n)}\right|_{s=\lambda_i}. \tag{1.25}$$

Once the residues have been calculated, the same relationship between the Laplace transform of the exponential function and its inverse, given by (1. 15), can be used in (1.24) to obtain the inverse transform of the rational function:

$$f(t) = \mathscr{L}^{-1}[F(s)] = \sum_{i=1}^{n} k_i e^{\lambda_i t}. \tag{1.26}$$

Note that each exponent in (1.26) contains the product of one root of $D(s)$ by variable t. The sign of the exponent depends on the convention adopted for the first-order polynomials, since $(s - \lambda_i)$ and $(s + \lambda_i)$ will have exponential solutions in time $e^{\lambda_i t}$ and $e^{-\lambda_i t}$, respectively.

In the case of the conjugated complex poles λ_i and λ_i^*, conjugated complex residues k_i and k_i^* are respectively obtained; therefore, calculation of only one of them is sufficient.

The procedure to calculate the inverse transform of complex poles is shown below:

$$f(t) = ...k_i e^{\lambda_i t} + k_i^* e^{\lambda_i^* t}. \tag{1.27}$$

If $k_i = K_i \angle \phi_i$ and $\lambda_i = \sigma_i + j\omega_i$, the complex poles can be grouped into:

$$f(t) = ...K_i e^{j\phi_i} e^{(\sigma_i + j\omega_i)t} + K_i e^{-j\phi_i} e^{(\sigma_i - j\omega_i)t}, \tag{1.28}$$

or,

$$f(t) = ...K_i e^{\sigma_i t}\left[e^{j(\omega_i t + \phi_i)} + e^{-j(\omega_i t + \phi_i)}\right]. \tag{1.29}$$

The following compact expression will be used in the inverse transform of conjugated complex poles:

$$f(t) = ...2K_i e^{\sigma_i t}\cos(\omega_i t + \phi_i). \tag{1.30}$$

1.3 The Convolution Integral

For causal systems, the convolution of two functions $u(t)$ and $g(t)$ can be defined by the integral:

$$u(t)*g(t) = \int_0^t u(\tau)g(t-\tau)d\tau, \tag{1.31}$$

since the integrand is null at $\tau < 0$ and $\tau > t$.

An important property in the application of the Laplace transform to the convolution of two functions $u(t)$ and $g(t)$, whose transforms are $U(s)$ and $G(s)$ is presented below.

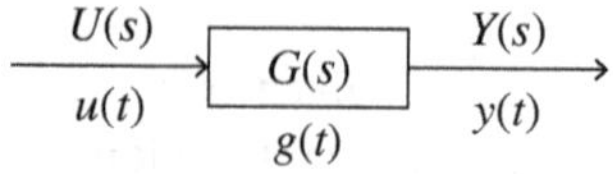

Figure 1.1 Excitation of the linear system.

$$\mathscr{L}[u(t)*g(t)] = U(s)G(s). \tag{1.32}$$

This property is particularly significant for the analysis of linear systems, when the input signal $[u(t), U(s)]$ is applied to a system whose transfer function is designated by $[g(t), G(s)]$, where the impulse response of the system is given by $g(t)$, resulting in the output $[y(t), Y(s)]$, according to Figure 1.1:

In the time domain, the output $y(t)$ is found by convolving the input $u(t)$ with $g(t)$, which can be complex as shown in Eq. (1.31). In the frequency domain, however, the response is just the product of the input $U(s)$ and the system's transfer function $G(s)$. In steady-state power system analysis, it becomes evident that all products involving current or voltage phasors with complex impedance or admittance are based on this property.

So far, the fundamental concepts of the Laplace transform were presented, and are best applied to the study of certain simple networks, starting with the RL circuit.

1.4 RL Circuit

The RL circuit has several applications and one of them is in the calculation of fault overcurrents in power systems. Assume a series RL circuit fed by a generic voltage source:

$$e(t) = Ri(t) + L\frac{di(t)}{dt}. \tag{1.33}$$

Applying the Laplace transform:

$$E(s) = RI(s) + sLI(s) - Li_0. \tag{1.34}$$

The complex impedance of the circuit can be written as:

$$Z(s) = (R + sL),$$

or,

$$Z(s) = L\left(s + \frac{R}{L}\right), \tag{1.35}$$

thus,

$$E(s) = Z(s)I(s) - Li_0. \tag{1.36}$$

Calculating the current:

$$I(s) = \frac{E(s)}{Z(s)} + \frac{Li_0}{Z(s)}. \tag{1.37}$$

The inverse of $Z(s)$ is the complex admittance of the circuit, $G(s) = Z(s)^{-1}$, which operates as a transfer function (network function), where $E(s)$ and $I(s)$ are input and output of the system, respectively.

1.4.1 RL Circuit with Sinusoidal Voltage Source

The study of the RL circuit with a sinusoidal voltage source is especially interesting as it is very common in power systems.

$$e(t) = E_0 \sin(\omega t + \theta). \tag{1.38}$$

using the trigonometric identity:

$$\sin(\omega t + \theta) = \sin(\omega t)\cos(\theta) + \sin(\theta)\cos(\omega t),$$

$$\mathscr{L}[E_0 \sin(\omega t + \theta)] = E_0 \frac{\omega\cos(\theta) + s\sin(\theta)}{s^2 + \omega^2}. \tag{1.39}$$

From (1.35) and (1.39), the curent in the circuit can be written in (1.40). The initial condition, $i_0 = 0$, is assumed for simplicity, since its contribution will be seen later in (1.53):

$$I(s) = \frac{E_0}{L} \frac{\omega\cos(\theta) + s\sin(\theta)}{(s - j\omega)(s + j\omega)\left(s + \frac{R}{L}\right)}. \tag{1.40}$$

The denominator is already factored, and the roots are:

$$\lambda_{1,2} = \pm j\omega, \quad \lambda_3 = -\frac{R}{L}.$$

where $\lambda_{1,2}$ and λ_3 are the excitation and network poles, respectively.

Making the expansion into partial fractions:

$$I(s) = \frac{k_1}{(s - j\omega)} + \frac{k_1^*}{(s + j\omega)} + \frac{k_2}{s + \frac{R}{L}}. \tag{1.41}$$

The calculation of k_1 and k_1^* is related to the steady-state response of the circuit:

$$k_1 = E_0 \left. \frac{\omega\cos(\theta) + s\sin(\theta)}{L\left(s + \frac{R}{L}\right)(s + j\omega)} \right|_{s = j\omega}. \tag{1.42}$$

The impedance in steady state of the circuit is a complex number, calculated at the same excitation frequency as the source,

$$Z\angle\varphi = R + j\omega L, \tag{1.43}$$

$$\tan(\varphi) = \frac{\omega L}{R}, \tag{1.44}$$

thus:

$$k_1 = E_0 \frac{\omega\cos(\theta) + j\omega\sin(\theta)}{L\left(j\omega + \frac{R}{L}\right)2j\omega}.$$

$$k_1 = \frac{E_0 e^{j\theta}}{j2Z\angle\varphi}. \tag{1.45}$$

With (1.45), the time response associated with the pair of complex conjugated poles is shown in (1.46):

$$\frac{E_0}{Z}\cos(\omega t + \theta - \varphi - \pi/2) = \frac{E_0}{Z}\sin(\omega t + \theta - \varphi). \tag{1.46}$$

The calculation of k_2 is related to the transient response of the circuit:

$$k_2 = \frac{E_0}{L}\frac{\omega\cos(\theta) + s\sin(\theta)}{(s - j\omega)(s + j\omega)}\bigg|_{s=-\frac{R}{L}},$$

$$k_2 = E_0 \frac{R\left(\frac{\omega L}{R}\cos(\theta) - \sin(\theta)\right)}{R^2\left(1 + \frac{(\omega L)^2}{R^2}\right)}, \tag{1.47}$$

$$k_2 = \frac{E_0}{R}\left(\frac{\frac{\sin(\varphi)\cos(\theta) - \sin(\theta)\cos(\varphi)}{\cos(\varphi)}}{1 + tg^2(\varphi)}\right), \tag{1.48}$$

since:

$$1 + \tan^2(\varphi) = \sec^2(\varphi),$$

thus:

$$R = Z\cos(\varphi),$$

rearranging:

$$k_2 = \frac{E_0}{Z}\sin(\varphi - \theta). \tag{1.49}$$

The transient response, associated with k_2, is given by:

$$i_t(t) = \frac{E_0}{Z}\sin(\varphi - \theta)e^{\frac{-t}{\tau}}. \tag{1.50}$$

With Eqs. (1.46) and (1.50), the time response of the RL circuit, which contains steady-state and transient components, is given by (1.51):

$$i\,(t) = \frac{E_0}{Z}\left[\sin\,(\omega\,t + \theta - \varphi) - \sin\,(\theta - \varphi)\,e^{-t/\tau}\right]. \tag{1.51}$$

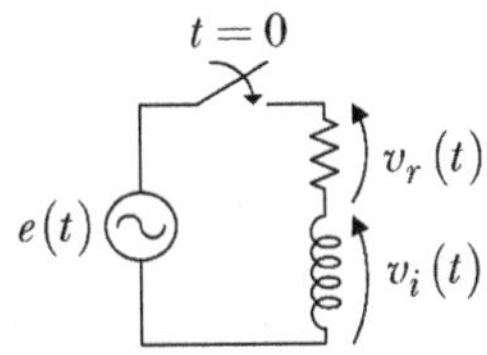

Figure 1.2 RL circuit with sinusoidal voltage source.

The inductance of the RL circuit does not allow rapid current variations, and consequently, low frequencies are expected in the transient current; thus, even large systems can be represented by their Thévenin equivalents calculated at 60 Hz. Therefore, the circuit in Figure 1.2 can be used to calculate the transient fault current in power systems. If the magnetization branch is considered linear, the circuit can give an initial idea of the critical conditions for the energization of reactors and transformers.

The RL circuit is used to calculate the maximum asymmetrical fault current and the damping of the DC component. The steady-state component is designated as the symmetrical fault current. The total current, calculated with the sum of steady-state and transient components is the asymmetrical fault current. The maximum fault current amplitude is calculated from the asymmetrical current. These currents are required to specify all power system equipment in a substation. For circuit breakers, in addition to the maximum current value, it is necessary to analyze the DC offset and damping.

A brief analysis of the current asymmetry is presented below.

Example 1.2 Calculate the transient current in the RL circuit for the angle θ (0°, 30°, 60°, and 90°), given that $R = 2\ \Omega$, $L = 30$ mH, $e(t) = E_0\sin(\omega\,t + \theta)$, and ($E_0 = 1.0$ V). Analyze the current asymmetry in the circuit as in Figure 1.2.

a) Applying expression (1.51), the time responses are calculated in Figure 1.3.
b) Analysis of the current asymmetry.

It can be observed in Figure 1.3 that the amplitude of the current is the highest at $\theta = 0°$, where the DC component has its greatest amplitude.

In order to highlight the main components of the current in Eq. (1.51), a practically pure inductive circuit is considered to calculate the steady-state component ($\theta = 0°$ and $\varphi \simeq 90°$), but a low resistance is used in the time constant for the DC component.

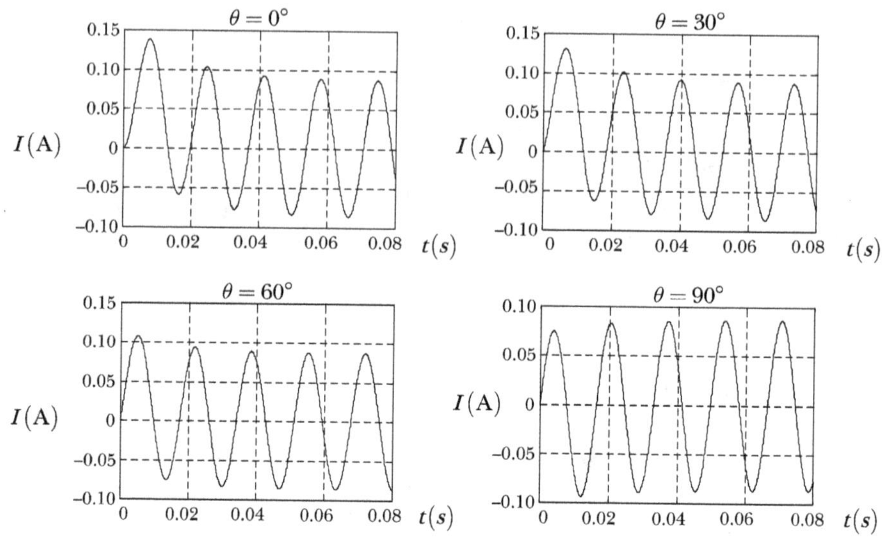

Figure 1.3 Time response of the RL circuit.

Equations (1.38) and (1.51) lead to:

$$e(t) = \frac{E_0}{Z}\sin(\omega t).$$

$$i(t) = \frac{E_0}{Z}\left(-\cos(\omega t) + e^{-\frac{R}{L}t}\right).$$

The current initially is zero and since the voltage is zero at the moment that the switch is closed, that is, when the steady-state current is expected to reach its peak value, a DC component arises so that there is no instantaneous current variation through the inductance. The peak value of the current obtained with this simplified formula is $I_{max} = 0.1375$ A, which occurs around half a cycle after the closing of the switch, that is, at $= 8.33$ ms if the excitation frequency is 60 Hz.

In fact, the maximum amplitude of the current depends on two variables (t, θ), which must be calculated iteratively in Eq. (1.51) to rigorously analyze the current asymmetry. Taking $\varphi = 79.9716°$, the maximum current amplitude is calculated to be $I_{max} = 0.1383$ A, at $t = 7.565$ ms and $\theta = 0.5713°$.

The simplified equation shows good accuracy, about 0.1%, in the predominant inductive RL circuit analyzed.

On the other hand, assuming symmetric current, that is:

$$\theta = 90° \text{ and } \varphi \simeq 90°,$$

which when substituted into (1.38) and (1.51), leads to:

$$e(t) = E_0 \cos(\omega t),$$
$$i(t) = \frac{E_0}{Z} \sin(\omega t).$$

In this case, at $t = 0$, the fault is applied when the excitation voltage is maximum and the current in the steady state is expected to be zero; the DC component does not exist, since there is no instantaneous current variation through the inductance. Figure 1.3 shows that the lowest current amplitude is obtained at $\theta = 90°$, when the amplitude of the DC component is at its lowest value. Evidently, since the system is linear, the values calculated with 1.0 V voltage can be scaled to any more realistic voltage range.

As an observation, the previous results were calculated for a unit source (1.0 V). Due to the linearity of the circuit, all results can be scaled for a real case. For example, if a 230 kV, 60 Hz substation has the same Thévenin equivalent for the positive sequence ($Z_t = 2 + j11.31\ \Omega$), the factor to be applied to the previous results would be:

$$\sqrt{2}\frac{230}{\sqrt{3}} \times 10^3 = 187.79 \times 10^3,$$

Therefore, the maximum current of 0.1375 A would become equal to 25.82 kA (187.79 × 0.1375).

1.4.2 RL Circuit with DC Voltage Source

The Laplace transform of the step function in Figure 1.4a is given by:

$$E(s) = \frac{E_0}{s}. \tag{1.52}$$

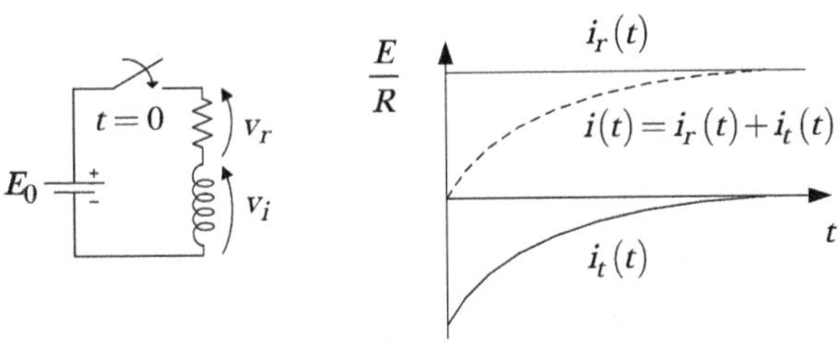

(a) (b)

Figure 1.4 RL circuit with DC excitation. (a) RL circuit. (b) Time response.

Note that the DC (direct current) source can be treated as a special case of the AC (alternating current) source by setting $\omega = 0$ and $\theta = \pi/2$ in (1.39).

Substituting (1.52) into (1.37) yields (1.53):

$$I(s) = \frac{E_0}{L}\frac{1}{s\left(s+\dfrac{R}{L}\right)} + \frac{i_0}{s+\dfrac{R}{L}}, \tag{1.53}$$

where i_0 is the initial current in the inductance, $i_0 = i(0^-)$, which will be considered to illustrate the calculation with the initial condition.

The expansion into partial fractions of the function in (1.53) needs the residues k_1 and k_2, given by:

$$k_1 = \left.\frac{1}{s+\frac{R}{L}}\right|_{s=0} = \frac{L}{R}, \quad k_2 = \left.\frac{1}{s}\right|_{s=-\frac{R}{L}} = -\frac{L}{R}$$

thus,

$$I(s) = \frac{E_0}{L}\left[\frac{L}{sR} - \frac{L}{R\left(s+\frac{R}{L}\right)}\right] + \frac{i_0}{s+\frac{R}{L}}$$

$$I(s) = \frac{E_0}{Rs} - \frac{E_0}{R\left(s+\frac{R}{L}\right)} + \frac{i_0}{s+\frac{R}{L}}. \tag{1.54}$$

Defining τ as the time constant:

$$\tau = \frac{L}{R},$$

and applying the inverse transform:

$$i(t) = \frac{E_0}{R}\left(1-e^{-\frac{t}{\tau}}\right) + i_0e^{-\frac{t}{\tau}}, \tag{1.55}$$

the steady state may be written,

$$i_r(t) = \frac{E_0}{R},$$

and the transient component is given by:

$$i_t(t) = \left(i_0 - \frac{E_0}{R}\right)e^{-t/\tau}.$$

The total current in the circuit is obtained as the sum of both components:

$$i(t) = i_r(t) + i_t(t). \tag{1.56}$$

The current i_0 changes only the transient component, which can be easily incorporated into the previous expressions. Figure 1.4b presents the time response of the RL circuit.

1.5 Series RLC Circuit

By incorporating capacitance, the analysis possibilities of the RLC circuit are expanded compared to the RL circuit. Its oscillatory behavior serves multiple purposes in the analysis of transients in power systems. Interestingly, the RLC circuit approaches the RL circuit when the capacitance tends to infinity, behaving like a short circuit. Conversely, when the capacitance is small and behaves like a high impedance, the circuit demonstrates how even small currents can result in high voltages across the capacitors.

The current and the voltage across the capacitor will be the variables in the equations, as in [4].

1.5.1 RLC Circuit with Sinusoidal Voltage Source

$$e\,(t) = E_0\ \sin\ (\omega\, t + \theta)$$

Applying Kirchhoff's second law (Figure 1.5):

$$e(t) = v_r(t) + v_i(t) + v_c(t). \tag{1.57}$$

- Calculation of current with AC source
  The current is solved writing:

$$e(t) = Ri(t) + L\frac{di(t)}{dt} + \frac{1}{C}\int_0^t i(t)dt.$$

Transforming:

$$E(s) = RI(s) + L[sI(s) - i_0] + \frac{1}{sC}[I(s) + Q_0],$$

and as $v_0 = \frac{Q_0}{C}$,

$$E(s) = \left(R + sL + \frac{1}{sC}\right)I(s) - Li_0 + \frac{v_0}{s}. \tag{1.58}$$

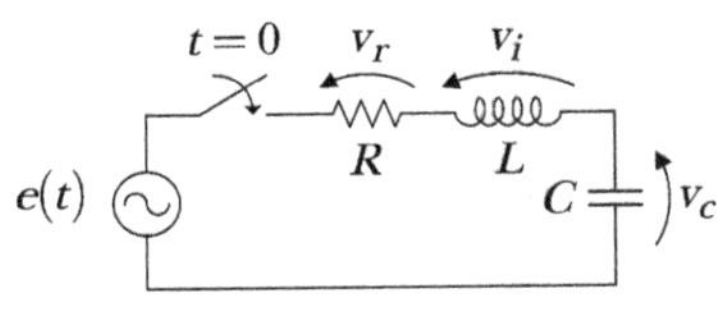

Figure 1.5 Series RLC circuit with sinusoidal voltage source.

The complex impedance of the circuit in (1.58) may be rewritten as:

$$Z(s) = \frac{L}{s}\left(s^2 + s\frac{R}{L} + \frac{1}{LC}\right). \tag{1.59}$$

Dividing (1.58) by $Z(s)$:

$$I(s) = \frac{E(s)}{Z(s)} + \frac{Li_0}{Z(s)} - \frac{v_0}{sZ(s)}. \tag{1.60}$$

The system transfer function is the complex admittance of the circuit, $G(s) = Z(s)^{-1}$, which has one zero (which is equal to zero) and two poles, which are the roots of the polynomial:

$$P(s) = s^2 + s\frac{R}{L} + \frac{1}{LC}.$$

In (1.60) $E(s)$ and $I(s)$ are the input and output of the system, respectively.

With null initial condition, $E(s)$ from (1.39) and $Z(s)$ from (1.59), the Laplace transform of the current is shown in (1.61):

$$I(s) = \frac{sE_0[\omega\cos(\theta) + s\sin(\theta)]}{L(s^2 + \omega^2)\left(s^2 + \frac{R}{L}s + \frac{1}{LC}\right)}. \tag{1.61}$$

The expansion into partial fractions of the equation for the current is as follows:

$$I_c(s) = \frac{k_1}{(s - j\omega)} + \frac{k_1^*}{(s + j\omega)} + \frac{k_2}{(s - \sigma + j\omega_d)} + \frac{k_2^*}{(s - \sigma - j\omega_d)}. \tag{1.62}$$

- Calculation of current with DC excitation
  Taking $E(s)$ from (1.52),

$$I(s) = \frac{E_0}{L\left(s^2 + \frac{R}{L}s + \frac{1}{LC}\right)}. \tag{1.63}$$

- Calculation of the capacitor voltage with AC excitation

  For the capacitor:

$$i(t) = C\frac{dv_c(t)}{dt}.$$

  Applying the Laplace transform:

$$I(s) = sCV_c(s) - Cv_c(0). \tag{1.64}$$

Designating:

$$v_c(0) = v_0,$$

Substituting (1.64) in (1.60),

$$sCV_c(s) = \frac{E(s)}{Z(s)} + \frac{Li_0}{Z(s)} - \frac{v_0}{sZ(s)} + Cv_0,$$

$$V_c(s) = \frac{E(s)}{sCZ(s)} + \frac{Li_0}{SCZ(s)} + \frac{v_0}{s}\left(1 - \frac{1}{sCZ(s)}\right). \tag{1.65}$$

Assuming quiescent initial conditions and substituting $Z(s)$ from (1.60):

$$V_c(s) = \frac{E_0[\omega\cos(\theta) + s\mathrm{sen}(\theta)]}{LC(s^2+\omega^2)\left(s^2 + \frac{R}{L}s + \frac{1}{LC}\right)}, \tag{1.66}$$

that can be rearranged into (1.67):

$$V_c(s) = \frac{E_0(\omega\cos\theta + s\sin\theta)}{(s^2+\omega^2)sC\left(R + sL + \frac{1}{sC}\right)}, \tag{1.67}$$

where,

$$V_c(s) = \frac{\frac{1}{sC}}{\left(R + sL + \frac{1}{sC}\right)}E(s),$$

is clearly a voltage divider equation.

The roots of the denominator in (1.61) and in (1.66) are the same and comprise two pairs of complex conjugate poles – one related to the sinusoidal voltage source $\lambda_{1,2} = \pm j\omega$, and the other one, $\lambda_{3,4} = \sigma \pm j\omega_d$, related to the polynomial $P(s)$.

$$\lambda_{3,4} = -\frac{R}{2L} \pm \sqrt{\left(\frac{R}{2L}\right)^2 - \frac{1}{LC}}.$$

The roots $\lambda_{3,4}$ meet one of three conditions:

a) $\frac{R}{2L} > \frac{1}{\sqrt{LC}}$ overdamped circuit with real roots.

b) $\frac{R}{2L} = \frac{1}{\sqrt{LC}}$ critically damped circuit with equal roots.

c) $\frac{R}{2L} < \frac{1}{\sqrt{LC}}$ underdamped circuit with complex roots.

The inverse transform process is the same for cases (a), (b), and (c), that is, expansion of (1.66) into partial fractions.

In Chapter 12, the case of an overdamped, parallel RLC circuit is addressed, with real roots and its respective application to power systems for the transient analysis of circuit-breaker openings. Here, the oscillatory case of a series RLC circuit is examined, with several applications to power systems, assuming complex roots. Regarding the approach, there is no concern about obtaining an algebraic expression for the solution of this example of a simple network. The focus is on the numerical procedure used to perform the calculation of an elementary case, and in the identification of the poles due to the excitation source, which are related to the steady-state response, and in the identification of the damped poles of the passive network, which are related to the transient response. The procedure used in this simple example can be extended to more complex networks.

Designating for the RLC circuit:

$$\tau = \frac{2L}{R}, \tag{1.68}$$

$$\omega_0 = \frac{1}{\sqrt{LC}}, \tag{1.69}$$

$$\sigma = -\frac{R}{2L}, \tag{1.70}$$

$$\omega_d = \sqrt{\frac{1}{LC} - \left(\frac{R}{2L}\right)^2}, \tag{1.71}$$

where:

τ: time constant of the RLC circuit.
ω_0: natural frequency of oscillation.
σ: real part of pole λ ($\sigma < 0$, related to the attenuation constant, $|\sigma| = \tau^{-1}$).
ω_d: imaginary part of pole λ (damped frequency of oscillation).

The damped and the natural frequencies of oscillation are related by:

$$\omega_d^2 = \omega_0^2 - \sigma^2.$$

The expansion into partial fractions of the equation for the voltage across the capacitor uses Eq. (1.66):

For the steady-state-response, residue k_1 is calculated:

$$k_1 = \left.\frac{E_0(\omega\cos\theta + s\text{sen}\theta)}{(s + j\omega)sC\left(R + sL + \frac{1}{sC}\right)}\right|_{s=j\omega},$$

$$k_1 = -\frac{E_0 e^{j\theta}}{2\omega C\left(R + j\omega L - \frac{j}{\omega C}\right)}.$$

The complex impedance of the RLC circuit at nominal frequency,

$$Z\angle\varphi = R + j\left(\omega L - \frac{1}{\omega C}\right),$$

has magnitude Z and phase angle φ:

$$Z = \sqrt{R^2 + \left(\omega L - \frac{1}{\omega C}\right)^2},$$

$$\tan\varphi = \frac{\omega L - \frac{1}{\omega C}}{R}.$$

The residue k_1 can be rewritten as:

$$k_1 = \frac{E_0}{2\omega CZ}\angle\theta - \varphi - \pi. \tag{1.72}$$

For the transient response, residue k_2 is calculated:

$$k_2 = \left.\frac{E_0(\omega\cos\theta + s\sin\theta)}{LC(s + j\omega)(s - j\omega)(s - \sigma + j\omega_d)}\right|_{s = \sigma + j\omega_d},$$

$$k_2 = \frac{E_0\omega_0^2(\omega\cos\theta + \sigma\sin\theta + j\omega_d\sin\theta)}{2j\omega_d[\sigma + j(\omega + \omega_d)][\sigma + j(\omega_d - \omega)]}, \tag{1.73}$$

the residue k_2 is defined as $k_2 = K_2 \angle \phi_2$.

Based on the above results, the time response obtained is:

$$v_c(t) = -\frac{E_0}{\omega CZ}\cos(\omega t + \theta - \varphi) + 2K_2 e^{\sigma t}\cos(\omega_d t + \phi_2). \tag{1.74}$$

The steady-state solution for the RLC circuit is analyzed next. The amplitude of the phasor of the voltage across the capacitor is obtained through the voltage divider equation, changing the sine function with phase θ to an equivalent cosine function with phase $\theta - \pi/2$:

$$V_{cr} = \frac{1}{j\omega CZ\angle\varphi}E_0\angle\left(\theta - \frac{\pi}{2}\right),$$

and in time domain,

$$v_{cr}(t) = -\frac{E_0}{\omega CZ}\cos(\omega t + \theta - \varphi),$$

which coincides with the steady-state response in time domain using the residue k_1 in (1.72).

The amplitude of current in the steady state can also be obtained:

$$I_r = \frac{E_0\angle\left(\theta - \frac{\pi}{2}\right)}{Z\angle\varphi},$$

with the current delayed by φ with respect to the phase angle of the source, θ:

$$i(t) = \frac{E_0}{Z}\cos\left(\omega t + \theta - \varphi - \frac{\pi}{2}\right) = \frac{E_0}{Z}\sin(\omega t + \theta - \varphi). \tag{1.75}$$

1.5.2 LC Circuit

The LC circuit equation is obtained from Eq. (1.59), neglecting the resistance and using the natural frequency defined in (1.69):

$$Z(s) = \frac{L}{s}\left(s^2 + \omega_0^2\right). \tag{1.76}$$

- Calculation of current and capacitor voltage with AC source

$$I(s) = \frac{sE_0[\omega\cos(\theta) + s\sin(\theta)]}{L(s^2 + \omega^2)(s^2 + \omega_0^2)} \tag{1.77}$$

$$V_c(s) = \frac{E_0[\omega\cos(\theta) + s\sin(\theta)]}{LC(s^2 + \omega^2)(s^2 + \omega_0^2)}. \tag{1.78}$$

- Calculation of current with DC source and residual voltage at the capacitor
  Excitation with a DC voltage source can be useful to approximate a cosine voltage by a constant value over the analyzed time interval, thus simplifying the analysis of the circuit. This is a basic circuit for analyzing capacitor bank switching (Figure 1.6).

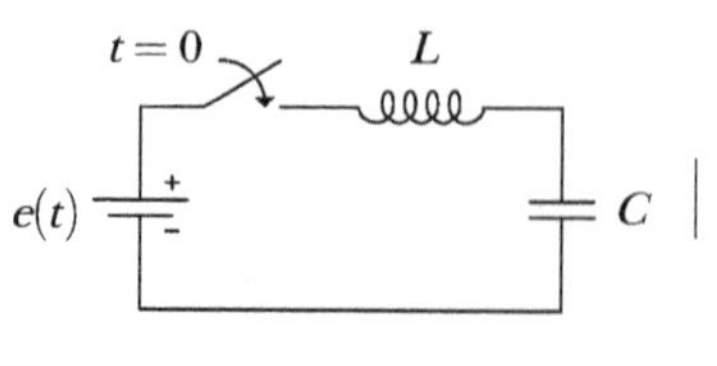

Figure 1.6 LC circuit with DC voltage source.

In this case, the initial current is assumed zero, but it is worth considering an initial voltage on the capacitor. Thus, from (1.60):

$$I(s) = \frac{E_0}{sZ(s)} - \frac{v_0}{sZ(s)},$$

$$I(s) = \frac{E_0 - v_0}{L} \frac{1}{s^2 + \omega_0^2},$$

multiplying numerator and denominator by ω_0,

$$I(s) = \frac{E_0 - v_0}{L\frac{1}{\sqrt{LC}}} \frac{\omega_0}{s^2 + \omega_0^2}.$$

Applying the inverse transform,

$$i(t) = \frac{E_0 - v_0}{\sqrt{L/C}} \sin \omega_0 t. \tag{1.79}$$

It is known that the maximum current occurs when $v_0 = -E_0$, that is, the voltage across the capacitor has the same excitation voltage magnitude but opposite in sign. On the other hand, the impedance $\sqrt{L/C}$, which is also designated as the surge impedance of the circuit, places limits on the current.

If $v_0 = 0$,

$$i(t) = \frac{E_0}{\sqrt{L/C}} \sin \omega_0 t. \tag{1.80}$$

- LC free response with initial conditions i_0 and v_0.

  The equation for the *LC* circuit is now examined from the point of view of the voltages developed across the capacitor. Initial conditions for current and voltage are also assumed:

  If excitation is eliminated in (1.65), making $E(s) = 0$, and taking $Z(s)$ from (1.76),

$$V_c(s) = \frac{i_0}{C} \frac{1}{s^2 + \omega_0^2} + \frac{s v_0}{s^2 + \omega_0^2},$$

Applying the inverse transform,

$$v_c(t) = i_0 \sqrt{\frac{L}{C}} \sin(\omega_0 t) + v_0 \cos(\omega_0 t). \tag{1.81}$$

The constant M is defined as:

$$M = \sqrt{i_0^2 \frac{L}{C} + v_0^2}.$$

Multiplying and dividing (1.81) by M and defining:

$$\sin(\varphi) = \frac{v_0}{M},$$

$$\cos(\varphi) = \frac{i_0\sqrt{\frac{L}{C}}}{M},$$

and using these relations in (1.81), the following equation is obtained:

$$v_c(t) = M[\sin(\omega_0 t)\cos(\varphi) + \sin(\varphi)\cos(\omega_0 t)],$$

which can be compressed into:

$$v_c(t) = \sqrt{v_0^2 + i_0^2\frac{L}{C}}\sin(\omega_0 t + \varphi). \tag{1.82}$$

One application of this equation is in the de-energization of a reactor with a model consisting of a high inductance in parallel with a stray capacitance, when analyzing the interruptions in small inductive currents, where the magnitude of the overvoltage depends on the initial values of voltage and current.

Example 1.3 Calculate the transient current for the circuit in Figure 1.5, using the Laplace transform.

Data

$e(t) = E_0\sin(\omega t + \theta)$, $E_0 = 1$, $\theta = 0°, 30°, 60°$, and $90°$.
$R = 1\ \Omega$, $L = 0.1$ H, $C = 10\ \mu$F.
Assume quiescent initial conditions.

Solution

From (1.61)

$$I(s) = \frac{sE_0[\omega\cos(\theta) + s\sin(\theta)]}{L(s^2+\omega^2)\left(s^2 + \frac{R}{L}s + \frac{1}{LC}\right)}.$$

The roots of the denominator, considering the oscillatory response and previous expressions (1.69) and (1.70):

$$\lambda_{1,2} = \pm j\omega,$$

$$\lambda_{3,4} = \sigma \pm j\omega_d.$$

Calculating the complex residues k_i using (1.61) and (1.62):

$$k_1 = \left.\frac{E_0 s(\omega\cos\theta + s\sin\theta)}{L(s+j\omega)(s-\sigma-j\omega_d)(s-\sigma+j\omega_d)}\right|_{s=j\omega},$$

$$k_1 = \frac{E_0\omega e^{j\theta}}{2L[-\sigma + j(\omega-\omega_d)][-\sigma + j(\omega+\omega_d)]},$$

$$k_2 = \frac{E_0}{L} \frac{(\sigma + j\omega_d)[\omega \cos(\theta) + (\sigma + j\omega_d) \sin(\theta)]}{[\sigma + j(\omega + \omega_d)]\,[\sigma + j(\omega_d - \omega)]2j\omega_d}.$$

The time response can be written as:

$$i(t) = 2K_1 \cos(\omega t + \phi_1) + 2K_2 e^{\sigma t} \cos(\omega_d t + \phi_2).$$

Assigning numerical values to the above expressions, the following results are obtained, where the phase angle of the residues is given in radians.

$$\theta = 0^\circ$$

$$\sigma = -5.0\,\mathrm{s}^{-1}$$

$$\omega = 377 \ \mathrm{rad/s}$$

$$\omega_d = 1000 \ \mathrm{rad/s}$$

$$k_1 = 2.1972 \times 10^{-3} E_0 \angle -0.0044$$

$$k_2 = 2.1972 \times 10^{-3} E_0 \angle 3.1349$$

$$i(t) = 2 \times 10^{-3} E_0 \begin{bmatrix} 2.1972 \cos(377t - 0.0044) \\ + 2.1972 e^{-5t} \cos(1000t + 3.1349) \end{bmatrix}$$

$$\theta = 30^\circ$$

$$i(t) = 2 \times 10^{-3} E_0 \begin{bmatrix} 2.1972 \cos(377t + 0.5192) \\ + 3.4724 e^{-5t} \cos(1000t - 2.1524) \end{bmatrix}$$

$$\theta = 60^\circ$$

$$i(t) = 2 \times 10^{-3} E_0 \begin{bmatrix} 2.1972 \cos(377t + 1.0428) \\ + 5.1603 e^{-5t} \cos(1000t - 1.7870) \end{bmatrix}$$

$$\theta = 90^\circ$$

$$i(t) = 2 \times 10^{-3} E_0 \begin{bmatrix} 2.1972 \cos(377t + 1.5664) \\ + 5.8284 e^{-5t} \cos(1000t - 1.5725) \end{bmatrix}.$$

In this example, the RLC circuit was studied with different voltages applied at $t = 0$ as a function of the phase angle of the sinusoidal voltage source. It can be observed that the steady-state component given by the residue k_1 varies only in phase angle, keeping the absolute value constant. On the other hand, the transient component given by the residue k_2 is altered in modulus and phase to ensure zero current at zero time.

It can also be shown that the transient component oscillates at the damped frequency of the network ω_d. The decay rate depends on the attenuation constant, which is the inverse of the time constant (Figure 1.7).

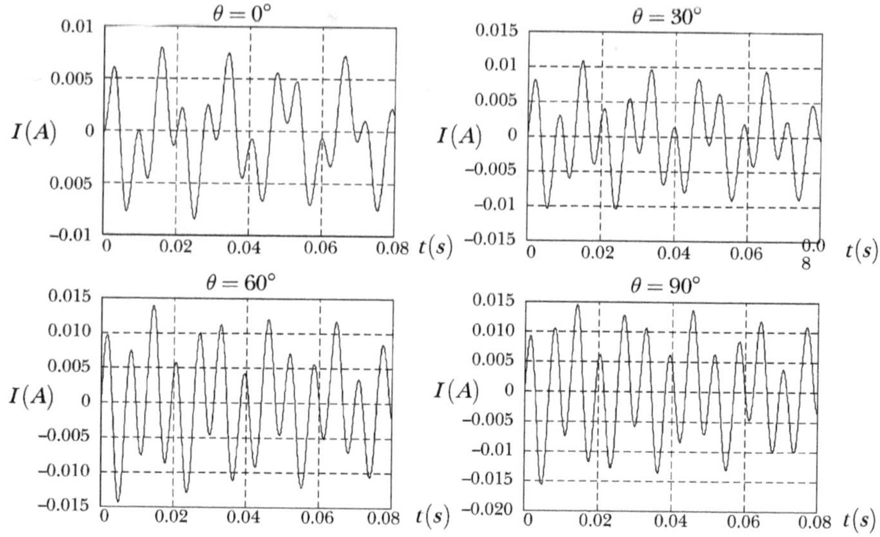

Figure 1.7 Time response of the current in the RLC circuit.

Example 1.4 Assuming a steady-state short-circuit in a transmission line, calculate the voltage across the circuit breaker terminals when the fault is cleared, for the circuit of Figure 1.8a.

Data

$$R = 2\,\Omega, L = 0.016\ \mathrm{H}, C = 0.2\,\mu\mathrm{F}.$$

The voltage at the generator is designated as $e(t) = E_0 \sin(\omega t + \theta)$.

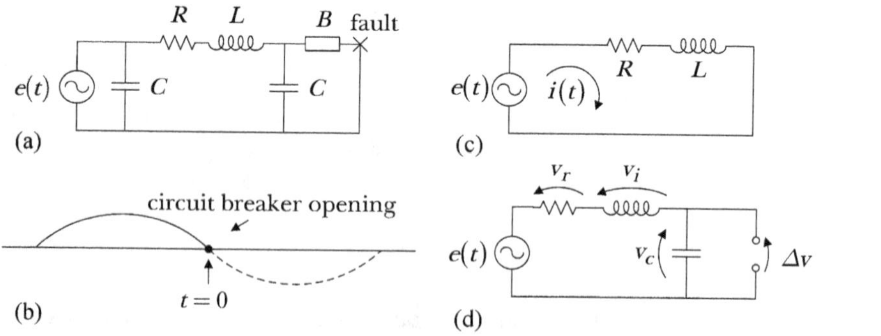

Figure 1.8 Opening of the circuit breaker. (a) Circuit for Transmission line fault. (b) Current zero crossing. (c) Equivalent circuit with fault current. (d) Circuit after current interruption.

Solution

The voltage across the capacitor is solved in this example.

Opening of the circuit breaker occurs at the point of the current's zero crossing, as in Figure 1.8b. The process of interrupting alternating currents through circuit breakers is more thoroughly discussed in Chapter 12. For the moment, it may be assumed that, from the thermal point of view, this is an appropriate point in time for the electric arc in the circuit breaker to be extinguished. From an electrical point of view, this is also an interesting point in time for current interruption, if the circuit is predominantly inductive, as is most commonly the case, since there is no instantaneous variation of current through the inductance.

The capacitor closest to the source can be neglected since the sending end voltage does not change with or without the capacitor. It can also be shown that the Thévenin equivalent of the voltage source in parallel with the capacitor is equal to that of an ideal voltage source.

a) Analysis of a circuit during a fault.

The circuit during the fault is given in Figure 1.8c. The voltages and currents in the RL circuit need to be calculated, as these are the initial conditions for the RLC circuit after opening the circuit breaker.

Assuming that the transient component of the fault has been damped, only the steady-state component of the RL circuit is present at the instant of opening:

$$i(t) = I \sin\ (\omega t + \theta - \varphi_1),$$

where $\varphi_1 = \arctan\left(\frac{\omega L}{R}\right)$ is the angle associated with the complex impedance $Ze^{j\varphi_1}$.

$$I = \frac{E_0}{|Z|}.$$

The circuit breaker is considered open only when the current is interrupted at its zero crossing, i.e., at $t = 0$:

$$0 = i_0 = I \sin(\theta - \varphi_1),$$

therefore, $\theta = \varphi_1$.

Substituting the numerical values:

$$\theta = \varphi_1 = \arctan\left(\frac{16 \times 10^{-3} \times 377}{2}\right) = 71.66^\circ.$$

For purposes of this calculation, the value of θ was chosen so that the current crosses zero at $t = 0$. If the phase angle θ differed in the example, the time when the current crosses zero would have to be calculated, to proceed with the analysis from that point on.

b) Analysis of the circuit after interruption of the current.

The circuit after interruption of the current is depicted in Figure 1.8d, where the generator voltage may be assumed as:

$$e(t) = E_0 \sin(\omega t + 71.66^\circ).$$

The voltage across the circuit breaker-terminals is equal to the voltage across the capacitor $v_c = (t)$. Given that the initial voltage across the capacitor is zero, since it is bypassed by the circuit-breaker and the fault, $v_0 = 0$, and that the current is also zero at the time of opening, that is, $i_0 = 0$, the initial conditions are assumed to be zero. An oscillatory circuit is also assumed.

Thus, from (1.66):

$$V_c(s) = \frac{E_0}{LC} \frac{(\omega \cos\theta + s \sin\theta)}{(s^2 + \omega^2)\left(s^2 + \frac{R}{L}s + \frac{1}{LC}\right)},$$

$$\lambda_{1,2} = \pm j\omega, \ \lambda_{3,4} = \sigma \pm j\omega_d.$$

Substituting the numerical values and assuming $E_0 = 1$:

$$\omega = 377 \ \text{rad/s}$$

$$\sigma = -\frac{R}{2L} = -62.5\,\text{s}^{-1},$$

$$\omega_d = \sqrt{\frac{1}{LC} - \left(\frac{R}{2L}\right)^2} = 17677.6 \ \text{rad/s},$$

$$k_1 = 0.5e^{-j18.35^\circ},$$

$$k_2 = 0.4748e^{-j180.6^\circ}.$$

The inverse Laplace transform is given by:

$$v_c(t) \cong \cos(377t - 0.1\pi) + 0.95e^{-62.5t}\cos(17678t - 1.003\pi).$$

The solution of the differential equation for this circuit has a steady-state and a transient response.

Regarding the transient opening of the circuit breaker, the result in Figure 1.9 provides a qualitative idea of the phenomenon, since single-phase faults unbalance the three-phase system, usually involving the zero-sequence network. In addition, it is important to examine whether the line model, with the equivalent π, is reasonable at the frequencies involved. Otherwise, the line model should be improved using a larger number of π units to represent higher frequencies or using a distributed parameter representation, including traveling waves along a

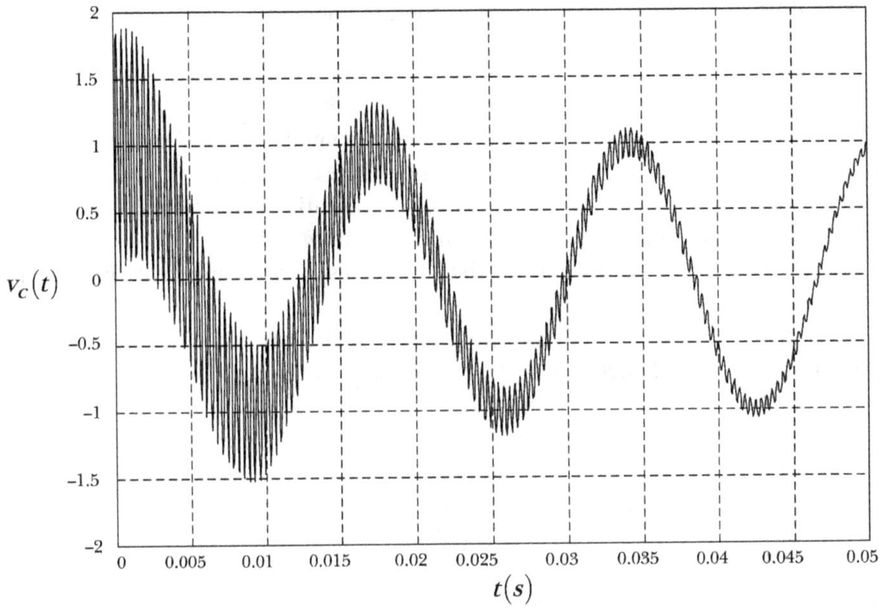

Figure 1.9 Voltage $v_c(t)$.

transmission line, which is normally the representation utilized in electromagnetic transient programs. It can be anticipated that a more accurate transient analysis would employ a model with distributed parameters to represent the transmission line, as will be seen in Chapter 12.

1.6 Resonance at the Nominal Frequency

In power systems, resonant conditions at the nominal frequency can be found, when the network has a natural frequency of oscillation coincident with the nominal excitation frequency.

Assume the elementary LC circuit whose natural frequency is equal to the nominal frequency of a sinusoidal source,

$$e(t) = E_0 \sin(\omega t).$$

Substituting ω_0 by ω and setting $\theta = 0$ in (1.77), the equation for the current is given below:

$$I(s) = \frac{E_0}{L} \frac{s\omega}{(s^2 + \omega^2)^2}. \tag{1.83}$$

Recalling Example 1.1, the current in the time domain is:

$$i(t) = \frac{E_0}{2L} t \sin(\omega t), \tag{1.84}$$

that is, an unstable response is obtained increasing with time and tending toward infinity. This fact is created by the coincidence of the complex pole of the excitation voltage with the complex pole of the network, generating a double pole, and by an inverse transform that increases with time to infinity.

In power systems, incorporating the losses, resonances are more frequently observed in RLC circuits.

Assuming an oscillatory circuit, the poles of the source and of the circuit were previously discussed in Section 1.5.1, $\lambda_{1,2} = \pm j\omega$, and $\lambda_{3,4} = \sigma \pm j\omega_d$, and therefore, perfect coincidence does not occur.

However, despite the limitations introduced by the resistance, there are resonant conditions that imply high values of voltages or currents.

For steady-state analysis, assuming the excitation voltage:

$$e(t) = E_0 \cos(\omega t),$$

the phasor for the current is given as:

$$I = \frac{\frac{E_0}{\sqrt{2}}}{R + j\left(\omega L - \frac{1}{\omega C}\right)}. \tag{1.85}$$

In the resonance condition:

$$\omega L = \frac{1}{\omega C} \quad \text{or} \quad \omega = \frac{1}{\sqrt{LC}}, \tag{1.86}$$

and therefore:

$$I = \frac{E_0}{\sqrt{2}R}. \tag{1.87}$$

1.7 Analysis of Simple Networks with More Than One Loop

In the study of simple networks with more than one loop, or even in more complex networks, the application of nodal analysis is the most appropriate procedure to organize the equations. When the network has several nodes, formulation of the equations using the nodal admittance matrix is one of the most efficient techniques that can be employed; however, loop analysis, through the loop impedance matrix, can also be considered complementary.

Thus, this item presents the study of transients in simple networks with more than one loop with the application of the Laplace transform, using the nodal admittance matrix and the loop impedance matrix.

1.7.1 Inductive and Capacitive Elements with Initial Conditions

Capacitor and inductor models suitable for analysis of the nodal admittance matrix and loop impedance matrix, including the initial conditions, using the Laplace transform, are described below.

- MODEL FOR INDUCTORS

  Equation for the inductance (Figure 1.10a):

$$v(t) = L\frac{di(t)}{dt}.$$

  Impedance model (Figure 1.10b):

$$V(s) = sLI(s) - Li_0. \tag{1.88}$$

  Admittance model (Figure 1.10c):

$$I(s) = \frac{V(s)}{sL} + \frac{i_0}{s}. \tag{1.89}$$

- MODEL FOR CAPACITORS

  Equation for the capacitance (Figure 1.11a):

$$v(t) = \frac{1}{C}\int i(t)dt.$$

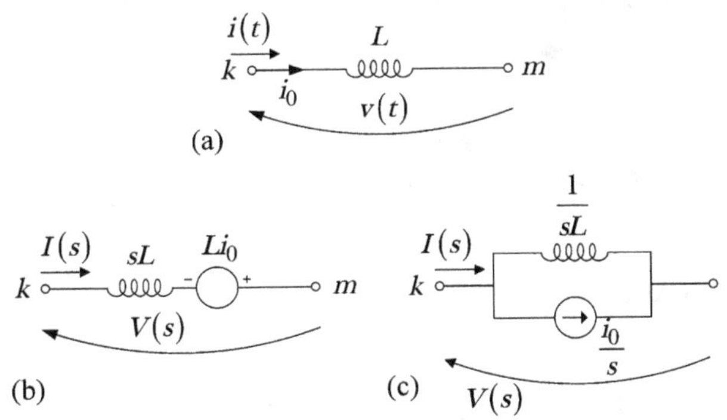

Figure 1.10 Inductance models. (a) Inductance. (b) Impedance model. (c) Admittance model.

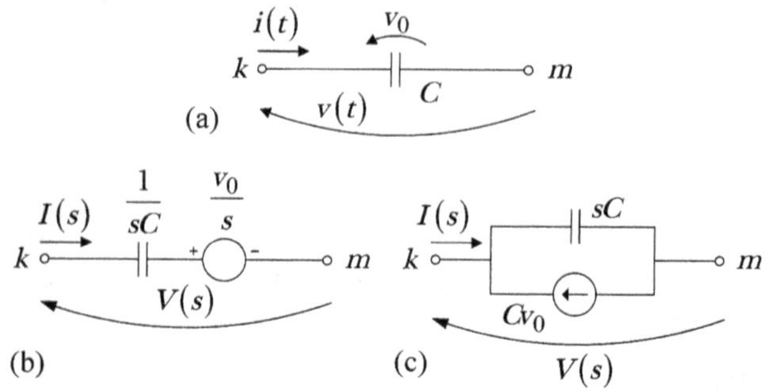

Figure 1.11 Capacitance models. (a) Capacitance. (b) Impedance model. (c) Admittance model.

Impedance model (Figure 1.11b):

$$V(s) = \frac{1}{sC}I(s) + \frac{v_0}{s}. \tag{1.90}$$

Admittance model (Figure 1.11c):

$$I(s) = sCV(s) - Cv_0. \tag{1.91}$$

1.7.2 Network Analysis

In the nodal analysis, the elements are represented as admittances and in the loop analysis, they are represented as impedances. A few examples with simplified values for the parameters are presented below.

Example 1.5 Examine the circuit in Figure 1.12a using the nodal admittance matrix.

Data

$$C_1 = C_2 = 1 \text{ F}, R_1 = 1\ \Omega, L_1 = 1 \text{ H}, e(t) = H(t).$$

Solution

Nodal analysis solution (Figure 1.12b).

$$\begin{bmatrix} \frac{1}{s} & -\frac{1}{s} & 0 \\ -\frac{1}{s} & s + \frac{1}{s} + 1 & -1 \\ 0 & -1 & s+1 \end{bmatrix} \times \begin{bmatrix} E_1(s) \\ V_2(s) \\ V_3(s) \end{bmatrix} = \begin{bmatrix} I_1(s) - \frac{i_0}{s} \\ \frac{i_0}{s} + C_1 v_{01} \\ C_2 v_{02} \end{bmatrix}.$$

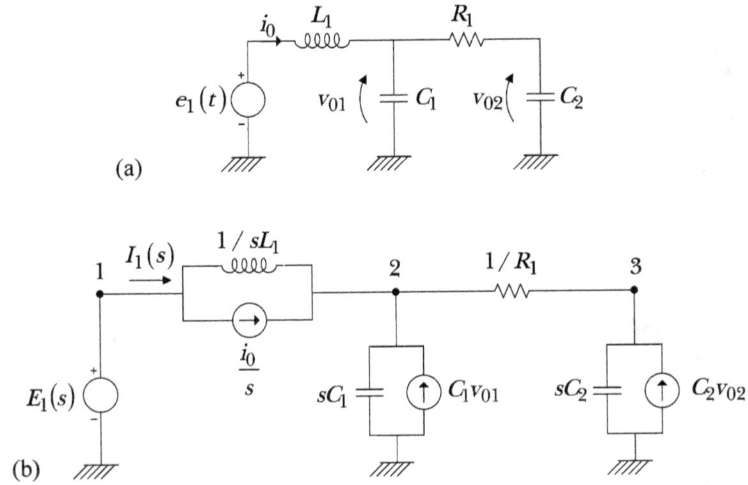

Figure 1.12 Electric network and equivalent models. (a) Electric circuit. (b) Equivalent model.

Since the voltage at node 1 is known, only rows 2 and 3 are used in the solution:

$$\begin{bmatrix} -\dfrac{1}{s} \\ 0 \end{bmatrix} E_1(s) + \begin{bmatrix} \dfrac{s^2+s+1}{s} & -1 \\ -1 & s+1 \end{bmatrix} \times \begin{bmatrix} V_2(s) \\ V_3(s) \end{bmatrix} = \begin{bmatrix} \dfrac{i_0}{s} + C_1v_{01} \\ C_2v_{02} \end{bmatrix},$$

defining,

$$A(s) = \begin{bmatrix} \dfrac{s^2+s+1}{s} & -1 \\ -1 & s+1 \end{bmatrix}.$$

Assuming quiescent initial conditions, to simplify calculation, the solution is obtained from the inverse of A(s):

$$\begin{bmatrix} V_2(s) \\ V_3(s) \end{bmatrix} = \frac{1}{\det A(s)} [\text{cof}A(s)]^t \begin{bmatrix} \dfrac{1}{s} \\ 0 \end{bmatrix} E_1(s).$$

$[\text{cof}A(s)]^t$ is the transpose of the cofactor matrix of $A(s)$:

$$\det A(s) = \frac{s^3+2s^2+s+1}{s},$$

$$\begin{bmatrix} V_2(s) \\ V_3(s) \end{bmatrix} = \frac{s}{s^3+2s^2+s+1} \begin{bmatrix} s+1 & 1 \\ 1 & \dfrac{s^2+s+1}{s} \end{bmatrix} \times \begin{bmatrix} \dfrac{1}{s} \\ 0 \end{bmatrix} E_1(s),$$

$$\begin{bmatrix} V_2(s) \\ V_3(s) \end{bmatrix} = \frac{s}{s^3 + 2s^2 + s + 1} \begin{bmatrix} \frac{s+1}{s} \\ \frac{1}{s} \end{bmatrix} E_1(s),$$

$$V_2(s) = \frac{s+1}{(s^3 + 2s^2 + s + 1)} \cdot \frac{1}{s},$$

$$V_3(s) = \frac{1}{(s^3 + 2s^2 + s + 1)} \cdot \frac{1}{s}.$$

The roots of the denominator are:

$$\lambda_1 = 0, \lambda_2 = -1.7549, \lambda_{3,4} = -0.1226 \pm j0.7449,$$

$$V_2(s) = \frac{k_1}{s} + \frac{k_2}{(s+1.7549)} + \frac{k_3}{(s+0.1226-j0.7449)} + \frac{k_3^*}{(s+0.1226+j0.7449)},$$

$$k_1 = 1, k_2 = 0.1336, k_3 = 0.5704\angle -173.54°,$$

$$v_2(t) = 1 + 0.1336e^{-1.7549\,t} + 1.1409e^{-0.1226\,t}\cos(0.7449t - 173.54°).$$

Analogously, for$e_3(t)$:

$$v_3(t) = 1 - 0.1770e^{-1.7549t} + 1.0e^{-0.1226t}\cos\,(0.7449\,t + 146.3°).$$

Example 1.6 Obtain the current in the network in Figure 1.13a, assuming the initial conditions of the capacitors. This circuit is used in Chapter 10 for the analysis of short-circuit currents in substations with shunt capacitors.

Data

$$C_1 = 1\,\text{F}, C_2 = 0.5\,\text{F}, L_1 = 2\,\text{H}, L_2 = 1\,\text{H}, L_3 = 1\,\text{H}, v_{01} = v_{02} = 10\,\text{V}.$$

Solution

Applying the Laplace transform, the loop impedance model is obtained in Figure 1.13b:

Using the loop impedance matrix:

$$\begin{bmatrix} 3s + \frac{3}{s} & -s - \frac{2}{s} \\ -s - \frac{2}{s} & 2s + \frac{2}{s} \end{bmatrix} \times \begin{bmatrix} I_1(s) \\ I_2(s) \end{bmatrix} = \begin{bmatrix} \frac{v_{01}}{s} - \frac{v_{02}}{s} \\ \frac{v_{02}}{s} \end{bmatrix},$$

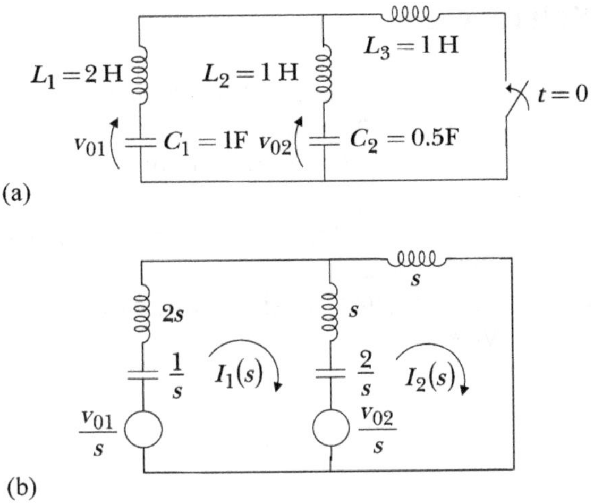

Figure 1.13 Electric network and equivalent model. (a) Electric circuit. (b) Equivalent model.

$$\begin{bmatrix} I_1(s) \\ I_2(s) \end{bmatrix} = 1/\det \begin{bmatrix} \frac{2s^2+2}{s} & \frac{s^2+2}{s} \\ \frac{s^2+2}{s} & \frac{3s^2+3}{s} \end{bmatrix} \times \begin{bmatrix} 0 \\ \frac{10}{s} \end{bmatrix}, \quad \det = \frac{5s^4+8s^2+2}{s^2},$$

$$\begin{bmatrix} I_1(s) \\ I_2(s) \end{bmatrix} = \frac{10}{(5s^4+8s^2+2)} \begin{bmatrix} s^2+2 \\ 3s^2+3 \end{bmatrix},$$

$$I_1(s) = \frac{10(s^2+2)}{(5s^4+8s^2+2)}, I_2(s) = \frac{30(s^2+1)}{(5s^4+8s^2+2)},$$

$$\lambda_{1,2} = \pm j0.5569, \lambda_{3,4} = \pm j1.1357.$$

The inverse transform of $I_1(s)$ is:

$$k_1 = \left.\frac{10(s^2+2)}{(s+j1.1357)(s^2+0.3101)}\right|_{s=j1.1357}, \quad k_1 = j0.6381,$$

$$k_2 = \left.\frac{10(s^2+2)}{(s^2+1.2899)(s+j0.5569)}\right|_{s=0.5569}, k_2 = -j3.0972,$$

$$i_1(t) = 6.1944\sin(0.5569t) - 1.2763\sin(1357t).$$

Analogously, for $I_2(s)$:

$$i_2(t) = 1.5631\sin(1.1357t) - 7.5866\sin(0.5569t).$$

The data in the example are hypothetical in order to simplify the presentation. With real data, the frequencies would be much higher, but the procedure is the same.

References

1 Debnath, L., & Bhatta, D. (2015). *Integral transforms and their applications* (3rd ed.). Boca Raton, FL: CRC Press.

2 Papoulis, A. (1962). *The Fourier integral and its applications*. New York: McGraw-Hill College.

3 Goldman, S. (1966). *Laplace transform theory and electrical transients*. New York: Dover Publications.

4 Greenwood, A. (1991). *Electrical transients in power systems* (2nd ed.). New York: John Wiley & Sons.

2

Traveling Waves in Single-Phase Lines

2.1 Introduction

In this chapter, electromagnetic transient phenomena are interpreted, taking into consideration the traveling waves in aerial transmission lines.

Although a basic treatment of propagation of electromagnetic waves is accommodated by the Maxwell equations, a simplified procedure is adopted here to approach this phenomenon.

The transmission line will be represented by a model with distributed parameters per unit length, in which an inductance L' is associated with the energy stored in the magnetic field and a capacitance C' is associated with the energy stored in the electric field. The elementary equations for voltage and current variations in the distributed model, incorporating second-order derivatives, lead to the derivation of wave equations.

The single-phase transmission line is analyzed, and the results can help in the understanding of three-phase lines.

The calculation of transmission line parameters is reviewed in some of its basic aspects, but the approach can be found in condensed form in Chapter 5 and in the specific literature. The elementary case of a lossless transmission line with a conductor of infinite length, radius r, and height h from the ground is initially considered (Figure 2.1a). The calculus of parameters for the transmission line is implemented by the method of images, assuming the magnetic permeability and permittivity of air to be the same as those in a vacuum.

Magnetic permeability in a vacuum: $\mu_0 = 4\pi \times 10^{-4}$ H/km.

Permittivity in a vacuum is obtained from the expression for the light velocity:

$$c^2 = \frac{1}{\mu_0 \varepsilon_0}.$$

Understanding Electromagnetic Transients in Power Systems, First Edition. Luiz Cera Zanetta Jr.

Published 2025 by John Wiley & Sons, Inc.

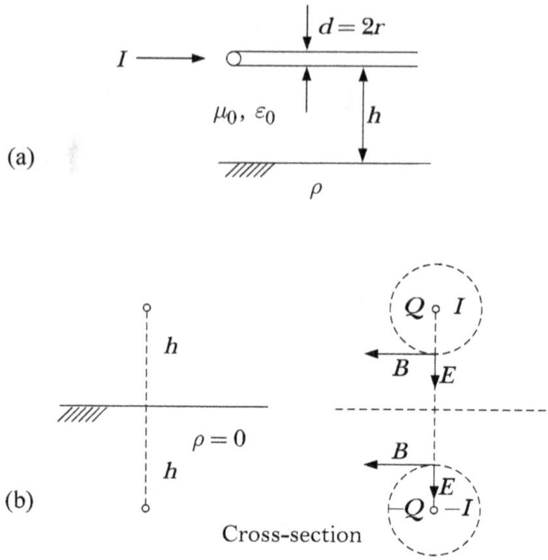

Figure 2.1 Single-phase transmission line. (a) Aerial conductor. (b) Method of Images

Assuming the light velocity, $c = 299792.5$ km/s, the value for the constant below is calculated to be:

$$\frac{1}{2\pi\varepsilon_0} = 17.975109 10^6 \text{ km}/\mu\text{F}.$$

Assuming a uniform and lossless transmission line and the soil to be a perfect conductor of null resistivity, the distributed inductances and capacitances of the transmission line can be calculated by the formulas.

$$L' = \frac{\mu_o}{2\pi} \ln \frac{2h}{r} \text{ H/m,} \tag{2.1}$$

$$C' = \frac{2\pi\varepsilon_0}{\ln \frac{2h}{r}} \text{ F/m.} \tag{2.2}$$

The electric and magnetic fields at a distance x from the conductor in Figure 2.1b are given by the following expressions, which are easily obtained by applying Ampere's and Gauss' laws [1].

$$B = \frac{\mu_0 I}{2\pi x}, E = \frac{Q}{2\pi\varepsilon_0 x}.$$

In order to obtain the external flux linkage due to current I in the conductor, on the rectangular surface between the conductor and ground in one unit length, the following integral is evaluated:

$$\phi_1 = \int_r^h B dx = \int_r^h \frac{u_o I}{2\pi x} dx = \frac{u_{oI}}{2\pi} \ln \frac{h}{r}. \tag{2.3}$$

Calculating the flux on the same surface, created by the conductor's image, using current − I:

$$\phi_2 = -\int_{2h}^h \frac{u_o I}{2\pi x} dx = \frac{u_{oI}}{2\pi} \ln \frac{2h}{h}.$$

Superimposing the linked fluxes due to the conductor and its image:

$$\phi_t = \phi_1 + \phi_2 = \frac{u_{oI}}{2\pi} \ln \frac{2h}{r}. \tag{2.4}$$

The internal flux of the conductor is small for DC current and at high frequencies, considering the skin effect, the internal flux is reduced even more and thus it can be neglected in this simplified approach.

Therefore, the inductance L', per unit length, presented in Eq. (2.1) may be obtained, since $\emptyset = LI$. Note that the same inductance may also be obtained by calculating the flux linked with the conductor in a group of conductors, according to [1].

In order to calculate the voltage between the conductor and ground, the integral in (2.5) is evaluated:

$$\Delta V_1 = \int_r^h \frac{Q}{2\pi\varepsilon_0 x} dx = \frac{Q}{2\pi\varepsilon_o} \ln \frac{h}{r}. \tag{2.5}$$

Repeating the calculus for image − Q, the total electric voltage difference is obtained:

$$\Delta V_2 = -\int_{2h}^h \frac{Q}{2\pi\varepsilon_0 x} dx = \frac{Q}{2\pi\varepsilon_o} \ln \frac{2h}{h}$$

$$\Delta V_t = \Delta V_1 + \Delta V_2 = \frac{Q}{2\pi\varepsilon_o} \ln \frac{2h}{r}. \tag{2.6}$$

The capacitance C' per unit length in Eq. (2.2) can then be associated with the circuit, by means of Eq. (2.6), given that $Q = C' \times \Delta V_t$.

To address traveling waves in single-phase transmission lines, certain simplifications are made to emphasize the fundamental concepts. Complementary information on this subject may be found in Chapters 3 and 5, where multiphase transmission lines are analyzed, using a formulation that is more appropriate for implementation in computer programs.

2.2 Basic Equations

2.2.1 Transmission Line with Losses

The classical formulation is described in several references [2–5] and is also presented here (Figure 2.2).

The transmission line is initially considered to have distributed parameters, R', L', and C' per unit length. For aerial transmission lines, the dielectric losses are very low and the conductance G' may be neglected. In insulated cables, this term needs to be included, but the formal development of the equations is the same.

Given the parameters for a small segment of length Δx: $R'\Delta x$, $L'\Delta x$, $G'\Delta x$, and $C'\Delta x$, the voltage drop across this segment may be written as:

$$v(x,t) - v(x+\Delta x,t) = R'\Delta x\, i(x,t) + L'\Delta x \frac{\partial i}{\partial t}(x,t). \tag{2.7}$$

Similarly, for current variation $i(x, t)$ in the same line segment:

$$i(x,t) - i(x+\Delta x,t) = G'\Delta x \cdot v(x+\Delta x,t) + C'\Delta x \frac{\partial v(x+\Delta x,t)}{\partial t}. \tag{2.8}$$

Calculating the partial derivatives in (2.7) and (2.8) by letting Δx approach zero, the derivative of voltage and current functions with respect to variable x are obtained:

$$\frac{\partial v(x,t)}{\partial x} = -R'i(x,t) - L'\frac{\partial i(x,t)}{\partial t}, \tag{2.9}$$

$$\frac{\partial i(x,t)}{\partial x} = -G'v(x,t) - C'\frac{\partial v(x,t)}{\partial t}. \tag{2.10}$$

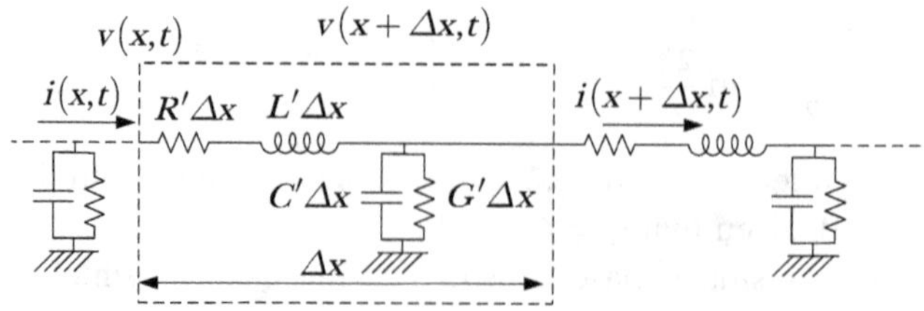

Figure 2.2 Distributed parameters.

The basic differential equations for currents and voltages are of the same type; therefore, the voltage equations are preferably developed, and only a few current equations will be elaborated as they are dual.

Deriving (2.9) with respect to x:

$$\frac{\partial^2 v(x,t)}{\partial^2 x} = -R'\frac{\partial i(x,t)}{\partial x} - L'\frac{\partial^2 i(x,t)}{\partial x \partial t}, \tag{2.11}$$

Deriving (2.10) with respect to t:

$$\frac{\partial^2 i(x,t)}{\partial x \partial t} = -G'\frac{\partial v(x,t)}{\partial t} - C'\frac{\partial^2 v(x,t)}{\partial t^2}. \tag{2.12}$$

Substituting (2.12) and (2.10) in (2.11):

$$\frac{\partial^2 v(x,t)}{\partial^2 x} = R'G'v(x,t) + (R'C' + L'G')\frac{\partial v(x,t)}{\partial t} + L'C'\frac{\partial^2 v(x,t)}{\partial t^2}. \tag{2.13}$$

And similarly for current:

$$\frac{\partial^2 i(x,t)}{\partial^2 x} = R'G'i(x,t) + (R'C' + L'G')\frac{\partial i(x,t)}{\partial t} + L'C'\frac{\partial^2 i(x,t)}{\partial t^2}. \tag{2.14}$$

These equations have partial derivatives with respect to variables with different dimensions, and the treatment becomes simpler if the Laplace transform is applied, given the initial conditions specified below,

$$v(x,0) = f(x), \tag{2.15}$$

the following ordinary differential equation needs to be solved:

$$v'(x,0) = \left.\frac{\partial v(x,t)}{\partial t}\right|_{t=0}, \tag{2.16}$$

$$\begin{aligned}\frac{d^2 V(x,s)}{d^2 x} &= (R' + sL')(G' + sC')V(x,s) \\ &+ L'C'\left\{\left[s + \left(\frac{R'}{L'} + \frac{G'}{C'}\right)\right]v(x,0) + v'(x,0)\right\}.\end{aligned} \tag{2.17}$$

The particular solution of (2.17) depends on the excitation, and the solution of the homogeneous equation is only possible under special conditions of the parameters, which are considered fixed. Therefore, there is limited interest in further detailing (2.16), despite its formal significance.

To provide some insight into traveling waves in transmission lines, a lossless transmission line at rest, excited by known functions at certain points of the line, is considered, which will be discussed in the next section.

2.2.2 Lossless Transmission Line

Assuming the lossless line and designating the traveling speed:

$$L'C' = \frac{1}{\nu^2}, \quad \text{or} \quad \nu = \frac{1}{\sqrt{L'C'}}. \tag{2.18}$$

For the lossless line, the wave travels at the speed of light denoted by c (note that for the lossless line $\nu = c$, which can be verified by substituting (2.1) and (2.2) into (2.18)). However, to examine more general cases, this velocity will be denoted by ν.

The equation of a lossless line is obtained from (2.13) with $R' = 0$ and $G' = 0$:

$$\frac{\partial^2 v(x,t)}{\partial^2 x} = \frac{1}{\nu^2}\frac{\partial^2 v(x,t)}{\partial t^2}, \tag{2.19}$$

whose solution is given by the equation:

$$v(x,t) = S(x - \nu t) + T(x + \nu t), \tag{2.20}$$

In this equation, S and T are generic functions of two variables to be determined by the initial and boundary conditions. These functions represent traveling waves moving in the positive and negative directions of the x-axis.

Assuming that the initial conditions can be represented by two functions $f(x)$ and $g(x)$ (Cauchy conditions in (2.15) and (2.16)):

$$f(x) = v(x,0) \text{ and } g(x) = v'(x,0),$$

the formal result of (2.19) subject to (2.15) is known as the D'Alembert solution, whose derivation can be found in [6]:

$$v(x,t) = \frac{f(x - \nu t) + f(x + \nu t)}{2} + \frac{1}{2\nu}\int_{x-\nu t}^{x+\nu t} g(\xi)d\xi. \tag{2.21}$$

It is observed that $v(x, t)$ depends only on points in the interval between $x - \nu t$ and $x + \nu t$, referred to as the domain of dependence of the variables (x, t).

Despite the conceptual interest in (2.21), the application of the Laplace transform is convenient and will be the subject of the following analysis. For this, a lossless line with zero initial conditions is considered, simplifying (2.16):

$$\frac{d^2V(x,s)}{dx^2} = s^2L'C'V(x,s), \tag{2.22}$$

The solution of the ordinary differential equation in (2.22) can be composed of two types of traveling waves, examined separately in (2.23) and (2.24),

$$V(x,s) = V^+(0,s)e^{-\gamma(s)x}, \tag{2.23}$$

$$V(x,s) = V^-(0,s)e^{+\gamma(s)x}, \tag{2.24}$$

in which the propagation constant below is employed:

$$\gamma(s) = \sqrt{s^2L'C'} = s\sqrt{L'C'}, \tag{2.25}$$

or, according to (2.18),

$$\gamma(s) = \frac{s}{\nu}. \tag{2.26}$$

The functions $V^+(0, s)$ and $V^-(0, s)$ are the Laplace transforms of the generic functions in time $v^+(t)$ and $v^-(t)$ and are referenced from the same origin, $x = 0$. However, the same reference point is not mandatory, as their existence or even their superposition depends on the boundary conditions.

To obtain the solutions in the time domain, the inverse Laplace transform is applied, considering the first shifting theorem:

$$F(s)e^{sa} \leftrightarrow f(t+a), \tag{2.27}$$

the inverse transforms of (2.23) and (2.24) are given in (2.28) and (2.29), respectively:

$$v(x,t) = v^+\left(0, t - \frac{x}{\nu}\right), \tag{2.28}$$

$$v(x,t) = v^-\left(0, t + \frac{x}{\nu}\right), \tag{2.29}$$

where v^+ is the traveling wave in the positive direction on the x-axis and v^- is the traveling wave in the opposite direction on the x-axis.

The exponential functions in (2.23) and (2.24) are useful in the development of complex transmission line models that involve the frequency dependence of parameters. For now, it is sufficient to accept that these operators displace the functions v^+ and v^- in space with specific time delays.

Analogously, for the currents, the following equations can be written:

$$I(x,s) = I^+(0,s)\,e^{-\gamma(s)x} \tag{2.30}$$

and

$$I(x,s) = I^-(0,s)\,e^{+\gamma(s)x} \tag{2.31}$$

If the arguments in (2.28) and (2.29) are held constant, the same values of these functions are obtained over the transmission line.

For v^+, in (2.28), assuming:

$$t - \frac{x}{\nu} = \text{const}, \tag{2.32}$$

the derivative with respect to time in (2.32) furnishes a traveling wave with positive velocity in the direction of positive x-axis:

$$\frac{dx}{dt} = \nu. \tag{2.33}$$

Assume a generator that applies a voltage, as a function of time $f(t)$, at the origin ($x = 0$) on a semi-infinite transmission line, as in Figure 2.3a.

For time $t = 0$.

At the origin, for $(x, t) = (0, 0) \rightarrow v(0, 0) = v^+(0, 0) = f(0)$.

At $x = \ell$, $(x, t) = (\ell, 0) \rightarrow v(\ell, 0) = 0$.

At time $t = \tau$,

At the origin, $(x, t) = (0, \tau) \rightarrow v(0, \tau) = f(\tau)$,

At $x = \ell$, the traveling wave arrives at this point at $t = \tau$ ($\ell = \nu\tau$), thus:

$$(x, t) = (\ell, \tau) \rightarrow v(\ell, \tau) = v^+ \left(0, \tau - \frac{\ell}{\nu}\right) = v^+ (0, 0) = f(0).$$

and $v^+ (0, 0) = f(0)$ is detected at point $x = \ell$.

And it repeats for any $t > \tau$, such that $f(t)$ applied at the origin is perfectly reproduced at a distance ℓ as $f(t - \tau)$, with a time delay τ (Figure 2.4).

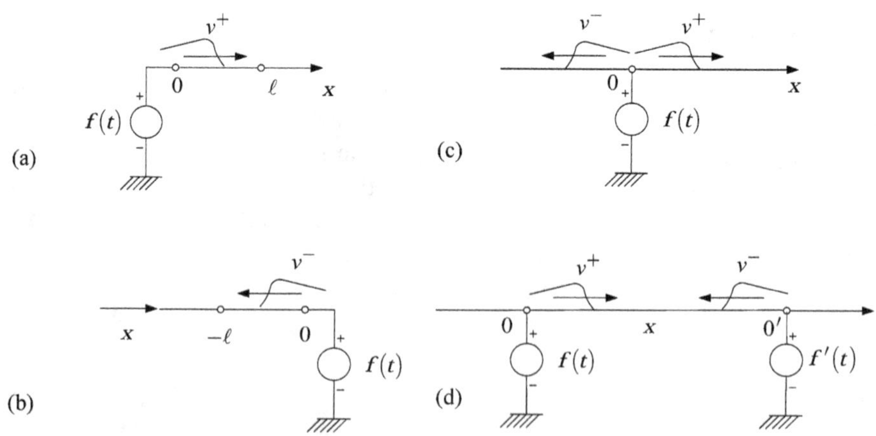

Figure 2.3 Transmission line excitation. (a) Semi-infinite line with progressive wave. (b) Semi-infinite line with regressive wave. (c) Infinite-length line with progressive and regressive waves. (d) Excitation with two generators.

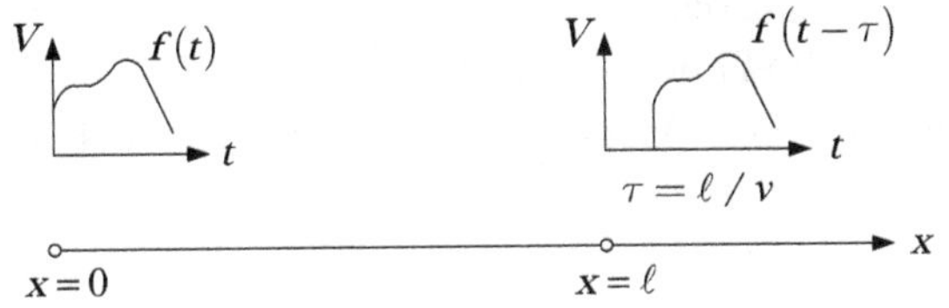

Figure 2.4 Delay of $f(t)$.

Note that when losses are taken into account, it can be demonstrated that the regressive component does not exist in the case of Figure 2.3a, as the modulus of the exponential function $\exp[+\gamma(s)x]$ tends to infinity as x approaches infinity, since $\text{Re}[+\gamma(s)x] > 0$, which is physically unrealistic. Therefore, only the progressive component can exist, and it should tend to zero as x approaches infinity.

Much like the case of v^+, if the argument of the regressive wave v^- in (2.29) is held constant,

$$t + \frac{x}{\nu} = \text{const}, \tag{2.34}$$

its derivative with respect to t is given by,

$$\frac{dx}{dt} = -\nu, \tag{2.35}$$

and v^- travels in the negative direction on the x-axis.

In Figure 2.3b, the same experiment may be repeated with v^- to conclude that at $t = \tau$, $v^-(0, 0)$ is observed at $x = -\ell$, as:

$$v(-\ell, \tau) = v^-\left(0, \tau + \frac{-\ell}{\nu}\right) = v^-(0, 0). \tag{2.36}$$

Consider the same generator applying $f(t)$ at $x = 0$ on an infinite transmission line in Figure 2.3c. Using the continuity principle, the same voltage wave will travel on opposite sides of the line and will never superpose.

The general solution considering the superposition of progressive waves in (2.23) and regressive waves in (2.24), as a result of exciting transmission lines at rest by generators, can be exemplified with a second generator applying voltage at $x = \ell$ in Figure 2.3d. The voltage at a point x for $0 \leq x \leq \ell$, could be the result of superposition.

$$V(x, s) = V^+(0, s)e^{-\gamma(s)x} + V^-(\ell, s)e^{+\gamma(s)(x-\ell)}. \tag{2.37}$$

However, the boundary conditions change significantly because there is now a finite length of line between the generators with reflections. To avoid reflections, voltage generators should be replaced by current generators to maintain the conditions of an infinitely long line without discontinuities.

The superposition of progressive and regressive waves is of greater interest in lines with finite length, when there are waves reflected at the ends of the transmission lines that travel in the opposite direction to the incident waves, which will be analyzed later.

2.3 Voltage and Current Relations and Surge Impedance of a Lossless Transmission Line

Using the model of lossless transmission line, it is possible to perform a simplified analysis of the relationship between voltages and currents.

From Eq. (2.9):

$$\frac{\partial V(x,s)}{\partial x} = -sL'I(x,s). \tag{2.38}$$

For progressive waves, substituting the derivative of (2.23) with respect to x into (2.38),

$$I(x,s) = -\frac{1}{sL'}\left[-\gamma(s)V^{+}(0,s)e^{-\gamma(s)x}\right], \tag{2.39}$$

or:

$$I(x,s) = \frac{\gamma(s)}{sL'}V^{+}(0,s)e^{-\gamma(s)x}. \tag{2.40}$$

With (2.25) in (2.40), and designating the surge impedance of the transmission line as:

$$Z = \sqrt{\frac{L'}{C'}}, \tag{2.41}$$

the relation between voltage and current is achieved:

$$I(x,s) = \frac{1}{Z}V^{+}(0,s)e^{-\gamma(s)x}. \tag{2.42}$$

For regressive waves, substituting the derivative of (2.24) into Eq. (2.38):

$$I(x,s) = -\frac{1}{Z}V^{-}(0,s)e^{+\gamma(s)x}, \tag{2.43}$$

Equations (2.42) and (2.43) relate voltages and currents that move in progressive and regressive directions along the x-axis, respectively.

The surge impedance given in (2.41), for the lossless transmission line, is a real number and shows the behavior of a resistance; thus, voltage and current waves

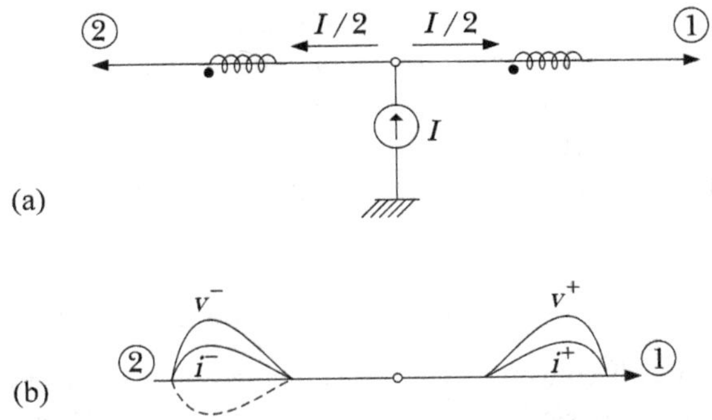

Figure 2.5 Current and voltage waveforms and measurement. (a) Applied current. (b) Measurements.

traveling in the same direction are in phase, both showing the same waveform and related by a proportionality factor. This fact deserves attention because, in several situations, the line can be modeled as a resistance, which maintains the same behavior along the transmission line with respect to voltage and current.

In the positive direction on the x-axis, voltage and current have the same sign and in the negative direction on the x-axis, they have opposite signs, since the convention in equations is that positive current moves in the positive direction on the x-axis. The following comments are illustrative [5].

A current generator is applied to the transmission line according to Figure 2.5a.

Due to the symmetry condition, the charge making up the current wave is divided into two equal parts on both sides, and the same relationship between voltage and current exists, given by Z.

The voltages and charges associated with the current should split in agreement with the continuous lines in Figure 2.5b. The voltage on both sides has the same polarity and the current measured on side 1 is positive, in accord with the convention that positive current flows in the positive direction on the x-axis. The sign of the current aligns with the positive current convention of the current transformer in Figure 2.5a. The measured current on side 2 is negative, contradicting the positive current convention of the current transformer, as represented by the dashed line in Figure 2.5b, where the voltage and current have opposite signs.

2.4 Traveling Waves in Discontinuities – Reflected and Refracted Waves

The behavior of voltage and current waves is analyzed at points of discontinuity in the transmission line. It should be noted that this account places greater emphasis on the voltage waves, which are responsible for the overvoltages in power systems;

however, the calculation of voltages and currents can be considered as dual calculations, according to the equations previously presented.

These discontinuities may exist in circuits with lumped parameters such as resistors, inductors, capacitors, or even in the form of distributed parameters, such as transmission lines with different surge impedances.

2.4.1 A Generic Impedance at the Line Terminal

Some discontinuities at the transmission line terminal are examined next, assuming a generic load at the discontinuity point (Figure 2.6).

Z_1: surge impedance of the lossless transmission line on side 1.

$Z_2(s)$: complex impedance at line terminal on side 2.

Initially, there is a traveling wave v_1^+ in transmission line 1 and, when it reaches the point of discontinuity, at $x = \ell$, the reflected and refracted waves need to be calculated.

As the wave equations in the time and frequency domain are fundamentally the same, in line 1 before reaching point $x = \ell$ currents and voltages have the same sign and the same waveform, written in the frequency domain:

$$I_1^+ (s) = \frac{V_1^+ (s)}{Z_1}. \tag{2.44}$$

For the reflected wave traveling in the negative direction on the x-axis, the following can be written:

$$I_1^- (s) = -\frac{V_1^- (s)}{Z_1}. \tag{2.45}$$

The case of a generic impedance on side 2 can be analyzed witting:

$$I_2(s) = \frac{V_2(s)}{Z_2(s)}. \tag{2.46}$$

Currents and voltages on sides 1 and 2 can be equated using the principle of continuity. At the point where the two lines are connected, the voltage on side 1 is the sum of the incident plus the reflected wave, which is equal to the voltage on side 2, thus:

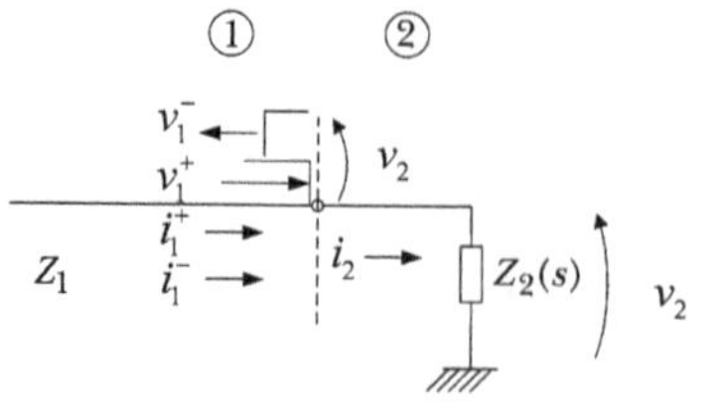

Figure 2.6 Complex impedance at the line terminal $Z_2(s)$.

$$V_1^+ (s) + V_1^- (s) = V_2(s). \tag{2.47}$$

The same equality is valid for the current:

$$I_1^+ (s) + I_1^- (s) = I_2(s). \tag{2.48}$$

The application of this principle can follow different orientations for the reflected and refracted currents. In (2.48), all currents are oriented in the same direction as the x-axis .

The Eqs. (2.44) and (2.45) are substituted into the left side of (2.48), just as (2.46) and (2.47) are substituted into the right side of (2.48):

$$\frac{V_1^+(s)}{Z_1} - \frac{V_1^-(s)}{Z_1} = \frac{V_1^+(s) + V_1^-(s)}{Z_2(s)}, \tag{2.49}$$

thus:

$$V_1^-(s) = \frac{Z_2(s) - Z_1}{Z_2(s) + Z_1} V_1^+(s). \tag{2.50}$$

Let Γ_r be the reflection coefficient for the voltage wave:

$$\Gamma_r(s) = \frac{Z_2(s) - Z_1}{Z_2(s) + Z_1}, \tag{2.51}$$

which relates to the incident and reflected waves.

For the reflected wave, Eq. (2.52) is written:

$$V_1^-(s) = \Gamma_r(s)\, V_1^+(s) \tag{2.52}$$

Substituting (2.52) into (2.47), although medium 2 is generic and not necessarily a transmission line with distributed parameters, a refraction coefficient can be defined, thus eliminating $V_1^-(s)$ from the equations:

$$V_2(s) = [1 + \Gamma_r(s)]V_1^+(s), \tag{2.53}$$

$$V_2(s) = \Gamma_t(s)\, V_1^+(s), \tag{2.54}$$

where:

$$\Gamma_t(s) = 1 + \Gamma_r(s) = \frac{2Z_2(s)}{Z_1 + Z_2(s)}. \tag{2.55}$$

The current waves can be written as follows. From (2.45) and (2.52):

$$I_1^-(s) = -\frac{V_1^-(s)}{Z_1} = -\frac{\Gamma_r(s)V_1^+(s)}{Z_1}, \tag{2.56}$$

from (2. 44):

$$I_1^-(s) = -\Gamma_r(s)I_1^+(s). \tag{2.57}$$

The reflection coefficient for the current is defined as $\Gamma_{(i)r}$:

$$\Gamma_{(i)r}(s) = -\Gamma_r(s) = \frac{Z_1 - Z_2(s)}{Z_1 + Z_2(s)}, \tag{2.58}$$

In other words, the reflection coefficients for current and voltage have the same absolute value, but they are opposite in sign.

For the refracted current:

$$I_2^+(s) = (1 + \Gamma_{(i)r}(s))I_1^+(s), \tag{2.59}$$

$$\Gamma_{(i)t}(s) = 1 + \Gamma_{(i)r}(s). \tag{2.60}$$

The refraction coefficient for the current wave is obtained as:

$$\Gamma_{(i)t}(s) = \frac{2Z_1}{Z_1 + Z_2(s)}. \tag{2.61}$$

From the equations for a generic impedance terminal, some special cases are examined next.

- Lines with Different Surge Impedances

  In this case, a voltage wave propagates along line 2, which also has distributed parameters. Since they are ideal lines, the analysis is simpler in the time domain, so it suffices to replace the complex impedance $Z_2(s)$ with the surge impedance Z_2 and all previous equations remain valid (Figure 2.7).

$$\Gamma_r = \frac{Z_2 - Z_1}{Z_2 + Z_1}, \tag{2.62}$$

$$\Gamma_t = \frac{2Z_2}{Z_1 + Z_2}. \tag{2.63}$$

Throughout the text, the following terms will also be used:

i_i, v_i: incident waves at the connection point (i_1^+, v_1^+).
i_r, v_r: reflected waves at the connection point (i_1^-, v_1^-).
i_t, v_t: refracted waves at the connection point (i_2, v_2).

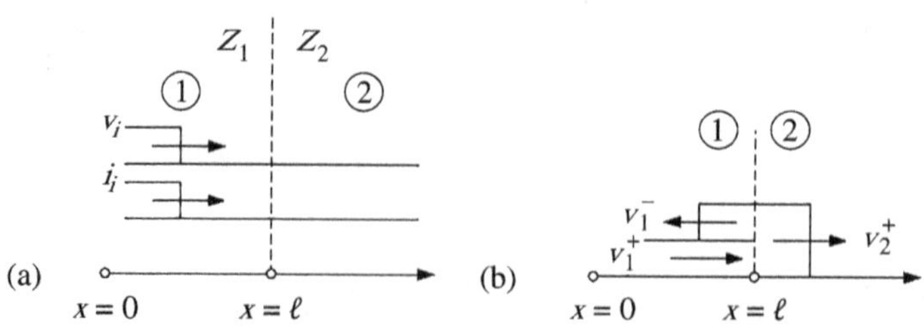

Figure 2.7 Transmission lines with different surge impedances. (a) Incident waves. (b) Incident, refracted and reflected voltage waves.

The refracted waves are denoted with the subscript t because the subscript r is used for reflected waves. There is an association with transmitted waves, but the definition is not exactly the same.

Example 2.1 A voltage step of 100 V is traveling in transmission line 1, with a surge impedance of 200 Ω, when it reaches the series connection to transmission line 2, with a surge impedance of 300 Ω. Calculate the reflected and refracted waves.

Solution

Applying (2.62) and (2.63) Figure 2.8 is obtained:

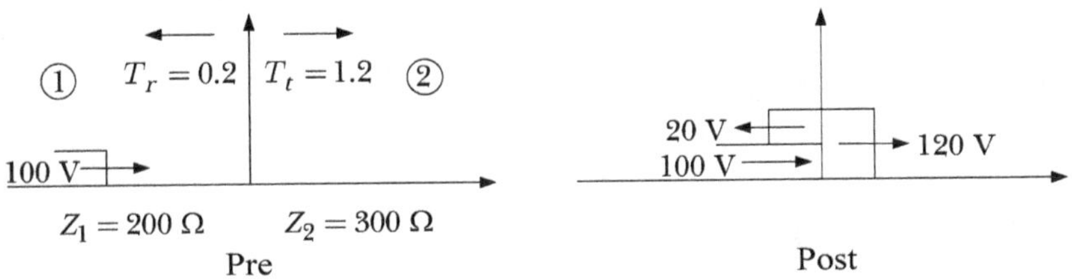

Figure 2.8 Refracted and reflected voltages.

- Open Line
  The open line can be analyzed as a particular case of the previous formulation.
  Just making $Z_2(s) \to \infty$ in expressions (2.51) and (2.55), expressions (2.64) and (2.65) are obtained:

$$\Gamma_r = \lim_{Z_2 \to \infty} \frac{Z_2 - Z_1}{Z_1 + Z_2} = 1. \tag{2.64}$$

$$\Gamma_t = \lim_{Z_2 \to \infty} \frac{2Z_2}{Z_1 + Z_2} = 2, \tag{2.65}$$

- Line with Short-Circuited Terminal
  In this case, to calculate the transmission and reflection coefficients, it is sufficient to let $Z_2(s) = 0$.

$$\Gamma_t = 0. \tag{2.66}$$

$$\Gamma_r = -1, \tag{2.67}$$

- Resistance at the Line Terminal
  As in item 3.3, in a lossless line, the surge impedance behaves as a resistance, since the voltage and current are in phase. In the same way, it can also be supposed that the resistance exhibits the behavior of a transmission line of infinite length, with the waves traveling in one direction only. Thus, the equations for the resistance are obtained in the time domain assuming $Z_2(s) = R$ in Figure 2.6.

- Inductance at the Line Terminal
  In this case, substituting $Z_2(s) = sL$ in (2.55), the behavior of a voltage step reaching the line terminal connected to an inductance is examined:

$$V_2(s) = \frac{2sL}{Z_1 + sL}\frac{E_0}{s}, \tag{2.68}$$

where $\frac{E_0}{s}$ is the Laplace transform of the incident wave.

$$V_2(s) = \frac{2E_0}{s + \frac{Z_1}{L}}. \tag{2.69}$$

Inversion of the transform back to the time domain yields an expression for the voltage at the line terminal:

$$v_2(t) = 2E_0 e^{-\frac{Z_1}{L}t}. \tag{2.70}$$

At zero time, $t \to 0$, the voltage at the line terminal is $v_2(0) = 2E_0$, and the behavior is that of an open line. For $t \to \infty$, that voltage is reduced to zero, and the behavior is the same as that of the line at a short-circuited terminal. The result can be explained by appreciating that the inductance, when excited by a voltage step, initially behaves as an open circuit and as time increases, as a short circuit.

The reflected wave is obtained from:

$$v_1^+ + v_1^- = v_2, \tag{2.71}$$

$$v_1^- = v_2 - v_1^+. \tag{2.72}$$

Because $v_1^+ = E_0$, thus:

$$v_1^-(t) = E_0\left(2e^{-\frac{Z_1}{L}t} - 1\right). \tag{2.73}$$

The time response of the voltage at the terminal is shown in Figure 2.9.
Initially $v_1^- = v_1^+$ is obtained and after a long period of time $v_1^- = -v_1^+$.

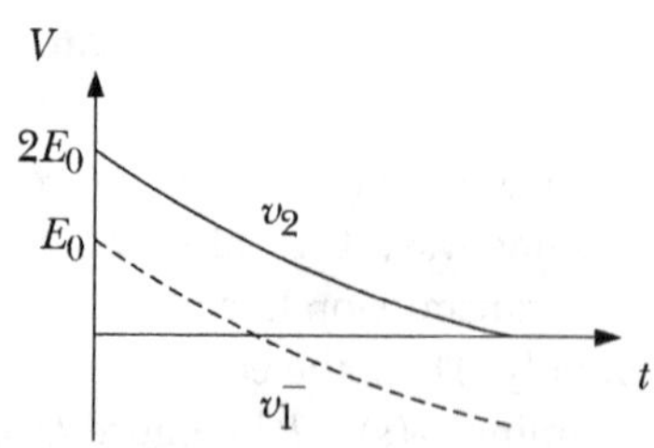

Figure 2.9 Response to step excitation of an inductive terminal.

Example 2.2 A traveling cosine wave reaches a reactor connected to the receiving end of a transmission line, at point F. Determine the equation for the voltage, assuming that the transmission line is long enough to disregard reflections from the sending end. The data are presented in Figure 2.10.

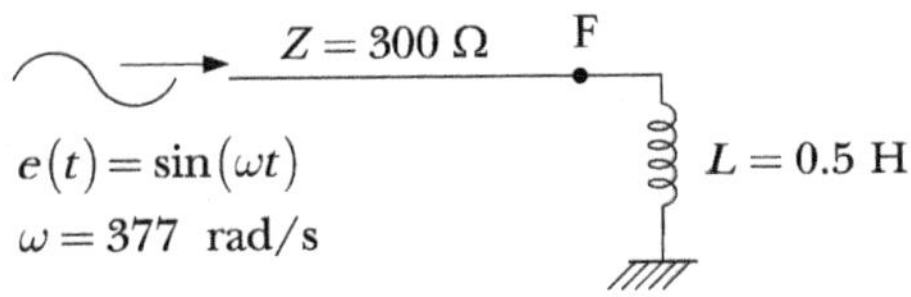

Figure 2.10 Transmission line excited by a sinusoidal voltage.

Solution

Voltage at the receiving end:

$$V_f(s) = \Gamma_t(s)E(s),$$

$$\Gamma_t(s) = \frac{2sL}{Z + sL},$$

$$E(s) = \frac{\omega}{s^2 + \omega^2},$$

$$V_f(s) = \frac{2\omega s}{(s^2 + \omega^2)\left(s + \frac{Z}{L}\right)}.$$

Expanding into partial fractions:

$$V_f(s) = \frac{k_0}{s + 600} + \frac{k_1}{s - j377} + \frac{k_1^*}{s + j377},$$

$$k_0 = \left.\frac{2\omega s}{s^2 + \omega^2}\right|_{s=-600} = -0.901,$$

$$k_1 = \left.\frac{2\omega s}{(s + j377)(s + 600)}\right|_{s=j377} = 0.5320\angle -32.14°.$$

Assuming the initial time to be $t = 0$, when the wave reaches the terminal, the time response of the voltage is given by:

$$v_f(t) = -0.901e^{-600t} + 1.064\cos(377t - 32.14°).$$

- Capacitance at the Line Terminal
  The impedance of the capacitor is given by:

$$Z_2(s) = \frac{1}{sC}, \tag{2.74}$$

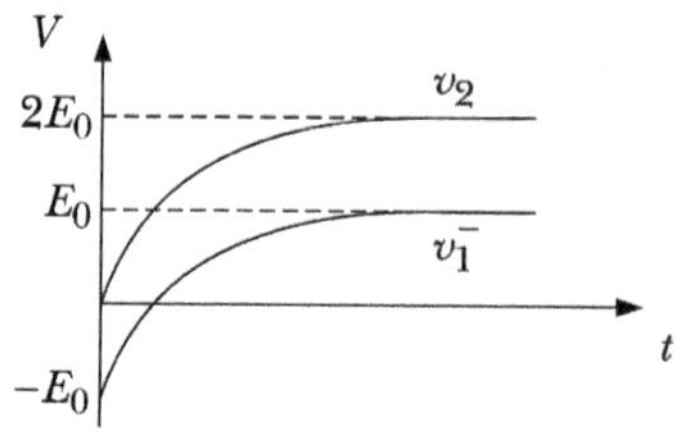

Figure 2.11 Step response of a capacitive line terminal.

The behavior of a step wave that reaches a line terminal connected to a capacitance is now analyzed in Figure 2.11, substituting the new impedance $Z_2(s)$ in (2.55):

$$V_2(s) = \frac{2\frac{1}{sC}}{Z_1 + \frac{1}{sC}} \frac{E_0}{s}, \tag{2.75}$$

$$V_2(s) = \frac{2E_0}{CZ_1 s\left(s + \frac{1}{CZ_1}\right)}, \tag{2.76}$$

$$V_2(s) = 2E_0\left[\frac{1}{s} - \frac{1}{s + \frac{1}{CZ_1}}\right], \tag{2.77}$$

$$v_2(t) = 2E_0\left[1 - e^{-\frac{t}{CZ_1}}\right]. \tag{2.78}$$

At the first instant, the capacitor behaves as a short-circuit, with null voltage at the line terminal, and thus the reflected wave is:

$$v_1^- = -v_1^+. \tag{2.79}$$

After a long period of time, the capacitor behaves as an open circuit, resulting in $v_1^- = v_1^+$.

For the reflected voltage wave:

$$v_1^+ + v_1^- = v_2 \tag{2.80}$$

or:

$$v_1^-(t) = 2E_0\left(1 - e^{-\frac{t}{Z_1C}}\right) - E_0, \tag{2.81}$$

$$v_1^-(t) = E_0\left(1 - 2e^{-\frac{t}{Z_1C}}\right). \tag{2.82}$$

This case is interesting because the transformer is represented predominantly as a capacitance for surges with fast wavefronts, and many times, only the transformer's equivalent stray capacitance is used.

- Coupling Point with Several Lines

  For the case in Figure 2.12a, it is sufficient to obtain the equivalent impedance of all lines that are on side 2 (Figure 2.12b):

$$Z_2 = Z_2^{(1)} \parallel Z_2^{(2)} \cdots \parallel Z_2^{(n)}, \tag{2.83}$$

and use the same previous expressions.

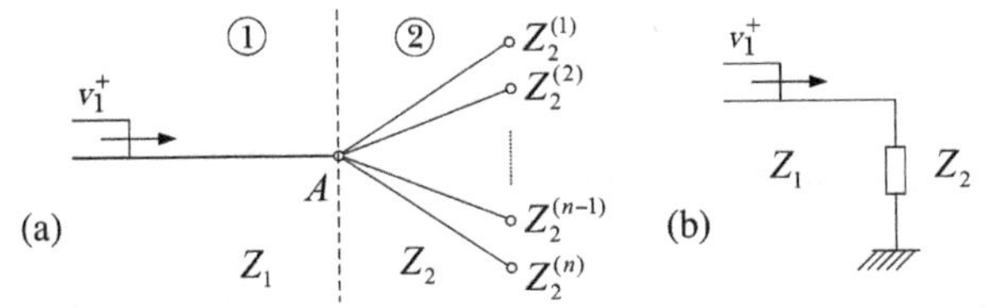

Figure 2.12 Several lines at the same coupling point. (a) Incident wave. (b) Equivalent circuit.

It is noted that the same voltage levels are refracted to all lines $Z_2^{(i)}$, according to the continuity principle. Thus, the current splits so as to maintain the voltage at the same value in all lines.

- Lumped Impedance Between Two Transmission Lines

  The case of two transmission lines with surge impedances Z_1 and Z_2, connected by a generic lumped impedance $Z(s)$ is analyzed (Figure 2.13). For a traveling wave on transmission line 1, the reflection and refraction coefficients in relation to line 2 are obtained.

  To calculate the voltage propagating on line 2, it is necessary to subtract the voltage drop across the bipole $Z(s)$. The equation becomes straightforward by applying the voltage divider rule, modeling line 2 as a resistance.

  The first step is to calculate the reflection and refraction coefficients at point 1. Therefore, the sum of the resistance, representing transmission line 2, and the series lumped impedance results in an equivalent lumped impedance connected to line 1 (Figure 2.14).

  The equivalent impedance is calculated as:

$$Z_{eq}(s) = Z(s) + Z_2, \tag{2.84}$$

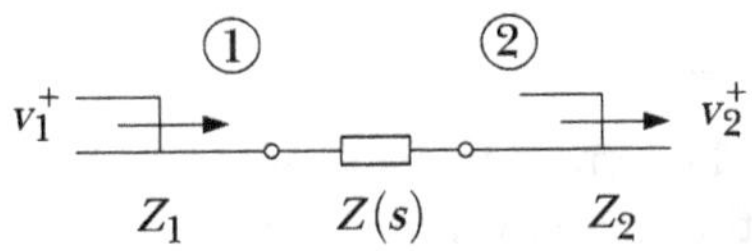

Figure 2.13 Lumped impedance in series with the transmission lines.

Thus, the refracted and reflected coefficients at point 1 are attained:

$$\Gamma_{t1} = \frac{2Z_{eq}(s)}{Z_1 + Z_{eq}(s)}, \tag{2.85}$$

$$\Gamma_{r1} = \frac{Z_{eq}(s) - Z_1}{Z_1 + Z_{eq}(s)}. \tag{2.86}$$

The second step is to calculate the wave refracted to line 2 using the voltage divider equation:

$$V_2^+(s) = \frac{2Z_{eq}(s)}{Z_1 + Z_{eq}(s)} V_1^+(s) \frac{Z_2}{Z_{eq}(s)}, \tag{2.87}$$

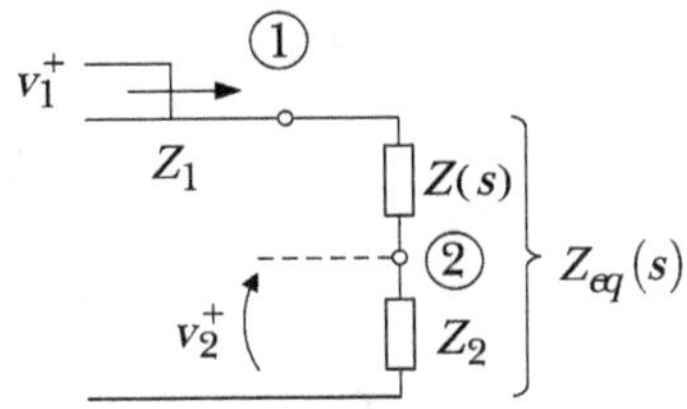

Figure 2.14 Equivalent model.

$$V_2^+(s) = \frac{2Z_2}{Z_1 + Z(s) + Z_2} V_1^+(s). \tag{2.88}$$

The refraction coefficient at point 2 is given by:

$$\Gamma_{t2} = \frac{2Z_2}{Z_1 + Z(s) + Z_2}. \tag{2.89}$$

Regarding elements in series with the AC (alternating current) transmission lines, current limiting reactors can be found in special applications, but they are frequent in DC (direct current) transmission lines. However, modeling these reactors at high frequencies may need to account for stray capacitances in parallel. Similarly, series capacitors in AC transmission lines reduce the total reactance of the line. Power line carriers can still be found in some communication systems.

The case of a series LC (inductor-capacitor) circuit is presented below, as an example of a generic series impedance in the transmission line.

Example 2.3 A voltage step is traveling from line 1 to line 2. Calculate the time response of the voltage at the terminal of line 2 (at point F). Assume lines with the same parameters and sufficient length to disregard reflections at point 2, traveling back to point F (Figure 2.15).

Data

$$Z_1 = Z_2 = 100\ \Omega; L = 0.1\ \text{H}; C = 10.4167\ \mu\text{F}; e(t) = 1H(t).$$

Solution

The refraction coefficient is calculated at point 2, according to (2.89), considering the complex impedance of the LC series circuit.

$$Z(s) = sL + \frac{1}{sC},$$

$$\Gamma_{t2}(s) = \frac{2Z_2}{Z_1 + Z_2 + Z(s)}.$$

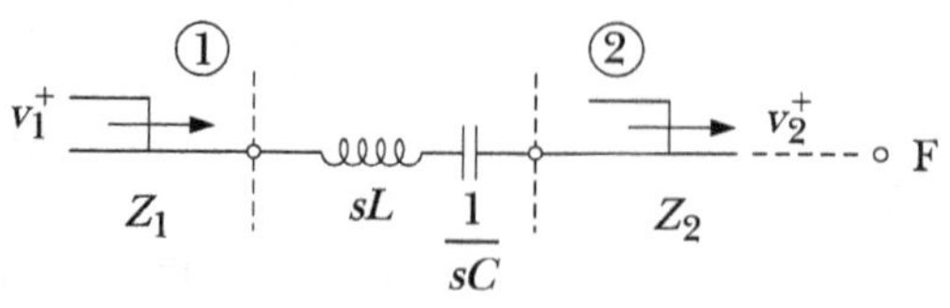

Figure 2.15 Lines connected by an LC circuit.

At the terminal of line 2, the refraction coefficient is:

$$\Gamma_{t_f}(s) = 2.$$

Thus:

$$V_f(s) = \Gamma_{t_2}(s)\Gamma_{t_f}(s)\frac{1}{s},$$

$$V_f(s) = \frac{4Z_2}{2Z_2 + sL + \frac{1}{sC}}\frac{1}{s},$$

$$V_f(s) = \frac{4Z_2 sC}{s^2LC + 2Z_2Cs + 1}\frac{1}{s},$$

or:

$$V_f(s) = \frac{4Z_2}{L}\frac{1}{s^2 + \frac{2Z_2}{L}s + \frac{1}{LC}}.$$

- Poles:

$$s_{1,2} = -\frac{Z_2}{L} \pm \sqrt{\left(\frac{Z_2}{L}\right)^2 - \frac{1}{LC}},$$

$$s_{1,2} = -1000 \pm \sqrt{1000^2 - 96 \times 10^4},$$

$$s_{1,2} = -1000 \pm 200,$$

$$s_1 = -800, s_2 = -1200.$$

- Residues:

$$V_f(s) = \frac{4000}{(s+800)(s+1200)},$$

$$k_1 = \left.\frac{4000}{s+1200}\right|_{s=-800} = 10,$$

$$k_2 = \left.\frac{4000}{s+800}\right|_{s=-1200} = -10.$$

- Time response:

$$v_f(t) = 10\left(e^{-800t} - e^{-1200t}\right).$$

2.4.2 Analysis of Discontinuities Using the Thévenin Equivalent

It is of interest to analyze reflected and refracted waves with the aid of Thévenin equivalent circuit, which is a useful concept in the management of nonlinearities

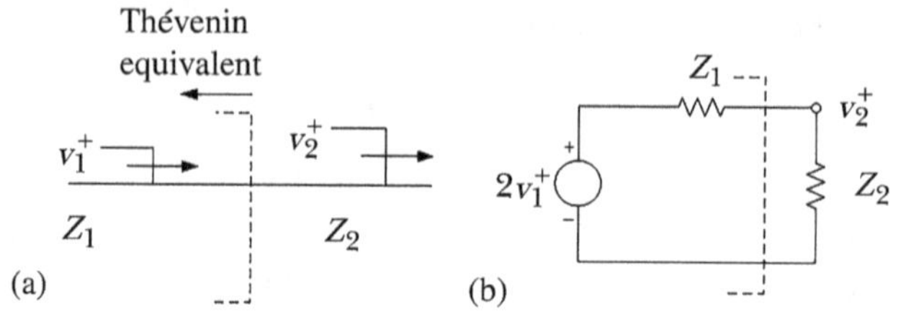

Figure 2.16 Thévenin equivalent of the transmission line. (a) Incident wave. (b) Thévenin equivalent circuit.

in electromagnetic transient programs [7]. This method is general, but it is initially applied to the linear case of two transmission lines, studied above, aiming to demonstrate its validity using a well-known case. Transmission lines 1 and 2 are modeled by their surge impedances, Z_1 and Z_2, respectively, in Figure 2.16a.

At the open line-1 terminal, the Thévenin equivalent voltage is obtained from:

$$e_{th} = 2v_1^+, \tag{2.90}$$

because the voltage doubles when it reaches the open line terminal.

The Thévenin equivalent impedance for traveling waves is a resistance equal to the surge impedance of the transmission line, given by Z_1. The concept is the same as in steady-state analysis, since the applied voltage at the line terminal divided by the current gives the surge impedance Z_1. Note that far generators don't need to be short-circuited, since at the first moment, this terminal is decoupled by the line time delay from the rest of the system. The Thévenin equivalent circuit with surge impedances Z_1 connected to the surge impedance Z_2 is presented in Figure 2.16b:

The voltage at the line terminal is given by the voltage divider equation:

$$v_2^+ = 2v_1^+ \frac{Z_2}{Z_1 + Z_2} = v_1^+ \frac{2Z_2}{Z_1 + Z_2}, \tag{2.91}$$

that is,

$$v_2^+ = \Gamma_t v_1^+, \tag{2.92}$$

The result of expression (2.91), applying the Thévenin equivalent, is the same as Eq. (2.63), applying the refraction coefficient for a traveling wave.

- Internal Impedances of Generators

  So far, excitation of the transmission line has been implemented with ideal generators. Next, the internal impedances of the generators are introduced, allowing the elaboration of more detailed models of the excitation sources. It should be noted that the applied voltages should not be calculated with refraction coefficients because the generator is not represented using distributed

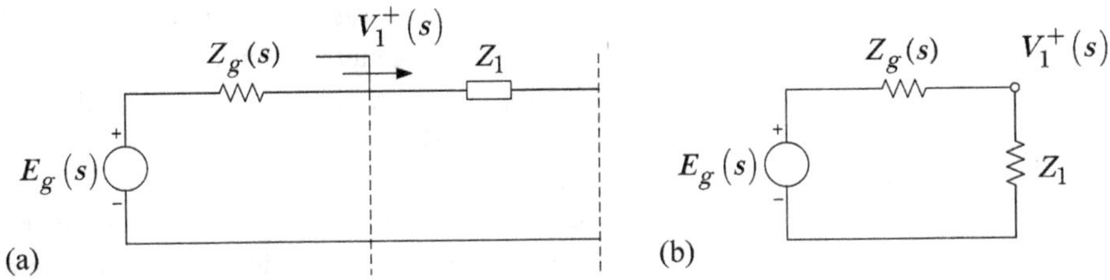

Figure 2.17 Generator internal impedance and equivalent circuit. (a) Applied wave. (b) Equivalent circuit.

parameters, at least in routine analyses, and therefore, there are no internal traveling waves in the generator to be refracted to the transmission line. This impedance can also represent the Thévenin equivalent of a network, but further analysis is needed in this case.

The voltage applied by a generator to a transmission line is studied considering the circuit of Figure 2.17a. The voltage is calculated by the voltage divider equation, and for this purpose, the transmission line is modeled as a resistance, Figure 2.17b.

With the equivalent circuit:

$$V_1^+(s) = \frac{Z_1}{Z_g(s) + Z_1} E_g(s). \tag{2.93}$$

For the regressive waves that move from the transmission line and reach the generator, since they are traveling waves, the calculations use the reflection and refraction coefficients and take into consideration the internal impedance of the generator connected to ground, by the same approach as an impedance connected at the line terminal.

The synchronous generator model is often presented as an internal voltage and an equivalent inductance related to the windings. It is also possible to include a resistance to represent losses, by an analogous approach.

Assuming e_g to be a voltage step of amplitude E_0, the equation for the voltage applied to the transmission line is given by:

$$V_1^+(s) = \frac{E_0}{s} \frac{Z_1}{Z_1 + sL_{int}}. \tag{2.94}$$

The inversion to the time domain gives:

$$v_1^+(t) = E_0\left(1 - e^{-\frac{Z_1}{L_{int}}t}\right), \tag{2.95}$$

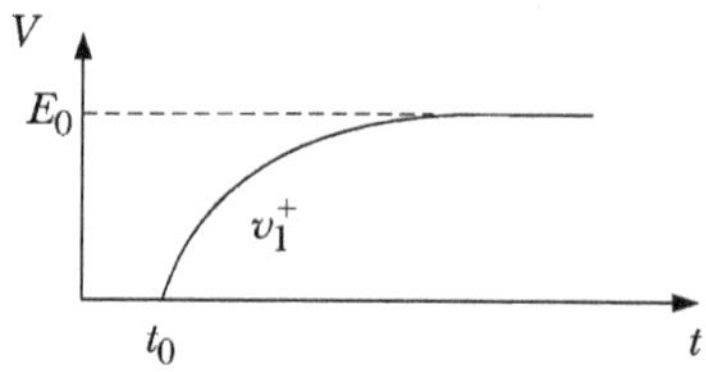

Figure 2.18 Increase in the voltage applied at time t_0.

thus, as can be seen, although the internal voltage of the generator has a constant value, the applied wave shows a (1 − exp) behavior (Figure 2.18).

A simplified analysis using a voltage step is illustrative and Eq. (2.94) can easily be transformed to a sinusoidal excitation. As a matter of fact, the sinusoidal excitation can be decomposed into a superposition of steps, and in certain situations, a step function can represent the cosine function.

Power systems with a high short-circuit power and small L values have small time constants, causing the waveform to approach a step voltage; hence, the exponential behavior is more pronounced for small values of the short-circuit power with large inductances.

Example 2.4 A transmission line with a surge impedance of 250Ω is energized by a system whose equivalent impedance is inductive. The nominal line voltage of the system is 500 kV and the frequency is 60 Hz. Calculate the time constant of the wave applied to the transmission line, assuming that the short-circuit power of the system is 25 000 or 2500 MVA.

Solution

a) Inductance for a short-circuit power of 25 000 MVA.

$$X = \frac{500^2}{25000} = 10\ \Omega,$$

$$\tau = \frac{10}{377 \times 250} = 106\ \mu\text{s},$$

The value of the inductance is 26.5 mH and the time constant is 106 μs.

b) Inductance for a short-circuit power of 2500 MVA

For a short-circuit power ten times lower, the time constant is 10 times higher, about 1 ms, and in this case, the effect of the inductance is more evident.

2.5 Nonlinear Elements

In this item, the behavior of resistive nonlinear elements connected to the network and their effect on the traveling waves will be analyzed.

So far, the behavior of linear elements, such as resistors, inductors, and capacitors has been examined. The analysis of nonlinear resistors is of great

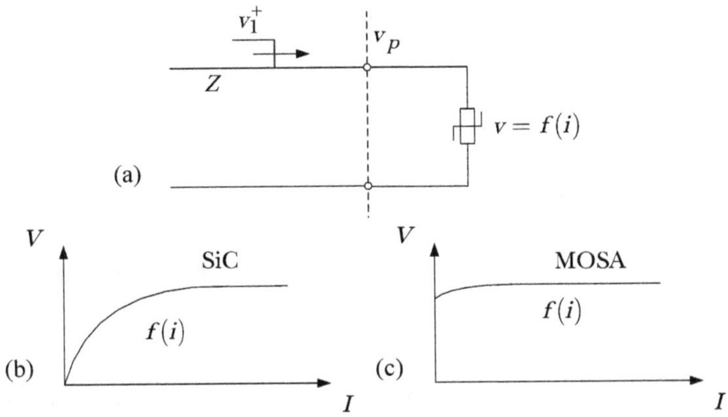

Figure 2.19 Line terminal with surge arrester. (a) Applied wave. (b) SiC arrester. (c) MOSA

interest for the simulation of surge arresters applied to control the overvoltages in power systems.

The main advantage of the method lies in its simplicity due to the use of the Thévenin equivalent. In the first step, the Thévenin equivalent of the network is obtained, without regard to the presence of the nonlinear element. Subsequently, a simultaneous solution of the equation for the Thévenin equivalent and the equation for the nonlinear element is performed.

A transmission line with a nonlinear element at its terminal is presented in Figure 2.19a. The nonlinear element may be a surge arrester.

In Figure 2.19a, i is the current in the surge arrester, and the voltage is given by the residual voltage, which is a nonlinear function, $f(i)$. Further details about surge arresters may be found in Chapter 13.

Figure 2.19b illustrates a typical residual curve for a SiC (silicon sarbide) surge arrester, and Figure 2.19c shows the residual curve for a MOSA (metal oxide surge arrester), which has a more pronounced nonlinear characteristic. When calculating transients with SiC arresters, spark gaps are usually considered, but for simplicity, they're treated as a permanent part of the network here.

As seen in item 2.4.1, the voltage doubles at the open-end line terminal, also designated as prospective voltage, resulting in the Thévenin circuit connected with the nonlinear element in Figure 2.20a:

This system of equations is obtained:

$$v_p = 2v_1^+ - Z_1\, i, \tag{2.96}$$

$$v_p = f(i). \tag{2.97}$$

The equation system can be solved either by graphical or iterative procedures, as in Figure 2.20b.

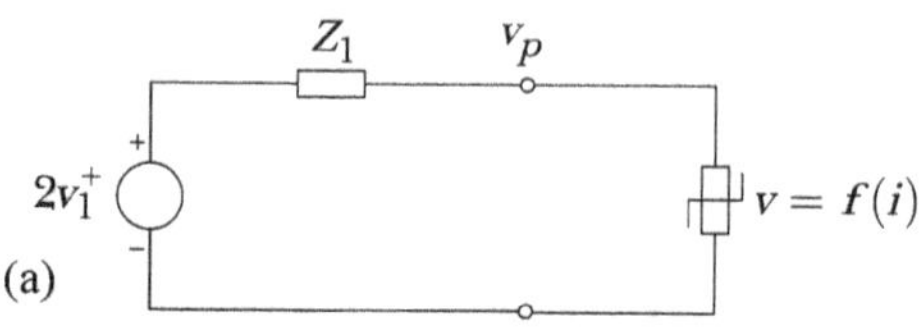

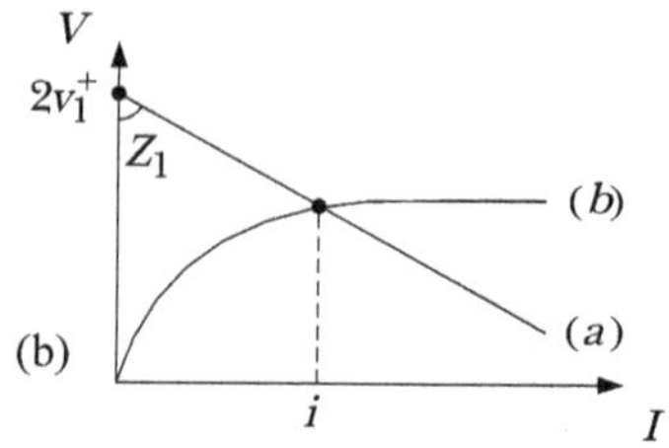

Figure 2.20 Transmission line and arrester. (a) Thévenin equivalent circuit and nonlinear element. (b) Graphical solution.

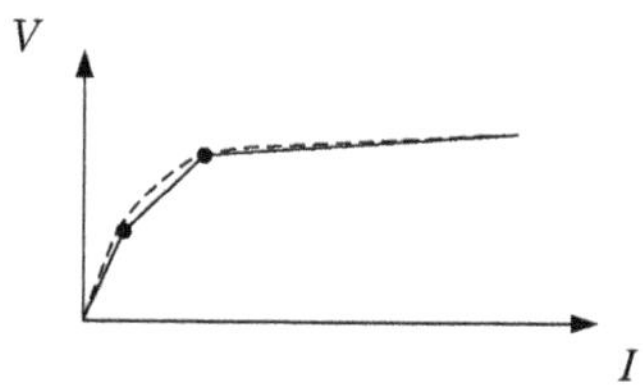

Figure 2.21 Piecewise representation of the nonlinear element.

There is more than one way to solve this nonlinear problem. One of the procedures adopted in computer programs is piecewise representation of the residual voltage $V \times I$ of the nonlinear element, which significantly simplifies the solution (Figure 2.21).

With the nonlinear curve subdivided into straight-line segments, the solution is obtained at each time step, with an algorithm identifying the correct segments for the transient solution.

The following example is illustrative of the solution of traveling waves with piecewise representation of the nonlinear element. A simplified SiC surge arrester with only two line segments is assumed connected to the transmission line.

Example 2.5 A voltage step of 1000 kV is traveling toward the transmission line terminal, which is protected by a surge arrester, Figure 2.22a. The surge impedance of the line is 400 Ω and the arrester residual voltage is presented in Figure 2.22b.

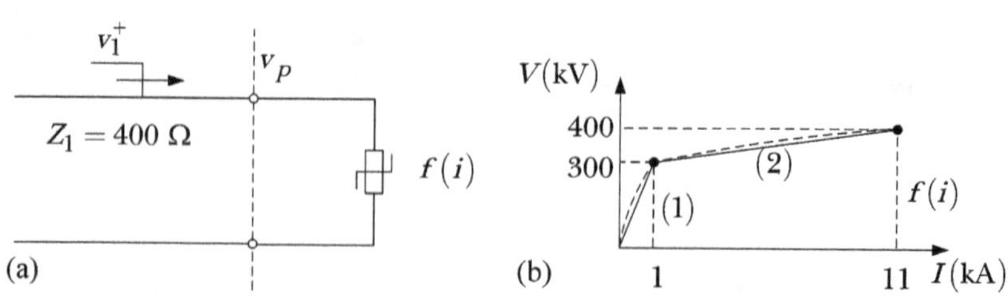

Figure 2.22 Transmission line and surge arrester. (a) Incident wave at the line terminal. (b) Arrester residual curve and data.

Solution

According to (2.96) and (2.97), the equations for the line terminal are:

$$v_p = 2v_1^+ - Z_1 i,$$

$$v_p = f(i).$$

Substituting the numerical values, the equation for the Thévenin equivalent is given by:

$$v_p = 2000 - 400i.$$

The algorithm should sequentially test the line segments. If the solution applies to segment 1, the following equations are obtained:

For $i < 1$ kA, the line equation is $f(i) = 300i$, resulting in:

$$i = \frac{2000}{700} = 2.85\text{ kA}.$$

However, the solution is outside of the interval $i < 1$ kA.

Assuming the solution to be in segment 2, 1 kA $< i <$ 11 kA, the equation for this line segment is obtained by setting:

$$300 = a + b, a = 10,$$

$$400 = 11a + b, b = 290,$$

Therefore, the equation for the segment 2 is given by:

$$f(i) = 10i + 290.$$

And the solution is obtained by setting:

$$2000 - 400i = 10i + 290,$$

$$i = \frac{1710}{410} = 4,17\text{ kA},$$

now 4.17 kA is a valid solution since it lies within the 1 kA $< i <$ 11 kA interval.

Note that if there is a spark gap, the nonlinear resistor must be inserted into the circuit as soon as the Thévenin equivalent voltage exceeds the spark level. After insertion of the resistor, the solution would be the same, with or without a spark gap.

The surge arrester lowers the prospective voltage of 2000 kV to the following value:

$$v_p = 2000 - 400i = 2000 - 400 \times 4.17,$$

$$v_p = 331.7\,\text{kV}.$$

To calculate the reflected wave, the following equations are used:

$$v_1^+ + v_1^- = v_f,$$

$$1000 + v_1^- = 331,7,$$

$$v_1^- = -668,3\,\text{kV},$$

which has a negative sign.

Example 2.6 Calculate the voltage for a voltage step of 2000 kV striking a substation with three transmission lines (Figure 2.23a).

The approximate residual curve of the MOSA resistor is given by the equations:

$$f(i): \text{if } v_p \leq 1000\,\text{kV}, \qquad i = 0;$$

$$\text{if } v_p > 1000\,\text{kV}, \qquad v_p = 10i + 1000.$$

Solution

The procedure is the same as that for the previous example, and the Thévenin equivalent has a surge impedance given by:

$$Z_{th} = Z_1 \parallel Z_2 \parallel Z_3; Z_{th} = 133\,\Omega.$$

As previously discussed, the Thévenin equivalent voltage e_{th} is the prospective voltage that would exist in the absence of a surge arrester (Figure 2.23b).

$$e_{th} = v_1^+ \Gamma_t, \text{where } \Gamma_t = \frac{2Z_2 \parallel Z_3}{Z_1 + Z_2 \parallel Z_3}.$$

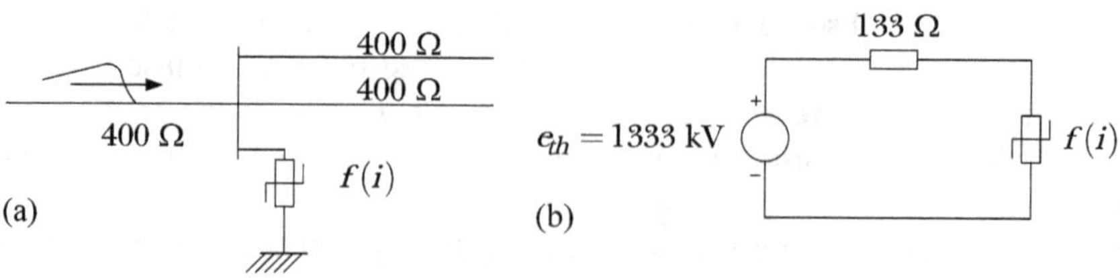

Figure 2.23 Voltage step striking the substation. (a) Incident wave. (b) Thévenin equivalent circuit.

$$\Gamma_t = \frac{2}{3}; e_{th} = \frac{2}{3} \times 2000 = 1333\,\text{kV}.$$

Thus:

$$1333 - 133i = 1000 + 10i,$$

$$i = \frac{333}{143} = 2.33\,\text{kA},$$

$$v_p = 1023.3\,\text{kV}.$$

2.6 Lattice Diagram

So far, refraction and reflection coefficients in transmission lines of infinite or semi-infinite length were considered, and now transmission lines of finite length will be considered. Thus, a lossless transmission line model is established based on its surge impedance and its travel time.

The reflection and refraction coefficients at both extremities and also the time delay are first calculated for a generic segment of a transmission line.

The diagram shows the axes for voltage and time, Figure (2.24). The expressions for the voltage at terminals 1 and 2 are written:

$$v_1(t) = v(t) + \Gamma_{t_1}\Gamma_{r_2}v(t-2\tau) + \Gamma_{t_1}\Gamma_{r_2}^2\Gamma_{r_1}v(t-4\tau) + \dots, \tag{2.98}$$

$$v_2(t) = \Gamma_{t_2}v(t-\tau) + \Gamma_{t_2}\Gamma_{r_2}\Gamma_{r_1}v(t-3\tau) + \dots. \tag{2.99}$$

The method is very useful for interpreting traveling waves in discontinuities in transmission lines. This technique is suitable for simple networks because otherwise it is more convenient to use the algorithms from EMTP (electromagntic transient program)-type, such as the method of characteristics, discussed in Chapter 4.

Example 2.7 Analyze the voltages in the circuit shown in Figure 2.25 by applying the lattice diagram.

$$\tau_2 = 1.5\tau_1, \tau_3 >> \tau_1, e_g = 200\,\text{V}.$$

$$Z_1 = Z_3 = 200\,\Omega,\ Z_2 = 100\,\Omega,\ Z_\ell = 500\,\Omega.$$

Solution

- Refraction and reflection coefficients:

$$\Gamma_{t_{1-23}} = \frac{2Z_2 \parallel Z_3}{Z_1 + Z_2 \parallel Z_3} = \frac{2 \times 66.67}{266.67} = 0.5,\ \Gamma_{r_{1-23}} = 0.5 - 1 = -0.5,$$

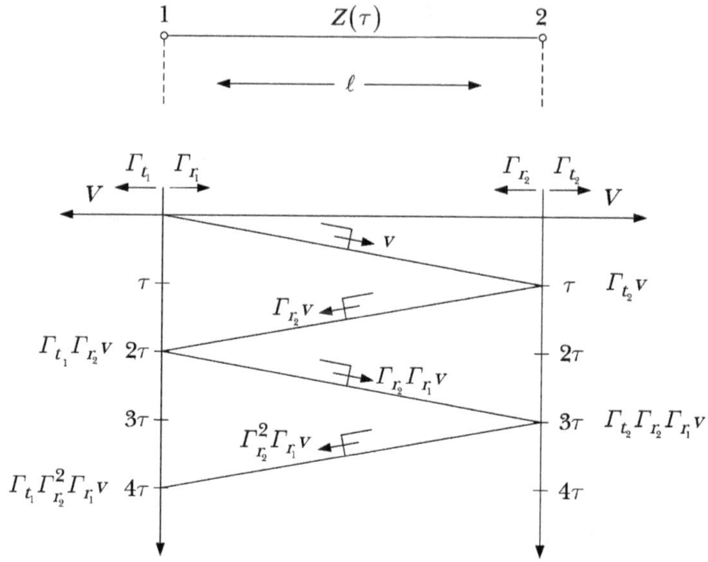

Figure 2.24 Refracted and reflected waves in the lattice diagram.

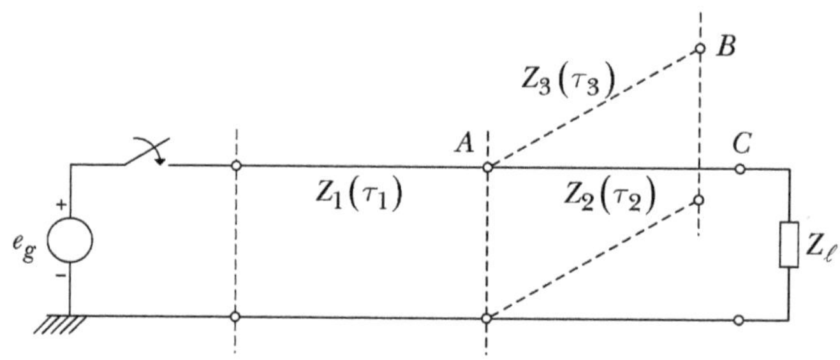

Figure 2.25 Transmission line with bifurcation.

$$\Gamma_{t_2-13} = \frac{2Z_1 \parallel Z_3}{Z_2 + Z_1 \parallel Z_3} = \frac{2 \times 100}{100 + 100} = 1, \Gamma_{r_2-13} = 0,$$

$$\Gamma_{t_2-\ell} = \frac{2 \times 500}{500 + 100} = 1.67, \Gamma_{r_2-\ell} = 1.67 - 1 = 0.67.$$

The refracted wave that travels toward B is equal to the voltage at node A and since τ_3 is much greater than τ_1 and τ_2, no reflection is assumed to come back from line segment A–B, which behaves as an infinite-length line (Figure 2.26).

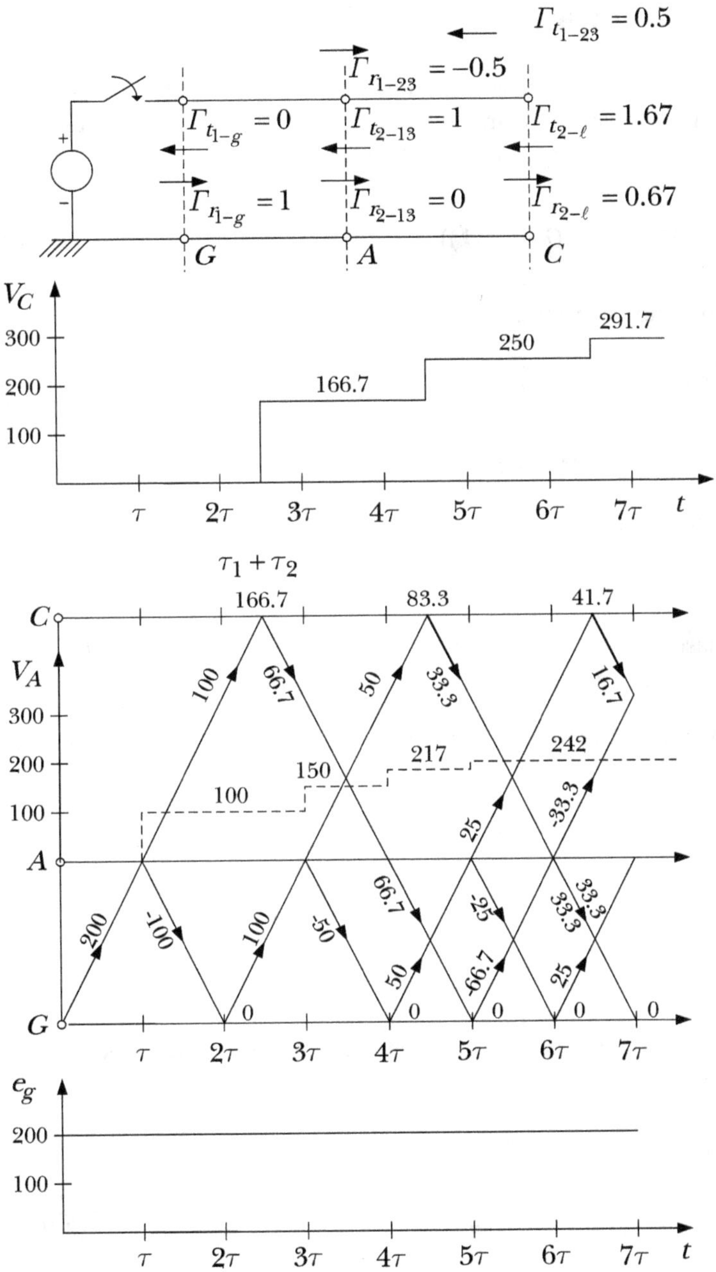

Figure 2.26 Data and traveling waves.

2.7 Sine Voltage Waves

Assuming a semi-infinite line, the voltage at point x can be calculated for a specific frequency with the sinusoidal source applied at the origin. Under steady-state sinusoidal conditions, Eq. (2.16) simplifies to a function of x:

$$\frac{d^2V(x)}{dx^2} = (R' + j\omega L')(G' + j\omega C')V(x), \tag{2.100}$$

With the propagation constant:

$$\gamma = \sqrt{(R' + j\omega L')(G' + j\omega C')}, \tag{2.101}$$

the solution of (2.100) can be written as:

$$V(x) = V(0)e^{-\gamma x},$$

where:

$V(0)$: voltage phasor applied by the generator at the origin, $V(0) = V_0 \angle \theta$.
$V(x)$: voltage phasor at point x along the transmission line.

For frequency $s = j\omega$ the propagation constant has a real part and an imaginary part,

$$\gamma = \alpha + j\beta,$$

where:

α: damping factor.
β: phase factor.

The voltage in the time domain can be obtained at point x from the voltage phasor at point $x = 0$,

$$v(x,t) = \text{Re}\left[\sqrt{2}V_0 e^{j\theta} e^{-(\alpha + j\beta)x} e^{j\omega t}\right],$$

$$v(x,t) = \sqrt{2}V_0 e^{-\alpha x}\cos(\omega t + \theta - \beta x). \tag{2.102}$$

Given the line boundary conditions, the steady-state model of the finite-length transmission line, either through two-port networks or π-equivalent circuits, can be found in [1].

References

1 Stevenson, W. D. Jr. (1975). *Elements of power systems analysis* (3rd ed.). New York: McGraw-Hill.
2 Johnson, W. C. (1950). *Transmission lines and networks.* New York: McGraw-Hill.
3 Bewley, L. V. (1963). *Travelling waves on transmission systems* (2nd ed.). New York: Dover.
4 Mariotto, P. A. (1981). *Ondas e linhas.* Rio de Janeiro: Guanabara Dois.
5 Greenwood, A. (1991). *Electrical transients in power systems* (2nd ed.). New York: John Wiley & Sons.
6 Stephenson, G. (1970) *An introduction to partial differential equations for science students.* Longman.
7 *ATP: Alternative Transients Program: Rule book.* (1987). Heverlee (BE): Leuven EMTP Center.

3

Traveling Waves in Multiphase Lines

3.1 Introduction

The traveling waves in multiphase transmission lines can be analyzed using an approach like the one employed in Chapter 2, through extension of the scalar equations for single-phase lines to matrix equations for multiphase lines.

This chapter provides a brief review of some concepts of matrix algebra useful in the study of traveling waves in multiphase lines, without intending to cover the theory in depth. The generic case of a non-transposed transmission line is analyzed using the Laplace transform to solve differential equations containing partial derivatives.

The relationship between multiphase voltage and current waves is initially analyzed in phase domain as in [1, 2]. Then, the advantages introduced into the calculations by modal transformations are discussed, which are applied to diagonalize the system of equations, allowing significant simplification through modal domain [3–5]. These transformations are applied to transposed and non-transposed transmission lines.

The equations for traveling waves in the time domain are simplified by assuming lossless transmission lines, which is useful for describing basic models in electromagnetic transient (EMT) programs. Chapter 5 supplements this with detailed models for lossy transmission lines, including the frequency dependence of their parameters.

Phase and modal traveling waves are exemplified in simple cases, and the energization of a three-phase transmission line with sequential breaker pole closing is analyzed using a streamlined approach to facilitate understanding of a very complex transient.

Understanding Electromagnetic Transients in Power Systems, First Edition. Luiz Cera Zanetta Jr.

Published 2025 by John Wiley & Sons, Inc.

3.2 Elements of Matrix Algebra

3.2.1 Calculation of the Exponential Matrix e^{Ax}

The mathematical notation e^{Ax} is used to represent an exponential matrix, which can be obtained by the expansion of the matrix into an infinite series of matrices, identical to the expansion of the exponential function e^{ax} into a series.

$$e^{ax} = 1 + a\frac{x}{1!} + a^2\frac{x^2}{2!} + a^3\frac{x^3}{3!} + \ldots, \tag{3.1}$$

$$e^{Ax} = I + A\frac{x}{1!} + A^2\frac{x^2}{2!} + A^3\frac{x^3}{3!} + \ldots \tag{3.2}$$

Analogously to (3.2):

$$e^{-Ax} = I - A\frac{x}{1!} + A^2\frac{x^2}{2!} - A^3\frac{x^3}{3!} + \ldots \tag{3.3}$$

Taking the derivative of (3.2) with respect to x:

$$\frac{d}{dx}\, e^{Ax} = A\, e^{Ax}, \tag{3.4}$$

demonstrating the same derivative properties as the exponential function e^{ax}.

Example 3.1 Given the matrices:

$$A = \begin{bmatrix} 1 & 0 \\ 1 & 1 \end{bmatrix};\ A^2 = \begin{bmatrix} 1 & 0 \\ 2 & 1 \end{bmatrix};\ A^3 = \begin{bmatrix} 1 & 0 \\ 3 & 1 \end{bmatrix} \text{ etc,}$$

calculate the matrix e^{Ax} by means of its series expansion.

Applying (3.2):

$$e^{Ax} = \begin{bmatrix} 1 & 0 \\ 0 & 1 \end{bmatrix} + \begin{bmatrix} 1 & 0 \\ 1 & 1 \end{bmatrix}\frac{x}{1} + \begin{bmatrix} 1 & 0 \\ 2 & 1 \end{bmatrix}\frac{x^2}{2} + \begin{bmatrix} 1 & 0 \\ 3 & 1 \end{bmatrix}\frac{x^3}{6} + \ldots$$

or:

$$e^{Ax} = \begin{bmatrix} 1 + x + \dfrac{x^2}{2} + \dfrac{x^3}{6} + \ldots & 0 \\ x + \dfrac{2x^2}{2} + \dfrac{3x^3}{6} + \ldots & 1 + x + \dfrac{x^2}{2} + \dfrac{x^3}{6} + \ldots \end{bmatrix},$$

therefore:

$$e^{Ax} = \begin{bmatrix} e^x & 0 \\ xe^x & e^x \end{bmatrix}.$$

3.2.2 Modal Decomposition

The convenience of working with modal decomposition is briefly reviewed.

- Diagonalization
  Let A be a real or complex matrix, diagonalizable by another matrix P, also real or complex. The following can be written:

$$\overline{A} = P^{-1}AP, \tag{3.5}$$

$$A = P\overline{A}P^{-1}, \tag{3.6}$$

$$\overline{A} = \text{diag}(\lambda_i), \quad 1 \leq i \leq n. \tag{3.7}$$

$\overline{A}$ is a diagonal matrix, where λ_i are the eigenvalues of A.
It is also known that:

$$\det(\lambda_i I - A) = 0. \tag{3.8}$$

Matrix P is composed of columns that are the eigenvectors associated with the eigenvalues of A, obtained from the solution of the homogeneous system of equations,

$$A\underset{\sim}{p_i} = \lambda_i \underset{\sim}{p_i}, \tag{3.9}$$

if $\underset{\sim}{p_i}$ is the eigenvector associated with the eigenvalue λ_i, also designated as right eigenvector of matrix A.

Matrix P can be written as follows:

$$P = \left[\underset{\sim}{p_1} \middle| \underset{\sim}{p_2} \middle| \cdots \middle| \underset{\sim}{p_n} \right]. \tag{3.10}$$

The eigenvectors in (3.10) need to be linearly independent. It can be demonstrated that, using diagonalization, any function $f(A)$ can be calculated from the following equation:

$$f(A) = Pf(\overline{A})P^{-1}, \tag{3.11}$$

In a very simple way, this can be verified for the exponential matrix, since in the series expansion of (3.2), each exponential A^n can be decomposed in the product of matrix A written as (3.6) n times:

$$A^n = P\overline{A}P^{-1} \times P\overline{A}P^{-1} \times \cdots \times P\overline{A}P^{-1}, \tag{3.12}$$

which may be rewritten as:

$$A^n = P\overline{A}^n P^{-1}, \tag{3.13}$$

as all internal multiplications $P^{-1} \times P$ are equal to the identity matrix. Summing all the terms in the series in (3.3), the exponential function of matrix A can be calculated as:

$$e^{Ax} = Pe^{\overline{A}x}P^{-1}. \tag{3.14}$$

Designating $R = P^{-1}$ as the matrix of left eigenvectors, each row of R can be written as:

$\underset{\sim}{r}_i^t$: left eigenvector of λ_i.

The matrix e^{Ax} can also be expressed in a special format, and for that, the function $e^{\bar{A}x}$ is first calculated:

$$e^{\bar{A}x} = \text{diag}\left(e^{\lambda_i x}\right), \quad 1 \le i \le n. \tag{3.15}$$

Substituting (3.15) into (3.14), the following matrix product is obtained:

$$e^{Ax} = \begin{bmatrix} \underset{\sim}{p}_1 & \underset{\sim}{p}_2 & \cdots & \underset{\sim}{p}_n \end{bmatrix} \begin{bmatrix} e^{\lambda_1 x} & & & \\ & e^{\lambda_2 x} & & \\ & & \ddots & \\ & & & e^{\lambda_n x} \end{bmatrix} \times \begin{bmatrix} \underset{\sim}{r}_1^t \\ \underset{\sim}{r}_2^t \\ \vdots \\ \underset{\sim}{r}_n^t \end{bmatrix}. \tag{3.16}$$

The eigenvectors in (3.16) will be assumed orthonormal (with unit Euclidean norm) and multiplying first and second matrices in the right-hand side, it can be reduced to:

$$e^{Ax} = \begin{bmatrix} e^{\lambda_1 x}\underset{\sim}{p}_1 | e^{\lambda_2 x}\underset{\sim}{p}_2 | \ldots | e^{\lambda_n x}\underset{\sim}{p}_n \end{bmatrix} \times \begin{bmatrix} \underset{\sim}{r}_1^t \\ \underset{\sim}{r}_2^t \\ \vdots \\ \underset{\sim}{r}_n^t \end{bmatrix}. \tag{3.17}$$

The matrix product in (3.17) can be expressed as a sum of products between columns of P by rows of R. However, to clarify further, the first matrix is expressed as a sum of submatrices, each with only i-th column nonzero and all other elements zero.

$$e^{Ax} = \left\{ \begin{bmatrix} e^{\lambda_1 x}\underset{\sim}{p}_1 | \underset{\sim}{0} | \ldots | \underset{\sim}{0} \end{bmatrix} + \ldots \begin{bmatrix} \underset{\sim}{0} | \ldots e^{\lambda_i x}\underset{\sim}{p}_i | \ldots | \underset{\sim}{0} \end{bmatrix} + \ldots \begin{bmatrix} \underset{\sim}{0} | \underset{\sim}{0} | \ldots | e^{\lambda_n x}\underset{\sim}{p}_n \end{bmatrix} \right\} \times \begin{bmatrix} \underset{\sim}{r}_1^t \\ \underset{\sim}{r}_i^t \\ \vdots \\ \underset{\sim}{r}_n^t \end{bmatrix}, \tag{3.18}$$

where $\underset{\sim}{0}$ indicates a null column vector.

Multiplying each submatrix by R, only the products of i-th column with the correspondent i-th row are nonzero.

$$e^{Ax} = \left[e^{\lambda_1 x} \underset{\sim}{p}_1 \underset{\sim}{r}_1^t\right] + \dots \left[e^{\lambda_i x} \underset{\sim}{p}_i \underset{\sim}{r}_i^t\right] + \dots \left[e^{\lambda_n x} \underset{\sim}{p}_n \underset{\sim}{r}_n^t\right],$$

This equation can be written in a compact form, which is referred to as the spectral decomposition of the matrix e^{Ax}:

$$e^{Ax} = \sum_{i=1}^{n} e^{\lambda_i x} \underset{\sim}{p}_i \underset{\sim}{r}_i^t. \tag{3.19}$$

3.2.3 Properties of Symmetric and Balanced Matrices

A symmetrical and balanced matrix A, of order 3 is assumed. Note that the matrix has equal terms on the diagonal as well as equal terms outside the diagonal, presenting the following structure:

$$A = \begin{bmatrix} a & b & b \\ b & a & b \\ b & b & a \end{bmatrix}. \tag{3.20}$$

The eigenvalues λ_i are obtained using (3.8), and the solution is as follows:

$$\begin{aligned} \lambda_0 &= a + 2b, \\ \lambda_1 &= a - b, \\ \lambda_2 &= a - b. \end{aligned} \tag{3.21}$$

In (3.20) and (3.21), the similarity to the application of symmetrical components can be observed.

An example of matrix P is shown for balanced matrices [5], designated as Karrembauer's matrix of order n.

$$P = \begin{bmatrix} 1 & 1 & 1 & \cdots & 1 \\ 1 & 1-n & 1 & \cdots & 1 \\ 1 & 1 & 1-n & \cdots & 1 \\ \vdots & \vdots & \vdots & \ddots & \vdots \\ 1 & 1 & 1 & \cdots & 1-n \end{bmatrix}. \tag{3.22}$$

3.2.4 Diagonalization of the Product of Symmetrical Matrices

In the study of traveling waves in transmission lines, it is necessary to work with equations involving matrix products. For this reason, it is of interest to understand how to deal with these products, since the elements in each matrix consist of transmission line parameters. It is also known that the matrices are complex and symmetric.

Consider Z_f and Y_f, complex matrices for series impedance and shunt admittance in phase domain, respectively. These matrices contain the parameters per unit length of the transmission line and are symmetric; thus, $Z_f^t = Z_f, Y_f^t = Y_f$.

Consider the products defined in (3.23) and (3.24) below:

$$A = Z_f Y_f, \tag{3.23}$$

$$B = Y_f Z_f. \tag{3.24}$$

It is known that matrix A and its transpose have the same eigenvalues; thus:

$$\det(\lambda_i I - A) = \det(\lambda_i I - A^t). \tag{3.25}$$

In the case of symmetric matrices:

$$A^t = B, \tag{3.26}$$

therefore, A and B have the same eigenvalues, that is:

$$\overline{A} = \overline{B}. \tag{3.27}$$

However, the eigenvectors for A and B are different, but since both matrices are symmetric, there is a correspondence between the eigenvectors.

Designating P and Q as the diagonalization matrices of A and B, respectively, composed of eigenvectors that have unitary Euclidian norm, Eqs. (3.28) and (3.29) follow:

$$\overline{A} = P^{-1}AP, \tag{3.28}$$

$$\overline{B} = Q^{-1}BQ, \tag{3.29}$$

since $\overline{A} = \overline{B}$:

$$P^{-1}AP = Q^{-1}BQ. \tag{3.30}$$

As the matrices in (3.27) are diagonal, they can be transposed on the right side of (3.30), and equality is maintained:

$$P^{-1}AP = Q^t B^t \left(Q^{-1}\right)^t, \tag{3.31}$$

therefore, the relation between the eigenvectors is given by (3.32):

$$P^{-1} = Q^t. \tag{3.32}$$

That is, the eigenvectors of the two matrix products are related through a relatively simple equation, such as (3.32). It can also be demonstrated that Z_f and Y_f can be diagonalized by setting [2]:

$$\overline{Z}_f = P^{-1} Z_f Q, \tag{3.33}$$

$$\overline{Y}_f = Q^{-1} Y_f P. \tag{3.34}$$

If Eqs. (3.33) and (3.34) are multiplied term by term, $\overline{A}$ appears on the left-hand side, that is,

$$\overline{A} = \overline{Z}_f \overline{Y}_f, \tag{3.35}$$

and $P^{-1}AP$ appears on the right-hand side since:

$$P^{-1} Z_f Q Q^{-1} Y_f P = P^{-1} Z_f Y_f P.$$

In the special case of balanced matrices, with the same structure as in (3.20), the products in (3.23) and (3.24) are equal, that is $A = B$, and it can be concluded that the eigenvectors are the same; thus $P = Q$.

Based on this preliminary information, an analysis of wave propagation in multiphase transmission lines is performed in the next section.

3.3 Phase Domain

3.3.1 Multiphase Line

Multiphase lines will be analyzed, often exemplified by three-phase lines (Figure 3.1).

Distributed resistance and inductance matrices are considered in the differential equations for the variation of phase voltages along the x-axis: The single-phase equations from Chapter 2 are generalized as follows.

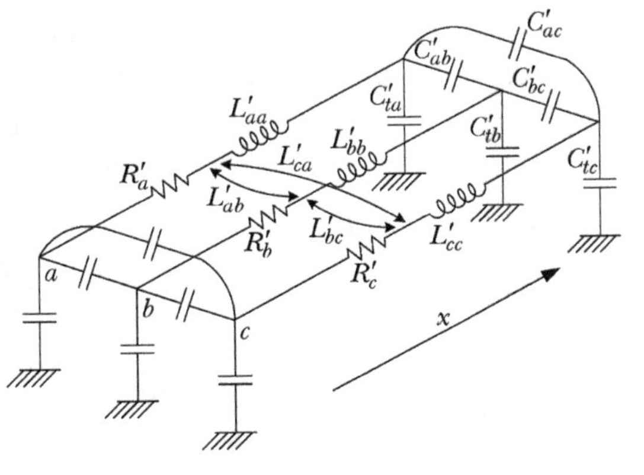

Figure 3.1 Three-phase transmission line with distributed parameters.

$$-\frac{\partial}{\partial x}\begin{bmatrix} v_a \\ v_b \\ v_c \end{bmatrix} = \begin{bmatrix} r'_{aa} & r'_{ab} & r'_{ac} \\ r'_{ba} & r'_{bb} & r'_{bc} \\ r'_{ca} & r'_{cb} & r'_{cc} \end{bmatrix} \times \begin{bmatrix} i_a \\ i_b \\ i_c \end{bmatrix} + \frac{\partial}{\partial t}\begin{bmatrix} l'_{aa} & l'_{ab} & l'_{ac} \\ l'_{ba} & l'_{bb} & l'_{bc} \\ l'_{ca} & l'_{cb} & l'_{cc} \end{bmatrix} \times \begin{bmatrix} i_a \\ i_b \\ i_c \end{bmatrix}. \tag{3.36}$$

$$-\frac{\partial}{\partial x}\begin{bmatrix} i_a \\ i_b \\ i_c \end{bmatrix} = \begin{bmatrix} g'_{aa} & g'_{ab} & g'_{ac} \\ g'_{ba} & g'_{bb} & g'_{bc} \\ g'_{ca} & g'_{cb} & g'_{cc} \end{bmatrix} \begin{bmatrix} v_a \\ v_b \\ v_c \end{bmatrix} + \frac{\partial}{\partial t}\begin{bmatrix} c'_{aa} & c'_{ab} & c'_{ac} \\ c'_{ba} & c'_{bb} & c'_{cc} \\ c'_{ca} & c'_{cb} & c'_{cc} \end{bmatrix} \begin{bmatrix} v_a \\ v_b \\ v_c \end{bmatrix} \tag{3.37}$$

The matrices of conductance and capacitance have the structure of a nodal admittance matrix, and the following notation is assumed:

- Voltage and current vectors

$$\underset{\sim}{v} = \begin{bmatrix} v_a \\ v_b \\ v_c \end{bmatrix}, \ \underset{\sim}{i} = \begin{bmatrix} i_a \\ i_b \\ i_c \end{bmatrix}. \tag{3.38}$$

- Matrices of parameters per unit length of the line.

$$R'_f = \begin{bmatrix} r'_{aa} & r'_{ab} & r'_{ac} \\ r'_{ba} & r'_{bb} & r'_{bc} \\ r'_{ca} & r'_{cb} & r'_{cc} \end{bmatrix}, L'_f = \begin{bmatrix} l'_{aa} & l'_{ab} & l'_{ac} \\ l'_{ba} & l'_{bb} & l'_{bc} \\ l'_{ca} & l'_{cb} & l'_{cc} \end{bmatrix} \text{ e } C'_f = \begin{bmatrix} c'_{aa} & c'_{ab} & c'_{ac} \\ c'_{ba} & c'_{bb} & c'_{bc} \\ c'_{ca} & c'_{cb} & c'_{cc} \end{bmatrix}.$$

$$G'_f = \begin{bmatrix} g'_{aa} & g'_{ab} & g'_{ac} \\ g'_{ba} & g'_{bb} & g'_{bc} \\ g'_{ca} & g'_{cb} & g'_{cc} \end{bmatrix} \tag{3.39}$$

The matrix notation for distributed parameters can be simplified by setting $R_f = R'_f, L_f = L'_f,\ G_f = G'_f$, and $C_f = C'_f$.

Applying the Laplace transform with the assumptions discussed in Chapter 2, the following expressions can be written:

$$Z_f(s) = R_f + sL_f, \tag{3.40}$$

$$Y_f(s) = G_f + sC_f \tag{3.41}$$

Rewriting (3.36) and (3.37):

$$-\frac{d}{dx}\underset{\sim}{V}(x,s) = Z_f(s)\underset{\sim}{I}(x,s), \tag{3.42}$$

$$-\frac{d}{dx}\underset{\sim}{I}(x,s) = Y_f(s)\underset{\sim}{V}(x,s). \tag{3.43}$$

Taking the second derivative of (3.42) with respect to x and substituting (3.43):

$$\frac{d^2}{dx^2}\underset{\sim}{V}(x,s) = Z_f(s)Y_f(s)\underset{\sim}{V}(x,s). \tag{3.44}$$

Similarly, for the current:

$$\frac{d^2}{dx^2}\underset{\sim}{I}(x,s) = Y_f(s)Z_f(s)\underset{\sim}{I}(x,s). \tag{3.45}$$

Assuming the general case of untransposed lines with symmetrical but unbalanced matrices, the matrix products in (3.44) and (3.45), although different, are related by:

$$\left[Z_f(s)Y_f(s)\right]^t = Y_f(s)Z_f(s).$$

In the case of transposed lines with symmetrical and balanced matrices, these matrix products are equal. In fact, the idea of perfectly transposed lines is assumed; however, this is a topic to be discussed later in Section 3.4.3. Furthermore, for lossless lines, they are not only equal but also diagonal matrices.

Designating:

$$[\gamma_v(s)] = \left[Z_f(s)Y_f(s)\right]^{1/2}, \tag{3.46}$$

$$[\gamma_i(s)] = \left[Y_f(s)Z_f(s)\right]^{1/2}. \tag{3.47}$$

Note that, by transposing (3.46), it is immediate that:

$$[\gamma_v(s)] = [\gamma_i(s)]^t \tag{3.48}$$

The following solutions can be proposed for Eqs. (3.44) and (3.45), in the form of exponential matrices associated with progressive waves.

$$\underset{\sim}{V}^+(x,s) = e^{-[\gamma_v(s)]x}\underset{\sim}{V}^+(0,s), \tag{3.49}$$

$$\underset{\sim}{I}^+(x,s) = e^{-[\gamma_i(s)]x}\underset{\sim}{I}^+(0,s), \tag{3.50}$$

where $\underset{\sim}{V}(0,s)$ and $\underset{\sim}{I}(0,s)$ are defined by the boundary conditions at point $x = 0$. The same comments on boundary conditions in Chapter 2 remain valid in (3.49) and (3.50).

Much like the progressive waves, the regressive waves can be written as:

$$\underset{\sim}{V}^-(x,s) = e^{[\gamma_v(s)]x}\underset{\sim}{V}^-(0,s), \underset{\sim}{I}^-(x,s) = e^{[\gamma_i(s)]x}\underset{\sim}{V}^-(0,s).$$

The inversion of (3.49) and (3.50) to the time domain is of practical interest for modeling the transmission line in EMT programs, considering the dependence of the parameters on frequency, which is addressed in Chapter 5. The following equations are developed with some simplification, considering the parameters calculated at a single frequency, and some basic elements are also presented through lossless lines.

3.3.2 Relationship Between Voltages and Currents – Matrix of Characteristic Impedances

To obtain the relationship between voltages and currents in the transmission line, the same considerations for traveling waves in the positive and negative direction on the *x-axis* are valid as in the single-phase case. Only the forward waves in the positive direction on the *x-axis* are considered.

By substituting the progressive wave from Eqs. (3.49) and (3.50) into Eq. (3.42),

$$[\gamma_v(s)]e^{-[\gamma_v(s)]x}\underset{\sim}{V}^+(0,s) = Z_f(s)e^{-[\gamma_i(s)]x}\underset{\sim}{I}^+(0,s), \tag{3.51}$$

$$e^{-[\gamma_v(s)]x}\underset{\sim}{V}^+(0,s) = [\gamma_v(s)]^{-1}Z_f(s)e^{-[\gamma_i(s)]x}\underset{\sim}{I}^+(0,s), \tag{3.52}$$

but it can be demonstrated that:

$$[\gamma_v(s)]^{-1}Z_f(s) = \left(Y_f(s)^{-1}Z_f(s)\right)^{\frac{1}{2}},$$

and designating the matrix of characteristic impedances of the transmission line as:

$$Z(s) = \left(Y_f(s)^{-1}Z_f(s)\right)^{\frac{1}{2}}, \tag{3.53}$$

thus,

$$e^{-[\gamma_v(s)]x}\underset{\sim}{V}^+(0,s) = Z(s)e^{-[\gamma_i(s)]x}\underset{\sim}{I}^+(0,s). \tag{3.54}$$

Recalling (3.49) and (3.50),

$$\underset{\sim}{V}^+(x,s) = Z(s)\underset{\sim}{I}^+(x,s), \tag{3.55}$$

where $Z(s)$ is composed of elements that are the characteristic impedances of the phases (self-impedances) on the diagonal, and the mutual characteristic impedances off the diagonal. Voltages and currents can be related using the matrix of characteristic impedances.

Note that the presence of $[\gamma_v(s)]$ and $[\gamma_i(s)]$ on the left and right sides of equation (3.54), respectively, does not indicate an inconsistency, as the equality is ensured by the symmetry of the matrices. For instance, transposing the matrix product on the right side maintains the equality, making it clear that:

$$\left[Z(s)e^{-[\gamma_i(s)]x}\right]^t = e^{-[\gamma_v(s)]x}Z(s), \tag{3.56}$$

with the equality being verified in (3.55), since:

$$\underset{\sim}{V}^+(0,s) = Z(s)\underset{\sim}{I}^+(0,s).$$

The matrix of characteristic impedances in the frequency domain is calculated using the transmission line parameters. This matrix relates voltage and current waves in phase domain. The dual expression for the matrix of characteristic admittances can be analogously obtained.

For the regressive wave, a similar equation to (3.55) is obtained, with a negative sign.

$$\underset{\sim}{V}^-(x,s) = -Z(s)\underset{\sim}{I}^-(x,s)$$

3.3.3 Lossless Transmission Line

Although simplified, the model of lossless transmission lines has several applications in transient studies, and therefore, the topic is addressed below.

The traveling waves in a lossless multiphase transmission line, assuming voltage waves, are calculated from (4.49) with the simplification of equations (4.46) and (4.53), resulting, for the progressive waves, in:

$$\underset{\sim}{V}^+(x,s) = e^{-[\gamma_v(s)]x}\underset{\sim}{V}^+(0,s), \tag{3.57}$$

$$[\gamma_v(s)] = s(L_f C_f)^{1/2}, \tag{3.58}$$

$$Z = \left[(C_f)^{-1}L_f\right]^{1/2} \tag{3.59}$$

The parameters of the transmission line are calculated, using the following expressions for inductances and capacitances (Figure 3.2):

$$L_f = \frac{\mu_0}{2\pi}M, \tag{3.60}$$

$$C_f = 2\pi\varepsilon_0 M^{-1}. \tag{3.61}$$

With the elements in matrix M given by:

$$m_{ii} = \ln\left(\frac{2h_i}{r_i}\right), \tag{3.62}$$

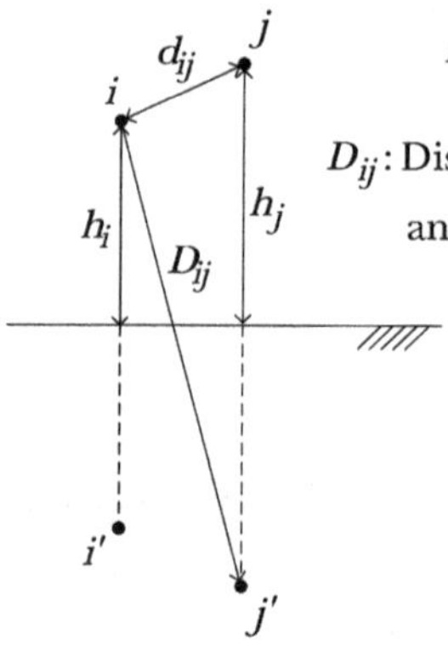

Figure 3.2 Ideal conductors and images.

$$m_{ij} = \ln\left(\frac{D_{ij}}{d_{ij}}\right) \; (i \neq j), \tag{3.63}$$

By substituting (3.60) and (3.61) into (3.58), Eq. (3.64) is obtained:

$$[\gamma_v] = s\sqrt{\mu_0 \varepsilon_0} I_d, \tag{3.64}$$

where I_d is the identity matrix, and in this case, the propagation matrices for voltages and currents are equal, that is, $[\gamma] = [\gamma_v] = [\gamma_i]$.

The model is defined in the phase domain, where the waves propagate at the speed of light in the three phases, and therefore, when it is necessary to differentiate the speeds in the phases, it is more convenient to work in the modal domain, as will be seen later.

The characteristic impedances may be designated surge impedance; thus, the matrix of the surge impedances is calculated, substituting (3.60) and (3.61) into (3.59):

$$Z = \frac{1}{2\pi\varepsilon_0} M (\mu_0 \varepsilon_0)^{\frac{1}{2}}, \tag{3.65}$$

or:

$$Z = \frac{1}{2\pi}\sqrt{\frac{\mu_0}{\varepsilon_0}} M \tag{3.66}$$

thus:

$$Z = 60M. \tag{3.67}$$

Matrix Z has the same structure as matrix M, and its elements are surge impedances that behave like resistances.

3.3.4 Traveling Waves in Multiphase Lines with Discontinuities

- Discontinuities in Transmission Lines

To simplify notation, multiphase waves traveling from line 1 to line 2 are analyzed in the time domain, though all results derived using the Laplace transform remain valid. Here, the case of two transmission lines with different characteristic impedances is examined, extending the terminology and equations from Chapter 2.

i_i, v_i: incident waves at the connection point (i_1^+, v_1^+).
i_r, v_r : reflected waves at the connection point (i_1^-, v_1^-).
i_t, v_t : refracted waves at the connection point (i_2, v_2).
Z_1: matrix of surge impedances of line 1.
Z_2: matrix of surge impedances of line 2.

Recalling the equations from (2.44) to (2.46) in Chapter 2, generalized for the multiphase case:

$$\underset{\sim}{i}_i = (Z_1)^{-1}\underset{\sim}{v}_i,\ \underset{\sim}{i}_r = -(Z_1)^{-1}\underset{\sim}{v}_r,\ \underset{\sim}{i}_t = (Z_2)^{-1}\underset{\sim}{v}_t.$$

With the condition of continuity in voltage and current in (3.47) and (3.48), respectively:

$$\underset{\sim}{v}_i + \underset{\sim}{v}_r = \underset{\sim}{v}_t,$$

$$\underset{\sim}{i}_i + \underset{\sim}{i}_r = \underset{\sim}{i}_t$$

the equation (3.49) for currents can be rewritten as,

$$(Z_1)^{-1}\left(\underset{\sim}{v}_i - \underset{\sim}{v}_r\right) = (Z_2)^{-1}\left(\underset{\sim}{v}_i + \underset{\sim}{v}_r\right),$$

and then Eq. (3.68) is derived:

$$\underset{\sim}{v}_r = (Z_2 - Z_1)(Z_1 + Z_2)^{-1}\underset{\sim}{v}_i. \tag{3.68}$$

From (3.68), the matrix of reflection coefficients for incident and reflected voltage vectors can be written in (3.69), almost like the single-phase approach in (2.50), Chapter 2.

$$\Gamma_r = (Z_2 - Z_1)(Z_1 + Z_2)^{-1}. \tag{3.69}$$

Transposing the product on the right side in (3.69), which is composed of symmetrical matrices, it can be shown that:

$$(Z_2 - Z_1)(Z_1 + Z_2)^{-1} = (Z_1 + Z_2)^{-1}(Z_2 - Z_1),$$

thus, Γ_r can be determined using either expression.

The boundary condition $\underset{\sim}{v}_i + \underset{\sim}{v}_r = \underset{\sim}{v}_t$ can be rewritten as, $(I_d + \Gamma_r)\underset{\sim}{v}_i = \Gamma_t \underset{\sim}{v}_i$ thus, the matrix of refraction coefficients can be calculated from:

$$\Gamma_t = I_d + \Gamma_r \text{ or } \Gamma_t = 2Z_2(Z_1 + Z_2)^{-1} \tag{3.70}$$

I_d: identity matrix.

Example 3.2 Calculate the refracted traveling waves from line 1 to line 2.

Data

The incident voltage vector and the surge impedances of lines 1 and 2 are:

$$\underset{\sim}{v}_i = \begin{bmatrix} 100 \\ 20 \\ 80 \end{bmatrix}, \quad Z_1 = \begin{bmatrix} 450 & 110 & 90 \\ 110 & 390 & 110 \\ 90 & 110 & 450 \end{bmatrix}, \quad Z_2 = \begin{bmatrix} 350 & 80 & 80 \\ 80 & 350 & 80 \\ 80 & 80 & 350 \end{bmatrix}.$$

Solution

Calculating Γ_r using (3.69):

$$\Gamma_r = \begin{bmatrix} -0.1256 & -0.0127 & 0.0172 \\ -0.0226 & -0.0424 & -0.0226 \\ 0.0172 & -0.0127 & -0.1256 \end{bmatrix}.$$

Calculating Γ_t using (3.70):

$$\Gamma_t = \begin{bmatrix} 0.8744 & -0.0127 & 0.0172 \\ -0.0226 & 0.9576 & -0.0226 \\ 0.0172 & -0.0127 & 0.8744 \end{bmatrix},$$

therefore, $\underset{\sim}{v}_t = \Gamma_t \underset{\sim}{v}_i$:

$$\underset{\sim}{v}_t = \begin{bmatrix} 88.5590 \\ 15.0805 \\ 71.4162 \end{bmatrix},$$

$$\underset{\sim}{v}_r = \Gamma_r \underset{\sim}{v}_i,$$

- Voltages Applied by Generators

The voltages in transmission lines, driven by multiphase generators (or Théve-nin equivalents) with internal impedances, are analyzed in the time domain below (Figure 3.3). This analysis assumes a lossless transmission line without any loss of generality. The matrix representing the generator's internal impedances consists of lumped elements, while the lossless transmission line is represented by its matrix of surge impedances. Although a matrix of internal resistances is assumed, using generic impedances would yield the same equations in the frequency domain.

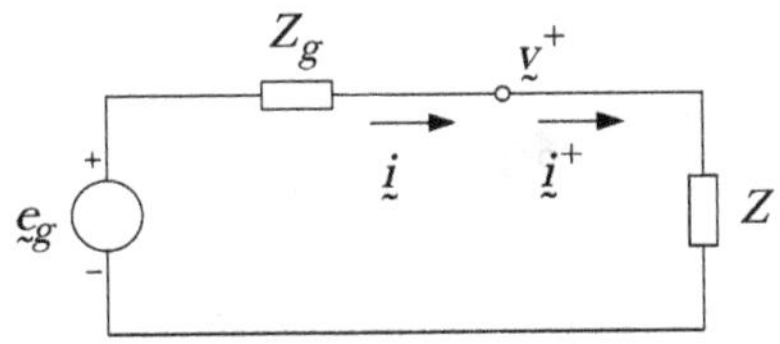

Figure 3.3 Voltages applied by a generator.

The vector of the forward waves in the transmission line is designated $\underset{\sim}{v}^+$. Since the applied voltages are not traveling waves inside the generator, the voltage across the internal resistance of the generator is obtained from:

$$\underset{\sim}{e}_g - \underset{\sim}{v}^+ = Z_g \underset{\sim}{i}, \tag{3.71}$$

In the transmission line, voltages and currents are related by the matrix of surge impedances:

$$\underset{\sim}{i}^+ = Z^{-1} \underset{\sim}{v}^+ \tag{3.72}$$

The matrix of surge admittances is calculated from $Y = Z^{-1}$, and the current is given by:

$$\underset{\sim}{i}^+ = Y \underset{\sim}{v}^+, \tag{3.73}$$

Substituting (3.73) into (3.71), knowing that $i = i^+$:

$$\underset{\sim}{e}_g - \underset{\sim}{v}^+ = Z_g Y \underset{\sim}{v}^+, \tag{3.74}$$

hence,

$$\underset{\sim}{v}^+ = (I_d + Z_g Y)^{-1} \underset{\sim}{e}_g. \tag{3.75}$$

The relationship between the internal voltages of the generator and the voltages applied to the transmission line is given by the inverse of $(I_d + Z_g Y)$, which is analogous to the voltage divider rule in the single-phase case.

3.3.5 Thévenin Equivalent in Multiphase Circuits

The Thévenin equivalent of single-phase circuits, covered in Chapter 2, can be extended to multiphase networks. The part of the multiphase network to be inserted into the equivalent circuit is considered to be disconnected from the rest

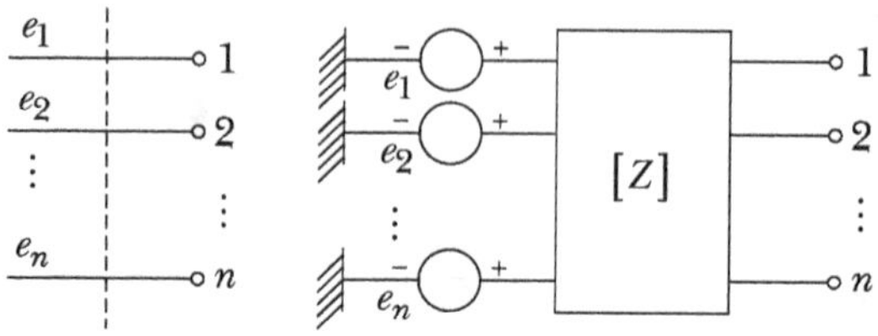

Figure 3.4 Thévenin equivalent circuit.

of the system at the boundary nodes to calculate the Thévenin generator, impedance, and voltage.

The impedance matrix for the Thévenin equivalent is calculated by the same method as single-phase circuits. For transmission lines, the matrix of the impedances in the Thévenin equivalent is the matrix of characteristic-impedances, Z, for the line.

Based on voltages e_1, e_2, ..., e_n calculated at a certain time at the interface points, the Thévenin equivalent in Figure 3.4 can be assumed to model the multiphase network.

3.4 Modal Domain

3.4.1 Modal Analysis

Traveling waves are analyzed in the complex domain through the application of the Laplace transform. The modal transformation that diagonalizes the system of equations introduces an important simplification in the analysis of traveling waves. Nontransposed transmission lines are considered initially and then transposed lines.

If P and Q are matrices that diagonalize the system of equations in (3.44) and (3.45), respectively, the transformed variables can be written as:

$$\underset{\sim}{V}(x,s) = P\underset{\sim}{\overline{V}}(x,s), \tag{3.76}$$

$$\underset{\sim}{I}(x,s) = Q\underset{\sim}{\overline{I}}(x,s). \tag{3.77}$$

where:

$\underset{\sim}{\overline{V}}, \underset{\sim}{\overline{I}}$: modal voltage and modal current vectors.

For the case of a three-phase transmission line:

$$\underset{\sim}{\overline{V}}(x,s) = \begin{bmatrix} V_0(x,s) \\ V_1(x,s) \\ V_2(x,s) \end{bmatrix}, \quad \underset{\sim}{\overline{I}}(x,s) = \begin{bmatrix} I_0(x,s) \\ I_1(x,s) \\ I_2(x,s) \end{bmatrix}. \tag{3.78}$$

Equation (3.76) is substituted into (3.44), and the ordinary differential equation for voltages is written as:

$$\frac{d^2}{dx^2} P\underset{\sim}{\overline{V}}(x,s) = (Z_f Y_f) P\underset{\sim}{\overline{V}}(x,s). \tag{3.79}$$

Multiplying (3.79) by P^{-1} and recalling that $A = Z_f Y_f$, equations (3.80) and (3.81) can be written:

$$\overline{A} = P^{-1}(Z_f Y_f)P, \tag{3.80}$$

$$\frac{d^2}{dx^2} \underset{\sim}{\overline{V}}(x,s) = \overline{A}\,\underset{\sim}{\overline{V}}(x,s). \tag{3.81}$$

Analogously for the currents, given that $B = Y_f Z_f$ and using (3.77) and (3.45):

$$\overline{B} = Q^{-1}(Y_f Z_f)Q, \tag{3.82}$$

$$\frac{d^2}{dx^2} \underset{\sim}{\overline{I}}(x,s) = \overline{B}\,\underset{\sim}{\overline{I}}(x,s). \tag{3.83}$$

For a given frequency $s = j\omega$, assuming different eigenvalues and eigenvectors, the numerical values for the above expressions can easily be calculated with numerical computation software.

The diagonal matrix $\overline{A}$ for a n-phase transmission line is written as:

$$\overline{A} = \text{diag}(a_j), 0 \leq j \leq n-1. \tag{3.84}$$

The matrix of modal propagation constants is obtained from:

$$[\overline{\gamma}] = \overline{A}^{1/2}. \tag{3.85}$$

$$[\overline{\gamma}] = \text{diag}(\gamma_j), 0 \leq j \leq n-1, \tag{3.86}$$

where $\gamma_j = \sqrt{a_j}$.

The equations in modal domain are the same as those in Section 3.3, with the added simplification that the system is diagonal. Considering only the progressive waves, the solution for (3.81), using the matrix calculated in (3.86), is given by:

$$\underset{\sim}{\overline{V}}(x,s) = e^{-[\overline{\gamma}(s)]x}\underset{\sim}{\overline{V}}^{+}(0,s), \tag{3.87}$$

where the exponential matrix $e^{-[\overline{\gamma}(s)]x}$ is diagonal since it is a function of a diagonal matrix.

$\underset{\sim}{\overline{V}}^{+}(0,s)$ is the vector of modal components defined by the boundary conditions at $x = 0$:

$$\underset{\sim}{\overline{V}}^{+}(0,s) = P^{-1}\underset{\sim}{V}^{+}(0,s).$$

As A and B have the same eigenvalues, only one propagation matrix $[\overline{\gamma}]$ is assumed for voltages and currents. Thus, for (3.83):

$$\underset{\sim}{\overline{I}}(x,s) = e^{-[\overline{\gamma}(s)]x}\underset{\sim}{\overline{I}}^{+}(0,s). \tag{3.88}$$

It is noted that expressions (3.87) and (3.88) could be obtained from (3.49) and (3.50), by applying the corresponding transformation matrices P and Q, and recalling the diagonalization property of matrix functions.

For any given propagation mode j, the following expression holds:

$$V_j(x,s) = e^{-\gamma_j(s)x}V_j^{+}(0,s). \tag{3.89}$$

Similarly for the current:

$$I_j(x,s) = e^{-\gamma_j(s)x}I_j^{+}(0,s) \tag{3.90}$$

Therefore, all the properties for single-phase transmission lines, as presented in Chapter 2, can be applied to a given propagation mode j.

For frequency $s = j\omega$, $\gamma_j = \alpha_j + j\beta_j$ is obtained (do not confuse j indicating an imaginary number with j indicating a modal index):

α_j: $\mathrm{Re}(\gamma_j)$, modal attenuation constant.
β_j: $\mathrm{Im}(\gamma_j)$, modal phase constant.
$\frac{\omega}{\beta_j}$: modal propagation speed
$\frac{2\pi}{\beta_j}$: modal wavelength.

The matrix of modal characteristic impedances is calculated from:

$$\overline{Z} = \left(\overline{Y}_f^{-1}\overline{Z}_f\right)^{1/2}. \tag{3.91}$$

Matrix $[\overline{\gamma}]$ is unique, and matrices $\overline{Z}, \overline{Z}_f$ e $\overline{Y}_f$ only have meaning when joined by their corresponding transformation matrices.

3.4.2 Analysis of the Propagation Modes

The diagonalization process is like the use of symmetrical components. The columns in the modal transformation matrix are the eigenvectors associated with the eigenvalues of the matrix.

For a three-phase transmission line, matrix P can be presented by its columns:

$$P = \left[\underset{\sim}{p_1}\middle|\underset{\sim}{p_2}\middle|\underset{\sim}{p_3}\right], \tag{3.92}$$

where each column $\underset{\sim}{p_j}$, is associated with a propagation mode j.

For a generic variable, current or voltage, phase, and modal components are related by:

$$\begin{bmatrix} f_a \\ f_b \\ f_c \end{bmatrix} = \begin{bmatrix} p_{11} & p_{12} & p_{13} \\ p_{21} & p_{22} & p_{23} \\ p_{31} & p_{32} & p_{33} \end{bmatrix} \times \begin{bmatrix} f_0 \\ f_1 \\ f_2 \end{bmatrix}. \tag{3.93}$$

Next, an example of a three-phase line excited by a function composed only of mode 1 is shown. The other modes are null [5].

$$\begin{bmatrix} f_a \\ f_b \\ f_c \end{bmatrix} = \begin{bmatrix} x & -0.3 & x \\ x & 1.0 & x \\ x & -0.7 & x \end{bmatrix} \times \begin{bmatrix} 0 \\ f_1 \\ 0 \end{bmatrix}, \tag{3.94}$$

The information given by the eigenvector is that for a unitary applied signal, having only the mode 1 component, 100% of the signal is observed to flow through phase b, with 30% returning through phase a and 70% returning through phase c. The sum of the phase components for this aerial mode is zero.

Analogously, applying a positive excitation sequence to a transmission line:

$$\begin{bmatrix} F_a \\ F_b \\ F_c \end{bmatrix} = \begin{bmatrix} 1 & 1 & 1 \\ 1 & \alpha^2 & \alpha \\ 1 & \alpha & \alpha^2 \end{bmatrix} \times \begin{bmatrix} 0 \\ F_1 \\ 0 \end{bmatrix}, \qquad \alpha = 1\angle 120^\circ. \tag{3.95}$$

computing the product:

$$\begin{bmatrix} F_a \\ F_b \\ F_c \end{bmatrix} = \begin{bmatrix} 1 \\ \alpha^2 \\ \alpha \end{bmatrix} F_1, \tag{3.96}$$

the three phasors in a positive sequence, shifted by 120°, are obtained. The sum of the components is also zero.

Assume a voltage vector applied to the transmission line at point $x = 0$ proportional to one of the eigenvectors of matrix P. Signal propagation can be analyzed by the spectral decomposition, as in Eq. (3.19).

Let $\underset{\sim}{e}_0$ be an excitation at the origin, parallel to an eigenvector $\underset{\sim}{p}_m$ and h a real constant:

$$\underset{\sim}{e}_0 = h\underset{\sim}{p}_m. \tag{3.97}$$

$\underset{\sim}{p}_m$: column m of matrix P, right eigenvector associated with mode m.

$\underset{\sim}{p}_j$: column j of matrix P, right eigenvector associated with mode j.

$\underset{\sim}{r}_j^t$: row j of matrix $R = P^{-1}$, left eigenvector associated with mode j.

The vector for the voltages in the transmission line can be obtained by:

$$\underset{\sim}{v} = \sum_{j=1}^{n} e^{-\gamma_j x} \underset{\sim}{p}_j \underset{\sim}{r}_j^t \underset{\sim}{e}_0. \tag{3.98}$$

Replacing (3.97) in (3.98):

$$\underset{\sim}{v} = \sum_{j=1}^{n} e^{-\gamma_j x} \underset{\sim}{p}_j \underset{\sim}{r}_j^t \eta \underset{\sim}{p}_m, \tag{3.99}$$

as product $\underset{\sim}{r}_j^t \underset{\sim}{p}_m$ is not null and equal to 1 only for $j = m$, the following result is obtained:

$$\underset{\sim}{v} = \eta e^{-\gamma_m x} \underset{\sim}{p}_m, \tag{3.100}$$

that is, if the transmission line is excited by a voltage vector proportional to $\underset{\sim}{p}_m$, only propagation in mode m is observed in the phases of the transmission line.

Example 3.3 This example shows the calculation of the modal transformation matrix, the surge impedance, the propagation speed, and the attenuation constant at frequencies of 60 Hz, 50 kHz, and 1 MHz, given a typical nontransposed transmission line of 230 kV. The results were calculated using the routines in reference [6].

Results

Matrix of modal current transformations:

$$Q(60\text{ Hz}) = \begin{bmatrix} 1.000 + j0 & 0.500 + j0 & -0.3440 - j0.0003 \\ 0.9190 + j0.009 & 0 + j0 & 0.6839 - j0.0034 \\ 1.000 + j0 & -0.500 + j0 & -0.3440 - j0.0003 \end{bmatrix},$$

$$Q(50\text{ kHz}) = \begin{bmatrix} 1.000 + j0 & -0.500 + j0 & -0.3431 - j0.0003 \\ 0.8950 + j0.0079 & 0 + j0 & 0.6929 - j0.0030 \\ 1.000 + j0 & -0.500 + j0 & -0.3431 - j0.0003 \end{bmatrix},$$

$$Q(1\text{ MHz}) = \begin{bmatrix} 1.000 + j0 & -0.9136 + j0.0196 & -0.1882 - j0.0043 \\ 0.9174 + j0.0132 & 0 + j0 & 0.3745 - j0.0053 \\ 1.000 + j0 & 0.9136 - j0.0196 & -0.1882 - j0.0043 \end{bmatrix}.$$

The real part of matrix Q dominates at all frequencies and only a slight change in the imaginary part is observed. However, there may be cases with more significant variations depending on the geometries of transmission lines, as will be discussed in Chapter 5.

Some modal line parameters are calculated as a function of the frequency and are presented below.

Modal values.

Frequency (Hz)	Mode	Characteristic impedance (Ω)	Propagation speed (km/s) ×10^6	Attenuation (Np/km)
60	0	883.6−j114.9	0.19079	0.25694×10^{-3}
60	1	410.3−j36.1	0.29211	0.11357×10^{-3}
60	2	357.8−j36.1	0.29333	0.12968×10^{-3}
50×10^3	0	624.6−j20.1	0.26946	0.37543×10^{-1}
50×10^3	1	402.3−j0.7	0.29788	0.18261×10^{-2}
50×10^3	2	350.8−j0.5	0.29930	0.16305×10^{-2}
10^6	0	588.9−j15.5	0.28626	0.57683
10^6	1	401.3−j0.6	0.29862	0.30394×10^{-1}
10^6	2	350.2−j0.1	0.29966	0.75669×10^{-2}

With an increase in frequency, damping increases in all modes, with a particularly pronounced effect in mode zero.

Example 3.4 This example shows the calculation of traveling waves in a lossy transmission line. The results are obtained in phase and modal domain using the routines in [7]. The parameters are calculated at 60 Hz using routine Line Constants [6].

Applying (3.46) and (3.53), calculate the characteristic impedance and propagation matrices. Assuming a semi-infinite transmission line, excited at the sending end, calculate the voltages at a distance $x = 100$ km from the origin. The applied voltages are sinusoidal, and the calculations are performed using phasor values (Section 2.7 in Chapter 2).

$$\underset{\sim}{E_0} = \begin{bmatrix} 1\angle 10^\circ \\ 1\angle -100^\circ \\ 1\angle +100^\circ \end{bmatrix}.$$

For the 60 Hz frequency, the following data are furnished:

$$Z_f = \begin{bmatrix} 0.2100 + j0.9000 & 0.1200 + j0.4300 & 0.1200 + j0.3800 \\ 0.1200 + j0.4300 & 0.2200 + j0.9000 & 0.1200 + j0.4300 \\ 0.1200 + j0.3800 & 0.1200 + j0.4300 & 0.2100 + j0.9000 \end{bmatrix} \Omega/\text{km},$$

$$Y_f = \begin{bmatrix} 0 + j0.2900 & 0 - j0.0500 & 0 - j0.0200 \\ 0 - j0.0500 & 0 + j0.3000 & 0 - j0.0500 \\ 0 - j0.0200 & 0 - j0.0500 & 0 + j0.2900 \end{bmatrix} \times 10^{-5}\ \text{S/km}.$$

Solution

a) Calculation in phase domain $A = Z_f Y_f$.

$$Z_f Y_f = \begin{bmatrix} -0.2319 + j0.0525 & -0.0650 + j0.0195 & -0.0707 + j0.0246 \\ -0.0711 + j0.0214 & -0.2270 + j0.0540 & -0.0711 + j0.0214 \\ -0.0707 + j0.0246 & -0.0650 + j0.0195 & -0.2319 + j0.0525 \end{bmatrix} \times 10^{-5},$$

recalling that $B = (Z_f Y_f)^t$.

Matrix of propagation constants:

$$[\gamma_v] = \begin{bmatrix} 0.1616 + j1.5005 & 0.0360 + j0.2066 & 0.0522 + j0.2261 \\ 0.0396 + j0.2260 & 0.1707 + j1.4861 & 0.0396 + j0.2260 \\ 0.0522 + j0.2261 & 0.0360 + j0.2066 & 0.1616 + j1.5005 \end{bmatrix} \times 10^{-3},$$

noting that $[\gamma_i] = [\gamma_v]^t$.

Matrix of characteristic impedances:

$$Z = \begin{bmatrix} 5.5991 - j0.6223 & 1.8700 - j0.2684 & 1.4883 - j0.2691 \\ 1.8700 - j0.2684 & 5.5771 - j0.6585 & 1.8700 - j0.2684 \\ 1.4883 - j0.2691 & 1.8700 - j0.2684 & 5.5991 - j0.6223 \end{bmatrix} \times 10^2.$$

Exponential matrix and voltages in the phase domain at $x = 100$ km:

$$e^{-[\gamma_v]x} = \begin{bmatrix} 0.9725 - j0.1468 & -0.0067 - j0.0194 & -0.0086 - j0.0211 \\ -0.0074 - j0.0213 & 0.9718 - j0.1453 & -0.0074 - j0.0213 \\ -0.0086 - j0.0211 & -0.0067 - j0.0194 & 0.9725 - j0.1468 \end{bmatrix},$$

$$\underset{\sim}{V}(x) = e^{-[\gamma_v]x}\underset{\sim}{E}(0) = \begin{bmatrix} 0.9768 - j0.1416 \\ -0.2970 - j0.9567 \\ -0.0508 - j0.9721 \end{bmatrix} \quad (x = 100).$$

a) Calculation in modal domain.

$$P = \begin{bmatrix} 0.5757 - j0.0164 & -0.7071 + j0.0012 & 0.3856 + j0.0120 \\ 0.5800 - j0.0158 & 0.0000 + j0.0000 & -0.8377 - j0.0244 \\ 0.5757 - j0.0164 & 0.7071 - j0.0012 & 0.3856 + j0.0120 \end{bmatrix},$$

$$\overline{A} = P^{-1}AP,$$

$$\overline{A} = \begin{bmatrix} -0.3681 + j0.0967 & 0 & 0 \\ 0 & -0.1612 + j0.0279 & 0 \\ 0 & 0 & -0.1615 + j0.0344 \end{bmatrix}.$$

Matrix of propagation constants in modal domain.

$$[\bar{\gamma}] = \begin{bmatrix} 0.2498 + j1.9348 & 0 & 0 \\ 0 & 0.1095 + j1.2744 & 0 \\ 0 & 0 & 0.1347 + j1.2799 \end{bmatrix} \times 10^{-3}.$$

Exponential matrix in modal domain at $x = 100$ km:

$$e^{-[\bar{\gamma}]x} = \begin{bmatrix} 0.9571 - j0.1875 & 0 & 0 \\ 0 & 0.9811 - j0.1257 & 0 \\ 0 & 0 & 0.9786 - j0.1257 \end{bmatrix}.$$

Transforming to phase domain, the propagation matrix is given by:

$$Pe^{-[\bar{\gamma}]x}P^{-1} = \begin{bmatrix} 0.9725 - j0.1468 & -0.0067 - j0.0194 & -0.0086 - j0.0211 \\ -0.0074 - j0.0213 & 0.9718 - j0.1453 & -0.0074 - j0.0213 \\ -0.0086 - j0.0211 & -0.0067 - j0.0194 & 0.9725 - j0.1468 \end{bmatrix}.$$

This result is identical to the previous calculation in phase domain.
Modal voltages at the origin, $x = 0$ km:

$$\underset{\sim}{\overline{E}}(0) = P^{-1}\underset{\sim}{E}(0) = \begin{bmatrix} 0.3951 + j0.0570 \\ -0.8310 + j0.6950 \\ 0.5165 + j1.1926 \end{bmatrix}.$$

Modal voltages at $x = 100$ km:

$$\underset{\sim}{\overline{V}}(x) = e^{-[\bar{\gamma}_v]x}\underset{\sim}{\overline{E}}(0) = \begin{bmatrix} 0.3888 - j0.0196 \\ -0.7279 + j0.7863 \\ 0.6554 + j1.1021 \end{bmatrix}.$$

Note the higher attenuation in mode zero at 100 km.
Calculating the voltages in phase domain:

$$\underset{\sim}{V}(x) = P\underset{\sim}{\overline{V}}(x) = \begin{bmatrix} 0.9768 - j0.1416 \\ -0.2970 - j0.9567 \\ -0.0508 - j0.9721 \end{bmatrix}.$$

The result is the same as item *a*.

3.4.3 Basic Models in the Modal Domain

This topic analyses models of the transmission line used in electromagnetic transient programs that consider the specific time delays for each mode due to the different propagation speeds of the modes.

Simplified but very efficient models of transmission lines represent losses as equivalent lumped resistances at a few points along the line, calculated at a given frequency (see Chapter 4 and reference [6]). The models address the traveling waves as lossless in the segments between these equivalent resistances and assume different time delays in the aerial and ground modes, according to the calculation of parameters for a lossy line.

- Non-Transposed Lines

The traveling waves in the lossless segments of a nontransposed three-phase transmission line are calculated using propagation modes. Each transmission line has its own modal transformation matrix P to obtain the parameters for the propagation modes. Equation (3.101) exemplifies a three-phase line, but the number of phases and modes can be higher than three.

$$\overline{L_f} = \begin{bmatrix} l'_0 & 0 & 0 \\ 0 & l'_1 & 0 \\ 0 & 0 & l'_2 \end{bmatrix}, \quad \overline{C_f} = \begin{bmatrix} c'_0 & 0 & 0 \\ 0 & c'_1 & 0 \\ 0 & 0 & c'_2 \end{bmatrix}. \tag{3.101}$$

In the modal equations, each mode is evaluated independently of the others, as in (3.89) and (3.90), and it is possible to apply the same results to the single-phase line presented in Chapter 2.

- Transposed Lines

For transposed transmission lines, a single modal transformation matrix for voltages and currents is used for all transmission lines in the network, regardless of their design. Note that the model assumes perfectly transposed lines, where transposition is idealized as occurring in very small segments along the line, rather than only at designated points. If this assumption is not accurate enough for the desired calculation, non-transposed line segments should be used with auxiliary transposition at designated points.

For a three-phase transmission line, the following matrices can be used for modal transformation, as in Eq. (3.22) with $n = 3$:

$$P = \begin{bmatrix} 1 & 1 & 1 \\ 1 & -2 & 1 \\ 1 & 1 & -2 \end{bmatrix}, P^{-1} = \frac{1}{3}\begin{bmatrix} 1 & 1 & 1 \\ 1 & -1 & 0 \\ 1 & 0 & -1 \end{bmatrix}. \tag{3.102}$$

The parameters are given in per unit length of the line, and matrix elements in phase domain are given by:

L': inductance of the phase (self-inductance).
L'_m: mutual inductance between phases.
C'_t: capacitance to ground.
C'_m: coupling capacitance between phases,

thus:

$$L_f = \begin{bmatrix} l' & l'_m & l'_m \\ l'_m & l' & l'_m \\ l'_m & l'_m & l' \end{bmatrix}, \quad C_f = \begin{bmatrix} c' & c'_m & c'_m \\ c'_m & c' & c'_m \\ c'_m & c'_m & c' \end{bmatrix}.$$

where:

$l' = L'$,
$l'_m = L'_m$,
$c' = C'_t + 2C'_m$,
$c'_m = -C'_m$.

The elements of modal matrices in (3.101) are given by:

$l'_0 = L' + 2L'_m$: inductance of mode 0 (zero sequence inductance).
$l'_1 = L' - L'_m$: inductance of mode 1 (positive sequence inductance).
$l'_2 = l'_1$: inductance of mode 2 (negative sequence inductance).
$c'_0 = C'_t$: capacitance of mode 0 (zero sequence capacitance).
$c'_1 = C'_t + 3C'_m$: capacitance of mode 1 (positive sequence capacitance).
$c'_2 = c'_1$: capacitance of mode 2 (negative sequence capacitance).

The modal propagation speeds are:

$$v_0 = \frac{1}{\sqrt{l'_0 c'_0}}, v_1 = \frac{1}{\sqrt{l'_1 c'_1}}, v_2 = v_1 \tag{3.103}$$

The line segments contain the three propagation modes $j = (0, 1, 2)$. Modes 1 and 2 have identical characteristics which differ from those of mode 0.

3.4.4 Traveling Waves in Discontinuities

The same analysis from item 3.3.4 is repeated below in the modal domain, for the case of two connected transmission lines. The matrices of surge impedances Z_1 and Z_2, and matrices for modal transformation P_1 and P_2, are considered for lines 1 and 2, respectively.

The equation of voltage continuity (2.47) in Chapter 2, extended to multiphase lines,

$$\underset{\sim}{v_i} + \underset{\sim}{v_r} = \underset{\sim}{v_t},$$

can be rewritten to modal domain,

$$P_1 \underset{\sim}{\overline{v}_i} + P_1 \underset{\sim}{\overline{v}_r} = P_2 \underset{\sim}{\overline{v}_t}$$

which premultiplied by $(P_1)^{-1}$ gives:

$$\underset{\sim i}{\overline{v}} + \underset{\sim r}{\overline{v}} = (P_1)^{-1} P_2 \underset{\sim t}{\overline{v}}. \tag{3.104}$$

It is observed that a specific propagation mode in line 1 interacts with other propagation modes in line 2, due to the fact that:

$$(P_1)^{-1} P_2 \neq I_d, \text{(where } I_d \text{ is the identity matrix)},$$

and with the interaction between modes, one of the benefits of modal decoupling is lost.

However, if the transformation matrices are the same for lines 1 and 2 and are simply denoted by P, then the product $P^{-1}P$ equals the identity matrix. The equation now shows the modes independently on both sides, which is the case of transposed lines and simplifies the analysis.

$$\underset{\sim i}{\overline{v}} + \underset{\sim r}{\overline{v}} = \underset{\sim t}{\overline{v}} \tag{3.105}$$

Assuming transposed lines, applying the same modal transformation matrix to the incident, reflected, and refracted voltage phase vectors ($\underset{\sim}{v}_t = \Gamma_t \underset{\sim}{v}_i$), from (3.69) and (3.70) the matrices corresponding to refraction and reflection coefficients are determined in the modal domain:

$$P\underset{\sim}{\overline{v}}_t = \Gamma_t P \underset{\sim}{\overline{v}}_i,$$

thus:

$$\overline{\Gamma}_t = P^{-1} \Gamma_t P. \tag{3.106}$$

Substituting (3.70) in (3.106) and inserting PP^{-1} into the matrix product:

$$\overline{\Gamma}_t = 2P^{-1} Z_2 P P^{-1} (Z_1 + Z_2)^{-1} P,$$

it is concluded that:

$$\overline{\Gamma}_t = 2\overline{Z}_2 \left(\overline{Z}_1 + \overline{Z}_2\right)^{-1}. \tag{3.107}$$

Matrix $\overline{\Gamma}_t$ is diagonal, and no interaction between modes occurs. The same is true for the matrix of reflection coefficients.

$$\overline{\Gamma}_r = \left(\overline{Z}_2 - \overline{Z}_1\right)\left(\overline{Z}_1 + \overline{Z}_2\right)^{-1}, \tag{3.108}$$

Thus, if the lines are transposed, a significant simplification in the calculation is achieved, as the same transformation matrix can be assumed for every transmission line.

For non-transposed lines, the main advantage of modal models in calculating modal propagation velocities remains, but they offer no benefit for calculations at discontinuities. It is also worth mentioning that in EMT programs, the interaction of traveling waves at discontinuity points is handled somewhat differently, without using the matrices Γ_t and Γ_r. Instead, a transformation from the modal domain to the phase domain is performed to interact with the network, and then the results are recalculated into modal domain to analyze the traveling waves along the transmission line.

The following examples, in reference to items 3.3 and 3.4, show the application of modal transformations to transmission lines and their correspondence to the calculation in phase domain.

Example 3.5 shows an analysis of the surge impedances in the modal and phase domain for two cables in a transmission line. Detailed calculations of the eigenvalues, eigenvectors, and modal transformations are presented. Example 3.6 illustrates the calculation of traveling waves in the time domain for three-phase energization of a transmission line.

Example 3.5 Obtain the modal parameters for a transmission line, assuming the shield wire and just one phase conductor (Figure 3.5).

Data

Conductor 1 represents the shield wire and conductor 2 represents the bundle of cables of phase a. The model assumes a non-transposed and lossless transmission line, and phases b and c are ignored.

$$h_1 = 17\text{ m},\ h_2 = 13\text{ m},\ d = 5.0\text{ m},\ d_{12} = 6.4\text{ m},$$
$$D_{12} = 30.41\text{ m},\ r_1 = 1.13\text{ cm},\ r_2 = 13.32\text{ cm}.$$

h_1 and h_2 are the average distances from the conductors to the ground.
r_2 is the equivalent radius of the bundle of conductors of phase a.

Solution

This model can be used to calculate the induced voltage on a phase conductor from a lightning strike to the shield wire.

The surge impedance matrix in the phase domain is built with the aid of (3.67):

$$Z = 60 \begin{bmatrix} \ln\left(\dfrac{2h_1}{r_1}\right) & \ln\left(\dfrac{D_{12}}{d_{12}}\right) \\ \ln\left(\dfrac{D_{12}}{d_{12}}\right) & \ln\left(\dfrac{2h_2}{r_2}\right) \end{bmatrix}.$$

In this example, conductors 1 and 2 are treated as a two-phase, nontransposed, and lossless transmission line.

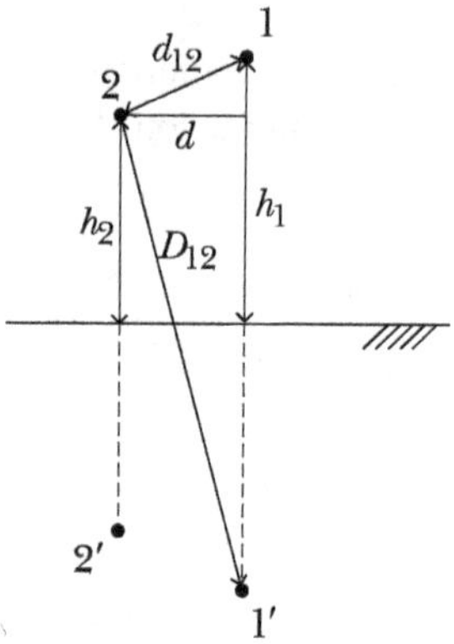

Figure 3.5 Shield wire, phase conductor, and images.

The following surge impedance matrix is calculated:

$$Z = \begin{bmatrix} 480.6 & 93.5 \\ 93.5 & 316.4 \end{bmatrix}.$$

- Calculation of the eigenvalues.

$$\det(\lambda I_d - Z) = 0,$$

where I_d is the identity matrix.

$$\det\begin{bmatrix} \lambda - 480.6 & -93.5 \\ -93.5 & \lambda - 316.4 \end{bmatrix} = 0,$$

$$(\lambda - 480.6)(\lambda - 316.4) - 93.5^2 = 0,$$

$$\lambda_1 = 522.93 = Z_0$$

$$\lambda_2 = 274.07 = Z_1 : \text{modal surge impedances.}$$

- Calculation of the eigenvectors.

From equation (4.9), the homogeneous system of equations needs to be solved,

$$(\lambda_i I_d - A)\underset{\sim}{p_i} = \underset{\sim}{0},$$

and to arrive at a non-trivial solution $\left(\underset{\sim}{p_i} \neq 0\right)$, it is necessary and sufficient that the determinant of $(\lambda_i I_d - A)$ is equal to zero, which is verified because λ_i is an eigenvalue of the system, as in (3.8).

In summary, the following steps are performed: (a) identification of dependent equations and reduction of the system through rank analysis; (b) assignment of a value to the free variable for one of the components of the eigenvector and expression of the remaining components in terms of that free variable.

With respect to the eigenvectors of matrix Z, for λ_1,

$$(\lambda_1 I_d - Z)\underset{\sim}{x} = 0,$$

$$\begin{bmatrix} 42.33 & -93.50 \\ -93.50 & 206.53 \end{bmatrix} \times \begin{bmatrix} x_1 \\ y_1 \end{bmatrix} = 0.$$

The two lines of this matrix are linearly dependent, and by taking the first one and assuming $x_1 = 1$:

$$x_1 = 2.2088 y_1 \Rightarrow \underset{\sim}{p_1} = \begin{bmatrix} 1 \\ 0.4527 \end{bmatrix}.$$

Using the same procedure for λ_2 and assuming $y_2 = 1$:

$$\begin{bmatrix} -206.53 & -93.50 \\ -93.50 & -42.33 \end{bmatrix} \times \begin{bmatrix} x_2 \\ y_2 \end{bmatrix} = \begin{bmatrix} 0 \\ 0 \end{bmatrix},$$

$$x_2 = -0.4527 y_2 \Rightarrow \underset{\sim}{p_2} = \begin{bmatrix} -0.4527 \\ 1 \end{bmatrix}.$$

Normalizing the eigenvectors:

$$\left|\underset{\sim}{p_1}\right| = \left|\underset{\sim}{p_2}\right| = \sqrt{1^2 + 0.4527^2},$$

$$P = \begin{bmatrix} 0.9110 & -0.4124 \\ 0.4124 & 0.9110 \end{bmatrix}, P = \left[\underset{\sim}{p_1} \ \underset{\sim}{p_2}\right].$$

It can also be shown that:

$$\overline{Z} = P^{-1} Z P$$

$$\overline{Z} = \begin{bmatrix} 522.93 & 0 \\ 0 & 274.07 \end{bmatrix},$$

where $\overline{Z}$ is the matrix of modal surge impedances, for mode 0 and mode 1.

Example 3.6 Analyze the modal traveling waves in the energization of a three-phase transmission line with the sequential closing of the breaker poles, in Figure 3.6a.

Data

- Three-phase generator as in Figure 3.6b:

$$e_a(t) = \cos(\omega t),$$
$$e_b(t) = \cos(\omega t - 120°),$$
$$e_c(t) = \cos(\omega t + 120°).$$

  Sinusoidal voltages of unitary amplitude are applied; however, it is assumed that during the brief analytical period, the generator voltages remain constant.

- Time delays:
  Assuming that mode 0 has a different propagation velocity from modes 1 and 2, the following time delays are proposed:
  mode 1, τ_1; mode 2, $\tau_2 = \tau_1$; mode 0, $\tau_0 = 1.5\tau_1$.

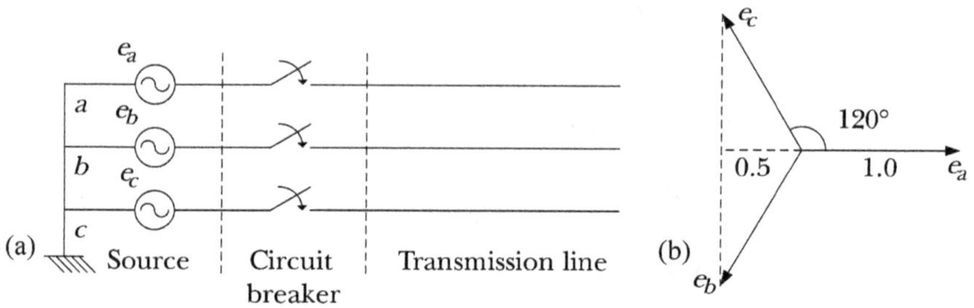

Figure 3.6 Energization of the transmission line. (a) Closing of the circuit-breaker poles. (b) Three-phase generator voltages.

- Surge impedances in modal domain:

$$Z_0 = 400\,\Omega$$

$$Z_1 = Z_2 = 200\,\Omega.$$

- Surge impedances in phase domain:
- $Z_m = 66.67\,\Omega$, mutual surge impedance,
- $Z = 400 - 2 \times 66.67 = 266.67\,\Omega$, phase surge impedance (self-impedance).
- Circuit breaker:

  The circuit breaker is represented by switches, whose three-phase closing is analyzed under two conditions:

Phase switch	Condition I $t = 0$	condition II $t = \tau_1$
a	Closed	Closed
b	Open	Closed
c	Open	Closed

Under condition I, for $t = 0$, the switch for phase a is closed.

Under condition II, for $t = \tau_1$, the switches for phases b and c are simultaneously closed.

Under conditions I and II, the generator voltages are assumed to remain constant, keeping their initial value at $t = 0$.

$$e_a = 1.0, e_b = -0.5 \text{ and } e_c = -0.5$$

Solution

Balanced matrices are assumed for the network, with the transformation matrices in (3.102):

(a) Condition I

Closing of the phase a switch.

(a.1) Calculation of the voltages applied to the line.
To calculate the phase voltages applied to the line, since only phase a is connected and the currents in phases b and c are zero, one can write:

$$\begin{bmatrix} 1 \\ e_b \\ e_c \end{bmatrix} = \begin{bmatrix} Z & Z_m & Z_m \\ Z_m & Z & Z_m \\ Z_m & Z_m & Z \end{bmatrix} \times \begin{bmatrix} i_a \\ 0 \\ 0 \end{bmatrix}.$$

It can be shown that:

$$i_a = \frac{1}{Z}, e_a = 1, e_b = \frac{Z_m}{Z}, e_c = \frac{Z_m}{Z}.$$

Vector of voltages applied to the line:

$$\underset{\sim}{e}_{g1} = \begin{bmatrix} 1 \\ 0.25 \\ 0.25 \end{bmatrix}.$$

The vector of modal voltages is calculated.

$$\underset{\sim}{\overline{e}}_{g1} = \frac{1}{3} \begin{bmatrix} 1 & 1 & 1 \\ 1 & -1 & 0 \\ 1 & 0 & -1 \end{bmatrix} \times \begin{bmatrix} 1 \\ 0.25 \\ 0.25 \end{bmatrix} = \begin{bmatrix} 0.50 \\ 0.25 \\ 0.25 \end{bmatrix}.$$

(a.2) Calculation of the matrices $\overline{\Gamma}_{ri}$ of reflection coefficients in modal domain at the sending end of the line (point i), in condition I.
The calculation of Γ_{ri} is performed in phase domain at the sending end by applying (3.70):

$$\Gamma_{ri} = (Z_2 - Z_1)(Z_1 + Z_2)^{-1},$$

where Z_2 is the impedance matrix of the generator, assuming high resistances in phases b and c simulating open phases.

$$Z_2 = \begin{bmatrix} 0 & 0 & 0 \\ 0 & r & 0 \\ 0 & 0 & r \end{bmatrix} : \text{side 2 – generator}.$$

$$Z_1 = \begin{bmatrix} Z & Z_m & Z_m \\ Z_m & Z & Z_m \\ Z_m & Z_m & Z \end{bmatrix} \text{ : side 1 – transmission line.}$$

$$Z_1 + Z_2 = \begin{bmatrix} Z & Z_m & Z_m \\ Z_m & r + Z & Z_m \\ Z_m & Z_m & r + Z \end{bmatrix}.$$

Performing the calculation by adopting, for instance, a resistance of $r = 10^6\ \Omega$, the following is obtained:

$$\Gamma_{ri} = \begin{bmatrix} -1 & 0 & 0 \\ -0.5 & 1 & 0 \\ -0.5 & 0 & 1 \end{bmatrix}.$$

The reflection coefficient matrix in phase domain and subsequent transformation matrices for modal domain can be calculated. However, in this example, the reflection coefficient matrix is calculated in modal domain:

$$\overline{\Gamma}_{ri} = P^{-1}\Gamma_{ri}P.$$

Substituting the numerical values:

$$\overline{\Gamma}_{ri} = \begin{bmatrix} 0 & -1 & -1 \\ -0.5 & 0.5 & -0.5 \\ -0.5 & -0.5 & -0.5 \end{bmatrix}.$$

As matrix Z_2 is not balanced, elements outside the diagonal are found in $\overline{\Gamma}_{ri}$.

(a.3) Calculation of the matrix of reflection coefficients $\overline{\Gamma}_{rf}$ on the receiving end, side 1, under condition I.

At the line terminal, the resistances on side 2 are assumed very high for the open line, $r \rightarrow \infty$:

$$Z_2 = \overline{Z}_2 = \begin{bmatrix} r & 0 & 0 \\ 0 & r & 0 \\ 0 & 0 & r \end{bmatrix} \text{, side 2 – open – ended line :}$$

and also:

$$\overline{Z}_1 = \begin{bmatrix} Z_0 & 0 & 0 \\ 0 & Z_1 & 0 \\ 0 & 0 & Z_2 \end{bmatrix} \text{, side 1 – transmission line.}$$

In this case, where there are diagonal matrices, the phase and modal components are coincident. Hence, the matrix of the modal reflection coefficients is calculated using Eq. (3.108).

$$\overline{\Gamma}_{rf} = \left(\overline{Z}_2 - \overline{Z}_1\right)\left(\overline{Z}_1 + \overline{Z}_2\right)^{-1}$$

Calculating the products and the limit as $r \to \infty$:

$$\overline{\Gamma}_{rf} = I_d,$$

where I_d is the identity matrix.

(b) Condition II.

Closing of switches b and c (all switches closed).

(b.1) Calculation of the voltage applied by the generator to the line.

With the three phases of the ideal generator closed, Z_g is composed of null elements, thus:

$$\left(I_d + Z_g Y\right)^{-1} = I_d.$$

The phase voltages applied by the generator are:

$$\underset{\sim}{e}_{g2} = \begin{bmatrix} 1 \\ -0.5 \\ -0.5 \end{bmatrix}.$$

The vector of modal voltages for $t = \tau_1$ is calculated when switches b and c are closed.

$$\underset{\sim}{\overline{e}}_{g2} = \frac{1}{3}\begin{bmatrix} 1 & 1 & 1 \\ 1 & -1 & 0 \\ 1 & 0 & -1 \end{bmatrix} \times \begin{bmatrix} 1 \\ -0.5 \\ -0.5 \end{bmatrix}, \underset{\sim}{\overline{e}}_{g2} = \begin{bmatrix} 0 \\ 0.5 \\ 0.5 \end{bmatrix}.$$

For condition II, it is necessary to calculate the vector of applied voltage variations in the transmission line, since there were already voltages applied to the sending end given by $\underset{\sim}{\overline{e}}_{g1}$.

$$\Delta \underset{\sim}{\overline{e}}_{g} = \begin{bmatrix} 0 \\ 0.5 \\ 0.5 \end{bmatrix} - \begin{bmatrix} 0.5 \\ 0.25 \\ 0.25 \end{bmatrix} = \begin{bmatrix} -0.5 \\ 0.25 \\ 0.25 \end{bmatrix}.$$

(b.2) Calculation of the matrix of reflection coefficients at the sending end of the transmission line.

Since Z_g is null, by applying (3.108), the matrix of modal reflection coefficients is obtained, given by $-I_d$, where I_d is the identity matrix.

(b.3) Calculation of the matrix of reflection coefficients at the receiving end

The conditions at the receiving end do not change and the identity matrix remains.

(c) Numerical results.

(c.1) Closing of phase a switch.

The modal voltages $\underset{\sim}{\overline{e}}_{g1}$ are applied to the transmission line, according to item a.1:

$$\underset{\sim}{\overline{e}}_{g1} = \begin{bmatrix} 0.5 \\ 0.25 \\ 0.25 \end{bmatrix}.$$

Figure 3.7 presents the traveling waves calculated in modal domain, according to the lattice diagram.

- For $t = \tau_1$ the arrival of modal components v_1 and v_2 at the receiving end is observed, with the same reflected waves, since $\overline{\Gamma}_{rf}$ is the identity matrix.
- For $t = \tau_{0,}$ the arrival of mode v_0 at the receiving and its reflection are observed.
- For $t = 2\tau_1$ the modes v_1 and v_2 return to the sending end and the reflection of mode v_0 is observed.

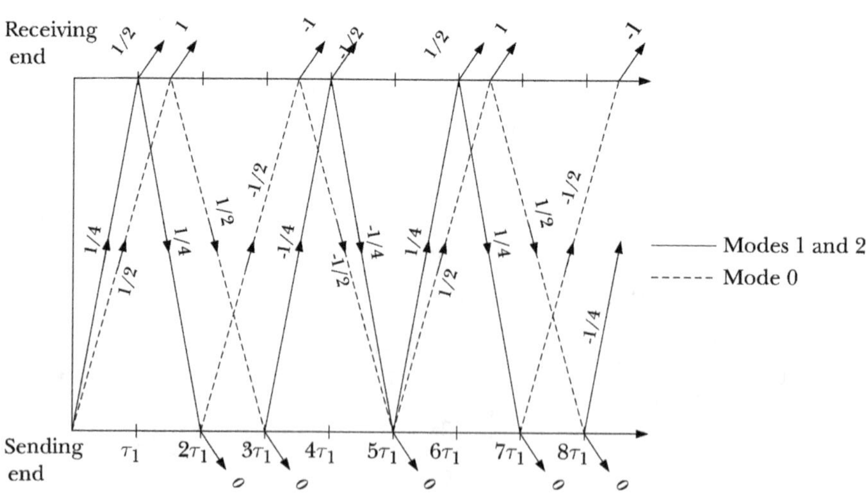

Figure 3.7 Mode propagation with the switch of one phase closed.

$$\underset{\sim}{v}_r = \begin{bmatrix} 0 & -1 & -1 \\ -0.5 & 0.5 & -0.5 \\ -0.5 & -0.5 & 0.5 \end{bmatrix} \times \begin{bmatrix} 0 \\ 0.25 \\ 0.25 \end{bmatrix} = \begin{bmatrix} -0.5 \\ 0 \\ 0 \end{bmatrix}$$

For $t = 2\tau_0$ the arrival of v_0 at the sending end of the line, and the reflection of the coordinates in modes 1 and 2 are observed:

$$\underset{\sim}{v}_r = \begin{bmatrix} 0 & -1 & -1 \\ -0.5 & 0.5 & -0.5 \\ -0.5 & -0.5 & 0.5 \end{bmatrix} \times \begin{bmatrix} 0.5 \\ 0 \\ 0 \end{bmatrix} = \begin{bmatrix} 0 \\ -0.25 \\ -0.25 \end{bmatrix}$$

An alternation in the modes for the reflected waves at the sending end of the transmission line can be observed, with mode zero creating reflected waves in modes 1 and 2 and modes 1 and 2 creating reflected waves in mode zero.

The voltage in phase a is:

$$v_a = v_0 + v_1 + v_2.$$

The voltages in phases b and c can also be calculated from:

$$v_b = v_0 - 2v_1 + v_2,$$

$$v_c = v_0 + v_1 - 2v_2,$$

$$\text{as } v_1 = v_2, \text{thus :}$$

$$v_b = v_c = v_0 - v_1,$$

in other words, as the line is perfectly transposed, when phase a is energized, equal voltages are induced in phases b and c.

Figure 3.8 shows the phase components at the end of the line.

The phase voltages at the sending end of the transmission line are obtained with the modal components that return to this terminal, according to Figure 3.7. These modal components are presented in the following table:

Voltage \ Time	$2\tau_1$	$3\tau_1$	$5\tau_1$	$7\tau_1$
v_0	0	0.5000	−0.5000	0
v_1	0.2500	0	−0.2500	0.2500
v_2	0.2500	0	−0.2500	0.2500

The transmitted values are calculated by transforming modal to phase components by means of the matrix P and then multiplying the result by the matrix of

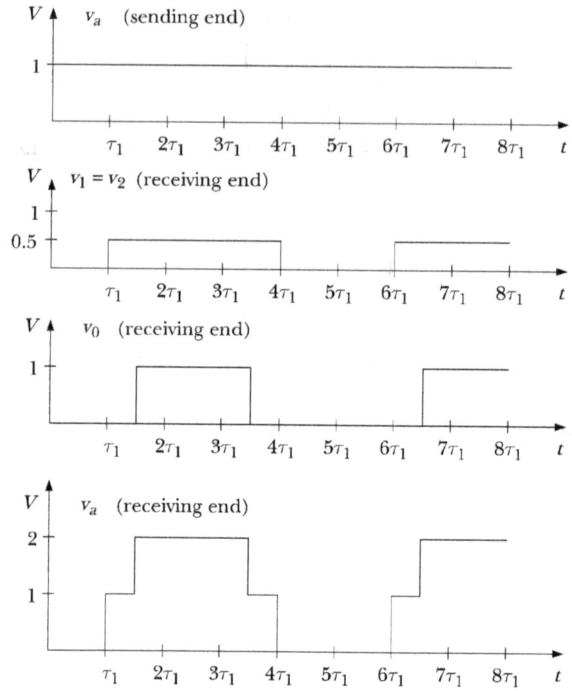

Figure 3.8 Modal and phase components at the receiving end, with one phase switch closed.

transmission coefficients Γ_{ti}. These two operations can be summarized as the product of the matrices $T = \Gamma_{ti}P$. Matrix T is given by:

$$T = \begin{bmatrix} 0 & 0 & 0 \\ 1.5 & -4.5 & 1.5 \\ 1.5 & 1.5 & -4.5 \end{bmatrix}.$$

Voltages calculated by application of matrix T to the previous modal variables are presented in the following table:

Voltage \ Time	$2\tau_1$	$3\tau_1$	$5\tau_1$	$7\tau_1$
v_a	0	0	0	0
v_b	−0.7500	0.7500	0	−0.7500
v_c	−0.7500	0.7500	0	−0.7500

The phase voltages at the sending end of the transmission line are calculated by superposing the voltages previously applied to the line in item a.1.

Voltage \ Time	0	$2\tau_1$	$3\tau_1$	$5\tau_1$	$7\tau_1$
v_a	1.0000	1.0000	1.0000	1.0000	1.0000
v_b	0.2500	−0.5000	1.0000	0.2500	−0.5000
v_c	0.2500	−0.5000	1.0000	0.2500	−0.5000

(c.2) Closing of b and c switches.

The connection of the switches in phases b and c is examined.

Now all modes are uncoupled and therefore for one incident mode, reflected and transmitted waves refer to the same mode. The variations in modal voltages calculated in item (b.1) are applied to the sending end.

Figures 3.9 and 3.10 present the modal variables according to the lattice diagram. For $0 < t < \tau_1$, matrices for condition I are applied. For $t > \tau_1$, matrices for condition I are replaced by matrices for condition II.

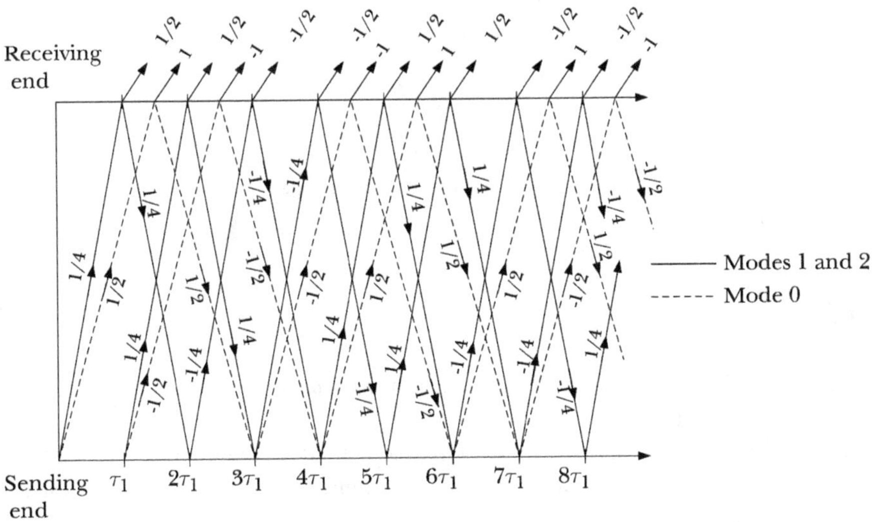

Figure 3.9 All switches closed – traveling waves in modal components.

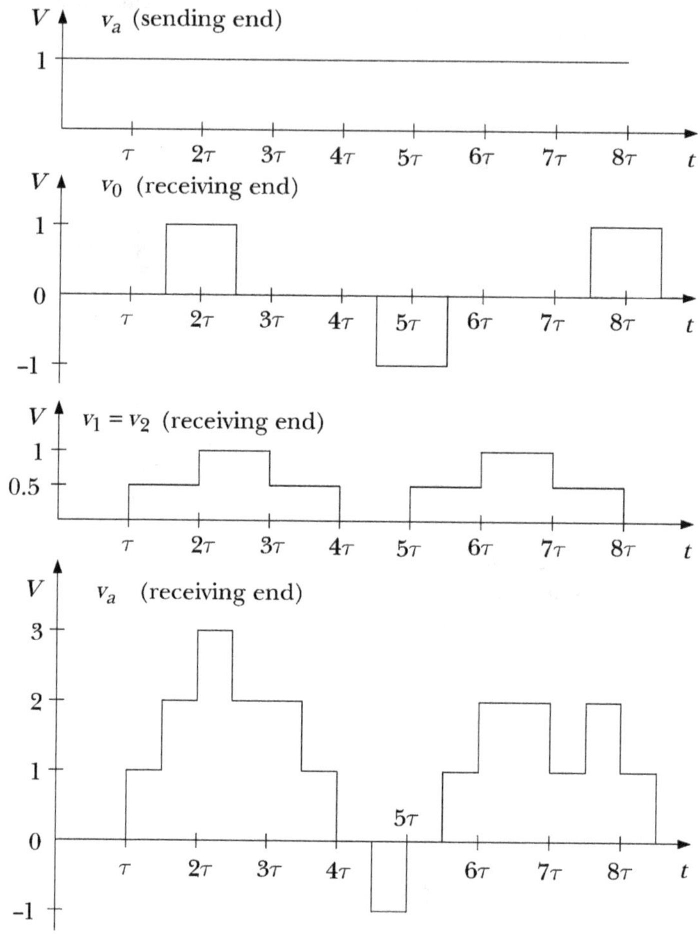

Figure 3.10 All switches closed – modal and phase voltages.

References

1 Bewley, L. V. (1963). *Travelling waves on transmission systems* (2nd ed.). New York: Dover.

2 Wedepohl, L. M. (1963). Application of matrix methods to the solution of travelling-wave phenomena in polyphase systems. *Proceedings of the Institution of Electrical Engineers*, 110(12), 2200–2212.

3 Hedman, D. E. (1965). Propagation on overhead transmission lines. I: Theory of modal analysis. *IEEE Transactions on Power Apparatus and Systems*, 84(3), 200–205.

4 Hedman, D. E. (1965). Propagation on overhead transmission lines. II: Earth-conduction effects and practical results. *IEEE Transactions on Power Apparatus and Systems*, 84(3), 205–211.

5 Dommel, H. W. (1986). *Electromagnetic transients program reference manual: (EMTP) theory book*. Portland: Bonneville Power Administration.

6 Leuven EMTP Center. *ATP: Alternative Transients Program: Rule book*. (1987). Heverlee (BE): Leuven EMTP Center.

7 *MATLAB: high-performance numeric computation and visualization software: Reference guide*. (1992). Natick (MA): Math Works.

8 Kailath, T. (1980). *Linear systems*. Englewood Cliffs (NJ): Prentice-Hall.

4

Numerical Solution of Electromagnetic Transients

4.1 Introduction

This chapter presents a numerical solution for electromagnetic transients in power systems using the trapezoidal integration method. This technique was introduced by H. Dommel [1–4], and allowed the development of efficient programs for the calculation of electromagnetic transients by means of nodal analysis.

Nowadays, one can find different EMT programs, but the original development of the EMTP program was sponsored by Bonneville Power Administration (BPA). These programs today contain several models of power system equipment, such as generators, transformers, transmission lines, lightning arresters, circuit breakers, power electronic equipment, and the possibility of simulation of control systems. These models accurately represent the complexity of transient phenomena in power systems.

This chapter is not intended to reproduce in detail all the capabilities of EMT programs, but only to introduce the theoretical basis for understanding certain elementary models.

Basic elements, such as resistors, inductors, and capacitors, are treated as lumped parameters, and the transmission line, as distributed parameters.

Initially, the models are presented on a single-phase basis, for later extension to three-phase cases, including interactions between phases. Some elements of the treatment of nonlinearities are also presented.

Understanding Electromagnetic Transients in Power Systems, First Edition. Luiz Cera Zanetta Jr.

Published 2025 by John Wiley & Sons, Inc.

4.2 Single-Phase Models

4.2.1 Inductance Model

An inductance with a branch current $i_{km}(t)$, in Figure 4.1a, having terminals at nodes k and m, yields the following first-order differential equation:

$$v_k(t) - v_m(t) = L\frac{d}{dt}i_{km}(t). \tag{4.1}$$

For the numerical integration, it is necessary to perform a temporal discretization, at regular Δt intervals, thus:

$$i_{km}(t) = i_{km}(t-\Delta t) + \frac{1}{L}\int_{t-\Delta t}^{t}[v_k(t) - v_m(t)]dt. \tag{4.2}$$

The graphic interpretation of the integral function is presented in Figure 4.2:

With the application of the trapezoidal method of integration, the following is obtained:

$$i_{km}(t) = i_{km}(t-\Delta t) + \frac{\Delta t}{2L}([v_k(t) - v_m(t)] + [v_k(t-\Delta t) - v_m(t-\Delta t)]). \tag{4.3}$$

In (4.3), note that it is possible to obtain the current at time t as a function of previously calculated values at time $(t - \Delta t)$, and therefore the following expression can be written:

$$i_{km}(t) = \frac{\Delta t}{2L}[v_k(t) - v_m(t)] + I_{km}(t-\Delta t), \tag{4.4}$$

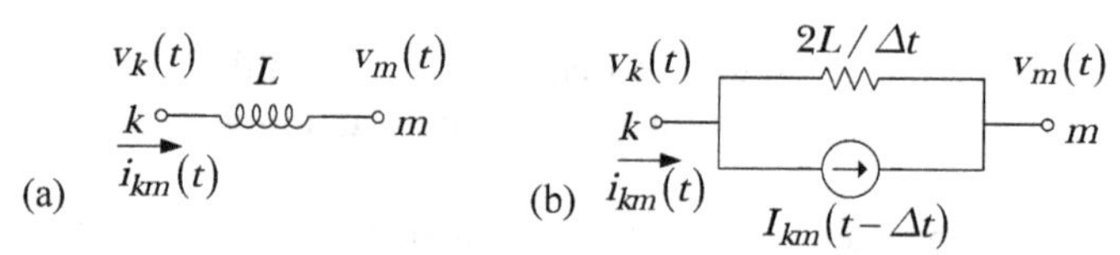

Figure 4.1 Inductance model. (a) Inductance. (b) Equivalent model.

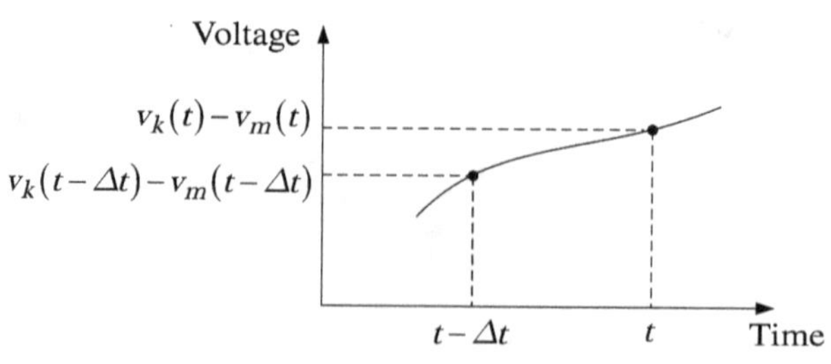

Figure 4.2 Integration in the Δt interval.

where:

$$I_{km}(t-\Delta t) = i_{km}(t-\Delta t) + \frac{\Delta t}{2L}[v_k(t-\Delta t) - v_m(t-\Delta t)]. \tag{4.5}$$

From Eq. (4.4), an equivalent model for the inductor is established and its transient behavior is calculated by the trapezoidal rule, Figure 4.1b.

The model consists of a resistance of value $R = 2L/\Delta t$, whose corresponding conductance is $G = \Delta t/2L$, and a current generator $I_{km}(t-\Delta t)$ that contains all the past information needed to update voltages and currents at time t. Thus, this current generator is related to past information and is also designated as a history term in the equations.

4.2.2 Capacitance Model

The capacitance model is built in a manner like the inductance model (Figure 4.3a).

$$v_k(t) - v_m(t) = v_k(t-\Delta t) - v_m(t-\Delta t) + \frac{1}{C}\int_{t-\Delta t}^{t} i_{km}(t)dt. \tag{4.6}$$

Applying the trapezoidal method of integration:

$$v_k(t) - v_m(t) = v_k(t-\Delta t) - v_m(t-\Delta t) + \frac{\Delta t}{2C}[i_{km}(t) + i_{km}(t-\Delta t)], \tag{4.7}$$

And the terms at time t and $t-\Delta t$ can be separated:

$$i_{km}(t) = \frac{2C}{\Delta t}[v_k(t) - v_m(t)] + I_{km}(t-\Delta t), \tag{4.8}$$

where:

$$I_{km}(t-\Delta t) = -i_{km}(t-\Delta t) - \frac{2C}{\Delta t}[v_k(t-\Delta t) - v_m(t-\Delta t)]. \tag{4.9}$$

The equivalent circuit for (4.8) may be seen in Figure 4.3b.

Again, the branch current $i_{km}(t)$ is composed of a current flowing through a resistance $R = \Delta t/2C$, whose corresponding conductance is $G = 2C/\Delta t$, in parallel with a current generator $I_{km}(t-\Delta t)$ that contains the past information.

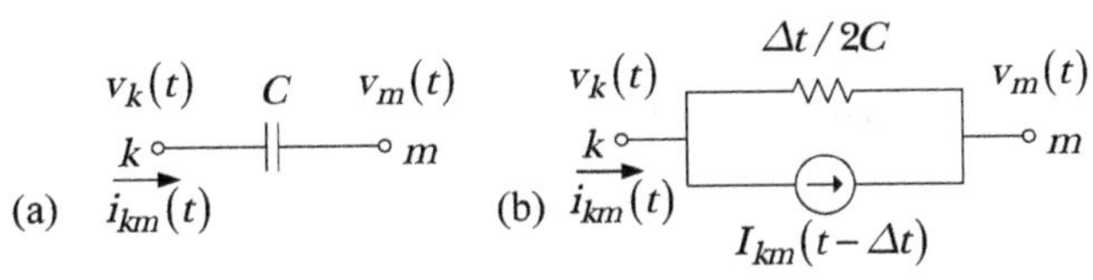

Figure 4.3 Capacitance model. (a) Capacitance. (b) Equivalent model.

4.2.3 Resistance Model

The relationship between the voltage and current in the resistance does not involve differential equations, and therefore, it does not require the application of the trapezoidal rule; Eq. (4.10) follows immediately:

$$i_{km}(t) = \frac{v_k(t) - v_m(t)}{R}. \tag{4.10}$$

4.2.4 RL Circuit

The equation for the k–m branch of the RL circuit is written as (Figure 4.4):

$$v_k(t) - v_m(t) = Ri_{km}(t) + L\frac{d}{dt}i_{km}(t), \tag{4.11}$$

integrating:

$$\int_{t-\Delta t}^{t} [v_k(t) - v_m(t)]dt = R\int_{t-\Delta t}^{t} i_{km}(t)dt + L\int_{t-\Delta t}^{t} \frac{d}{dt}i_{km}(t)dt, \tag{4.12}$$

thus:

$$\frac{\Delta t}{2}[v_k(t) - v_m(t) + v_k(t-\Delta t) - v_m(t-\Delta t)] = R\frac{\Delta t}{2}[i_{km}(t) + i_{km}(t-\Delta t)] + \\ + L[i_{km}(t) - i_{km}(t-\Delta t)].$$

Let:

$$R_{eq} = R + \frac{2L}{\Delta t},$$

thus

$$i_{km}(t) = R_{eq}^{-1}[v_k(t) - v_m(t)] + I_{km}(t-\Delta t), \tag{4.13}$$

the history term is given by:

$$I_{km}(t-\Delta t) = R_{eq}^{-1}\left[v_k(t-\Delta t) - v_m(t-\Delta t) + \left(\frac{2L}{\Delta t} - R\right)i_{km}(t-\Delta t)\right]. \tag{4.14}$$

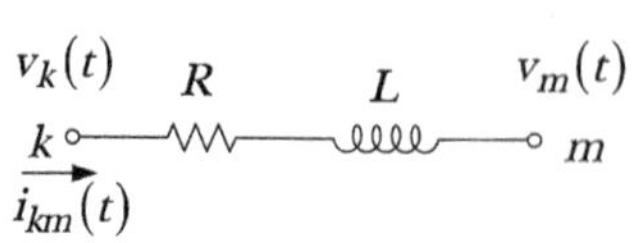

Figure 4.4 Single-phase RL circuit.

4.2.5 Single-Phase Transmission Line Models

- Lossless Transmission Lines – Method of Characteristics

For a lossless transmission line, the following variables are defined:

Z: surge impedance,
v: propagation velocity,
τ: time delay in the transmission line.

The method of characteristics represents a significant advancement in the inclusion of efficient transmission line models in EMT-type programs, overcoming the limitations of lattice diagram-based programs. It will be shown below that the principle is based on Thévenin or Norton equivalents, as discussed in Chapters 2 and 3. This will first be demonstrated for a lossless single-phase line, and its generalization to multiphase lines with losses will be addressed later in Section 4.6.4.

The initial concept of the method was based on a voltage function known as a forward wave, traveling in the transmission line from node k to node m.

$$f_k^{(v)}(t) = v_k(t) + Zi_{km}(t). \tag{4.15}$$

Voltage and current can be assumed as progressive waves with the same signal:

$$v_k(t) = Zi_{km}(t),$$

and it can be concluded that the forward voltage wave can be written as:

$$f_k^{(v)}(t) = 2v_k(t). \tag{4.16}$$

The forward wave that departs from k at $t - \tau$ and reaches m at time t is termed as the backward wave at m, as in Figure 4.5a.

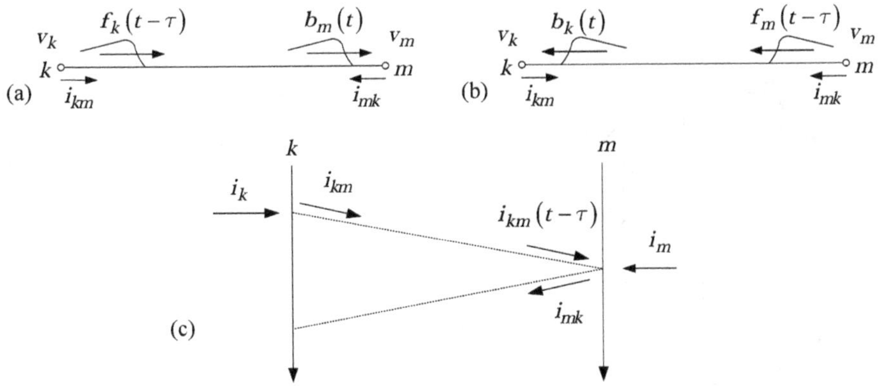

Figure 4.5 Wave displacement according to time delay τ. (a) Forward and backward waves from k to m. (b) Forward and backward waves from m to k. (c) Reflected-current convention.

$$b_m^{(v)}(t) = f_k^{(v)}(t - \tau). \tag{4.17}$$

The backward wave to be used at node m, at time t, is pre-determined and designated as a history term, since it was previously calculated at node k, at time $t - \tau$. Note that the backward wave at node m is actually the voltage of the Thévenin equivalent model of the transmission line (or $2v_k^+$), as seen in Section 2.4.2 in Chapter 2, and presented in Figure 4.6a.

The forward wave f_k can also be expressed in terms of currents multiplying (4.15) by $Y = Z^{-1}$, and the same is valid for node m.

$$\begin{aligned} f_k(t) &= Yv_k(t) + i_{km}(t), \\ f_m(t) &= Yv_m(t) + i_{mk}(t), \end{aligned} \tag{4.18}$$

where $f = Yf^{(v)}$ and analogously to (4.16), it can be concluded for nodes k and m that:

$$\begin{aligned} f_k(t) &= 2i_{km}(t), \\ f_m(t) &= 2i_{mk}(t), \end{aligned} \tag{4.19}$$

which means that the backward wave at m is equal to twice the departed current from k at $t - \tau$. This is in accordance with the Norton equivalent at node m, where the current arriving at this point doubles if the line is short-circuited, thus, $b_m(t)$ can also be written as:

$$b_m(t) = 2i_{km}(t - \tau), \tag{4.20}$$

and a similar Norton equivalent can be obtained at node k.

$$b_k(t) = 2i_{mk}(t - \tau). \tag{4.21}$$

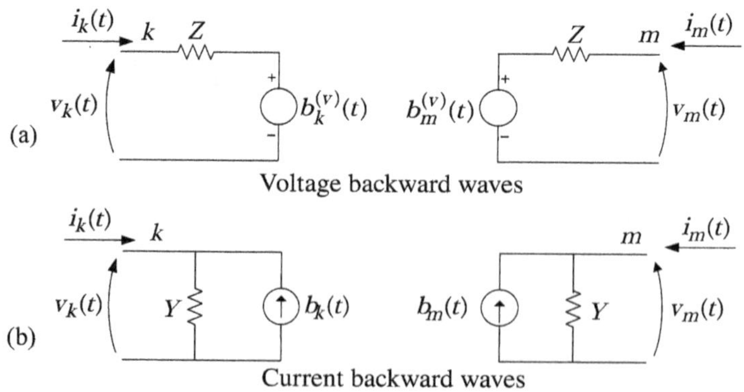

Figure 4.6 Transmission line equivalent model. (a) Thévenin equivalent model. (b) Norton equivalent model.

The Thévenin equivalent (for voltages) in Figure 4.6a can be transformed into the Norton equivalent (for currents) in Figure 4.6b.

The equations that use currents, with the convention adopted in Figures 4.5 and 4.6, are convenient for nodal analysis and the solution of the circuit in Figure 4.6b can be written as:

$$i_k(t) = Yv_k(t) - b_k(t), \tag{4.22}$$

$$i_m(t) = Yv_m(t) - b_m(t), \tag{4.23}$$

Note that in Chapter 2, the incident and reflected currents have the same direction, both entering the discontinuity node. However, using the convention from Figure 4.5c [5] (detailed for node m), the incident current enters the node, and the reflected current exits the node.

To complete the solution, it is necessary to calculate the reflected currents $i_{km}(t)$ and $i_{mk}(t)$ that will actually flow through the line, using Figure 4.5c. As shown in this figure, $i_{mk}(t-\tau)$ and $i_{km}(t-\tau)$ are the incident currents at nodes k and m at time t, respectively, thus:

$$i_{km}(t) = i_k(t) + i_{mk}(t-\tau), \tag{4.24}$$

and,

$$i_{mk}(t) = i_m(t) + i_{km}(t-\tau). \tag{4.25}$$

Example 4.1 For a lossless and no-load transmission line, with surge impedance 10 Ω apply a 10 V step at the sending end and obtain the voltages at the receiving end using the method of characteristics. Consider Figure 4.6 with $k = 1$ as sending end and $m = 2$ as receiving.

A time step Δt equal to the time delay τ of the transmission line is assumed. Thus, $t = n\Delta t = n\tau$. The following expressions are written for a generic step n:

Solution using the Norton equivalent.

$$\begin{aligned}
b_1(n) &= 2i_{21}(n-1),\\
b_2(n) &= 2i_{12}(n-1),\\
i_1(n) &= Yv_1(n) - 2i_{21}(n-1),\\
i_2(n) &= Yv_2(n) - 2i_{12}(n-1),\\
v_1(n) &= 10.
\end{aligned}$$

Terminal 2 is open, thus, $i_2(n) = 0$:

$$\begin{aligned}
&0 = Yv_2(n) - b_2(n), \text{or } v_2(n) = 2i_{12}(n-1) \times Z.\\
&i_{12}(n) = i_1(n) + i_{21}(n-1)\\
&i_{21}(n) = i_1(n) + i_{12}(n-1)
\end{aligned}$$

All the following values are calculated in volts and amperes.

(a) for $t = 0$, $n = 0$:

$b_1(0) = 2i_{21}(-1) = 0,$

$b_2(0) = 2i_{12}(-1) = 0,$

$e_1(0) = 10V, v_2(0) = 0,$

$i_1(0) = Yv_1(0) - 2i_{21}(-1),$

$i_1(0) = 0.1 \times 10 + 0 = 1,$

$i_2(0) = Yv_2(0) - 2i_{12}(-1),$

$i_2(0) = 0,$

$i_{12}(0) = i_1(0) + i_{21}(-1) = 1$

$i_{21}(0) = i_2(0) + i_{12}(-1) = 0$

(b) for $t = \tau$, $n = 1$:

$b_1(1) = 2i_{21}(0) = 0,$

$b_2(1) = 2i_{12}(0) = 2,$

$v_1(1) = 10,$

$v_2(1) = 2i_{12}(0) \times Z, v_2(1) = 2 \times 10 = 20,$

$i_1(1) = Yv_1(1) - 2i_{21}(0),$

$i_1(1) = 0.1 \times 10 - 2 \times 0 = 1,$

$i_2(1) = Yv_2(1) - 2i_{12}(0),$

$i_2(1) = 0.1 \times 20 - 2 \times 1 = 0,$

$i_{12}(1) = i_1(1) + i_{21}(0) = 1$

$i_{21}(1) = i_2(1) + i_{12}(0) = 1$

(c) for $t = 2\tau$, $n = 2$:

$b_1(2) = 2i_{21}(1) = 2,$

$b_2(2) = 2i_{12}(1) = 2,$

$v_1(2) = 10,$

$v_2(2) = 2i_{12}(1) \times Z, v_2(2) = 2 \times 10 = 20,$

$i_1(2) = Yv_1(2) - 2i_{21}(1),$

$i_1(2) = 0.1 \times 10 - 2 \times 1 = -1,$

$i_2(2) = Yv_2(2) - 2i_{12}(1),$

$i_2(2) = 0.1 \times 20 - 2 \times 1 = 0,$

$i_{12}(2) = i_1(2) + i_{21}(1) = 1,$

$i_{21}(2) = i_2(2) + i_{12}(1) = 1,$

and so on.

Table 4.1 present some of the calculated steps.

The figure 4.7 shows the Norton equivalents and reflected currents used at some steps in the calculation. Figure 4.7a presents incident and reflected current waves (with the convention of Figure 4.5c) at nodes 1 and 2. Figures 4.7c to 4.7e present

Table 4.1 Calculation steps.

$t = n\tau$	v_1(V)	v_2(V)	i_1(A)	i_2(A)	i_{12}(A)	i_{21}(A)
0^+	10	0	1	0	+1	0
	10	20	1	0	+1	+1
2	10	20	−1	0	0	+1
3	10	0	−1	0	0	0
4	10	0	1	0	+1	0
5	10	20	1	0	+1	+1

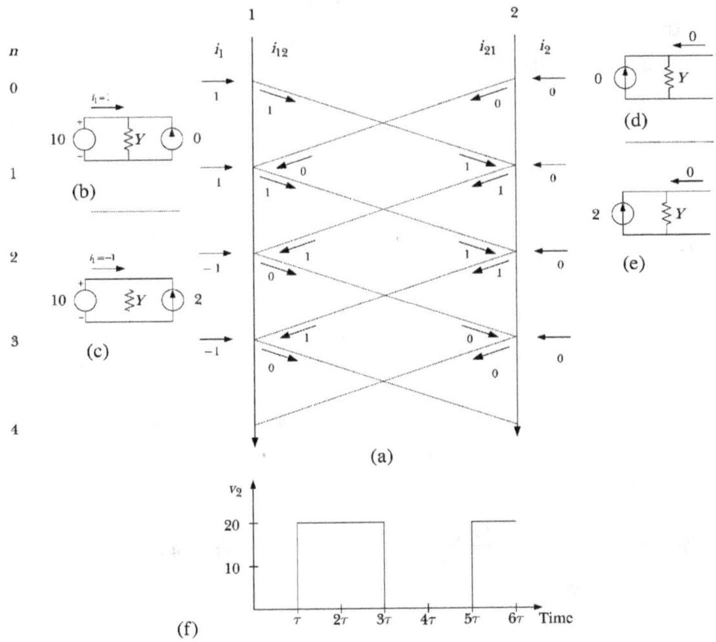

Figure 4.7 Voltage at the open terminal of the transmission line. (a) Incident and reflected current waves. (b) Norton equivalent at node 1 for steps 0 and 1. (c) Norton equivalent at node 1 for steps 2 and 3. (d) Norton equivalent at node 2 for step 0. (e) Norton equivalent at node 2 for steps 1 and 2. (f) Voltage time response at node 2.

Norton equivalent for some steps. Figure 4.7b for steps 0 and 1 at node 1. Figure 4.7c for steps 2, 3 at node 1. Figure 4.7d for step 0 at node 2. Figure 4.7e for steps 1, 2 at node 2. Figure 4.7f presents the time response of voltage v_2.

This example shows that the model permits an accurate representation of traveling waves and eliminates the use of refraction and reflection coefficients, thereby allowing an approach to currents and voltages based on nodal analysis.

It was also assumed that the time delays along the lines are integer multiples of the integration steps; however, in more general cases, it does not occur, which will be addressed later.

- Approximation for Lossy Transmission Lines

  The lossy line model assumes that a distributed series resistance can be lumped at some of the points connecting lossless line segments.

  - Two lumped resistances.

    $R = R'\ell$: total resistance of the transmission line.

    A simple equivalent model of lossy lines incorporating lumped resistances inserted at the ends of the segment of lossless line is presented below.

    $R/2$: half of the lumped resistance at the line segment terminals.

    The following can be written:

$$v_k'(t) = v_k(t) - \frac{R}{2} i_{km}(t), \tag{4.26}$$

$$v_m'(t) = v_m(t) - \frac{R}{2} i_{mk}(t), \tag{4.27}$$

$$i_{km}(t) = \frac{v_k'(t)}{Z} - b_k(t), \tag{4.28}$$

$$b_k(t) = f_m(t-\tau) = \left(\frac{v_m'(t-\tau)}{Z} + i_{mk}(t-\tau)\right). \tag{4.29}$$

In the first step, Eq. (4.27) calculated at time $t-\tau$ is substituted into (4.29). After that, the modified equation along with (4.26) are substituted into Eq. (4.28), resulting in Eq. (4.30):

$$i_{km}(t) = \frac{1}{Z} v_k(t) - \frac{R}{2Z} i_{km}(t) - \frac{v_m(t-\tau)}{Z} + \frac{R}{2Z} i_{mk}(t-\tau) - i_{mk}(t-\tau), \tag{4.30}$$

thus,

$$i_{km}(t) = \frac{1}{Z + \frac{R}{2}} v_k(t) - b_k(t), \tag{4.31}$$

$$b_k(t) = \left(\frac{v_m(t-\tau)}{Z+R/2} + \frac{Z-R/2}{Z+R/2} i_{mk}(t-\tau)\right). \tag{4.32}$$

Setting:

$$Z_{eq} = Z + R/2, \tag{4.33}$$

$$h = \frac{Z-R/2}{Z+R/2}, \tag{4.34}$$

the equivalent model is obtained:

$$i_{km}(t) = \frac{v_k(t)}{Z_{eq}} - b_k(t), \tag{4.35}$$

$$b_k(t) = f_m(t-\tau) = \left[\frac{v_m(t-\tau)}{Z_{eq}} + h i_{mk}(t-\tau)\right]. \tag{4.36}$$

Although the model in Figure 4.8a is not widely used, it is quite simple and illustrative of how the equivalent model can be obtained.

- Three lumped resistances.

In Figure 4.8b, one more lumped resistance is inserted in the middle of the line model [1, 6], which improves the model accuracy. In reality, it is as if the line were divided in half, with each half represented by the previous model.

By following a procedure similar to the previous one, it can be demonstrated that the equivalent model is composed of the following equations:

$$Z_{eq} = Z + R/4, \tag{4.37}$$

$$i_k(t) = \frac{v_k(t)}{Z_{eq}} - b_k(t), \tag{4.38}$$

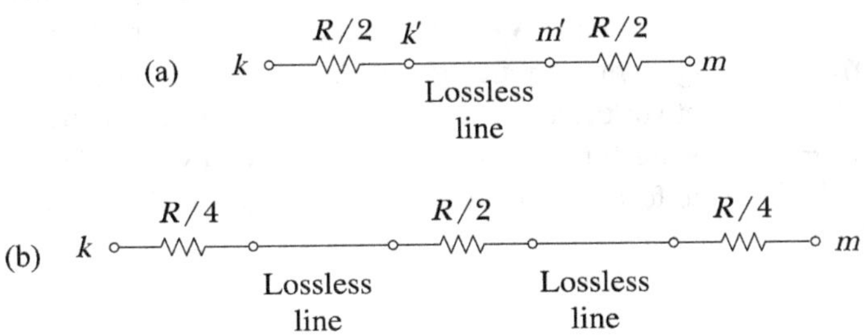

Figure 4.8 Lossy transmission line. (a) Two lumped resistances. (b) Three lumped resistances.

$$b_k(t) = \frac{1+h}{2}\left[\frac{v_m(t-\tau)}{Z_{eq}} + hi_{mk}(t-\tau)\right] + \left(\frac{1-h}{2}\right)\left[\frac{v_k(t-\tau)}{Z_{eq}} + hi_{km}(t-\tau)\right], \tag{4.39}$$

with:

$$h = \frac{Z - R/4}{Z + R/4}. \tag{4.40}$$

The model in Figure 4.8b is part of the embedded models in EMT-type programs and is an essential tool for analyzing networks with multiple transmission lines, providing relatively accurate results. In general, calculating the resistance at the nominal frequency yields conservative results, which is often already important information. A useful trick is to calculate the resistance at the dominant transient frequency, which can improve the accuracy of the results.

4.3 Transient Solution Using Nodal Analysis

In the calculation of electromagnetic transients, it is convenient to apply nodal analysis. For node k, all currents entering the node are added up according to Kirchhoff's first law (Figure 4.9).

$$i_k = (v_k - v_1)Y_{k1} + (v_k - v_2)Y_{k2} + v_k Y_{kk} + (v_k - v_m)Y_{km}, \tag{4.41}$$

or, rearranging:

$$i_k = -v_1 Y_{k1} - v_2 Y_{k2} + \dots + (Y_{k1} + Y_{k2} + Y_{kk} + \dots + Y_{km})v_k + \dots - v_m Y_{km}. \tag{4.42}$$

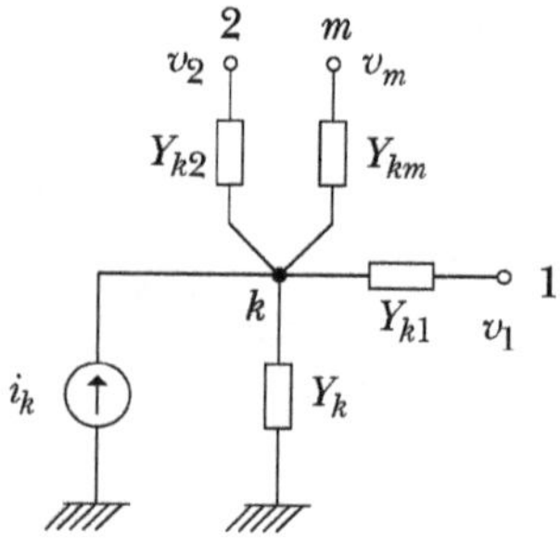

Figure 4.9 Elements connected to node k.

Equation (4.42), when applied to all nodes in the network, is formulated using the nodal admittance matrix, which relates nodal currents and nodal voltages throughout the entire system. A single-phase network is addressed in (4.42), but mutual elements in the three-phase networks are easily included in the formulation, as shown in a later section.

$$\underset{\sim}{i} = Y\underset{\sim}{v}, \tag{4.43}$$

$\underset{\sim}{i}$: vector of nodal currents.

$\underset{\sim}{v}$: vector of nodal voltages.

A generic matrix line k from Eq. (4.43) is illustrated,

$$\begin{bmatrix} \vdots \\ \vdots \\ i_k \\ \vdots \\ \vdots \end{bmatrix} = \begin{bmatrix} \cdots & & & & \\ \cdots & & & & \\ y_{k1} & \cdots & y_{kk} & \cdots & y_{km} \\ \cdots & & & & \\ \cdots & & & & \end{bmatrix} \times \begin{bmatrix} v_1 \\ \vdots \\ v_k \\ \vdots \\ v_m \end{bmatrix}, \tag{4.44}$$

where:

$$y_{k1} = -Y_{k1}, y_{k2} = -Y_{k2'} \ldots y_{kk} = +(Y_{k1} + Y_{k2} + \ldots + Y_{km}), y_{km} = -Y_{km}.$$

The structure of the system of nodal equations for a simplified network, considering the models developed so far, is exemplified in Figure 4.10a with $k = 1$.

Replacing the lumped elements R, L, C and the transmission line of surge impedance Z, by their respective equivalents, Figure 4.10b is obtained. The convenience of the transmission line model for nodal analysis can be appreciated in this figure.

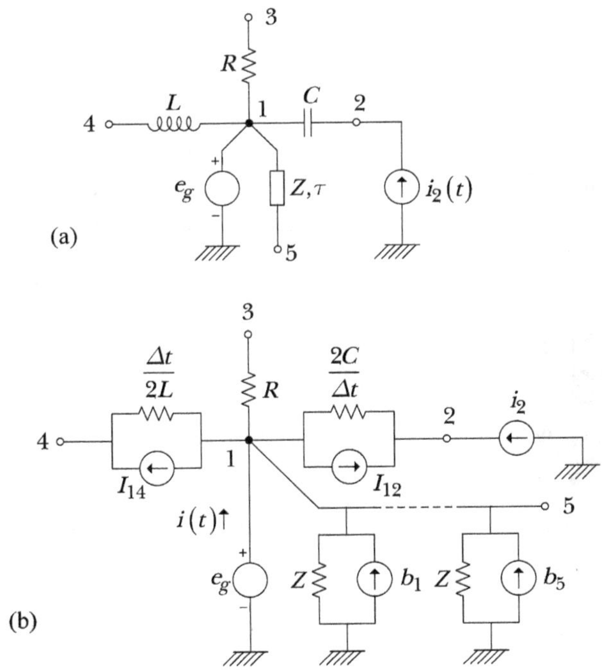

Figure 4.10 Generic network. (a) Elements connected to node 1. (b) Equivalent model.

The system of equations is formulated based on nodal analysis:

$$\underset{\sim}{i}(t) + \underset{\sim}{I}(t) = Y\underset{\sim}{v}(t), \tag{4.45}$$

Y: nodal admittance matrix.
$\underset{\sim}{i}(t)$: known currents applied by generators.
$\underset{\sim}{I}(t)$: known current generators calculated in previous time steps.
$\underset{\sim}{v}(t)$: nodal voltages.

The exemplified case of Figure 4.10b is written in expression (4.46):

$$\begin{bmatrix} b_1(t) - I_{14}(t-\Delta t) \\ -I_{12}(t-\Delta t) \\ \hdashline I_{12}(t-\Delta t) + i_2(t) \\ \hdashline 0 \\ \hdashline +I_{14}(t-\Delta t) \\ \hdashline +b_5(t) \end{bmatrix} = \begin{bmatrix} \frac{2C}{\Delta t} + \frac{1}{R} + \frac{\Delta t}{2L} + \frac{1}{Z} & -\frac{2C}{\Delta t} & -\frac{1}{R} & -\frac{\Delta t}{2L} & 0 \\ -\frac{2C}{\Delta t} & \frac{2C}{\Delta t} & 0 & 0 & 0 \\ -\frac{1}{R} & 0 & \frac{1}{R} & 0 & 0 \\ -\frac{\Delta t}{2L} & 0 & 0 & \frac{\Delta t}{2L} & 0 \\ 0 & 0 & 0 & 0 & \frac{1}{Z} \end{bmatrix} \times \begin{bmatrix} v_1 = e_g \\ \hdashline v_2 \\ \hdashline v_3 \\ \hdashline v_4 \\ \hdashline v_5 \end{bmatrix} \tag{4.46}$$

The nodal voltages are grouped into two sets. The first set contains the known voltages imposed by voltage generators and the second set contains the unknown voltages to be calculated at each time step:

$$\begin{bmatrix} \underset{\sim g}{i} \\ \hdashline \underset{\sim u}{i} \end{bmatrix} = \begin{bmatrix} Y_{gg} & Y_{gu} \\ \hdashline Y_{ug} & Y_{uu} \end{bmatrix} \times \begin{bmatrix} \underset{\sim g}{e} \\ \hdashline \underset{\sim u}{v} \end{bmatrix}. \tag{4.47}$$

g: nodes of known voltage (generators).
u: nodes of unknown voltage.

The calculation of the unknown voltages for each time step is performed as follows:

$$\underset{\sim u}{i} = Y_{ug}\underset{\sim g}{e} + Y_{uu}\underset{\sim u}{v}, \tag{4.48}$$

$$Y_{uu}\underset{\sim u}{v} = \left[\underset{\sim u}{i} - Y_{ug}\underset{\sim g}{e}\right], \tag{4.49}$$

$$\underset{\sim}{v}_u = [Y_{uu}]^{-1}\left[\underset{\sim}{i}_u - Y_{ug}\underset{\sim}{e}_g\right]. \tag{4.50}$$

Vector $\underset{\sim}{i}_u$ is related to unknown voltages, but composed of current generators, established by current sources or history information, and therefore, totally determined. The vector of voltage generators $\underset{\sim}{e}_g$ is also known.

Matrices Y_{uu} and Y_{ug} do not change until a topological modification of the network occurs and the time step Δt remains constant. At each time step, the system of equations in (4.48) is solved as follows:

(a) Instead of inverting Y_{uu} in (4.50), the augmented matrix $[Y_{uu}|Y_{ug}]$ is triangularized before entering the integration loop and being stored in the table of factors [7].

(b) Procedure for each time step:

- **(b.1)** Vectors $\underset{\sim}{i}_u$ and $Y_{ug}\underset{\sim}{e}_g$ are calculated and the system of equations is solved by applying the table of factors with back-substitution, thus obtaining the unknown voltages$\underset{\sim}{v}_u$.
- **(b.2)** Branch currents are obtained after the calculation of unknown voltages$\underset{\sim}{v}_u$.
- **(b.3)** Vector $\underset{\sim}{I}$ is calculated and stored at time $t - \Delta t$ to be used at time t.

Note that matrices Y_{uu} and Y_{ug} are sparse, and only the non-null elements in the matrix are stored following calculation on the basis of optimally ordered triangular factorization.

Example 4.2 Solve the following RL circuit by the trapezoidal integration method (Figure 4.11):

- Selection of the integration step.

In this example, a time step five times smaller than the time constant of the circuit is adopted just to exemplify the first few steps of the calculation, although a value 10 times smaller than the time constant would give more accurate results:

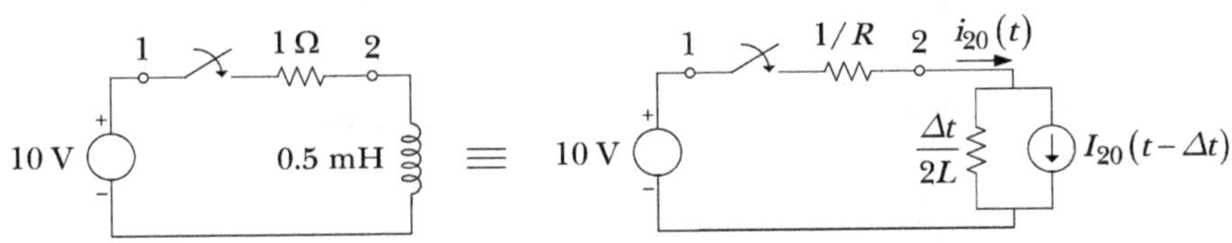

Figure 4.11 RL circuit.

$$\tau = \frac{L}{R} = \frac{0.5 \times 10^{-3}}{1} = 0.5 \text{ ms}, \ \ \Delta t = 0.1 \text{ ms}.$$

Writing the equation:

$$I_{20}(t-\Delta t) = [v_2(t-\Delta t) - v_0(t-\Delta t)]\frac{\Delta t}{2L} + i_{20}(t-\Delta t),$$

the voltage at node zero is null, since it is grounded, thus:

$$I_{20}(t-\Delta t) = v_2(t-\Delta t)\frac{\Delta t}{2L} + i_{20}(t-\Delta t),$$

$$\frac{\Delta t}{2L} = \frac{1}{10},$$

$$I_{20}(t-\Delta t) = \frac{v_2(t-\Delta t)}{10} + i_{20}(t-\Delta t),$$

and the nodal equation is:

$$\begin{bmatrix} I_1 \\ I_{20}(t-\Delta t) \end{bmatrix} = \begin{bmatrix} 1 & -1 \\ -1 & 11/10 \end{bmatrix} \times \begin{bmatrix} e_1(t) \\ v_2(t) \end{bmatrix}.$$

Equation (4.50) is used to facilitate the solution of this simple example, obtained from inversion of Y_{gg}. As previously noted, the algorithm is more efficient when triangularization and back-substitution are applied.

$$\underset{\sim}{v}_u = [Y_{uu}]^{-1}\left[\underset{\sim}{i}_u - Y_{ug}\underset{\sim}{e}_g\right].$$

(1) Calculation of the unknown voltage, at time t.

$$v_2(n) = \frac{10}{11}[-I_{20}(n-1) + e_1(n)].$$

(2) Solution of the network. After finding the unknown voltage, the current in the branch is calculated.

$$i_{20}(n) = \frac{v_2(n)}{10} + I_{20}(n-1).$$

(3) Updating the history current at step n to be used in the following step, $n+1$.

$$I_{20}(n) = \frac{v_2(n)}{10} + i_{20}(n).$$

Solution

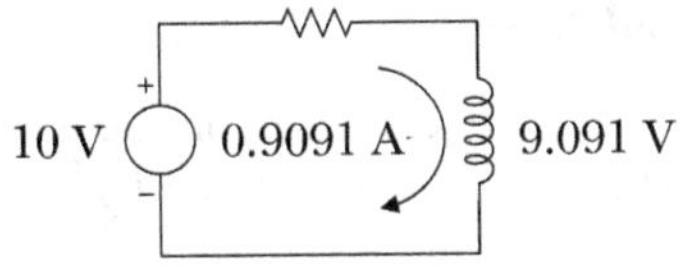

Figure 4.12 Voltages and currents for Δt = 0.1 ms.

a) for $n = 0$.
All variables in the network are assumed null at time zero, except the generator voltage:

$$e_1(0) = 10,\ i_{12}(0) = 0,$$
$$v_2(0) = 0,\ i_{20}(0) = 0.$$

b) for $n = 1$ (Closing the switch at $t = \Delta t = 0.1$ ms):

1) $v_2(1) = \dfrac{10 \times 10}{11} = 9.091\text{ V},$
2) $i_{20}(1) = \dfrac{9.091}{10} + 0 = 0.9091\text{ A},$
3) $I_{20}(1) = \dfrac{9.091}{10} + 0.9091 = 1.8182\text{ A}$ (Figure 4.12).

c) for $n = 2$.

1) $v_2(2) = \dfrac{10}{11}(-1.8182 + 10) = 7.438\text{ V},$
2) $i_{20}(2) = \dfrac{7.438}{10} + 1.818 = 2.562\text{ A},$
3) $I_{20}(2) = 0.7438 + 2.562 = 3.306\text{ A}.$

d) for $n = 3$.

1) $v_2(3) = \dfrac{10}{11}(-3.306 + 10) = 6.085\text{ V},$
2) $i_{20}(3) = \dfrac{6.085}{10} + 3.306 = 3.915\text{ A},$
3) $I_{20}(3) = \dfrac{6.085}{10} + 3.915 = 4.524\text{ A},$

and so on.

Example 4.3 The following circuit represents the energization of a transmission line. Calculate the transient using the trapezoidal method (Figure 4.13).

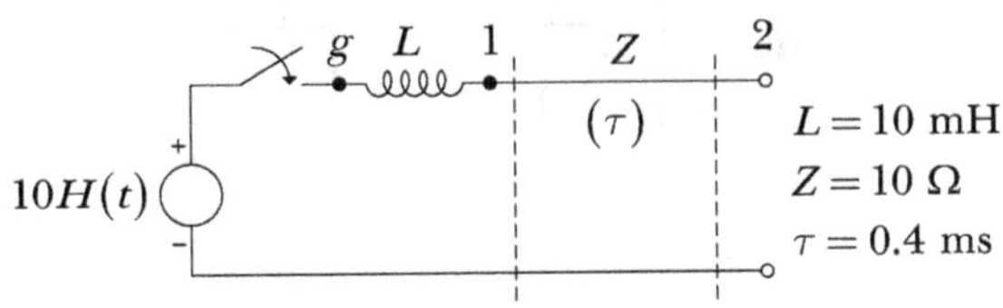

Figure 4.13 Line energization.

Solution

The time constant of the applied wavefront, as discussed in Example 2.4 of Chapter 2, is calculated by:

$$\tau = \frac{L}{Z} = \frac{10 \times 10^{-3}}{10} = 1 \text{ ms}.$$

This example assumes a time step $\Delta t = 0.2$ ms, one-fifth of the time constant of the applied voltage at the sending end, to speed up the calculation, but a shorter time step would give more accurate results.

For a generic network with transmission lines, the time step should be smaller than the period corresponding to the maximum frequency to be analyzed and smaller than the time delay of the shortest transmission line. One-tenth of the time delay is usually considered a reasonable value.

- Equivalent circuit for integration by the trapezoidal rule (Figure 4.14):
- System of nodal equations.

$$\begin{bmatrix} -I_{g1}(t-\Delta t) \\ I_{g1}(t-\Delta t) + b_1(t) \\ + b_2(t) \end{bmatrix} = \begin{bmatrix} 0.01 & -0.01 & 0 \\ -0.01 & 0.11 & 0 \\ 0 & 0 & 0.10 \end{bmatrix} \times \begin{bmatrix} 10 \\ v_1(t) \\ v_2(t) \end{bmatrix},$$

Since the system is very small, it will be solved by matrix inversion, although triangularization and back substitution are more appropriate.

$$\underset{\sim}{v}_u = [Y_{uu}]^{-1}\left[\underset{\sim}{i}_u - Y_{ug}\underset{\sim}{e}_g\right],$$

where:

$$Y_{ug} = \begin{bmatrix} -0.01 \\ 0 \end{bmatrix}, \; e_g = 10,$$

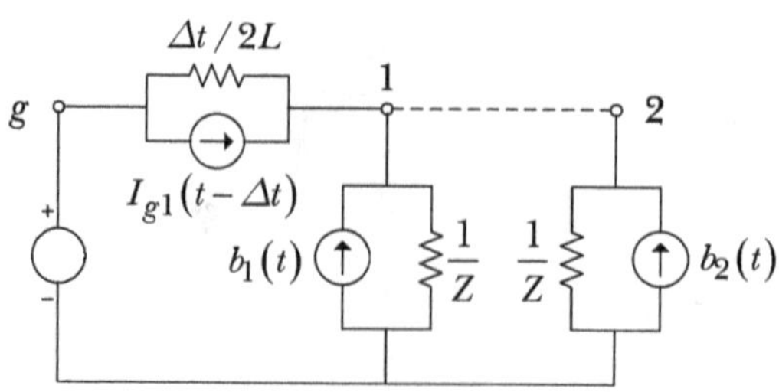

Figure 4.14 Equivalent circuit.

$$Y_{uu}^{-1} = \begin{bmatrix} 9.0909 & 0 \\ 0 & 10 \end{bmatrix}, \ \underset{\sim}{i}_u = \begin{bmatrix} I_{g1}(t-\Delta t) + b_1(t) \\ b_2(t) \end{bmatrix}.$$

Since $\tau = 2\Delta t$ and by setting $t = n\Delta t$, the following can be written for step number n:

$$\begin{bmatrix} v_1(n) \\ v_2(n) \end{bmatrix} = \begin{bmatrix} 9.0909 & 0 \\ 0 & 10 \end{bmatrix} \times \left(\begin{bmatrix} I_{g1}(n-1) + b_1(n) \\ b_2(n) \end{bmatrix} - \begin{bmatrix} -0.1 \\ 0 \end{bmatrix} \right),$$

$$v_1(n) = 9.0909\left[I_{g1}(n-1) + b_1(n) + 0.1\right],$$

$$v_2(n) = 10b_2(n).$$

- Transmission line equations:

$$b_1(n) = 2i_{21}(n-2),$$
$$b_2(n) = 2i_{12}(n-2),$$
$$i_1(n) = Yv_1(n) - b_1(n),$$
$$i_2(n) = Yv_2(n) - b_2(n) = 0 \rightarrow v_2(n) = 10 \times 2i_{12}(n-2).$$

- Inductor equations.

  The calculation of the current is as follows:

$$i_{g1}(n) = 0.01\left[e_g(n) - v_1(n)\right] + I_{g1}(n-1),$$

  history term update:

$$I_{g1}(n) = i_{g1}(n) + 0.01\left[e_g(n) - v_1(n)\right].$$

  The following results are in volts and amperes.

(a) At $t = 0$, $n = 0$.

$$e_g(0) = 10,$$

$$b_1(0) = 2i_{21}(-2) = 0,$$

$$b_2(0) = 2i_1(-2) = 0,$$

$$I_{g1}(-1) = 0.$$

(a.1) Calculation for unknown voltages.

$$v_1(0) = 9.0909 \times 0.1 = 0.9091,$$
$$v_2(0) = 0.$$

(a.2) Solution of the network.

$$i_{g1}(0) = 0.01 \times (10 - 0.9091) + 0 = 0.0909,$$
$$i_{12}(0) = 0.01 \times 0.9091 - 0 = 0.0909,$$
$$i_{21}(0) = 0.$$

(a.3) Current generator update.

$$I_{g1}(0) = 0.0909 + 0.01 \times (10 - 0.9091) = 0.1818,$$

(b) For $t = \Delta t = 0.2$ ms, $n = 1$.

$$e_g(1) = 10,$$
$$b_1(1) = 2i_{21}(-1) = 0,$$
$$b_2(1) = 2i_{12}(-1) = 0,$$
$$I_{g1}(0) = 0.1818.$$

(b.1) $v_1(1) = 9.0909 \times (0.1818 + 0 + 0.1) = 2.5618,$
$v_2(1) = 10 \times 0 = 0.$

(b.2) $i_{g1}(1) = 0.01 \times (10 - 2.5618) + 0.1818 = 0.2562,$

$i_1(1) = 0.01 \times 2.5618 - 0 = 0.2562,$

$i_2(1) = 0.$

(b.3) $I_{g1}(1) = 0.2562 + 0.01 \times (10 - 2.5618) = 0.3306,$
$i_{21}(1) = 0.1 \times 0 + 0 = 0,$

$i_{12}(1) = 0.1 \times 2.5618 + 0.2562() = 0.5124,$

and so on.

In this example, the time delay is an integer multiple of the adopted time step, which makes the solution simpler. However, it does not happen, in general, and this numerical issue is addressed in item 4.7.

4.4 Nonlinear Elements

4.4.1 Resistive Elements

The solution of a network containing several nonlinear elements can be obtained by means of different iterative methods.

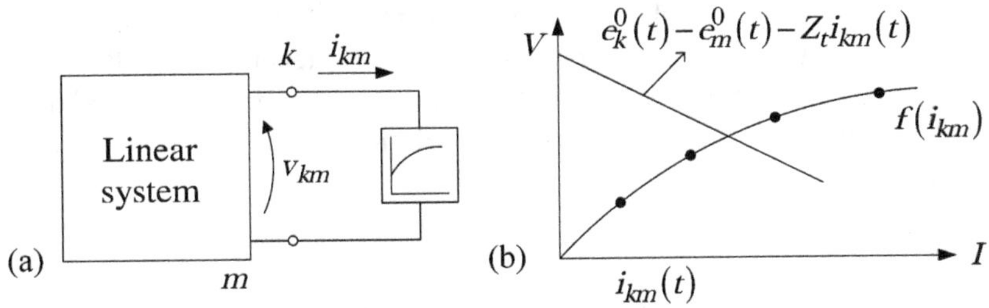

Figure 4.15 Linear system connected to the nonlinear resistive element. (a) Thévenin equivalent (b) Graphical solution.

An introduction to the treatment of nonlinearities will be presented, but with the simplification that only one nonlinear element in the electric network will be taken into consideration, Figure 4.15a.

As seen in Chapter 2, when there is just one nonlinear element, the problem can be addressed by simultaneously solving the linear system and the equation for the nonlinear element.

The solution is also known as the compensation method and the equation for the linear system is found with the aid of the Thévenin equivalent.

$$v_k(t) - v_m(t) = e_k^0(t) - e_m^0(t) - Z_t i_{km}(t). \tag{4.51}$$

For the linear network at time t, voltages $e_k^0(t)$ and $e_m^0(t)$ are calculated with the nonlinear element disconnected, representing the Thévenin equivalent voltage at nodes k and m, and Z_t is the Thévenin equivalent impedance between nodes k and m.

The nonlinear element is described by the nonlinear equation:

$$v_k(t) - v_m(t) = f(i_{km}), \tag{4.52}$$

The solution of (4.51) and (4.52) can be obtained by any numerical method.

As will be seen in Chapter13, arresters can have their nonlinear behavior represented by a residual voltage, like the $V \times I$ curve in Figure 4.15b. Simultaneous solution of Eqs. (4.51) and (4.52) leads to the result as in same figure.

The residual voltage can be represented by a nonlinear fitted function or by linearized segments in a piecewise model. If the arrester has a spark gap, insertion of the nonlinear element in the network only occurs after the voltage exceeds the sparkover level, that is, $i_{km}(t) = 0$ until $e_k^0(t) - e_m^0(t)$ reaches the sparkover level.

When there is more than one nonlinear element, the previously described algorithm can still be used, provided that the arresters are present in topologically disconnected sub-networks, each one with just one arrester.

The transmission line decouples both terminals in time. In this case, the solution of only one arrester at each line terminal can be treated independently. Otherwise, iterative procedures become necessary for the solution when there are several non-linear elements operating simultaneously. The solution for the more general case is presented in [2, 3].

$$\begin{bmatrix} I & 0 & 0 & 0 \\ 0 & II & 0 & 0 \\ 0 & 0 & III & 0 \\ 0 & 0 & 0 & IV \end{bmatrix} : \text{disconnected sub-networks } I, II, III, IV.$$

Thévenin equivalent impedance, Z_t.

For the linear part of the network, the nodal equation is written as follows:

$$\underset{\sim}{i} = Y\underset{\sim}{v}. \tag{4.53}$$

The Thévenin equivalent impedance between nodes k and m, can be obtained from the inverse of matrix Y, by applying unitary currents at these nodes, but the equations will not be detailed here.

$$Z = Y^{-1}. \tag{4.54}$$

So far, the problem has been approached by the compensation method, based on the solution of (4.51) and (4.52). Another solution, still based on the piecewise model, is presented below. The resistive element, represented by line segments, is given in Figure 4.16.

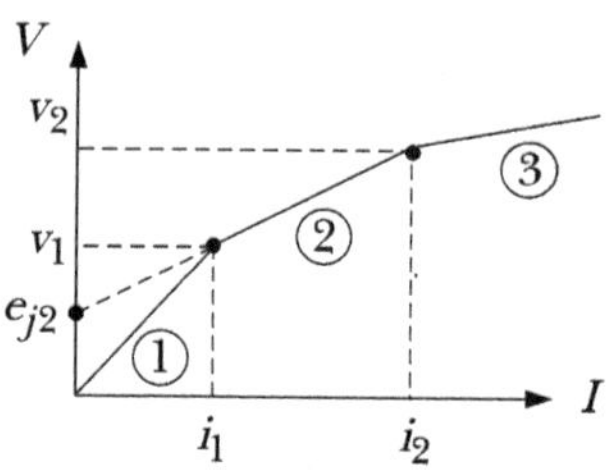

Figure 4.16 Piecewise representation of the nonlinear element.

Taking segment 2, for instance:

For $i_1 \le i \le i_2$ and $v_1 \le v \le v_2$, the resistance is given by:

$$R_2 = \frac{v_2 - v_1}{i_2 - i_1}. \tag{4.55}$$

In this line segment, the nonlinear element can be represented by the following voltage-resistance equivalent or by its Norton equivalent, Figures 4.17a and b, respectively.

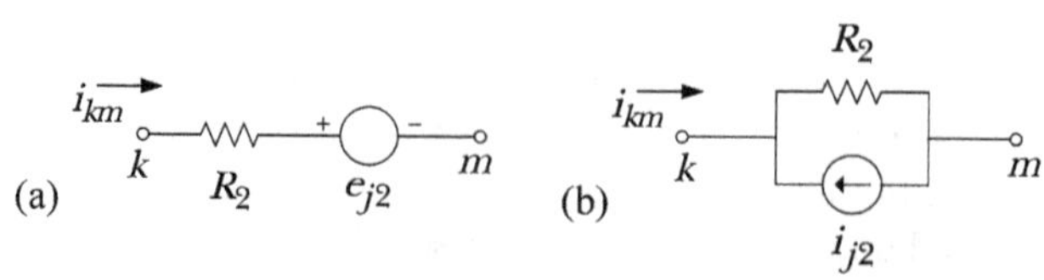

Figure 4.17 Nonlinear element model. (a) Voltage-resistance equivalent. (b) Norton equivalent.

$$i_{j2} = \frac{e_{j2}}{R_2}. \tag{4.56}$$

In this way, equivalent models are introduced into the admittance matrix according to the valid intervals of the equivalents. When the current is outside the defined interval in the model, an update is made to the matrix of nodal admittances and to the current generators.

Example 4.4 Analyze the voltages and currents in an arrester located at the receiving end of a no-load transmission line. Assume a surge wave represented by a voltage step of 1000 kV.

Data

Residual voltage of the arrester:

segment 1 for $i < 0.5$ kA, $v = 600i$;
segment 2 for $0.5 < i < 2.5$ kA, $v = 50i + 275$.

Surge impedance of the transmission line: $Z = 300\ \Omega$.

Solution

Figure 4.18 shows the Thévenin equivalent circuit associated with the traveling wave reaching the receiving end of the transmission line. The model of the arrester is assumed for segment 2.

$$i = \frac{2000 - 275}{350} = 4.93 \text{ kA},$$

$$v = 521.5 \text{ kV}.$$

4.4.2 Inductive Elements

In electric networks, nonlinear inductive elements are mainly associated with the magnetization branches of transformers and reactors.

A simple case with just one nonlinear inductive element in the network will be attempted (Figure 4.19). Cases with more than one nonlinear inductive element

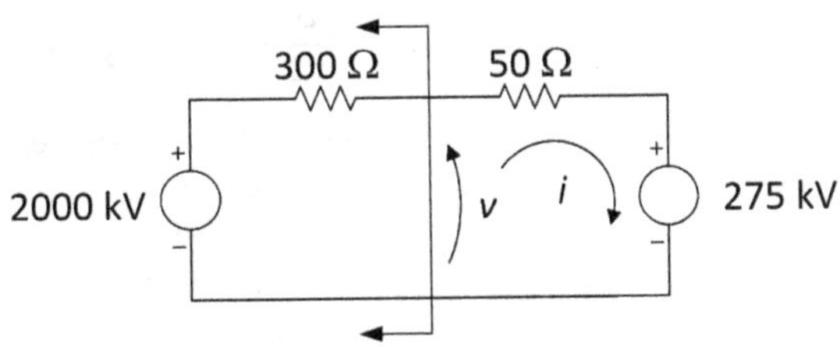

Figure 4.18 Thévenin equivalent circuit.

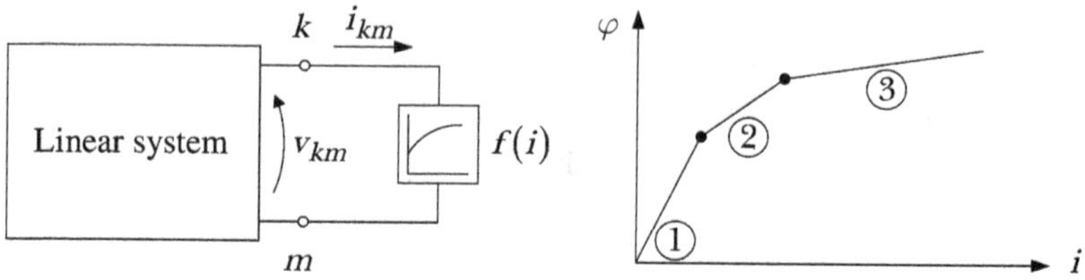

Figure 4.19 Electric system and nonlinear inductive element.

can be solved by iterative methods, such as the Newton–Raphson method, for instance [2, 3].

The solution presented previously for nonlinear resistive elements employs Eqs. (4.51) and (4.52), based on a nonlinear relationship between voltage and current. In the case of inductive elements, a known nonlinear relationship between flux and current is assumed. Thus, the variables need to be preprocessed so that a nonlinear relationship between voltage and current can be derived, allowing the approach outlined in (4.51) and (4.52) to be applied.

The equation that relates the flux and current in the nonlinear element is given by (4.57):

$$\varphi = f(i). \tag{4.57}$$

The equation that relates flux and voltage is given by the integral:

$$\varphi(t) = \varphi(t-\Delta t) + \int_{t-\Delta t}^{t} [v_k(t) - v_m(t)]dt. \tag{4.58}$$

Applying the trapezoidal rule in (4.58):

$$\varphi(t) = \varphi(t-\Delta t) + \frac{\Delta t}{2}[v_k(t) - v_m(t) + v_k(t-\Delta t) - v_m(t-\Delta t)],$$

or:

$$\varphi(t) = \frac{\Delta t}{2}[v_k(t) - v_m(t)] + h(t-\Delta t), \tag{4.59}$$

with the history term:

$$h(t-\Delta t) = \varphi(t-\Delta t) + \frac{\Delta t}{2}[v_k(t-\Delta t) - v_m(t-\Delta t)].$$

Substituting (4.57) into (4.59), the equation relating voltage and current in the nonlinear inductance is obtained, similarly to the resistive case in (4.52):

$$v_k(t) - v_m(t) = \frac{2}{\Delta t}[f(i) - h(t-\Delta t)]. \tag{4.60}$$

Let:

$$f_1(i) = \frac{2}{\Delta t}[f(i) - h(t - \Delta t)]. \tag{4.61}$$

At each time step, the relationship between voltage and current is obtained simply by shifting the function describing the flux by the history term $h(t - \Delta t)$ and multiplying the result by $2/\Delta t$. Therefore, the solution can be treated analogously to the resistive case, previously discussed.

Example 4.5 Analyze the current in the nonlinear inductance for a sinusoidal applied voltage.

Data

$$\begin{aligned}
&e(t) = 100\cos(\omega t),\\
&\varphi = f(i) \text{ (piecewise model)}\\
&\text{Segment 1}:\\
&0 < i < 1.81581 \Rightarrow \varphi = 0.1i.\\
&\text{Segment 2}:\\
&1.81581 \leq i \Rightarrow \varphi = 0.02i + 0.14526,\\
&\text{Frequency}: 60\text{ Hz}.
\end{aligned}$$

Solution

A known voltage is applied by the generator to the nonlinear element, and this simplifies the solution of this example; however, the voltage and impedance of the Thévenin equivalent are also known at the nonlinear element nodes, making it feasible to solve a generic circuit using the same procedure.

For segment 1:

$$L = \frac{\lambda_1 - \lambda_0}{i_1 - i_0}, L = 0.1\text{ H},$$

$$e(t) = 100\cos(377t), i(t) = \frac{1}{L}\int e(t)dt = 10\frac{100}{377}\sin(377t).$$

The following table shows the calculations (Table 4.2).

For $t < 2$ ms, since $i < 1.81581$ A, the first segment of the inductance is used. For $t > 2$ ms, there is a change in the slope of the piecewise nonlinear inductance.

Assuming that $\Delta t = 0.25$ ms, the Thévenin equivalent voltage should be calculated at $t = 2.25$ ms in a generic case, but here, the generator voltage, 66.12969 V, is available.

For segment 2:

$$\varphi(i) = 0.02i + 0.14526.$$

Table 4.2 Flux and current time response.

t(ms)	*e(t)* V	$\varphi(t)$ Wb	*i(t)* A
0	100	0	0
1	92.9773	0.09765	0.9765
2	72.8956	0.181581	1.81581

Thus:

$$h(t-\Delta t) = h(2.00),$$

$$h(2.00) = \varphi(2.00) + \frac{\Delta t}{2}e(2.00) = 0.181581 + \frac{\Delta t}{2}72.8956 = 0.190693,$$

$$f_1(i) = [\varphi(i) - h(2.00)]\frac{2}{\Delta t},$$

$$f_1(i) = (0.02i + 0.14526 - 0.190693)\frac{2}{\Delta t} = 160i - 363.464.$$

After this calculation, the solution can be found immediately:

$$66.12969 = 160i - 363.464,\ i = 2.684961\ \text{A}.$$

In this simple case, it is possible to extract an exact result that allows the integration error to be calculated.

$$\varphi \text{ for } t = 2.25\ \text{ms},\ \varphi(2.25) = 0.198972\ \text{Wb}.$$

Thus:

$$0.198972 = 0.02i + 0.14526,\ i = 2.68559\ \text{A},$$

which is very close to the calculated value using the trapezoidal rule. The error can be reduced by reducing the time step Δt.

4.4.3 Conversion of the Saturation Curve

The data for the magnetization branch are usually supplied in rms units, obtained from calculations or experiments. For analyses in the time domain, it is necessary to convert voltage × current saturation curve in rms values ($V \times I$) to flux × current ($\varphi \times i$) in peak values.

In Figure 4.20, curves A and B show two different slopes, α_1 and α_2, in the saturated region of the magnetization branch of the transformer.

Slope α_0 corresponds to the segment with a linear relation between flux and current.

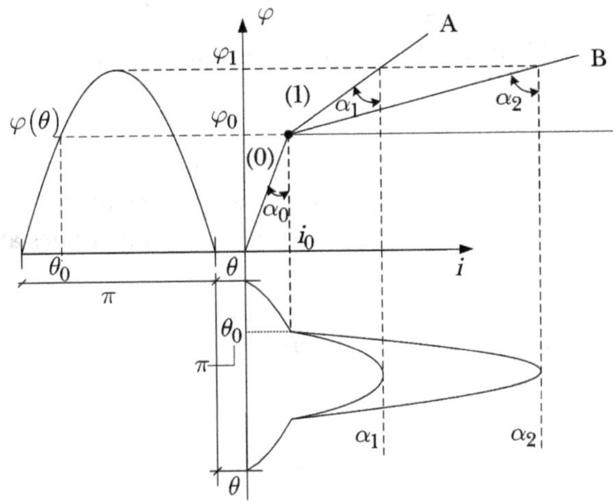

Figure 4.20 Flux and current at intervals (0) and (1).

Assuming the application of a sinusoidal voltage to the inductor:

$$v(t) = \sqrt{2}V\sin(\omega t),$$

which corresponds to the flux:

$$\varphi(t) = -\frac{\sqrt{2}V}{\omega}\cos(\omega t).$$

The peak value of the flux is obtained from the rms value of the voltage using the following relation:

$$\varphi_m = \frac{\sqrt{2}V}{\omega}, \tag{4.62}$$

where φ_m is the amplitude of the sinusoidal flux.

For the current in interval (0), corresponding to the first segment, the relationship between the peak value and rms value is $\sqrt{2}$, since it is in the linear region.

The relationship between the currents and fluxes is examined in interval (1), when the second straight line segment introduces a nonlinearity, and the current consists of sinusoidal arcs.

The relationship between flux and current is established using the line segments:

- Interval (0) $i = f_0(\varphi)$ $0 < i \leq i_0$:
  In this interval, the coefficient α_0 allows the current to be obtained from the flux using the linear equation:

$$i = \alpha_0\varphi.$$

- Interval (1) $i = f_1(\varphi)$ $i_0 < i$:
  In this interval, using coefficient α_1, the current is obtained from:

$$i = i_0 + \alpha_1(\varphi - \varphi_0).$$

Assume that the $\varphi \times i$ curve is given by points φ_0, i_0, φ_1, i_1 and the corresponding $V \times I$ curve (V_0, I_0, V_1, I_1) needs to be calculated. Points V_0, I_0, V_1 are obtained by means of simple relationships between amplitudes and rms values for sinusoidal functions.

Taking advantage of the current symmetry, the rms value in interval $[0, \pi/2]$, can be calculated using the definition:

$$I_1^2 = \frac{2}{\pi}\int_0^{\frac{\pi}{2}} i^2 d\theta. \tag{4.63}$$

Hence, the current I_1 in interval (1) can be easily calculated as the rms value.

The inverse problem is much more common when V_0, I_0, V_1, I_1 are supplied as rms values and the corresponding φ_1, i_1 values must be obtained.

From the previous relationships, φ_0, i_0, φ_1 are immediately obtained and i_1 must be calculated. The problem is to obtain the slope α_1 of the line segment in interval (1), so that the calculated current i_1 can be converted to the rms value I_1 provided (see Figure 4.21).

In order to make this conversion, reference [8] adopts the trapezoidal rule using a recursive approach. However, the problem is solved here by applying the definition of the rms value, taking the slope, α_1, as the unknown variable.

The saturation curve can be composed of several straight-line segments, but only two segments are assumed to exemplify the method, using an applied flux of amplitude φ_1, within the limits of interval (1):

$$\varphi = \varphi_1 \sin\theta.$$

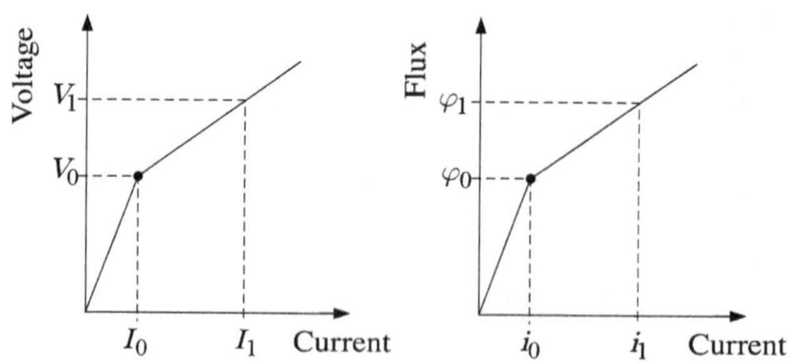

Figure 4.21 Conversion of $V \times I$ to $\varphi \times i$ curve.

The slope α_0 is calculated from φ_0, i_0, and the equation for the current in interval (0) is given by:

$$i = \alpha_0 \varphi_1 \sin\theta. \tag{4.64}$$

For interval (1), the expression for the current is:

$$i = i_0 + \alpha_1(\varphi_1 \sin\theta - \varphi_0). \tag{4.65}$$

Applying the definition of the rms value in the two intervals, (0) and (1),

$$I_1^2 = \frac{2}{\pi}\left[\int_0^{\theta_0} i^2 d\theta + \int_{\theta_0}^{\pi/2} i^2 d\theta\right], \tag{4.66}$$

where θ_0 delimits the two integration intervals and is given by:

$$\theta_0 = \arcsin\left(\frac{\varphi_0}{\varphi_1}\right).$$

Within interval $[0, \theta_0]$, the expression for the square of the current is:

$$i^2 = \alpha_0^2 \varphi_1^2 \sin^2\theta,$$

and analogously in interval $[\theta_0, \pi/2]$:

$$i^2 = i_0^2 + 2i_0\alpha_1(\varphi_1 \sin\theta - \varphi_0) + \alpha_1^2\left(\varphi_1^2 \sin^2\theta - 2\varphi_0\varphi_1 \sin\theta + \varphi_0^2\right).$$

The value of α_1 is determined from the solution to the second-degree equation:

$$x\alpha_1^2 + y\alpha_1 + z = 0 \tag{4.67}$$

with the coefficients indicated below:

$$x = \int_{\theta_0}^{\pi/2} \left(\varphi_1^2 \sin^2\theta - 2\varphi_0\varphi_1 \sin\theta + \varphi_0^2\right) d\theta,$$

$$y = 2i_0 \int_{\theta_0}^{\pi/2} (\varphi_1 \sin\theta - \varphi_0) d\theta,$$

$$z = \alpha_0^2\varphi_1^2 \int_0^{\theta_0} \sin^2\theta d\theta + i_0^2 \int_{\theta_0}^{\pi/2} d\theta - \frac{\pi I_1^2}{2}.$$

Coefficients x, y, z are calculated, given that:

$$\int \sin^2(\theta) d\theta = \frac{1}{2}\left(\theta - \frac{\sin 2\theta}{2}\right).$$

For a $\varphi \times i$ curve with more saturation points, the previous results can easily be extended to other line segments.

Example 4.6 For the $V \times I$ saturation curve plotted in rms values, obtain the corresponding $\varphi \times i$ curve as peak values. The curve can be approximated by two straight-line segments [8].

The assumed base values are:

$$V_b = 635.1\ \text{kV and } S_b = 50\ \text{MVA}.$$

$V \times I$ curve in per unit (pu):

$$V_0 = 0.9; I_0 = 0.0056; V_1 = 1.0; I_1 = 0.015.$$

Thus:

$$\varphi_0 = \frac{\sqrt{2}V_0}{\omega} = 2144.22\ \text{Vs}.$$

$$\varphi_1 = \frac{\sqrt{2}V_1}{\omega} = 2382.46\ \text{Vs},$$

$$i_0 = 0.6235\ \text{A}, \left(i_0 = \sqrt{2}I_0 \frac{50 \times 10^6}{635.1 \times 10^3} \right).$$

Applying the previous expressions:

$$\theta_0 = 1.1198\ \text{rad},$$

$$\alpha_0 = 2.90779 \times 10^{-4},$$

$$x = 1.359 \times 10^4,\ y = 89.025,\ z = -1.841,$$

resulting in:

$$\alpha_1 = 0.008815.$$

Applying (4.65) at the peak value, $\theta = \pi/2$:

$$i_1 = i_0 + \alpha_1(\varphi_1 - \varphi_0) \text{ or } i_1 = 2.72372\ \text{A}.$$

4.5 Representation of Switches

Transients due to changes in network topology can be simulated by altering the switch status of the modeled network. One possible way to simulate the closing and opening of circuit breakers is by means of switches that are represented in the network as resistances, which would ideally assume a null value when closed and an infinite value when open. In practical computational terms, very small or very large values can be assumed.

Another possibility is to use the node elimination technique. The nodes at switch terminals are arranged so that they are located at the end of the matrix and only the upper nodes without switches are subjected to triangular factorization. When

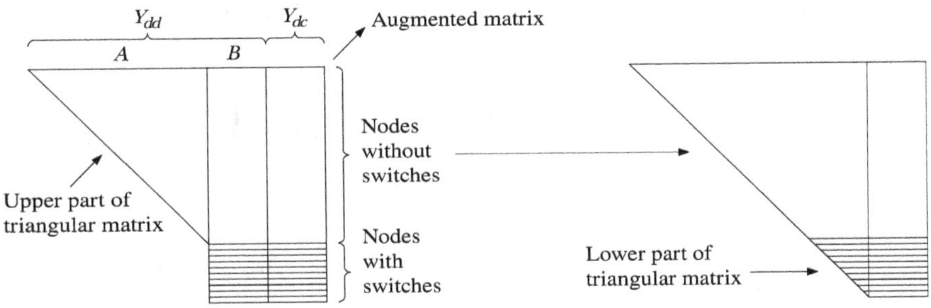

Figure 4.22 Switched network.

the status of a switch changes, this alteration is updated in the lower part of the matrix. If the switch closes, lines and columns are added and a single node is stored. Subsequently, the method of ordered triangular factorization is performed (Figure 4.22).

When a switch closes between nodes k and m, if node k is maintained and node m is eliminated, the sum of nodal currents $i_k + i_m$ is obtained at node k, and the same voltage $v_k = v_m$ is maintained at node k. This operation corresponds to the sum of rows and columns k and m, maintaining row and column k and eliminating row and column m.

4.6 Multiphase Models

In the previous sections, several single-phase basic elements of electric networks were presented.

Transient analysis of power systems requires a three-phase representation, since there are interactions between phases in transmission lines, as well as in other power system components, such as transformers, for instance.

Initially, the basic treatment of coupled circuits with lumped parameters, consisting of inductors, capacitors, and resistors will be presented and, subsequently, coupled circuits with distributed parameters will be addressed.

4.6.1 Three-Phase Inductive Circuit with Mutual Inductances

The inductive circuit is represented in the matrix formulation (Figure 4.23):

$$\begin{bmatrix} v_k^a(t) \\ v_k^b(t) \\ v_k^c(t) \end{bmatrix} - \begin{bmatrix} v_m^a(t) \\ v_m^b(t) \\ v_m^c(t) \end{bmatrix} = \begin{bmatrix} L_a & L_{ab} & L_{ac} \\ L_{ab} & L_b & L_{bc} \\ L_{ac} & L_{bc} & L_c \end{bmatrix} \times \frac{d}{dt} \begin{bmatrix} i_{km}^a(t) \\ i_{km}^b(t) \\ i_{km}^c(t) \end{bmatrix}, \tag{4.68}$$

where the inductance matrix L is assumed to be symmetrical.

$$\frac{d}{dt}\begin{bmatrix} i_{km}^{a}(t) \\ i_{km}^{b}(t) \\ i_{km}^{c}(t) \end{bmatrix} = \begin{bmatrix} & L & \end{bmatrix}^{-1} \times \left\{ \begin{bmatrix} v_{k}^{a}(t) \\ v_{k}^{b}(t) \\ v_{k}^{c}(t) \end{bmatrix} - \begin{bmatrix} v_{m}^{a}(t) \\ v_{m}^{b}(t) \\ v_{m}^{c}(t) \end{bmatrix} \right\}. \tag{4.69}$$

Defining the vectors:

$$\underset{\sim}{i}_{km}(t) = \begin{bmatrix} i_{km}^{a}(t) \\ i_{km}^{b}(t) \\ i_{km}^{c}(t) \end{bmatrix}, \; \underset{\sim}{v}_{k}(t) = \begin{bmatrix} v_{k}^{a}(t) \\ v_{k}^{b}(t) \\ v_{k}^{c}(t) \end{bmatrix}, \; \underset{\sim}{v}_{m}(t) = \begin{bmatrix} v_{m}^{a}(t) \\ v_{m}^{b}(t) \\ v_{m}^{c}(t) \end{bmatrix}. \tag{4.70}$$

Much like the single-phase case in Eq. (4.4), Eq. (4.69) is integrated:

$$\underset{\sim}{i}_{km}(t) = \frac{\Delta t}{2} L^{-1} \times \left[\underset{\sim}{v}_{k}(t) - \underset{\sim}{v}_{m}(t) \right] + \underset{\sim}{I}_{km}(t - \Delta t), \tag{4.71}$$

where:

$$\underset{\sim}{I}_{km}(t - \Delta t) = \underset{\sim}{i}_{km}(t - \Delta t) + \frac{\Delta t}{2} L^{-1} \times \left[\underset{\sim}{v}_{k}(t - \Delta t) - \underset{\sim}{v}_{m}(t - \Delta t) \right]. \tag{4.72}$$

The conductance matrix can be defined as:

$$G_L = \frac{\Delta t}{2} L^{-1}. \tag{4.73}$$

The conductance matrix is inserted into the nodal matrix, in a three-phase arrangement of phases a, b, c, analogous to the single-phase case.

To exemplify the three-phase nodal equation, it is assumed that nodes $k_{a,\,b,\,c}$ and $m_{a,\,b,\,c}$ have been sequentially ordered in the admittance matrix.

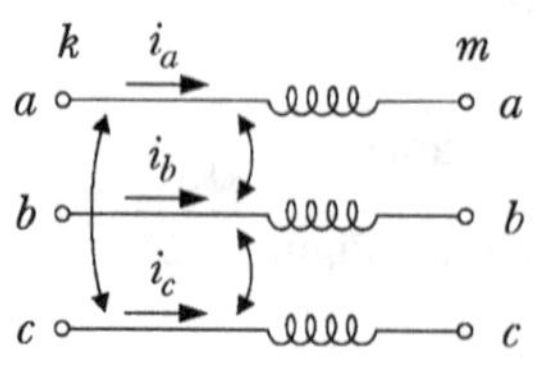

Figure 4.23 Circuit with mutual inductances.

$$\underset{\sim}{i} - \begin{bmatrix} I_{k}^{a}(t-\Delta t) \\ I_{k}^{b}(t-\Delta t) \\ I_{k}^{c}(t-\Delta t) \\ I_{m}^{a}(t-\Delta t) \\ I_{m}^{b}(t-\Delta t) \\ I_{m}^{c}(t-\Delta t) \end{bmatrix} = \left[\begin{array}{c|c} [G_L] & -[G_L] \\ \hline -[G_L] & [G_L] \end{array} \right] \times \begin{bmatrix} v_{k}^{a}(t) \\ v_{k}^{b}(t) \\ v_{k}^{c}(t) \\ v_{m}^{a}(t) \\ v_{m}^{b}(t) \\ v_{m}^{c}(t) \end{bmatrix} \tag{4.74}$$

Current generator ($\underset{\sim}{i}$); History values ($I(t-\Delta t)$ vector)

In the nodal admittance matrix, Y, of a generic system, submatrices $[G_L]$ are inserted as follows:

$$Y = \begin{array}{c} \\ k_a \\ k_b \\ k_c \\ m_a \\ m_b \\ m_c \end{array} \begin{array}{c} \begin{array}{cccccc} k_a & k_b & k_c & m_a & m_b & m_c \end{array} \\ \left[\begin{array}{c|c} +[G_L] & -[G_L] \\ \hline -[G_L] & +[G_L] \end{array}\right] \end{array}. \tag{4.75}$$

4.6.2 Three-Phase Circuit with Resistances and Inductances

Simplifying the matrix notation, for a three-phase RL circuit, the following can be written:

$$\underset{\sim k}{v}(t) - \underset{\sim m}{v}(t) = R\underset{\sim km}{i}(t) + L\frac{d}{dt}\underset{\sim km}{i}(t), \tag{4.76}$$

integrating:

$$\int_{t-\Delta t}^{t}\left[\underset{\sim k}{v}(t) - \underset{\sim m}{v}(t)\right]dt = R\int_{t-\Delta t}^{t}\underset{\sim km}{i}(t)dt + L\int_{t-\Delta t}^{t}\frac{d}{dt}\underset{\sim km}{i}(t)dt, \tag{4.77}$$

Similarly to Section 4.2.4:

$$Z = R + \frac{2}{\Delta t}L, \tag{4.78}$$

thus:

$$\underset{\sim km}{i}(t) = Z^{-1}\left[\underset{\sim k}{v}(t) - \underset{\sim m}{v}(t)\right] + \underset{\sim km}{I}(t-\Delta t), \tag{4.79}$$

the history term is given by:

$$\underset{\sim km}{I}(t-\Delta t) = Z^{-1}\left[\underset{\sim k}{v}(t-\Delta t) - \underset{\sim m}{v}(t-\Delta t) + \left(\frac{2}{\Delta t}L - R\right)\underset{\sim km}{i}(t-\Delta t)\right]. \tag{4.80}$$

Matrix Z^{-1} is inserted into the nodal admittance matrix analogously to item 4.6.1. For a more compact formulation of previous equations, see [3].

The coupled inductances are more common in π-equivalent circuits of transmission lines and in transformer matrices. Network Thévenin equivalents are frequently represented with this model.

4.6.3 Three-Phase Capacitive Circuit

The most usual application is to represent π-circuits in three-phase transmission lines. The matrices of capacitances on both sides of the π-circuit are included in the nodal admittance matrix of the system.

This matrix of capacitances is symmetric, where the diagonal elements contain the sum of all capacitances connected to the nodes, and the off-diagonal elements contain the negative value of the capacitances between the associated nodes (Figure 4.24).

$$C = \begin{bmatrix} c_a & c_{ab} & c_{ac} \\ c_{ab} & c_b & c_{bc} \\ c_{ac} & c_{bc} & c_c \end{bmatrix}, \quad \begin{matrix} c_a = C_{at} + C_{ab} + C_{ac} \\ c_{ab} = -C_{ab}. \end{matrix}$$

The following three-phase equation considers the vector notation of (4.70). For this circuit, the nodal equation of currents and voltages is given as:

$$\underset{\sim}{i}_k(t) = C\frac{d}{dt}\underset{\sim}{v}_k(t). \tag{4.81}$$

It should be recalled that the matrix C on each side of the π-circuit represents half of the total capacitance of the transmission line.

The trapezoidal rule is applied analogously to Eq. (4.8), and simplifying the notation for matrix C, Eq. (4.82) is given as:

$$\underset{\sim}{i}_k(t) = \frac{2}{\Delta t}C\underset{\sim}{v}_k(t) + \underset{\sim}{I}_k(t-\Delta t) \tag{4.82}$$

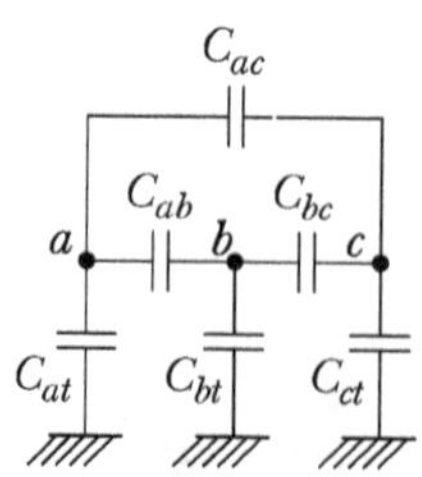

Figure 4.24 Transmission line capacitances.

and

$$\underset{\sim}{I}_k(t-\Delta t) = -\underset{\sim}{i}_k(t-\Delta t) - \frac{2}{\Delta t}C\underset{\sim}{v}_k(t-\Delta t), \tag{4.83}$$

Let:

$$G_c = \frac{2}{\Delta t}C.$$

In the nodal admittance matrix Y of a generic system, submatrices $[G_c]$ of the transmission line between nodes k and m are inserted analogously to (4.75):

4.6.4 Three-Phase Transmission Lines

For practical reasons, some models in EMT-type programs address attenuation in a simplified yet sufficiently accurate manner, enabling their widespread use in the study of large electrical networks with many transmission lines, then providing an extraordinary saving of computational resources.

The design of these simpler models, with resistances interleaved with lossless line segments, has already been discussed in item 4.2.5 and illustrated in Figure 4.8. The following presents the models of the multiphase line segments, considered lossless.

The generalization of the method of characteristics from the single to the three-phase transmission lines is trivial, requiring only a rewriting of the current Eqs. (4.19)–(4.25) in vector form. Below only Eqs. (4.22) and (4.21) are exemplified in vector notation.

$$\underset{\sim k}{i}(t) = Y\underset{\sim k}{v}(t) - \underset{\sim k}{b}(t), \tag{4.84}$$

where

$$\underset{\sim k}{b}(t) = 2\underset{\sim mk}{i}(t-\tau).$$

Equation (4.84) is incorporated into the nodal admittance matrix of the electrical network and solved in the phase domain. At each time step, all variables are computed in the phase domain.

$$\begin{bmatrix} \cdots & \cdots & \cdots & \cdots \\ \cdots & Y & 0 & \cdots \\ \cdots & 0 & Y & \cdots \\ \cdots & \cdots & \cdots & \cdots \end{bmatrix} \times \begin{bmatrix} \cdots \\ \underset{\sim k}{v}(n) \\ \underset{\sim m}{v}(n) \\ \cdots \end{bmatrix} = \begin{bmatrix} \cdots \\ \underset{\sim k}{b}(n) \\ \underset{\sim m}{b}(n) \\ \cdots \end{bmatrix} + \left[\underset{\sim g}{i}(n)\right],$$

where $\underset{\sim g}{i}(n)$ represents other history terms and generators.

However, the traveling waves related to currents are calculated in the modal domain.

$$\underset{\sim k}{\overline{b}}(t) = 2\underset{\sim mk}{\overline{i}}(t-\tau). \tag{4.85}$$

$$\underset{\sim mk}{\overline{i}}(t) = \underset{\sim m}{\overline{i}}(t) + \underset{\sim km}{\overline{i}}(t-\tau) \tag{4.86}$$

Equations (4.85) and (4.86) accurately describe the time delays τ for the propagation modes. Modal equations enable the calculation of each propagation mode in a manner similar to the single-phase case discussed in Section 4.2.5. Details on modal transformation in lines with and without transposition can be found in

Chapter 3. Next, the case of a transposed transmission line is considered, which has some very useful particularities for common applications in EMT-type programs.

Matrices P and P^{-1} are used to interchange modal and phase variables:

$$\underset{\sim}{\overline{v}} = P^{-1}\underset{\sim}{v}, \text{ and } \underset{\sim}{\overline{i}} = P^{-1}\underset{\sim}{i}.$$

Thus, the history term in phase domain is calculated to be used in (4.84) as:

$$\underset{\sim k}{b} = P\underset{\sim k}{\overline{b}}$$

The matrices of inductances and capacitances per unit length are exemplified in phase domain:

$$L_f = \begin{bmatrix} l' & l'_m & l'_m \\ l'_m & l' & l'_m \\ l'_m & l'_m & l' \end{bmatrix}, \; C_f = \begin{bmatrix} c' & c'_m & c'_m \\ c'_m & c' & c'_m \\ c'_m & c'_m & c' \end{bmatrix},$$

by applying a modal transformation to L_f and C_f:

$$\overline{L_f} = \begin{bmatrix} l'_0 & 0 & 0 \\ 0 & l'_1 & 0 \\ 0 & 0 & l'_2 \end{bmatrix}, \; \overline{C_f} = \begin{bmatrix} c'_0 & 0 & 0 \\ 0 & c'_1 & 0 \\ 0 & 0 & c'_2 \end{bmatrix},$$

where:

$$\overline{L_f} = P^{-1}L_fP, \; \overline{C_f} = P^{-1}C_fP.$$

$$l'_0 = l' + 2l'_m, \quad c'_0 = c' + 2c'_m,$$

$$l'_1 = l' - l'_m, \quad c'_1 = c' - c'_m.$$

In the case of transposed lines, modes 1 and 2 are identical, as $l'_1 = l'_2$, $c'_1 = c'_2$. The modal surge impedances and surge admittances are given by:

$$\overline{Z} = \begin{bmatrix} Z_0 & 0 & 0 \\ 0 & Z_1 & 0 \\ 0 & 0 & Z_2 \end{bmatrix}, \; \overline{Y} = \begin{bmatrix} 1/Z_0 & 0 & 0 \\ 0 & 1/Z_1 & 0 \\ 0 & 0 & 1/Z_2 \end{bmatrix}.$$

For each propagation mode j, the following parameters are calculated:

$$Z_j = \sqrt{\frac{l'_j}{c'_j}}, \; v_j = \frac{1}{\sqrt{l'_jc'_j}}, \; \tau_j = \frac{\ell}{v_j}, \; j = 0, 1, 2,$$

where ℓ is the length of the lossless line segment.

The calculation is exemplified in a transmission line with terminals k and m (Figure 4.25).

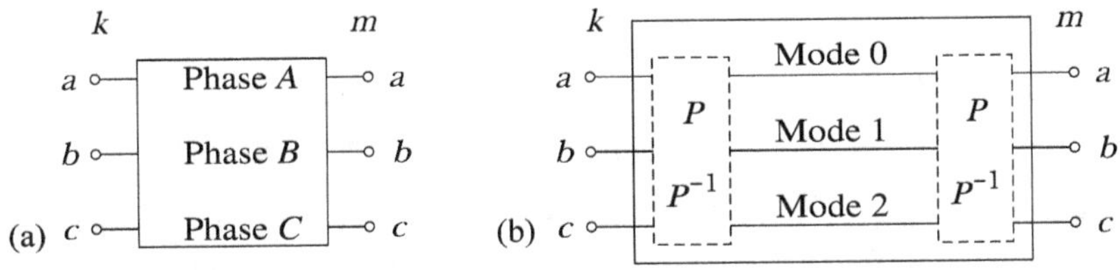

Figure 4.25 Transmission line. (a) Phase domain. (b) Modal domain.

Each mode j is calculated separately, as in Figure 4.6.

The nodal solution for transients in power systems was presented in Section 4.3 for the single-phase case. The same procedure is applied to the three-phase system with the assembly of the nodal admittance matrix in phase domain, including the transmission lines.

Although the following example does not qualify as a complete presentation of the numerical calculation, it is intended to give an overview of modal propagation in transmission lines, illustrating the procedure to be adopted in more complex networks.

Example 4.7 This example refers to the energization of a three-phase transmission line. Example 3.6 in Chapter 3 analyzed the sequential closing of switches connecting ideal voltage generators to the phases of the transmission line. The same case is repeated here but focuses only on the closing of the first phase, with the equations based on modal propagation. The transmission line is presented in Figure 4.26a:

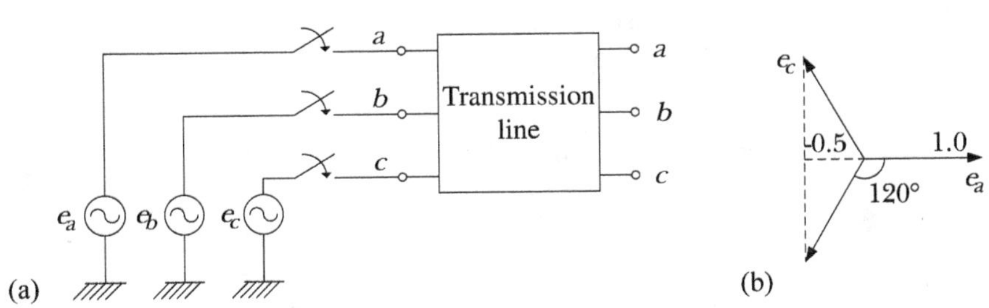

Figure 4.26 Transmission line energization. (a) Three-phase and lossless transmission line. (b) Three-phase source voltages at $t = 0$.

Data

The three-phase voltages of the generator are given by:

$$e_a = \cos(\omega t), e_b = \cos(\omega t - 120°), e_c = \cos(\omega t + 120°).$$

The switch for phase a is closed at time zero ($t = 0$) when phase a is at its maximum value, ($e_a = 1.0$). The transmission line is assumed to be perfectly transposed and the data are extracted from Example 3.6 in Chapter 3:

$$Z_0 = 400\,\Omega, Z_1 = 200\,\Omega, \tau_0 = 1.5\tau_1, \tau_1 = \tau_2.$$

To simplify the solution, the voltages applied by the generators can be assumed to be constant throughout the analysis period, as in Figure 4.26b.

Since phases b and c remain open, the currents in phases a, b, and c at the receiving end, and in phases b and c at the sending end, are zero.

Solution

Only the calculation steps are provided for node $k = 1$ with the corresponding equations to be used, but the numerical results are not shown. The equations for node $m = 2$ are similar.

The modal transformation matrices, as in Section 3.2.3 in Chapter 3, can be written for $n = 3$:

$$P = \begin{bmatrix} 1 & 1 & 1 \\ 1 & -2 & 1 \\ 1 & 1 & -2 \end{bmatrix}, P^{-1} = \frac{1}{3}\begin{bmatrix} 1 & 1 & 1 \\ 1 & -1 & 0 \\ 1 & 0 & -1 \end{bmatrix}.$$

The surge admittance matrix data are furnished in modal values:

$$\overline{Y} = \frac{1}{400}\begin{bmatrix} 1 & 0 & 0 \\ 0 & 2 & 0 \\ 0 & 0 & 2 \end{bmatrix},$$

and they need to be transformed to phase domain:

$$Y = P\overline{Y}P^{-1}.$$

The surge admittance matrix in phase domain is calculated as:

$$Y = \frac{1}{3} \times 10^{-3}\begin{bmatrix} 12.5 & -2.5 & -2.5 \\ -2.5 & 12.5 & -2.5 \\ -2.5 & -2.5 & 12.5 \end{bmatrix}.$$

Time step as in Chapter 3: $\Delta t = \frac{\tau_1}{2}$.

The backward vector in terms of modal coordinates is calculated according to Eq. (4.85):

$$\overline{\underset{\sim 1}{b}}(t) = 2\overline{\underset{\sim 21}{i}}(t-\tau).$$

The calculated vector $\overline{\underset{\sim 1}{b}}$ considers the specific time delay in each mode, and then the corresponding phase values are retrieved

$$\underset{\sim 1}{b} = P\overline{\underset{\sim 1}{b}}.$$

In this case, with just one transmission line, the nodal matrix of the system has only the sub-matrices containing the surge admittances of the transmission line. The insertion of the submatrices of the three-phase transmission line into the network's nodal admittance matrix is similar to that presented in Eq. (4.75), except that the transmission line matrix is block diagonal, as nodes k and m are decoupled. The nodal equations for the network are written in phase domain (as in the second equation after (4.84)):

$$\begin{bmatrix} Y & 0 \\ 0 & Y \end{bmatrix} \times \begin{bmatrix} \underset{\sim 1}{v}(n) \\ \underset{\sim 2}{v}(n) \end{bmatrix} = \begin{bmatrix} \underset{\sim 1}{b}(n) \\ \underset{\sim 2}{b}(n) \end{bmatrix}.$$

The first line is eliminated, since it contains the known voltage applied by phase a of the generator $v_1^a = 1.0\,\text{pu}$. After the calculation of the unknown voltages in $\underset{\sim}{v_1}$ and $\underset{\sim 2}{v}$, a transformation to modal domain is performed,

$$\overline{\underset{\sim 1}{v}} = P^{-1}\underset{\sim 1}{v} \text{ and } \overline{\underset{\sim 2}{v}} = P^{-1}\underset{\sim 2}{v},$$

The next step is to calculate the reflected currents and backward waves to solve again the nodal equations.

$$\overline{\underset{\sim 12}{i}}(t) = \overline{\underset{\sim 1}{i}}(t) + \overline{\underset{\sim 21}{i}}(t-\tau), \text{and } \overline{\underset{\sim 21}{i}}(t) = \overline{\underset{\sim 1}{i}}(t) + \overline{\underset{\sim 12}{i}}(t-\tau).$$

The results are identical to those obtained in Figure 3.8 in Chapter 3, and therefore are not reproduced here.

4.7 Comments on Numerical Errors

The numeric integration method for differential equations is subject to errors that depend on the integration step used. The trapezoidal integration method is also subject to numerical errors, although it shows good numeric stability, Figure 4.27a.

A rigorous analysis of numerical oscillations in discrete systems can be performed by means of the Z transform, as in [3]; however, this is not the purpose of this chapter.

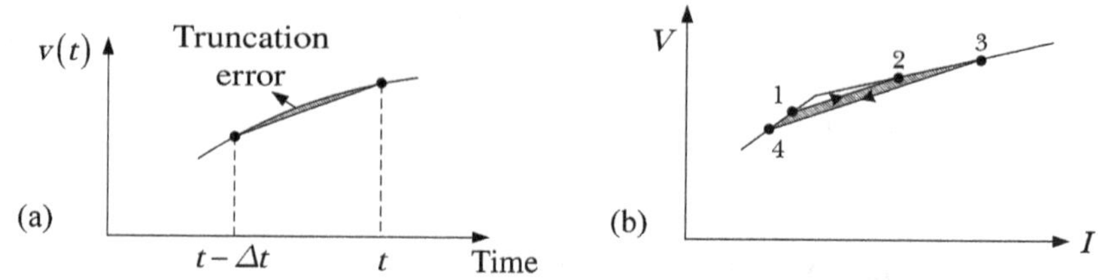

Figure 4.27 Trapezoidal integration error. (a) Truncation error. (b) Non-linear element error.

In the specific case of electromagnetic transient programs, especially EMT-type programs, errors can occur in the nodal admittance matrix. A frequent problem occurs when very different numerical values of the admittance are summed at the same node. Hence, the program verifies the permissible values of the admittance within a tolerance range.

Some nonlinear elements can also introduce numerical errors. As discretized intervals are used, there is no guarantee that the forward and backward paths are the same, if an effect like hysteresis occurs, as in Figure 4.27b.

Numerical methods can show inaccuracy when the derivative of a function changes abruptly. This is a particularly true for sudden variations both current through inductances and voltages across capacitances. Next, the problem of sudden current variations through an inductance becomes evident when current interruption by a switch is analyzed. This is the most common situation in power systems since it is often necessary to deal with current interruptions due to the opening of circuit breakers.

The voltage variation across a capacitance can be understood by the duality with the inductive case analyzed here.

A lossless inductance is initially assumed, with interruption of the sinusoidal current occurring exactly at the zero-crossing point, which would be a rare coincidence in a discrete calculation (Figure 4.28a).

The solution of the circuit in Figure 4.28b is the same as that of item 4.2.1, and applying the trapezoidal rule, Eq. (4.87) is written:

$$i(t) = i(t - \Delta t) + \frac{\Delta t}{2L}(e(t) + e(t - \Delta t)). \tag{4.87}$$

After the switch opens, the current is always zero:

$$i(t) = i(t - \Delta t) = 0, \tag{4.88}$$

therefore:

$$e(t) = -e(t - \Delta t), \tag{4.89}$$

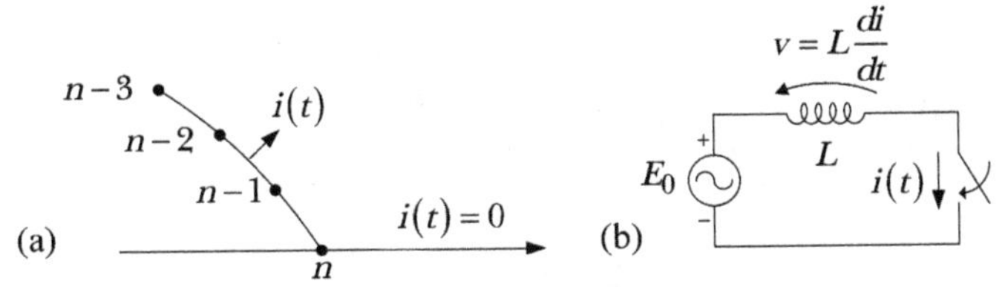

Figure 4.28 Current interruption in an inductive circuit. (a) Zero-crossing of the current at step *n*. (b) Switch opening at step *n*.

which implies an oscillatory voltage in the subsequent time steps (Figure 4.29).

In this case, a numerical oscillation in the voltage is observed due to the sudden variation in the derivative of the current.

Zero-crossing of a current occurs with inversion of its sign, and a residual current remains stored in the inductance. Yet another problem arises due to instantaneous variations in the flux through the inductance.

In this case, the residual energy stored in the inductance due to the residual current i_n in Figure 4.30 cannot be released through any current path, and the oscillations can get even worse.

A commonly used technique to damp the oscillations is to insert a large resistance in parallel with the inductance, to interfere as little as possible in the overall results of the network and to create a path to discharge the energy stored in the inductance, thus attenuating the numerical oscillations (Figure 4.31a).

Models from items 4.2.1 and 4.2.3 are applied in Figure 4.31b.

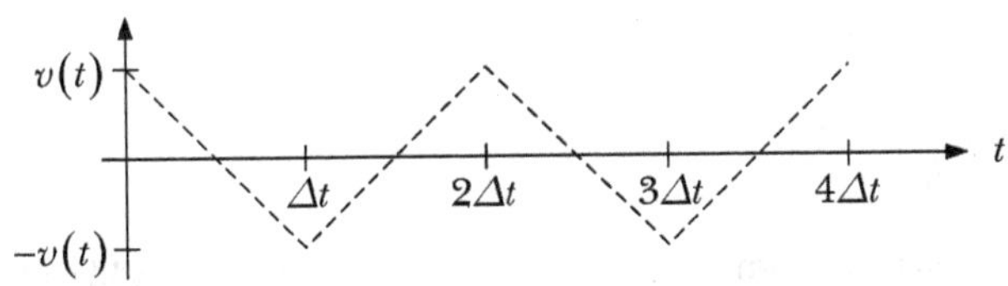

Figure 4.29 Numerical oscillation in the voltage.

Figure 4.30 Change in the sign of the current.

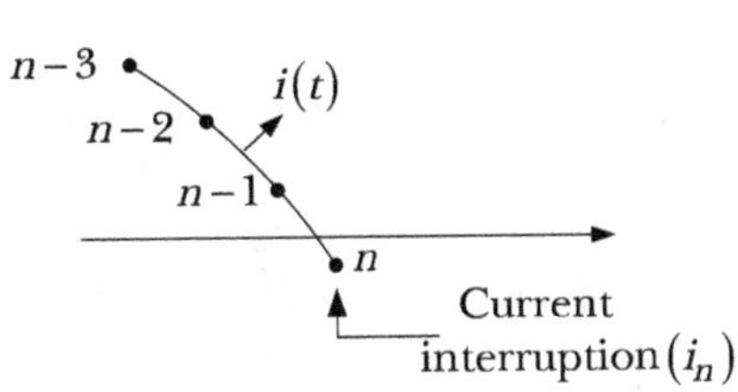

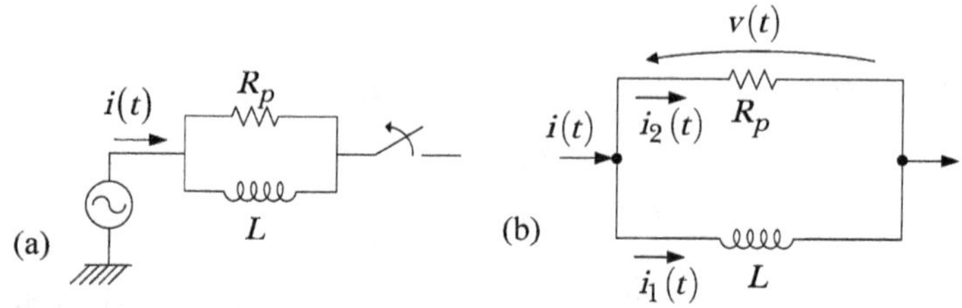

Figure 4.31 Current interruption. (a) Large resistance in parallel with an inductance. (b) Current division in resistance and inductance.

$$i(t) = i_1(t) + i_2(t), \tag{4.90}$$

with:

$$i_1(t) = i_1(t-\Delta t) + \frac{\Delta t}{2L}(e(t) + e(t-\Delta t), \tag{4.91}$$

$$i_2(t) = \frac{1}{R_p}v(t). \tag{4.92}$$

Adding (4.91) and (4.92):

$$i(t) = \left[\frac{\Delta t}{2L} + \frac{1}{R^p}\right] v(t) + i_1(t-\Delta t) + \frac{\Delta t}{2L}.v(t-\Delta t). \tag{4.93}$$

For time $t-\Delta t$:

$$i_1(t-\Delta t) = i(t-\Delta t) - i_2(t-\Delta t), \tag{4.94}$$

$$i_2(t-\Delta t) = v(t-\Delta t)\frac{1}{R_p}. \tag{4.95}$$

Substituting (4.95) into (4.94) and combining the result with (4.93), Eq. (4.96) is obtained:

$$i(t) = \left[\frac{\Delta t}{2L} + \frac{1}{R^p}\right] v(t) + i_1(t-\Delta t) - \left[\frac{1}{R_p} - \frac{\Delta t}{2L}\right] v(t-\Delta t), \tag{4.96}$$

Rewriting (4.96) after interruption of the current, when $i(t) = i(t-\Delta\, t) = 0$:

$$v(t) = -\frac{R_p - \dfrac{2L}{\Delta t}}{R_p + \dfrac{2L}{\Delta t}} v(t-\Delta t), \tag{4.97}$$

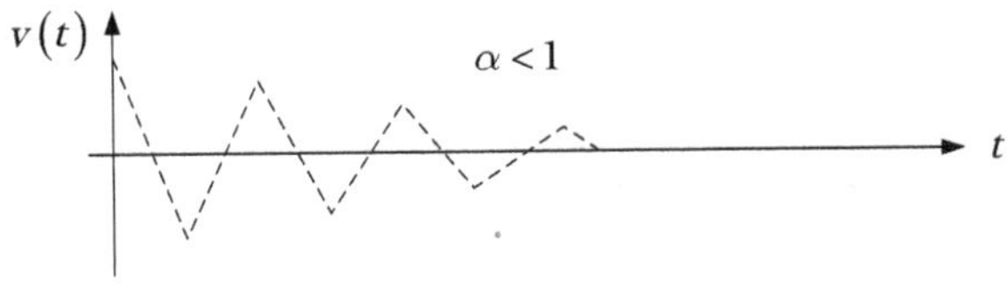

Figure 4.32 Oscillations with a resistance in parallel.

setting:

$$\alpha = \frac{R_p - \dfrac{2L}{\Delta t}}{R_p + \dfrac{2L}{\Delta t}}. \tag{4.98}$$

Factor α is observed to be a positive number smaller than 1, indicating the stability of the solution (Figure 4.32):

$$v(t) = -\alpha v(t - \Delta t). \tag{4.99}$$

Factor $\alpha = 0$ can be considered, which attenuates the oscillations very quickly, setting:

$$R_p = \frac{2L}{\Delta t}. \tag{4.100}$$

Practical results, however, demonstrated that larger values may be more suitable because they have a minimal impact on the network solution and on the steady-state calculation, even though the oscillations are damped more slowly.

Reasonable values for R_p can be found in the range indicated in [3]:

$$5\frac{2L}{\Delta t} \le R_p \le 9\frac{2L}{\Delta t}. \tag{4.101}$$

A resistance in parallel with an inductance can be assumed to somehow contribute to an improved physical representation of the circuit, since lossless inductances are not found on the network.

While damping of the oscillations by means of resistances may be an interesting trick, computing an interpolated value of current at exact zero to open the switch is an elegant solution used in EMTP-type programs.

An example of interpolation is presented below, but in the case of nonmultiplicity of the time delay of transmission lines in relation to the integration step.

With respect to transmission lines, it should be recalled that the simple cases analyzed previously considered the time delay τ as an integer multiple of the time step Δt. In practice, this rarely occurs, and the solution is to interpolate between calculated values. The interpolation of the history terms $I(t - \tau)$ can be done according to Figure 4.33.

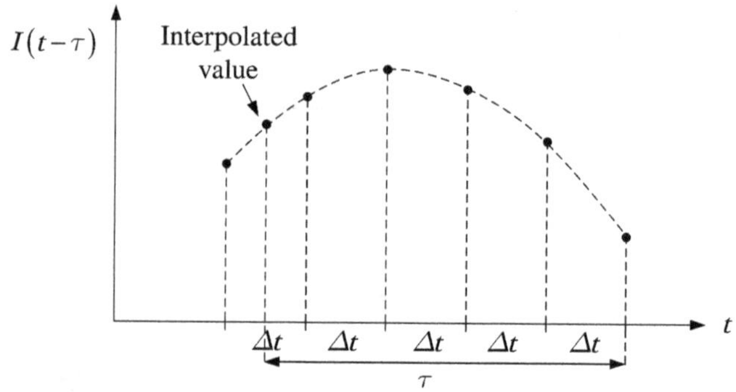

Figure 4.33 Interpolation of the history terms.

References

1 Dommel, H. W. (1969). Digital computer solution of electromagnetic transients in single-and multiphase networks. *IEEE Transactions on Power Apparatus and Systems*, PAS-88(4), 388–399.

2 Dommel, H. W. (1971). Nonlinear and time-varying elements in digital simulation of electromagnetic transients. *IEEE Transactions on Power Apparatus and Systems*, PAS-90(6), 2561–2567.

3 Dommel, H. W. (1986). *Electromagnetic transients program reference manual: (EMTP) theory book*. Portland: Bonneville Power Administration.

4 Dommel, H. W., & Meyer, W. S. (1974). Computation of electromagnetic transients. *Proceedings of the IEEE*, 62(7), 983–993.

5 Gustavsen, B., Sletbak, J., Henriksen, T. (1995). Calculation of electromagnetic transients in transmission cables and lines taking frequency dependent effects accurately into account IEEE. *Transactions on Power Delivery*, 10(2), 1076–1084.

6 *ATP: Alternative Transients Program: Rule book*. (1987). Heverlee (BE): Leuven EMTP Center.

7 Tinney, W. F., & Walker, J. W. (1967). Direct solutions of sparse network equations by optimally ordered triangular factorization. *Proceedings of the IEEE*, 55(11), 1801–1809.

8 Dommel, H. W., Yan, A., & Wei, S. (1986). Harmonics from transformer saturation. *IEEE Transactions on Power Delivery*, 1(2), 209–215.

5

Electrical Parameters Dependence on Frequency

5.1 Introduction

In this chapter, the traveling waves in transmission lines will be analyzed, considering the frequency dependence of the electrical parameters.

Various components of a power system have parameters that depend on frequency, with transmission lines being the most significant due to their frequent involvement in basic transient studies. Other equipment, such as transformers, may also require this level of detail, but it can be said that this is more relevant in certain specialized analyses.

In aerial transmission lines or in isolated cables, calculation of the electrical parameters per unit length, including the skin effect and Carson's corrections [1, 2], shows that the resistance increases with frequency, while the inductance decreases with frequency, and the effect is much more pronounced in the ground mode [3]. The shunt parameters of aerial transmission lines can be assumed constant but their dependence on frequency should be taken into consideration in isolated cables due to the frequency dependence of the permittivity of the insulation. This chapter does not cover the basic theory for calculating frequency-dependent transmission line parameters, offering only a brief summary of the main formulation, given the extensive literature on the topic.

Electric transient phenomena can be studied with reasonable accuracy by means of the Fourier transform, since the series parameters dependence on frequency, such as resistances and inductances, can be accurately calculated in the frequency domain.

The need for a model suitable for implementation in computer programs has motivated the investigations of several researchers since the 1970s [4–7].

Since it is not feasible to accurately analyze the transmission line only in the time domain, and that frequency domain is the most suitable approach, it is necessary to perform efficient inversions of the equations from the frequency domain to the time domain. This calculation is performed using the method of

Understanding Electromagnetic Transients in Power Systems, First Edition. Luiz Cera Zanetta Jr.

Published 2025 by John Wiley & Sons, Inc.

characteristics with rational functions fitted in the frequency domain, which allows efficient recursive convolutions in the time domain.

To incorporate the parameters dependence on frequency, different methods were proposed, such as the weighting functions [6], exponential recursive convolutions [7], Z transform [8], etc.

The approach proposed in reference [9] has proven suitable for efficient incorporation into electromagnetic transient programs. It consists of fitting transfer functions that approximate the impulsive response of the transmission line, including the parameter's dependence on frequency. The general solution applies recursive formulas much like the approach in [7]. The fitting of rational functions is done by means of asymptotic approximations assuming real poles. Each mode is fitted independently of the others, with proper adjustment for the time delay. An approximation is made by assuming a modal transformation matrix that is constant.

At low frequencies, in lines with large asymmetries in their conductors, or in cases of parallel circuits, the assumption of a constant transformation matrix may not be accurate. However, the assumption of constant transformation matrices in most cases is accurate and increases the speed of the calculations.

For certain transmission-line configurations, the need for transformation matrices dependent on frequency may be necessary, and this subject is of current interest [10–20].

The universal line model (ULM) proposed in [21, 22] is formulated in the phase domain and can be applied to any type of line configuration, even when the line has a transformation matrix that is highly dependent on frequency and when the modal time delays vary widely. It is the most general model proposed so far that is capable of accurately representing both asymmetrical overhead lines and underground cables; however, the use of the model implies higher processing times. Rational functions are fitted with the aid of the vector fitting method, assuming real and complex poles.

FDNEs (frequency-dependent network equivalents) enlarge the dimensions of the matrices, when there are several connection ports between the network and the equivalent, which may contain transmission lines and transformers. The methodology is basically the same as for transmission lines, although there may be a greater number of elements to be fitted by rational functions.

Although the items presented below do not represent an exhaustive treatment of the subject, they include the major topics associated with numeric calculation of electromagnetic transients with frequency-dependent parameters.

5.2 Elements for Mathematical Modeling

This section covers two auxiliary calculations that facilitate the use of transmission line models with frequency-dependent parameters. The first topic addresses the adjustment of transmission line and equipment frequency responses to fit a rational function, enabling the application of linear system properties. By representing transmission line functions with rational functions, the calculation of convolutions with voltages or currents becomes significantly more efficient using recursive methods. Consequently, this second topic is also discussed here.

5.2.1 Fitting of Rational Functions

The mathematical fitting of the frequency response of the transfer function $G(s)$, known only by its samples $G(s_k)$, is performed by a rational function $G_f(s)$, as in (5.1).

$$G_f(s) = \frac{N_f(s)}{D_f(s)}, \tag{5.1}$$

given by the ratio of two polynomials. The function can be improper, proper, or strictly proper, depending on the orders n_1 and n of the numerator and denominator, respectively.

The solution is achieved by minimizing the Euclidian norm $\|G(s) - G_f(s)\|$, for m sampled points $s_k = j\omega_k$, $1 \leq k \leq m$, and $m > n$ calculating the polynomial coefficients, a_i and b_i, in:

$$\min \sum_{k=1}^{m} \left| G(s_k) - \frac{b_0 + b_1(s_k) + \ldots + b_{n_1}(s_k)^{n_1}}{a_0 + a_1(s_k) + \ldots + a_n(s_k)^n} \right|^2. \tag{5.2}$$

This formulation is efficient when it is linearized; the differences in (5.2) are multiplied by the denominator of $G_f(s)$ and the least-squares problem is subsequently solved for the overdetermined system.

This approximation can be rewritten in a compact form (where the symbol $\approx$ indicates an approximation in the least-squares sense),

$$N_f(s) \approx D_f(s)G(s),$$

which remains valid when both sides are divided by a known polynomial $D(s)$:

$$\frac{N_f(s)}{D(s)} \approx \frac{D_f(s)}{D(s)} G(s), \tag{5.3}$$

Reference [22] addresses the minimization problem presented in (5.2) using the approach described in (5.3). While this seemingly complicates the formulation, it enables the expansion of rational functions into partial fractions.

The fitting is executed in two steps. In the first step, a least squares formulation is established to iteratively calculate the poles of $G_f(s)$. In the second step, with the poles already defined, a new least squares problem is formulated, aiming solely at calculating the residues of the fitted function. Within the iterative process, the first step is initiated with poles, λ_i, estimated over the frequency range of interest.

The algorithm is implemented with two auxiliary functions, $\sigma G(s)$ and $\sigma(s)$, defined as partial fractions and with the same denominator (assumed determined). The rational functions are also indicated in brackets for reference only.

$$\sigma G(s) = \sum_{i=1}^{n} \frac{\rho_i}{s-\lambda_i} + d \leftrightarrow \left[\frac{N_z(s)}{D(s)}\right], \tag{5.4}$$

$$\sigma(s) = 1 + \sum_{i=1}^{n} \frac{\tilde{\rho}_i}{s-\lambda_i} \leftrightarrow \left[\frac{N_p(s)}{D(s)}\right], \tag{5.5}$$

It can be anticipated that the most interesting results in (5.4) and (5.5) are the residues $\tilde{\rho}_i$, related to the zeros in $N_p(s)$.

The following approximation is proposed, analogous to (5.2):

$$\sigma G(s) \approx \sigma(s) G(s), \tag{5.6}$$

which can be interpreted as:

$$\left[\frac{N_z(s)}{D(s)}\right] \approx \left[\frac{N_p(s)}{D(s)}\right] G(s).$$

It is clear that $N_z(s)$ approximates the zeros while $N_p(s)$ approximates the poles of the fitted function.

Equation (5.6) expands into Eq. (5.7):

$$\left[\sum_{i=1}^{n} \frac{\rho_i}{s-\lambda_i} + d\right] \approx \left[1 + \sum_{i=1}^{n} \frac{\tilde{\rho}_i}{s-\lambda_i}\right] G(s). \tag{5.7}$$

For the solution of (5.7), the expressions in (5.8) provide further details on the equations of the least squares method, particularly in constructing a generic row k of the matrix A and the main vectors used. The following coefficient is defined $\alpha_{ik}=(s_k-\lambda_i)$.

$$A_k^t x = b_k; x = [\rho_1 \ \dots \ \rho_n \ d \ \tilde{\rho}_1 \ \dots \ \tilde{\rho}_n]; \ b_k = G(s_k); \ Ax = b;$$

$$A_k^t = \left[\frac{1}{\alpha_{1k}} \cdots \frac{1}{\alpha_{nk}} \ 1 \ \frac{-G(s_k)}{\alpha_{1k}} \cdots \frac{-G(s_k)}{\alpha_{nk}}\right]; \ A = \begin{bmatrix} A_1^t \\ \vdots \\ A_m^t \end{bmatrix}; \ b = \begin{bmatrix} b_1 \\ \vdots \\ b_m \end{bmatrix}. \tag{5.8}$$

By solving the overdetermined system the residues $\tilde{\rho}_i$ are obtained, and as a consequense the zeros of $\sigma(s)$ need to be calculated.

An efficient calculation of these zeros, which are the approximated poles λ_i of $G_f(s)$, can be extracted from the eigenvalues of Eq. (5.9), as demonstrated in [23].

$$\underset{\sim}{z} = \mathrm{eig}\left(D - \underset{\sim}{b}\underset{\sim}{c}^t\right), \tag{5.9}$$

where:

$$D = \begin{bmatrix} \lambda_1 & \cdots & 0 \\ \vdots & \ddots & \vdots \\ 0 & \cdots & \lambda_n \end{bmatrix}, \quad \underset{\sim}{b} = \begin{bmatrix} 1 \\ \vdots \\ 1 \end{bmatrix}, \quad \underset{\sim}{c}^t = [\widetilde{\rho}_1 \ \cdots \ \widetilde{\rho}_n] \tag{5.10}$$

The poles used in each iteraction will be updated with the results from (5.9), and a new iteration is carried out, if necessary. In the second step, once the approximated poles λ_i have been achieved, the fit is completed by calculating the residues of $G_f(s)$ through a new least squares formulation.

Two significant improvements were introduced in the method, and one of them is given in reference [24], which accelerates the convergence in the pole calculation and reduces the importance of specification of the initial poles, especially in the case of fitting responses with noise contamination. The second modification, proposed in [25], takes advantage of matrix sparsity in the iterative process to calculate the poles by means of the QR transformation.

One of the greatest difficulties in fitting models is passivity enforcement, and some implementations have been undertaken for this purpose, as in reference [26]. The constraints of symmetry, causality, and stability are complementary issues to be discussed. Only an overview of the core of the method has been presented above, and several additional aspects require discussion, such as the distribution of initial poles, fitting order, complex poles, matrix fitting, and so on. However, these are topics for further exploration in the related references. Rational function fitting is an essential tool in EMT-program applications, such as in [27] and its family of related programs, not only for transmission line but also for network equivalent modeling.

5.2.2 Convolution Integral by the Recursive Method

Next, it will be shown how to obtain numerically the convolution integral for a generic input function in a system with an impulse response of the exponential type. Initially, a first-order differential equation is analyzed for a continuous system.

$$\dot{x} = -ax + bu. \tag{5.11}$$

Equation (5.11) represents a first-order system with the following transfer function:

$$X(s) = \frac{b}{s+a} U(s). \tag{5.12}$$

The solution in time domain is given by:

$$x(t) = e^{-at}x_0 + b\int_0^{\infty} u(t-\tau)e^{-a\tau}d\tau. \tag{5.13}$$

The analysis is like that of [7], but assuming (5.14) calculated from (5.13),

$$x(t) = e^{-a\Delta t}x(t-\Delta t) + b\int_{t-\Delta t}^{t} e^{-a\tau}u(t-\tau)d\tau, \tag{5.14}$$

in which $x(t-\Delta t)$ is known and the integration is carried out in one interval Δt. Note that (5.13) turns discrete in (5.14).

Equation (5.14) can be calculated recursively by making a linear approximation of the input function, which has been shown to be very efficient for this purpose (Figure 5.1).

$u(t-\tau)$ can be obtained by linear interpolation.

$$u(t-\tau) = \frac{u(t-\Delta t)-u(t)}{\Delta t}\tau + u(t). \tag{5.15}$$

Substituting (5.15) into (5.14), the integral can be rewritten using a simplified notation:

$$u(t) = u_n,\ u(t-\Delta t) = u_{n-1},\ x(t) = x_n,$$

Calculating the integrals, the equation for x_n may be written as:

$$x_n = \alpha x_{n-1} + \beta u_{n-1} + \gamma u_n, \tag{5.16}$$

with constants α, β and γ calculated by:

$$\alpha = e^{-a\Delta t}, \tag{5.17}$$

$$\beta = \frac{b}{a}\left(-\alpha + \frac{1-\alpha}{a\Delta t}\right), \tag{5.18}$$

$$\gamma = \frac{b}{a}\left(1 - \frac{1-\alpha}{a\Delta t}\right). \tag{5.19}$$

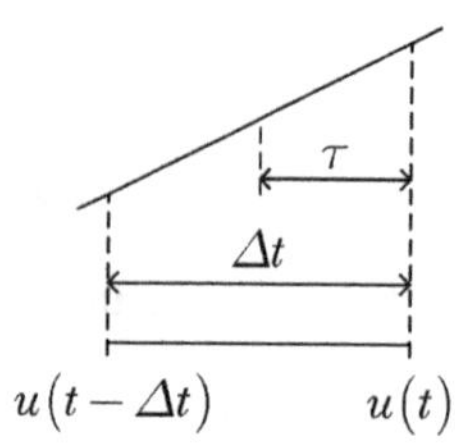

Figure 5.1 Linear interpolation.

Alternative recursive formulas can be obtained more simply [9] when the trapezoidal method of integration is applied to Eq. (5.11).

Applying the trapezoidal rule, which implicitly assumes a linear approximation for $u(t)$, and simplifying the notation, for $t = n\Delta t$, the following result is obtained,

$$x_n = \frac{2 - a\Delta t}{2 + a\Delta t} x_{n-1} + \frac{b\Delta t}{2 + a\Delta t} u_{n-1} + \frac{b\Delta t}{2 + a\Delta t} u_n. \tag{5.20}$$

Setting:

$$\alpha_t = \frac{2 - a\Delta t}{2 + a\Delta t}, \ \beta_t = \frac{b\Delta t}{2 + a\Delta t}, \tag{5.21}$$

thus:

$$x_n = \alpha_t x_{n-1} + \beta_t u_{n-1} + \beta_t u_n, \tag{5.22}$$

which provides results equivalent to those of expression (5.16).

In the case of a second-order system with complex poles, Eq. (5.12) may be rewritten:

$$X(s) = \left[\frac{r}{s + \lambda} + \frac{r^*}{s + \lambda^*}\right] U(s), \tag{5.23}$$

in which:

$$\lambda = \sigma + j\omega,$$

$$r = \delta + j\upsilon.$$

Applying (5.20) to the two parcels associated with λ and λ^* in (5.23), and adding the complex conjugates, it is found that:

$$\alpha_t = \frac{2\left[4 - (\sigma\Delta t)^2\right] - (\omega\Delta t)^2}{(2 + \sigma\Delta t)^2 + (\omega\Delta t)^2}, \tag{5.24}$$

$$\beta_t = \frac{2\Delta t[\delta(2 + \sigma\Delta t) + \omega\Delta t]}{(2 + \sigma\Delta t)^2 + (\omega\Delta t)^2}, \tag{5.25}$$

$$x_n = \alpha_t x_{n-1} + \beta_t u_{n-1} + \beta_t u_n. \tag{5.26}$$

This recursive method to calculate the convolution integral is applicable to systems that have impulse responses composed of exponentials, as in the case of electrical systems. The method, which was demonstrated for first or second-order systems, may be applicable to an nth-order system, when expanded into partial fractions.

Example 5.1 A unitary step voltage is applied to an RL circuit with $R = 1\ \Omega$, $L = 1\text{H}$. Recursively calculate the current at 1.05s knowing its value at 1.00s.

Solution

The time response of the current is known (Chapter 1, Section 1.4.2):

$$i(t) = 1 - e^{-t}:$$

With reference to (5.13) $a = b = 1$, and the current values at 1.0s and 1.05s are calculated:

$$i(1) = 0.63212 \text{ A}, i(1.05) = 0.65006 \text{ A}.$$

The current i_n, $(t = 1.05s)$, is recursively obtained from i_{n-1}, $(t = 1.00s)$, applying (5.17)–(5.19).

Since $\Delta t = 0.05$ s, $a = 1$, $b = 1$; coefficients α, β and γ are calculated:

$$\alpha = e^{-\Delta t} = 0.9512,$$

$$\gamma = 1 - \frac{1-\alpha}{\Delta t} = 0.02459,$$

$$\beta = -\alpha + \frac{1-\alpha}{\Delta t} = 0.02418.$$

For a step excitation: $u_{n-1} = u_n = 1$.

$$i_n = \alpha i_{n-1} + \beta u_{n-1} + \gamma u_n,$$

thus:

$$i_n = 0.65006 \text{ A}.$$

5.3 Modal Domain Approach

In the following treatment of multiphase lines, the propagation matrix will be referred to as $H(\omega)$ for simplicity,

$$H(\omega) = e^{-[\gamma\ell]} \text{ or } H(\omega) = e^{-\gamma\ell}, \tag{5.27}$$

meaning that in the multiphase case, $H(\omega)$ has dimensions $(n \times n)$, and if it is assumed for current propagation, then its transpose $[H(\omega)]^t$ should be used for voltage propagation. In the modal approach, $H(\omega)$ is a scalar designated for a specific propagation mode, and it is the same for voltage and current.

The advantage of the modal domain is that the propagation delays are well defined, allowing for efficient implementation of the line models. The modal transformations facilitate the solution by decoupling the propagation equations into independent modes.

An approximation can be made by considering the modal transformation matrix real and constant across the frequency band, as seen in Example 3.3 in Chapter 3. This assumption allows accurate and faster calculations in most cases of overhead transmission lines [9].

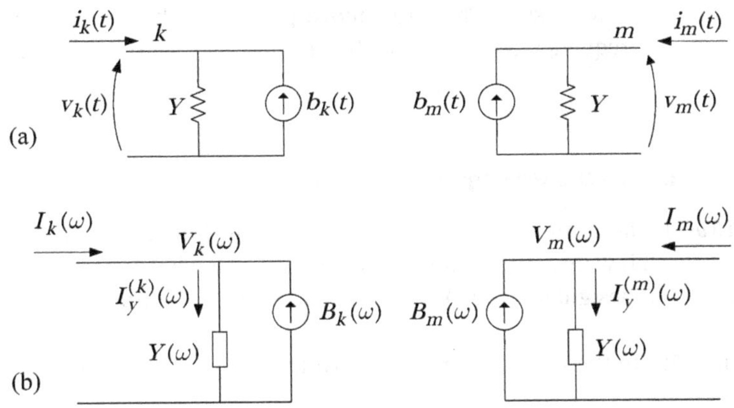

Figure 5.2 Equivalent model. (a) Norton equivalent in time domain. (b) Norton equivalent in frequency domain.

Analogous to Chapter 4, the equivalent circuit is as follows in Figure 5.2a:

In the case of multiphase lines, each mode is treated separately, and the analysis is similar to that of a single-phase line. Practically, only the zero mode, which involves the ground and is the most significant, is typically considered, saving computational time.

The calculation is presented only for node k, since, after substituting k by m, the equations for node m are just the same.

For node k, in Figure 5.2b, there are two convolutions to be calculated. The first refers to the calculation of $B_k(\omega)$, given by the convolution of $H(\omega)$ with a forward wave that propagates from m to k, designated as $F_m(\omega)$:

$$B_k(\omega) = H(\omega)F_m(\omega) \tag{5.28}$$

Note that here $H(\omega)$ is a scalar function.

In reality, in terms of currents and applying Norton's equivalent, Eq. (5.28) should be written as:

$$B_k(\omega) = 2H(\omega)I_{mk}(\omega),$$

however, the form described in (5.28) will be used for simplification.

The second convolution is related to the calculation of the current $I_y^{(k)}$, at node k, given by the product $Y(\omega)$ (characteristic admmitance) times $V_k(\omega)$, as in Figure 5.2b.

$$I_y^{(k)}(\omega) = Y(\omega)V_k(\omega), \tag{5.29}$$

The functions $H(\omega)$ and $Y(\omega)$ have well-established frequency responses; however, the functions $I_{mk}(\omega)$ and $V_k(\omega)$ do not, with their values needing to be calculated at each integration step. Thus, the above products, which would be straightforward in the frequency domain, must be resolved by integration in the time domain.

5.3.1 Convolution with the Propagation Function

The function $H(\omega)$ is the impulse response at node k for an impulse excitation applied at node m, as if it were a semi-infinite transmission line. The inverse Fourier transform of $H(\omega)$ is designated as $h(t)$, whose typical response in time is shown in Figure 5.3.

To return Eq. (5.28) to the time domain, the convolution theorem is applied,

$$b_k(t) = \int_0^t f_m(t-u)h(u)du, \tag{5.30}$$

in which the lower limit in the integral can be replaced by τ, since $h(t)$ is null for $t < \tau$. The constant τ represents the time delay between nodes k and m.

- Rational function fitting

  To perform the calculation in (5.30) efficiently, in the first step, the function $H(\omega)$ is approximated by a rational function $H_f(s)$ with constant coefficients and order n_h, with poles $-\lambda_i$, $1 \le i \le n_h$. The poles can be real or complex variables and assumed real in the following for simplicity.

  The rational function $H_f(s)$ is written below as a strictly proper function, without any loss of generality, and in the expanded form of partial fractions.

$$H_f(s) = \sum_{i=1}^{n_h} \frac{\rho_i}{s + \lambda_i} \tag{5.31}$$

To reproduce the response in Figure 5.3, it is necessary to include the time delay τ, denoting the approximation with this delay as:

$$H_{fd}(s) = H_f(s)e^{-s\tau}. \tag{5.32}$$

Since the parameters are now constant, the first shifting theorem can be applied, and in this format, it is possible to use recursive convolution calculations. The first-order polynomials are written in line with the recursive method from Section 5.2.2.

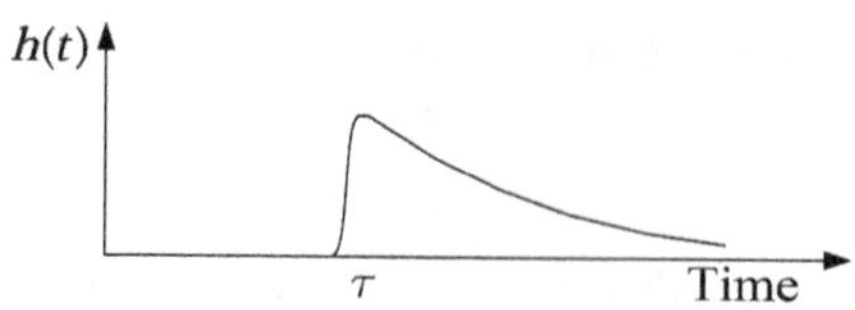

Figure 5.3 Inverse transform of $H(\omega)$.

The time delay τ depends on frequency, and is assumed for the fast components of the propagation. In the case of ref. [9], which involves real poles, only the magnitude of the function $H(\omega)$ is considered in the fitting process, with the phase angle derived as a consequence of the magnitude for the minimum phase system. The time delay can be determined by comparing the phase angle of the propagation function with that of the rational approximation.

$$\tau = \frac{\angle H_f(\omega) - \angle H(\omega)}{\omega}. \tag{5.33}$$

However, when complex poles are involved, both the real and imaginary parts of $H(\omega)$ are used in the fitting process. Therefore, it is recommended to remove the time delay before performing the fitting (a process known as backwinding), achieved by multiplying $H(\omega)$ by $\exp(j\omega\tau)$ and discussed in [17]. The optimized calculation of this time delay, based on Bode's theorem, is proposed in [28].

The backward wave at node k is rewritten with the commutative property of convolution.

$$B_k(s) = H_f(s)F_m(s)e^{-s\tau}, \tag{5.34}$$

which can be rewritten, substituting (5.31) in (5.34):

$$B_k(s) = \left[\sum_{i=1}^{n_h} \frac{\rho_i}{s+\lambda_i}\right] F_m(s)e^{-s\tau}. \tag{5.35}$$

The number of poles n_h in (5.35) depends on the characteristics of the transmission line, the propagation mode analyzed, and the accuracy of the fit to the function. A generic term for the convolution in (5.35) is shown below:

$$B_{ki}(s) = \frac{\rho_i}{s+\lambda_i} F_m(s)e^{-s\tau}. \tag{5.36}$$

- Recursive calculation

  Recalling Eqs. (5.17)–(5.19) constants α_i, β_i, γ_i are calculated and each term $b_{ki}(t)$ is written as:

  $$b_{ki}(t) = \alpha_i b_{ki}(t-\Delta t) + \beta_i f_m(t-\tau-\Delta t) + \gamma_i f_m(t-\tau) \tag{5.37}$$

  Summing all terms at node k:

  $$b_k(t) = \sum_{i=1}^{n_h} b_{ki}(t) \tag{5.38}$$

This current generator, in time domain, is the sum of precalculated values, also denoted as history terms.

In the following example, only the propagation function $H(\omega)$, in a semi-infinite transmission line will be considered.

Example 5.2 For a semi-infinite line, apply a unitary step at the sending end (node m) and calculate the voltage at 200 km from that point (node k), assuming that the parameters are frequency dependent.

Data

$V_m(t) = 1.0\,\text{V}$

Time delay $\tau = 0.70$ ms.

Fitted propagation function $H_{fd}(s)$ [27]:

$$H_{fd}(s) = \sum_{i=1}^{n_h} \frac{\rho_i}{s+\lambda_i} e^{-s\tau}, n_h = 9.$$

The residues ρ_i are the elements of vector $[\rho]$. The poles are the elements of vector $[\lambda]$ with a negative sign.

$$[r] = \begin{bmatrix} 2.75183 \\ 0.855246 \times 10^3 \\ 0.204996 \times 10^8 \\ 19.0354 \\ 0.131952 \times 10^4 \\ 0.408980 \times 10^9 \\ 386.985 \\ 0.186253 \times 10^5 \\ 0.429501 \times 10^9 \end{bmatrix} \quad [l] = \begin{bmatrix} 0.451391 \times 10^3 \\ 0.613296 \times 10^4 \\ 0.532257 \times 10^5 \\ 0.8606804 \times 10^3 \\ 0.8211034 \times 10^4 \\ 0.5213520 \times 10^5 \\ 0.4709777 \times 10^4 \\ 0.2337699 \times 10^5 \\ 0.5218734 \times 10^5 \end{bmatrix}.$$

Note that the negative sign of the last residue may have been caused by the reduction of the original 14 poles to 9 poles to simplify the example.

The poles are given by the corresponding elements in the columns of matrix $[\lambda]$, with a negative sign.

$$[\lambda] = \begin{bmatrix} 0.451391 \times 10^3 & 0.8606804 \times 10^3 & 0.4709777 \times 10^4 \\ 0.613296 \times 10^4 & 0.8211034 \times 10^4 & 0.2337699 \times 10^5 \\ 0.532257 \times 10^5 & 0.5213520 \times 10^5 & 0.5218734 \times 10^5 \end{bmatrix}.$$

Solution

Initially, some calculations will be made with the provided data for ρ and λ.

The impulse response of the rational function $H_f(s)$ is written below and presented in Figure 5.4a:

$$h_f(t) = \sum_{i=1}^{9} \rho_i e^{-\lambda_i t}$$

For the semi-infinite line, the voltage at node k can be calculated from $V_k(\omega) = H(\omega)V_m(\omega)$, (as well as with $B_k(\omega) = H(\omega)F_m(\omega)$, since it is easy to show that $V_k(\omega) = B_k(\omega)/2$ holds for the propagating voltages in the semi-infinite transmission line). The impulse response at node k is not shown, but it is easily attained shifting $h_f(t)$ by τ,

$$h_{fd}(t) = \sum_{i=1}^{9} \rho_i e^{-\lambda_i (t-\tau)}.$$

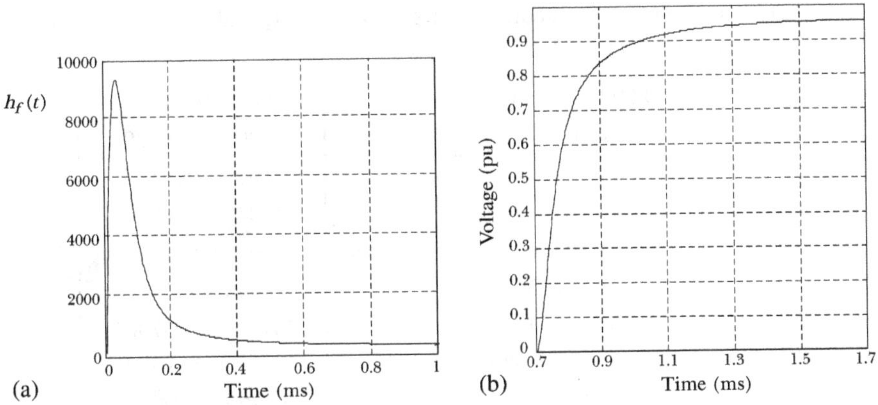

Figure 5.4 Impulse and step response. (a) Impulse response $h_f(t)$ without the time delay. (b) Step response at point m.

With the application of a unitary step at node m, that is, $v_m(t) = 1.0$ V, the voltage at node k is calculated by integrating the impulse response and applying the time delay, in Figure 5.4b:

$$v_k(t) = \sum_{i=1}^{9} \frac{\rho_i}{\lambda_i}\left(1 - e^{-\lambda_i(t-\tau)}\right)$$

For node m, voltages at two consecutive time steps $t_1 = 0.7070$ and $t_2 = 0.7072$ are calculated as $v_1 = 0.5219$ V and $v_2 = 0.5241$ V, respectively. These values were also verified using the EMT program from [27].

Next, the calculation of voltage v_2 is repeated using the recursive formulas, assuming the pre-calculated voltage componentes of v_1 at t_1.

$$v_{ki}(t) = \alpha_i v_{ki}(t - \Delta t) + \beta_i v_m(t - \tau - \Delta t) + \gamma_i v_m(t - \tau),$$

where α_i, β_i and γ_i are calculated from (5.17)–(5.19).

The vectors $\underset{\sim}{\alpha}, \underset{\sim}{\beta}$, and $\underset{\sim}{\gamma}$ have components α_i, β_i, and γ_i for all poles, $1 \le i \le 9$.

$$\underset{\sim}{\alpha} = \begin{bmatrix} 0.99909762538507 \\ 0.99828011989178 \\ 0.99062467102983 \\ 0.98780899976393 \\ 0.98371203905014 \\ 0.95432215104777 \\ 0.89901873941026 \\ 0.90098163904488 \\ 0.90088768957817 \end{bmatrix}, \underset{\sim}{\beta} = \begin{bmatrix} 0.000002750174 \\ 0.000019013570 \\ 0.000384563412 \\ 0.000848284478 \\ 0.001305162410 \\ 0.018054814811 \\ 19.101254774988 \\ 381.631717843013 \\ -400.75287171561 \end{bmatrix}, \underset{\sim}{\gamma} = \begin{bmatrix} 0.000002751002 \\ 0.000019024482 \\ 0.000385772781 \\ 0.000851759904 \\ 0.001312326475 \\ 0.018338392360 \\ 19.791149638395 \\ 395.128115502314 \\ -414.93990815182 \end{bmatrix}.$$

For node k, the vectors with components v_{ki} at times t_1 and t_2 are designated as $\underset{\sim}{v}_k(t_1)$ and $\underset{\sim}{v}_k(t_2)$, respectively.

$$\underset{\sim}{v}_k(t_1) = \begin{bmatrix} 0.00018961663 \\ 0.00129313278 \\ 0.02307635339 \\ 0.04867390774 \\ 0.07025310629 \\ 0.64162643304 \\ 375.86515314163 \\ 7640.60460157905 \\ -8016.74295256299 \end{bmatrix}, \underset{\sim}{v}_k(t_1) = \begin{bmatrix} 0.00019494670 \\ 0.00132894680 \\ 0.02363034118 \\ 0.04978056850 \\ 0.07172631532 \\ 0.64871152492 \\ 376.80222057902 \\ 7660.80429056987 \\ -8037.87781634397 \end{bmatrix}.$$

The elements of vector $\underset{\sim}{v}_k(t_1)$ are the history terms in $t - \Delta t$.

Applying the recursive formula: $\underset{\sim}{\alpha}. \times \underset{\sim}{v}_k(t_1) + \underset{\sim}{\beta} + \underset{\sim}{\gamma}$, where $\underset{\sim}{\alpha}. \times \underset{\sim}{v}_k(t_1)$ denotes the individual products, element-by-element, vector $\underset{\sim}{v}_k(t_2)$ is calculated and presented above. Note that $v_m(t_1 - \tau)$ and $v_m(t_2 - \tau)$ are unitary.

Summing all elements in $\underset{\sim}{v}_k(t_2)$, the voltage at node k is calculated to be:

$$v_k(t_2) = 0.5241 \text{ V}.$$

5.3.2 Convolution with the Characteristic Admittance

The second convolution to be calculated refers to the current at the node k, denoted by $I_y^{(k)}(s)$ in Figure 5.2b, given by the product $Y(\omega)V_k(\omega)$ in (5.29).

$$I_y^{(k)}(\omega) = Y(\omega)V_k(\omega). \tag{5.39}$$

- Fitting of the Characteristic Admittance

  The fitting of the characteristic admittance is proposed as:

$$Y_f(s) = Y_0 + \sum_{j=1}^{n_y} \frac{\rho_j}{s + \lambda_j}. \tag{5.40}$$

n_y: number of poles for Y.
Substituting (5.40) in (5.39):

$$I_y^{(k)}(s) = \left(\sum_{j=1}^{n_y} \frac{\rho_j}{s + \lambda_j} \right) V_k(s) + Y_0 V_k(s) \tag{5.41}$$

- Recursive Calculation
  The convolution of a generic term j in the sum above is performed by means of the same recursive procedure used from (5.35) to (5.38).

$$i_{yj}^{(k)}(t) = \alpha_j i_{yj}^{(k)}(t - \Delta t) + \beta_j v_k(t - \Delta t) + \gamma_j v_k(t). \tag{5.42}$$

The sum of γ_j in (5.42) with Y_0 in (5.41), form an equivalent admittance to be introduced into the nodal admittance matrix.

$$Y_e = Y_0 + \sum_{j=1}^{n_y} \gamma_j. \tag{5.43}$$

The sum of history terms in (5.42) is written as:

$$b_y^{(k)}(t) = \sum_{j=1}^{n_y} \alpha_j i_{yj}^{(k)}(t - \Delta t) + \beta_j v_k(t - \Delta t).$$

The history term $b_y^{(k)}(t)$ above is subtracted from $b_k(t)$ in (5.38) to form the equivalent current generator in Figure 5.5 (note the currents $i_y^{(k)}$, and b_k have opposite signs at node k).

$$b_{ek}(t) = b_k(t) - b_y^{(k)}(t). \tag{5.44}$$

The Norton's equivalents are presented in Figure 5.5 for nodes k and m:

After the nodal solution, the reflected current $i_{km}(t)$, that forms the forward wave to be used at node m is calculated as in Chapter 4, Eq. 4.24.

$$i_{km}(t) = i_k(t) + i_{mk}(t - \tau), \tag{5.45}$$

thus, the information for node k is complete, and by analogy, so is the information for node m.

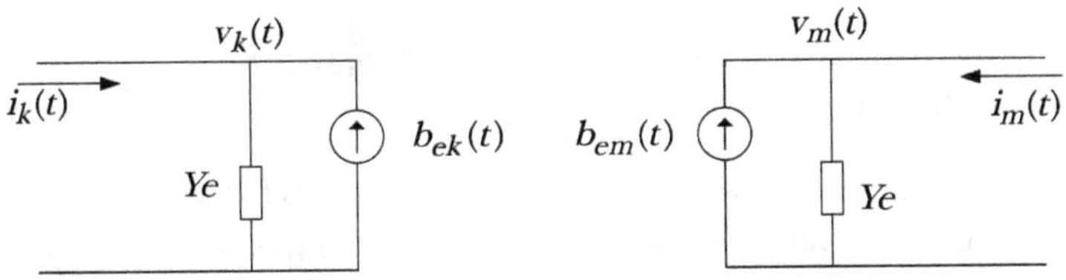

Figure 5.5 Equivalent model.

5.4 Frequency-Dependent Transformation Matrix

For multi-circuit transmission lines with large conductor asymmetry, as well as underground cables, the assumption of a constant transformation matrix may not be accurate, and the use of a frequency-dependent transformation matrix becomes necessary.

Modal analysis becomes more complex when the transformation matrices depend on frequency. This requires a greater number of convolutions and increases the difficulty of fitting rational functions, such as when eigenvalues cross in some frequency range, which could lead to ill-conditioned matrices [11, 13–15].

Let the propagation matrix be $H(\omega)$ of dimension $(n \times n)$ in the phase domain, with the corresponding frequency-dependent modal transformation,

$$H(\omega) = Q(\omega)\overline{H}(\omega)R(\omega), \tag{5.46}$$

and recalling Chapter 3, $Q(\omega)$ is the matrix of right eigenvectors with columns $\underset{\sim\mu}{q}(\omega)$, related to mode μ.

$R(\omega)$: matrix of left eigenvectors with rows $\underset{\sim\mu}{r^t}(\omega)$, related to mode μ, $R(\omega) = Q^{-1}(\omega)$.

$\overline{H}(\omega)$: matrix of modal propagation functions ($\overline{H}(\omega) = \text{diag}[H_\mu(\omega)]$ with elements $H_\mu(\omega) = e^{-\gamma_\mu(\omega)\ell}$).

Spectral decomposition is the key element for accurately incorporating modal velocities into the phase domain approach (idempotent matrices), and for understanding modal mixing in the phases [12, 16].

Equation (5.46) is rewritten in (5.47), as in Eq. (3.19) in Chapter 3.

$$H(\omega) = \sum_{\mu=1}^{n} e^{-\gamma_\mu(\omega)x} M_\mu(\omega), \tag{5.47}$$

where,

$$M_\mu(\omega) = \underset{\sim\mu}{q}(\omega).\underset{\sim\mu}{r^t}(\omega), \tag{5.48}$$

$M_\mu(\omega)$: residue matrix associated with propagation mode μ, $(n \times n)$.
n: number of propagation modes.

In the modal domain, the fit is done on the right side of (5.47) for the functions $e^{-\gamma_\mu(\omega)}$ and $M_\mu(\omega)$, and in the phase domain, the fit is done on the left side for the function $H(\omega)$.

The fitting in the phase domain is analyzed in references [15, 16]. The ARMA model proposed in [16] highlights the advantages of fitting the propagation matrix as a sum of modal contributions directly in the phase domain, using the Z-transform. Later, the model is improved in several aspects in [29], including the interpolations. The discussion on interpolations using the Z-transform is also addressed in reference [30].

The method in reference [21, 22] is designated as the ULM. This model is considered the most efficient and will be analyzed in more detail below.

For a given mode μ in (5.47),

$$H_\mu(\omega) = e^{-\gamma_\mu(\omega)\ell} M_\mu(\omega), \tag{5.49}$$

in which the propagation function is approximated by a rational function:

$$e^{-\gamma_\mu(\omega)\ell} \approx G_f^{(\mu)}(s)e^{-s\tau_\mu}, \text{ recalling that } G_f^{(\mu)}(s)$$

(or $G_{fd}^{(\mu)}(s)$) designates the fitted rational function for mode μ.

Two simplifying assumptions are considered. The first step assumes that the fitting of each propagation function $e^{-\gamma_\mu(\omega)\ell}$ is done using a constant and real modal transformation matrix, calculated at a frequency of 1 MHz, which avoids the issue of artificial mode crossover mentioned earlier. The fitted poles for mode μ are designated as $-\lambda_{\mu m}$. The second step assumes that the matrix $H(\omega)$ in the phase domain can be accurately fitted using the same previously calculated poles $-\lambda_{\mu m}$, which introduces a significant simplification to the approach.

For each frequency, the element h_{ij} of the matrix of propagation functions can be calculated as in (5.50):

$$h_{ij}(\omega) = \sum_{\mu=1}^{n_\mu}\left[\sum_{m=1}^{n_m}\left(\frac{\rho_{\mu m}^{ij}}{j\omega + \lambda_{\mu m}}\right)\right]e^{-j\omega\tau_\mu}, \tag{5.50}$$

in which:

n_μ: number of modes.
n_m: number of poles for each mode.
$\rho_{\mu m}^{ij}$: mth residue of mode μ, for element h_{ij} in H.
$-\lambda_{\mu m}$: mth pole of mode μ.

The Eq. (5.50) is linear with respect to the residues, and an overdetermined linear matrix equation for several frequencies can be written as:

$$AX = B, \tag{5.51}$$

which is solved with the least-squares method.

The following matrices exemplify the simple case for only two modes, one with two poles and the other with three poles, with the elements h_{ij} to be fitted, $1 \le i$, $j \le 2$, in the frequency range $\omega_{min} \le \omega \le \omega_{max}$.

Assuming that the fitting of the propagation functions $e^{-\gamma_1(\omega)\ell}$ and $e^{-\gamma_2(\omega)\ell}$ are given by the rational functions $G_{fd}^{(1)}(s)$ and $G_{fd}^{(2)}(s)$, with their respective time delays.

$$G_{fd}^{(1)}(s) = \left[\frac{\rho_{11}}{s+\lambda_{11}} + \frac{\rho_{12}}{s+\lambda_{12}}\right]e^{-s\tau_1}; G_{fd}^{(2)}(s) = \left[\frac{\rho_{21}}{s+\lambda_{21}} + \frac{\rho_{22}}{s+\lambda_{22}} + \frac{\rho_{23}}{s+\lambda_{23}}\right]e^{-s\tau_2},$$

applying (5.46),

$$\begin{bmatrix} h_{11} & h_{12} \\ h_{21} & h_{22} \end{bmatrix} = \begin{bmatrix} q_{11} & q_{12} \\ q_{21} & q_{22} \end{bmatrix} \times \begin{bmatrix} G_{fd}^{(1)}(s) & 0 \\ 0 & G_{fd}^{(2)}(s) \end{bmatrix} \times \begin{bmatrix} r_{11} & r_{12} \\ r_{21} & r_{22} \end{bmatrix}.$$

By multiplying the matrices or using (5.47),

$$H = \begin{bmatrix} q_{11}r_{11}G_{fd}^{(1)}(s) + q_{12}r_{21}G_{fd}^{(2)}(s) & q_{11}r_{12}G_{fd}^{(1)}(s) + q_{12}r_{22}G_{fd}^{(2)}(s) \\ q_{21}r_{11}G_{fd}^{(1)}(s) + q_{22}r_{21}G_{fd}^{(2)}(s) & q_{21}r_{12}G_{fd}^{(1)}(s) + q_{22}r_{22}G_{fd}^{(2)}(s) \end{bmatrix},$$

it is clear that each element i,j of the matrix H in phase domain is a combination of all propagation modes.

$$h_{ij}(s) = \left[\frac{\rho_{11}^{ij}}{s+\lambda_{11}} + \frac{\rho_{12}^{ij}}{s+\lambda_{12}}\right]e^{-s\tau_1} + \left[\frac{\rho_{21}^{ij}}{s+\lambda_{21}} + \frac{\rho_{22}^{ij}}{s+\lambda_{22}} + \frac{\rho_{23}^{ij}}{s+\lambda_{23}}\right]e^{-s\tau_2} \quad (5.52)$$

In (5.52), it is important to note that the poles are known while the residues are the unknowns to be determined. This allows writing the matrices A and B for calculating the residues using the least squares method.

$$A = \begin{bmatrix} \frac{e^{-j\omega_l\tau_1}}{j\omega_l+\lambda_{11}} & \frac{e^{-j\omega_l\tau_1}}{j\omega_l+\lambda_{12}} & \frac{e^{-j\omega_l\tau_2}}{j\omega_l+\lambda_{21}} & \frac{e^{-j\omega_l\tau_2}}{j\omega_l+\lambda_{22}} & \frac{e^{-j\omega_l\tau_2}}{j\omega_l+\lambda_{23}} \\ \vdots & \vdots & \vdots & \vdots & \vdots \\ \frac{e^{-j\omega\tau_1}}{j\omega+\lambda_{11}} & \frac{e^{-j\omega\tau_1}}{j\omega+\lambda_{12}} & \frac{e^{-j\omega\tau_2}}{j\omega+\lambda_{21}} & \frac{e^{-j\omega\tau_2}}{j\omega+\lambda_{22}} & \frac{e^{-j\omega\tau_2}}{j\omega+\lambda_{23}} \\ \vdots & \vdots & \vdots & \vdots & \vdots \\ \frac{e^{-j\omega_h\tau_1}}{j\omega_h+\lambda_{11}} & \frac{e^{-j\omega_h\tau_1}}{j\omega_h+\lambda_{12}} & \frac{e^{-j\omega_h\tau_2}}{j\omega_h+\lambda_{21}} & \frac{e^{-j\omega_h\tau_2}}{j\omega_h+\lambda_{22}} & \frac{e^{-j\omega_h\tau_2}}{j\omega_h+\lambda_{23}} \end{bmatrix} \quad (5.53)$$

$$X = \begin{bmatrix} \rho_{11}^{11} & \rho_{11}^{12} & \rho_{11}^{21} & \rho_{11}^{22} \\ \rho_{12}^{11} & \rho_{12}^{12} & \rho_{12}^{21} & \rho_{12}^{22} \\ \rho_{21}^{11} & \rho_{21}^{12} & \rho_{21}^{21} & \rho_{21}^{22} \\ \rho_{22}^{11} & \rho_{22}^{12} & \rho_{22}^{21} & \rho_{22}^{22} \\ \rho_{23}^{11} & \rho_{23}^{12} & \rho_{23}^{21} & \rho_{23}^{22} \end{bmatrix} \quad (5.54)$$

$$B = \begin{bmatrix} h_{11}(j\omega_{\min}) & h_{12}(j\omega_{\min}) & h_{21}(j\omega_{\min}) & h_{22}(j\omega_{\min}) \\ \vdots & \vdots & \vdots & \vdots \\ h_{11}(j\omega) & h_{12}(j\omega) & h_{21}(j\omega) & h_{22}(j\omega) \\ \vdots & \vdots & \vdots & \vdots \\ h_{11}(j\omega_{\max}) & h_{12}(j\omega_{\max}) & h_{21}(j\omega_{\max}) & h_{22}(j\omega_{\max}) \end{bmatrix} \tag{5.55}$$

As pointed out in [21, 22], all poles $-\lambda_{\mu m}$ are present in all elements h_{ij} of the fitted matrix H, which allows a column-wise realization of H. Each row in A and B is written for a given frequency and each column in X and B is related to an element of H.

5.5 Model of the Transmission Line with the Nodal Admittance Matrix

The line models with frequency-dependent parameters are based on the method of characteristics, since the functions to be fitted are relatively smooth, both for the propagation matrix and for the characteristic admittance matrix. The method has an advantage over formulations using the nodal admittance matrix because it allows the fitting of rational functions with a smaller number of poles and thus with a lower number of convolutions. Therefore, the approach to the transmission line via the nodal admittance matrix was substituted by the method of characteristics; however, interest in the calculation via the nodal admittance matrix persists. One of the motivations to pursue this modeling is to obtain benchmarks, without the intermediate steps of fitting rational functions.

5.5.1 Inverse Fourier Transform

The following is an approach based on application of the inverse Fourier transform to the calculation of transients using the nodal admittance matrix.

Applying the inverse transform to the nodal admittance matrix for a three-phase transmission line, Y, the nodal currents are calculated for a given time t:

$$i_{ij}(t) = \frac{2}{\pi} \int_0^{\infty} \mathrm{Re}\left[Y_{ij}(\omega)V_j(\omega)\right] \cos(\omega t)d\omega, \tag{5.56}$$

$$i_i(t) = \sum_{j=1}^{n} i_{ij}(t), \tag{5.57}$$

in the previous equations:

Y_{ij}: element i, j of the nodal admittance matrix Y.
V_j: Fourier transform of the voltage applied at node j.

i_{ij}: components of the nodal current at node i.
i_i: nodal current at node i.
n: dimension of nodal admittance matrix Y.

Matrix Y could represent a higher-order system of n nodes or a single three-phase transmission line when $n = 6$.

The use of this formulation implies that the entire network be represented in the frequency domain. If the nodal currents are known, an analogous formulation based on the matrix of nodal impedances may be used to calculate the nodal voltages.

The calculation times associated with this procedure are prolonged due to the need for small frequency increments to ensure precise resolution around resonance points.

The modified (inverse) Fourier transform (MFT) can speed up the calculations [6, 31, 32], although processing times remain high. In this same context of inverting functions from the frequency domaim to the time domain the numerical Laplace transform can be included, discussed in references, such as [33–35].

The following expression gives the MFT with real constant a to shift the eigenvalues and smooth the frequency response, allowing a constant integration step. An upper limit for the frequency Ω and a sigma function to damp Gibbs' oscillations are also considered.

$$i_i(t) = \frac{2e^{at}}{\pi} \int_0^{+\infty} \mathrm{Re}\left\{ \sum_{j=1}^{n} Y_{ij}(a + j\omega) V_j(a + j\omega) \right\} \cos(\omega t) d\omega. \tag{5.58}$$

Calculation of the network parameters displaced by a real constant in the frequency domain requires dedicated routines for this purpose, and that is in addition to some difficulties, such as optimizing the real constant a, the integration step $\Delta\omega$ and the upper limit of integration Ω.

If the voltage in (5.58) is sinusoidal, with a nominal frequency of 60 Hz, the frequency response of the YV product becomes markedly attenuated at high frequencies, allowing a reduction in the upper limit of integration Ω, without significant error.

If the voltage is unknown, the convolution of the impulse response of the admittance matrix, with the voltage calculated at each time step is needed, which adds complexity to the calculation.

A refined alternative method in [36], calculates the Fourier integral in Eq. 5.58 (with $a = 0$) using uneven spacing to reduce the number of samples, while the time-domain response is computed using a semianalytical approach based on the adaptive Simpson's method.

The IFFT (inverse fast Fourier transform) is presented in [35, 37].

5.5.2 State-Space Model of the Transmission Line

A computer code for the rational approximation of the frequency-dependent Y matrix for the transmission line is detailed in [38]. For ill-conditioned situation, the subdivision of the nodal admittance matrix into two submatrices is recommended, considering the short-circuit and open-circuit representation, known as the folded line equivalent [39].

The model of the nodal admittance matrix can be written with spectral decomposition, according to Eq. (5.59), as shown in [19].

$$Y(s) = \sum_{j=1}^{n_y} \frac{1}{s + \lambda_j} R_j + D. \tag{5.59}$$

In (5.59):

$-\lambda_j$: pole of the nodal admittance matrix fitting.
R_j: matrix of residues.
n: dimension of the matrices Y, R, and D, $(n \times n)$.
n_y: number of poles fitting the rational function.

Represented as a state-space model, the matrix of transfer functions of the nodal admittance matrix is written as:

$$Y(s) = C(sI - A)^{-1}B + D. \tag{5.60}$$

The transformation from (5.59) to (5.60) can be understood if the terms of the pole-residue model in (5.59) are rewritten as in Figure 5.6:

$$\frac{1}{s + \lambda_j} R_j = R_j \left(\frac{1}{s + \lambda_j} I \right) I, \tag{5.61}$$

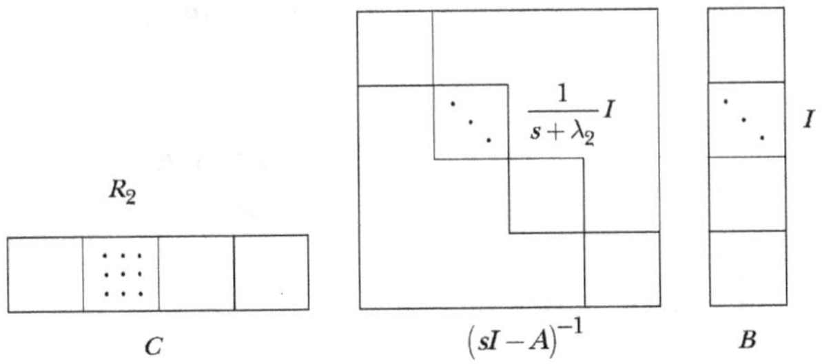

Figure 5.6 State space model.

in which I is the identity matrix and has the same dimensions as matrix R_j. Comparison of (5.60) with (5.61) allows construction of matrices A, B, and C. Note that the sign convention of matrix A in (5.60), which is the most usual, would require that the denominator polynomials in (5.61) should be written as $(s - \lambda_j)$. However, this rigor is irrelevant assuming the eigenvalues in (5.61) as $-\lambda_j$.

In Figure 5.6, an example of the construction of these matrices with real poles is presented, assuming $j = 2$, $n = 3$ and $n_y = 4$; thus, R_2 is the residue matrix of pole λ_2 (note that for a three-phase line, the order is $n = 2 \times 3$). Matrix $(sI - A)^{-1}$ is diagonal and contains the poles repeated n times in the same order as the nodal admittance matrix. Matrix C contains n_y residue matrices and matrix B is composed of n_y identity matrices.

Transformation of the nodal admittance matrix into a state-space model when poles and residues are complex is also discussed in [19].

The MIMO state-space model for $Y(s)$ presented in (5.62) and (5.63) is already in diagonal form, which dispenses with the need for a modal transformation or distinguishing between A and $\overline{A}$. This transmission line representation, accounting for frequency-dependent parameters, represents an elegant approach in state-space systems.

$$\dot{\underset{\sim}{x}} = A\underset{\sim}{x} + B\underset{\sim}{u}, \tag{5.62}$$

$$\underset{\sim}{y} = C\underset{\sim}{x} + D\underset{\sim}{u}, \tag{5.63}$$

5.5.3 Norton's Equivalent

Integrating the linear system in (5.62) from $t - \Delta t$ to t,

$$\underset{\sim}{x}(t) = \underset{\sim}{x}(t - \Delta t) + \int_{t-\Delta t}^{t} [Ax(\zeta) + Bu(\zeta)]\, d\zeta, \tag{5.64}$$

and applying the trapezoidal rule,

$$x(t) = \underset{\sim}{x}(t - \Delta t) + \frac{\Delta t}{2}\{A[x(t - \Delta t) + x(t)] + B[u(t - \Delta t) + u(t)]\}, \tag{5.65}$$

$$\left[I - \frac{\Delta t}{2}A\right] \underset{\sim}{x}(t) = \left[I + \frac{\Delta t}{2}A\right] \underset{\sim}{x}(t - \Delta t) + \frac{\Delta t}{2}B[u(t - \Delta t) + u(t)]. \tag{5.66}$$

Rewriting (5.66) and (5.63) in discrete variables:

$$\underset{\sim k}{x} = \left[I - \frac{\Delta t}{2}A\right]^{-1}\left[I + \frac{\Delta t}{2}A\right] \underset{\sim k-1}{x} + \frac{\Delta t}{2}\left[I - \frac{\Delta t}{2}A\right]^{-1} B \left[\underset{\sim k}{u} + \underset{\sim k-1}{u}\right]. \tag{5.67}$$

$$\underset{\sim k}{y} = C\underset{\sim k}{x} + D\underset{\sim k}{u}, \tag{5.68}$$

The Eq. (5.67) can be expressed in the following form:

$$\underset{\sim k}{x} = \alpha \underset{\sim k-1}{x} + \lambda B \underset{\sim k}{u} + \lambda B \underset{\sim k-1}{u}, \tag{5.69}$$

in which matrices α end λ are defined in (5.70) and (5.71), respectively:

$$\alpha = \left[I - \frac{\Delta t}{2}A\right]^{-1}\left[I + \frac{\Delta t}{2}A\right], \tag{5.70}$$

$$\lambda = \frac{\Delta t}{2}\left[I - \frac{\Delta t}{2}A\right]^{-1}, \tag{5.71}$$

recalling that A, α, and λ are diagonal matrices.

The following substitution of variable $\underset{\sim k}{x}$ by a new variable $\underset{\sim k}{z}$ may be proposed in (5.72):

$$\underset{\sim k}{x} = (\alpha\lambda + \lambda)\underset{\sim k}{z} + \lambda B \underset{\sim k}{u}. \tag{5.72}$$

Substituting (5.72) in (5.69),

$$(\alpha\lambda + \lambda)\underset{\sim k}{z} + \lambda B \underset{\sim k}{u} = \alpha\left[(\alpha\lambda + \lambda)\underset{\sim k-1}{z} + \lambda B \underset{\sim k-1}{u}\right] + \lambda B \underset{\sim k}{u} + \lambda B \underset{\sim k-1}{u}, \tag{5.73}$$

$$(\alpha\lambda + \lambda)\underset{\sim k}{z} = \alpha(\alpha\lambda + \lambda)\underset{\sim k-1}{z} + (\alpha\lambda + \lambda)B\underset{\sim k-1}{u}. \tag{5.74}$$

Since the matrices are diagonal, Eq. (5.74) can be simplified to:

$$\underset{\sim k}{z} = \alpha \underset{\sim k-1}{z} + B \underset{\sim k-1}{u}. \tag{5.75}$$

Analogously, for the output:

$$\underset{\sim k}{y} = C\left((\alpha\lambda + \lambda)\underset{\sim k}{z} + \lambda B \underset{\sim k}{u}\right) + D\underset{\sim k}{u}, \tag{5.76}$$

$$\underset{\sim k}{y} = C(\alpha\lambda + \lambda)\underset{\sim k}{z} + (C\lambda B + D)\underset{\sim k}{u}. \tag{5.77}$$

Designating:

$$S = C(\alpha\lambda + \lambda), \tag{5.78}$$

$$T = (C\lambda B + D). \tag{5.79}$$

The output can be obtained:

$$\underset{\sim k}{y} = S\,\underset{\sim k}{z} + T\,\underset{\sim k}{u} \tag{5.80}$$

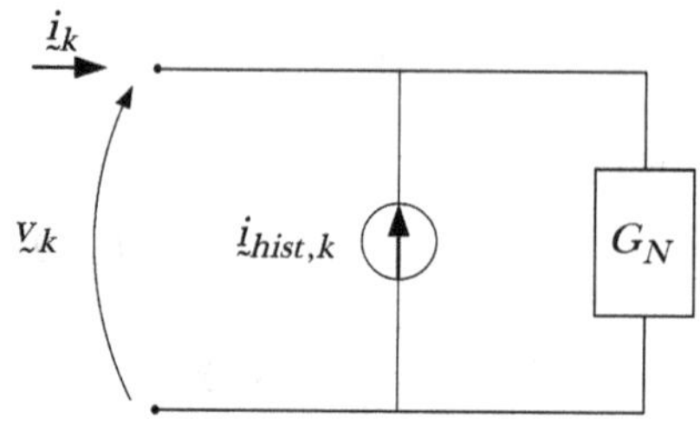

Figure 5.7 Norton's equivalent.

Equation (5.80) can be represented by means of Norton's equivalent, according to Figure 5.7.

$$\underset{\sim}{i}_{hist,k} = -S\underset{\sim}{z}_k. \tag{5.81}$$

The history term $\underset{\sim}{i}_{hist,k}$ is calculated by substituting (5.75) in (5.81).

The conductance matrix G_N for step k is equal to matrix T in (5.79), current $\underset{\sim}{i}_k$ is the output $\underset{\sim}{y}_k$ and voltage $\underset{\sim}{v}_k$ is the input $\underset{\sim}{u}_k$ [40].

$$G_N = T \tag{5.82}$$

This transmission line model can be implemented in electromagnetic transient programs, since it is consistent with the formulation presented in Chapter 4 that describes the application of the trapezoidal rule to the calculation of transients.

5.5.4 Calculation of the Nodal Admittance Matrix in Frequency Domain

The nodal admittance matrix of the transmission line relates the nodal voltages to the nodal currents at the two terminals of the line, with its elements dependent on frequency. For an n-phase transmission line, the nodal admittance matrix dimension is $2n$.

To calculate the frequency response of the nodal admittance matrix, an efficient numerical method is required because the number of calculated points is high.

Calculation of the nodal admittance matrix in the frequency domain can be accomplished using the formulations presented in [36, 41]. The advantage of the method in reference [36] is its greater speed because it avoids hyperbolic expressions.

5.5.5 Frequency-Dependent Network Equivalents-FDNEs

Power systems usually contain a very large number of components, such as transmission lines, transformers, and other equipment. Thus, there is a need to represent parts of the network that are not the focus of the analysis, through equivalent networks that consider the dependence of the parameters on frequency. The frequency-dependent equivalents of these sub-networks may contain transformers and transmission lines. Frequency sweep measurements and frequency scan calculations are used to obtain rational models of power transformers and subnetworks with transmission lines, respectively.

Models obtained by rational functions fitting can be interconnected with EMT-type programs through equivalent circuits (RLC networks). These circuits were, for a long time, the viable alternative, from studies in TNAs to digital studies. However, there are currently more efficient techniques to calculate the multi-port equivalents [38, 40, 41], and Section 5.2.1 introduced the technique for the fitting of rational functions of transmission lines, which can be accurately extended to multiport equivalents, and part of this analysis was seen in Sections 5.5.2 and 5.5.3.

The FDNEs can be calculated analogously, with the augmented state-space system composed of transmission lines and other equipment, like transformers, fitted by rational functions.

The most efficient way to implement the equivalent would be to apply the convolutions directly to the fitted rational function, as in transmission line models with frequency-dependent parameters. However, the EMT-program may not have a user-friendly interface for this purpose.

5.6 Transmission Line Parameters

5.6.1 Internal Impedance of the Conductor

Calculation of the internal impedance of a tubular conductor, presented below, employs well-known skin-effect formulae, with modified Bessel functions. The internal resistance calculation is more significant than the internal inductance, since the internal inductance represents a small portion of the total inductance. Stranded cables as the steel-reinforced aluminum conductor, ACSR, are approximated by a tubular conductor with the same cross-sectional geometry. Solid conductors can be treated as a particular instance of tubular conductors.

q: inner radius of the conductor.
r: outer radius of the conductor.
m: inverse of the complex penetration depth.

$$m = \sqrt{j\omega\mu\sigma}. \tag{5.83}$$

Let $x = mr$, then, the solution to the differential equation for current density is a zero-order modified Bessel equation. For a tubular conductor, the internal and external radii, given by q_i and r_i, respectively, correspond to $x_i = m_i q_i$ and $x_e = m_i r_i$; after variable substitutions, the internal impedance of the conductor is given by the expression:

$$z_{i-\text{int}} = r_{i-\text{int}} + jx_{i-\text{int}}(\Omega/\text{km.}). \tag{5.84}$$

$r_{i-\text{int}}$: conductor internal resistance.
$x_{i-\text{int}}$: conductor internal reactance.

$$z_{i-\text{int}} = \frac{j\omega\mu}{2\pi x_e}\left[\frac{I_0(x_e)K_1(x_i) + I_1(x_i)K_0(x_e)}{I_1(x_e)K_1(x_i) - I_1(x_i)K_1(x_e)}\right]. \tag{5.85}$$

For a solid conductor, if $x_i = 0$ in (5.85), then $I_1(0) = 0$. Although $K_1(0)$ tends to infinity, the numerator and denominator cancel, giving:

$$z_{i-\text{int}} = \frac{j\omega\mu}{2\pi x_e}\frac{I_0(x_e)}{I_1(x_e)}. \tag{5.86}$$

ρ_i: conductor resistivity Ω. km.
σ_i: conductor conductivity S/km.
μ_i: conductor relative permeability.
$I_0(x)$: modified Bessel function of the first kind and order zero.
$I_1(x)$: modified Bessel function of the first kind and order one.
$K_0(x)$: modified Bessel function of the second kind and order zero.
$K_1(x)$: modified Bessel function of the second kind and order one.

5.6.2 Matrix of Series Impedance with Carson's Corrections

In the calculation of the series impedance matrix with frequency-dependent parameters, in addition to the skin effect, the effect of frequency on return of the current through earth is also considered with the aid of Carson's corrections [1].

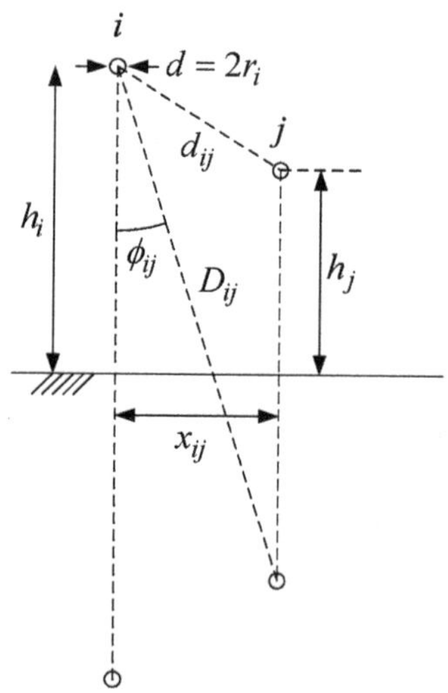

Figure 5.8 Aerial conductors and method of images.

h_i: average height of conductor i in relation to the ground.
d_{ij}: distance between conductors i and j.
D_{ij}: distance between conductor i and the image of conductor j.
ϕ_{ij}: angle between the side and diagonal of the trapezoid in Figure 5.8.
x_{ij}: horizontal distance between conductors i and j.

$$\omega = 2\pi f \text{ rad/s}.$$

Calculation of the series reactance of a transmission line in the first phase assumes the earth to be a perfect conductor for purposes of applying the method of images. In the second phase, Carson's corrections and the skin effect are introduced [2].

The series impedance matrix consists of the following elements in Ω/km.

Self-reactance:

$$z_{ii} = \left(r_{i-int.} + \Delta r_{ii}^{c}\right) + j\left(\omega \times 2 \times 10^{-4} \ln \frac{2h_i}{r_i} + x_{i-int.} + \Delta x_{ii}^{c}\right). \tag{5.87}$$

Mutual-Reactance:

$$z_{ij} = \Delta r_{ij}^{c} + j\left(\omega \times 2 \times 10^{-4} \ln \frac{D_{ij}}{d_{ij}} + \Delta x_{ij}^{c}\right). \tag{5.88}$$

Δr^c: Carson's correction for the resistance.
Δx^c: Carson's correction for the reactance.

Carson's correction terms, Δr^c and Δx^c, are developed in infinite series, arranged for easier programming, and can be introduced into the formulation below, depending on the value of parameter a.

$$a = 4\pi\sqrt{5} \times 10^{-4} D \sqrt{\frac{f}{\rho}}. \tag{5.89}$$

f: frequency (Hz).
ρ: earth resistivity (Ωm).
D: distance equal to $2h_i$ for the impedance of the phase and D_{ij} for the mutual impedance.

The correction terms for the resistance Δr^c and reactance Δx^c are developed for low frequency $a \leq 5$, and high frequency $a > 5$, which can be found in [2], in Ω/km.

5.6.3 Matrix of Series Impedance with a Complex Ground Return Plane

Reference [42] models the impedance to the return current through homogeneous soil, introducing the concept of the complex ground return plane. This plane of perfect conductibility replaces the real earth and is situated at a complex distance δ below the earth's surface; the impedance results from the following equations.

$$\delta = \frac{1}{\sqrt{j\omega\mu_0\sigma}}. \tag{5.90}$$

Terms in the matrix diagonal are:

$$z_{ii} = r_{i-\text{int}} + j\left(\omega \frac{\mu_0}{2\pi} \ln \frac{2(h_i + \delta)}{r_i} + x_{i-\text{int}}\right). \tag{5.91}$$

Terms off the matrix diagonal are:

$$z_{ij} = j\omega \frac{\mu_0}{2\pi} \ln\left(\frac{\sqrt{(h_i + h_j + 2\delta)^2 + x_{ij}^2}}{d_{ij}}\right). \tag{5.92}$$

The advantages of this formulation are simplicity and precision, and it also avoids the discontinuities arising in Carson's series at low and high frequencies.

5.6.4 Matrix of Capacitances

The conductance to earth is neglected in the calculation of the matrix of capacitances. In the calculation of the matrix of Maxwell's potential coefficients, the earth is assumed to be a perfect conductor at zero voltage.

Terms in the matrix diagonal:

$$p_{mii} = \frac{1}{2\pi\varepsilon_0} \ln \frac{2h_i}{r_i}, \tag{5.93}$$

Terms off the matrix diagonal:

$$p_{mij} = \frac{1}{2\pi\varepsilon_0} \ln \frac{D_{ij}}{d_{ij}}. \tag{5.94}$$

$$[P_m] = \left[p_{mij}\right], \quad 1 \leq i, j \leq n, \tag{5.95}$$

in which n is the number of conductors.

Matrix C is obtained by inversion of P_m:

$$C = P_m^{-1}. \tag{5.96}$$

The susceptance matrix is calculated by multiplying C times $j\omega$.

References

1 Carson, J. R. (1926). Wave propagation in overhead wires with ground return. *The Bell System Technical Journal*, 5(4), 539–554.

2 Dommel, H. W. (1986). *Electromagnetic transients program reference manual: (EMTP) theory book*. Portland: Bonneville Power Administration.

3 CIGRE Working Group 13.05. (1979). The Calculation of Switching Surges III: transmission line representation for energization and re-energization studies with complex feeding networks. *Electra*(62), 45–78.

4 Budner, A. (1970). Introduction of frequency-dependent line parameters into an electromagnetic transients program. *IEEE Transactions on Power Apparatus and Systems*, PAS-89(1), 88–97.

5 Snelson, J. K. (1972). Propagation of travelling waves on transmission lines: frequency dependent parameters. *IEEE Transactions on Power Apparatus and Systems*, PAS-91(1), 85–91.

6 Meyer, W. S., & Dommel, H. W. (1974). Numerical modelling of frequency-dependent transmission-line parameters in an electromagnetic transients program. *IEEE Transactions on Power Apparatus and Systems*, PAS-93(5), 1401–1409.

7 Semlyen, A., & Dabuleanu, A. (1975). Fast and accurate switching transient calculations on transmission lines with ground return using recursive convolutions. *IEEE Transactions on Power Apparatus and Systems*, 94(2), 561–571.

8 Humpage, W. D., Wong, K. P., Nguyen, T. T., & Sutanto, D. (1980). Z-transform electromagnetic transient analysis in power systems. *IEE Proceedings C (Generation, Transmission and Distribution)*, 127(6), 370–378.

9 Marti, J. R. (1982). Accurate modelling of frequency-dependent transmission lines in electromagnetic transient simulations. *IEEE Transactions on Power Apparatus and Systems*, PAS-101(1), 147–157.

10 Ametani, A. (1973). Refraction coefficient method for switching-surge calculations on untransposed transmission-lines (accurate and approximate inclusion of frequency dependency) [abstract C 73 444-7]. In: IEEE PES Summer Power Meeting; 1973 Jul 15-20; Vancouver. *IEEE Transactions on Power Apparatus and Systems, PA92*(6), 1809.

11 Marti, L. (1988). Simulation of transients in underground cables with frequency-dependent modal transformation matrices. *IEEE Transactions on Power Delivery*, 3(3), 1099–1110.

12 Marcano, F. J., & Marti, J. R. (1997). Idempotent line model: case studies. In *Proceedings of IPST'97 International Conference on Power Systems Transients;* 1997 Jun 22–26; Seattle (pp. 67–72).

13 Brandao Faria, J. A. (1988). Overhead three-phase transmission lines-nondiagonalizable situations. *IEEE Transactions on Power Delivery*, 3(4), 1348–1355.

14 Wedepohl, L. M., Nguyen, H. V., & Irwin, G. D. (1996). Frequency-dependent transformation matrices for untransposed transmission lines using Newton-Raphson method. *IEEE Transactions on Power Systems*, 11(3), 1538–1546.

15 Angelidis, G., & Semlyen, A. (1995). Direct phase-domain calculation of transmission line transients using two-sided recursions. *IEEE Transactions on Power Delivery*, 10(2), 941–949.

16 Noda, T., Nagaoka, N., & Ametani, A. (1996). Phase domain modeling of frequency-dependent transmission lines by means of an ARMA model. *IEEE Transactions on Power Delivery*, 11(1), 401–411.

17 Gustavsen, B., & Semlyen, A. (1998). Simulation of transmission line transients using vector fitting and modal decomposition. *IEEE Transactions on Power Delivery*, 13(2), 605–614.

18 Gustavsen, B., Irwin, G., Mangelrod, R., Brandt, D., & Kent, K. (1999). Transmission line models for the simulation of interaction phenomena between parallel AC and

DC overhead lines. In: *Proceedings of the International Conference on Power Systems Transients (IPST)*; 1999 Jun 20–24; Budapest, Hungary. (pp. 61–68). Budapest: IPST.

19 Semlyen, A., & Gustavsen, B. (2009). A half-size singularity test matrix for fast and reliable passivity assessment of rational models. *IEEE Transactions on Power Delivery*, 24(1), 345–351.

20 Gustavsen, B. (2012). Modal domain-based modeling of parallel transmission lines with emphasis on accurate representation of mutual coupling effects. *IEEE Transactions on Power Delivery*, 27(4), 2159–2167.

21 Morched, A., Gustavsen, B., & Tartibi, M. (1999). A universal model for accurate calculation of electromagnetic transients on overhead lines and underground cables. *IEEE Transactions on Power Delivery*, 14(3), 1032–1038.

22 Gustavsen, B., & Semlyen, A. (1998). Application of vector fitting to state equation representation of transformers for simulation of electromagnetic transients. *IEEE Transactions on Power Delivery*, 13(3), 834–842.

23 Kailath, T. (1980). *Linear systems*. Englewood Cliffs (NJ): Prentice-Hall.

24 Gustavsen, B. (2006). Improving the pole relocating properties of vector fitting. *IEEE Transactions on Power Delivery*, 21(3), 1587–1592.

25 Deschrijver, D., Mrozowski, M., Dhaene, T., & De Zutter, D. (2008). Macromodeling of multiport systems using a fast implementation of the vector fitting method. *IEEE Microwave and Wireless Components Letters*, 18(6), 383–385.

26 Gustavsen, B. (2008). Fast passivity enforcement for pole-residue models by perturbation of residue matrix eigenvalues. *IEEE Transactions on Power Delivery*, 23(4), 2278–2285.

27 *ATP: Alternative Transients Program: Rule book*. (1987). Heverlee (BE): Leuven EMTP Center.

28 Gustavsen, B. (2017). Optimal time delay extraction for transmission line modeling. *IEEE Transactions on Power Delivery*, 32(1), 45–54.

29 Noda, T., Nagaoka, N., & Ametani, A. (1997). Further improvements to a phase-domain ARMA line model in terms of convolution, steady-state initialization, and stability. *IEEE Transactions on Power Delivery*, 12(3), 1327–1334.

30 Humpage, W. D., Wong, K. P., & Nguyen, T. T. (1982). Z-plane synthesis of response functions and interpolators in z-transform electromagnetic transient analysis in power systems. *IEE Proceedings C (Generation, Transmission and Distribution)*, 129(3), 104–110.

31 Bickford, J. P., Mullineux, N., & Reed, J. R. (1976). *Computation of power system transients*. London: Peter Peregrinus.

32 Papoulis, A. (1962). *The Fourier Integral and its applications*. New York: McGraw-Hill College.

33 Wilcox, D. J. (1978). Numerical Laplace transformation and inversion. *International Journal of Electrical Engineering & Education*, 15(3), 247–265.

34 Gómez, P., & Uribe, F. A. (2009). The numerical Laplace transform: an accurate technique for analyzing electromagnetic transients on power system devices. *International Journal of Electrical Power & Energy Systems*, 31(2–3), 116–123.

35 Moreno, P., & Ramirez, A. (2008). Implementation of the numerical Laplace transform: A review task force on frequency domain methods for EMT studies, working group on modeling and analysis of system transients using digital simulation, general systems subcommittee, IEEE Power Engineering Society. *IEEE Transactions on Power Delivery*, 23(4), 2599–2609.

36 Gustavsen, B. (2005). Validation of frequency-dependent transmission line models. *IEEE Transactions on Power Delivery*, 20(2), 925–933.

37 Ametani, A. (1973). The application of the fast Fourier transform to electrical transient phenomena. *International Journal of Electrical Engineering & Education*, 10(4), 277–287.

38 Gustavsen, B. (2002). Computer code for rational approximation of frequency dependent admittance matrices. *IEEE Transactions on Power Delivery*, 17(4), 1093–1098.

39 Gustavsen, B., & Semlyen, A. (2009). Admittance-based modeling of transmission lines by a folded line equivalent. *IEEE Transactions on Power Delivery*, 24(1), 231–239.

40 Gustavsen, B., & De Silva, H. M. J. (2013). Inclusion of rational models in an electromagnetic transients program: Y-parameters, Z-parameters, S-parameters, transfer functions. *IEEE Transactions on Power Delivery*, 28(2), 1164–1174.

41 Morched, A. S., Ottevangers, J. H., & Marti, L. (1993). Multi-port frequency dependent network equivalents for the EMTP. *IEEE Transactions on Power Delivery*, 8(3), 1402–1412.

42 Deri, A., Tevan, G., Semlyen, A., & Castanheira, A. (1981). The complex ground return plane a simplified model for homogeneous and multi-layer earth return. *IEEE Transactions on Power Apparatus and Systems*, PAS-100(8), 3686–3693.

6

Elements of Power Electronics

6.1 Introduction

The use of power electronics in alternating current networks was initially restricted to industrial rectifiers, and the first effective application in power transmission occurred in 1954 in the Gotland system [1], and since then high voltage direct current (HVDC) applications have become a frequent solution in large long-distance power transmission projects. At the same time, semiconductor switches began to be used to control reactive power, as in static voltage compensators (SVCs), in addition to conventional shunt capacitors and reactors.

The continuous evolution in power electronics has led to the development of static switches that can precisely regulate the onset and cessation of current flow, supporting increasingly higher voltages and currents. Consequently, new reactive compensation equipment based on voltage source converters (VSCs) has emerged alongside significant applications in direct current (DC) links, thereby broadening the possibilities in power transmission. Several benefits can be offered by flexible AC transmission systems (FACTS) devices, such as the ability to control voltage and redirect power flows very quickly. Examples of FACTS equipment include the static compensator (STATCOM) for shunt converter applications, and the static synchronous series compensator (SSSC) for series converter applications. Within this group of equipment, the unified power flow controller (UPFC) is included, as well as other equipment such as the interline power flow controller (IPFC) and the generalized interline power flow controller (GIPFC). A broad description of FACTS equipment is presented in [2–5].

Recently, renewable generators based on inverters (inverter-based generator – IBG) have assumed a relevant share in power system generation, whose dynamic performance deserves special attention [6]. It then became necessary to develop accurate models of IBG, which can no longer be represented simply as negative charges.

Understanding Electromagnetic Transients in Power Systems, First Edition. Luiz Cera Zanetta Jr.

Published 2025 by John Wiley & Sons, Inc.

The IBGs are normally connected to the grid through power converters and have control functions that operate at both the individual and systemic levels, coordinated with the other generators in operation. The generation possibilities with IBGs can form a very large set of small units distributed on the kW scale or even composed of a set of larger units on the MW scale in the same point of common coupling (PCC). It still represents a major challenge to consolidate an aggregation methodology for IBGs, specifically distributed photovoltaic installations, which are not adequately defined at the moment.

A recurring aspect in all power electronics applications is related to the ratio between the short circuit power of the network and the nominal power of the equipment at the point of common coupling, designated as short-circuit ratio (SCR), as a higher ratio simplifies the implementation of power converters, reducing technical issues and the need for additional control functions.

The objective of this chapter is to provide an introduction in the application of static switches in modern power systems, without delving into static switching technology. Finally, the subject is very broad and only a few basic topics are briefly discussed here; hence, students encountering the topic for the first time might naturally sense a dearth of supplementary information, which is extensively covered in the literature on this topic.

6.2 LCC – Line Commutated Converters

In this section, the principles of switching that occur in converters with thyristors will be discussed, as well as some characteristics of the valves and the basic equations for calculating the electrical quantities on the AC and DC sides will be developed.

Accurate switch models are part of a very specialized area of power electronics and are not considered here. Only the simplest models of ideal switches are needed for this introduction.

The diode is the simplest switch, which only conducts when the voltage at the anode is greater than that at the cathode, that is, when this voltage difference is positive.

The current flows from the anode to the cathode in the semiconductor and the opposite only occurs in abnormal conditions in which the diode cannot withstand the applied reverse voltage. Reverse voltage is an important design feature, as it is the same one that allows the switching process from one valve to another.

The diode only interrupts the current at zero crossing; thus, the switching conditions between valves are imposed by the voltages of the electrical network.

Thyristors have the same characteristics as diodes, but they require a gate signal to start conduction, which lasts until the current crosses zero. Thus, unlike diodes, it is possible to control the DC voltage level with the gate signal.

6.2.1 Rectifier Bridge without Commutation Angle

An introduction to line commutated converters (LCCs) is presented in an elementary case with thyristor valves. A six-pulse bridge is assumed, with no commutation inductance and the triggering angle α is initially zero. In this case, the bridge operates without control, like the case of diode bridges. In this simplified bridge, the direct current is assumed constant and ripple-free. Commutation between phases is instantaneous and without overlap and the smoothing reactor maintains the current constant and without ripple. The current pulses are shown in Figure 6.1a.

This figure gives a summary of the current pulses in the upper and lower valves as a function of time. Each upper valve conducts current when the alternating

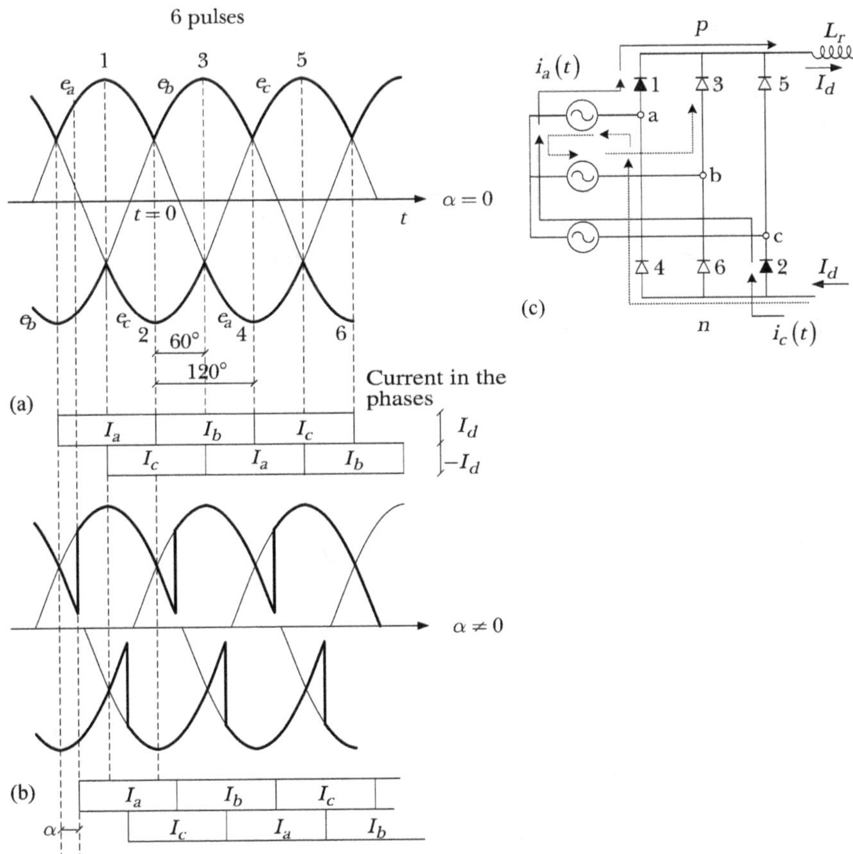

Figure 6.1 Rectifier bridge waveforms. (a) Firing angle α = 0. (b) Firing angle $\alpha \neq 0$. (c) Current path in fired valves. *Source:* Kimbark [1]/John Wiley & Sons.

voltage at the corresponding anode is greater than the other voltages, ensuring a positive polarity between anode and cathode. On the other hand, each lower valve conducts when its phase voltage is at a minimum, and positive polarity between anode and cathode is ensured.

In the upper row, if valves 1 and 2 are switched on, the valves 3 and 5, in phases b and c, respectively, cannot be switched on, as they have reverse voltage: $V_{ap} = 0$, $V_{bp} = V_{ba} < 0$, and $V_{cp} = V_{ca} < 0$.

Assuming that valves 1 and 2 are closed before $t = 0$, the switching from valve 1 to valve 3 occurs at $t = 0$ in Figure 6.1a, when the voltage at phase b exceeds the voltage in phase a. Then the voltage in valve 3 becomes direct and in valve 1, it becomes reverse: $V_{bp} = 0$, $V_{ap} < 0$, $V_{cp} < 0$, and the process is repeated in subsequent cycles.

As a rule, the valve that conducts in the upper line (1,3,5) is the one with the highest algebraic voltage level at the anode and the valve that conducts in the lower line (4,6,2) is the one with the lowest algebraic voltage level at the cathode. Applying this rule, it is possible to establish the diode bridge switching sequence as: (1,2); (3,2); (3,4); (5,4); (5,6); (1,6) and so on.

In Figure 6.1b, the firing of the thyristor bridges with $\alpha \neq 0$ is observed, the effect of which is to delay the initiation of valve commutation, which allows to control the voltage on the DC side. Note that the currents in each phase are delayed by α with respect to the voltage and this is the main reason for the consumption of reactive power by the converter.

Figure 6.2 illustrates the current in phase a of the controlled bridge. The current paths in phase a are delineated in Figure 6.1c, with the paths in fired valves (1,2)

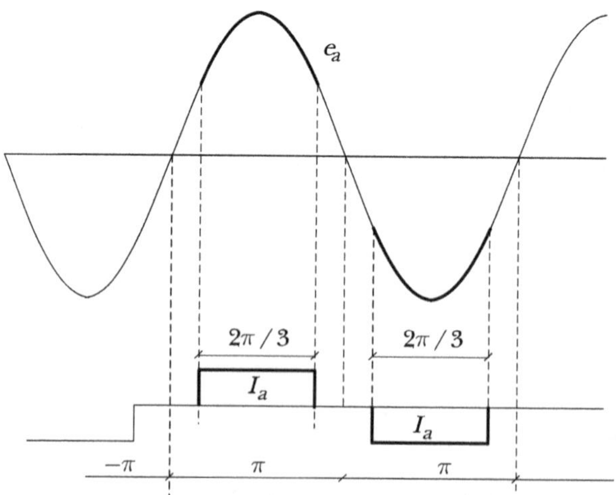

Figure 6.2 Current in phase *a*. *Source:* Kimbark [1]/John Wiley & Sons.

depicted with a continuous line, corresponding to the positive semi-cycle. Conversely, the paths in fired valves (3,4) are represented with a dashed line, corresponding to the negative semi-cycle.

The Fourier transform of a rectangular function, like the rectangular current pulses on the AC side of the bridge, is presented in Example 6.1.

6.2.2 Rectifier Bridge with Commutation Angle

The AC source side has an inductance, mainly due to the converter transformer, and its effect needs to be analyzed since the current switching between valves does not happen immediately. The commutation time required for the current corresponds to the overlap angle μ [1].

The overlap angle in normal operation is assumed less than 60° (ensuring that the number of valves conducting is less than or equal to 3) and typical values for steady-state operation range from 18° to 25°.

- Current Commutation

The current commutation from valve 1 to valve 3 is analyzed by means of the equivalent circuit in Figure 6.3.

On the basic assumption of constant direct current, ripple-free even during the commutation interval, and applying Kirchhoff's first law to node P, Eq. (6.1) can be written:

$$i_1 + i_3 = I. \tag{6.1}$$

At zero time in Figure 6.1a, Eq. (6.2) is given by:

$$e_b - e_a = e_{ba} = \sqrt{3}E_0 \sin \omega t, \tag{6.2}$$

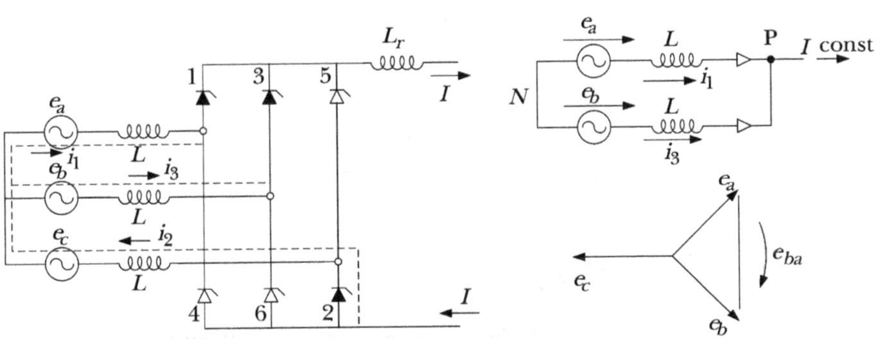

Figure 6.3 Equivalent circuit. *Source:* Kimbark [1]/John Wiley & Sons.

where E_0 is the amplitude of the phase-to-earth voltage. Applying Kirchhoff's second law to the loop formed by phases a and b:

$$\sqrt{3}E_0 \sin \omega t = Li_3' - Li_1'. \tag{6.3}$$

Deriving (6.1) with respect to time, the following relation between the derivatives of the currents can be shown:

$$i_1' = -i_3', \tag{6.4}$$

Thus, Eq. (6.3) is rewritten:

$$\sqrt{3}E_0 \sin(\omega t) = 2Li_3'. \tag{6.5}$$

Integrating (6.5) in the interval α to ωt:

$$i_3(t) = \frac{\sqrt{3}E_0}{2\omega L}[\cos(\alpha) - \cos(\omega t)], \tag{6.6}$$

where

$$I_{s2} = \frac{\sqrt{3}E_0}{2\omega L},$$

corresponds to the current amplitude of the phase-to-phase fault on the AC side of the converter transformer. The current commutation is illustrated in Figure 6.4:

At the end of the commutation, for $\omega t = \alpha + \mu$, the following equation is obtained:

$$I = \frac{\sqrt{3}E_0}{2\omega L}[\cos(\alpha) - \cos(\alpha + \mu)]. \tag{6.7}$$

Thus, the current can be expressed in the following intervals:

- Interval $[\alpha, \alpha + \mu]$:

$$i_3(t) = I\left[\frac{\cos(\alpha) - \cos(\omega t)}{\cos(\alpha) - \cos(\alpha + \mu)}\right], \tag{6.8}$$

$$i_1(t) = I - i_3(t). \tag{6.9}$$

- Interval $\left[\alpha + \mu, \alpha + \frac{2\pi}{3}\right]$:

$$i_3(t) = I. \tag{6.10}$$

- Interval $\left[\alpha + \frac{2\pi}{3}, \alpha + \frac{2\pi}{3} + \mu\right]$:

Within this interval, current $i_3(t)$ is equal to current $i_1(t)$ in the interval $[\alpha, \alpha + \mu]$, shifted by $2\pi/3$.

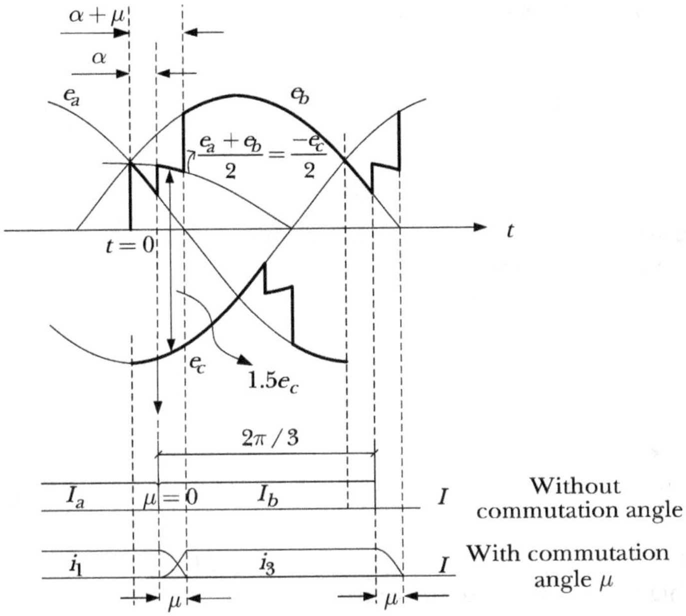

Figure 6.4 Current commutation angle. *Source:* Kimbark [1]/John Wiley & Sons.

$$i_3(t) = I - I\left[\frac{\cos(\alpha) - \cos\left(\omega t - \frac{2\pi}{3}\right)}{\cos(\alpha) - \cos(\alpha + \mu)}\right]. \tag{6.11}$$

The current waveform is defined by these three intervals. The harmonic content on the AC side is calculated using the Fourier series, and the same waveform is addressed in Example 6.1. The sinusoidal arcs introduced by the commutation inductances make the behavior of the current smoother than that of rectangular pulses, lowering the harmonic distortion.

In a 12-pulse bridge, the converters are connected to transformers with isolated star and delta connections, and this angular displacement leads to a reduction in the harmonic content of the current.

- Average DC voltages
  With respect to voltages at converter terminals:
  Calculating the average voltage in Figure 6.1a, in an interval $\pi/3$, with $\alpha = 0$ and $\mu = 0$.

$$V_d = \frac{3}{\pi}\int_{\frac{\pi}{3}}^{\frac{2\pi}{3}} \sqrt{3}\, E_0 \,\text{sen}\theta \, d\theta = \frac{3}{\pi}\sqrt{3}\, E_0. \tag{6.12}$$

Assuming $\alpha \neq 0$ and $\mu = 0$, as in Figure 6.1b

$$V_d = \frac{3}{\pi} \int_{\frac{\pi}{3}+\alpha}^{\frac{2\pi}{3}+\alpha} \sqrt{3}\, E_0 \,\text{sen}\theta \, d\theta = \frac{3}{\pi}\sqrt{3}\, E_0[-\cos\theta]_{\frac{\pi}{3}+\alpha}^{\frac{2\pi}{3}+\alpha}. \tag{6.13}$$

$$V_d = V_{do} \cos\alpha. \tag{6.14}$$

Analogously, the average voltage is easily evaluated with the area under the arcs of sinusoids in Figure 6.4, $\alpha \neq 0$ and $\mu \neq 0$ [1]:

$$V_d = \frac{V_{d0}}{2}(\cos\alpha + \cos(\alpha + u)). \tag{6.15}$$

Equation (6.15) may also be written as:

$$V_d = V_{d0}\cos\alpha - \frac{3}{\pi} X_c I_d, \tag{6.16}$$

where X_c is equal to ωL, see Eq. (6.6). This reactance refers basically to the transformer, assuming high SCR and sinusoidal voltage at the AC side.

6.2.3 Inverter Bridge

For the inverter, the equations are basically with the firing angle α greater than 90°, but an extinction angle γ will be introduced, defined as follows:

$$\gamma = \pi - (\alpha + u). \tag{6.17}$$

The angle γ corresponds to the complement of $\alpha + u$ in an interval π, that is, once the commutation is complete γ is the angle that remains for the voltage e_{ba} to become negative. Switching between phase a and phase b is only possible as long as there is positive voltage on valve 3, that is, as long as the voltage e_{ba} is greater than zero.

The inverter usually works with the angle γ which represents the margin that exists after switching so that the valve recovers its capacity to withstand the reverse voltage after π.

If α is too large or if μ is greater than expected in a switching period, the angle γ may be lower than the minimum allowable and the valve may not recover, resulting in commutation failure.

From (6.17), the inverter equation is obtained with the following substitutions:

$$\alpha = \pi - (\gamma + u),$$

$$(\alpha + u) = \pi - \gamma.$$

Replacing α and $\alpha + \mu$ in the rectifier Eq. (6.15):

$$V_d = -\frac{V_{d0}}{2}[\cos\gamma + \cos(\gamma + u)]. \tag{6.18}$$

and analogously to (6.16), it can be demonstrated that:

$$V_d = -\left(V_{d0}\cos\gamma - \frac{3}{\pi}X_c I_d\right), \tag{6.19}$$

where X_c refers to the inverter side, but the definition is the same as in (6.16).

This equation was derived for a rectifier bridge with the position indicated on the left side in Figure 6.5a for both angles α and γ. With α between $\pi/2 - \pi$ (γ between $0 - \pi/2$) the rectifier voltage is negative. The same rectifier bridge can operate as an inverter if it is flipped and connected to the opposite DC link terminal, changing the sign at the right side in (6.19). It now takes on the role of an inverter, complementing the monopolar arrangement of the DC link with ground return (Figure 6.5a).

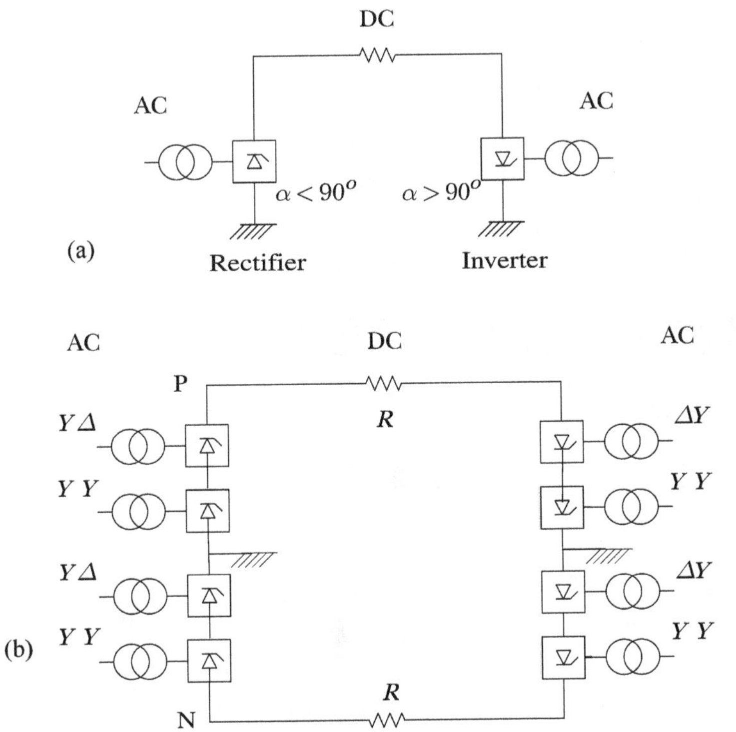

Figure 6.5 DC links. (a) Monopolar 6 pulses DC link. (b) Bipolar 12 pulses DC llink.

The rectifier voltage (controlled by α) is kept slightly above the voltage in the inverter, which now operates as a DC voltage (controlled by γ) in opposition to the rectifier voltage, which allows controlling the current in the link and enabling the operating mode at constant current. This is the basic control, but many other functions are added, which are not detailed here.

Figure 6.5b shows the basic configuration of a DC link bipole, with the positive and negative poles.

6.2.4 Fourier Analysis of Current in Six-Pulse Bridges

The definition of the complex Fourier coefficient and the fundamental time-shift property are presented below.

Given a periodic function $f(t)$ of period T,

$$f(t) = f(t + T),$$

the function $f(t)$ can be decomposed into Fourier complex series by calculating:

$$f(t) = \sum_{h=-\infty}^{\infty} c_h e^{jh\omega t}, \tag{6.20}$$

$$c_h = \frac{1}{T} \int_{t_0}^{t_0+T} f(t) e^{-jh\omega t} dt, \tag{6.21}$$

where $\omega = 2\pi/T$ is the fundamental angular frequency of the signal, h representing the harmonic order under examination.

If the complex coefficients of a given function $f(t)$ are known, the coefficients of the same function with a time delay τ may be written as follows:

$$f(t-\tau) = \sum_{h=-\infty}^{\infty} c_h e^{jh\omega(t-\tau)},$$

or:

$$f(t-\tau) = \sum_{h=-\infty}^{\infty} \bar{c}_h e^{jh\omega t},$$

where:

$$\bar{c}_h = c_h e^{-jh\omega\tau}. \tag{6.22}$$

The calculation of the complex Fourier coefficients for a pulse waveform is an example of periodic pulses decomposition with Fourier complex series, (Figure 6.6).

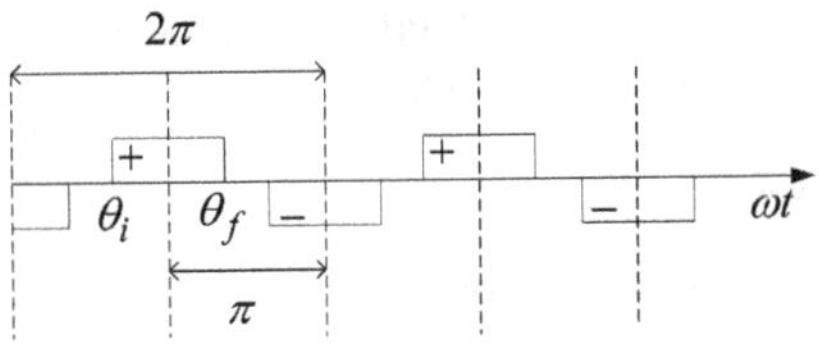

Figure 6.6 Periodic pulses.

Initially, in (6.21), only a unitary function and positive polarity pulses are considered, with the following substitution of variables,

$$\omega t = \theta, \ \ dt = \frac{d\theta}{\omega}, \ (\omega T = 2\pi).$$

thus, Eq. (6.21) is rewritten as:

$$c_h^+ = \frac{1}{2\pi} \int_{\theta_0}^{\theta_0+2\pi} e^{-jh\theta} d\theta \tag{6.23}$$

The complex coefficient of the positive pulse defined on the generic interval $[\theta_i, \theta_f]$ is indicated below:

$$c_h^+ = \frac{1}{2\pi} \int_{\theta_i}^{\theta_f} e^{-jh\theta} d\theta = \frac{1}{2\pi} \frac{e^{-jh\theta_f} - e^{-jh\theta_i}}{-jh}. \tag{6.24}$$

The complex coefficients c_h^-, referring to negative polarity pulses, are calculated simply by changing the sign of c_h^+ and applying a delay, as in (6.22). The functions to be analyzed have a period 2π and the negative half-cycles are behind π in relation to the positive half-cycle and therefore c_h^- can be written as:

$$c_h^- = -c_h^+ e^{-jh\pi}.$$

The complex Fourier coefficients of the full wave, composed of positive and negative half-cycles, are given by the sum of c_h^+ and c_h^- indicated below:

$$c_h = c_h^+ \left(1 - e^{-jh\pi}\right), \tag{6.25}$$

The composition of anti-symmetrical positive and negative pulses demonstrates some interesting properties, typical of three-phase AC systems. The factor $(1 - e^{-jh\pi})$ is null for even values of h; in other words, constant term and even harmonics are null. For the fundamental and odd harmonics, the factor is equal to 2, thus, $c_h = 2c_h^+$. As the magnitude of a harmonic component is twice that of the complex Fourier coefficient, this magnitude can be simply calculated by $4c_h^+$.

Example 6.1 Analyze the harmonic components of the alternating current in a six-pulse rectifier bridge.

Solution

The complex coefficients are initially calculated considering a current pulse of amplitude I, without commutation angle μ, as in Figure 6.2, with period 2π, positive polarity, and centered in the interval $\theta_i = -\pi/3$ to $\theta_f = +\pi/3$. Thus, from (6.24):

$$c_h^+ = \frac{I}{2\pi} \frac{e^{-jh\pi/3} - e^{+jh\pi/3}}{-jh},$$

or,

$$c_h^+ = \frac{I}{\pi h} \sin\left(\frac{h\pi}{3}\right).$$

There are no harmonic components in multiples of three (3, 6, 9, 12, etc.), since, for these harmonics, the sinusoidal value is null.

Thus, the odd harmonic components remain (5, 7, 11, 13, etc.), including the fundamental component. The harmonic order h for the current generated in a six-pulse converter can be obtained by increasing n in the expression $h = 6n \pm 1$.

The amplitude of the fundamental component of the Fourier series can be calculated with $4c_h^+$:

$$I_1 = \frac{2\sqrt{3}}{\pi} I = 1.103I.$$

The magnitudes of the other harmonics are given by:

$$I_h = \frac{I_1}{h}.$$

In order to address the generic case, the thyristor firing angle and the commutation reactance should be included by introducing sinusoidal arcs both at the beginning and at the end of the current flowing through the thyristor valve; however, the analytical procedure is very much the same.

The equations for the current are detailed in Eqs. (6.8)–(6.11).

The current waveform during the operation of the rectifier bridge is presented in Figure 6.7.

The Fourier complex coefficients for the positive pulses are calculated below. Let:

$$k = \frac{1}{\cos(\alpha) - \cos(\alpha + \mu)}, \text{ and } \theta = \omega t, \text{ then,}$$

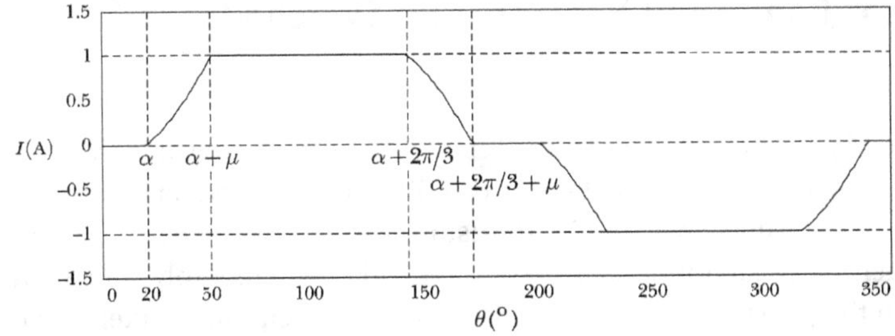

Figure 6.7 Current in one phase, for α = 20° and μ = 30°.

$$c_h^+ = \frac{I}{2\pi}\left[\begin{array}{l} \displaystyle\int_{\alpha}^{\alpha+\mu} k[\cos(\alpha) - \cos(\theta)]e^{-jh\theta}d\theta + \int_{\alpha+\mu}^{\alpha+\frac{2\pi}{3}} e^{-jh\theta}d\theta + \\ \displaystyle + \int_{\alpha+\frac{2\pi}{3}}^{\alpha+\mu+\frac{2\pi}{3}} \left[1 - k\left(\cos(\alpha) - \cos\left(\theta - \frac{2\pi}{3}\right)\right)\right]e^{-jh\theta}d\theta \end{array}\right].$$

The expression for these coefficients can be calculated, given that:

$$\int \cos(\theta)e^{-jh\theta}d\theta = e^{-jh\theta}\left(\frac{jh\cos(\theta) - \sin(\theta)}{h^2 - 1}\right),$$

Figure 6.8 shows the fifth-order current harmonic as a percentage of the fundamental component for a few values of the angles α and μ, calculated with the aid of a computer program.

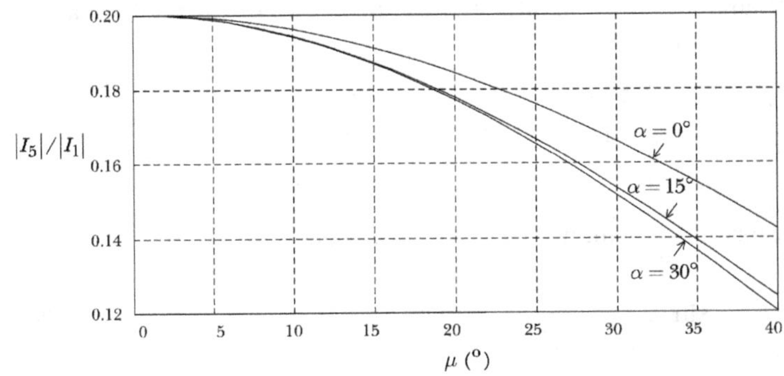

Figure 6.8 Fifth-order harmonic in pu of the current at fundamental frequency.

6.3 Thyristor Controlled Reactors and Switched Capacitors

Within the scope of power electronics applied in the AC system, controllable reactive compensation is part of the early stage of FACTS devices, based on controlled reactors and capacitor banks switched by thyristors.

Equipment such as capacitors and reactors are switched smoothly and with great precision in the phase angle using thyristors, replacing circuit breakers, providing a reduction in voltage and current transients. The inductors can be controlled continuously, and the capacitors are switched at certain times, so that there is no sudden variation in voltage.

- Controlled reactor

  A significant aspect concerning thyristor bridges is that the reactor can be controlled continuously, varying the current conduction interval with a delay in the firing angle α in the thyristors, connected in antiparallel as in the thyristor-controlled reactor (TCR) in Figure 6.9a. The current in the reactor, with this delay, is no longer sinusoidal, Figure 6.9b, however, with the application of the Fourier series, its fundamental component presents a lower value than that obtained without this delay, which has the effect of increasing the reactance (or reducing the susceptance) at the fundamental frequency. Thus, a continuously variable reactor is obtained, which depends on α.

  The equation for the reactor is written as:

$$v(t) = L\frac{di(t)}{dt}, \tag{6.26}$$

with a voltage source,

$$v(t) = E_0 \cos(\omega t),$$

where $E_0 = \sqrt{2}V$, and V is the magnitude of the phasor of the applied voltage.

Integrating (6.26) in the interval $[\alpha, \omega t]$:

$$i(t) = \frac{\sqrt{2}V}{\omega L}(\sin \omega t - \sin \alpha). \tag{6.27}$$

In (6.27) the constant $\frac{\sqrt{2}V}{\omega L}$ multiplies a periodic function $f(t)$, designated as:

$$f(t) = \sin \omega t - \sin \alpha. \tag{6.28}$$

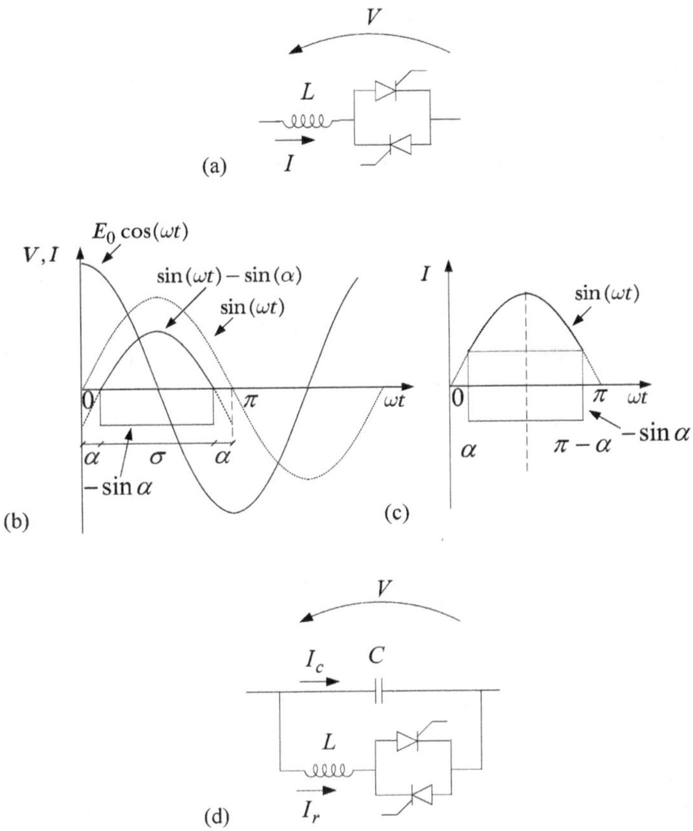

Figure 6.9 TCR and TCSC, voltage, and current. (a) TCR. (b) Voltage and current in the reactor. (c) Current components. (d) TCSC.

This function is composed of a sinusoidal function subtracted from a rectangular pulse with amplitude sinα, Figure 6.9b. The Fourier transform of $f(t)$ is presented in [3, 7], but it can also be calculated by adding the complex Fourier coefficients of the sinusoidal function (c_{hs}) and the pulse function (c_{hp}) in a half-cycle.

$$c_h^+ = c_{hs}^+ + c_{hp}^+ \tag{6.29}$$

The complex coefficients of the sine function are calculated for a half-cycle in Figure 6.9c, in the interval $[\alpha, \pi - \alpha]$, using its complex form.

$$c_{hs}^{+} = \frac{1}{2\pi}\int_{\alpha}^{\pi-\alpha}\frac{e^{+j\theta}-e^{-jh\theta}}{2j}e^{-jh\theta}d\theta \tag{6.30}$$

$$c_{hs}^{+} = \frac{1}{j4\pi}\left[\frac{e^{+j(1-h)(\pi-\alpha)}-e^{+j(1-h)\alpha}}{j(1-h)}-\frac{e^{-j(1+h)(\pi-\alpha)}-e^{-j(1+h)\alpha}}{-j(1+h)}\right] \tag{6.31}$$

The coefficient of the pulse function with amplitude sinα, c_{hp}^{+}, is calculated in the same interval, initially as if the pulse had the same polarity as the sinusoidal function, taking advantage of Eq. (6.24), but in fact, it has opposite polarity and must be subtracted in (6.29).

The complex coefficient of the pulse function with amplitude sinα is given by:

$$c_{hp}^{+} = \frac{\sin\alpha}{2\pi}\left[\frac{e^{-jh(\pi-\alpha)}-e^{-jh\alpha}}{-jh}\right] \tag{6.32}$$

Adding (6.32) with its polarity corrected in (6.31), the coefficient c_h^{+} of $f(t)$ may be obtained in (6.33).

$$c_{h}^{+} = \frac{1}{j4\pi}\begin{bmatrix}\dfrac{e^{+j(1-h)(\pi-\alpha)}-e^{+j(1-h)\alpha}}{j(1-h)}-\dfrac{e^{-j(1+h)(\pi-\alpha)}-e^{-j(1+h)\alpha}}{-j(1+h)}\\ -2\sin\alpha\left[\dfrac{e^{-jh(\pi-\alpha)}-e^{-jh\alpha}}{-h}\right]\end{bmatrix} \tag{6.33}$$

It is interesting to obtain the fundamental component of the current for $h = 1$. In this case, there is an indeterminacy in the denominator in the first term of the bracket in (6.33), which can be overcome by applying L'Hôpital's rule. However, the calculation of c_1^{+}, ($h = 1$), in (6.30) is immediate, without indeterminacies and leads to the same result:

$$c_1^{+} = \frac{1}{j4\pi}[(\pi-2\alpha)+\sin 2\alpha-4\sin\alpha\cos\alpha]$$

$$c_1^{+} = \frac{1}{j4\pi}[(\pi-2\alpha)-\sin 2\alpha] \tag{6.34}$$

The value of the fundamental current is calculated multiplying Eq. (6.34) by the factor 4 and by the constant that multiplies the function $f(t)$ in (6.27). To calculate the current in rms the factor $\sqrt{2}$ is omitted. Note that the same procedure is adopted to calculate the harmonics using (6.33).

$$I_1 = \frac{V}{j\pi\omega L}[(\pi - 2\alpha) - \sin 2\alpha] \tag{6.35}$$

The susceptance of the reactor may be written in function of the firing angle α [3, 7]:

$$B(\alpha) = B_m\left[1 - \frac{2\alpha}{\pi} - \frac{\sin(2\alpha)}{\pi}\right], \tag{6.36}$$

B_m: maximum susceptance of the reactor ($1/\omega L$), when $\alpha = 0$.

Current and voltage in rms are related by $B(\alpha)$:

$$I_1 = B(\alpha)V. \tag{6.37}$$

The harmonics that result from nonsinusoidal current waveforms are attenuated by filters. In SVCs, the reactors and thyristors are connected to the grid through transformers, at a voltage level much lower than that of the grid. The delta connection allows the circulation of third harmonics only in the secondary of the transformers.

The variable reactor can be connected in parallel with other reactors or capacitors, offering a continuous variation of reactive power to the network, as in SVCs. The TCR-thyristor-controlled reactor, can complement reactive compensation equipment, avoiding switching transients.

The association of a TCR in parallel with a capacitor has application in the series compensation of transmission lines, as in the TCSC-thyristor-controlled series capacitor, Figure 6.9d.

Adding the susceptances of the capacitor and of the variable reactor, the equivalent susceptance of the TCSC is also variable, given by:

$$B_e(\alpha) = \omega C - B(\alpha) \tag{6.38}$$

In (6.38), the susceptance of the reactor varies within the range 0 to B_m, when α is equal to $\pi/2$ and 0, respectively. At a resonance point, in which the firing angle is designated as α_r, a zero equivalent susceptance (infinite impedance) is found, which makes operation in this point and its surroundings unfeasible. Assuming α_{min} above this prohibited range, the permissible operating range lies between α_{min} and $\pi/2$, in which the equivalent susceptance is capacitive, since $\omega C > B(\alpha)$. Therefore, the control must limit the operating range to avoid possible resonance between the reactor and the capacitor.

6.4 Power Electronics – with VSC

6.4.1 Voltage Source Converters – VSC in Transmission Systems

Unlike thyristors, a new generation of switches now governs both the initiation and cessation of current flow. This evolution represents a paradigm shift in reactive compensation and HVDC systems. These converters initially used GTOs (gate turn-off transistors), but the most current technology is based on IGBTs (insulated gate bipolar transistor); however, it may be just one of the stages in the evolution of switches.

The basic concept is to create a three-phase system of positive sequence voltages from a DC voltage source that may be supplied by a capacitor. If the three-phase voltages are connected to the grid, there is a possibility of exchanging reactive power with the grid, and in this case, the capacitor voltage remains constant.

The second possibility is the exchange of active power with the network; however, in this case, the capacitor needs to be connected to an energy source, so that the voltage remains constant on the DC bus. As an example, in the HVDC system, the transmission line supplies the DC voltage to the inverter, playing the role of the capacitor, which is charged by the generation side and discharged by the load side.

Through the VSC converter, a family of new FACTS equipment was developed, such as the STATCOM (static shunt synchronous compensator), the SSSC (static synchronous series compensator), and the UPFC (unified power flow controller), whose analysis is the subject of several papers, as in the references [4, 5]. The STATCOM is an evolution of the SVC, with the advantage of eliminating physical equipment, such as capacitors and shunt reactors. In this shunt connection, the STATCOM operates with the current leading or lagging 90 degrees of the system voltage, that is, as a shunt capacitor bank or as a shunt reactor.

In addition to the shunt connection, it is possible to make the series connection of the converter in applications such as the SSSC, which operates as a series capacitor bank. These electronic components are connected to the network through transformers, which reduces the voltage level of the electronic equipment and increases protection against disturbances in the electrical network. Due to the need for a transformer connected in series with the transmission line, over an isolated platform, this may represent a limitation for high voltages and powers; however, new technologies, such as MMC (Modular multilevel converters), to be presented in Section 6.6, can overcome this limitation, inserting the series voltage without the transformer.

The basic topology of the UPFC results from the union of a STATCOM with an SSSC, as shown in Figure 6.10 [8, 9].

The functional principle is that the STATCOM controls the bus voltage, operating with the current in quadrature with the voltage. Note that there is a coupling

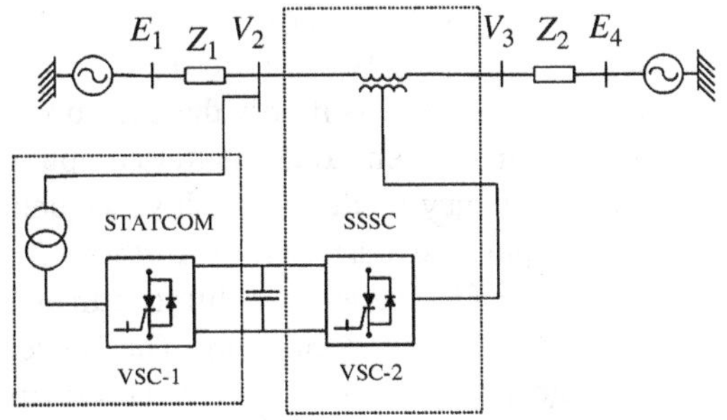

Figure 6.10 UPFC – STATCOM/SSSC.

inductance between the converter voltage and the network voltage, which may be the transformer's leakage reactance. The control sets the two voltages in phase, so that the reactive power (inductive) flows from the highest to the lowest voltage level, that is, if the converter voltage is higher than that of the network, the STATCOM operates as a capacitor and otherwise as a reactor.

As for the SSSC, there are two possibilities. If the current is in quadrature with the voltage, the series element helps to adjust the voltage at the line terminal, only exchanging reactive power with the network or simply serving the role of a series capacitor. On the other hand, if the control shifts the phase angle of the voltage, the series element starts to operate as a phase-shifting transformer and now the UPFC can more effectively control the flow of active power in the line. In this operating mode, a fraction of the shunt current in the STATCOM charges the DC link capacitor with active power, which returns to the network through the SSSC, shifting the SSSC series voltage angle. Note that this active power flow can also be reversed from the SSSC to the STATCOM. The remaining fraction of the current is associated with the reactive power compensation.

The model code of the UPFC in EMT programs, based on GTO bridges, is presented in reference [10]. In addition to transmission lines, other applications can be visualized with shunt and series converters, and this versatility can also be explored in the operation of autotransformers assisted by the UPFC in [11].

The UPFC operating principles are analyzed below with an elementary steady-state model, representing the STATCOM by a shunt current source and the SSSC by a series voltage source. The shunt current source has two components. One is orthogonal to the bus voltage and controls the exchange of reactive power with the system and, consequently, the bus voltage. The other one, which is in phase with the bus voltage angle, controls the exchange of active power with the coupling capacitor, keeping the voltage on the capacitor constant. The SSSC is represented by a voltage source in series with the network, with a given amplitude.

If the internal losses of the converter are neglected, it is assumed that the active power removed from the network by the shunt converter returns to the network through the series converter in such a way that the net balance of active power exchanged with the network is zero.

The model is simple as it only depends upon a quadratic equation whose solution is the shunt current to thereafter compute the remaining variables. The fundamental frequency model for positive sequence, shown in Figure 6.11a, will be used for this purpose. Basically, the forthcoming equations resulted from the independent effect of the series and shunt sources (characterizing the inverters) upon the system (superposition theorem). The Thévenin equivalents, the series compensator voltage, and the shunt reactive power must be furnished to the model.

To simplify the equations coming forth, the resistances are neglected, and the total reactance is designated as:

$$X_t = X_1 + X_2$$

Without the shunt current generator (STATCOM) in Figure 6.11a, the current in the series circuit, including the series converter, is calculated in real and imaginary coordinates (r–i components).

$$I = i_r + ji_i; V_c = v_{cr} + jv_{ci};\ E_1 = e_{1r} + je_{1i};\ E_4 = e_{4r} + je_{4i}.$$

$$i_r + ji_i = \frac{(e_{1r} + v_{cr} - e_{4r}) + j(e_{1i} + v_{ci} - e_{4i})}{jX_t}. \tag{6.39}$$

Designating:

$$\Delta e_r = e_{1r} + v_{cr} - e_{4r};\ \Delta e_i = e_{1i} + v_{ci} - e_{4i}.$$

The components i_r and i_i from (6.39) are rewritten:

$$i_r = \frac{\Delta e_i}{X_t},\ i_i = -\frac{\Delta e_r}{X_t}, \tag{6.40}$$

and from Figure 6.11b, the current I_{sh} can be written as:

$$i_{sh1} = I_{sh}\frac{X_2}{X_t};\ i_{sh2} = I_{sh}\frac{X_1}{X_t}. \tag{6.41}$$

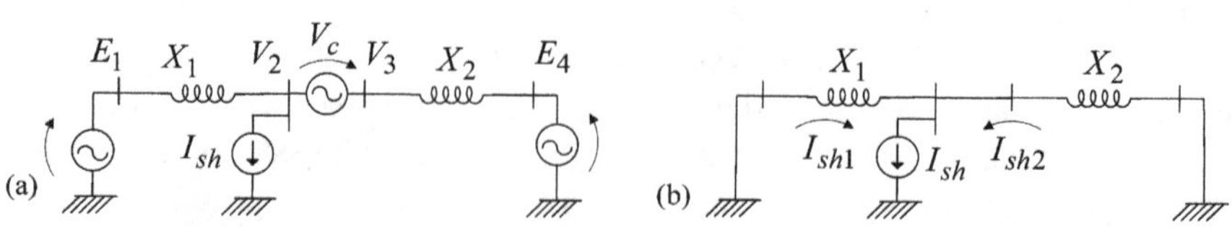

Figure 6.11 Elementary system with a UPFC. (a) Equivalent circuit. (b) Effect of the shunt current generator.

The total current on sides 1 and 2 can be calculated with the superposition theorem, adding (6.39) and (6.40):

$$i_{1r} = \frac{\Delta e_i + X_2 i_{shr}}{X_t}; i_{1i} = \frac{-\Delta e_r + X_2 i_{shi}}{X_t}, \tag{6.42}$$

$$i_{2r} = \frac{\Delta e_i - X_1 i_{shr}}{X_t}; i_{2i} = \frac{-\Delta e_r - X_1 i_{shi}}{X_t}. \tag{6.43}$$

The voltage at bus 2 on r–i components is given by:

$$v_{2r} = e_{1r} + X_1 i_{1i}; v_{2i} = e_{1i} - X_1 i_{1r}. \tag{6.44}$$

Designating,

$$X_p = \frac{X_1 X_2}{X_t}. \tag{6.45}$$

As the r-i components, i_{1r} and i_{1i}, were already obtained in (6.42):

$$v_{2r} = e_{1r} - \frac{X_1}{X_t}\Delta e_r + X_p i_{shi};\ v_{2i} = e_{1i} - \frac{X_1}{X_t}\Delta e_i - X_p i_{shr}.$$

Designating:

$$v_x = e_{1r} - \frac{X_1}{X_t}\Delta e_r;\ v_y = e_{1i} - \frac{X_1}{X_t}\Delta e_i,$$

thus,

$$v_{2r} = v_x + X_p i_{shi}; v_{2i} = v_y - X_p i_{shr} \tag{6.46}$$

The active power, P_{se}, at the output of the series converter:

$$P_{se} = (v_{cr} i_{2r} + v_{ci} i_{2i}).$$

As the r-i components, i_{2r} and i_{2i}, were previously calculated in (6.43):

$$P_{se} = \left[\frac{v_{cr}\Delta e_i + v_{ci}\Delta e_r}{X_t} - \frac{X_1}{X_t}(v_{cr} i_{shr} + v_{ci} i_{shi})\right]. \tag{6.47}$$

Similarly, the active power for the shunt converter can be calculated,

$$P_{sh} = (v_{2r} i_{shr} + v_{2i} i_{shi}),$$

$$P_{sh} = (v_x + X_p i_{shi}) i_{shr} + (v_y - X_p i_{shr})\, i_{shi},$$

$$P_{sh} = v_x i_{shr} + v_y\, i_{shi}. \tag{6.48}$$

If the inverter losses are neglected, the equality between the shunt and series active powers in the dc-link can be written as:

$$P_{sh} = P_{se}. \tag{6.49}$$

$$\frac{v_{cr}\Delta e_i + v_{ci}\Delta e_r}{X_t} - \frac{X_1}{X_t}(v_{cr}i_{shr} + v_{ci}i_{shi}) = v_x i_{shr} + v_y i_{shi}.$$

$$\frac{v_{cr}\Delta e_i + v_{ci}\Delta e_r}{X_t} - \left(v_x + \frac{X_1 v_{cr}}{X_t}\right)i_{shr} - \left(v_y + \frac{X_1 v_{ci}}{X_t}\right)i_{shi} = 0. \tag{6.50}$$

And the following coefficients can be defined:

$$c = \frac{v_{cr}\Delta e_i + v_{ci}\Delta e_r}{X_t}; b = v_x + \frac{X_1 v_{cr}}{X_t}; a = v_y + \frac{X_1 v_{ci}}{X_t}.$$

The simplified form for Eq. (6.50) is given in (6.51):

$$c - bi_{shr} - ai_{shi} = 0, \tag{6.51}$$

thus,

$$i_{shi} = \frac{c - bi_{shr}}{a}. \tag{6.52}$$

The reactive power in the shunt inverter may be written as:

$$Q_{sh} = (v_{2q}i_{shr} - v_{2d}i_{shi}).$$

From (6.46):

$$Q_{sh} = (v_y - X_p i_{shr})i_{shr} - (v_x + X_p i_{shi})i_{shi}.$$

$$Q_{sh} = v_y i_{shr} - v_x i_{shi} - X_p i_{shr}^2 - X_p i_{shi}^2. \tag{6.53}$$

Substituting (6.52) in (6.53) and after some algebraic manipulation, the coefficients can be defined:

$$A = X_p\left(1 + \frac{b^2}{a^2}\right),$$

$$B = -\left(v_y + \frac{b}{a}v_x + X_p\frac{2bc}{a^2}\right),$$

$$C = \left(Q_{sh} + X_p\frac{c^2}{a^2} + v_x\frac{c}{a}\right).$$

The second-order equation provides a direct solution, thus, avoiding the iterative calculation.

$$Ai_{shr}^{\ 2} + Bi_{shr} + C = 0. \tag{6.54}$$

Once computed i_{shr} and i_{shi}, currents i_{2r} and i_{2i} can be obtained in (6.43). Finally, the power at the receiving-end of the line can be computed through (6.55) and (6.56):

$$P_4 = e_{4r}i_{2r} + e_{4i}i_{2i}, \tag{6.55}$$

$$Q_4 = e_{4i}i_{2r} - e_{4r}i_{2i} \tag{6.56}$$

The P–Q results obtained with this simple model and those in [12] are the same.

One of the main advantages related to the installation and operation of these FACTS devices refer to the instantaneous control of active and reactive power flow in multiline systems, yielding substantial benefits such as increasing the transmission capacity of existing lines and postponing the construction of new transmission lines, control flexibility and reduction of transmission losses by redirecting the power flow, dampening system oscillations, among others.

The UPFC model used in available power flow programs is based on the power injection model [13]; however, other models can be proposed and those based on current injection can be useful mainly in optimization analysis [14].

There are proposals to increase operational flexibility in transmission systems, through the IPFC [15], and the GIPFC, whose steady-state analysis can be found in [12].

Power flow studies are necessary and involve long-, medium-, and short-term analyses. Dynamic studies analyze the system's response to frequency deviations, large voltage deviations of short duration, generally after fault elimination, and transient stability to large disturbances. Interactions between controls need to be investigated, as this possibility increases with the greater penetration of FACTS devices. From the point of view of overvoltage analysis, aimed at the insulation coordination, significant dynamic studies are related to the load rejection, more specifically those referring to the performance of FACTS equipment in controlling temporary overvoltages.

6.4.2 Application of VSC in Renewable Generation

Renewable generation presents notable growth in energy supply in power systems, with a tendency to surpass electrical energy generation based on conventional sources and become the main form of electrical energy generation [6].

This growth in renewable generation was favored by the evolution of VSCs, which are the interface between the primary energy source and the AC transmission system, giving rise to the so-called IBGs. The IBGs are supplied by different primary energy sources; however, the sources currently used on a larger scale are photovoltaic and wind.

Unlike synchronous generators, IBGs do not have inertia and do not present internal fluxes or voltages and their contribution to the short-circuit current is limited by the control to around 1 pu.

IBGs are synchronized with the AC network through control, which provides them with an inherent synchronization characteristic, except in cases where a significant voltage sag occurs in the network.

In wind generation, inverter control considers the mechanical drive of the turbines, which contains a certain inertia. Photovoltaic plants have no inertia and do not contribute to supporting the system in case of under frequencies unless there is energy storage. The absence of inertia is still a difficult problem to overcome during disturbances in the system and the implementation of synthetic inertia via control is not yet completely equivalent to the physical inertia of synchronous machines, but on the other hand, the inverter control response is fast and flexible.

For system analysis, when studying short-term dynamic performance and electromagnetic transients, it is reasonable to simplify the IBG model, which can only be composed of an inverter connected to a stiff DC source, especially if there is an energy store system.

6.5 VSC Elements

Figure 6.12 illustrates the basic topology of connecting a VSC to an AC system.

The development of VSCs was based on two main techniques for sine wave generation. The first consists of the generation of positive and negative rectangular pulses (or a composition of pulses) at the same network frequency, and the second is the generation of successive pulses at a high frequency, according to the pulse width modulation (PWM) technique. By applying the Fourier analysis to the voltage pulses generated in both cases, the fundamental wave has the same frequency of the network, and the harmonic components are neutralized using different techniques.

Coupling with the grid, in the case of FACTS equipment and HVDC systems, may be done through transformers. The transformer allows the use of electronic devices at a voltage level lower than that of the network and its leakage reactance limits current peaks and protects electronic devices from transients.

6.5.1 Converter Bridges

The principles of converting a DC voltage, supplied by a charged capacitor, into a sequence of rectangular pulses will be examined in elementary converter topologies that are useful in understanding those more complex. A review on converter bridge topology is presented in [16–18].

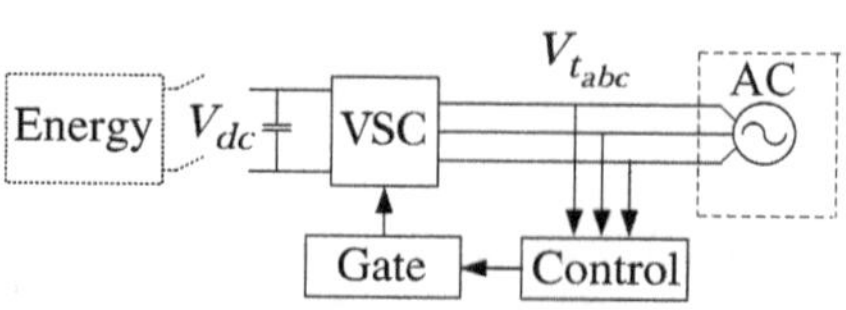

Figure 6.12 VSC elementary topology.

The choice of the topology depends on the specific VSC requirements and desired operational performance. The most appropriate configuration considers factors such as cost, nominal voltage, ripple, current levels, etc.

- Half-bridge submodule

  The half-bridge is attractive for its simplicity and reliability and normally constitutes a submodule of a more complex configuration, as will be seen in Section 6.7.

  Figure 6.13a shows the half-bridge single-bus topology composed of switches sw1 (s1,d1) and sw2 (s2,d2).

  Each switch has an IGBT (s) and a diode (d) connected in antiparallel, enabling current conduction in both directions of the switch.

  The capacitor is connected between terminals 1 and 2 when sw1 is closed (on) and sw2 is open (off), as in Figure 6.13b and c, the submodule voltage is $+V_{dc}$, and the current path is bidirectional through the IGBT or the diode. In the opposite situation, when sw1 is off and sw2 is on, the voltage applied to the AC side is zero (Figure 6.13d and e).

  The operation of switches sw1 and sw2 is mutually exclusive, that is, when sw1 is closed, sw2 must be open, and vice-versa.

  In short:

- Capacitor on (sw1 on): (s1 or d1) on, and (s2 and d2) off.
- Capacitor off (sw1 off): (s1 and d1) off and (s2 or d2) on.

To connect with the AC systems, another submodule is required to create a symmetrical state as in Section 6.7.

- Half-bridge with two levels

  The half-bridge with DC-bus split topology, in Figure 6.13f, has two identical capacitive units connected in series, dividing the DC voltage into $+V_{dc}/2$ and $-V_{dc}/2$. This bridge is not commonly used as a submodule, as in Section 6.6, but offers two voltage levels and can be connected to the AC system.

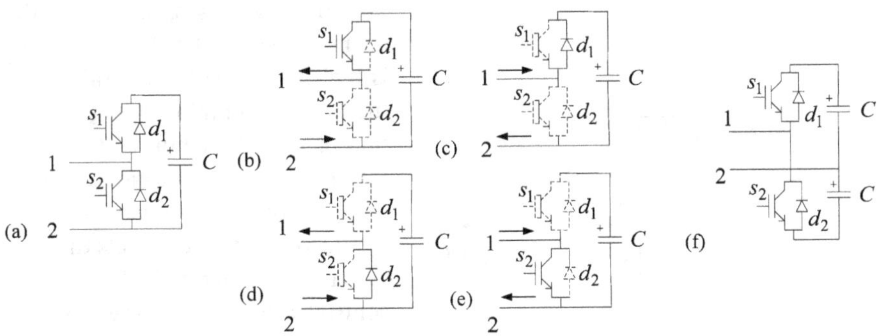

Figure 6.13 Half-bridge operation. (a) Half bridge topology. (b) Switch s1 on. (c) Diode d1 on. (d) Diode d2 on. (e) Switch s2 on. (f) Half-bridge topology with two levels.

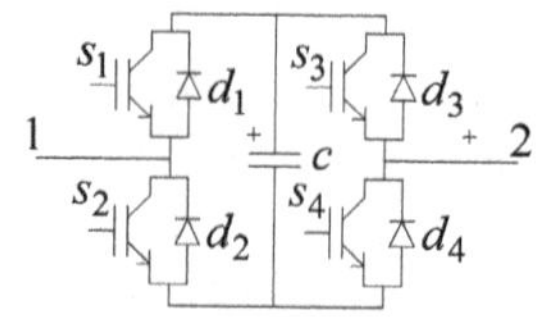

Figure 6.14 Full-bridge.

- Full-bridge
  The full-bridge topology contains four switches and a capacitor according to the arrangement in Figure 6.14. To simplify, only the on and off modes of the switches will be analyzed, without detailing the direction of the current.

Note that some states are not allowed since they short-circuit the capacitor, such as the switch conditions: (sw1 and sw2 on) or (sw3 and sw4 on).

The switches (sw1 and sw4) in the on state apply $+V_{dc}$ between terminals 1 and 2.

The switches (sw1 and sw3) or (sw2 and sw4) in the on state apply zero voltage between the terminals.

The switches (sw2 and sw3) in the on state apply $-V_{dc}$ between terminals 1 and 2.

Other bridge topologies may be proposed but will not be analyzed here. The two main types of gate drivers are presented below.

6.5.2 Gate Drivers

- Low-frequency gate drivers
  Gate drivers at the grid frequency (low frequency) were used in a family of converters having two or three voltage levels. For example, the two pulses in Figure 6.15 could be obtained very simply with a half-bridge connected to a split DC bus.

As the voltage pulses are at the network frequency, the fundamental component of the Fourier series will be a sinusoidal function, at that same frequency. This technique was most used in high-voltage systems, whose advantage is to present low losses; however, the harmonic content at low frequency is high and must be neutralized by filters. It was employed in a generation of VSCs based on GTOs.

Converters with two levels essentially apply a square wave through a transformer that connects the converter to the network. Due to the controlled switching of the capacitor on the secondary of the transformer, the voltage waveform on the primary side approaches a sine wave.

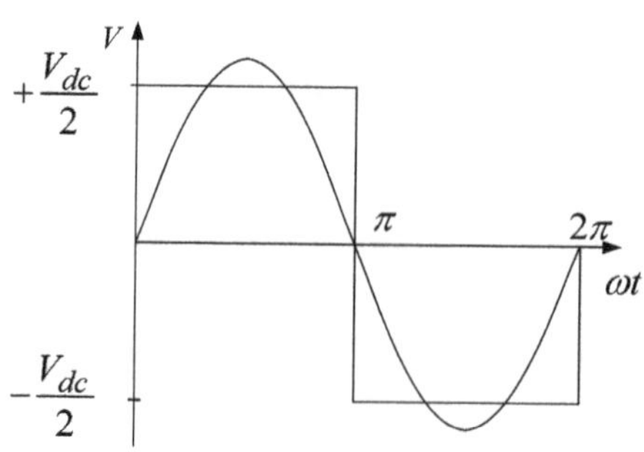

Figure 6.15 Low-frequency gate driver.

Three-level converters can generate waveforms closer to sinusoidal, with lower harmonic content, using a relatively low switching frequency and which can be used in high voltage applications due to the possibility of connecting GTOs in series. The disadvantage of this configuration in relation to the two-level configuration is the greater complexity of the arrangement and the additional cost of diodes.

To obtain a voltage close to sine wave, in addition to the number of levels, the magnetic circuit is used by phase-shifting the converters through transformer connections with star-star and star-delta, increasing the number of pulses. The zigzag secondary can also be used, allowing more effective cancellation of harmonics with a magnetic circuit [10].

- Pulse Width Modulation – PWM

The PWM technique uses a frequency much higher than the fundamental one, above 2 kHz, resulting in a series of pulses with different durations, as shown in Figure 6.16

The gate system uses two signals. The first is a sinusoidal reference signal, at the grid frequency (modulating wave), to be reproduced by the bridge in the

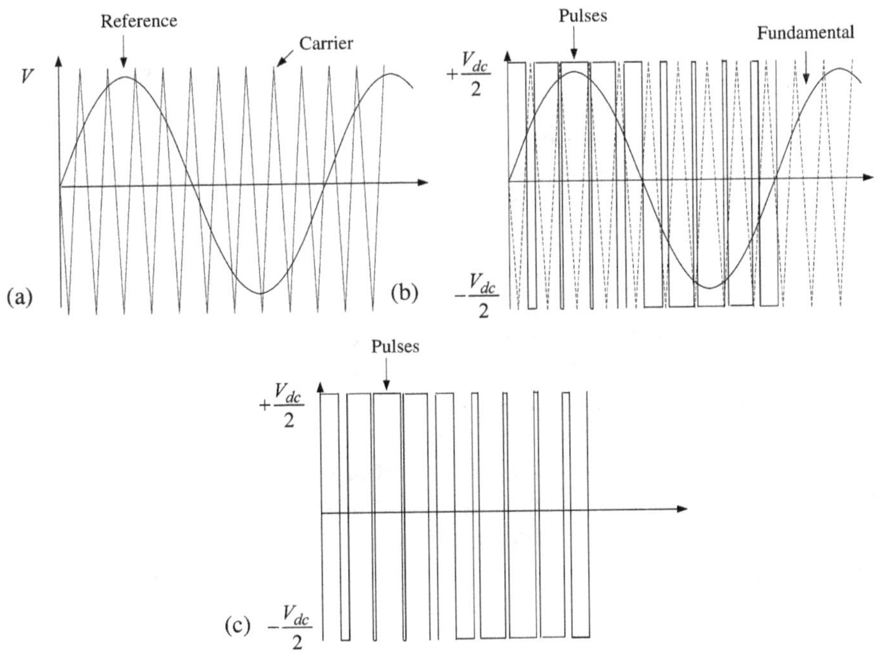

Figure 6.16 Voltage waveform generated with PWM. (a) Modulation and carrier voltages. (b) Pulses and fundamental (modulation) signal. (c) Generated pulses.

main circuit. The second uses a high-frequency, triangular-shaped carrier signal. When comparing signals, during the time interval in which the modulating wave is algebraically greater than the carrier wave, the gate signal is defined in the on (positive) state and otherwise in the off (negative) state, as indicated in Figure 6.16b (as for the half-bridge with two levels in Figure 6.13f). This PWM signal is then used to drive the bridge switches in the power circuit to obtain the desired sinusoidal voltage.

In Figure 6.16a, for each pair of pulses, when the width of the positive pulse exceeds that of the subsequent negative pulse, it contributes to an increase in the average voltage value. Conversely, if the width of the negative pulse is greater, it leads to a decrease in the average voltage value, leading to the desired waveform modulation. The decomposition of all pulses in Figure 6.16c into Fourier series presents a fundamental wave that reproduces the reference signal, with the harmonic spectrum concentrated in the high-frequency range [3]. A review on PWM techniques can be found in [17].

6.6 MMC – Modular Multilevel Converter

The recent evolution of the modular multilevel converter (MMC) has enabled several applications in HVDC, FACTS, and IBGs. This technology has been very successful for several reasons, including low switching frequency and, consequently, loss reduction. In addition to the several practical results already achieved, the technology is very promising in new conceptions of equipment to be applied in power systems. The modular application allows high voltage and power levels to be achieved without the need to use coupling transformers or even harmonic filters.

Figure 6.17 shows a three-phase five-level converter, with only four submodules (SM) per leg and two SM per arm, for illustration purpose [19]. The figure details one SM and the current paths are shown in Figure 6.13.

The DC bus capacitor serves several purposes, allowing temporary storage of energy during transient conditions or fluctuations in the power system, which increases the stability and reliability of the converter and improves voltage regulation. In both DC links and IBGs, the converter capacitor is connected to a primary energy source, such as a renewable source or a DC transmission line. This source keeps the capacitor charged during active power supply. Capacitors are necessary in IBGs for voltage stability but are not essential in DC links. For reactive power control, the capacitor is crucial for maintaining stable DC bus voltage, even without the primary energy source.

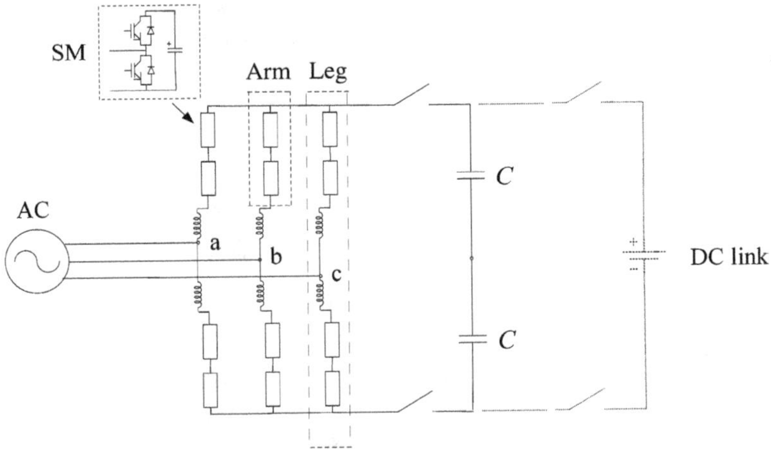

Figure 6.17 Three-phase five-level MMC.

The legs of a three-phase converter are identical and each of them is associated with a phase of the AC voltage. These legs contain SM connected in series and distributed over two arms. Each SM has two levels, $+V_{dc}$ when the SM capacitor is inserted or zero when this capacitor is off. The SM capacitor sizing is derived in [20, 21] and to be able to alternate the polarity of the pulses, two arms are needed, upper and lower.

The arm reactors in Figure 6.17 serve to mitigate circulating currents between converter legs and minimize the impact of internal and external faults on the converter. The voltage drops across the reactor are small at low frequency, and the reactor derivation may be found in [20, 22].

An SM trigger sequence with four steps is set up by the control and shown in Figure 6.18, with capacitors on and off in the leg of phase a.

The half-bridge rules were presented in Section 6.5.1, but in summary, switches sw1 and sw2, when turned on or off, put the capacitor in on or off modes, respectively.

Note that the submodule states of the lower arms are the opposite of the upper arms.

Steps 1, 2, 3, and 4, in table 6.1, are cyclically repeated in steps 5, 6, 7, and 8 and so on. The control adjusts the pulse durations, which in this simplified case are indicated on the arm voltage E_a^+ . In phases b and c the sequences are the same, delayed by 120° and 240°, respectively. Voltages in upper and lower arms are represented in Figure 6.19.

It is possible to calculate the line voltages (phase-to-phase) of a three-phase system with the upper arm voltages,

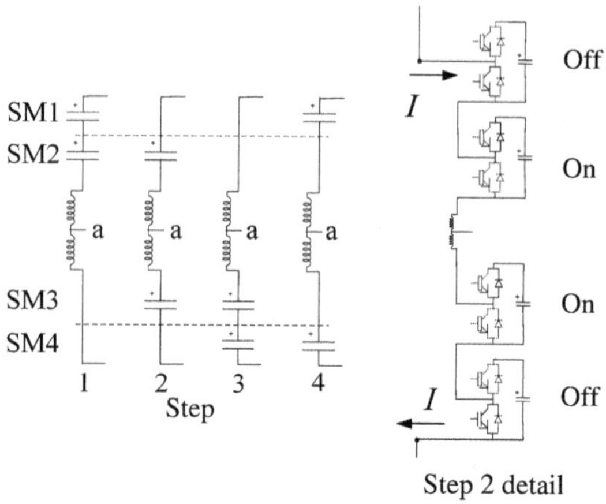

Figure 6.18 Triggering sequence with four steps.

Table 6.1 Submodule states.

	Upper SMs		Lower SMs	
Step	**1**	**2**	**3**	**4**
1	on	off	off	on
2	on	on	off	off
3	off	on	on	off
4	off	off	on	on

$$V_{ab} = E_b^+ - E_a^+ ; V_{bc} = E_c^+ - E_b^+ ; V_{ca} = E_a^+ - E_c^+ , \tag{6.57}$$

as in Figure 6.20.

But the line voltages can also be calculated with the lower arm voltages:

$$V_{ab} = E_a^- - E_b^- ; V_{bc} = E_b^- - E_c^- ; V_{ca} = E_c^- - E_a^- . \tag{6.58}$$

The line voltages in an MMC (three-wire) converter can interface with a grid that includes a neutral connection. From the phase-to-phase voltages, the phase-to-neutral voltages applied to a balanced three-phase star-connected load with a neutral point can be calculated with a loop analysis, but this very simple circuit will be solved with the superposition principle (Figure 6.21a).

When voltage V_{ab} is applied and V_{bc} is by-passed, V_{an} is equal to:

$$V_{an}^{(1)} = \frac{2}{3} V_{ab}$$

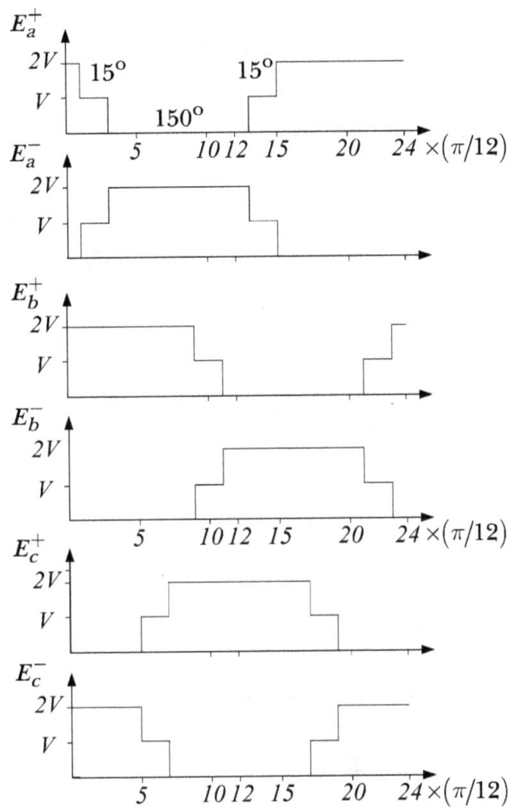

Figure 6.19 Voltages in upper and lower arms.

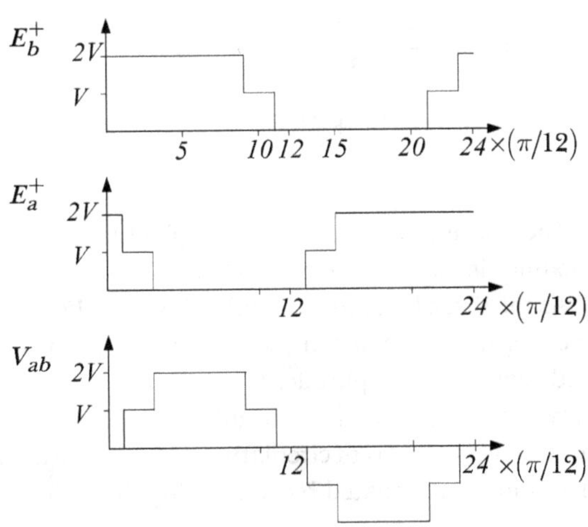

Figure 6.20 V_{ab} line-to-line voltage.

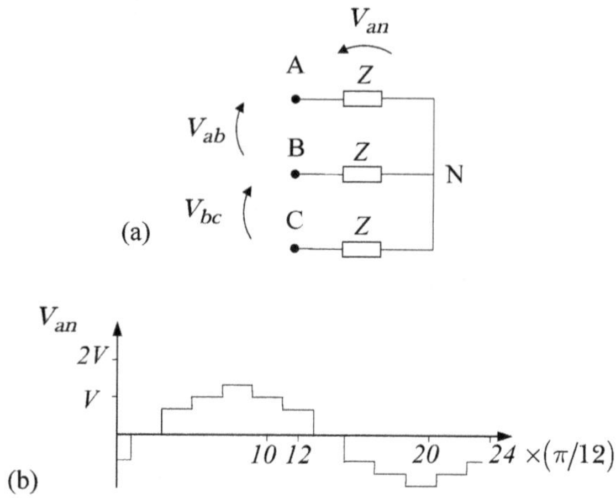

Figure 6.21 Phase-to-neutral voltage. (a) Star connected load. (b) Phase-to-neutral voltage with four sub modules.

When V_{bc} is applied and V_{ab} is by-passed V_{an} is given by:

$$V_{an}^{(2)} = \frac{1}{3} V_{bc}$$

Summing up the two parcels:

$$V_{an} = \frac{2V_{ab} + V_{bc}}{3}. \tag{6.59}$$

And doing the same for phases b and c:

$$V_{bn} = \frac{-V_{ab} + V_{bc}}{3};$$

$$V_{cn} = -\frac{V_{ab} + 2V_{bc}}{3};$$

The phase-to-neutral voltage applied to a load by the converter with four sub-modules is shown in Figure 6.21b.

As the three legs are in parallel (Figure 6.18), the number of modules inserted in each leg must be constant, so that there is no current circulation between the arms with significant amplitude. A circulating current exists, given the small imbalance between the legs, but if the number of modules is different, the imbalance becomes prohibitive in terms of circulating current. The circulating current, although small, needs to be attenuated by controlling the triggering of the IGBTs.

6.7 Converter Control

Converter control ensures efficient and reliable operation in various applications such as HVDC systems and IBGs. The control strategy involves several aspects such as current control, voltage balancing, real-time protection, network synchronization, etc. Effective MMC control strategies are essential for optimizing converter performance in diverse applications.

Space-vector control is the designation for converter control, which will be introduced in its basic elements, focusing only on the transformation of variables and synchronization functions. The aim is to provide a basic overview, without delving into the intricate electronic details of all control functions. A review of control methods may be found in [18].

6.7.1 Transformation abc/$\alpha\beta$ and $\alpha\beta$/dq

The transformations of variables used in the VSC control and electrical machines are quite similar [23].

In the stator of a synchronous generator, the three sinusoidal fluxes ϕ_a, ϕ_b and ϕ_c, related to three sinusoidal currents delayed by 120° in time, i_a, i_b, i_c feed three windings displaced by 120° in the space. The instantaneous composition of the fluxes creates a rotating magnetic field in space, with a constant magnitude whose frequency ω is equal to the frequency of the sinusoidal currents.

Thus, using an analogy with synchronous generators, only the basic properties of space vectors are presented.

a) Space vector in R2
   Initially, a fixed reference frame is defined in a plane (R2) with three fixed axes, $\vec{a}$, $\vec{b}$, $\vec{c}$, as shown in Figure 6.22a displaced in space by 120°.
   The complex number α is defined as:

$$\alpha = 1\angle 120°.$$

In the counterclockwise positive convention, axis $\vec{b}$ leads axis $\vec{a}$ by 120° while axis $\vec{c}$ lags behind axis $\vec{a}$ by 120°, thus

$$\vec{b} = \alpha\vec{a}, \vec{c} = \alpha^2\vec{a}.$$

Three-phase voltages with a positive sequence are assumed in which phase b now lags behind phase a by 120°, and phase c leads phase a by 120°.

A rotating vector, whose position depends on time, is defined with angular velocity in the counterclockwise-positive convention (Figure 6.22a). A stationary

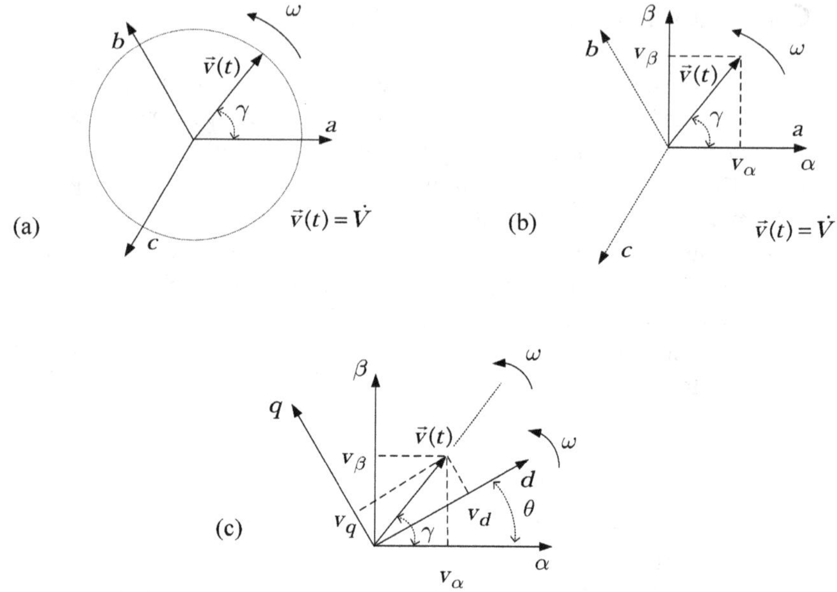

Figure 6.22 Reference frames abc/$\alpha\beta$/dq (R2). (a) Reference frame abc. (b) Transformation abc/$\alpha\beta$. (c) Transformation $\alpha\beta$/dq.

voltage phasor is designated in module and phase angle as $V\angle\gamma$ (rms) and the voltage phasors of the three-phase system are given by:

$$V_a = V\angle\gamma, V_b = \alpha^2 V\angle\gamma, V_c = \alpha V\angle\gamma.$$

To simplify the notation, let the rotating vector be designated as $\dot{V}$ in (6.60).

$$\dot{V} = \sqrt{2}V\angle\gamma e^{j\omega t}, \tag{6.60}$$

The voltage amplitudes over time in the three-phase system abc can be espressed as a function of this rotating vector:

$$v_a(t) = \text{Re}(\dot{V});\ v_b(t) = \text{Re}(\alpha^2\dot{V});\ v_c(t) = \text{Re}(\alpha\dot{V}),$$

which can be rewritten with Euler's formula:

$$v_a(t) = \frac{\dot{V} + (\dot{V})^*}{2};\ v_b(t) = \frac{\alpha^2\dot{V} + (\alpha^2\dot{V})^*}{2};\ v_c(t) = \frac{\alpha\dot{V} + (\alpha\dot{V})^*}{2}. \tag{6.61}$$

Placing voltages abc in the spatial reference frame, the vectors have variable amplitudes in time, but fixed directions in space:

$$\vec{v}_a(t) = v_a(t)\vec{a};\ \vec{v}_b(t) = v_b(t)\vec{b};\ \vec{v}_c(t) = v_c(t)\vec{c}$$

The space vector $\vec{v}(t)$ is defined as the sum of the three vectors above, according to Eq. (6.62).

$$\vec{v}(t) = \frac{2}{3}\left(v_a(t)\vec{a} + v_b(t)\vec{b} + v_c(t)\vec{c}\right). \tag{6.62}$$

Thus, substituting (6.61) into (6.62), Eq. (6.63) is achieved:

$$\vec{v}(t) = \frac{2}{3}\left(\frac{\dot{V} + (\dot{V})^*}{2} + \frac{\dot{V}\alpha^2 + (\dot{V}\alpha^2)^*}{2}\alpha + \frac{\dot{V}\alpha + (\dot{V}\alpha)^*}{2}\alpha^2\right)\vec{a}. \tag{6.63}$$

After some algebraic manipulation and knowing that $\vec{a} = 1\angle 0^\circ$, the result of the sum in (6.63) gives the rotating vector (6.64):

$$\vec{v}(t) = \frac{2}{3}\left(\frac{3}{2}\dot{V}\right) = \dot{V}, \tag{6.64}$$

which is the same as in (6.60). Thus, vector $\vec{v}(t)$ rotates with angular velocity ω counterclockwise, describing a circle with radius $\sqrt{2}V$ as in Figure 6.22a.

Note that the factor 2/3 allows the amplitude of $\vec{v}(t)$ to be equal to the amplitude of the balanced three-phase voltages.

- Transform abc/$\alpha\beta$

  The vector $\vec{v}(t)$ is transformed to another fixed base, designated as $\alpha\beta$, defined by an axis $\vec{\alpha}$ parallel to the axis $\vec{a}$ and by a second axis $\vec{\beta}$, which leads $\vec{\alpha}$ by 90°, as shown in Figure 6.22b. The designation $\alpha\beta$ is the same as that used in [24] and has no relation to the complex number α defined previously.

  From Figure 6.22b, the transformation matrix in (6.65) may be written:

$$\begin{bmatrix} v_\alpha \\ v_\beta \end{bmatrix} = \frac{2}{3}\begin{bmatrix} 1 & -\frac{1}{2} & -\frac{1}{2} \\ 0 & \frac{\sqrt{3}}{2} & -\frac{\sqrt{3}}{2} \end{bmatrix}\begin{bmatrix} v_a \\ v_b \\ v_c \end{bmatrix}. \tag{6.65}$$

In (6.65) the first column is the projection of vector $v_a\vec{a}$ in the direction $\vec{\alpha}$; the second column contains the projection of vector $v_b\vec{b}$ in the directions $\vec{\alpha}$ and $\vec{\beta}$, with cos($2\pi/3$) and sin($2\pi/3$), respectively. The third column has analogous projection. Note that both v_α and v_β are time-dependent variables.

- Transform $\alpha\beta$/dq

  After obtaining the rotating vector $\vec{v}(t)$ in fixed coordinates $\alpha\beta$, a new reference frame is defined by the orthogonal vectors $\vec{d}$ and $\vec{q}$, ($\vec{q}$ leading $\vec{d}$ by 90°) rotating with respect to axis $\vec{\alpha}$ with angular velocity ω in a counterclockwise-positive convention. If ω is the same angular velocity in (6.60), the rotating vector $\vec{v}(t)$ is stationary in the new reference frame dq and voltage components v_d and v_q are constant, as in Figure 6.22c.

Assuming θ the angle between the reference frames $\alpha\beta$ and dq,

$$\theta(t) = \theta_0 + \omega t. \tag{6.66}$$

From Figure 6.22c, the transformation $\alpha\beta/\,dq$ is given in (6.67).

$$\begin{bmatrix} v_d \\ v_q \end{bmatrix} = \begin{bmatrix} \cos\theta & \sin\theta \\ -\sin\theta & \cos\theta \end{bmatrix} \cdot \begin{bmatrix} v_\alpha \\ v_\beta \end{bmatrix}, \tag{6.67}$$

and for the inverse relationship, simply change the signs of function $\sin\theta$ in (6.67).

Only a brief introduction to the R2 system was presented, covering the fundamental aspects of space vectors based on the positive sequence. This analysis can be expanded in the case of a generic three-phase system, using the Fourier series. The harmonic components also describe circles, but with angular velocities that are multiples of the fundamental. The analysis of the negative and zero sequence, as well as the harmonic components, is important for the stability of the control system, as well as for the effectiveness of power conditioners and filters.

b) Space vector in R3

The space vector $\vec{v}(t)$ is defined in an orthonormal basis $\vec{a}, \vec{b}, \vec{c}$, in R3, in a similar way to (6.62), but without the factor 2/3. This reference frame can be visualized as the three unitary and converging edges of a transparent cube, as in Figure 6.23a.

$$\vec{v}(t) = \left(v_a(t)\vec{a} + v_b(t)\vec{b} + v_c(t)\vec{c}\right). \tag{6.68}$$

The geometric locus equidistant from the three axes $\vec{a}, \vec{b}, \vec{c}$ is in the main diagonal of the cube, whose length is $\sqrt{3}$. Thus, the unitary vector in the diagonal direction is defined as:

$$\vec{o} = \frac{\vec{a} + \vec{b} + \vec{c}}{\sqrt{3}}. \tag{6.69}$$

Assume the lower vertex "o" of the cube placed on a plane designated as $\alpha\beta$, with the vector $\vec{o}$ (parallel to the main diagonal) perpendicular to this $\alpha\beta$ plane (Figure 6.23a).

The upper view of the cube in Figure 6.23b shows only the projections of edges $\left(\vec{a}, \vec{b}, \vec{c}\right)$, indicated as $\left(\vec{a}', \vec{b}', \vec{c}'\right)$, dividing the plane $\alpha\beta$ into three sectors of 120°, but their magnitudes are lower than one (0.82).

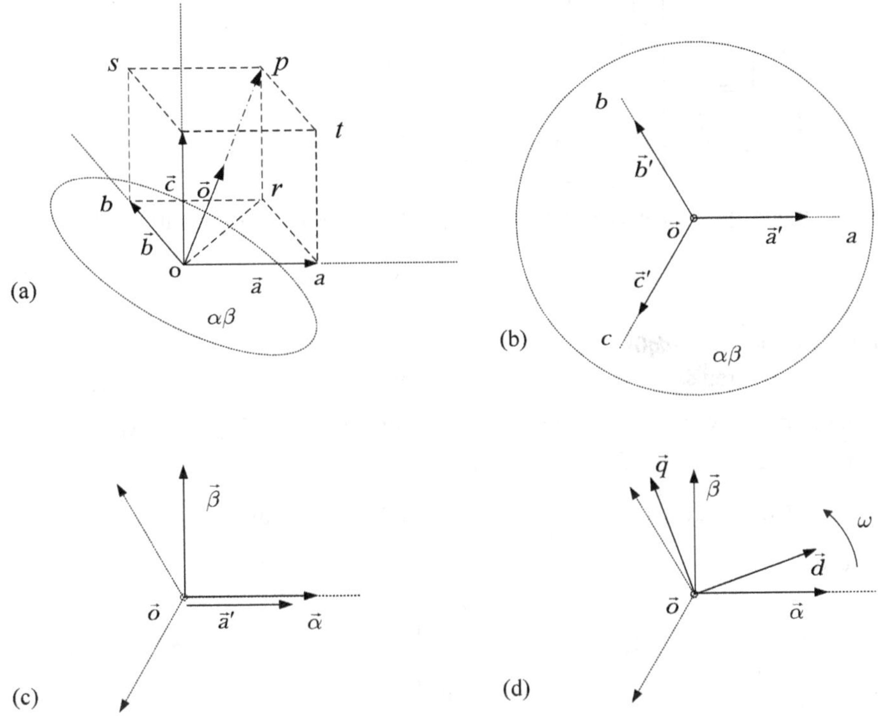

Figure 6.23 Geometric representation of reference frames abc/$\alpha\beta$0/dq0 (R3). (a) Reference frame abc in R3. (b) Projections of abc in $\alpha\beta$. (c) Upper view of reference frame $\alpha\beta$0. (d) Upper view of reference frame dq0.

For any space vector $\vec{v}(t)$, defined as in (6.68), the $\alpha\beta$ plane is the geometric locus for zero sequence null; thus, if axes $\left(\vec{a}', \vec{b}', \vec{c}'\right)$ are assumed with unitary amplitudes, the same reference frame $\left(\vec{a}, \vec{b}, \vec{c}\right)$ in R2 is achieved.

For the general case, the zero sequence in space is placed on the axis $\vec{o}$, assuring the same projections of the zero sequence on axes $(\vec{a}, \vec{b}, \vec{c})$.

- **Reference frame ($\alpha\beta$0)**

As a second step, unitary axes $\vec{\alpha}$ and $\vec{\beta}$ are placed in the plane $\alpha\beta$, with $\vec{\alpha}$ parallel to $\vec{a'}$ and $\vec{\beta}$ leading $\vec{\alpha}$ by 90°, which gives rise to the $\alpha\beta$0 reference frame, as in Figure 6.23c.

The transformation of the space vector $\vec{v}(t)$ from abc to $\alpha\beta$0 reference frame is given by Eq. (6.70), [25, 26].

$$\begin{bmatrix} v_0 \\ v_\alpha \\ v_\beta \end{bmatrix} = \sqrt{\frac{2}{3}} \begin{bmatrix} \frac{1}{\sqrt{2}} & \frac{1}{\sqrt{2}} & \frac{1}{\sqrt{2}} \\ 1 & -\frac{1}{2} & -\frac{1}{2} \\ 0 & \frac{\sqrt{3}}{2} & -\frac{\sqrt{3}}{2} \end{bmatrix} \cdot \begin{bmatrix} v_a \\ v_b \\ v_c \end{bmatrix}. \tag{6.70}$$

The v_0 component is calculated from the first line in (6.70), as in (6.71):

$$v_0 = \frac{v_a + v_b + v_c}{\sqrt{3}}. \tag{6.71}$$

- **Reference frame ($dq0$)**
  The third step is to make the transformation from $\alpha\beta 0$ to the $dq0$ reference frame, simply extending equation (6.67), in which the axes $\vec{d}, \vec{q}$ rotate with angular velocity ω around axis $\vec{o}$, which is fixed both in the $\alpha\beta 0$ and in the dq0 reference frames. Note that $\left(\vec{a}', \vec{b}', \vec{c}'\right)$, $\left(\vec{\alpha}, \vec{\beta}\right)$ and $\left(\vec{d}, \vec{q}\right)$ are all in plane $\alpha\beta$, in Figure 6.23d.

$$\begin{bmatrix} v_0 \\ v_d \\ v_q \end{bmatrix} = \begin{bmatrix} 1 & 0 & 0 \\ 0 & \cos\theta & \sin\theta \\ 0 & -\sin\theta & \cos\theta \end{bmatrix} \begin{bmatrix} v_0 \\ v_\alpha \\ v_\beta \end{bmatrix} \tag{6.72}$$

The transformation from abc to dq0 reference frame can be done directly using (6.73).

$$\begin{bmatrix} v_0 \\ v_d \\ v_q \end{bmatrix} = \frac{\sqrt{2}}{\sqrt{3}} \begin{bmatrix} \frac{1}{\sqrt{2}} & \frac{1}{\sqrt{2}} & \frac{1}{\sqrt{2}} \\ \cos\theta & \cos\left(\theta - \frac{2\pi}{3}\right) & \cos\left(\theta + \frac{2\pi}{3}\right) \\ -\sin\theta & -\sin\left(\theta - \frac{2\pi}{3}\right) & -\sin\left(\theta + \frac{2\pi}{3}\right) \end{bmatrix} \cdot \begin{bmatrix} v_a \\ v_b \\ v_c \end{bmatrix} \tag{6.73}$$

For the inverse transformation, from dq0 to abc, the terms in the third line in (6.73) have positive signs.

In Figure 6.24, space vector $\vec{v}(t)$ is represented by components (v_d, v_q, v_0) in axes dq0.

6.7.2 PLL – Phase-Locked Loop

The AC transmission system, in its essence, has a fundamental dependence on synchronism. The operation of connecting a synchronous generator to the grid uses

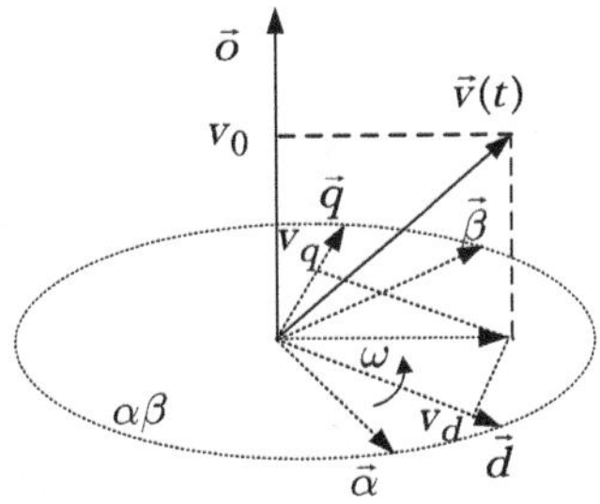

Figure 6.24 Space vector $\vec{v}(t)$ and its components (v_d, v_q, v_0) in axes dq0.

the synchronoscope, which adjusts the exact moment to parallel the generator with the grid. After synchronization, on a stable equilibrium point, the system operates continuously, and in the face of stable disturbances, thanks to the presence of synchronizing torques that restore equilibrium points in a steady state.

To exchange controlled active or reactive power with the grid, all equipment must comply with the conditions imposed by synchronism.

With thyristors, the LCC control regulates the gate signals based on the three-phase voltages supplied by the network to the equipment. Note that in this case, there is only one three-phase voltage system.

For VSC converters, which synthesize three-phase voltages from DC sources, it is necessary to establish synchronism between the three-phase sinusoidal voltages generated by the equipment with the three-phase voltages of the network, noting that in this case, there are two three-phase voltage systems.

The PLL has its origin in telecom systems and plays a fundamental role in synchronizing the inverter and grid voltages. In this case, synchronism must be maintained continuously by the control, as there are no synchronizing torques due to the absence of electromagnetic torques in the VSCs.

The basic principle is to place the PLL's space vector orthogonal or parallel to the grid's space vector.

The synchronous reference frame (SFR-PLL) is the most elementary PLL implementation, which assumes high strength and low distortion level for the network. As this is not always the case, several strategies seek to mitigate the distortion by changing the PLL bandwidth with filters, in such a way that the response time delay is as little as possible.

- PLL based on the dot product between space vectors
  In Figure 6.25a:

  u_d: comparison between signals u_{in} and u_{out}.
  u_f: filtered signal.
  PD: phase detector.
  LF: loop filter.
  VCO (voltage-controlled oscillator).

As shown in this figure, the procedure is to calculate the dot product of the space vectors u_{in} and u_{out} in the PD block whose output u_d can be understood as an error signal. The error approaches zero when the dot product also approaches zero in

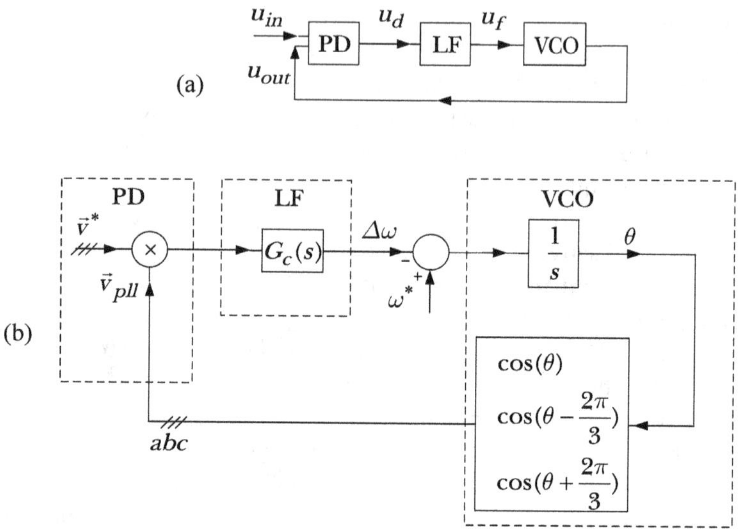

Figure 6.25 PLL – phased locked loop in abc reference frame. (a) PLL basic topology. (b) Space vectors dot product in a closed loop similar to the basic topology.

the closed loop, that is, when u_{in} is orthogonal to u_{out}. The second block, LF, has a PI-type controller and the third block, VCO has u_f as input, whose amplitude controls the frequency of u_{out}, which closes the feedback loop.

Figure 6.25b is intended to reproduce the PLL (in the abc domain) of a VSC with the same topology as Figure 6.25a.

In Figure 6.25b, the following variables are defined:

$\vec{v}^*(t)$: space vector of instantaneous grid voltages.
$\vec{v}_{pll}(t)$: PLL space vector.
ω^*: grid frequency.
$\Delta\omega$: frequency error (output of the PI controller).
ω_{pll}: PLL frequency.
θ: angle for transformation $\alpha\beta/dq$.
$G_c(s)$: controller transfer function.

In the PD block, when dot product $\vec{v}^*(t).\vec{v}_{pll}(t)$ is null, the spatial vectors $\vec{v}_{pll}(t)$ and $\vec{v}^*(t)$ are orthogonal, which is the fundamental information obtained by the PLL. Therefore, the control can set any arbitrary reference frame; for instance, the three-phase voltages of the converter may be advanced by 90°, setting its spatial vector parallel to $\vec{v}^*(t)$.

- PLL with transform abc/dq0

However, it is common to apply the abc/$\alpha\beta$ and $\alpha\beta/dq$ transforms, as in Figure 6.26a, and in the PLL loop, the input of the PI controller can be v_d or v_q.

The PLL sets $\vec{v}_{pll}(t)$ parallel to axis $\vec{d}$. In the closed loop, the PI controller input tends to zero in any case. If it is imposed $v_d = 0$, the space vector $\vec{v}^*(t)$ will be orthogonal to $\vec{d}$ (the inner product $v_d = \vec{v}^*(t).\vec{d}$ is null). However, if it is imposed $v_q = 0$, the space vector will be parallel to $\vec{d}$ (the inner product $v_q = \vec{v}^*(t).\vec{q}$ is null). For this reason, reference [27] argues that the similarity among the different PLL propositions is tied to the zero inner product between grid and PLL space vectors.

In Figure 6.26b, the controller imposes a null frequency error ($\Delta\omega$) and thus the derivative of the angle θ is the network frequency ω^* (rd/s). The function that describes the angle θ in time has a triangular shape that resets at 2π.

In steady state, the displacement between $\vec{v}^*(t)$ and $\vec{v}_{pll}(t)$ is zero, indicated by δ in Figure 6.26b, thus, $\vec{v}^*(t)$ and $\vec{d}$ are aligned. This alignment can also be admitted for analysis of stability, as the PLL response is much faster than electromechanical oscillations. For electromagnetic transients, there may be some change in the relative position of the vectors, but the PLL response is also very fast.

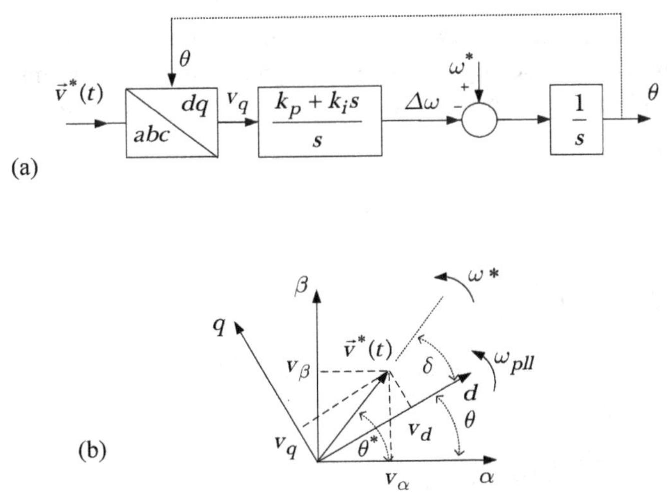

Figure 6.26 PLL in dq0 reference frame. (a) PLL closed loop to make $v_q$zero. (b) Reference frame dq.

There are other propositions for controllers, in real or pu quantities as well as in relation to feedback signals and axes orientation of the reference system, but this should be clarified in the specific PLL design.

6.7.3 Elementary Control

The purpose of VSC control is to generate symmetrical and balanced three-phase voltages, which can be adjusted in amplitude and phase, so that the device performs certain functions in synchronization with the network.

A simple topology in Figure 6.27a is composed of a feedback loop in the abc phase domain, with the comparison between reference and measured current. However, VSC control uses transformation from the abc to dq (or dq0) reference frames, as in Figure 6.27b,

Where:

T^{dq}_{abc}: Transformation from abc to dq domain.
T^{abc}_{dq}: Transformation from dq to abc domain.
θ: angle for the reference frame transformation (PLL).

The control topology in Figure 6.28 comprises a faster internal current loop (fast regulation) complemented by an external loop that modulates the current references and provides voltage control. The two loops give the VSC the attribute of a voltage generator with controlled current, making the VSC robust against short circuits.

Conventionally, inverter-based generators (IBGs) operate in GFL mode (grid-following mode). However, for IBGs operating in microgrids or islanded systems, recent discussions indicate a transition toward GFM (grid-forming mode)

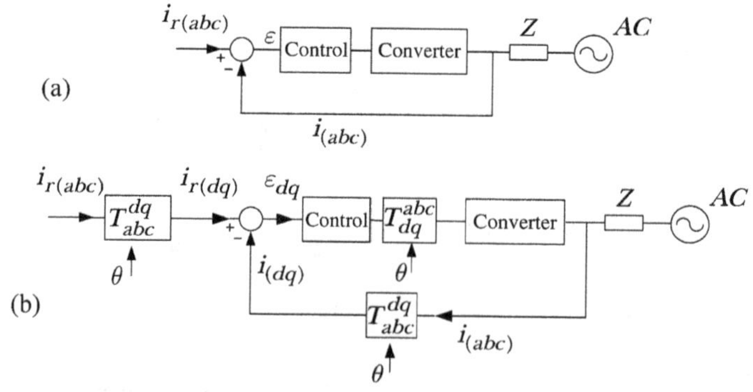

Figure 6.27 Current control. (a) Current control in abc reference frame. (b) Current control in dq reference frame.

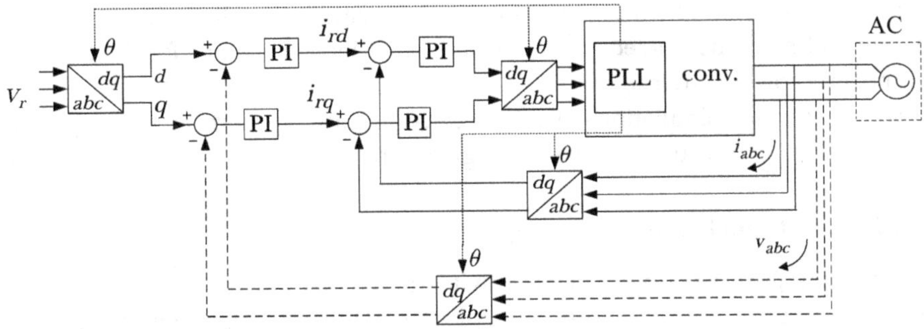

Figure 6.28 Current control with an external loop.

operation, as illustrated in the topology in Figure 6.28. This GFM concept may also have relevance in larger interconnected transmission networks, leading to discussions regarding the control structures.

In this figure, the loops are very simple with PI-type controllers that seek to bring the current and voltage errors close to zero. The feedback with sinusoidal abc variables might not perform adequately. Therefore, it is convenient to transform variables in such a way that the control operates with smoother variations, as with the transform abc/dq0.

Other control layers may be inserted into more external loops with network-level functions such as frequency control, active and reactive power, voltage, unbalances, protection, etc.

6.8 VSC Models

The models may be divided into two main families – the first focused on dynamic studies in the phasor domain, designated as rated mean square (RMS) models. The simplification in the number of variables and differential equations allows for increasing the number of dynamic components represented at the expense of a simplification of the models. The second family refers to models in the time domain, with detailed representation of equipment, designated as switching models. These models are bigger and more accurate, but the amount of equipment in the analysis is more restricted to reduce the computational burden.

The difficulties of detailed representation of large models in EMT programs are well known. When detailed modeling is required, it is necessary to include parts of the external network in equivalents, preferably using frequency dependent network equivalents (FDNEs). However, with VSCs and IBGs, calculating the equivalent is more difficult, as the number of switches (nonlinearities) is much higher and more distributed in the network, creating a challenge for the analysis of modern power systems.

Some aspects of simulations of these devices in the time domain are presented below, focusing on studies of electromagnetic transients.

6.8.1 Switching Models

The switching models represent each switch individually (IGBT, diode, thyristor, etc.) that is part of the inverter. Auxiliary routines allow the representation of control and gate drivers, and computer codes were created specifically for this purpose. This is the most accurate method, suitable both for inverter design and for analysis of the equipment interaction with the electrical grid. There are specific models of equipment, such as HVDC links, SVCs, and STATCOM, that have been developed since EMT-type programs emerged. These models require skills in using EMT program routines, as well as familiarity with electronic devices. Due to the high number of functions, only a restricted number of models can be analyzed simultaneously; therefore, several researches seek to reduce these limitations. Another issue is that the models are generally not public due to contractual clauses with manufacturers.

6.8.2 Averaged Switch Models

An alternative solution to switch number restrictions is to use the averaged switch method when the total number of switches is excessive [6, 28]. For each switch state, the modules that are active at the inverter (turned on capacitors) are included in an equivalent circuit, calculated using the trapezoidal integration rule (see Chapter 4). The submodules are reduced into Norton (or Thévenin) equivalents, drastically reducing the number of nodes that interact with the network. As an example, the model of an MMC converter with several submodules is analyzed below.

The model is based on the following steps.

- Forward process.
  a) By-pass.
     At each step, the control determines which switches will be open or closed, and, therefore, which capacitors will be inserted in that time interval, as in the upper arm indicated in Figure 6.29a. In addition to capacitors, snubber circuits can also be considered.
  b) Equivalents obtained with the trapezoidal rule.
     In ref. [6] all capacitors are replaced by their Norton equivalents, which are associated to obtain an equivalent network of the submodule with series/parallel association of electrical circuits.

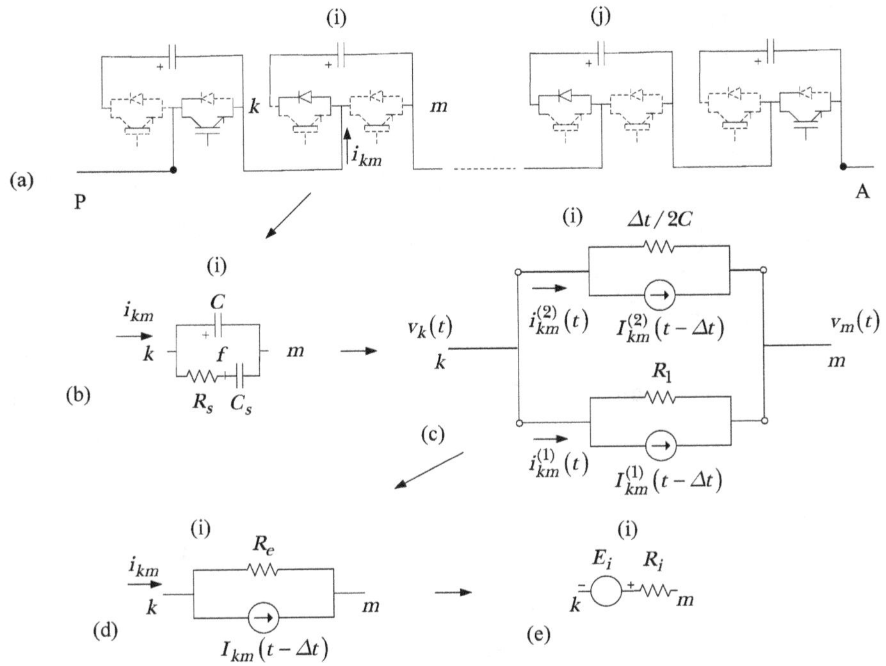

Figure 6.29 Submodule Thévenin equivalents. (a) Upper arm and active submodules. (b) Active *i*th submodule. (c) Equivalent circuit for the *i*th submodule. (d) Norton equivalent for the *i*th sub module. (e) Thévenin equivalent for the *i*th submodule.

Next, the second stage is detailed, modeling a submodule containing a snubber circuit (R_s, C_s) in parallel with the submodule's capacitor (C), as in Figure 6.29b. The submodule may be approached in different ways.

- SNUBBER CIRCUIT:

The snubber circuit may be needed in IBGs, as in [6], but they are not necessary in HVDC application [29]. In the following, the snubber circuit is integrated into each submodule. However, if its inclusion is deemed unnecessary, a more concise and simplified model is achieved.

The equations are valid for the ith module, although the index i is not highlighted in the equations so as not to overload the notation. The internal snubber node f will be eliminated, and the approach is analogous to the RL circuit in Chapter 4.

Assuming the current from node k to node m, for the capacitor C_s,

$$v_f(t) - v_m(t) = v_f(t-\Delta t) - v_m(t-\Delta t) + \frac{\Delta t}{2C_s}[i_{\text{km}}(t) + i_{\text{km}}(t-\Delta t)], \tag{6.74}$$

and for resistor R_s,

$$v_f(t) = v_k(t) - R_s i_{km}(t).$$

The RC circuit can be written only for nodes k and m.

$$v_k(t)-v_m(t)=[v_k(t-\Delta t)-v_m(t-\Delta t)]+\left(\frac{\Delta t}{2C_s}+R_s\right)i_{km}(t)+\left(\frac{\Delta t}{2C_s}-R_s\right)i_{km}(t-\Delta t),$$

$$v_k(t)-R_s i_{km}(t)-v_m(t)=v_k(t-\Delta t)-R_s i_{km}(t-\Delta t)-v_m(t-\Delta t)+ \\ +\frac{\Delta t}{2C_s}[i_{km}(t)+i_{km}(t-\Delta t)], \tag{6.75}$$

The current in the equations related to the snubber branch will be designated with the upper index 1,

$$i_{km}^{(1)}(t) = \frac{1}{\left(\frac{\Delta t}{2C_s}+R_s\right)}[v_k(t)-v_m(t)] + I_{km}^{(1)}(t-\Delta t), \tag{6.76}$$

$$I_{km}^{(1)}(t-\Delta t) = -\frac{\left(\frac{\Delta t}{2C_s}-R_s\right)}{\left(\frac{\Delta t}{2C_s}+R_s\right)}i_{km}^{(1)}(t-\Delta t)-\frac{1}{\left(\frac{\Delta t}{2C_s}+R_s\right)}[v_k(t-\Delta t)-v_m(t-\Delta t)], \tag{6.77}$$

whose equivalent circuit is at the bottom of Figure 6.29c, in which,

$$R_1 = \left(\frac{\Delta t}{2C_s}+R\right).$$

For capacitor C, simply repeat Eqs. (4.8) and (4.9) in Chapter 4 with upper index 2. The equivalent is at the top of Figure 6.29c:

$$i_{km}^{(2)}(t) = \frac{2C}{\Delta t}[v_k(t)-v_m(t)] + I_{km}^{(2)}(t-\Delta t), \tag{6.78}$$

where

$$I_{km}^{(2)}(t-\Delta t) = -i_{km}^{(2)}(t-\Delta t) - \frac{2C}{\Delta t}[v_k(t-\Delta t)-v_m(t-\Delta t)]. \tag{6.79}$$

The equivalent circuit of the submodule is shown in Figure 6.29d, in which:

$$R_e = \frac{R_1\frac{\Delta t}{2C}}{R_1+\frac{\Delta t}{2C}} \tag{6.80}$$

$$I_{km}(t-\Delta t) = I_{km}^{(1)}(t-\Delta t) + I_{km}^{(2)}(t-\Delta t) \tag{6.81}$$

c) Norton–Thévenin transformation

At this stage, Norton/Thévenin equivalent circuit transformations are performed, obtaining the Thévenin equivalent of each inserted submodule (i), as shown in Figure 6.29e.

Thévenin equivalent of submodule i:

$$R_i = R_e.$$

$$E_i = R_e I_{km}(t - \Delta t).$$

d) Sum of Thévenin equivalents

The equivalent of the converter arm is calculated with the sum of the partial equivalents of the submodules (Figure 6.30a), which can be transformed into a Norton equivalent for connection with the network, as shown in Figure 6.30b. As identical SMs are generally assumed, the sum becomes simpler, by multiplying by the number of active SMs.

The converter solution with the external network is calculated by inserting the Norton equivalents of all arms into the network's nodal admittance matrix, as shown in Figure 6.30c, at the coupling point. For this time t, the nodal voltages at points P, A, B, C, and N are calculated, as well as the currents in each branch of the equivalents.

- Reverse process

From the solution, at time t, the procedure is the same as that presented in Chapter 4, requiring a reverse process of calculating the voltages v_k and v_m in all submodules of the radial circuit in Figure 6.29a, which is facilitated because the current i_{km} is equal to solution i_{pa} in (current between nodes P and A) and is the same in all submodules in series. Consequently, the voltages v_k and v_m in each submodule can be obtained at the terminals of the Thévenin equivalents in Figure 6.29a. Only in step b, if the RC snubber is considered, there is a need to divide the current in both branches in Figure 6.29c, which is simple because the voltages v_k and v_m are known, as well as the hist terms. All hist terms are updated at time t, to be used in time $t + \Delta t$.

Examples of the averaged switch models' responses can be seen in reference [6].

It is also worth mentioning that there are alternative procedures, as the same calculation can be done with a submatrix of nodal admittances relative to the inverter arm, which is block diagonal. The solution of this subnetwork can be optimized as in references [28, 30], but the objective here is just to illustrate the averaged switch model, which is efficient because it reduces drastically the number of switches and nodes in the main network and is numerically stable.

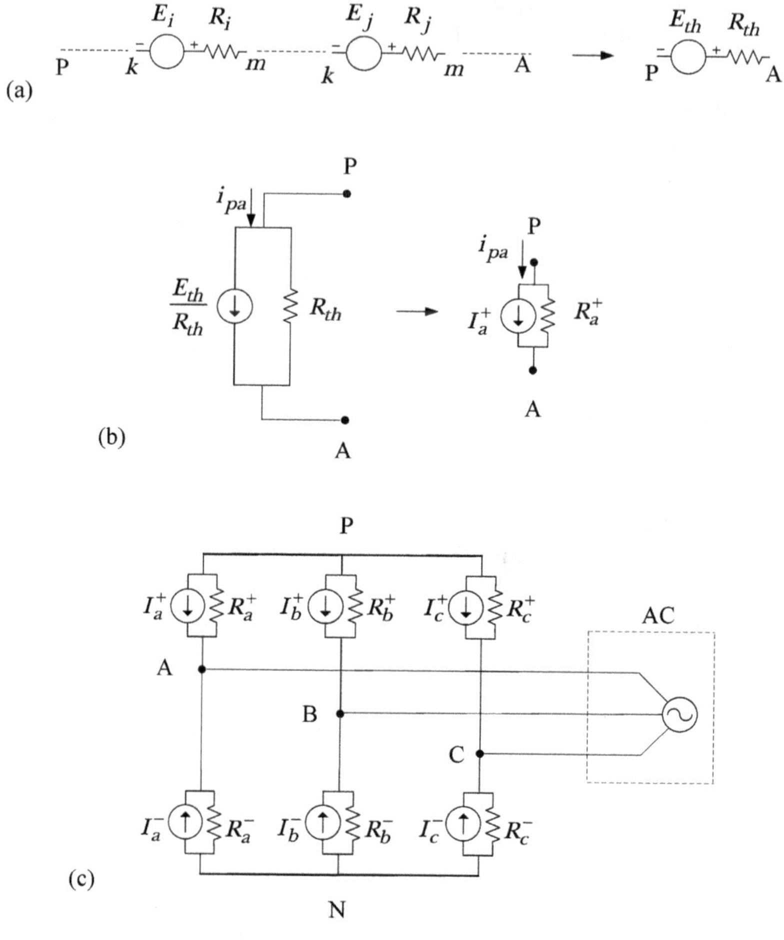

Figure 6.30 Arm equivalents. (a) Sum of series Thévenin equivalents. (b) Norton equivalent for the arm. (c) Norton equivalent for the MMC.

6.8.3 Simple Source Models

The model with sinusoidal current sources is based on a very simplified control structure that can adjust the angular reference and the amplitude of the current to be applied to the network. In general, the internal current control loop with an external voltage loop gives the model the characteristics of a voltage generator with limited current. This model does not generate harmonics and avoids commutation, but it must be used in a way that does not compromise more elaborate EMT studies.

References

1 Kimbark, E. W. (1971). *Direct current transmission*. New York: Wiley Interscience.

2 Hingorani, N. G. (1988). High power electronics and flexible AC transmission system. *IEEE Power Engineering Review*, 8(7), 3–4.

3 Hingorani, N. G., & Gyugyi, L. (2000). *Understanding FACTS: Concepts and technology of flexible AC transmission systems*. New York: IEEE Press.

4 Sen, K. K., & Sen, M. L. (2009). *Introduction to FACTS controller: Theory, modeling, and applications*. New Jersey: IEEE Press/Wiley.

5 Sen, K. K. (2015). Analysis of FACTS Controllers and their transient modelling techniques. In J. A. Martinez-Velasco (Ed.), *Transient analysis of power systems: Solution techniques, tools and applications* (pp. 195–247). Chichester: IEEE.

6 CIGRE. (2018). *Modelling of inverter-based generation for power system dynamic studies: Joint Working Group C4/C6.35/CIRED*. France: CIGRE. (Technical brochure, no. 727).

7 Eremia, M., Gole, A., & Toma, L. (2016). Static VAr Compensator (SVC). In M. Eremia, C.-C. Liu, & A.-A. Edris (Eds.), *Advanced solutions in power systems: HVDC, FACTS, and artificial intelligence* (pp. 271–338). New York: Wiley & IEEE Press.

8 Gyugyi, L. (1991). A unified flow control concept for flexible AC transmission systems. In *5th International Conference on AC and DC Power Transmission* (pp. 19–26). London: (IEE Conf. Pub., 345).

9 Schauder, C., Stacey, E., Lund, M., Gyugyi, L., Kovalsky, L., Keri, A., Mehraban, A., & Edris, A. (1998). AEP UPFC project: installation, commissioning and operation of the ±160 MVA STATCOM (phase I). *IEEE Transactions on Power Delivery*, 13(4), 1530–1535.

10 Sen, K. K., & Stacey, E. J. (1998). UPFC-unified power flow controller: theory, modeling, and applications. *IEEE Transactions on Power Delivery*, 13(4), 1453–1460.

11 Vasquez-Arnez, R. L., & Zanetta, L. C. (2005). Compensation strategy of autotransformers and parallel lines performance, assisted by the UPFC. *IEEE Transactions on Power Delivery*, 20(2), 1550–1557.

12 Vasquez-Arnez, R. L., & Zanetta, L. C. (2008). A novel approach for modeling the steady-state VSC-based multiline FACTS controllers and their operational constraints. *IEEE Transactions on Power Delivery*, 23(1), 457–464.

13 Noroozian, M., Angquist, L., Ghandhari, M., & Andersson, G. (1997). Use of UPFC for optimal power flow control. *IEEE Transactions on Power Delivery*, 12(4), 1629–1634.

14 Pereira, M., & Zanetta, L. C. (2013). A current based model for load flow studies with UPFC. *IEEE Transactions on Power Systems*, 28(2), 677–682.

15 Gyugyi, L., Sen, K. K., & Schauder, C. D. (1999). The interline power flow controller concept: a new approach to power flow management in transmission systems. *IEEE Transactions on Power Delivery*, 14(3), 1115–1123.

16 Nami, A., Liang, J., Dijkhuizen, F., & Demetriades, G. D. (2015). Modular multilevel converters for HVDC applications: Review on converter cells and functionalities. *IEEE Transactions on Power Electronics*, 30(1), 18–36.

17 Barros, L. A. M., Martins, A. P., & Pinto, J. G. (2022). A comprehensive review on modular multilevel converters, submodule topologies, and modulation techniques. *Energies*, 15(3), 1078.

18 Sun, P., Tian, Y., Pou, J., & Konstantinou, G. (2022). Beyond the MMC: Extended modular multilevel converter topologies and applications. *IEEE Open Journal of Power Electronics*, 3, 317–333.

19 Watanabe, E. H., Ferreira, J. R. B. L., & Portugal, P. M. M. (2016). Fundamentals on VSC. In *Jornada Técnica sobre Conversores Fonte de Tensão (VSC), CIGRE - B4, Rio de Janeiro*. Rio de Janeiro: COPPE.

20 Beddard, A. J. (2014). *Factors affecting the reliability of VSC-HVDC for the connection of offshore windfarms [Master's thesis]*. Manchester, UK: University of Manchester.

21 Marquardt, R., Lesnicar, A., & Hildinger, J. (2002). Modulares stromrichterkonzept für netzkupplungsanwendung bei hohen spannungen. *ETG-Fachbericht*, 88, 155–161.

22 Qingrui, T., Zheng, X., Huang, H., & Jing, Z. (2010). Parameter design principle of the arm inductor in modular multilevel converter based HVDC. In *International Conference on Power System Technology, Hangzhou* (pp. 1–6). Piscataway, NJ: IEEE.

23 Park, R. H. (1929). Two-reaction theory of synchronous machines generalized method of analysis-part I. *Transactions of the American Institute of Electrical Engineers*, 48(3), 716–727.

24 Clark, E. (1943). *Circuit analysis of A-C power systems: Vol. 1 – Symmetrical and related components*. New York: John Wiley.

25 Akagi, H., Ogasawara, S., & Hyosung, K. (1999). The theory of instantaneous power in three-phase four-wire systems: a comprehensive approach. In *Conference Record of the IEEE Industry Applications Conference. Thirty-Fourth IAS Annual Meeting* (Vol. 1, pp. 431–439). Piscataway: IEEE.

26 Akagi, H., Watanabe, E. H., & Aredes, M. (2007). *Instantaneous power theory and applications to power conditioning*. Piscataway, NJ: Wiley-IEEE Press.

27 Matakas, L., Komatsu, W., & Martinz, F. O. (2010). Positive sequence tracking Phase Locked Loops: A unified graphical explanation. In *Proceedings of the International Power Electronics Conference* (pp. 1273–1280). Tokyo: IEE Japan.

28 Gnanarathna, U. N., Gole, A. M., & Jayasinghe, R. P. (2011). Efficient modeling of modular multilevel HVDC converters (MMC) on electromagnetic transient simulation programs. *IEEE Transactions on Power Delivery*, 26(1), 316–324.

29 Davies, M., Dommaschk, M., Dorn, J., Lang, J., Retzmann, D., & Soerangr, D. (2008). *HVDC PLUS: Basics and principle of operation*. Erlangen: Siemens.

30 Le-Huy, P., Giroux, P., & Soumagne, J. C. (2011). Real-time simulation of modular multilevel converters for network integration studies. In *International Conference on Power Systems Transients (IPST2011) in Delft University of Technology, the Netherlands, 2011 June 14–17*.

7

Phasor Domain Analysis and Temporary Overvoltages

7.1 Introduction

This chapter addresses phasor domain analysis and the more common cases of temporary overvoltages in power systems.

These overvoltages last several cycles before system protection is activated, and they occur after the operation of a network switch or even after a fault in the system. Their effects on networks are analyzed in [1–4]. Temporary overvoltages frequently can be calculated based on RMS models, after the occurrence of some disturbance in the system. In some instances, these overvoltages may exhibit poorly damped time constants and long durations, as in cases of slow voltage regulators in generators that require additional evaluation of system dynamics.

In linear systems, after damping of the main transient components, as for example, after a switching surge, it is possible to find a solution for temporary overvoltages by simply using traditional power-system tools. However, when the phenomenon has a nonlinear behavior due to the presence of converter bridges or saturated transformers, simulations in the time domain by means of electromagnetic transient programs may be required. As the distorted waves usually contain harmonics in their spectrum, the natural oscillation frequencies of the electrical network should be properly represented, so that the time response is numerically accurate.

The computational methods are not analyzed extensively, since this chapter is focused only on the basic concepts underlying the most common temporary overvoltages.

7.2 Line Energization and Load Rejection

In this Section, a joint presentation of transmission line energization and load rejection will be made for phasor-based evaluations. Except for generator excitation levels, both maneuvers are quite similar.

Understanding Electromagnetic Transients in Power Systems, First Edition. Luiz Cera Zanetta Jr.

Published 2025 by John Wiley & Sons, Inc.

In the case of load rejection, transformer saturation may occur, as the excitation levels of generators are generally higher than is the case for line energization.

Energization and load rejection are presented below as linear phenomena, highlighting the most fundamental aspects of temporary overvoltages. Although nodal equations could be applied to calculate the phasor domain solution, the more intuitive two-port approach will be used.

Nonlinear phenomena, including saturation of transformers during temporary overvoltages are dealt with in Section 7.7.

7.2.1 Line Energization

- Two-Port Network

Elementary analysis of reactive compensation of transmission lines and the Ferranti effect are reviewed under this item.

In the energization of the transmission line, the sending-end is connected to the energized network and the receiving end remains unloaded.

Representation by means of a two-port network is executed as follows [5] (Figure 7.1).

The two-port network constants, determined at $s = j\omega$, for the network nominal frequency, are presented in Eq. (7.1):

$$\begin{bmatrix} V_1 \\ I_1 \end{bmatrix} = \begin{bmatrix} A & B \\ C_q & D \end{bmatrix} \times \begin{bmatrix} V_2 \\ I_2 \end{bmatrix}, \tag{7.1}$$

where C_q is the constant C for a two-port network, with the subscript q added to differentiate it from the line capacitance.

For the case of an unloaded transmission line $I_2 = 0$ and thus:

$$V_1 = AV_2. \tag{7.2}$$

Therefore, the two-port constant A relates the voltages at the sending and receiving ends of the transmission line. For ordinary lengths of transmission lines, constant A has a modulus smaller than 1, and therefore, the voltage at the receiving end is greater than that at the sending-end. Calculation of the voltage at the receiving end is important in connecting the load or even synchronizing with the electrical system on this side of the transmission line.

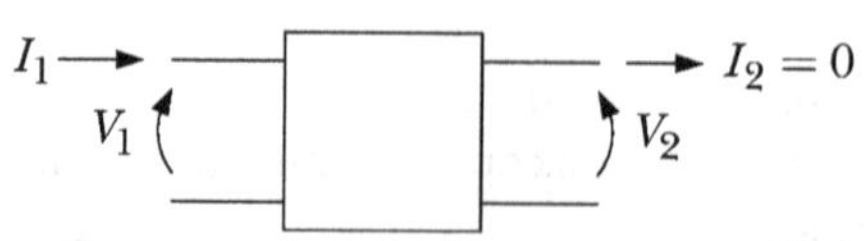

Figure 7.1 Two-port network.

When necessary, the transmission line capacitance is compensated by means of shunt reactors, which is a common practice employed to control temporary overvoltages in power systems [6].

Although the sizing of the reactive compensation in an electric network should consider other operative factors, this analysis focuses only on overvoltage control.

To approach the models through two-port networks and π circuits, initially, the simplified case of a lossless transmission line energized by an ideal voltage source is presented.

- Simple Cases

Example 7.1 Analyze the temporary overvoltages at the receiving end of a lossless, medium-length transmission line with the furnished data, assuming an ideal voltage source at the sending-end. Calculate also the reactive power absorbed by the transmission line.

Data

- $\ell = 150$ km.
- Positive-sequence parameters: $X' = 0.377$/km and $C' = 14$ nF/km.
- Nominal line voltage: $V = 500$ kV.

Solution

The solution is obtained with the aid of the equivalent two-port network of the line, making the association via the cascade of the two-port network of L and C lumped elements, as in Figure 7.2b.

Representation by the nominal – π circuit is assumed to be adequate, as the line is not too long (Figure 7.2a). A more accurate solution would require the equivalent – π of the transmission line, which will not be used for now.

$$\begin{bmatrix} A & B \\ C_q & D \end{bmatrix} = \begin{bmatrix} 1 & 0 \\ \dfrac{sC}{2} & 1 \end{bmatrix} \times \begin{bmatrix} 1 & sL \\ 0 & 1 \end{bmatrix} \times \begin{bmatrix} 1 & 0 \\ \dfrac{sC}{2} & 1 \end{bmatrix},$$

$$\begin{bmatrix} A & B \\ C_q & D \end{bmatrix} = \begin{bmatrix} 1 + s^2\dfrac{LC}{2} & sL \\ s^3\dfrac{LC^2}{4} + sC & 1 + s^2\dfrac{LC}{2} \end{bmatrix},$$

$$A = D = 1 + s^2\frac{LC}{2},$$

$$B = sL,$$

$$C_q = sC\left(1 + \frac{s^2LC}{4}\right).$$

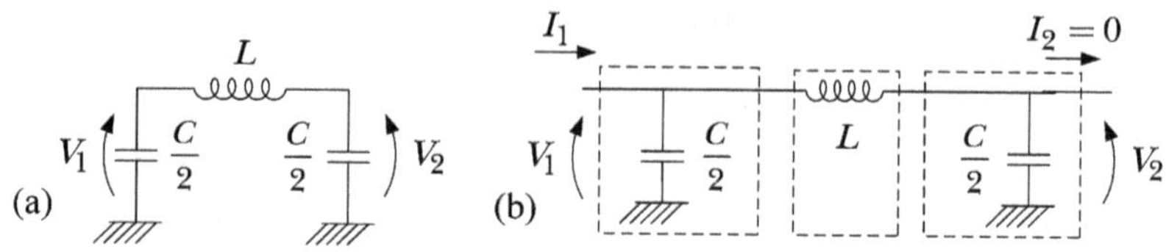

Figure 7.2 Energization of the transmission line. (a) Simplified π – nominal. (b) Cascade of two-port networks.

For the steady-state analysis at 60 Hz, constant A of the transmission line is calculated at $s = j\omega$.

$$s = j2 \times \pi \times 60 = j377.$$

In the case of a line of medium length, the total capacitance and inductance are calculated by applying the nominal – π circuit, an approximation that is acceptable:

$$L = \frac{0.377}{377} \times 150 = 0.15 \text{ H}$$

$$\frac{C}{2} = \frac{14 \times 150}{2} = 1050 \text{ nF or } \frac{C}{2} = 1.05 \text{ μF},$$

$$A = 1 + s^2 \frac{LC}{2}.$$

The same result would be obtained if the known expression ($A = 1 + ZY/2$) were applied to calculate the constant A of the transmission line, as shown in [5]. By setting:

$$Z = sL \text{ and } \frac{Y}{2} = \frac{sC}{2},$$

one obtains:

$$A = 1 + \frac{ZY}{2} = 1 + s^2 \frac{LC}{2},$$

$$A = 1 + (j377)^2 \times 0.15 \times 1.05 \times 10^{-6}, A = 0.978.$$

In this calculation for a lossless transmission line, it is clear that constant A for the line is less than 1, but even with the losses considered and more accurate models used, a constant A modulus less than 1 is the expected result for transmission lines of normal length.

Therefore, assuming $V_1 = 1.0$ pu:

$$V_2 = \frac{V_1}{A} = \frac{1.0}{0.978} = 1.022 \text{ pu}.$$

A small overvoltage is obtained at the receiving end of the line, since the line is relatively short. A voltage increase at the receiving end of an unloaded transmission line is known as the Ferranti effect.

- Reactive power absorbed by the transmission line during energization.
  The reactive power absorbed by the transmission line is calculated for one phase:

$$Q_c = V_1 I_1^*.$$

Current I_1 is calculated from:

$$I_1 = C_q V_2 + D I_2,$$

since $I_2 = 0$, I_1 is calculated from:

$$I_1 = C_q V_2,$$

$$C_q = j377 \times 2.1 \times 10^{-6}\left(1 - \frac{377 \times 2.01 \times 10^{-6} \times 377 \times 0.15}{4}\right),$$

$$C_q = j7.56 \times 10^{-4}\ \text{S}.$$

Assuming a 500-kV transmission line, the following results are obtained for the sending and receiving ends of the line:

$$V_1 = \frac{500}{\sqrt{3}} = 288.68\ \text{kV},\quad V_2 = 1.022 \times 288.68 = 295.03\ \text{kV}$$

$$Q_c = 288.68 \times 7.56 \times 10^{-4} \times 295.03 = 64.37\ \text{MVAr}.$$

That is, the three-phase reactive power absorbed by the transmission line of 150 km is 193.11 MVAr during the energization stage.

$$V_2 = 1.022'\ 288.68 = 295.03\ \text{kV}$$

A longer transmission line is analyzed next, including the reactive compensation to modify constant A of the transmission line.

Example 7.2 With the same data from Example 7.1, calculate the overvoltage at the receiving end of energization of a line of 400 km length. In addition, calculate the required reactive compensation for the voltages at the sending and receiving ends of the transmission line to maintain the same amplitude.

Solution

Constant A is initially calculated with the aid of an approximate model, using the nominal – π of the transmission line.

$$A = 1 + (j377)^2 \frac{0.377}{377} \times 400 \times \frac{400}{2} \times 14 \times 10^{-9},$$

The second term has a negative sign and it increases with the square of the line length. Thus, $A = 0.8408$, resulting in

$$V_2 = \frac{1}{0.8408} = 1.189 \text{ pu}.$$

In this case, a higher overvoltage is found. However, the result is somewhat approximate, as the nominal – π was used.

The exact calculation is performed as follows, applying the hyperbolic function [5]:

$$A = \cosh(\gamma \ell), \quad \gamma = \sqrt{jX' j\omega C'},$$

$$\gamma \ell = j\sqrt{0.377 \times 377 \times 14 \times 10^{-9}} \times 400 = j0.5642,$$

$$A = \frac{e^{-j0.5642} + e^{j0.5642}}{2},$$

$$A = \cos(0.5642) = 0.845,$$

$$V_2 = 1.183 \text{ pu}.$$

The results obtained with the parameters of the nominal – π and with the equivalent – π circuit do not differ substantially, but this is not true for higher line lengths.

The reactive compensation of the transmission line is calculated so that the nominal voltages at the sending and receiving terminals of the line are obtained. A reactor can be included at the end of the line, and the solution can be obtained via the cascade association of the two-port networks.

$$\begin{bmatrix} A & B \\ C_q & D \end{bmatrix} \times \begin{bmatrix} 1 & 0 \\ \frac{1}{sL_r} & 1 \end{bmatrix}.$$

- Line and shunt reactor (Figure 7.3):
  The association gives:

$$\begin{bmatrix} A + \frac{B}{sL_r} & B \\ C_q + \frac{D}{sL_r} & D \end{bmatrix} = \begin{bmatrix} A' & B' \\ C' & D' \end{bmatrix}$$

The new equivalent two-port network has:

$$A' = A + \frac{B}{sL_r} = A + \frac{sL}{sL_r} = A + \frac{L}{L_r}.$$

Figure 7.3 Cascade association of the reactor.

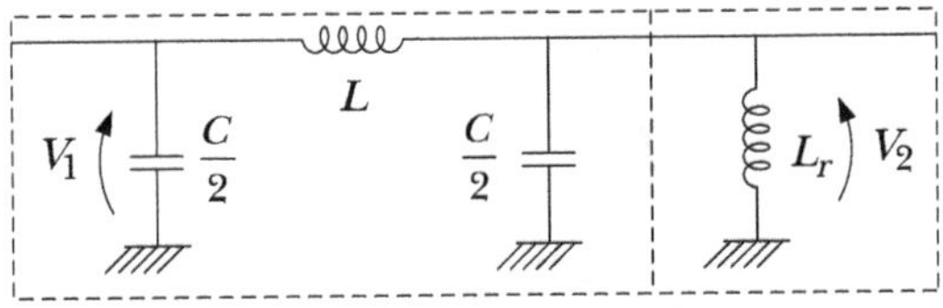

Since the line should have a unitary A', with the same voltage modules at both terminals, the inductance of the reactor is obtained from the following expression:

$$1 = 0.845 + \frac{0.4}{L_r},$$

Resulting in:

$$L_r = 2.58 \text{ H}.$$

In the previous examples, lossless transmission lines and two-port networks were used to introduce reactive compensation using very simple models. For transmission lines in which the resistance of the series impedance is well below the reactance, the approximations made are quite reasonable, as in the case of extra-high voltage (EHV) transmission lines. On the other hand, the equivalent $-\pi$ of the lossy line should be used in the calculation whenever possible to obtain accurate results.

It should be pointed out that the reactive compensation of large systems is performed using the nodal admittance matrix, but this subject is not addressed here.

- Voltage Divider

  In Example 7.2, it was seen how to provide efficient reactive compensation through an association involving more than one network; however, with another very simple analysis, one can understand the energization overvoltage at the receiving end of the transmission line.

  Assuming the transmission line without the reactor, in Figure 7.2a, voltage V_1 is set at 1.0 pu, and voltage V_2 can be obtained from the simple voltage divider equation, calculated at point $s = j\omega$:

$$V_2 = \frac{-\dfrac{j2}{\omega C}}{j\left(\omega L - \dfrac{2}{\omega C}\right)} V_1, \tag{7.3}$$

$$V_2 = \frac{\dfrac{2}{\omega C}}{\dfrac{2}{\omega C} - \omega L} V_1. \tag{7.4}$$

Since the denominator is smaller than the numerator $V_2 > V_1$.

The constant A for the transmission line is obtained by calculating V_1/V_2:

$$A = \frac{\dfrac{2}{\omega C}}{\dfrac{2}{\omega C} - \omega L}, \tag{7.5}$$

$$A = 1 - \frac{\omega^2 LC}{2}. \tag{7.6}$$

In the case of the nominal – π circuit, it is known that:

$$A = 1 + \frac{ZY}{2}, \quad \text{with } Z = j\omega L \quad \text{and} \quad \frac{Y}{2} = +j\omega\frac{C}{2}. \tag{7.7}$$

When analyzing long lines, the equivalent – π is assumed, and the previous formulas remain valid.

Still, in the case of reactive compensation, a simplified analysis can be performed by applying the voltage divider equation in Figure 7.2b:

Again, V_1 is adopted at 1.0 pu and voltage V_2 is calculated by:

$$V_2 = \frac{Z_{eq}}{j\omega L + Z_{eq}} V_1, \tag{7.8}$$

$$Z_{eq} = \frac{(j\omega L_r) \times \left(-\dfrac{j2}{\omega C}\right)}{j\omega L_r - \dfrac{j2}{\omega C}}, \tag{7.9}$$

$$Z_{eq} = j\frac{\dfrac{2L_r}{C}}{\dfrac{2}{\omega C} - \omega L_r}. \tag{7.10}$$

For the proposed condition of equal voltages at both transmission line terminals, that is, with $V_2 = V_1 = 1.0$ pu, $Z_{eq} \gg j\omega L$ should be imposed on (7.8), which means resonance between the impedance of the reactor and that of half of the total capacitance of the transmission line, that is, $Z_{eq} \to \infty$. Therefore, there is no current and no voltage drop through the inductance L in Figure 7.3 and $V_2 = V_1$. The reactor is given by:

$$\frac{2}{\omega C} - \omega L_r = 0 \to L_r = \frac{2}{\omega^2 C}. \tag{7.11}$$

In this case of an ideal source at the sending-end of the transmission line and the same modulus of the voltage at both terminals, the reactor compensates for half of the line capacitance. However, energization on the opposite side, as a matter of symmetry, would imply compensation for the other half of the capacitance, located at the sending-end, which would result in a compensation level of 100% of the transmission line capacitance. A compensation level of 100% is not realized for certain reasons to be explained later and, usually, long transmission lines operate at lower compensation levels, typically at 40–60% of the total capacitance.

- Effect of the Thévenin Equivalent of the Network

  In the previous Section, energization of the transmission line from the point of view of an infinite bus was considered. Situations can be found in which the electrical network on the side that energizes the transmission line can be considered strong or weak in terms of its short-circuit power. It is assumed that the system can be represented by the Thévenin equivalent in order to take its impedance into account.

  When the transmission line is energized, the following equivalent circuit may be proposed:

  - Z_t: Thévenin equivalent impedance.
  - Z: Series impedance of the transmission line
  - E: Thévenin equivalent voltage

The relationship between V_2 and V_1 is given by parameter A of the two-port network of the transmission line. However, the value for V_1 no longer corresponds to the Thévenin equivalent voltage E.

Making the two-port network cascade association in Figure 7.4a:

$$\begin{bmatrix} E \\ I_0 \end{bmatrix} = \begin{bmatrix} 1 & Z_t \\ 0 & 1 \end{bmatrix} \times \begin{bmatrix} A & B \\ C_q & D \end{bmatrix} \times \begin{bmatrix} V_2 \\ I_2 \end{bmatrix}, \tag{7.12}$$

$$\begin{bmatrix} E \\ I_0 \end{bmatrix} = \begin{bmatrix} A + Z_tC_q & B + Z_tD \\ C_q & D \end{bmatrix} \times \begin{bmatrix} V_2 \\ 0 \end{bmatrix}. \tag{7.13}$$

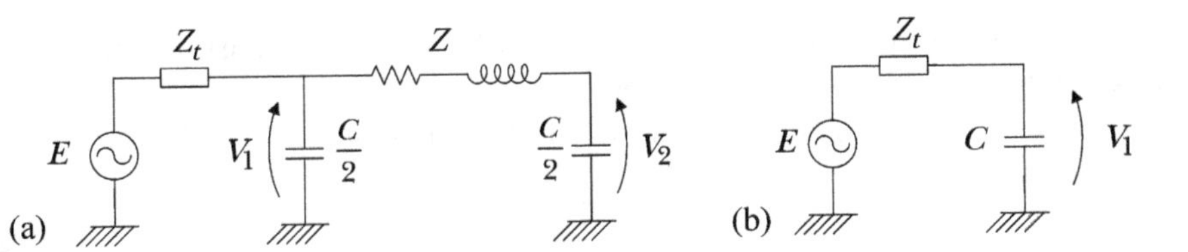

Figure 7.4 Energization of the transmission line. (a) Thévenin equivalent and transmission line. (b) Simplified circuit.

Voltage V_2 can be calculated from (7.14):

$$V_2 = \frac{E}{A + Z_t C_q}. \tag{7.14}$$

A simplified expression for the voltage at the sending-end of the transmission line V_1 is presented below. For this purpose, the equivalent impedance of the unloaded transmission line is calculated, as seen from the sending-end:

$$Z_l = -j\frac{2}{\omega C} \parallel \left(Z - j\frac{2}{\omega C}\right). \tag{7.15}$$

It can be assumed that:

$$|Z| \ll \left|-\frac{j2}{\omega C}\right|,$$

thus, the equivalent reactance of the transmission line is approximated only by its capacitive reactance:

$$X_c = \frac{1}{\omega C},$$

that is, the open line is simply represented by a large capacitor, assuming that the voltages at both terminals are approximately the same, if the Ferranti effect is omitted. The reactance at each line terminal is given by $2X_c$.

The simplified circuit in Figure 7.4b is proposed for line energization and Eq. (7.16) can be written assuming $Z_t \simeq jX_t$.

$$V_1 = \frac{-\dfrac{j}{\omega C}}{jX_t - \dfrac{j}{\omega C}} E = \frac{E}{1 - \omega C X_t}. \tag{7.16}$$

The three-phase short-circuit power of the network for a nominal line-to-line voltage V, assumed inductive, is given by:

$$S_{cc} \simeq \frac{V^2}{X_t}. \tag{7.17}$$

The capacitive three-phase power absorbed by the unloaded transmission line at the same nominal voltage V is given by:

$Q_c = V^2 \omega C$, where C is the total capacitance of the line.

Re-writing (7.16):

$$V_1 = \frac{E}{1 - \omega C X_t \dfrac{V^2}{V^2}}, \tag{7.18}$$

A first approach to calculating V_1 is obtained:

$$V_1 = \frac{E}{1 - \frac{Q_c}{S_{cc}}}. \tag{7.19}$$

In (7.19), the temporary overvoltage at the sending end of the line was shown to depend on the Q_c/S_{cc} ratio, and the worst-case scenario happens in long transmission lines with a large amount of reactive power Q_c and weak systems with little short-circuit powerS_{cc}.

If the transmission line is compensated using reactors, the formula is still valid since Q_c is considered to represent the net reactive power with the line compensation.

If accurate calculations for V_1 and V_2 are needed, the solution can be obtained either through the association of two-port networks or through nodal analysis.

In some cases, as in radial circuits, it may be advantageous to work with the cascade association of two-port networks. In Eq. (7.20), three transmission lines are series connected (cascaded) and the voltage V_r can be calculated at the end of the third line.

$$\begin{bmatrix} E \\ I \end{bmatrix} = \begin{bmatrix} 1 & Z_t \\ 0 & 1 \end{bmatrix} \times \begin{bmatrix} A_1 & B_1 \\ C_1 & D_1 \end{bmatrix} \times \begin{bmatrix} A_2 & B_2 \\ C_2 & D_2 \end{bmatrix} \times \begin{bmatrix} A_3 & B_3 \\ C_3 & D_3 \end{bmatrix} \times \begin{bmatrix} V_r \\ 0 \end{bmatrix}. \tag{7.20}$$

7.2.2 Load Rejection

- General Comments

  From the point of view of a phasor-based solution, the electric circuit for load rejection is very close to that of transmission line energization. The main difference in the two events is the excitation level of the generators or the Thévenin equivalent voltage, since in the case of load rejection, this level is considered to be higher.

  In Figure 7.5, the operating voltage at the load is assumed around 1.0 pu, and at the sending end, the voltage level is imposed by a generator voltage or an equivalent system, thus $V_1 = E$. The load is modeled as an equivalent impedance and both the load, and the transmission line are modeled in a very simplified way.

  It is known, however, that the load impedance is much smaller than the impedance corresponding to half the line capacitance, thus:

$$|Z_{load}| \ll \left| -\frac{j2}{\omega C} \right|. \tag{7.21}$$

The network in the steady state can be simplified even more (Figure 7.6a).

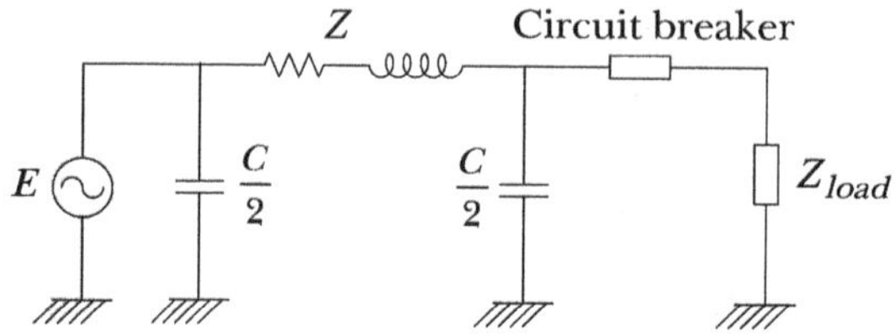

Figure 7.5 Transmission line with a load.

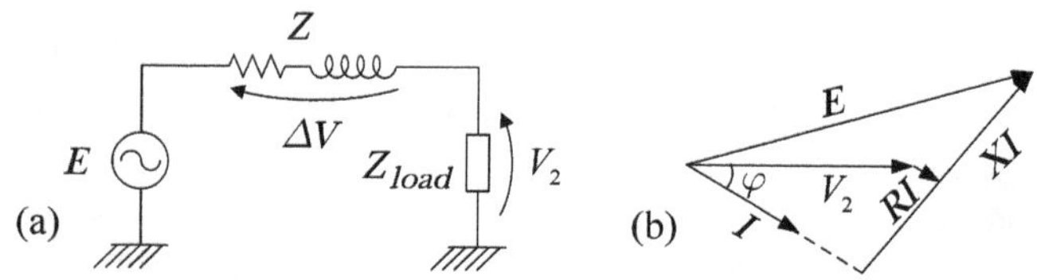

Figure 7.6 Transmission line with load. (a) Simplified circuit (b) Phasor diagram.

Usually, the load has an inductive power factor and the voltage at the receiving end is given by the voltage divider formula:

$$V_2 = \frac{Z_{load}}{Z + Z_{load}} E. \tag{7.22}$$

The phasor diagram in Figure 7.6b illustrates the behavior of voltages and currents for a load with a power factor, $\cos\varphi$.

Since there is a voltage drop through the transmission line $\Delta V = (R + jX)I$, the voltage E at the sending-end must be higher than the voltage at the load terminal V_2.

During load rejection (Figure 7.7a), on opening the switch, there is an overvoltage at the receiving end of the transmission line, resulting in a solution for the modified network according to the new phasor diagram in Figure 7.7b. Now the phase angle between voltage and current at the line terminal is $\varphi = 90°$ due to the line capacitance.

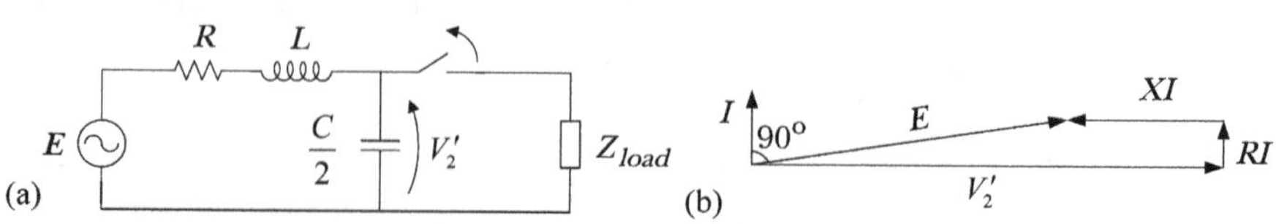

Figure 7.7 Load rejection. (a) Simplified circuit. (b) Phasor diagram.

In this case, the voltage at the sending-end E is higher than 1.0 pu according to Figure 7.6b and is assumed constant. In fact, this is a very reasonable assumption since the internal voltages of the generators are kept constant just after the opening of the circuit-breaker on the load side, but this will be detailed below.

Therefore, since the source voltages in Figures 7.6b and 7.7b are the same, voltage V_2' is observed to be even higher than E, as the phasor diagram in Figure 7.7b. The unloaded transmission line can also be analyzed using Eq. (7.14) to consider the Thévenin equivalent.

It is important to observe why the overvoltages associated with load rejection are higher than those associated with line energization, and the reason is the higher level of the source voltage at the sending-end of the transmission line in the case of load rejection.

In a more realistic analysis, if the Thévenin equivalent is included, a further increase in the temporary overvoltages is expected, according to Section 7.2.1.

The occurrence of an earth fault can increase the overvoltages due to load rejection, in the healthy phases. The combined event of load rejection due to an earth fault can occur if the breaker on the load side opens first, causing load rejection simultaneously with the fault, during the period that the breaker on the source side remains closed. On the other hand, if load rejection has very high overvoltages, it can cause an earth fault. These simultaneous events should be considered after a careful analysis of their probability so as not to overestimate the arrester ratings or insulation levels.

In order to ensure that temporary overvoltages due to load rejection are within the permissible ranges contained in the planning criteria, that is, below 1.4–1.5 pu, line reactive compensation studies are carried out with power-system software, in which the overvoltages are calculated for energization and load rejection under different operating conditions. With this information, the reactors permanently connected to the transmission lines are sized to reduce the overvoltages. The studies also determine the controlled reactive compensation and the sizing of the reactors and capacitor banks to be switched at the substation.

It is also appropriate to analyze the impact of temporary overvoltages, especially those of load rejection, on surge arresters and transformers.

The analysis of temporary overvoltages is a basic consideration in choosing the arrester rating, and this topic is discussed separately in Chapter 13.

Another important piece of equipment is the transformer, and it is necessary to check if over-excitation is acceptable, according to the curve of allowable overvoltages as a function of their duration.

Therefore, both for line arresters and for transformers, it is important to know the overvoltage level and the amount of time it may last.

A line energization overvoltage may last for a few minutes before the load is connected to the system or before synchronization with the system occurs.

Overvoltages due to load rejection should last for a shorter period, depending on the system protection time delay before the source side circuit breaker is tripped. Hence, when a circuit breaker opens on the load side, the one at the sending-end is expected to operate a few cycles later.

A conservative estimate of the duration of this event is around 500 ms, which would be the time available for the backup protection to trip the source side circuit breaker.

- Dynamic Behavior of Load Rejection

  Although some of the topics mentioned below are also valid for energization, their effects are more sensitive to load rejection.

  Chapter 8 covers switching surges, with a duration of a few cycles. For this maneuver, modeling of a synchronous machine with subtransient voltage, E''_d, and direct axis subtransient reactance, X''_d, is carried out, because the analysis resides in this period of machine behavior.

  Depending on the duration of load rejection, the transient period can be addressed, and, for that purpose, more complete models of the synchronous machine are necessary, with representation of winding fluxes, reactances, and time constants, on the direct axis and on the quadrature axis, by applying Park's transform.

  The models for electrical machines should be like those in the transient stability study, in which certain dynamic behaviors are evaluated in a period of one to two seconds. For transient stability analysis, the power grid is simulated using the parameters calculated at the fundamental frequency.

  The equations below can be used to represent the dynamic behavior of electrical machines [7]. Whether or not to consider any of the time constants will depend on the investigation that is being carried out.

  The application of Park's transform results in a set of equations on the direct axis D and on the quadrature axis Q, with currents, voltages, and fluxes represented in this new frame of reference. The machine has windings in the stator and in the rotor, represented on axes (D, Q), and the following relation can be obtained in per unit values. A complete description of these equations is found in the same reference:

$$
\begin{aligned}
\psi_d &= L_d i_d + L_{md} i_f + L_{md} i_{kd}, \\
v_f &= L_{md} p i_d + \left(R_f + L_{ff} p\right) i_f + L_{md} p i_{kd}, \\
0 &= L_{md} p i_d + L_{md} p i_f + (R_{kd} + L_{kkd} p) i_{kd}, \\
\psi_q &= L_q i_q + L_{mq} i_{kq}, \\
0 &= L_{mq} p i_q + \left(R_{kq} + L_{kkq} p\right) i_{kq},
\end{aligned}
\tag{7.23}
$$

where:

L_d, L_q, L_{kkd}, L_{kkq}, L_{ff}: inductances transformed to the direct and quadrature axis.

L_{md}, L_{mq}: mutual inductances transformed to the direct and quadrature axis.
ψ: flux linked to each winding.
f: field winding.
kd: damping winding on the direct axis.
kq: damping winding on the quadrature axis.
p: Heaviside operator (which has the same meaning as the complex variable s).

$$\begin{aligned}
v_d &= p\psi_d + \omega\psi_q + R_a i_d,\\
v_q &= p\psi_q - \omega\psi_d + R_a i_q,\\
T_e &= \frac{\omega_0}{2}\left(\psi_d i_q - \psi_q i_d\right),\\
\overline{\psi}_d &= \frac{X_d(p)}{\omega_0}\overline{i}_d + \frac{G_d(p)}{\omega_0}\overline{v}_f,\\
\overline{\psi}_q &= \frac{X_q(p)}{\omega_0}\overline{i}_q,
\end{aligned} \tag{7.24}$$

where:

$X_d(p)$ and $X_q(p)$: complex impedances (operational impedances on direct and quadrature axis).
T_e: electromagnetic torque.

Using the basic Eqs. (7.23) and (7.24), several models can be proposed for the representation of the synchronous machine. The dynamic behavior of the generator can be predicted, using these equations, with respect to currents, voltages and fluxes, and time constants.

Synchronous machines are equipped with automatic voltage regulators and speed regulators. In particular, the automatic voltage regulator is part of the excitation system of the machine (Figure 7.8), which acts to maintain the voltages at the terminals at a predetermined value in the range of 0.95–1.05 pu.

During load rejection, the voltage at the terminal of the generator rises since the line behaves as a capacitance, and without the voltage regulators, the voltage could become dangerously high. The voltage rise can be attributed to the same voltage divider effect observed when an inductive source supplies a capacitor, as in Eq. (7.19). With the voltage regulator in operation and considering the dynamic equations of the synchronous machine, Figure 7.9a shows a typical time response of the 60 Hz, temporary overvoltage behavior at the terminals of the generator.

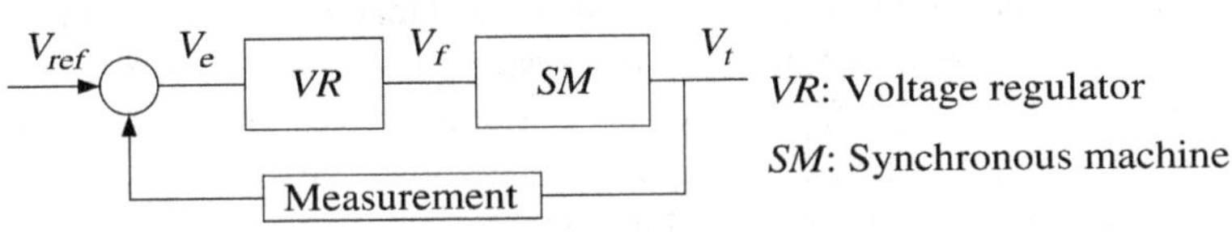

Figure 7.8 Excitation system.

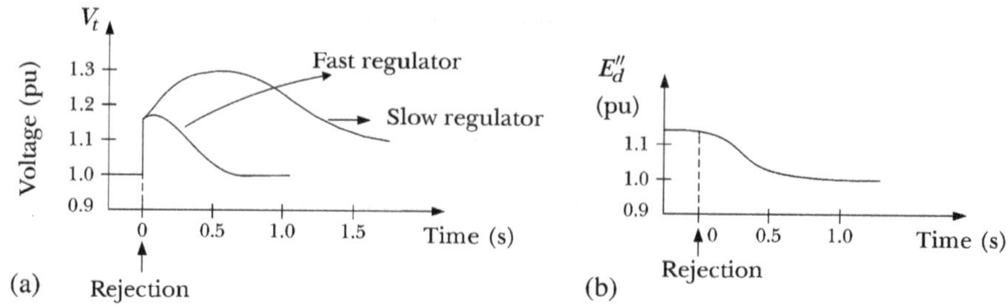

Figure 7.9 Time response of load rejection. (a) Voltage time response. (b) Subtransient voltage E''_d.

The time response is obtained using a transient stability program. Note that the voltage at the generator terminal exhibits a discontinuity at the instant that the load rejection occurs. This discontinuity is explained by a topology change in the network and by the fact that phasor equations are used in the calculations, and faster transients, such as switching surges, are neglected in the generator and transmission line model. However, even in stability programs, the internal voltage of the generator is a continuous function of time, since the internal fluxes do not change instantaneously. This internal voltage is the subtransient voltage E''_d.

The terms $p\psi_d$ and $p\psi_q$ in the Park equations in (7.24), which are omitted in stability programs, are included in electromagnetic transient programs, allowing the calculation of switching surges at the time they first begin.

Schematically, in very rudimentary terms, the excitation system of the synchronous machine can be represented by the macroblocks in Figure 7.8.

An error V_e is introduced into the voltage regulator, designated as automatic voltage regulator (AVR), generating a signal that modifies the field voltage V_f of the synchronous machine. Given the load conditions of the machine, the voltage at the terminal is used as a feedback signal acting on the field voltage, which does not change instantly due to delays in the voltage regulator and in the generator. Currently, as automatic voltage regulators have a fast response, much of the delay in the excitation loop is given by the time constant T'_{d0} of the field winding of the synchronous machine. These generator time constants and the time delay due to voltage regulators may explain the dynamic behavior of temporary responses, as in Figure 7.9a.

To conclude this introduction to the dynamic aspects of load rejection, it is helpful to recall that, in the representation of the system, only the low-frequency dynamic is considered and high-frequency effects are neglected. Thus, when the subtransient voltage $E''_d(t)$ is known, temporary overvoltages can be calculated on the busbars involved in load rejection, Figure 7.9b.

- Frequency Variations
  In load rejection, the mechanical torque surpasses the electromagnetic torque, which implies an increase in the electrical frequency.
  With a rise in electrical frequency, the series impedance of the transmission line $Z = R + j\omega L$ also increases, while the capacitive reactance of the line $X_c = 1/\omega C$ decreases.
  Neglecting the resistance and assuming that $R \ll j\omega L$, it turns out, according to Eq. (7.6), that the constant A decreases, whereas the voltage V_2 at the receiving end of the transmission line increases. It is known that an increase in the electrical frequency contributes to an increase in the load rejection overvoltage; however, this effect is not always very significant, mainly when protection is triggered quickly.
- Transformer Saturation
  Since both the internal voltage and frequency change, the generator dynamic should be modeled, so that temporary overvoltages may be correctly evaluated.
  The voltage rises and its waveform can be distorted due to the saturation of transformers with the introduction of harmonics into the electrical network. Determination of distorted waveforms is only possible by simulation in the time domain, with the saturation branch of the transformers adequately represented.
  Some topics related to overvoltages, including saturated transformers, which are quite common in load rejection analysis, are supplemented in Section 7.7.

7.3 Faults

The steady-state solution for the network, when a fault occurs, is performed using traditional power system software [8]. Symmetrical components are the proper theory with which to approach unbalanced conditions, such as the occurrence of a single-phase fault with network representation by means of sequence diagrams. As mentioned earlier, programs with three-phase representation of the network, such as the EMTP program, allow the analysis of unbalanced faults. One advantage of obtaining the three-phase admittance matrix of the network is the possibility of analyzing simultaneous faults, such as the opening of one phase and a fault in another phase. This flexibility is not normally possible in programs based on symmetric components unless the program employs the methodology in reference [9].

From the point of view of temporary overvoltages, it is convenient to analyze the single-phase fault, which can cause overvoltages in the healthy phases that are free of faults. Moreover, the single-phase fault has the highest probability of occurrence, according to several statistical analyses of EHV systems.

A rigorous analysis of system faults consists of a detailed simulation of the positive- and zero-sequence parameters of the network $R'_1, X'_1, C'_1, R'_0, X'_0, C'_0$, and it is possible to construct families of overvoltage curves by analyzing these values parametrically [3, 10]. Some simplifications are assumed below for the purpose of highlighting the most significant factors.

A system with ratios of $X_0/X_1 < 3$ and $R_0/X_1 < 1$ is considered grounded from the point of view of overvoltages, and in this case, the overvoltages at the healthy phases do not exceed values of 1.4–1.5 pu. Some unbalanced conditions in a lossless transmission system are examined, considering only the reactance in order to facilitate interpretation, designating:

$$k = \frac{X_0}{X_1} \tag{7.25}$$

Cases with a ratio of $k < 1$ are not analyzed here because they do not have overvoltages and are more important for the calculation of fault currents.

The formula for the overvoltage factor is shown in (7.26), for the case of a single-phase fault. This factor reflects the relationship between the voltages in the healthy phases divided by the prefault voltage at the busbar.

For the Thevenin equivalent in EHV systems, the resistance is usually much lower than the reactance, that is, $R \ll X$. In this case, the overvoltage factor for single-phase faults, given by the ratio between the overvoltage and the prefault voltage, can be obtained by means of a simplified formula:

$$f_{ov} = \sqrt{3}\frac{\sqrt{k^2 + k + 1}}{2 + k}. \tag{7.26}$$

Formula (7.26) reports exact values when the resistance is null.

The series connection of the sequence diagrams in Figure 7.10 is given by the zero-voltage boundary condition in phase a, ($V_a = 0$), and in that case $V_0 + V_1 + V_2 = 0$ [9, 11].

- Fault with a Ratio $k = 1$ ($X_0 = X_1$).
  A system busbar with a ratio of $X_0/X_1 = 1$ indicates that there is no coupling between phases. Therefore, one phase should have no influence on the other.

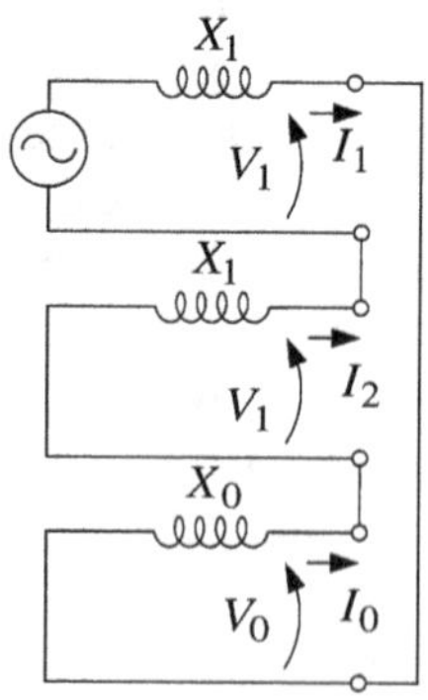

Figure 7.10 Series connection of the sequence diagrams for a single-phase fault.

Example 7.3 Calculate the voltages at the healthy phases on a system busbar with $X_0 = X_1$, in Figure 7.10, during a single-phase fault.

Thus,

$$I_0 = I_1 = I_2 = \frac{1}{3jX_1},$$

$$V_1 = 1 - \frac{jX_1}{3jX_1} = \frac{2}{3},$$

$$V_2 = -\frac{1}{3},$$

$$V_0 = -\frac{1}{3},$$

Calculating the phase components:

$$\begin{bmatrix} V_a \\ V_b \\ V_c \end{bmatrix} = \begin{bmatrix} 1 & 1 & 1 \\ 1 & \alpha^2 & \alpha \\ 1 & \alpha & \alpha^2 \end{bmatrix} \times \begin{bmatrix} V_0 \\ V_1 \\ V_2 \end{bmatrix},$$

where

$$\alpha = \angle 120°,$$
$$\alpha^2 = \angle -120°,$$
$$V_a = 0,$$
$$V_b = -\frac{1}{3} + \frac{2}{3}\angle -120° - \frac{1}{3}\angle 120° = 1\angle -120°,$$
$$V_c = -\frac{1}{3} + \frac{2}{3}\angle 120° - \frac{1}{3}\angle -120° = 1\angle 120°.$$

Applying (7.26), the overvoltage factor obtained is $f_{ov} = 1$.

Therefore, voltages V_b and V_c do not change, remaining the same in amplitude and phase.

- Single-Phase Fault in an Ungrounded System, $X_0 = \infty$
  For instance, given an unloaded transformer with a grounded star-delta connection, a single-phase fault is assumed on the delta side at phase a.
  By examining the phasors for the fault on the side of the delta connection, it is found that the overvoltage in the healthy phases b and c are $\sqrt{3}$ pu.
  When phase a is grounded, V_{ca} is obtained in phase c and $-V_{ab}$ in phase b. Both voltages have a modulus of $\sqrt{3}$ pu.

Example 7.4 Assuming the same case as Figure 7.11a, examine a single-phase fault on the delta side at phase a, by applying symmetrical components.

The solution is obtained using the sequence diagram in Figure 7.10, with $X_0 = \infty$.

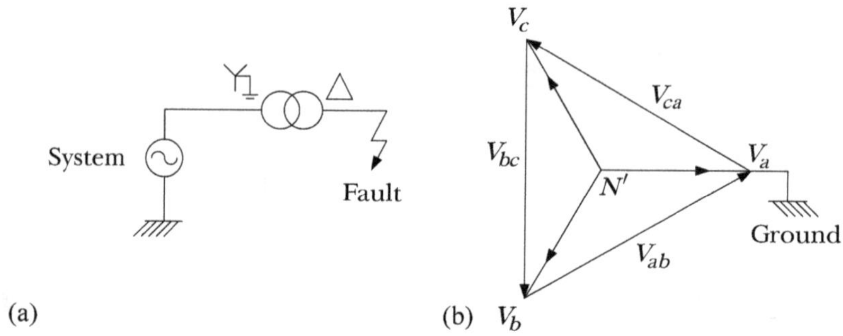

Figure 7.11 Single-phase fault in the delta-connected transformer. (a) Fault. (b) Three-phase voltages.

$$V_1 = 1.0,$$
$$V_2 = 0,$$
$$V_0 = -1.0.$$

thus

$$V_a = 0,$$
$$V_b = V_0 + \alpha^2 V_1 = -1 + 1\angle{-120^\circ} = 1.732\angle{-150^\circ},$$
$$V_c = V_0 + \alpha V_1 = -1 + 1\angle 120^\circ = 1.732\angle 150^\circ.$$

The voltages at the healthy phases have an overvoltage factor $f_{ov} = \sqrt{3}$. Note that phases b and c now have line voltages, since $V_b = -V_{ab}$ and $V_c = V_{ca}$ in Figure 7.11b. The same result is obtained by applying the overvoltage factor in (7.26) for $k = \infty$.

- Fault with a Ratio $k > 1$ ($X_0 > X_1$).
  The main case of a single-phase fault applied at the receiving end of a transmission line operating without a load is now analyzed in Figure 7.12a. Assume that the line is transposed and a ratio of $X_0/X_1 > 1$ exists at the receiving end.

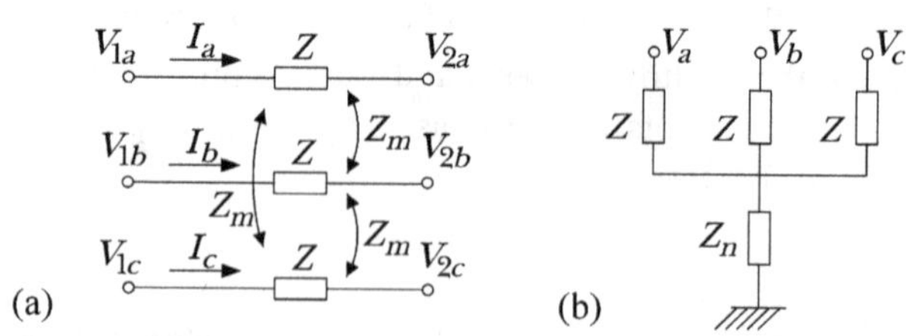

Figure 7.12 Three-phase transmission line. (a) Inductive mutual coupling. (b) Coupling introduced by an earthing impedance.

Disregarding the capacitances, the series impedance matrix of the line is given by:

$$[Z] = \begin{bmatrix} z & z_m & z_m \\ z_m & z & z_m \\ z_m & z_m & z \end{bmatrix}, \tag{7.27}$$

where

z: phase impedance.
z_m: mutual impedance between phases.
$z_0 = z + 2z_m$: zero-sequence impedance.
$z_1 = z_2 = z - z_m$: positive-sequence impedance.

The mutual coupling between phases gives rise to a positive-sequence impedance lower than the zero-sequence impedance, and interference between phases is observed. Disregarding losses, if the mutual reactance is zero, there is no interference from one phase to another, and the positive-sequence impedance is equal to the zero-sequence impedance, as in Example 7.3.

Another secondary effect of interference between phases, not due to inductive coupling, but rather to the coupling introduced by an impedance, can be seen in Figure 7.12b.

Figure 7.12b shows a balanced load with phase impedance Z, grounded by an impedance Z_n.

It is obvious that the imbalance in one phase, for example, a single-phase fault in phase a, modifies the voltages in phases b and c. It should then be expected that the positive-sequence impedance is different from the zero-sequence impedance.

In fact, the sequence impedances of the load are:

$$Z_1 = Z, \tag{7.28}$$

$$Z_0 = Z + 3Z_n. \tag{7.29}$$

The typical case of a fault at the receiving end of a transmission line, occurring during the energization or load rejection maneuvers, is considered in Example 7.5.

Example 7.5 A single-phase fault occurs at the receiving end of the transmission line without load. The nominal line voltage is 525 kV and the source is assumed to be an infinite bus. The data for the π-equivalent circuit, representing a line with a length of 300 km, are supplied below.

$$E = \frac{525}{\sqrt{3}} \text{ kV}, (V_1 = E),$$

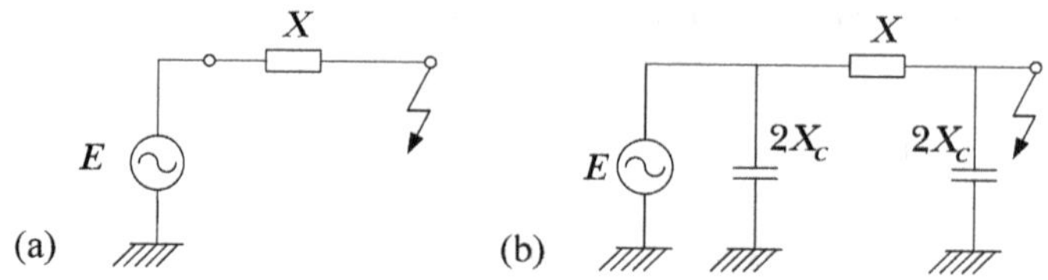

Figure 7.13 Single-phase fault. (a) Fault at the receiving end of the line. (b) Equivalent circuit.

X: series reactance of the transmission line,
$2X_c$: capacitive reactance of the transmission line at each terminal, corresponding to half of the total capacitance (Figure 7.13).

$$X_1 = j90\ \Omega;\ \ X_{c1} = -j1473.6\ \Omega.$$

$$X_0 = j420\ \Omega;\ \ X_{c0} = -j2210.4\ \Omega.$$

Calculate the voltages at the healthy phases using symmetrical components.

a) Solution, neglecting the line capacitance
For a 525 kV line, the phase-to-ground voltage is $E = 303$ kV. The sequence diagram corresponding to the single-phase fault is the same as in Figure 7.10:
Thus:

$$I_0 = I_1 = I_2 = \frac{303}{j600} = -j0.505\text{ kA},$$

$$V_1 = 303 - 0.505 \times 90 = 257.6\text{ kV},$$

$$V_2 = -45.4\text{ kV},$$

$$V_0 = -212.1\text{ kV}.$$

The phase components are obtained:

$$V_a = 0,$$

$$V_b = -212.1 + 257.6\angle -120^\circ - 45.4\angle 120^\circ = 412.4\angle -140.5^\circ\text{ kV},$$

$$V_c = -212.1 + 257.6\angle 120^\circ - 45.4\angle -120^\circ = 412.4\angle 140.5^\circ\text{ kV}.$$

The overvoltages at the receiving- end of the transmission line in phases b and c are 1.36 pu (412.4/303). If the line resistances are used, slightly different voltages for phases b and c are obtained.

Applying the overvoltage factor formula with $k = 4.666$ the result is the same, that is, $f_{ov} = 1.36$.

b) Solution including the line capacitance
The Thévenin equivalent circuit, as seen from the receiving end of the transmission line, is obtained using the circuit below (Figure 7.14):
The pre-fault voltage at the receiving end, which is the Thévenin equivalent voltage, is given by:

$$E_t = \frac{-j2X_{c1}}{j(X_1 - 2X_{c1})}E,$$

$$E_t = 1.031\,E.$$

The Thévenin equivalent impedances for positive and zero sequences are given by the parallel combination of the inductive reactance and capacitive reactance:

$$Z_{eq1} = j92.84\ \Omega,$$

$$Z_{eq0} = j464.09\ \Omega.$$

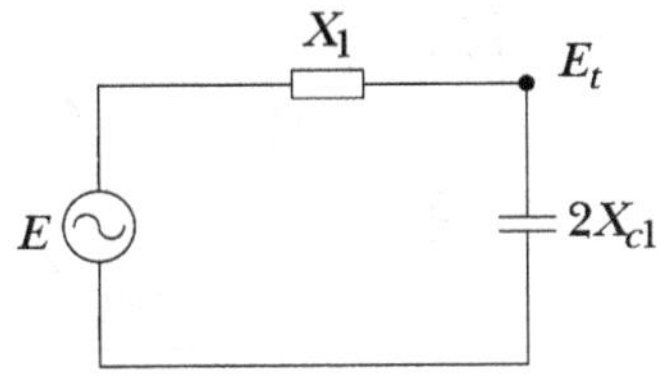

Figure 7.14 Circuit to calculate the Thévenin equivalent.

Recalculating according to item a.

$$I_0 = I_1 = I_2 = -j0.480\ \text{kA}.$$

$$V_a = 0,$$

$$V_b = 430.723\angle -141.05°\ \text{kV},$$

$$V_c = 430.723\angle 141.05°\ \text{kV}.$$

Applying the overvoltage factor formula for $k = 4.999$, the result is $f_{ov} = 1.378$, and the same value is obtained using the prefault voltage, with the quotient $430.723/(303 \times 1.031)$.

Fault analysis usually does not include the line capacitance when fault current levels are calculated, but consideration of the capacitance is necessary for overvoltage evaluations in single-phase faults for two reasons. One is related to the increase in the Thévenin equivalent voltage and the other is that the k ratio is also modified. In this example, the capacitance raised k from 4.666 to 4.999. However, the reactors usually provide compensation for long transmission lines, which can reduce the influence of the capacitance. The example provides a simplified overview of temporary overvoltages in long transmission lines due to the combined effect of single-phase faults with energization or load rejection.

More details on the analysis of unbalanced faults can be found in [9–11].

7.4 Open Phases in Transmission Lines

7.4.1 Introduction

Open-phase voltages during monopolar autoreclosure cannot be said to be the most significant from the point of view of the study of temporary overvoltages, although they can occur, especially in cases of resonance. However, the analyses

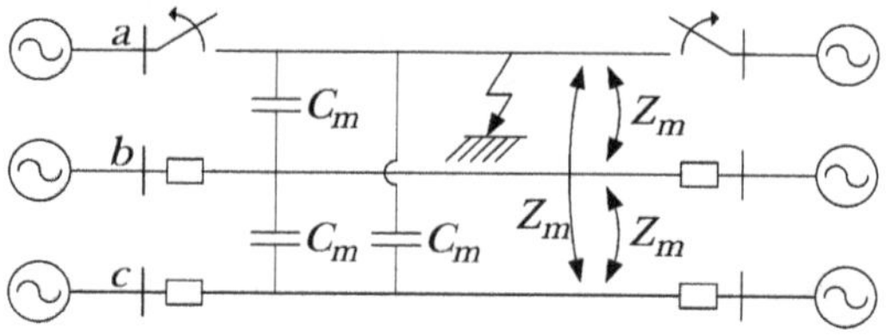

Figure 7.15 Monopolar opening.

that follow provide useful information on phase-to-phase coupling and line models, which contribute to the understanding of various phenomena connected with three-phase line imbalances and resonant conditions.

In single-phase autoreclosure analysis, the secondary arc current and the voltage at the open phase should be evaluated after the extinction of the secondary arc [12–14].

When the phase-to-ground fault occurs, the power arc, designated the primary arc, bypasses the insulator string.

For single-phase autoreclosure, the protection system selects only the pole of the faulted phase a, according to Figure 7.15.

After the opening of the poles of circuit breakers on both sides of the transmission line, current remains supplying the short circuit due to the capacitive and inductive coupling of the faulted phase with the healthy ones. This small current that persists, even after opening the circuit-breakers, is called the secondary arc current.

In order to analyze the extinction of the secondary arc, it is necessary to take into account the transient and steady-state conditions of the current and also the recovery voltages in the open phase after arc extinction.

The arc current and voltage in the open phase depend on the inductive and capacitive coupling of the phases. The influence of the capacitances between phases is preponderant, as discussed in more detail below. The influence of the inductive coupling depends on the current in the healthy phases, since the voltage drop through the open phase will be determined by the coupling and this current level. In general, a few simulations are sufficient to evaluate if the effect of the inductive coupling is low.

Another situation involving open phases is that of a stuck breaker at one terminal of the transmission line, when failure of one pole of the circuit breaker to open is possible.

In this case, as shown in Figure 7.16, since pole c of the circuit breaker failed in the three-phase opening of one terminal of the transmission line, a voltage can be induced in open phases a and b.

The models for all the open-phase studies mentioned above are similar and will be presented below.

Figure 7.16 Stuck breaker.

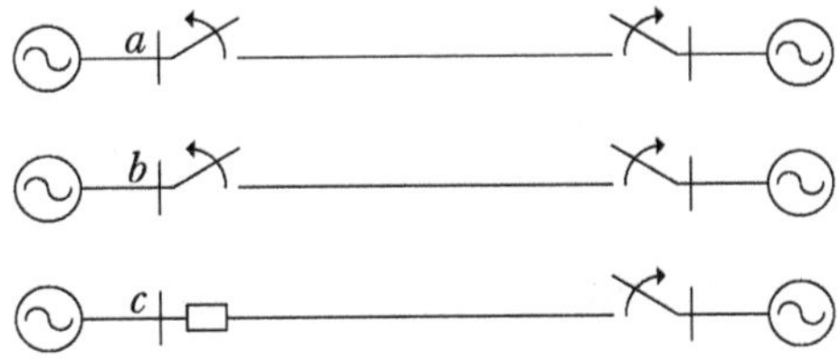

7.4.2 Network Modeling

- Phase-Domain Analysis

The analysis of open phases is presented with the three-phase admittance matrix of the transmission line representing the inductive and capacitive coupling, and, if necessary, other transmission lines in the system or equivalents representing the rest of the system can also be included. In the case of compensated lines, the reactor is represented by its nodal admittance matrix.

In general, long transmission lines are compensated through the installation of reactors. In the case of compensated transmission lines, it is necessary to include the nodal admittance matrix of the reactors in the nodal admittance matrix of the transmission line.

For compensated transmission lines, two cases are analyzed next:

a) Grounded single-phase reactors (Figure 7.17a).
   X: phase reactance.
   Admittance matrix:

$$Y_r = \begin{bmatrix} y & & \\ & y & \\ & & y \end{bmatrix}, \tag{7.30}$$

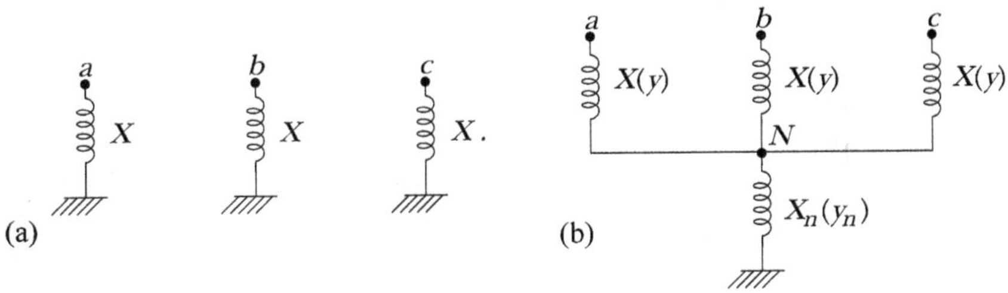

Figure 7.17 Three-phase reactors. (a) Solidly grounded. (b) Grounded by a neutral reactor.

where

$$y = -j\frac{1}{X}. \tag{7.31}$$

b) Single-phase reactors grounded by a neutral reactor (Figure 7.17b).

As will be seen further on in the analysis with symmetric components, this is a case in which $X_0 \neq X_1$, where$X$ is the phase reactance and X_n the neutral reactance. The sequence impedances are $X_1 = X$ and $X_0 = X + 3X_n$, which is like (7.28) and (7.29).

For the admittance, the following can be written:

$$y = -j\frac{1}{X}, \tag{7.32}$$

$$y_n = -j\frac{1}{X_n}. \tag{7.33}$$

The three-phase nodal admittance matrix of the reactor is assembled in (7.34):

$$Y = \begin{array}{c} \\ a \\ b \\ c \\ n \end{array} \begin{array}{c} \begin{array}{cccc} a & b & c & n \end{array} \\ \begin{bmatrix} y & & & -y \\ & y & & -y \\ & & y & -y \\ -y & -y & -y & 3y+y_n \end{bmatrix} \end{array}. \tag{7.34}$$

This representation accommodates the general case of unbalanced reactors, which can be used in nontransposed lines [13].

Node N is reduced, with the simple elimination of one nodal equation, which is also known as the Kron reduction. For the reader who is less familiar with this operation, in Section 7.5.3, Eqs. (7.70) and (7.71) are manipulated in a manner analogous to the Kron reduction. The reduced matrix will be designated as Y_r.

$$Y_r = \begin{array}{c} a \\ b \\ c \end{array} \begin{array}{c} \begin{array}{ccc} a & b & c \end{array} \\ \begin{bmatrix} y - \dfrac{y^2}{3y+y_n} & -\dfrac{y^2}{3y+y_n} & -\dfrac{y^2}{3y+y_n} \\ -\dfrac{y^2}{3y+y_n} & y - \dfrac{y^2}{3y+y_n} & -\dfrac{y^2}{3y+y_n} \\ -\dfrac{y^2}{3y+y_n} & -\dfrac{y^2}{3y+y_n} & y - \dfrac{y^2}{3y+y_n} \end{bmatrix} \end{array} \tag{7.35}$$

If the inverse of the matrix of series impedances is designated G_1, and half of the total susceptance of the line is designated $G2$, the nodal admittance matrix for the three-phase line can be written as follows:

$$Y_{line} = \begin{bmatrix} G_1 + G_2 & -G_1 \\ -G_1 & G_1 + G_2 \end{bmatrix}. \tag{7.36}$$

The reduced admittance matrix for the reactor Y_r can now be included in matrix Y_{line}, by inserting it at terminal 1, for instance:

$$Y_a = \begin{bmatrix} G_1 + G_2 + Y_r & -G_1 \\ -G_1 & G_1 + G_2 \end{bmatrix}. \tag{7.37}$$

This model allows for a very accurate representation of the transmission line. Frequently, the secondary arc can be calculated at both terminals of the transmission line, if internal faults will give intermediate results. The influence of the series impedance can be disregarded, that is, when the voltage drops due to line currents being small, the voltages at both terminals of the transmission line are assumed to be the same, and the capacitive model for the line is derived by doubling G_2. The whole transmission line can then be concentrated at a three-phase single node since the voltages at both line terminals are assumed to be approximately equal.

Thus, a more simplified transmission line model can be obtained (Figure 7.18).

In the following examples, the voltages at open phases will be calculated under different conditions, which have the same values as the steady-state recovery voltages, after secondary arc extinction.

These examples are intended to introduce an approach to unbalanced faults via the three-phase matrix of nodal admittances in simple cases. Other more complex cases must be analyzed using admittance matrices with more transmission lines in the system. It is still feasible to study several simultaneous faults by means of the admittance matrix. Specifically, for the analysis of open phases or the stuck breaker condition, the method can be very useful.

By imposing a null voltage on open phase a, the secondary arc current can also be calculated as a by-product of the analysis.

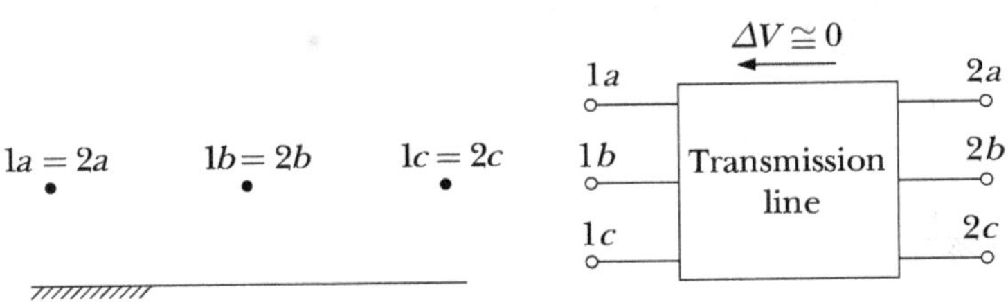

Figure 7.18 Equal voltages at the sending and receiving ends of the line.

Example 7.6 Calculate the voltages at the open-phase of a transmission line, with the three-phase admittance matrices G_1 and G_2 provided, and with a known transmission line length of 100 km. Assume that the nominal voltage of the transmission line is 230 kV, with a phase-to-ground value of $230/\sqrt{3} \cong 133.0\ \text{kV}$) (Figure 7.19).

In this simplified example, the voltages at the terminals of the healthy phases, which remain connected to the network, are assumed to be known and have nominal values.

$$
\begin{aligned}
E_{1b} &= 133\angle -120^\circ\ \text{kV},\\
E_{1c} &= 133\angle 120^\circ\ \text{kV},\\
E_{2b} &= 133\angle -130^\circ\ \text{kV},\\
E_{2c} &= 133\angle 130^\circ\ \text{kV}.
\end{aligned}
$$

The admittance matrix G_1 is calculated with the inversion of the series impedance matrix:

$$
G_1 = j\begin{bmatrix} -0.027778 & 0.011111 & 0.005555 \\ 0.0111111 & -0.031111 & 0.011111 \\ 0.005555 & 0.011111 & -0.027778 \end{bmatrix}\ \text{S}.
$$

Susceptance matrix for one side of the transmission line:

$$
G_2 = j\frac{\omega}{2}[C] = j\begin{bmatrix} 1.5 & -0.25 & -0.1 \\ -0.25 & 1.54 & -0.25 \\ -0.1 & -0.25 & 1.5 \end{bmatrix} \times 10^{-4}\ \text{S}.
$$

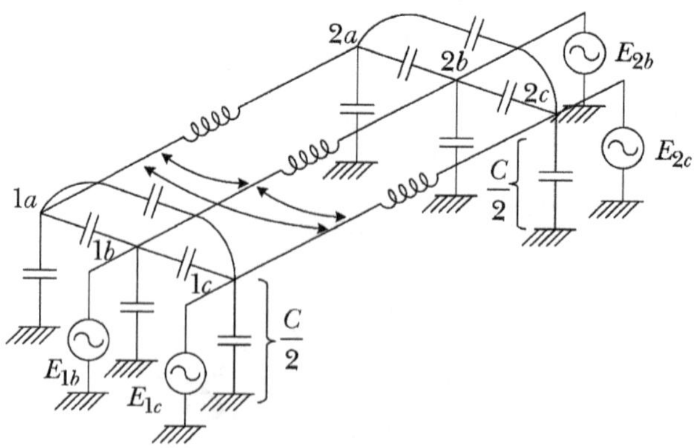

Figure 7.19 230 kV transmission line.

Admittance matrix for the three-phase transmission line, without the reactor:

$$Y_{line} = \begin{bmatrix} G_1 + G_2 & -G_1 \\ -G_1 & G_1 + G_2 \end{bmatrix}.$$

Writing the nodal equations,

$$\underset{\sim}{i} = Y_{line}\underset{\sim}{v}.$$

The voltages at nodes *1b, 1c* and *2b, 2c* are known.

$$\underset{\sim}{v} = \begin{bmatrix} V_{1a} \\ 133.0\angle -120° \\ 133.0\angle -120° \\ V_{2a} \\ 133.0\angle -130° \\ 133.0\angle 110° \end{bmatrix} \text{(voltage vector)}, \quad \underset{\sim}{i} = \begin{bmatrix} 0 \\ I_{1b} \\ I_{1c} \\ 0 \\ I_{2b} \\ I_{2c} \end{bmatrix} \text{(current vector)}.$$

Setting:

$$A = G_1 + G_2, \qquad A = \begin{bmatrix} a_{11} & a_{12} & a_{13} \\ a_{21} & a_{22} & a_{23} \\ a_{31} & a_{32} & a_{33} \end{bmatrix},$$

$$B = -G_1, \qquad B = \begin{bmatrix} b_{11} & b_{12} & b_{13} \\ b_{21} & b_{22} & b_{23} \\ b_{31} & b_{32} & b_{33} \end{bmatrix}.$$

The system of equations is assembled:

$$\begin{bmatrix} a_{11} & a_{12} & a_{13} & b_{11} & b_{12} & b_{13} \\ a_{21} & a_{22} & a_{23} & b_{21} & b_{22} & b_{23} \\ a_{31} & a_{32} & a_{33} & b_{31} & b_{32} & b_{33} \\ b_{11} & b_{12} & b_{13} & a_{11} & a_{12} & a_{13} \\ b_{21} & b_{22} & b_{23} & a_{21} & a_{22} & a_{23} \\ b_{31} & b_{32} & b_{33} & a_{31} & a_{32} & a_{33} \end{bmatrix} \times \begin{bmatrix} V_{1a} \\ E_{1b} \\ E_{1c} \\ V_{2a} \\ E_{2b} \\ E_{2c} \end{bmatrix} = \begin{bmatrix} 0 \\ I_{1b} \\ I_{2c} \\ 0 \\ I_{2b} \\ I_{2c} \end{bmatrix}.$$

Taking the first and the fourth rows, which have null nodal currents, and separating the known and unknown terms:

$$-\begin{bmatrix} a_{11} & b_{11} \\ b_{11} & a_{11} \end{bmatrix} \times \begin{bmatrix} V_{1a} \\ V_{2a} \end{bmatrix} = \begin{bmatrix} a_{12} & a_{13} & b_{12} & b_{13} \\ b_{12} & b_{13} & a_{12} & a_{13} \end{bmatrix} \times \begin{bmatrix} E_{1b} \\ E_{1c} \\ E_{2b} \\ E_{2c} \end{bmatrix},$$

thus:

$$\begin{bmatrix} V_{1a} \\ V_{2a} \end{bmatrix} = -\begin{bmatrix} a_{11} & b_{11} \\ b_{11} & a_{11} \end{bmatrix}^{-1} \times \begin{bmatrix} a_{12} & a_{13} & b_{12} & b_{13} \\ b_{12} & b_{13} & a_{12} & a_{13} \end{bmatrix} \times \begin{bmatrix} E_{1b} \\ E_{1c} \\ E_{2b} \\ E_{2c} \end{bmatrix}.$$

Substituting the numerical values:

$$\begin{aligned}
&a_{11} = -j0.027628, && a_{12} = j0.011086, && a_{31} = j0.005545,\\
&a_{21} = j0.011086, && a_{22} = -j0.030957, && a_{32} = j0.011086,\\
&b_{11} = j0.027778, && b_{12} = -j0.011111, && b_{13} = -j0.005555,\\
&b_{21} = -j0.011111, && b_{22} = j0.031111, && b_{23} = -j0.011111.\\
&E_{1b} = -66.5 - j115.2, && E_{2b} = -85.49 - j101.9, &&\\
&E_{1c} = -66.5 + j115, && E_{2c} = -45.49 + j125.0, &&
\end{aligned}$$

The voltages at both terminals of the open phase are:

$$\begin{aligned}
V_{1a} &= 20.114\angle -136.95^\circ \text{ kV},\\
V_{2a} &= 19.213\angle -160.42^\circ \text{ kV}.
\end{aligned}$$

Example 7.7 In Example 7.6, different phase angles were assumed for the voltages at the two terminals of the transmission line in order to observe the effect of the current on the healthy phases and their induced voltages on the open phase.

Using the data in Example 7.6, the calculation is performed assuming the same modulus and phase for the voltage at both line terminals. Now, since both terminals have the same voltage, the current can be considered null in the healthy phases, and the inductive effect can be neglected. In fact, this does not exactly occur in long lines, but an approximation is made here for a line of medium length. Thus, the induced voltage in phase a is caused only by the capacitiances between phases, and it is possible to work with the matrix of lumped capacitances (Figure 7.20).

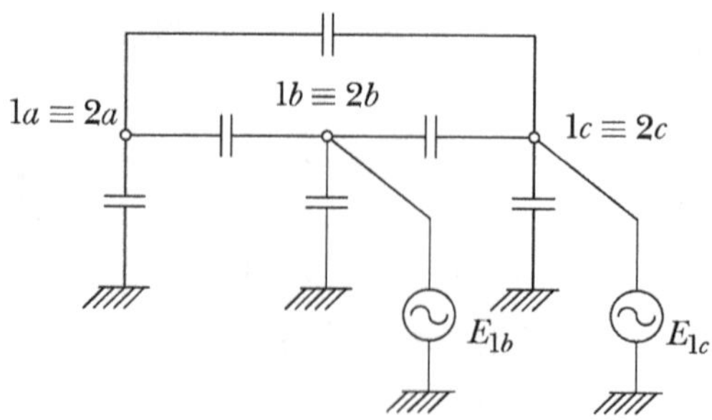

Figure 7.20 Capacitive model of the transmission line.

Solution

The nodal admittance matrix of the transmission line is given by:

$$Y_{line} = j\omega[C],$$

that is, $Y_{line} = 2G_2$.

Given that the known voltages at phases b and c are E_{1b} and E_{1c}, the induced voltage on the open phase a can be calculated from:

$$\begin{bmatrix} 0 \\ I_2 \\ I_3 \end{bmatrix} = j2\begin{bmatrix} 1.50 & -0.25 & -0.10 \\ -0.25 & 1.54 & -0.25 \\ -0.10 & -0.25 & 1.50 \end{bmatrix} \times 10^{-4} \begin{bmatrix} V_{1a} \\ E_{1b} \\ E_{1c} \end{bmatrix}.$$

Taking the first line with a nodal current of zero, the following result is obtained,

$$V_{1a} = -15.52 - j11.52.$$

In this case, with the same voltage values at both terminals, the results remain close to those obtained in Example 7.6, showing that the inductive effect is not significant in the case under consideration:

$$V_{1a} = V_{2a} = 19.33\angle -143.41^\circ \text{ kV}.$$

Example 7.8 Calculate the voltage in the open-phase a, assuming that the line in Example 7.7 is completely transposed and compensated by two reactors of 10 MVAr three-phase nominal power at the rated line voltage of 230 kV. The reactors are solidly grounded and Figures 7.21a and b present the transmission line without and with reactors, respectively.

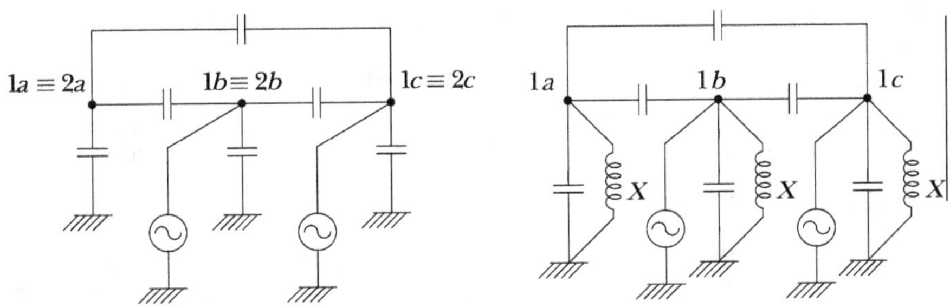

(a) Without reactor (b) With reactor

Figure 7.21 Feeding in two phases. (a) Without reactor. (b) With reactor.

Solution

A simplified solution is adopted by assuming equal voltages at both terminals of the transmission line and disregarding the series impedance.

Assuming a transposed line, a new nodal admittance matrix is built using the elements of the capacitance matrix of the nontransposed transmission line and placing on the diagonal the average value of all diagonal elements, and outside the diagonal, the average of the coupling elements.

Initially, the line without reactors is examined:

$$Y_{line} = j2\begin{bmatrix} 1.513 & -0.2 & -0.2 \\ -0.2 & 1.513 & -0.2 \\ -0.2 & -0.2 & 1.513 \end{bmatrix} \times 10^{-4}.$$

The applied voltages in the healthy phases are given by:

$$E_{1b} = -66.5 - j115.2;\ E_{1c} = -66.5 + j115.2;$$

$$\begin{bmatrix} 0 \\ I_2 \\ I_3 \end{bmatrix} = j2\begin{bmatrix} 1.513 & -0.2 & -0.2 \\ -0.2 & 1.513 & -0.2 \\ -0.2 & -0.2 & 1.513 \end{bmatrix} \times 10^{-4} \times \begin{bmatrix} V_{1a} \\ E_{1b} \\ E_{1c} \end{bmatrix}.$$

The induced voltage in the open phase a can be obtained from the first line of the nodal expression.

$$0 = 1.513V_{1a} - 0.2(-66.5 - j115.2) - 0.2(-66.5 + j115.2),$$

$$V_{1a} = -17.58\ \text{kV}.$$

The compensated line has 10 MVAr reactors, at both terminals; thus, the susceptance of each reactor is calculated to be:

$$y = -j\frac{10}{(230)^2} = -j1.89 \times 10^{-4}\ \text{S}.$$

The admittance matrix of the transmission line is calculated by summing the capacitance at both ends:

$$Y_{line} = 2G_2.$$

The nodal admittance matrix Y_a, including the reactors is calculated from:

$$Y_a = Y_{line} + 2Y_r,$$

$$Y_a = j2\begin{bmatrix} 1.513 & -0.2 & -0.2 \\ -0.2 & 1.513 & -0.2 \\ -0.2 & -0.2 & 1.513 \end{bmatrix} \times 10^{-4} - j2\begin{bmatrix} 1.89 & & \\ & 1.89 & \\ & & 1.89 \end{bmatrix} \times 10^{-4}.$$

Applying again the previous nodal equation to the first line of matrix Y_a:

$$0 = (1.513 - 1.89)V_{1a} - 0.4 \times 66.5,$$

$$V_{1a} = -70.56 \text{ kV}.$$

From this example, one can see that the reactors would create a resonant condition in the open phase, as $V_{1a} \rightarrow \infty$, if the susceptance were equal to 1.513×10^{-4} S, corresponding to a reactive three-phase power of $1.513 \times 10^{-4} \times 230^2 = 8$ MVar.

In this example, solidly grounded reactors were used. In the following example, the calculation will be repeated but with neutral reactors.

Example 7.9 With the same data from Example 7.8, calculate the voltage at the open-phase a considering both reactors to be grounded through neutral reactors of 1200 Ω each, corresponding to a quite common ratio of the zero- and positive-sequence reactances $X_0/X_1 \cong 1.7$ (see next Eq. (7.38)).

According to Eq. (7.35) and Figure 7.17b, Y_r matrix assembly with the elimination of the neutral node is shown:

$$y_n = -j8.33 \times 10^{-4} \text{ S},$$

$$Y_r = j\begin{bmatrix} -1.635 & 0.255 & 0.255 \\ 0.255 & -1.635 & 0.255 \\ 0.255 & 0.255 & -1.635 \end{bmatrix} \times 10^{-4},$$

$$Y_a = Y_{line} + 2Y_r,$$

$$Y_a = j2\begin{bmatrix} -0.122 & 0.055 & 0.055 \\ 0.055 & -0.122 & 0.055 \\ 0.055 & 0.055 & -0.122 \end{bmatrix} \times 10^{-4}.$$

The voltage at the open phase a is again obtained using the first line of the matrix:

$$0 = -0.122V_{1a} + 0.055 \times 2 \times 66.5,$$

$$V_{1a} = 59.96 \text{ kV}.$$

The cases presented in Examples 7.7–7.9 disregarded the series impedance and inductive coupling, since equal voltages were assumed for the line terminals. To take inductive coupling into consideration, it is convenient to use the three-phase equations of Example 7.6 as previously mentioned.

Although in the previous examples, only the voltages at the open phases were calculated, with small alterations in the equations, the secondary arc currents can easily be obtained, just by imposing a null voltage on the phase that has the fault.

Next, open-phase analysis will be approached using symmetrical components, assuming transposed transmission lines. The symmetrical component approach offers interesting models with which to understand the coupling between the phases of the transmission line. To simplify the equations, only the capacitances between phases were considered.

- Symmetrical Components

a) Transmission Line Model.
The following models describe a transposed transmission line, taking into consideration coupling by means of the positive sequence and the zero sequence. The total capacitance of the transmission line is assumed in Figure 7.22.
The behavior of the model is analyzed when positive- and zero-sequence voltages are applied to the transmission line model, as in Figures 7.23a and b, respectively.
For the positive sequence, the voltage at point N is null, and N can be connected to T, leading to Figure 7.24:
The resulting capacitance coincides with C_1, and the circuit responds correctly to the positive sequence.
When the zero-sequence voltages are applied to the line model, since the phase voltages are the same, only phase-to-earth currents are observed and

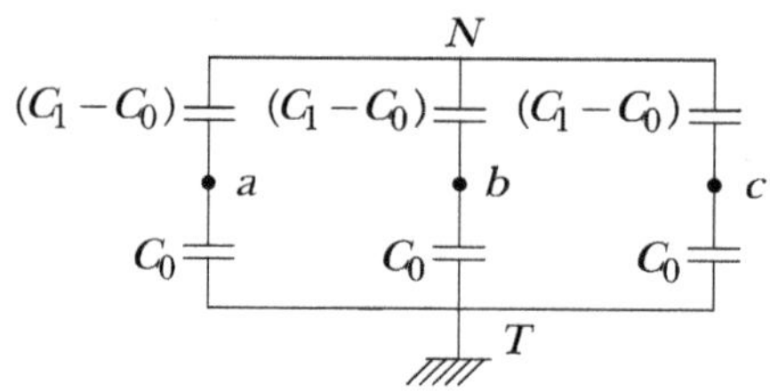

Figure 7.22 Capacitive model.

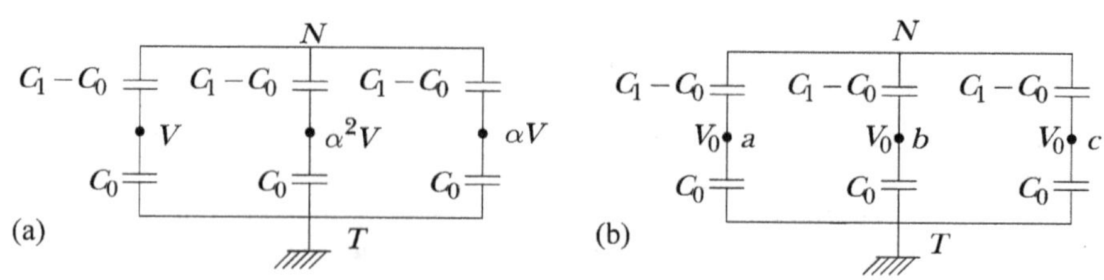

Figure 7.23 Capacitive model (a) Application of positive-sequence voltage. (b) Application of zero-sequence voltage.

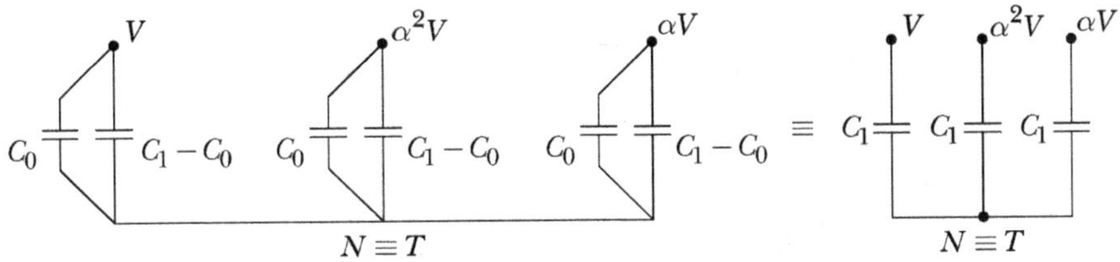

Figure 7.24 Positive-sequence capacitance.

therefore only the lower part containing capacitance C_0 is inserted into the network. The resulting capacitance, C_0, confirms the circuit's validity for the zero sequence.

b) Reactor Model.

The phase reactors can be grounded by reactors or by resistors. Grounding reactors, also known as Petersen coils, are considered to be more efficient for application in single-phase autoreclosure. Therefore, grounding reactors are analyzed, but the analysis can be extended to the case of resistors.

The following reactance are assigned to the reactor:

X: balanced phase reactance.
X_n: grounding reactor or neutral reactor.
$X_1 = X$: positive-sequence reactance.
$X_0 = X_1 + 3X_n$: zero-sequence reactance (see (7.29)).

The ratio between the zero-sequence reactance and positive-sequence reactance is given by:

$$k = \frac{X_0}{X_1}, \tag{7.38}$$

and the neutral reactance can be written:

$$X_n = \frac{X_0 - X_1}{3} = \frac{k-1}{3} X_1. \tag{7.39}$$

The reactor is represented, using the positive- and zero-sequence reactance values, by the model in Figure 7.25a.

Again, an interpretation of the model can be made with the application of positive- and zero-sequence voltages.

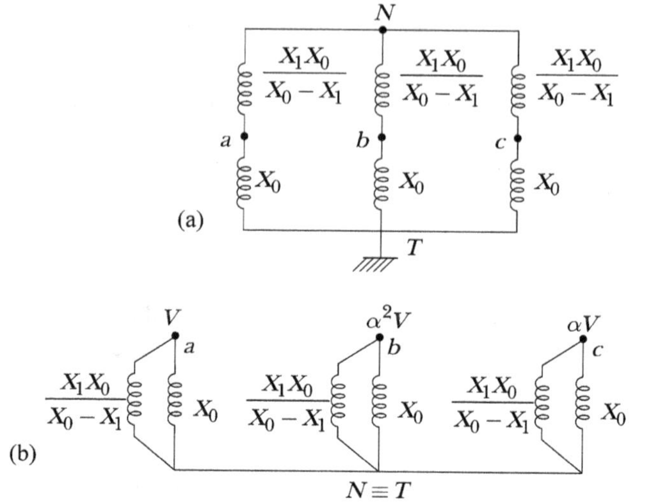

Figure 7.25 Reactor model. (a) Reactor model with positive and zero-sequence reactances. (b) Application of positive-sequence voltage.

Applying positive-sequence voltages to the phases, points N and T are at the same voltage level (which is zero) and therefore they can be connected.

Obtaining the equivalent reactance for each phase in Figure 7.25b:

$$X_1 = \frac{X_1X_0}{X_0 - X_1} \parallel X_0 = \frac{X_1X_0^2}{X_1X_0 + X_0^2 - X_1X_0}. \tag{7.40}$$

Therefore, the positive-sequence reactance $X_1$is obtained in each phase.

Applying zero-sequence voltages to the phases, the upper part of the circuit can again be disregarded and the zero-sequence reactance X_0 is obtained.

It can be shown that the model in Figure 7.25a correctly responds to the positive sequence and zero sequence.

c) Model of the Transmission Line Compensated by a Reactor.

Combining the models in Figures 7.22 and 7.25a, the final model of the compensated transmission line is presented in Figure 7.26.

It is noted that:

- When the reactor is solidly grounded, the upper part of the reactor model can be ignored, since $X_1 = X_0$.
- As previously discussed, the simplified model assumes that the series line impedance and network equivalent can be neglected.

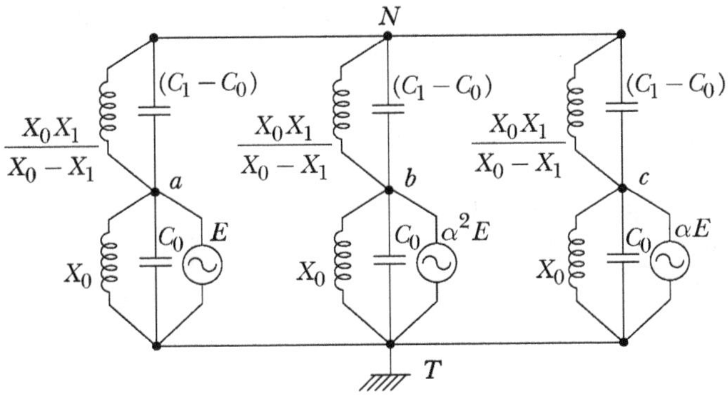

Figure 7.26 Model of the compensated transmission line in symmetrical components.

7.4.3 Model for Single-Phase Autoreclosure

The following models are suitable for monopolar autoreclosure analysis with symmetrical components.

- Voltage at the Open Phase

  Figure 7.27 shows the opening of phase a, after extinction of a secondary arc.

  The equivalent reactance in the upper part of the circuit is given by:

$$Z = -\frac{j}{\omega(C_1 - C_0)} \parallel j\frac{X_0X_1}{X_0 - X_1}, \tag{7.41}$$

and the equivalent reactance to ground is:

$$Z_0 = -\frac{j}{\omega C_0} \parallel jX_0. \tag{7.42}$$

The Thévenin equivalent, as seen from point A of Figure 7.27, is given by the open terminal voltage E_t in Figure 7.28 and the equivalent reactance Z_t.

$$E_t = -\frac{E}{2}, \; Z_t = \frac{Z}{2}. \tag{7.43}$$

The modulus of the Thévenin equivalent voltage is used in the simplified circuit in Figure 7.29a, which is equivalent to the reduced circuit in Figure 7.29b:

The voltage at open phase a is presented in Eq. (7.44), which is the recovery voltage at phase a, after secondary arc extinction, calculated in the steady state:

$$V_a = \frac{E}{2}\frac{Z_0}{1.5Z + Z_0}. \tag{7.44}$$

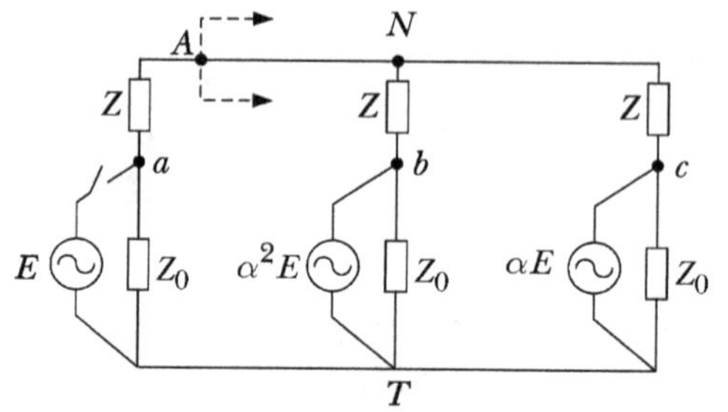

Figure 7.27 Opening of phase a.

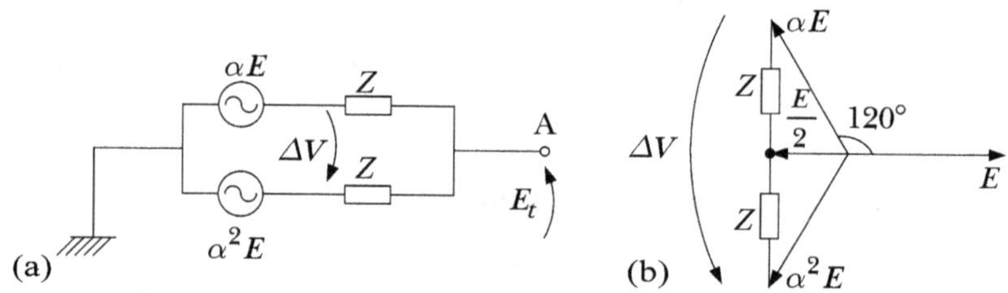

Figure 7.28 Equivalent circuit for phases *b* and *c*. (b) Phasor diagram

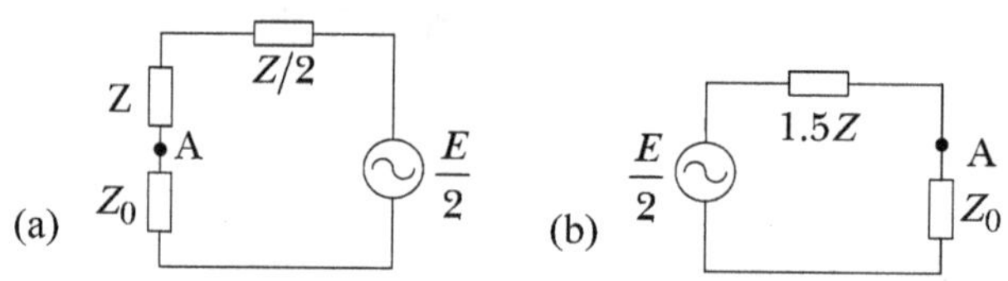

Figure 7.29 Simplified circuit. (a) Thévenin equivalent connected to the open phase. (b) Reduced circuit.

The resonance condition for (7.44) is given in (7.45) and can occur only in compensated lines:

$$Z_0 = -1.5Z. \tag{7.45}$$

Equation (7.46) shows the equivalent reactance of Eq. (7.41):

$$Z = \frac{\dfrac{1}{\omega(C_1 - C_0)} \dfrac{X_0X_1}{X_0 - X_1}}{-\dfrac{j}{\omega(C_1 - C_0)} + j\dfrac{X_0X_1}{X_0 - X_1}} = -j\frac{X_0X_1}{X_0X_1\omega(C_1 - C_0) + X_1 - X_0}. \tag{7.46}$$

By analogy, taking (7.42):

$$Z_o = \frac{-\frac{1}{\omega_0 C_0} jX_0}{-\frac{j}{\omega C_0} + jX_0} = -j\frac{X_0}{\omega C_0 X_0 - 1}, \tag{7.47}$$

Substituting (7.46) and (7.47) into (7.45), the resonant condition may be presented in Eq. (7.48):

$$\frac{X_0}{X_0 \omega C_0 - 1} = -\frac{1.5X_0X_1}{X_0X_1\omega(C_1 - C_0) + X_1 - X_0}. \tag{7.48}$$

Rearranging (7.48) and given that $X_0 = kX_1$, the following result is obtained:

$$X_1 = \frac{k + 0.5}{k\omega(C_1 + 0.5C_0)}. \tag{7.49}$$

Using the very simple expression in (7.49), the reactanceX_1that implies a resonant condition in the open phase is identified as a function of the transmission line capacitance and the k ratio.

It is worth noting that the transmission line and reactor parameters are calculated with an accuracy margin of about 5%. As a result the resonance points may shift slightly in their vicinity (See Section 7.5).

For reactance X_1, the three-phase reactive power of the reactor is given by (7.50), with the nominal line-to-line voltage V (kV).

$$Q_{r1} = \frac{V^2}{X_1}(\text{MVAr}), \tag{7.50}$$

where X_1 is the positive-sequence reactance of the reactor given in Ohm.

- Secondary Arc Current
  For the calculation of the secondary arc current, point A in Figure 7.29b is grounded, which results in Eq. (7.51):

$$I_s = \frac{E}{3Z}, \tag{7.51}$$

- where I_s is the secondary arc current.
  As $Z \rightarrow \infty$ in Eqs. (7.44) and (7.51), the voltage at the open phase and secondary arc current are both null, that is, the phase of the transmission line is uncoupled. Hence, the theoretical neutral reactor that imposes this condition can be obtained. This theoretical reactor furnishes important information concerning reduction of the secondary arc current. However, in practical terms, the secondary arc current and the voltage at the open phase do not need to be zero but

should have values within the extinction zone of the current. Therefore, it is not necessary to make such a drastic reduction in secondary current or recovery voltage, which could imply very high grounding impedances.

By setting the denominator of (7.46) equal to zero, the resonance condition that uncouples the phases is attained, resulting in a null voltage at the open phase (see Figure 7.26).

$$X_0 X_1 \omega (C_1 - C_0) + X_1 - X_0 = 0. \tag{7.52}$$

From the reactor model described previously, it is known that $X_0 = X_1 + 3X_n$:

$$(X_1 + 3X_n) X_1 \omega (C_1 - C_0) = 3X_n, \tag{7.53}$$

$$X_1^2 \omega (C_1 - C_0) = 3X_n (1 - X_1 \omega (C_1 - C_0)), \tag{7.54}$$

$$X_n = \frac{X_1^2 \omega (C_1 - C_0)}{3[1 - X_1 \omega (C_1 - C_0)]}. \tag{7.55}$$

Example 7.10 For the transposed transmission line with a 500 kV line voltage, calculate:

a) The voltage at the open phase assuming a 40% reactive compensation level.
b) The secondary arc current.
c) The voltage at the open phase, assuming a neutral reactor with a ratio of $k = 1.7$, at a reactive compensation level of 40%.
d) A theoretical grounding reactor that cancels the secondary arc current at 40% and 60% reactive compensation levels of the transmission line.

Data

$\ell = 200$ km (line length).
$C_1' = 13.3$ nF/km.
$C_0' = 8$ nF/km.

Solution

In the model of Figure 7.26, the reactors on both sides are assumed to be in parallel. Therefore, to obtain the reactance on each side, the calculated values in this example should be doubled.

To obtain an estimate in accord with the adopted hypotheses, the total capacitance of the nominal – π circuit is calculated.

$$C_1 = 13.3 \times 10^{-9} \times 200 = 2.66\ \mu\text{F}.$$

$$C_0 = 8 \times 10^{-9} \times 200 = 1.6\ \mu\text{F}.$$

Reactive power:

$$Q_{c1} = (500)^2 \times 377 \times 2.66 \times 10^{-6} = 250.7 \text{ MVAr}.$$

a) Compensation level of 40%, assuming that the reactor is solidly grounded:

$$Q_{r1} = 0.4 \times 250.7 = 100.28 \text{ MVAr},$$

$$X_1 = \frac{500^2}{100.28} = 2493\ \Omega.$$

Calculation of the equivalent reactance values, according to (7.46) and (7.47):

$$Z_0 = -j\frac{2493}{377 \times 1.6 \times 10^{-6} \times 2493 - 1} = -j4949$$

$$Z = -j\frac{(2493)^2}{(2493)^2 \times 377 \times 1.06 \times 10^{-6}} = \frac{-j}{377 \times 1.06 \times 10^{-6}} = -j2483\ \Omega.$$

As $X_1 = X_0$, there is only capacitive coupling, according to Figure 7.26, resulting in the following induced voltage:

$$V_a = \frac{500}{(\sqrt{3} \times 2)} \times \frac{4949}{(4949 + 1.5 \times 2483)} = 82.35 \text{ kV}.$$

b) Secondary arc current, by applying (7.51):

$$I_s = \frac{500}{\sqrt{3}}\frac{1}{3 \times 2483} = 38.8 \text{ A}.$$

c) With the neutral reactor, the zero-sequence reactance is obtained from:

$$X_0 = 1.7 \times 2438 = 4238\ \Omega.$$

Corresponding to a grounding reactor with$X_n = 581.7\ \Omega$ (calculated as parallel association of the reactances on both sides).

The induced voltage is recalculated:

$$Z_0 = -j\frac{4238}{377 \times 1.6 \times 10^{-6} \times 4238 - 1} = -j2723\ \Omega$$

$$Z = -j\frac{4238 \times 2493}{4238 \times 2493 \times 377(1.06 \times 10^{-6}) - 4238 + 2493} = -j4265\ \Omega$$

$$V_a = \frac{500}{(\sqrt{3} \times 2)} \times \frac{2723}{(2723 + 1.5 \times 4265)} = 43.1 \text{ kV}.$$

The induced voltage level is almost half with respect to the case of a grounded reactor.

Secondary arc current is also reduced:

$$I_s = \frac{500}{\sqrt{3}} \frac{1}{3 \times 4265} = 22.6 \text{ A}.$$

d) Theoretical grounding reactor, as in (7.55):

$$X_n = \frac{2493^2 \times 377 \times 1.06 \times 10^{-6}}{3(1 - 2493 \times 377 \times 1.06 \times 10^{-6})} = 220926\ \Omega.$$

For a compensation level of 40%, the theoretical grounding reactor calls for an isolated neutral of the main reactor to reduce the secondary arc current to zero, which is not advisable due to other factors, as for example, the high insulation level of the reactor neutral.

As the coupling impedance approaches infinity, that is, $Z \to \infty$, Eq. (7.55) may be applied, or alternatively, the parallel resonance condition may be imposed on the inductance and capacitance between the phases, as in Eq. (7.41) and Figure 7.26.

$$\frac{X_0 X_1}{X_0 - X_1} = \frac{1}{\omega(C_1 - C_0)}.$$

However, a high value for the positive-sequence reactance in this case is obtained:

$$X_1 \cong \frac{1}{\omega(C_1 - C_0)} \cong 2500\ \Omega,$$

and therefore, the resonance is only valid for $X_0 \gg X_1$, so that:

$$\frac{X_0 X_1}{X_0 - X_1} \cong X_1 \cong \frac{1}{\omega(C_1 - C_0)}.$$

This explains the extremely high value for the grounding reactor.

Assume that a 60% reactive compensation level of the transmission line is implemented.

$$Q_{r1} = 0.6 \times 270.51 = 150.43 \text{ MVAr},$$

$$X_1 = \frac{500^2}{150.43} = 1662\ \Omega.$$

Using (7.55) again:

$$X_n = \frac{(1662)^2 377(1.06 \times 10^{-6})}{3(1 - 1662 \times 377 \times 1.06 \times 10^{-6})} = 1096\ \Omega,$$

with a ratio of $k = 2.98$.

The theoretical reactor, in this case, is substantially lower, signifying far better conditions under which to reduce the current However, it is not advisable to adopt a theoretical reactor with $k = 2.98$, since k ratios lower than 2.0 lead to feasible solutions inside the arc extinction zone, where the secondary currents are low but not necessarily zero.

7.4.4 Model for Stuck Breaker Analysis

This model assumes that the end of one phase of the transmission line remains connected to the system, and therefore, is kept energized after breaker opening.

Assuming that the stuck breaker situation occurs in phase a, voltages at the open phases b and c can be calculated with the circuit in Figure 7.30a, which can be simplified in Figure 7.30b.

The voltage in phases b and c is calculated below:

$$V_b = V_c = E\frac{0.5Z_0}{1.5Z + 0.5Z_0}. \tag{7.56}$$

In the case of resonance, the denominator of (7.56) should be equal to zero.

$$1.5Z = -0.5Z_0. \tag{7.57}$$

Equation (7.58) is then obtained for the resonant condition:

$$X_1 = \frac{k + 0.5}{k\omega(C_0 + 0.5C_1)}. \tag{7.58}$$

7.4.5 Single-Phase Autoreclosure

For the extinction of the secondary arc currents and because recovery voltages in the open phases cannot be high, conditions close to resonance should be avoided.

Analysis of the secondary arc in the steady state should be supplemented with transient studies of EHV transmission lines, according to Chapter 8.

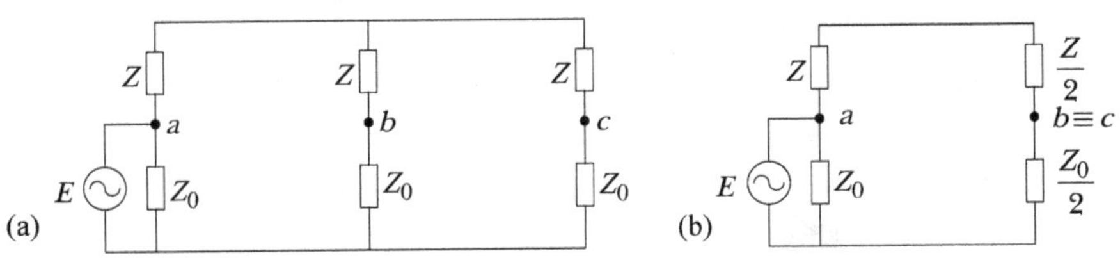

Figure 7.30 Analysis of stuck breaker. (a) Stuck breaker in phase a. (b) Reduced circuit.

7.5 Voltages Induced in Parallel Circuits

7.5.1 General Considerations

In power system applications, it is quite common to encounter transmission lines with parallel circuits. The number of parallel circuits depends on planning requirements, and, for economic reasons, the circuits are installed in the same tower or very close to each other, leading to capacitances and mutual inductances between the phases of the circuits. This item analyzes the voltages induced in a disconnected transmission line when a parallel line remains in operation.

In general, the system equivalent impedances at the terminals of the transmission lines, on sides 1 and 2, have much lower values than impedances Z_g and Z_m due to capacitances, where:

Z_m: impedance between the phases of the circuits.
Z_g: phase-to-ground impedance of one circuit.

The voltages in the phases of the operating transmission line are imposed by the system and can be assumed to be around 1.0 pu. Therefore, the voltage across the open circuit depends on Z_m and Z_g (Figure 7.31).

Two major simplifications are needed to build a model enabling the important aspects of the voltages induced in parallel circuits to be understood. By analogy with single-phase autoreclosure, only the dominant capacitive coupling is considered. Another simplification is to assume transposed transmission lines, which simplifies the presentation of the fundamental concepts, using symmetrical components.

Detailed calculations can be performed using the nodal admittance matrix for the three-phase system, considering the inductive effect and nontransposed transmission lines.

7.5.2 Model for the Capacitive Coupling Between Circuits

In Figure 7.32a, the capacitances related to phase a of circuit 1 are indicated.

Two parallel circuits and a possible transposition scheme are shown in Figure 7.32b.

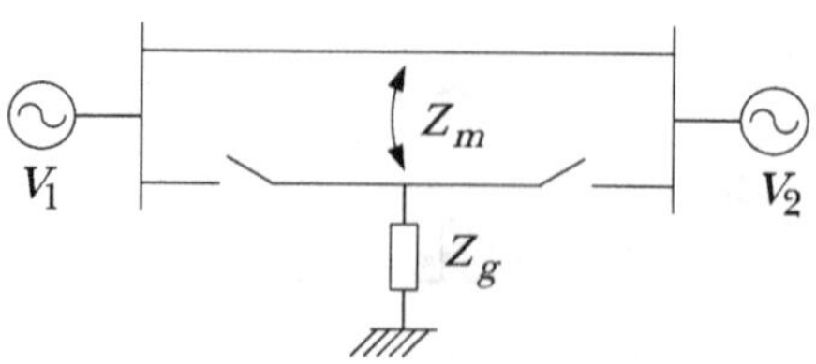

Figure 7.31 Parallel circuits.

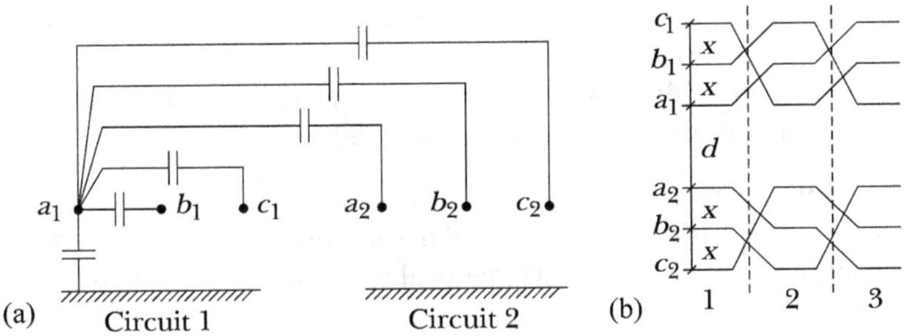

Figure 7.32 Capacitive model of the parallel circuits. (a) Capacitance related to phase a of circuit 1. (b) Example of transposition.

Sums of the distances between two phases of different circuits in segments 1, 2, and 3 along the line are examined for certain pairs of phases, without considering the distance d, which is the same in all cases.

$$a_1 - a_2 = 0 + 2x + 4x = 6x \qquad a_1 - c_2 = 2x + x + 3x = 6x$$
$$b_1 - b_2 = 2x + 4x + 0 = 6x \qquad a_1 - b_2 = x + 3x + 2x = 6x$$
$$c_1 - c_2 = 4x + 0 + 2x = 6x \qquad b_1 - c_2 = 3x + 2x + x = 6x$$

Other pairs of phases could be examined, using the same sum. Since the average distances are equal, equal coupling capacitances between circuits are obtained. For the phases of each circuit, transposition is also achieved.

The nodal susceptance matrix for both circuits is assembled in (7.59), considering the coupling in all phases.

$$\begin{bmatrix} i_{a1} \\ i_{b1} \\ i_{c1} \\ i_{a2} \\ i_{b2} \\ i_{c2} \end{bmatrix} = \begin{bmatrix} y_1 & y_{m1} & y_{m1} & y_{12} & y_{m12} & y_{m12} \\ & y_1 & y_{m1} & y_{m12} & y_{12} & y_{m12} \\ & & y_1 & y_{m12} & y_{m12} & y_{12} \\ & & & y_2 & y_{m2} & y_{m2} \\ & & & & y_2 & y_{m2} \\ & & & & & y_2 \end{bmatrix} \times \begin{bmatrix} v_{a1} \\ v_{b1} \\ v_{c1} \\ v_{a2} \\ v_{b2} \\ v_{c2} \end{bmatrix}, \tag{7.59}$$

Note that since the conductances are neglected, the admittance and susceptance matrices are the same. Furthermore, the capacitance and susceptance matrices are related by the factor $j\omega$.

For this transposition scheme, the following elements may be defined:

y_1, y_2: diagonal elements in susceptance matrices for circuits 1 and 2, respectively.

y_{m1}, y_{m2}: off-diagonal elements in susceptance matrices for circuits 1 and 2, respectively.

y_{12}, y_{m12}: diagonal and off-diagonal elements in the susceptance matrix for coupling elements between phases of circuits 1 and 2.

Assuming identical circuits, $y_1 = y_2$ and $y_{m1} = y_{m2}$.

In other words, symmetric and balanced submatrices are obtained by means of a suitable transposition. Applying the symmetrical components to the double circuit, the matrix is written as:

$$Y_{012} = \begin{bmatrix} T^{-1} & \\ & T^{-1} \end{bmatrix} \times \begin{bmatrix} Y_{11(abc)} & Y_{12(abc)} \\ Y_{21(abc)} & Y_{22(abc)} \end{bmatrix} \times \begin{bmatrix} T & \\ & T \end{bmatrix}, \tag{7.60}$$

where Y_{11}, Y_{12}, Y_{21}, Y_{22} are the submatrices indicated in (7.59), and $Y_{21} = Y_{12}$. Now the diagonal submatrices are obtained with the application of symmetrical components.

$$Y_{012} = \begin{bmatrix} Y_{11(012)} & Y_{12(012)} \\ Y_{21(012)} & Y_{22(012)} \end{bmatrix}, \tag{7.61}$$

where:

$$Y_{ij(012)} = T^{-1} Y_{ij(abc)} T \quad 1 \le i,j \le 2, \tag{7.62}$$

$$\begin{bmatrix} \underset{\sim}{i}_{1(012)} \\ \underset{\sim}{i}_{2(012)} \end{bmatrix} = [Y_{012}] \times \begin{bmatrix} \underset{\sim}{v}_{1(012)} \\ \underset{\sim}{v}_{2(012)} \end{bmatrix}, \tag{7.63}$$

The symmetrical components of the current in circuits 1 and 2 are given by:

$$\underset{\sim}{i}_{1(012)} = \begin{bmatrix} i_{10} \\ i_{11} \\ i_{12} \end{bmatrix} \text{ and } \underset{\sim}{i}_{2(012)} = \begin{bmatrix} i_{20} \\ i_{21} \\ i_{22} \end{bmatrix},$$

and matrix$Y_{012}$is also exemplified:

$$Y_{012} = \begin{bmatrix} y_{10} & 0 & 0 & y_{mo} & 0 & 0 \\ 0 & y_{11} & 0 & 0 & y_{m1} & 0 \\ 0 & 0 & y_{12} & 0 & 0 & y_{m2} \\ y_{mo} & 0 & 0 & y_{20} & 0 & 0 \\ 0 & y_{m1} & 0 & 0 & y_{21} & 0 \\ 0 & 0 & y_{m2} & 0 & 0 & y_{22} \end{bmatrix}, \tag{7.64}$$

where:

y_{ik}: i indicates the circuit and k indicates the sequence.

y_{mk}: m indicates the coupling between circuits 1 and 2 and k indicates the sequence.

Through the couplingl capacitance, one circuit can induce voltages in the other, as for example, when operating circuit 1 is fed with a positive sequence, the induced voltage in disconnected circuit 2 is also of positive sequence.

The capacitance values related to the susceptance in equation (7.64) are:

C_{m1}: coupling capacitance of positive sequence between circuits 1 and 2.
C_{11}, C_{21}: capacitance of positive sequence for circuits 1 and 2, separately.

The terms on the diagonal of the capacitance matrix, in lower case letters, are given by $c_{11} = C_{11} + C_{m1}$ and $c_{21} = C_{21} + C_{m1}$, which are the sum of the capacitance values connected to nodes 1 and 2, respectively. The coupling term is given by $c_{m1} = -C_{m1}$.

Assuming positive sequence in the source in Figure 7.33, the induced voltage V_{21} in circuit 2 is smaller than the operating voltage V_{11} in circuit 1.

The capacitive coupling was evident in the nodal matrices in phase domain, as in (7.59), and symmetrical components need not be invoked to show this effect. However, the main interest in symmetrical components relates to understanding the resonant circuits that may occur in compensated transmission lines, which is addressed in the next Section.

7.5.3 Circuits with Reactive Compensation

For compensated transmission lines, there are situations in which V_{21} is larger than V_{11} in Figure 7.34 because the presence of the reactor can change the impedance ratio of the voltage divider.

Assume a reactor with different reactance values for the positive and zero sequences, as for example, a three-phase reactor grounded through a neutral reactor in which $k \neq 1$,according to Eq. (7.38) and Figure 7.17b.

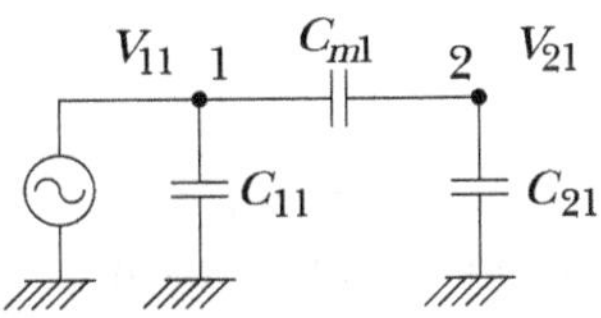

Figure 7.33 Positive sequence.

$X_1 = \omega L_1$: positive sequence.
$X_0 = \omega L_0$: zero sequence.
X_n: neutral reactor.

Including the positive-sequence inductance of the reactor in Figure 7.33, Figure 7.34 is obtained:

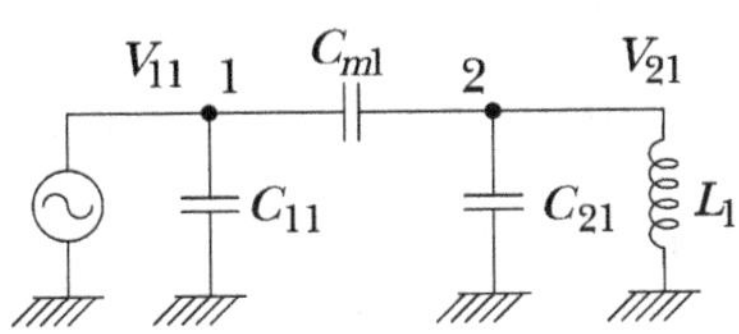

Figure 7.34 Circuit with reactor.

The positive-sequence parameters of circuit 2 alone, that is, the line capacitance and the reactor inductance can show a parallel resonance in lines that are 100%

compensated. However, for a coupled circuit, the resonant condition changes a little bit.

The small difference in the resonant condition, due to the coupling capacitance C_{m1}between circuits 1 and 2, can be seen, by including the reactor's nodal admittance matrix, which can be written in phase coordinates or in symmetrical components.

The admittance matrix of the reactor for sequences relating the nodal voltages and currents is written as:

$$\begin{bmatrix} I_{r0} \\ I_{r1} \\ I_{r2} \end{bmatrix} = \begin{bmatrix} y_{r0} & & \\ & y_{r1} & \\ & & y_{r2} \end{bmatrix} \times \begin{bmatrix} V_{r0} \\ V_{r1} \\ V_{r2} \end{bmatrix}. \tag{7.65}$$

Thus, with the reactor installed in circuit 2:

$$\begin{bmatrix} V_{r0} \\ V_{r1} \\ V_{r2} \end{bmatrix} = \begin{bmatrix} V_{20} \\ V_{21} \\ V_{22} \end{bmatrix}. \tag{7.66}$$

By introducing (7.65) into (7.63):

$$\begin{bmatrix} \underset{\sim}{i}_{1(012)} \\ \underset{\sim}{i}_{2(012)} + \underset{\sim}{i}_{r(012)} \end{bmatrix} = \begin{bmatrix} Y_{11(012)} & Y_{12(012)} \\ Y_{21(012)} & Y'_{22(012)} \end{bmatrix} \times \begin{bmatrix} \underset{\sim}{v}_{1(012)} \\ \underset{\sim}{v}_{2(012)} \end{bmatrix}, \tag{7.67}$$

where:

$$Y'_{22(012)} = \begin{bmatrix} y_{20} + y_{r0} & & \\ & y_{21} + y_{r1} & \\ & & y_{22} + y_{r2} \end{bmatrix}. \tag{7.68}$$

When Kirchhoff's first law is applied to circuit 2, the nodal current is zero, since there is no current source in circuit 2. Thus, Eq. (7.69) is obtained:

$$\underset{\sim}{i}_{2(012)} + \underset{\sim}{i}_{r(012)} = 0 \tag{7.69}$$

From (7.67) and (7.69), the following systems of linear equations are obtained.

$$\underset{\sim}{i}_{1(012)} = Y_{11}\underset{\sim}{v}_{1(012)} + Y_{12}\underset{\sim}{v}_{2(012)}, \tag{7.70}$$

$$\underset{\sim}{0} = Y_{21}\underset{\sim}{v}_{1(012)} + Y'_{22}\underset{\sim}{v}_{2(012)}. \tag{7.71}$$

From (7.71):

$$\underset{\sim}{v}_{2(012)} = -\left(Y'_{22}\right)^{-1} Y_{21}\underset{\sim}{v}_{1(012)}. \tag{7.72}$$

For the symmetrical components, since the matrices in (7.72) are diagonal, the product in (7.72) is easily obtained. Note that Y_{21} can be observed in (7.64).

$$\begin{bmatrix} V_{20} \\ V_{21} \\ V_{22} \end{bmatrix} = -\begin{bmatrix} \frac{y_{m0}}{y_{20}+y_{r0}} & & \\ & \frac{y_{m1}}{y_{21}+y_{r1}} & \\ & & \frac{y_{m2}}{y_{22}+y_{r2}} \end{bmatrix} \times \begin{bmatrix} V_{10} \\ V_{11} \\ V_{12} \end{bmatrix}. \quad (7.73)$$

The same case for positive-sequence resonance as in Figure 7.34 can be exemplified:

$$V_{21} = -\frac{y_{m1}}{y_{21}+y_{r1}} V_{11}. \quad (7.74)$$

For the resonant condition, the denominator of (7.74) is null.

$$y_{21} = -y_{r1}, \quad (7.75)$$

$$j\omega c_{21} = \frac{j}{\omega L_1}, \quad (7.76)$$

$$\omega = \frac{1}{\sqrt{c_{21} L_1}}, \quad (7.77)$$

where:

ω: nominal frequency of the network in rad/s.

Equation (7.77) shows that resonance occurs if the natural oscillation frequency of parameters c_{21} and L_1 is equal to the nominal frequency of the network. But, as previously seen in Section 7.5.2, c_{21} in the nodal admittance matrix is the sum of the capacitance values connected to node 2, rewritten in (7.78) as:

$$c_{21} = C_{m1} + C_{21}. \quad (7.78)$$

The resonance condition can be interpreted in another way, by rewriting the voltage divider equation in Figure 7.34:

$$Z_{eq} = \frac{j\omega L_1 \frac{-j}{\omega C_{21}}}{j\left(\omega L_1 - \frac{1}{\omega C_{21}}\right)}, \quad (7.79)$$

$$V_{21} = \frac{Z_{eq}}{-\frac{j}{\omega C_{m1}} + Z_{eq}} V_{11}. \quad (7.80)$$

Under the resonant condition:

$$Z_{eq} = \frac{j}{\omega C_{m1}}. \tag{7.81}$$

Equation (7.79) is substituted into (7.81):

$$\frac{j\omega L_1 \dfrac{-j}{\omega C_{21}}}{j\left(\omega L_1 - \dfrac{1}{\omega C_{21}}\right)} = \frac{j}{\omega C_{m1}}, \tag{7.82}$$

$$\frac{L_1/C_{21}}{j\left(\omega L_1 - \dfrac{1}{\omega C_{21}}\right)} = \frac{j}{\omega C_{m1}}, \tag{7.83}$$

$$\omega L_1 \frac{C_{m1}}{C_{21}} = -\omega L_1 + \frac{1}{\omega C_{21}}, \tag{7.84}$$

thus:

$$\omega L_1 \left(1 + \frac{C_{m1}}{C_{21}}\right) = \frac{1}{\omega C_{21}}, \tag{7.85}$$

and therefore:

$$\omega^2 L_1 (C_{21} + C_{m1}) = 1, \tag{7.86}$$

finally:

$$\omega = \frac{1}{\sqrt{L_1 (C_{21} + C_{m1})}}. \tag{7.87}$$

Equations (7.77) and (7.87) are equivalent, since $c_{21} = (C_{21} + C_{m1})$, as in (7.78).

It can be concluded that the positive-sequence resonance occurs between the inductance of reactor, L_1, and capacitance c_{21}.

However, in practical terms, $C_{m1} \ll C_{21}$ and therefore it can be assumed that $c_{21} \simeq C_{21}$ in most cases, but it is an approximation.

The result obtained for the positive sequence may be extended to the zero sequence.

A simplified case, which disregards the coupling capacitance C_{m1} and considers only the positive-sequence capacitance of circuit 2, given by C_{21}, is investigated below.

The total capacitance of the positive sequence and the total capacitance of the zero sequence are designated as C_1 and C_0, respectively. The three-phase reactive power of the transmission line and the reactor are calculated for the nominal line voltage V.

Three-phase reactive power of the transmission line:

$$Q_{c1} = V^2\omega C_1. \tag{7.88}$$

Three-phase reactive power of the reactor:

$$Q_{r1} = \frac{V^2}{\omega L_1}. \tag{7.89}$$

When $Q_{c1} = Q_{r1}$, in module, resonance of the positive sequence occurs, that is, when 100% of the capacitive power of a transmission line of positive sequence is compensated by a line reactor. More precisely, this condition would be close to the resonance point as previously discussed.

Hence, the drawback of 100% compensation for the positive-sequence capacitance of the transmission line becomes evident when there are parallel circuits. On the other hand, for long transmission lines, high compensation levels, from 80% to 90%, may be necessary.

The zero-sequence capacitance typically varies from around 60–70% of the positive-sequence capacitance, that is:

$$C_0 = 0.6C_1. \tag{7.90}$$

If the reactor compensates for 60% of the line's positive-sequence capacitance:

$$Q_{c1} = V^2\omega C_1, \tag{7.91}$$

$$Q_{r1} = \frac{V^2}{X_1} = 0.6Q_{c1}. \tag{7.92}$$

In cases without a neutral reactor ($X_n = 0$), and $X_{r1} = X_{r0}$:

$$Q_{r0} = Q_{r1} = 0.6Q_{c1}, \tag{7.93}$$

$$Q_{c0} = V^2\omega C_0 = 0.6V^2\omega C_1 = 0.6Q_{c1}. \tag{7.94}$$

Therefore, from (7.93) and (7.94), $Q_{c0} = Q_{r0}$ is obtained, which represents a resonant condition for the zero sequence.

In the steady state, reasonably balanced three-phase voltages are expected; therefore, the voltage component of the positive sequence of the operating circuit is much higher than that of the zero sequence, that is: $V_{11} \gg V_{10}$, thus, zero-sequence induced voltages in the de-energized circuit are usually lower than that of positive sequence.

Any relatively large imbalance noticeably increases the induced voltage in the zero sequence, as for example, the occurrence of a single-phase fault in the operating circuit. In this unbalanced condition, the zero-sequence voltage in the operating circuit is increased, raising the induced voltages in the open circuit.

So far, a simplified approach to transposed transmission lines was presented to analyze resonance conditions in parallel circuits by applying symmetric components.

7.5.4 Comments on Resonance Analysis in Parallel Circuits

In the case of nontransposed lines, there may be a third resonance point besides the two resonance conditions presented, for positive and zero-sequence parameters. An analysis of nontransposed lines can be found in [15, 16], and the use of computers in the calculation is advisable in this case. In Example 7.12, a nontransposed transmission line is analyzed, using the nodal capacitance matrix.

The previous simplifications are suitable for a preliminary analysis of the resonance range, when there is capacitive coupling. To obtain accurate numeric values, a detailed representation of the circuits and their transpositions should be made using power system software that takes inductive coupling into account.

It is necessary to take the accuracy of the line parameters in the positive and zero sequence into consideration, since the calculation is done using estimates of the average height of the conductors, soil resistivity, etc. The accuracy of reactor power is normally set at 5% and, in addition, the frequency of the system is sometimes subject to oscillations, especially when there are disturbances in the network.

Hence, it is convenient to study the resonance conditions by weighing all these factors. A procedure that can be adopted, only for analytical purposes, is to perform the studies with the average parameters of the line constants and to embed all the uncertainties in a larger tolerance margin for the reactor, as for example, to set the precision on the order of 10%.

Two main caveats should guide the resonance analysis. The first is to avoid positive-sequence resonance, as this is the normal operating condition of the system, and high voltages in the open circuit are likely to occur. The second caveat is to evaluate the resonance conditions for the zero sequence and to investigate the overvoltages during normal operation and under fault conditions. In the case of two parallel circuits, the probability that a fault in one of the circuits occurs when another is not in operation is low in EHV networks, but the consequences of such an event should be avoided.

In order to avoid a positive-sequence resonant condition, the nominal power of the reactors should be changed. In the case of the zero sequence, neutral reactors or resistors should be installed.

For transmission lines with single-phase autoreclosure, a complementary analysis should be performed, such as, for example, investigations on stuck breakers and open phases. In this case, the reactor should also be suitable for such conditions, which leads us to a more complex study of parallel circuits.

Example 7.11 For a 440-kV double circuit transmission line, calculate the voltages induced in open circuit 2 with circuit 1 operating. Calculate the induced positive-sequence voltages without reactive compensation. Determine the resonance conditions and calculate the induced voltages with a 60% reactive compensation level.

Data

a) The following capacitance matrix is assumed in a transposed line, like (7.59).

$$C = \begin{bmatrix} 12 & -2 & -2 & -0.8 & -0.4 & -0.4 \\ & 12 & -2 & & -0.8 & -0.4 \\ & & 12 & & & -0.8 \\ -0.8 & -0.4 & -0.4 & 12 & -2 & -2 \\ & -0.8 & -0.4 & & 12 & -2 \\ & & -0.8 & & & 12 \end{bmatrix} \text{nF/km}.$$

b) Positive and zero-sequence parameters.

Circuits 1 and 2:

$$C'_0 = 8 \text{ nF/km}, \quad C'_1 = 14 \text{ nF/km}.$$

Circuits 1 and 2, coupling capacitance:

$$C'_{m0} = 1.6 \text{ nF/km}, \quad C'_{m1} = 0.4 \text{ nF/km}.$$

c) Line length $\ell = 200$ km.

Solution

a) Calculation of the induced voltages without reactive compensation.

Since the case is in reference to a line of medium length, the total capacitance for the transmission line can be calculated in a simplified way. Note that for the voltage dividers the length and $j\omega$ cancel out.

$$C_1 = 14 \times 10^{-9} \times 200 = 2.8\ \mu\text{F},$$

$$C_0 = 8 \times 10^{-9} \times 200 = 1.6\ \mu\text{F},$$

$$C_{m0} = 1.6 \times 10^{-9} \times 200 = 0.32\ \mu\text{F},$$

$$C_{m1} = 0.4 \times 10^{-9} \times 200 = 0.08\ \mu\text{F}.$$

Positive-sequence voltage in circuit 2:

$$V_{11} = \frac{440}{\sqrt{3}} = 254\ \text{kV}.$$

Applying Eq. (7.74), and noting that the admittance $y_{r1} = 0$, without a reactor:

$$V_{21} = -\frac{-j\omega \times 0.4 \times 10^{-9} \times 200}{j\omega \times 14 \times 10^{-9} \times 200} \times 254 = \frac{0.4}{14} \times 254.$$

$$V_{21} = 7.26\ \text{kV}.$$

b) With reactive compensation.
Reactive power:

$$Q_{c1} = (440)^2 \times 377 \times 2.8 \times 10^{-6} = 204.36\ \text{MVAr},$$

$$Q_{c0} = (440)^2 \times 377 \times 1.6 \times 10^{-6} = 116.78\ \text{MVAr}.$$

Reactive compensation at 60%:

$$Q_{r1} = 0.6 \times 204.36\ \ \text{MVAr} = 122.62\ \text{MVAr},$$

$$y_{r1} = -j\frac{122,62}{440^2} = -j6,33 \times 10^{-4}\ \text{capacitive admittance.}$$

Induced positive-sequence voltage:
Applying (7.74):

$$V_{21} = -\frac{-j\omega \times 0.4 \times 10^{-9} \times 200}{+j1.056 \times 10^{-3} - j6.334 \times 10^{-4}} \times 254.$$

$$V_{21} = 18.13\ \text{kV}.$$

For the compensated transmission line, the induced positive-sequence voltage is not high, but the zero-sequence voltages can be, since the circuits are very close to the resonant condition of the zero sequence, with $Q_{r1} = Q_{r0}$=122.62 MVAr and Q_{c0} = 116.78 MVAr. A possible solution to avoid resonance would be to install a neutral reactor.

In the following example, the general case of a nontransposed line is analyzed.

Example 7.12

a) Calculate the induced voltages in a double circuit transmission line, similar to that of Example 7.11, now assumed to be nontransposed.

Use the following capacitance matrix:

$$[C'] = \begin{bmatrix} 11.6 & -1.8 & -2.0 & -0.8 & -0.25 & -0.6 \\ 1.8 & 12.5 & -2.1 & -0.6 & -0.3 & -1.3 \\ -2.0 & -2.1 & 12.3 & -0.25 & -0.1 & -0.3 \\ -0.8 & -0.6 & -0.25 & 11.6 & -2.0 & -1.8 \\ -0.25 & -0.3 & -0.1 & -2.0 & 12.3 & -2.1 \\ -0.6 & -1.3 & -0.3 & -1.8 & -2.1 & 12.5 \end{bmatrix} \text{nF/km.}$$

Solution

In cases of nontransposed lines, the following equation may be written:

$$\begin{bmatrix} I_{1a} \\ I_{1b} \\ I_{1c} \\ 0 \\ 0 \\ 0 \end{bmatrix} = \begin{bmatrix} Y_{11} & Y_{12} \\ Y_{21} & Y_{22} \end{bmatrix} \times \begin{bmatrix} 254\angle 0^\circ \\ 254\angle -120^\circ \\ 254\angle 120^\circ \\ V_{2a} \\ V_{2b} \\ V_{2c} \end{bmatrix},$$

where $Y = j\omega\ell[C']$.

Since the length and $j\omega$ cancel out, only the capacitance per unit length may be considered.

Taking the last three lines of the system of equations, the voltages in circuit 2 are obtained.

$$\begin{bmatrix} V_{2a} \\ V_{2b} \\ V_{2c} \end{bmatrix} = -\begin{bmatrix} 11.6 & -1.8 & -2.0 \\ -1.8 & 12.5 & -2.1 \\ -2.0 & -2.1 & 12.3 \end{bmatrix}^{-1} \times \begin{bmatrix} -0.8 & -0.25 & -0.6 \\ -0.6 & -0.3 & -1.3 \\ -0.25 & -0.1 & -0.3 \end{bmatrix} \times 254 \begin{bmatrix} 1 \\ \alpha^2 \\ \alpha \end{bmatrix},$$

$$\begin{bmatrix} V_{2a} \\ V_{2b} \\ V_{2c} \end{bmatrix} = \begin{bmatrix} 14.014\angle 54.47^\circ \\ 20.908\angle 97.06^\circ \\ 9.177\angle 77.94^\circ \end{bmatrix}.$$

b) Example 7.11 is solved again, on the assumption that the line is transposed and without reactive compensation, using the last three lines of the system of equations.

$$\begin{bmatrix} V_{2a} \\ V_{2b} \\ V_{2c} \end{bmatrix} = - \begin{bmatrix} 12.0 & -2.0 & -2.0 \\ -2.0 & 12.0 & -2.0 \\ -2.0 & -2.0 & 12.0 \end{bmatrix}^{-1} \times \begin{bmatrix} -0.8 & -0.4 & -0.4 \\ -0.4 & -0.8 & -0.4 \\ -0.4 & -0.4 & -0.8 \end{bmatrix} \times 254 \times \begin{bmatrix} 1 \\ \alpha^2 \\ \alpha \end{bmatrix},$$

$$\begin{bmatrix} V_{2a} \\ V_{2b} \\ V_{2c} \end{bmatrix} = 7.26 \times \begin{bmatrix} 1 \\ \alpha^2 \\ \alpha \end{bmatrix}.$$

7.6 Frequency Response Analysis

7.6.1 Introduction

Steady-state analysis of a nonsinusoidal wave in an electrical network can be performed using the Fourier series. The origin of the distortion may be due to saturation of transformers, rectifier circuits, static compensators, arc furnaces, etc.

After obtaining the harmonic components of the distorted wave at a certain node of the network, it is possible to calculate the behavior of each harmonic component in other nodes of the network.

For this purpose, transfer functions for the network in the frequency domain should be obtained, and the usual method for that is to set up the nodal admittance matrix for the harmonic frequencies to be analyzed.

7.6.2 Modeling the Network Elements

Frequency response analysis requires that all elements in the network are linear. The result of injected voltages or harmonic currents at one node is evaluated at other network nodes. A solution is obtained using methods for network analysis in the steady state.

The main components of power systems are transmission lines, generators, transformers, reactors, capacitors, which are represented by the respective impedances calculated at the frequency being analyzed. Only the linear branches are included in the models. In power systems, it is difficult to model the load dependence on frequency, and a precise model is always difficult to obtain.

When it is necessary to restrict the representation of the network through equivalents, the choice of boundary nodes, the degree of detail, and the number of nodes to be preserved are tasks that are not always easily solved, and familiarity with the network is always recommended.

The representation of the transmission lines should also be carefully performed at each harmonic frequency, due to the dependence of the series impedance on frequency, $Z(\omega) = R(\omega) + j\omega L(\omega)$, as shown in Chapter 5.

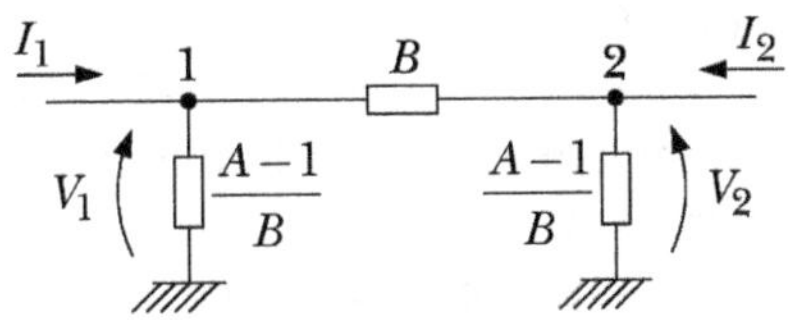

Figure 7.35 π-equivalent and two-port parameters.

The circuit in Figure 7.35 is used to obtain the admittance matrix of the transmission line, to be included in the nodal admittance matrix of the network.

$$\begin{bmatrix} I_1 \\ I_2 \end{bmatrix} = \begin{bmatrix} \frac{A-1}{B} + \frac{1}{B} & -\frac{1}{B} \\ -\frac{1}{B} & \frac{A-1}{B} + \frac{1}{B} \end{bmatrix} \times \begin{bmatrix} V_1 \\ V_2 \end{bmatrix}, \tag{7.95}$$

$$Y = \begin{bmatrix} \frac{A}{B} & -\frac{1}{B} \\ -\frac{1}{B} & \frac{A}{B} \end{bmatrix} = \frac{1}{B} \times \begin{bmatrix} A & -1 \\ -1 & A \end{bmatrix}, \tag{7.96}$$

$$Y = \frac{1}{Z_c \sinh(\gamma \ell)} \times \begin{bmatrix} \cosh(\gamma \ell) & -1 \\ -1 & \cosh(\gamma \ell) \end{bmatrix}. \tag{7.97}$$

The usual tasks in power systems deal only with the steady state at 60 Hz, and thus the behavior of a transmission line with respect to higher-order harmonics may be surprising. The following example seeks to illustrate this fact.

Example 7.13 For a harmonic voltage applied at the beginning of a no-load transmission line, obtain the voltage at the receiving end by calculation of the two-port constant A. Assume a lossless transmission line with a constant inductance.

$$C' = 14\,\text{nF/km},\ L' = 1\,\text{mH/km},\ \ell = 400\,\text{km}.$$

$$\gamma\,\ell = j\omega_h\sqrt{14 \times 10^{-9} \times 0.001} \times 400, \qquad \gamma\,\ell = j0.0015\omega_h.$$

The harmonic frequencies considered are $\omega_h = h\omega$, for h = {2, 3, 5, 7, 9, 11, 13}.

A	h
0.4281	2
–0.1215	3
–0.9491	5
–0.6909	7
0.3577	9
0.9971	11
0.4959	13

In the present case of a lossless line, it can be observed at certain frequencies that the signal can be considerably amplified as, for example, at $h = 3$. However, harmonic propagation in long lines is typically well attenuated when losses are considered.

The amplitude of the higher-order harmonics usually decreases with the order. For rectifier bridges, for instance, harmonic currents of order $h > 1$ have the following relationship with the fundamental frequency current [17, 18]:

$$I_h = K\frac{I_1}{h}. \tag{7.98}$$

The harmonic flow on the network is particularly important when there are rectifier bridges, as in the case of DC links, static compensators, rectifiers for industrial loads, arc furnaces, etc. Although higher-order harmonics have low amplitudes, as seen previously, this does not mean that the effects are negligible, since the network can amplify the voltage at certain nodes.

7.6.3 Harmonic Flow

Harmonics on transmission lines can be analyzed according to certain criteria, including the following:

- Total Harmonic Distortion of Voltage

  For a distorted signal decomposed into a Fourier series, where:

V_1: rms value of the fundamental,
V_h: rms value of the harmonic order ($h > 1$).

The total harmonic distortion is defined as:

$$THD = \sqrt{\sum_{h=2}^{50}\left(\frac{V_h}{V_1}\right)^2}. \tag{7.99}$$

- Telephone Interference Factor

  The harmonic flow in transmission lines, due to mutual coupling, can cause interference in the circuits of nearby telephone cables. Technological advances have reduced some risks, but harmonics remains an issue, especially in areas with old infrastructure.

  The induced voltage in the telephone cable parallel to the line can be calculated from:

$$V_h = I_h Z_m(h\omega), \tag{7.100}$$

where $Z_m(h\omega)$ is the mutual impedance between the telephone cable and the transmission line, calculated at frequency $h\omega$.

The criterion to be adopted for this type of interference is based on the telephone interference factor TIF.

$$TIF = \sqrt{\sum_{h=2}^{50} \left(\frac{V_h}{V_1} \cdot f_h\right)^2}, \tag{7.101}$$

where:

f_h: weighting factor for the harmonic h, according to international tables on the effect of this harmonic on the human ear.

The harmonics show a statistical behavior as a function of several factors that occur in a power system such as thyristor firing angles or network topology.

Two types of equipment related to harmonics in the network are presented below, in Sections 7.6.4 and 7.6.5, respectively.

7.6.4 Harmonics of Transformers

The transformer model is based on the T circuit, with a magnetizing branch to ground and reactances representing the leakage flux. The transformer model must accurately consider the magnetization branch, the saturation level, and the leakage reactance. This model makes some approximations regarding the precise level of excitation of the core, but, for several reasons, the model is widely applied in calculations that analyze the saturation effect on the network.

The elementary case of a single-phase, two windings, and unloaded transformer is assumed to describe the saturation effect without considering hysteresis, according to Figure 7.36. This model clarifies the origin of the harmonic currents generated in the magnetizing branch.

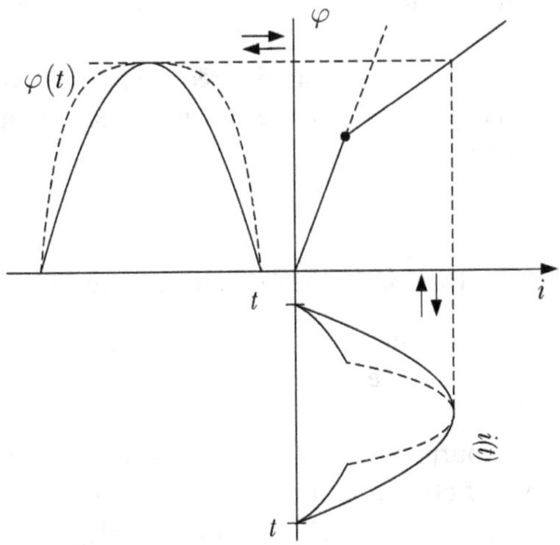

Figure 7.36 Relationship between fluxes, voltages, and currents.

Harmonic currents due to the magnetizing branch are of low amplitude when the transformer is operating near the rated voltage. However, the flow of harmonic currents, depending on the transformer winding connection, needs to be known, both for the system and the transformer design.

When a sinusoidal voltage is applied, that is, a sinusoidal flux in the transformer core, a nonsinusoidal magnetization current is generated (dashed line for the current in Figure 7.36), and there is only one nonsinusoidal current waveform that implies sinusoidal flux in the core. On the other hand, if a sinusoidal current is imposed, a nonsinusoidal flux will be generated, (dashed line for the flux in Figure 7.36).

The transformer saturation levels are related to economic considerations in core design, taking into account that the knee point of the saturation curve depends on the temporary overvoltages in the network.

This first single-phase analysis indicates that a sinusoidal flux is obtained with a distorted current and therefore with harmonic currents that, when decomposed into a Fourier series, have components of order 1, 3, 5, 7, 9, 11, 13, etc.

In three-phase systems, these harmonics require a path through the network. The most significant current harmonics in relation to transformers are of order 3, 5, and 7. The order-3 current harmonics show a behavior like the zero-sequence currents, requiring a return path to ground; current flow is favored in transformers with delta windings and grounded star windings.

Harmonics of order 5 and 7 have a behavior similar to those of negative- and positive-sequence currents, respectively, not requiring a return path to ground.

In order to analyze distorted time-domain voltages and model the transformer in electromagnetic transient programs, saturation data are generally available as rms values, and it is necessary to convert them to instantaneous values. The appropriate treatment can be found in [19] and in Section 4.4.2 in Chapter 4.

Another topic of interest concerns models for the transformer subjected to currents injected by other harmonic sources, as for example, rectifiers. In this case, the models for harmonic frequencies do not take saturation into account, but instead assume that the magnetization branch becomes linear with increasing frequency.

7.6.5 Harmonics of Converters and Filtering

The treatment of harmonic filters in power systems is quite an extensive topic that will not be detailed here. More information on these issues may be found in [17, 18, 20].

The purpose of including an analysis of harmonic filters is to show the importance of obtaining the frequency response of the network, for calculation of the system impedance $Z(\omega)$ and harmonic flow analyses of the network.

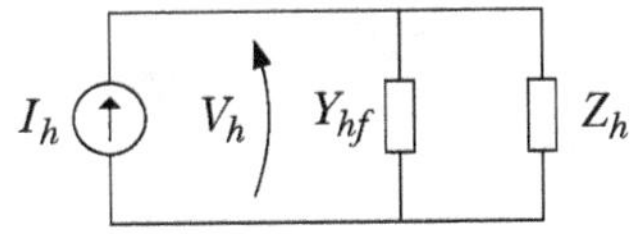

Figure 7.37 Electric network and filter.

For thyristor converters, the harmonic levels injected into the network can be reasonably estimated. This calculation will not be detailed here; however, as in the case of DC links, it is known that the injected harmonic currents behave like current generators applied at the coupling point.

Only the basic concepts of filter tuning are discussed below, and a rough filter analysis can be performed using a simplified network (Figure 7.37)

where:

Y_{hf}: filter admittance for harmonic h.
Z_h, Y_h: impedance and admittance of the system, respectively, for harmonic h.
I_h: injected harmonic current.
V_h: harmonic voltage.

At the system nominal frequency, the filter should have a very high impedance in order to interfere as little as possible with the system steady state. Application of harmonic current I_h to the filter impedance connected in parallel with the system impedance should result in an acceptable harmonic voltage at the system busbar.

$$V_h = I_h \frac{1}{Y_{hf} + Y_h}. \tag{7.102}$$

Installation of the filter is treated in a simplified manner above. Practical filter installations require a much more complex approach, as the following issues should be addressed [17]:

- Equation (7.102) can be generalized to nodal analysis in the case of multiple filter designs.
- The network has a variable topology that changes according to system operating requirements.
- Frequency-dependent models for equipment and transmission lines are well defined, but it is not simple to accurately model the dependence on frequency of the load.
- The impedances of the system for several load and topology scenarios are evaluated for each frequency approached. Each impedance is represented by a point leading to a cloud of points in the R–X plane. Geometrical figures like sectors, circles, or polygons are proposed to encompass the cloud to perform the filter design. The cloud is assumed to comprise eventual modeling errors.
- The characteristic and noncharacteristic harmonic currents are subject to random factors.

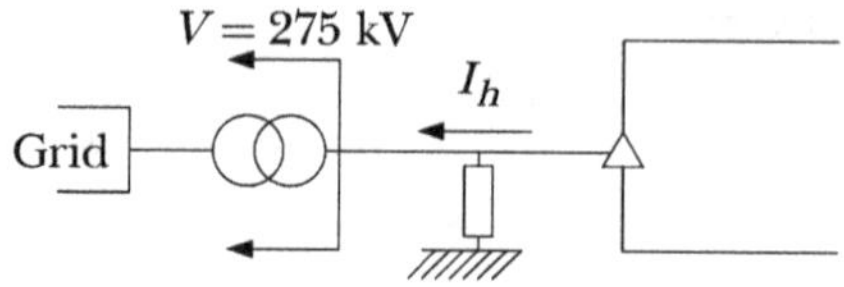

Figure 7.38 DC link in the steady state.

- Filter optimization and interaction with the network and other filters should also be addressed.

The purpose of the following example is to investigate, in a simple network, the tuning of a filter in the AC side of a DC link.

Example 7.14 The DC link bridge converter is assumed to be connected to a network operating in the steady state (Figure 7.38).

The filter must be tuned to limit the voltage of the eleventh harmonic, $h = 11$, generated by a bridge converter.

As mentioned previously, an economic filter design should consider several factors, such as the accuracy and quality factor of the components, in addition to network parameter uncertainties. This example presents only the basic idea of filter tuning without detailing the design of the components.

Data

$I_1 = 1.4$ kA, the fundamental current at 60 Hz.
$I_{11} = 4.6\%$ of the fundamental current.

The system impedance $Z(\omega)$ at the point of common coupling should be calculated using nodal analysis, and the methodology to build the nodal admittance matrix (see Chapter 4) or the nodal impedance matrix is presented in [8]. Figure 7.39 is hypothetical.

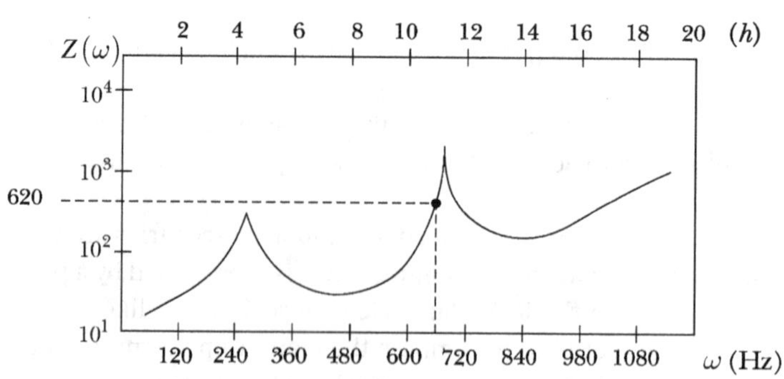

Figure 7.39 Frequency response of the system impedance $Z(\omega)$.

The amplitude of the nominal voltage, which is the same as that of the fundamental component, is given by:

$$V_1 = \sqrt{2}\frac{275}{\sqrt{3}} = 224.54 \text{ kV},$$

Assume a maximum voltage for the 11th harmonic of 2%.

$$\omega = 2\pi f = 377.$$

Solution

Assume that $Z(11 \times \omega) = 620 \angle 82°$.

The amplitude of the 11th harmonic of the current is:

$$I_{11} = 0.046 \times I_1 = 64.4 \text{ A}.$$

Without the filter, there would be a harmonic voltage modulus:

$$|V_{11}| = 0.0644 \times 620 = 39.93 \text{ kV}.$$

the percentage of the 11th harmonic voltage is given by:

$$\frac{39.93}{224.54} = 17.8\%.$$

The installed filter should reduce the 11th harmonic, below 2% of the fundamental, corresponding to $V_{11} = 4.49$ kV.

- Network admittance:

$$Y_{11} = \frac{1}{620\angle 82°} = 0.0016\angle -82°.$$

Applying (7.102):

$$V_{11} = I_{11}\frac{1}{Y_{11f} + Y_{11}},$$

$$V_{11} = 0.02 \times 224.54 = 4.49 \text{ kV},$$

$$4.49 = 0.0644\frac{1}{0.0016\angle -82° + Y_{11f}},$$

$$Y_{11f} + 0.0016\angle -82° = 0.0143,$$

$$Y_{11f} = 0.0142\angle 6.42°.$$

This example demonstrated a simplified approach to the filter tuning problem. HVDC converters are provided with Filters on the AC and DC sides. On the AC side, tunable filters are strictly adjusted to low harmonic orders, and high-pass filters are supposed to damp a broad band of frequencies (to low harmonic orders).

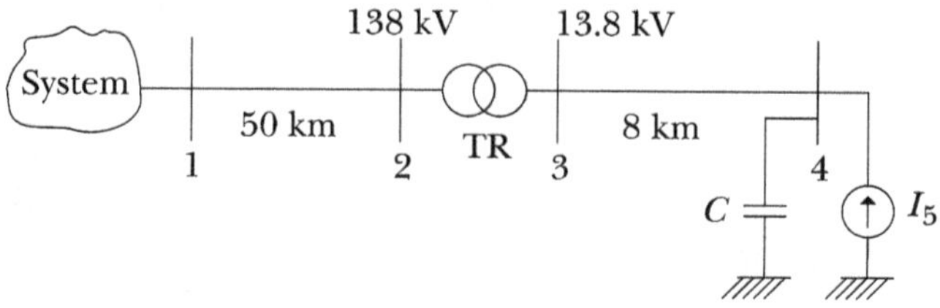

Figure 7.40 Analyzed system.

Example 7.15 Examine the flow of the 5th harmonic current into the network in Figure 7.40.

Data

a) The impedance of a feeding system is furnished for the 5th harmonic in pu with $S_b = 100$ MVA and the line voltage $V_b = 138$ kV:

$$z_{cc} = 0.037894 + j5 \times 0.433128 \text{ pu,}$$

Note that the load should be considered under the operating conditions of the system, that is, it should be included in the system equivalent circuit. The impedance at the fundamental frequency corrected to the 5th harmonic is a rather useful simplification.

b) Line 1–2, parameters already calculated for the 5th harmonic:

$$r' = 0.17124 \ \Omega/\text{km}.$$
$$x' = 2.42039 \ \Omega/\text{km}.$$
$$y' = 1.721 \times 10^{-5} \text{ S/km}.$$

c) Line 3–4, parameters for the 5th harmonic:

$$r' = 0.17404 \ \Omega/\text{km}.$$
$$x' = 2.01867 \ \Omega/\text{km}.$$
$$y' = 2.104 \text{ S/km}.$$

d) Transformer 2–3:

138/13.8 kV.

$r = 2\%$.

$$\frac{x}{r} = 4.5 \text{ constant.}$$

Nominal Power: 25 MVA.

e) Capacitor:

$$Q = 1.2\ \text{MVAr}.$$

f) Fundamental and 5th harmonic currents:

Fundamental component of the current, $I = 100$ A.
Fifth harmonic, $I_5 = 20\%$ of fundamental, phase $= 0°$.

Solution

For this simplified solution, the nominal – π model is adopted for the transmission line, applying per unit values.

1) Base values, $S_b = 100$ MVA.

$$Z_{b1} = \frac{138^2}{100} = 190.44\,\Omega, Z_{b2} = \frac{13.8^2}{100} = 1.9044\,\Omega.$$

2) Admittance values in pu.
   a) System.

$$z_{cc} = 0.037894 + j5 \times 0.433128 \text{ pu},$$

$$y_{cc} = \frac{1}{z_{cc}} = 0.008077 - j0.461616.$$

   b) Transmission Lines
   line 1–2

$$y_{12} = \frac{Z_{b1}}{Z_{12}} = \frac{190.44}{(0.17124 + j2.42039) \times 50},$$

$$y_{12} = 0.110778 - j1.565793 \text{ pu},$$

$$y_1 = y_2 = \frac{y'\ell Z_{b1}}{2},$$

$$y_1 = j8.1937 \times 10^{-2} \text{ pu}.$$

   line 3–4

$$y_{34} = \frac{Z_{b2}}{Z_{34}} = 1.0092 \times 10^{-2} - j0.117054 \text{ pu},$$

$$y_3 = y_4 = \frac{y'\ell Z_{b2}}{2},$$

$$y_3 = 1.6027 \times 10^{-4} \text{ pu}.$$

c) Transformer

$$z = (r + jhx)\frac{S_b}{S_n},$$

$$z = (0.02 + j5 \times 4.5 \times 0.02) \times \frac{100}{25},$$

$$y = \frac{1}{z} = 2.4643 \times 10^{-2} - j0.554460 \text{ pu}.$$

d) Capacitor

$$x_c = -j\frac{V^2}{Q}h\frac{1}{Z_{b2}}, \quad y_c = \frac{1}{x_c} = j0.06 \text{ pu}.$$

3) Admittances of circuit elements (Figure 7.41).
4) Nodal admittance matrix

	1	2	3	4
1	0.118855 − j 1.945472	− 0.110778 +j 1.565793	0	0
2		0.135421 − j 2.038316	− 0.024643 +j 0.554460	0
3			0.034735 − j 0.671354	− 0.010092 +j 0.117054
4	symmetric			0.010092 − j 0.056894

5) Solution for the network.

$$I_b = \frac{100 \times 10^3}{\sqrt{3} \times 13.8} = 4183.7 \text{ A}, \qquad I_5 = \frac{0.2 \times 100}{I_b} = 0.0048 \text{ pu},$$

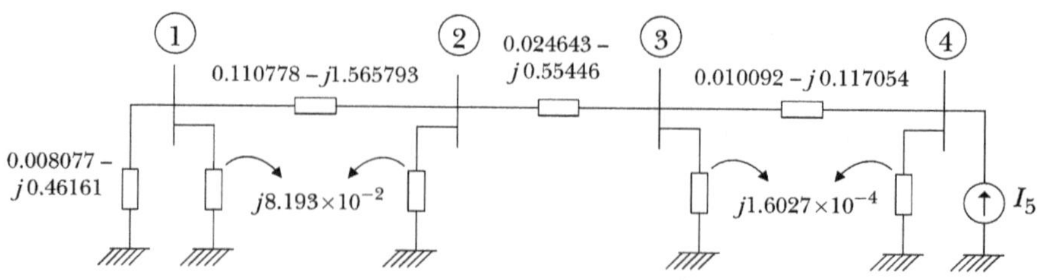

Figure 7.41 Admittances in per unit values.

$$\begin{bmatrix} 0 \\ 0 \\ 0 \\ 0.0048 \end{bmatrix} = [Y_a] \times \begin{bmatrix} V_1 \\ V_2 \\ V_3 \\ V_4 \end{bmatrix}, \quad \begin{bmatrix} V_1 \\ V_2 \\ V_3 \\ V_4 \end{bmatrix} = \begin{bmatrix} 0.0672 + j0.1166 \\ 0.0848 + j0.1440 \\ 0.1192 + j0.2021 \\ 0.2925 + j0.4690 \end{bmatrix}.$$

Three nodal voltages in pu were calculated for the fifth-order harmonic of the current. The high amplitudes of the nodal voltages indicate possible resonance.

7.7 Temporary Overvoltages with Transformers

When transformers are energized or load rejections occur, the nonlinear magnetizing branch of the transformer inserted into the network may result in distorted voltages. In this case, the traditional methods for solving linear systems under the steady state regime with phasors, which are basically adopted in this chapter, cannot be applied, and the time response for the transient and steady-state period needs to be simulated. The transient period is the subject of Chapter 8, which analyzes switching surges. After this initial period, there may be temporary overvoltages with harmonic distortions, which are studied in this chapter.

Some conditions involving transformers may give rise to nonlinear circuits that require simulations in the time domain, thereby furnishing important information on the behavior of the network, without, however, providing a comprehensive analysis of the phenomenon. Despite this, simulation is still a viable resource, as tools employing direct methods to identify unstable conditions in nonlinear networks with a certain degree of complexity are not yet available.

Although this Section does not intend to consider nonlinear phenomena analytically, it is convenient to analyze the stability of the state variables. Thus, the subject is treated in an introductory manner, without intending to solve the nonlinear differential equations that describe the ferroresonance. The comment in reference [21] is worthy of mention, as it characterizes ferroresonance as not fully understood, despite the interest aroused since the 1920s.

7.7.1 Transformer Energization and Load Rejection

In transformer energization, the transient flux can be high and saturate the core. The transient increase in the flux can be understood with the aid of the simple RL circuit in Chapter 1, since current and flux are proportional, and the unloaded transformer behaves much like a large inductance. In three-phase power systems, the phenomenon is even more complex due to the nonlinear saturation branch, the mutual inductance in the windings, and the influence of the residual flux that can elevate the transient flux. Energization of the transformer is expected to occur under favorable conditions; however, there are energization situations that can cause temporary overvoltages, as for example, in long radial systems or during restoration of the transmission system after a large disturbance.

When load rejection occurs, there is usually a rise in voltage in the part of the system that is disconnected from the load. The reasons for the voltage rise were presented assuming the simplified case of a transmission line, in item 7.2.2, which, in a way, is one of the most critical temporary overvoltage conditions in a weak system with a long transmission line and a transformer connected to the sending-end. Cases of transformers at the receiving end are even more critical. The transient switching surges that occur during load rejection are discussed in Chapter 8.

Temporary overvoltages can occur and last for many cycles during transformer energization and load rejection. In this case, the voltage rise creates a flux that saturates the transformer core and introduces harmonics into the network. The nonlinearity introduced by the transformer gives rise to distorted waveforms that are analyzed with simulations in the time domain, considering the presence of the saturation branch. If the saturation branch of the transformer is omitted, for instance, when the distortion level is low, the temporary overvoltage can be calculated using programs that work with phasors.

7.7.2 Ferroresonance

In the previous analysis, despite distorted voltage or current waveforms, it was assumed that the system maintains a stable behavior and that a steady state is reached.

The same behavior is not guaranteed in ferroresonance cases, in which a more complex time response of the nonlinear system is observed, including altered waveforms that may not be reproduced every cycle.

- Single-Phase RLC Circuit

  In Chapter 1, the origin of the resonance in networks was presented based on a simple LC circuit (in fact a RLC circuit with small resistance should be admitted to guarantee the steady-state). This circuit, containing a linear inductor, has a well-defined resonance frequency.

  While maintaining a simplified approach to this LC circuit, however assuming a nonlinear inductor, its phasor interpretation was initially proposed in [22].

  The equilibrium condition for this circuit is analyzed assuming a sinusoidal voltage source set to an rms value E, in series with a capacitance and with a nonlinear inductance, Figure 7.42a.

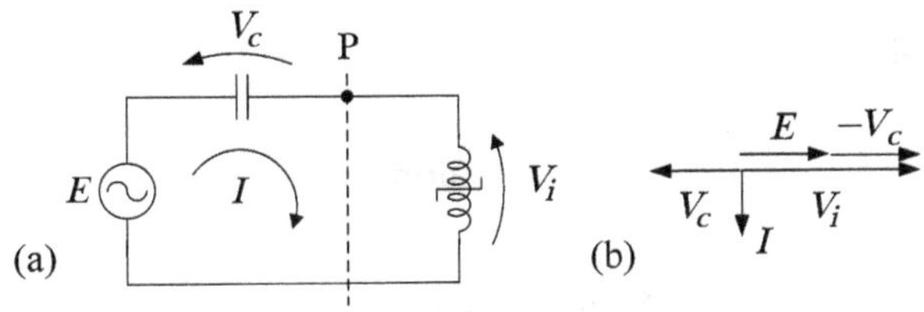

Figure 7.42 Analysis of equilibrium point. (a) Nonlinear LC circuit. (b) Phasor analysis.

The following equation is written :

$$E - V_c = V_i. \tag{7.103}$$

Assuming the current phasor as $I \angle -90^\circ$ in Figure 7.42b, the voltage drop in the capacitance is negative:

$$V_c = -X_c I. \tag{7.104}$$

Thus, the voltage at point P calculated from the source side gives the straight lines of Figure 7.43, with positive angular coefficient.

$$V_p = E + X_c I, \tag{7.105}$$

The voltage at point P is also equal to the voltage at the nonlinear inductor, obtained from the saturation curve of the nonlinear inductor, designated by $f(i)$ in Eq. (7.106).

$$V_i = f(I), \tag{7.106}$$

An interpretation of the circuit suggests the graphic representation in Figure 7.43: There are three possible solutions for the circuit in case A; two solutions are in the first quadrant and one in the third quadrant. In case B, there is only one solution, in the third quadrant.

In Figure 7.43, the source voltage in case B is higher than in case A; however, other solutions can be obtained by changing the value of X_c.

Point 1, in case A, represents an equilibrium point in the first quadrant because if the current increases in the vicinity of point 1, with a slight increase in current, ΔI, it is observed that the voltage ΔV_i (in the inductance) is greater than ΔV_p (provided by the source), which causes a reduction in the current and its return to point 1. On the other hand, if there is a reduction in the current, ΔV_i is less than ΔV_p, which causes a rise in current and again its return to point 1 (Figure 7.44).

Point 2 in case A does not correspond to an equilibrium condition. Point 3 in case A (and 3′ in case B), in the third quadrant, tends to exhibit behaviour similar to point 2, but using the phasor interpretation is even less suitable.

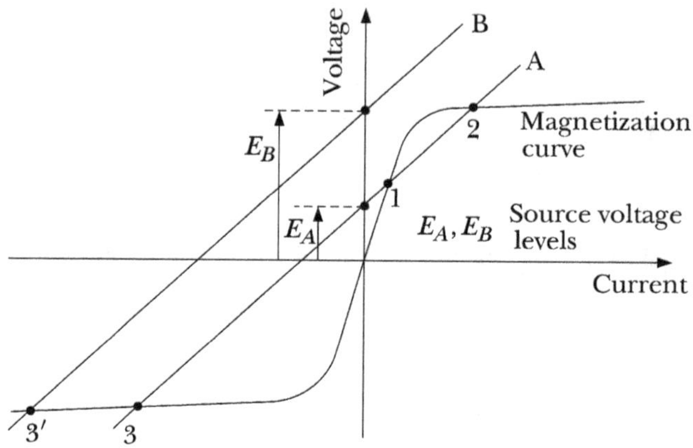

Figure 7.43 Graphic interpretation of the equivalent circuit.

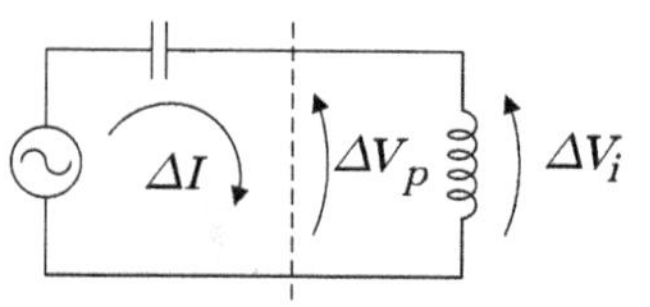

Figure 7.44 Current and voltage increments.

By analyzing the same circuit with the aid of simulators, reference [23] identifies conditions considered to represent ferroresonance, without a solution in the first quadrant; these are characterized by periodic and almost rectangular voltage waveforms with high current peaks (Figure 7.45).

Within interval 1, voltage $v_i(t)$ at the terminals of the inductor is integrated, giving rise to an increasing flux that will reach the saturation level. In intervals 1 and 2, the derivatives of the currents are small, but, since the inductance is high at a low current, the voltage across the inductor remains elevated. In intervals 1′ and 2′, there is a high current peak in the third quadrant and a rapid increase in the voltage across the capacitor. Within interval 1′, the derivative of the current is positive and initially high but decreases continuously to reach zero at the end of the interval. That is why the voltage at the inductor is initially positive and then rapidly decreases to become zero and reverse polarity. Interval 2′ is a mirror image of interval 1′, and the derivative of the current has a change of sign, which explains the polarity change of the inductor voltage$v_i(t)$. The inclusion of hysteresis can slightly alter the symmetry of intervals 1′ and 2′, but this does not substantially change the behavior of the voltages and currents.

The interpretation of the phenomenon presented in reference [24] is very interesting and worth knowing. Other ferroresonant conditions are characterized by voltage oscillograms without periodicity, and may be associated with chaotic situations, as shown below.

Figure 7.45 Current and voltage.

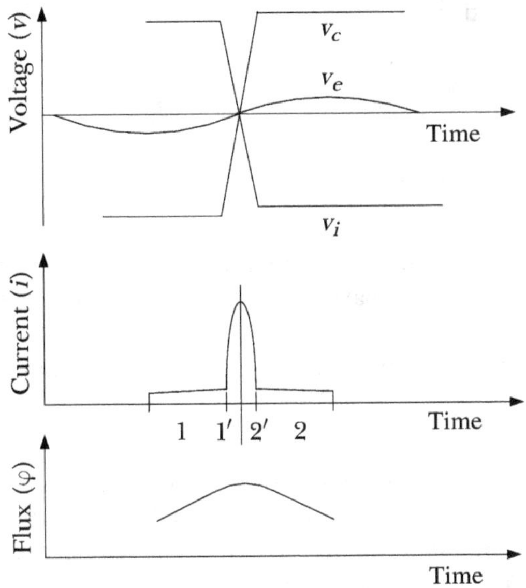

- Analysis Methods

  Second-order free systems are characterized by stable or unstable equilibrium points, while forced systems show periodic or nonperiodic trajectories. While linear systems have well-defined stability regions, nonlinear systems show a much more diverse behavior, with more than one equilibrium point and trajectories to regions of attraction. These regions are studied under the nomenclature of attractors. For second-order free systems, the attractors are equilibrium points or periodic trajectories. In the case of forced systems, there are, in addition, chaotic attractors and strange attractors [25].

  The state space equations of the RLC circuit are written for the case of a nonlinear inductor.

φ: flux linked with the nonlinear inductor.
v_c: voltage across the capacitor.
i: current in the circuit.

Current i is related to the flux through a nonlinear function:

$$i = i(\varphi),$$

recalling that in the linear case, this relation is given by $i = \varphi/L$.

From Figure 7.46, the state space Eqs. (7.107) are proposed:

$$\frac{d}{dt}\begin{bmatrix} v_c \\ \varphi \end{bmatrix} = \begin{bmatrix} C^{-1}i(\varphi) \\ -v_c - Ri(\varphi) \end{bmatrix} + \begin{bmatrix} 0 \\ e(t) \end{bmatrix}. \tag{7.107}$$

Setting:

$$\underset{\sim}{x} = \begin{bmatrix} v_c \\ \varphi \end{bmatrix}, \tag{7.108}$$

$$\frac{d}{dt}\left[\underset{\sim}{x}\right] = G\left(\underset{\sim}{x}, t\right). \tag{7.109}$$

A time domain solution to the above nonlinear equation is sought. There are numerical methods applicable to the solution of this nonlinear system, such as the harmonic balance method or Galerkin's method [26].

A procedure adopted in some references is to obtain the periodic solution of this system with the following boundary condition:

$$\underset{\sim}{x}(t) = \underset{\sim}{x}(t + T), \tag{7.110}$$

with T denoting the period of the excitation function $e(t)$.

Newton-Raphson's method and optimization methods are also used, in an attempt to minimize the error function derived from that boundary condition.

The search for a periodic solution, using numerical optimization methods, has not yet been completely successful for the analysis of nonlinear circuits, particularly those with ferroresonance [21].

A more comprehensive analysis of the problem perhaps would be based on nonlinear system theory and the study of bifurcation and chaos theory. The theory is described in various references, as for example, [25, 27].

In nonlinear system books, it is common to encounter discussions of the behavior of the solutions to the Duffing equation in the description of nonlinear oscillators with forced excitation, under different parameters and initial conditions. The Duffing equation describes a second-order mechanical spring-mass system with nonlinear behavior:

$$\frac{d^2}{dt^2}x + k\frac{d}{dt}x + x^3 = A\cos(t), \tag{7.111}$$

which may be put in the form of state variables:

$$\frac{d}{dt}\begin{bmatrix} x_1 \\ x_2 \end{bmatrix} = \begin{bmatrix} x_2 \\ -kx_1 - f(x_1) \end{bmatrix} + \begin{bmatrix} 0 \\ A\cos(t) \end{bmatrix}, \tag{7.112}$$

- with:

$$f(x) = x^3 \text{ and } \frac{d}{dt}x_1 = x_2\ (\dot{x}_1 = x_2). \tag{7.113}$$

Function $f(x)$ is cubic, while for the representation of $i(\varphi)$, different approximations can be assumed, using polynomials of the fifth, seventh, or ninth order or even approximations by means of a piecewise model. Although the description of the differential equations is not the same, it is possible to associate the solutions of the second-order system in Figure 7.46 with Duffing equation. In [25], there is an extensive discussion of the surprising solutions of Duffing equation and its varied behavior for different values of the initial conditions and parameters k and A.

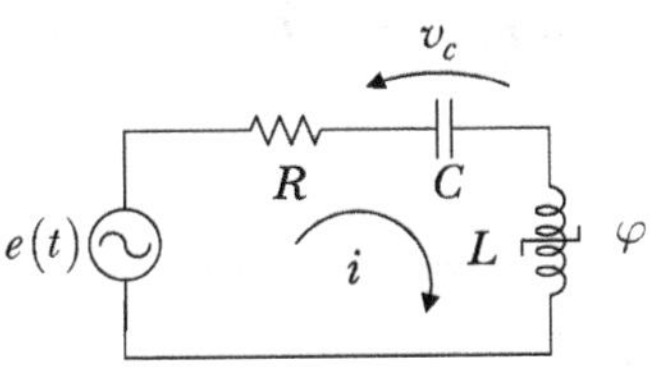

Figure 7.46 State variables.

Regarding the different forms of ferroresonance, reference [21] notes that, when a distorted periodic wave is obtained, yet characterized as ferroresonant, the frequency spectrum contains, besides the fundamental frequency, odd higher-order harmonics. The other form of ferroresonance, characterized by a chaotic response, contains a spectrum of harmonic and sub-harmonic frequencies.

A comprehensive analysis of the state-variables trajectories and possible solutions for the differential equations for the network in Figure 7.47 can be found in [28].

Because of its high value of 50 000 Ω, resistor R_2 in the circuit and in the tests performed has little meaning at low frequencies, behaving almost like an open circuit. Thus, the circuits in Figures 7.46 and 7.47 show a similar behavior.

In the latter, the inductor is linearized by segments, allowing each circuit to be described by linear differential equations, which are more easily solved.

By adopting numerical procedures that avoid the accumulation of errors, an analysis of the trajectories in the phase plane can be performed, revealing instances of periodic solutions that can be identified at points 1, 2, 3, and 3′ in Figure 7.43. Nonsymmetric solutions for flux and voltage are also examined, revealing conditions with subharmonic solutions. Complementing these results, an analysis is presented of chaotic trajectories of fluxes and voltages in the phase plane.

Through linear solutions, mapping of state variable trajectories is performed for each segment, analyzing stability by an approach like Poincaré mapping. In the analyses performed, that is, changing the amplitude of $e(t)$, this reference finds different behaviors of the time responses and their trajectories in the phase plane.

An example of a bifurcation study of a ferroresonant circuit can also be found in [21], using transient electromagnetic programs and time responses.

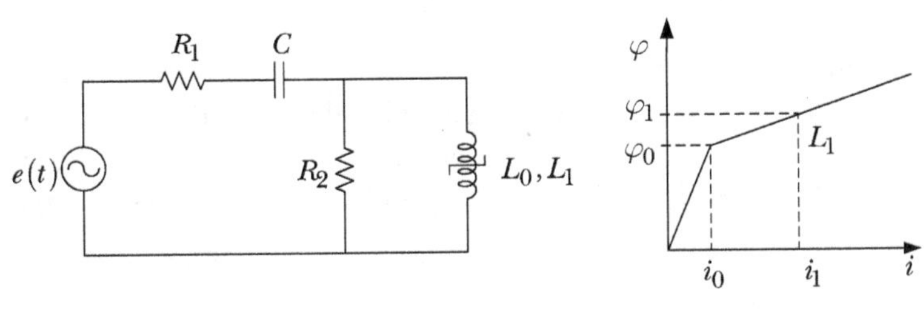

Figure 7.47 Nonlinear circuit and $\varphi \times i$ curve linearized by segments.

Example 7.16 In this example, time responses are analyzed in the circuit shown in Figure 7.46, when the excitation voltage is altered. The inductor is linearized by segments, according to Figure 7.47 using the data below:

$C = 8$ and $2\,\mu$F.

R = 0.01 Ω.

φ (Vs)	i (A)
0	0
0.075	0.028
0.100	0.280

The calculation was initiated with the voltage at its maximum value, that is, with null initial current, without introducing a DC component into the flux.

Simulation results are presented in Figure 7.48. In this figure, the phase-plane (v_c, φ) is presented on the left side and the time response $v_i(t)$ on the right side.

a) With a capacitance of 8 μF

The behavior of the system is stable, with a sinusoidal source voltage with an amplitude of up to 18 V (see Figure 7.48a and b).

Figure 7.48c and d describe the behavior of the system when the source voltage is increased from 18 to 19 V, achieving ferroresonant, chaotic behavior.

b) With a capacitance of 2 μF.

Up to 8.8 V, system behavior remains stable. The electrical system is poised at the threshold of the ferroresonant condition, at an excitation voltage of 9 V (Figure 7.48e and f).

When the capacitance is reduced, with respect to the 8 μF case, the Thévenin equivalent impedance rises, and at lower voltages, there is no solution in the first quadrant. In Figure 7.48f, the fifth-order subharmonic of 12 Hz is evident.

In Figure 7.48g and h, the source voltage was raised from 9 to 10 V, with the system losing the equilibrium point in the first quadrant and going to a ferroresonant chaotic behavior.

So far, some examples of ferroresonant circuits have been illustrated which, despite being simplified and reduced to the second order, are still difficult to interpret in the time domain. Thus, it is possible to appreciate the difficulty of analytically addressing three-phase circuits with multiple transformers in large power systems.

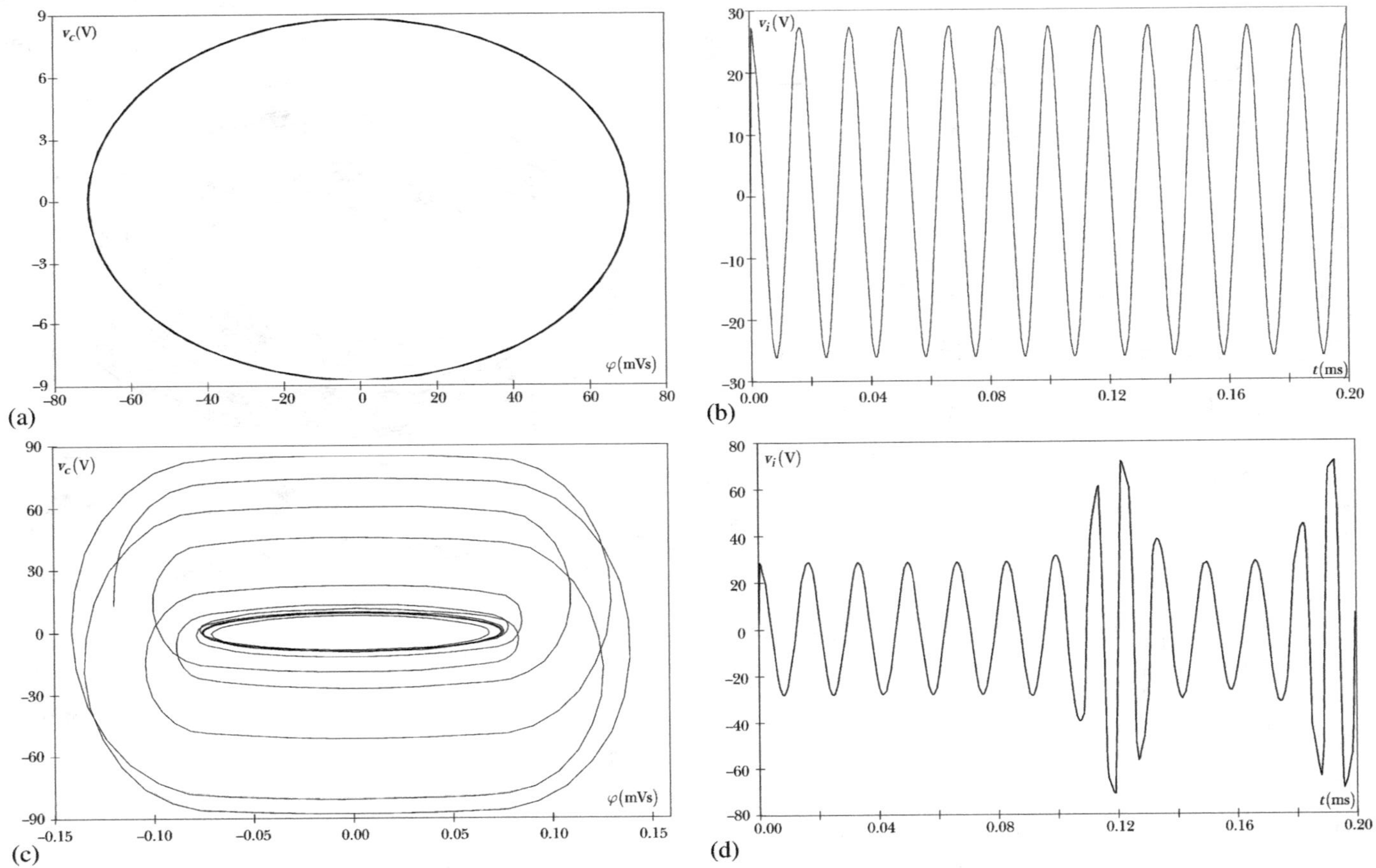

Figure 7.48 Phase-plane and time domain responses. (a) and (b) Capacitance 8 μF, source with 18 V. (c) and (d) Capacitance 8 μF, source with 18 V. (e) and (f) Capacitance 2 μF, source with 9 V. (g) and (h) Capacitance 2 μF, source with 10 V.

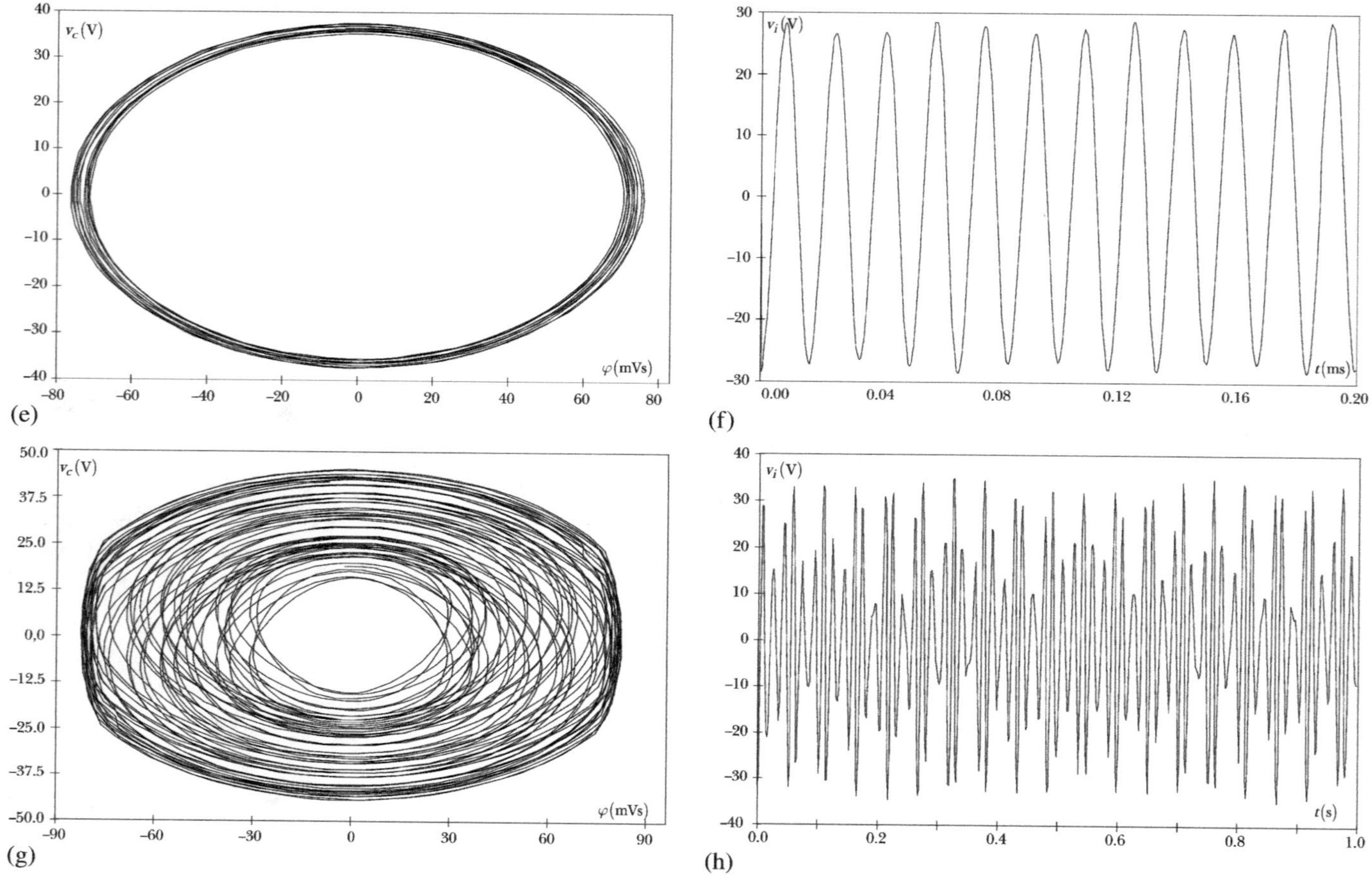

Figure 7.48 (Continued)

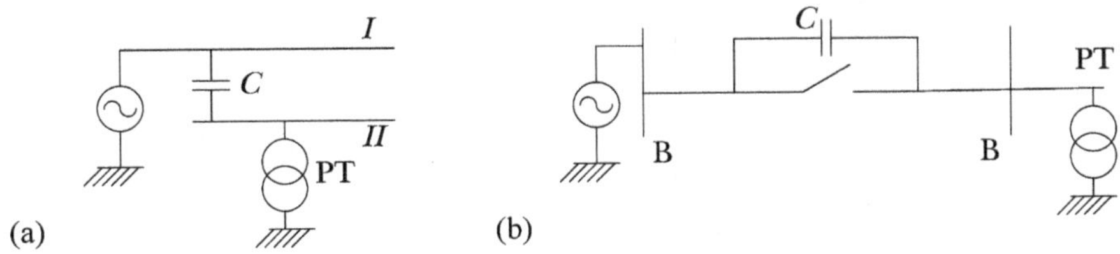

Figure 7.49 Potential transformer connection. (a) Double-circuit (b) Deenergized busbar.

However, ferroresonance occurs in very specific situations that are analyzed by time domain simulation. Some examples of three-phase circuits under potentially ferroresonant conditions are mentioned below.

- Ferroresonance in Three-Phase Circuits

  There are several conditions in which ferroresonance can be established in three-phase circuits [29], and some examples are presented below:

  a) Voltage transformers on transmission lines disconnected from the system and voltage transformers on deenergized busbars.

     An inductive voltage transformer may be connected to a deenergized circuit on a double-circuit transmission line. The capacitive coupling between parallel circuits, as shown in Section 7.5, is illustrated in Figure 7.49a, which depicts a series connection between the coupling capacitance between circuits and the nonlinear inductance of the PT.

     In Figure 7.49b, after the circuit-breaker opening, the inductive potential transformer remains connected to the grid by the potential equalizer capacitors of the circuit-breaker [30].

  b) Single-phase opening of inductive voltage transformer

     In medium voltage networks or industrial systems, the use of single-phase switches or fuses may result in conditions of one or more open phases.

     The single-phase opening of a cable connected to an inductive potential transformer is shown in Figure 7.50a.

     The simple case of a three-phase potential transformer with single-phase units is assumed. The explanation for the occurrence of ferroresonance may be found in [30, 31]. In a lossless single-phase circuit, the charge stored in the line capacitance oscillates with the transformer magnetization inductance, with a waveform that is approximately square. In each half cycle, during which the capacitor voltage is constant, the voltage at the inductor is integrated, raising the flux until a saturation point is reached. At that instant, a current peak occurs, much like the one in Figure 7.45, which discharges and inverts the polarity of the capacitor. The process is repeated indefinitely, at a constant frequency in the lossless system, Figure 7.50b.

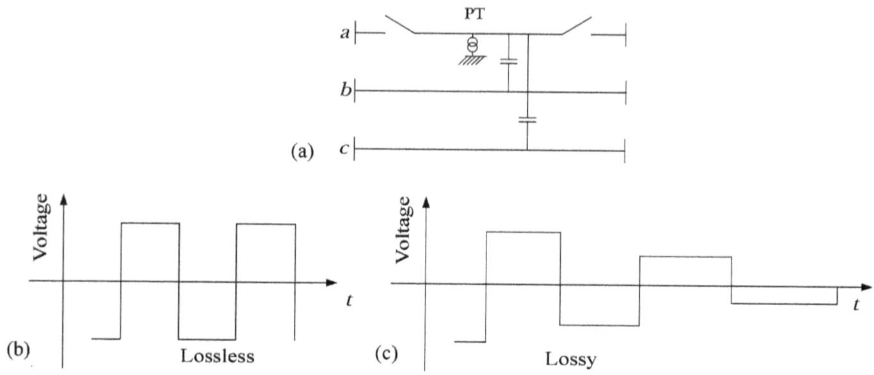

Figure 7.50 Potential transformer and single-phase opening. (a) Simplified circuit. (b) Lossless line. (c) Lossy line.

When losses are introduced, the voltage across the line capacitance is reduced, making a longer integration time necessary in order to reach the saturation point and causing a reduction in the oscillation frequency, Figure 7.50c. In the single-phase case, the circuit is damped and ferroresonance does not occur, as the oscillations are not sustainable.

Ferroresonance can occur in the three-phase circuit of Figure 7.50a, since the coupling to the other phases supplies the energy dissipated in the open phase, thus sustaining the oscillations. As the frequency changes, tuning at the sub-harmonic frequencies of the network can occur.

c) Single-phase opening of medium voltage circuits with ungrounded transformers
An analysis is undertaken of a network with feeding circuits containing capacitances and ungrounded transformers, in star or delta connection, as in the case of the underground networks fed by insulated cables. Assuming an unloaded three-phase transformer bank, composed of single-phase units, the following three-phase circuit is obtained (Figure 7.51).

X_c: capacitive reactance of the underground cable.
X_m: nonlinear magnetization branch of the transformer.

Note that, although the Thévenin equivalent requires a linear circuit, it is possible to derive the equivalent in Figures 7.52a and b, much like the equivalents in Figures 7.28 and 7.29a and b, solely to illustrate the series arrangement between the capacitor and the nonlinear inductor.

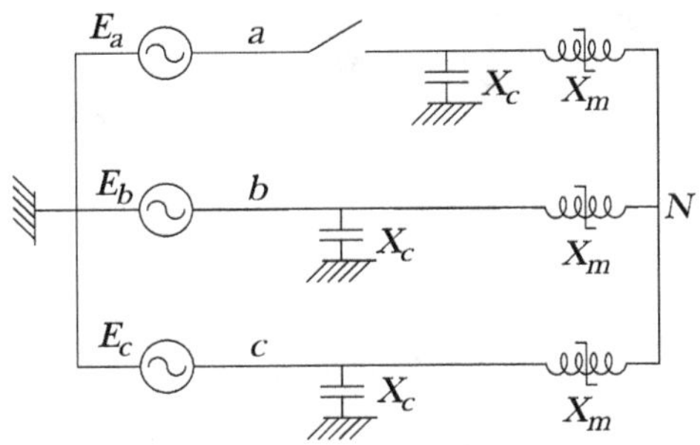

Figure 7.51 Three-phase circuit with one open phase.

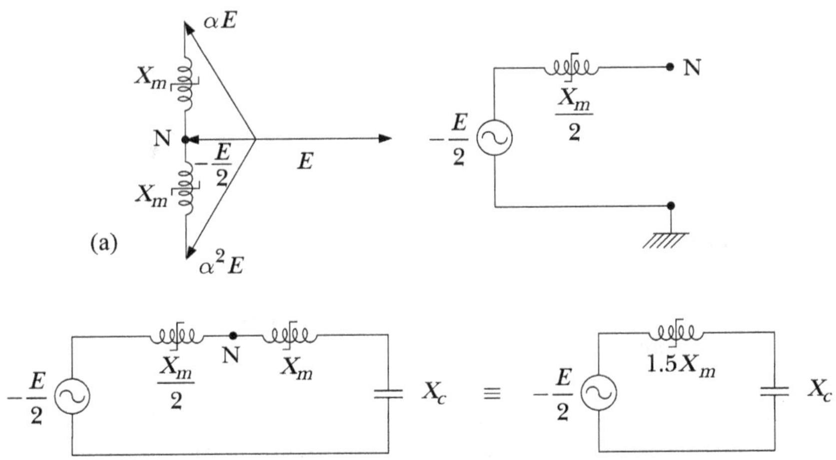

Figure 7.52 Equivalent circuit. (a) Equivalent circuit for phases *b* and *c*. (b) Reduced circuit.

In the equivalent circuit of Figure 7.52b, a capacitance in series with a nonlinear reactance is obtained, with the same behavior of the nonlinear RLC circuit as previously seen.

Reference [32] analyses other types of winding connections for transformers, which could create ferroresonance.

It was observed that the presence of a small load at the secondary coil of the transformer, at 2–5% of the nominal power, for instance, can be sufficient to eliminate the ferroresonant condition; however, it depends on each individual case.

Finally, Reference [33] offers a thorough review of the various approaches to this phenomenon and the conditions under which it can occur in power systems.

- Ferroresonance in Complex Networks and Transformer Models
  In the analysis of the RLC circuit with nonlinear inductors, a capacitive Thévenin equivalent was obtained, as seen from the connection point of the nonlinear inductor.
  There are ferroresonance conditions in which the Thévenin equivalent, as seen with the nonlinear inductor, is much more complex, with a frequency response that shows resonant peaks, thus complicating the analysis.
  Three-phase transformers may have magnetic coupling between the phases, making the models still more complex.
  In modeling saturation, there may be a certain degree of imprecision with respect to the distribution of leakage inductances between the windings, even in the simple case of a single-phase transformer. With respect to more complex arrangements, such as transformers with more than two windings, precise characterization of flux saturation is not trivial [34]. Besides, the $\varphi \times i$ characteristic may not exhibit its usual behavior in cases of high interwinding capacitances [21].

References

1 CIGRE Working Group 33.10. (1990). Temporary overvoltages: causes, effects and evaluation. In *International Conference on High Voltage Engineering* (Vol. 2, pp. 33.210/1–15). Paris: CIGRE.

2 Elovaara, J., Foremann, K., Schei, A., & Volcke, O. (1990). Metal oxide surge arresters in AC systems. Part III: Temporary overvoltages and their stresses on metal oxide surge arresters. *Electra* (129), 115–118.

3 *International Standard IEC 60071-2. Insulation co-ordination. Part 2: Application guidelines.* (2018). Geneva: International Electrotechnical Committee.

4 Silva, F. M. F., Velitsikakis, K., Lennerhag, O., Barakou, F., Cruz, A., Michel, J., Mills, D., Munji, K., Peng, J., Rahimi, I., Skovgaard, C., & Tannemaat, I. (2023). *Evaluation of temporary overvoltages in power systems due to low order harmonic resonances.* Paris: CIGRE.

5 Stevenson, W. D. Jr. (1975). *Elements of power systems analysis* (3rd ed.). New York: McGraw-Hill.

6 Miller, T. J. E. (1988). *Reactive power control in electric systems.* New York: John Wiley.

7 Adkins, B., & Harley, R. G. (1975). *The general theory of alternating current machines: Application to practical problems.* London: Chapman and Hall.

8 Stagg, G. H., & El-Abiad, A. H. (1968). *Computer methods in power system analysis.* New York: McGraw-Hill.

9 Anderson, P. M. (1973). *Analysis of faulted power systems*. Ames: Iowa State University Press.

10 Westinghouse Electric Corporation. (1964). *Electrical transmission and distribution reference book* (4th ed.). East Pittsburgh: Westinghouse Electric Corporation.

11 Clark, E. (1943). *Circuit analysis of A-C power systems: Vol. 1- Symmetrical and related components*. New York: John Wiley.

12 Kimbark, E. W. (1975). Charts of three quantities associated with single-pole switching. *IEEE Transactions on Power Apparatus and Systems*, 94(2), 388–395.

13 Shperling, B. R., Fakheri, A. J., Shih, C. H., & Ware, B. J. (1981). Analysis of single phase switching field tests on the AEP 765 KV system. *IEEE Transactions on Power Apparatus and Systems*, PAS-100(4), 1729–1735.

14 Single phase tripping and auto reclosing of transmission lines-IEEE Committee Report. (1992). *IEEE Transactions on Power Delivery*, 7(1), 182–192.

15 Electric Power Research Institute. (1982). *Transmission line reference book: 345 kV and above* (2nd ed.). Palo Alto: Electric Power Research Institute.

16 Priest, K. W., Ramirez, A., Nowak, H., & LaForest, J. J. (1972). Resonant voltages reactor compensated extra-high-voltage lines. *IEEE Transactions on Power Apparatus and Systems*, PAS-91(6), 2528–2536.

17 Kimbark, E. W. (1971). *Direct current transmission*. New York: Wiley Interscience.

18 Arrillaga, J., Bradley, D. A., & Bodger, P. S. (1985). *Power systems harmonics*. Chichester: John Wiley.

19 Dommel, H. W., Yan, A., & Wei, S. (1986). Harmonics from transformer saturation. *IEEE Transactions on Power Delivery*, 1(2), 209–215.

20 Institute of Electrical and Electronics Engineers. (1993). *IEEE Recommended Practices and Requirements for Harmonic Control in Electrical Power Systems (IEEE Std 519-1992)*. New York: Institute of Electrical and Electronics Engineers.

21 Iravani, M. R., Chaudhary, A. K. S., Giesbrecht, W. J., Hassan, I. E., Keri, A. J. F., Lee, K. C., Martinez, J. A., Morched, A. S., Mork, B. A., Parniani, M., Sharshar, A., Shirmohammadi, D., Walling, R. A., & Woodford, D. A. (2000). Modeling and analysis guidelines for slow transients. III. The study of ferroresonance. *IEEE Transactions on Power Delivery*, 15(1), 255–265.

22 Rudenberg, R. (1950). *Transient performance of electrical power systems*. New York: McGraw-Hill.

23 Kilgour, A. E., Harlow, J. H., & Phadke, A. G. (1968). Analysis and understanding ferroressonance on distribution systems. *Transmission & Distribution*, 20(6), 82–87.

24 Ferracci, P. (1986). *Ferroressonance*. Chichester: Cahier Technique 190.

25 Thompson, J. M. T., & Stewart, H. B. (1986). *Nonlinear dynamics and chaos: Geometrical methods for engineers and scientists*. Chichester: John Wiley.

26 Germond, A. J. (1975). Computation of ferroresonant overvoltages in actual power systems using Galerkin's method. In *IEEE Power Industry Computer Applications Conference, New Orleans* (pp. 127–135). New York: IEEE.

27 Guckenheimer, J., & Holmes, P. (1983). *Nonlinear oscillations, dynamical systems and bifurcation of vector fields.* New York: Springer.

28 Chua, L., Hasler, M., Neirynck, J., & Verburgh, P. (1982). Dynamics of a piecewise-linear resonant circuit. *IEEE Transactions on Circuits and Systems*, 29(8), 535–547.

29 Germay, M., Mastero, S., & Broman, J. (1974). Review of ferro-resonance phenomena in high voltage power systems and presentation of a voltage transformer model for predetermining them. In *International Conference on High Voltage Engineering, Paris.* (pp. 33–18). Paris: CIGRE.

30 Dolan, E. J., Gillies, D. A., & Kimbark, E. W. (1972). Ferroresonance in a transformer switched with an EHV line. *IEEE Transactions on Power Apparatus and Systems*, PAS-91(3), 1273–1280.

31 Sagardía, S. R., & Morched, A. (1997). Potential transformer failure due to ferroressonance. In *International Conference on Power Systems Transients, Seattle.* (pp. 191–196). IPST.

32 Hopkinson, R. H. (1968). Ferroresonant overvoltage control based on TNA tests on three-phase wye-delta transformer banks. *IEEE Transactions on Power Apparatus and Systems*, PAS-87(2), 352–361.

33 CIGRE Working Group C4-307 (2014). Resonance and ferroresonance in power networks. TB 569. CIGRE.

34 Dommel, H. W. (1986). *Electromagnetic transients program reference manual: (EMTP) theory book.* Portland: Bonneville Power Administration.

8

Switching Surges

8.1 Introduction

Switching surges are taken into consideration in the insulation coordination of extra-high voltage (EHV) and ultra-high voltage (UHV) transmission lines and substation equipment, and this chapter presents an overview of the most relevant overvoltages occurring in power systems.

This subject is very broad, and each aspect addressed here could be expanded to include a great deal of additional information. However, only the elementary topics that characterize these overvoltages will be pursued here.

Switching surges stem from topological changes in power systems due to circuit-breaker operation, short circuits, etc. During the accommodation period between two different topological configurations, there is a transient from one equilibrium point to the next whose duration depends on the system damping, but during the first few cycles of the transient period, high voltage levels may be encountered, which could damage the equipment's insulation.

Switching surges are classified as overvoltages with slow wavefronts. They are also known to be of internal origin since they are created by topological changes in the network.

Among the overvoltages of internal origin there is a special group that presents very high frequencies and wavefronts and is therefore designated as very fast transient overvoltages (VFTO). As the frequencies are very high, the models are close to those used in lightning surge studies. The need to maintain a balanced coverage of different topics leads to treating these overvoltages as a section within switching surges, given the space limitations.

Understanding Electromagnetic Transients in Power Systems, First Edition. Luiz Cera Zanetta Jr.

Published 2025 by John Wiley & Sons, Inc.

8.2 General Considerations

It is convenient to work with per unit values of a base voltage, that is, 1 pu means 100% of the adopted base. For three-phase lines with an rms line-to-line voltage, V, the base value assumed is

$$V_b = \frac{\sqrt{2}V}{\sqrt{3}},$$

which is the amplitude of the phase to ground voltage.

Switching surges are created by topological changes to the network, derived from the operation of circuit breakers or system faults. The switching surge waveform has a relatively slow wavefront period, compared to a lightning surge for instance (Figure 8.1), reaching values ranging from tens to hundreds of microseconds and containing frequencies from hundreds of Hz up to a few dozen kHz, except for the case of disconnector operation in SF_6 substations that is characterized by very fast wavefronts, according to Section 8.9 and belongs to another class of overvoltages.

Power system overvoltages are grouped into the following categories:

- Temporary overvoltages.
- Overvoltages with slow wavefronts.
- Overvoltages with fast wavefronts.

The behavior of an electrical network excited by sinusoidal sources can be represented by nonhomogeneous differential equations of order n. The solution is found based on the composition of the particular solution, which is the steady-state response, superimposed on the free transient response, obtained by solving the homogeneous equations.

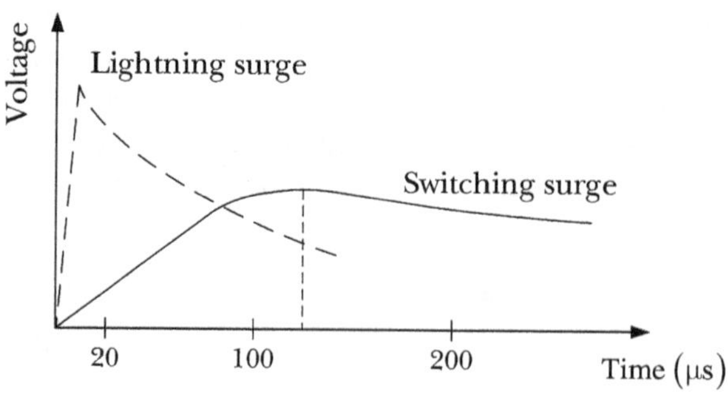

Figure 8.1 Switching surge and lightning surge.

An electric network, depending on its complexity, can generate higher-order differential equations for the solution of a transient period. However, independently of the order, a steady state and transient response will always be obtained. In Chapter 1, the solution to some differential equations associated with simple electric networks was presented using the Laplace transform. The solution of complex networks requires a broader analysis, and an approach using computational methods is necessary.

The presence of losses in the power system introduces a damping effect on the transient parcel of the response, allowing it to last for only a few cycles. The effect of network losses is often addressed from a negative point of view. On the other hand, the presence of losses is partly responsible for the damping of the transient response in several phenomena that may occur, allowing the existence of a steady state in a new operating situation. It is expected that a system without losses is not feasible to operate.

Since the switching surge is the result of the composition of two parcels, the transient response, and the steady-state response of the system, in some cases, it is not possible to significantly reduce the overvoltage when the steady-state parcel already has a high magnitude. That is, the procedures that attempt to control just the transient parcel are ineffective, since the problem is related to the steady state, as for example, in cases close to resonant conditions.

The representation of a network involves elements of two basic kinds. Elements such as capacitors, inductors, and resistors are treated as lumped parameters whose equations are based on electric circuit theory. Transmission lines, whenever possible, should be represented by distributed parameters, although in certain low-frequency simulations, lumped parameters can be used, as for example, in the analysis of trapped charges in transmission lines.

In the same way, equipment typically represented with lumped parameters may require treatment with distributed parameters, when the phenomenon contains high frequencies, as for example, in the case of electromagnetic wave propagation in the transformer windings.

When the frequencies are not too high, a simplified representation of the transmission line can be admitted, as in the early stage of research with TNAs, subdividing it into a finite number of π cells.

Currently, these overvoltages are analyzed using electromagnetic transient programs, such as EMT-type programs, and the basic modeling was presented in Chapter 4. In the calculation of these overvoltages, it becomes necessary to consider how representative and detailed the models are with respect to the frequencies present during the switching operation. There is an extensive literature available on the subject, and cited references as [1–8], representing pertinent contributions, serve the purpose of illustrating how knowledge in this area has evolved, without comprehensively covering all relevant contributions.

The most common switching surges in power systems are presented in the following, covering transmission lines and substation equipment.

8.3 Line Energization and Line Autoreclosure

Line energization and line autoreclosure have a close correspondence when it comes to the switching surge phenomenon. Some basic characteristics will be discussed here, and many complementary references as [5–8] give more details on this subject.

8.3.1 Energization

Energization of a transmission line causes a transient surge that occurs relatively frequently in a power system. This operation is carried out by closing the contacts of the circuit breaker at one line terminal.

A typical switching surge during energization of a transmission line can be seen in Figure 8.2.

In this oscillogram, as can be seen, the voltage is initially zero, since the line was deenergized, and reaches a maximum value during the transient period. If the measurement is at the receiving terminal, there is still a delay in which the voltage is zero, corresponding to the time delay for the traveling waves arriving at the open terminal.

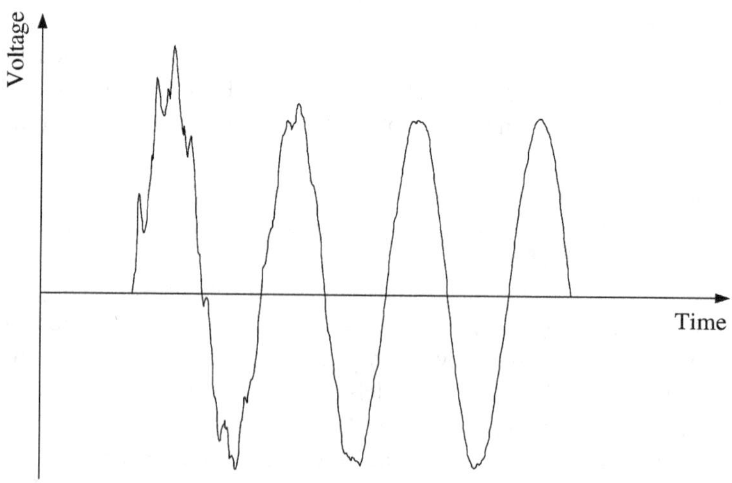

Figure 8.2 Typical energization switching surge.

After the initial transient, with a duration of a few cycles, the system may exhibit a period of temporary overvoltage before finally reaching a steady state.

A single-phase elementary model is initially analyzed using an ideal voltage generator (null internal impedance), which represents the excitation source, and the transmission line is represented by a very simple model consisting of a surge impedance and a time delay (travel time). A constant voltage generator is assumed to simplify the analysis even more, Figure 8.3a.

The traveling waves are analyzed using the lattice diagram in Figure 8.3b.

Chapter 2, dedicated to the analysis of traveling waves in single-phase lines, gives more details on the development of the lattice diagram.

The traveling wave takes a time delay τ to reach the open line terminal, and the oscillatory phenomenon has a period of 4τ.

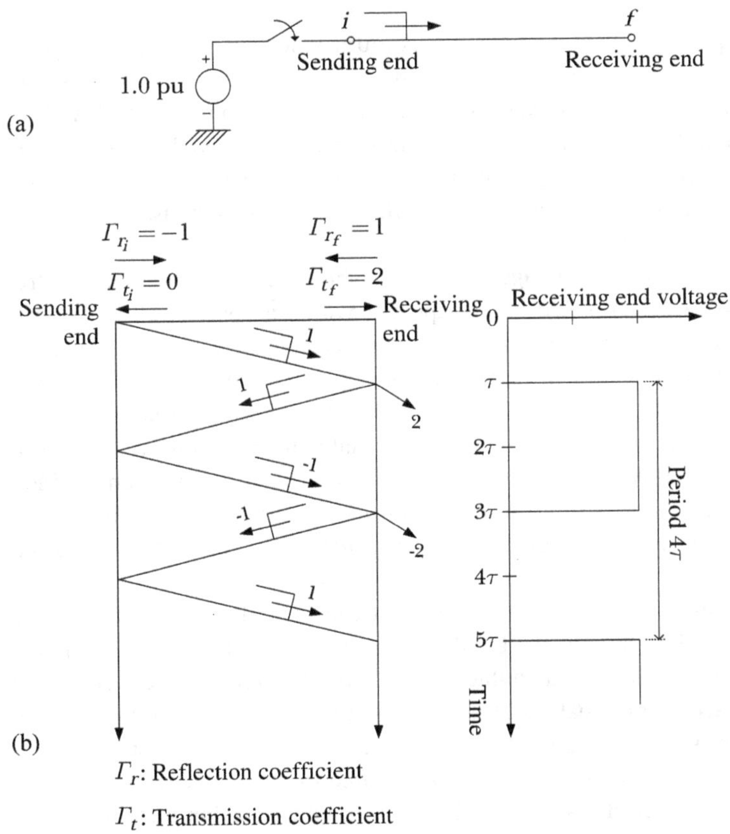

Figure 8.3 Transmission line energization. (a) Lossless line. (b) Lattice diagram.

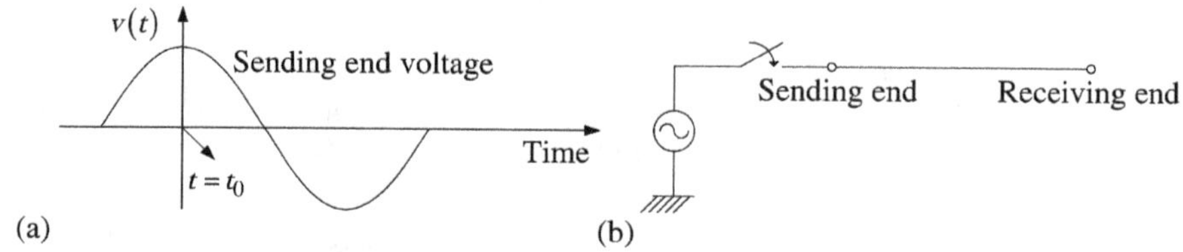

Figure 8.4 Energization with a sinusoidal generator. (a) Time of maximum generator voltage. (b) Lossless single-phase line.

In a lossless 300 km transmission line with waves traveling at the speed of light, a period of 4 ms is expected for the oscillatory phenomenon, corresponding to an oscillation frequency of 250 Hz in the transmission line.

In Figure 8.4, an ideal sine wave generator, corresponding to an infinite bus, energizes the transmission line. The voltage can be applied at different points on the sine wave.

When closing the switch at the time of maximum generator voltage, $t = t_0$, a critical case of overvoltage occurs because the traveling wave applied to the line is high in amplitude and these waves overlap with the sinusoidal voltage at 60 Hz. With representation of the losses, the transient component is damped, leaving only the sinusoidal response in the steady state. It is also possible to observe the importance of the closing instant of the circuit breaker with respect to the phase angle of the sine wave.

In the case of a sinusoidal voltage applied to a short transmission line, high frequencies due to the brief time delay are expected, with most of the transient occurring within a short time interval, without any appreciable alteration in the voltage level applied by the generator. This is even more evident if the sinusoidal voltage is applied to the line at its peak value, since the wave is almost flat around the peak. In this condition, the generator may be approximated by a constant voltage without significantly modifying the results, mainly for short and medium-length lines.

Actually, the energization of a transmission line is more complex and should be analyzed using three-phase models for generators, switches, and transmission lines.

The three poles of the circuit breaker do not close simultaneously, as there is mechanical variability among the breaker contacts; in addition, the electric arc established prior to mechanical contact depends on the voltage on the generator side, and ordinary circuit breakers cannot choose the appropriate points in time to close; thus, some scatter is expected for the breaker contacts. Only circuit breakers with controlled switching, see Section 8.9, are able to make electrical contact at favorable times. The electrical contact between the energization system and the transmission line can be understood as a random phenomenon for ordinary circuit breakers, according to the normal curves presented next.

Given:

t_m: average prestriking time.
σ: standard deviation.

Times t_a, t_b, and t_c represent possible random samples reflecting the electrical connection of the three poles of the circuit breaker with respect to the three-phase energization of the transmission line.

For these normal curves, an average closing time for the electrical contacts of the circuit breaker and its standard deviation, σ, are associated with each pole. The average is assumed to be the same for the three poles, and the maximum make-time scatter remains below a given limit, such as 5 ms. For example, in Figure 8.5, the difference $t_c - t_a$ shold be lower than 5 ms.

Another random variable is the average t_m with respect to time, since the precise instant the breaker closes with respect to the sinusoidal waveform depends on several factors that are not controlled by the switch (Figure 8.6).

The three-phase line shows inductive and capacitive coupling between phases and waves at different speeds in the conductors.

The details of calculation of polyphase traveling waves are discussed in Chapter 3, and the induced voltages and currents from one phase to another

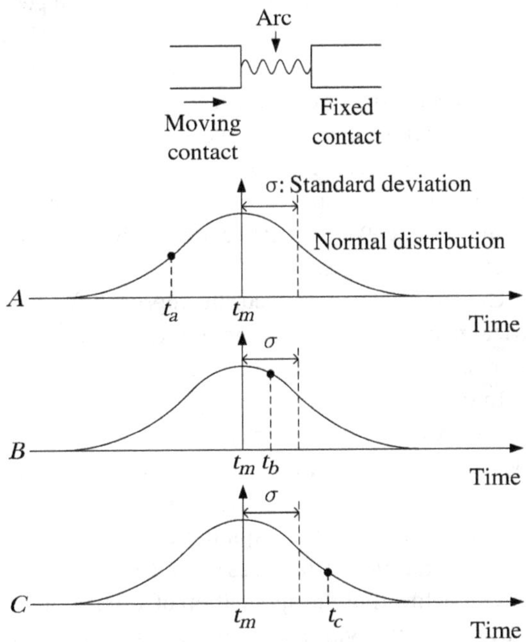

Figure 8.5 Random contact closure.

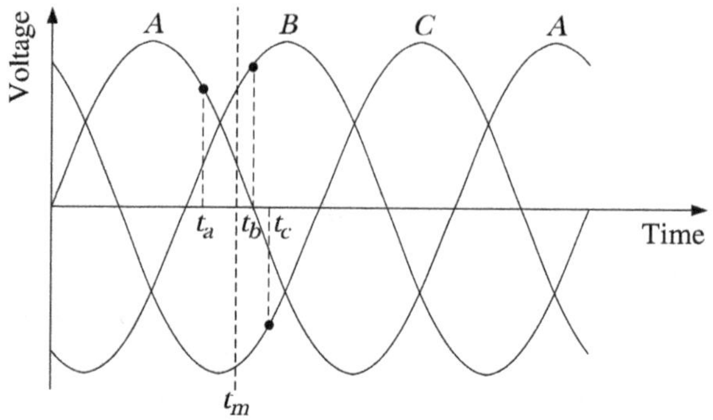

Figure 8.6 Random contact closure in the three-phase circuit.

should be considered in the transmission line model. Statistical calculations are performed, considering the random nature of breaker-making times, with the aid of digital programs such as EMTP (see Chapter 4). By taking random samples of switch closing times, the Monte Carlo simulation method becomes an efficient tool for the statistical analysis of energization overvoltages, provided that an adequate number of experiments are performed.

Having determined the overvoltages, they are statistically analyzed to obtain the distribution of maximum values during the transient period.

Several factors can influence the overvoltages during line energization, such as the short-circuit power of the source and its Thévenin equivalent. The line can be energized either by a generator connected to a step-up transformer or by a network with several transmission lines.

Representation of the network by its Thévenin equivalent at 60 Hz can be inappropriate [1, 7], and the accurate fitting of the equivalent should consider the frequency response of the network. Usually, representation by the Thévenin equivalent at 60 Hz leads to conservative results, which may be regarded as acceptable in many cases.

Currently, the most accurate approach for calculating the equivalent of an electrical network is through the frequency-dependent network equivalents (FDNEs), as discussed in Chapter 5.

In general, even at high frequencies, steps or abrupt discontinuities in the overvoltage waveforms are not common, even if a step voltage is applied by the excitation source. This may be explained by certain factors such as different modal propagation speeds (see Chapter 3), losses, dependence of parameters on frequency, etc. Part of this explanation is due to the inductive nature of the excitation sources and transformers, as will be examined next. It can be seen in the simplified circuit below that the application of a step voltage to a transmission line results in a traveling wave with slower wavefront (Figure 8.7).

In this interpretation, the lossless transmission line is represented by its surge impedance, as in Chapter 2.

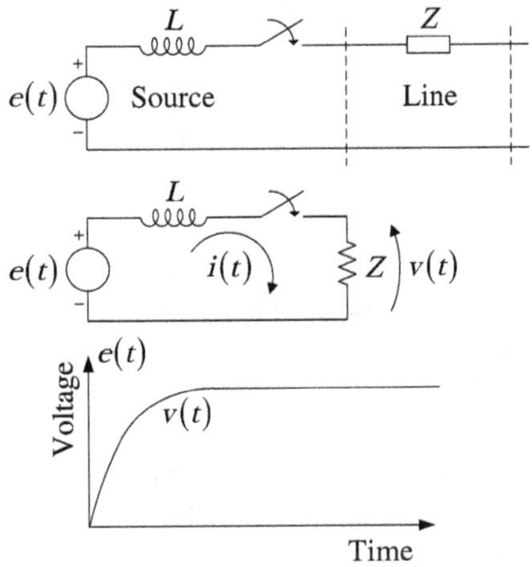

Figure 8.7 Effect of the internal inductance of the generator.

$$e(t) = L\frac{di(t)}{dt} + Zi(t), \tag{8.1}$$

$$v(t) = Zi(t). \tag{8.2}$$

The solution for the circuit shows that the step voltage applied by the source travels along the lossless transmission line with a 1 – exp function rather than as a step function.

When a transmission line energizes a second line in cascade, a reduction in the applied voltage occurs (considering the Thevenin equivalent with nominal voltage, in Figure 2.16, Chapter 2, and similar surge impedances on sides 1 and 2, a reduction in 50% is obtained).

Besides the closing time of the circuit breaker, other factors play important roles in the energization overvoltage amplitudes, such as the length of the transmission line, the reactive compensation level, preinsertion resistors, and surge arresters.

Most EHV transmission lines have surge arresters. In transmission lines without arresters, the most critical overvoltages occur at the end of the line. With arresters at the terminals, since the arresters clamp the overvoltage to its maximum protection level, the overvoltages occur at certain distances from the line terminal.

Controlled switching of circuit breakers at favorable times for the make current (to avoid overvoltages) is already a reality, and the prevalence of its use depends only on the cost of the control and measurement system. Controlled switching is the subject of Section 8.9.

8.3.2 Autoreclosure

- Tripolar Autoreclosure

  Tripolar autoreclosure is executed in the course of a given sequence of events following a fault. The circuit breakers open on both sides of the faulted section and wait for a period of time, called the dead time, to carry out an attempt to reclose the line. The purpose of this period, on the order of 500–700 ms, is to increase the probability of extinction of an electric arc for non-permanent faults. This maneuver is also referred to as fast reclosing.

When autoreclosure is performed, after the delay related to the dead time, there is a residual voltage due to the charge stored in the line capacitance, which has not yet been fully damped.

The basic difference between autoreclosure and line energization is the presence of this trapped charge as the initial condition in the transmission line.

In order to implement autoreclosure of the transmission lines, the systems at both transmission line terminals have to remain synchronized after the opening of the line, which requires a network stability analysis. In the case of a double-circuit interconnection between two systems, the link must remain stable during the tripolar opening of any of the circuits.

Autoreclosure can also be understood with the aid of a single-phase model of a line with a surge impedance Z and time delay τ. Much like the energization, a voltage step of 1.0 pu is applied to the sending-end, but now it is assumed that the transmission line has a constant residual voltage of −1.0 pu, Figure 8.8a.

In this case, a voltage step of 2.0 pu is applied to the transmission line if the reference is shifted from 0 to −1.0 pu. Note that, in reality, the line undergoes a voltage variation Of 2.0 pu, as in Figure 8.8b.

The voltage step travels to the receiving-end of the transmission line and doubles in value, reaching 4.0 pu relative to the new reference set to −1.0 pu, but the overvoltage reaches 3.0 pu with respect to the original reference, at zero voltage, Figure 8.8c.

In the same example, if a −1.0 pu voltage step were applied, there would be no voltage change at the sending-end and consequently no overvoltage along the transmission line. In the case of a sinusoidal voltage source, in a simplistic way, the task of the synchronization device would be to adjust the contact closure time when the voltage source reaches −1.0 pu in order to eliminate the overvoltage. More details about controlled switching are discussed in Section 8.9.

With this elementary model, adverse conditions can be observed upon applying an excitation voltage relative to the residual voltage on the line. This happens when the polarity of the applied voltage wave is opposite to that of the trapped charge.

The autoreclosure worst cases should exceed those of energization, because, as seen in the simple single-phase example, an increase from 2.0 pu for line energization to 3.0 pu for line autoreclosure is found.

When considering the statistical aspects of autoreclosure, there are cases in which the applied voltage is close to the residual voltage, and negligible

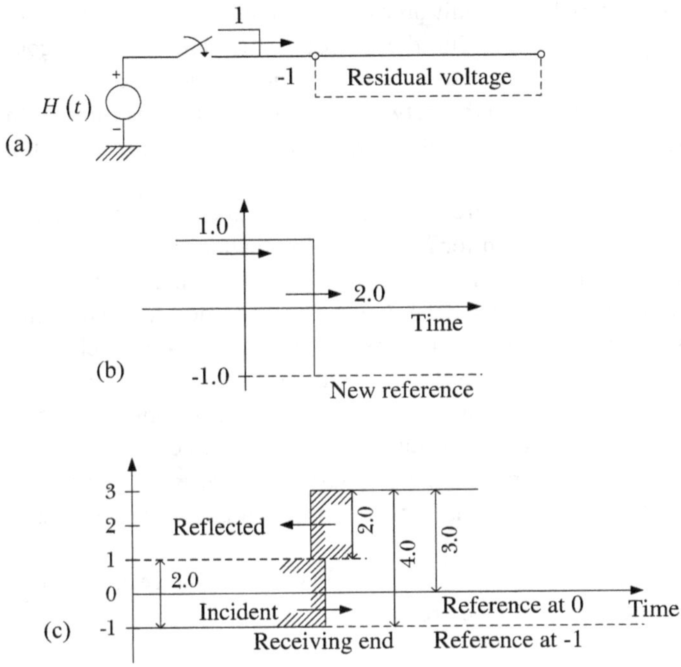

Figure 8.8 Auto reclosure of the transmission line. (a) Constant residual voltage. (b) Applied voltage variation of 2.0 pu. (c) Voltage at the receiving end.

overvoltage levels would thus be expected. Hence, the average levels for line energization and line autoreclosure observed in statistical distributions are quite similar.

- Single-Phase Autoreclosure
  As most of the faults are single-phase, single-phase fault clearing can be carried out only in the phase subject to the fault. In general terms, single-phase autoreclosure is a switching operation identical to tripolar autoreclosure, except that it is a single-phase application.

  Single-phase autoreclosure requires more sophisticated protection equipment, in order to identify the faulted phase and to perform single-phase opening. In this maneuver, synchronism problems are reduced because the healthy phases remain connected, which is helpful to the synchronization task. Thus, single-phase autoreclosure can be applied to weakly connected systems or even to single-circuit links.

  The electric arc goes through two stages during a single-phase opening. The first, known as the primary arc, occurs at the time of the fault with circuit breakers B_1 and B_2 closed. The second, called the secondary arc, occurs after the monopolar opening of these circuit-breakers, with inductive and capacitive coupling to the healthy phases feeding the arc. Capacitive and inductive coupling also occur in tripolar autoreclosure; however, the voltages on the healthy phases are no longer connected to the network and are lower in magnitude. In addition, the inductive effect is negligible since there is no current through the healthy phases.

Since the effect of the secondary arc is crucial in single-phase autoreclosure, it is necessary to check its extinction for a successful maneuver.

The success of single-phase autoreclosure in transmission lines is subject to extinction of the electric arc within a dead time of 0.5–1 seconds, following fault clearing. The extinction of the electric arc depends on several factors, such as arc length, dielectric renewal, time to zero current crossing, recovery voltage, etc.

Field and laboratory tests can establish reference values for secondary arc current and recovery voltage under both steady-state and transient conditions, indicating the successful extinction of the secondary arc. However, due to the complexity of the subject, the criteria for arc extinction should be addressed in specific literature.

Single-phase autoreclosure is performed with the residual voltage on the open phase, after extinction of the electric arc.

8.3.3 Residual Voltage for Tripolar Opening

In studies of tripolar autoreclosure, it is necessary to understand the behavior of the trapped charge on the transmission line.

- Line Without Reactive Compensation.
  The analysis of the single-phase circuit below clarifies the behavior of the trapped charge. In this simple circuit, since the frequencies involved are not high, the line is represented by its $\pi-$ circuit (Figure 8.9a).

  Assuming that the circuit breaker at the receiving end has already been opened, the behavior of the unloaded transmission line is equivalent to that of a capacitor because the frequencies are not high; thus:

$$|j\omega L| << \left| -\frac{j2}{\omega C} \right|. \tag{8.3}$$

The opening of the circuit breaker on the sending-end side occurs with the no-load line in a situation like the opening of a shunt capacitor. At the opening of a capacitor, the voltage and the current are in quadrature, and at the zero crossing of the current, the voltage reaches the maximum value, according to Figure 8.9b.

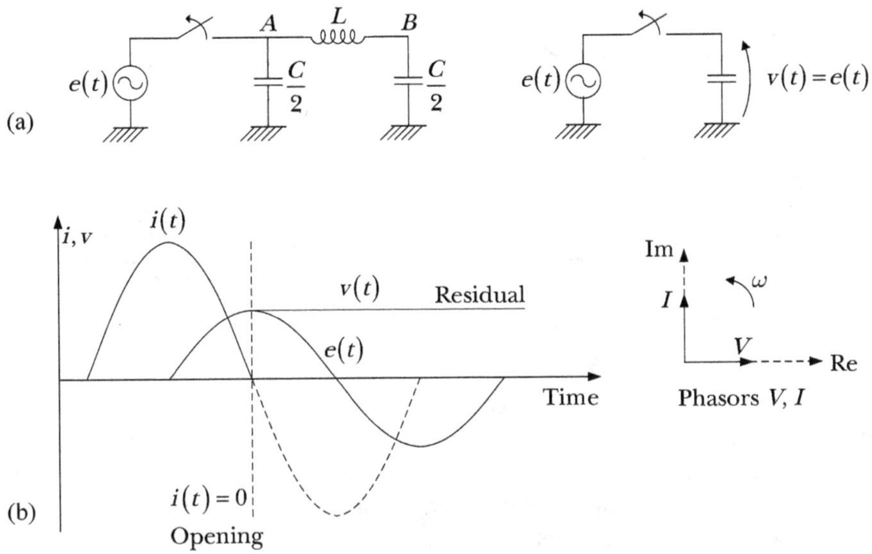

Figure 8.9 Transmission line without reactive compensation. (a) Transmission line model. (b) Residual voltage.

A DC residual voltage equal to the maximum value of the line operating voltage is obtained. A transmission line without reactive compensation has a practically constant residual voltage, despite the small oscillations between the series inductance and capacitance to ground and some initial traveling waves.

Since the switching is three-phase, the zero crossing of the current in the three phases is not simultaneous, as illustrated in Figure 8.10.

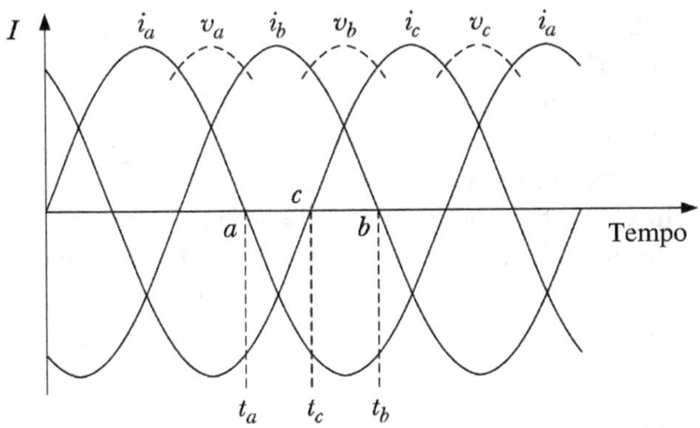

Figure 8.10 Three-phase opening.

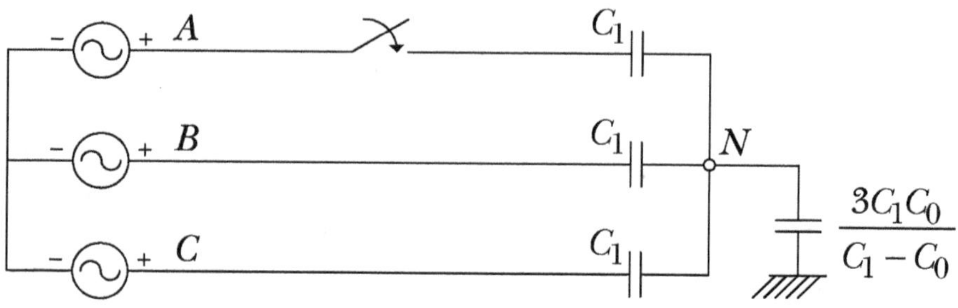

Figure 8.11 Capacitive coupling effect.

In Figure 8.10, a possible sequence for the opening of the phases would be: *a, c,* and *b,* occurring very close in time to the current zero crossing in a balanced system. After the opening of the first phase, the three-phase system becomes unbalanced, which has little impact on these time points, but in this case, it is assumed that the sequence t_a, t_c, t_b is not substantially affected.

In the case of unbalanced currents, all possible sequences should be taken into consideration, especially if the faulted phase is the first to open.

In any event, during the opening process, even for a healthy and balanced transmission line, after the opening of the first phase, a neutral shift occurs, altering the voltage at the other phases still connected to the system. A three-phase model for the line capacitances, including the positive sequence and the zero sequence is shown in Figure 8.11, as in TNA models. To check the validity of the model, simply apply the positive and zero sequences, as in Chapter 7, Section 7.4.2.

C_1: positive sequence capacitance.
C_0: zero sequence capacitance.

When phase *a* is open, a neutral shift is observed, which consequently modifies the voltages at phases *b* and *c*. Thus, different residual voltages are obtained in phases *a, b,* and *c,* even in cases without a fault.

In the case of single-phase fault clearing, after opening the first circuit-breaker in one line terminal, the second circuit-breaker to be opened in the opposite terminal operates under conditions in which the line is unloaded and has a single-phase fault. The imbalance is significant and increases the voltages at the healthy phases even more. It is also important to verify the opening sequence of the healthy phases, with respect to the faulted phase.

- Line with Reactive Compensation.

  The π circuit of the transmission line is again assumed, this time including the reactor used for the purpose of reactive compensation, Figure 8.12a.

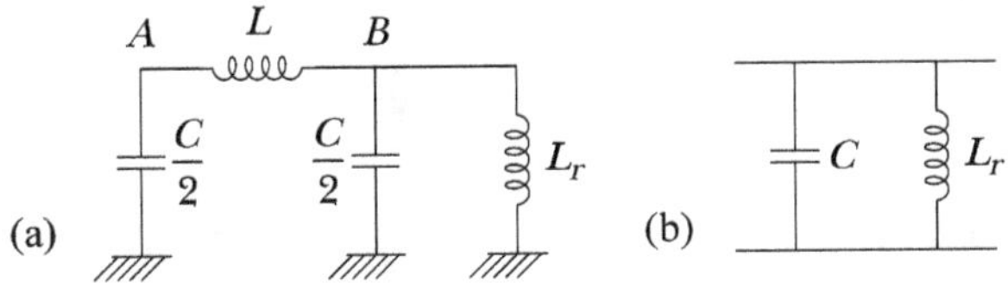

Figure 8.12 Transmission line with reactor. (a) Simplified model. (b) Reduced model.

The presence of the reactor at one or both ends of the transmission line does not substantially change the analysis. Considering, as in the previous case, that the frequencies involved are not high, the L series inductance of the line may be neglected, and the equivalent oscillatory circuit in Figure 8.12b may be considered.

The natural frequency of oscillation of the circuit is:

$$f = \frac{1}{2\pi\sqrt{L_r C}}. \tag{8.4}$$

The phases of the three-phase transmission line are coupled, showing positive and zero sequence parameters, and the reactor, depending on the design, with or without neutral reactors, or even depending on the type of core, can also be represented as positive sequence and zero sequence reactances.

Hence, the existence of two natural oscillation frequencies can be expected, one with positive sequence parameters and the other with zero sequence parameters. These natural frequencies can be calculated:

$$f_1 = \frac{1}{2\pi\sqrt{L_{r1} C_1}}, \tag{8.5}$$

$$f_0 = \frac{1}{2\pi\sqrt{L_{r0} C_0}}. \tag{8.6}$$

The reactive three-phase power of the transmission line and the reactive three-phase power of the reactor are calculated, taking V as the nominal line-to-line voltage:

$$Q_r = \frac{V^2}{\omega L_r}, \tag{8.7}$$

$$Q_c = V^2 \omega C_1. \tag{8.8}$$

The shunt compensation level of the transmission line for the positive sequence is given by:

$$k_1 = \frac{Q_r}{Q_c},$$

and the shunt compensation level for the zero sequence is given by:

$$k_0 = \frac{C_1}{C_0} k_1.$$

The following expressions for the positive and zero sequences can be written:

$$L_r = \frac{V^2}{\omega Q_r}, \tag{8.9}$$

$$C = \frac{Q_c}{\omega V^2}. \tag{8.10}$$

By making proper substitutions into (8.5) and (8.6), the following expressions for the positive and zero sequence frequencies are obtained:

$$f_1 = \frac{\omega}{2\pi} \sqrt{k_1} \tag{8.11}$$

assuming a system nominal frequency of 60 Hz:

$$f_1 = 60\sqrt{k_1}, \tag{8.12}$$

and similarly, for the zero sequence:

$$f_0 = 60\sqrt{k_0}. \tag{8.13}$$

The residual voltage shows oscillations composed of the frequencies in the positive and zero sequences. The waveform of an oscillatory residual voltage is illustrated in Figure 8.13.

Frequencies f_d and f_s correspond approximately to half the difference and half the sum of f_0 and f_1.

$$f_d = \frac{f_0 - f_1}{2}, \tag{8.14}$$

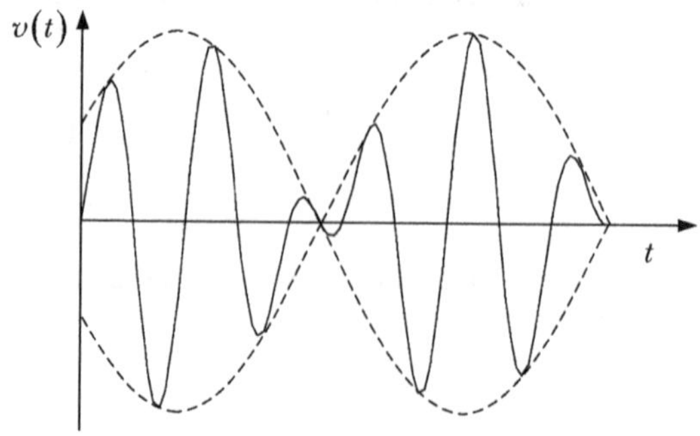

Figure 8.13 Oscillatory residual voltage.

Figure 8.14 Transmission line detailed model.

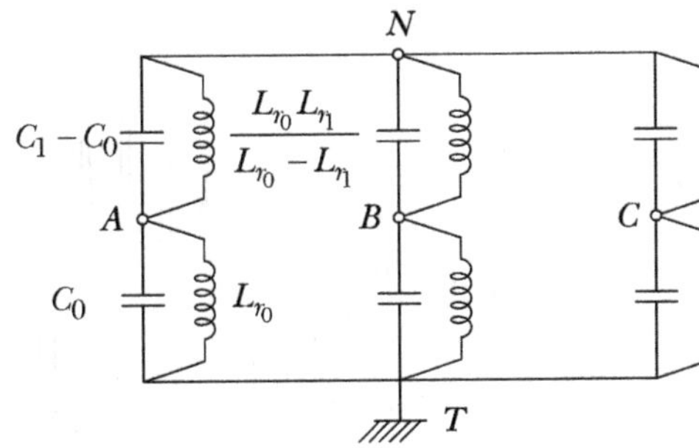

$$f_s = \frac{f_0 + f_1}{2}. \tag{8.15}$$

To an approximation, the composition of these two frequencies can be understood as the transformation of a sum of cosine terms, with equal amplitudes, as follows:

$$v(t) = \cos(\omega_0 t + \alpha) + \cos(\omega_1 t + \beta), \tag{8.16}$$

$$v(t) = \cos\left[\left(\frac{\omega_0 + \omega_1}{2}\right)t + \frac{\alpha + \beta}{2}\right] \cos\left[\left(\frac{\omega_0 - \omega_1}{2}\right)t + \frac{\alpha - \beta}{2}\right]. \tag{8.17}$$

However, despite good results in practical cases, this interpretation is only approximate because it assumes the same amplitude for the two sequences. For a more comprehensive understanding, one should analyze a more detailed circuit, containing the positive and zero sequences according to the model in Figure 8.14, as shown in Chapter 7, Section 7.4.2, Figure 7.26.

Example 8.1 Calculate the oscillation frequencies of the residual voltages in a 500 kV transmission line, with compensation for 50% of its reactive capacitive power.

Data

Line: length = 300 km, $C_1 = 13$ nF/km and $C_0 = 8$ nF/km.
Reactive compensation degree: $k_1 = 0.5$.

Solution

The approximate solution can be obtained from:

$$k_0 = \frac{13}{8} k_1,$$

$$f_1 = \sqrt{k_1} 60 = 42.7 \text{ Hz}.$$

Similarly, $f_0 = 54$ Hz.

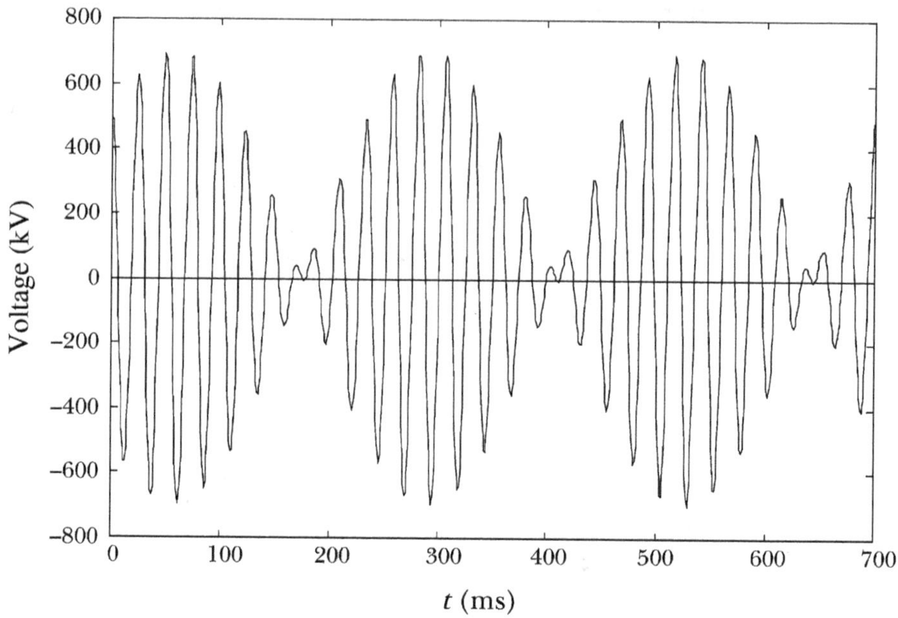

Figure 8.15 Residual voltage in phase *a* of a compensated transmission line. Single phase fault in phase c.

Thus, the following is obtained (Figure 8.15):

$$f_d = 5.8 \text{ Hz}, \qquad f_s = 48.2 \text{ Hz}.$$

8.3.4 Preinsertion Resistor

A preinsertion resistor is added to a circuit breaker in order to reduce the transients associated with energization and autoreclosure of transmission lines.

During displacement of the movable contact, a resistor is inserted transiently for an approximate period of 6–10 ms. This resistor is then short-circuited by connecting the movable contact to the fixed contact on the circuit breaker, Figure 8.16a.

The circuit breaker with a preinsertion resistor can be modeled as the sequential closing of switches, Figure 8.16b.

S_1 closes, simulating insertion of the resistor.
S_2 closes later, simulating by-pass of the resistor.

When the preinsertion resistor is inserted, the switching surge applied to the transmission line is reduced. This can be partly explained using an elementary model that assumes a line represented by its surge impedance, Figure 8.17a.

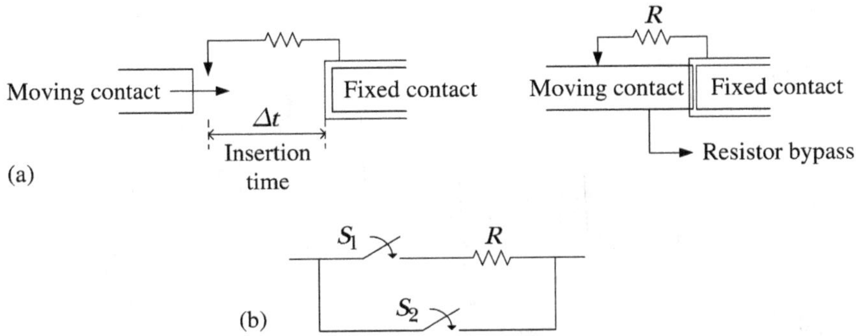

Figure 8.16 Preinsertion resistor. (a) Mechanical sketch. (b) Electrical model.

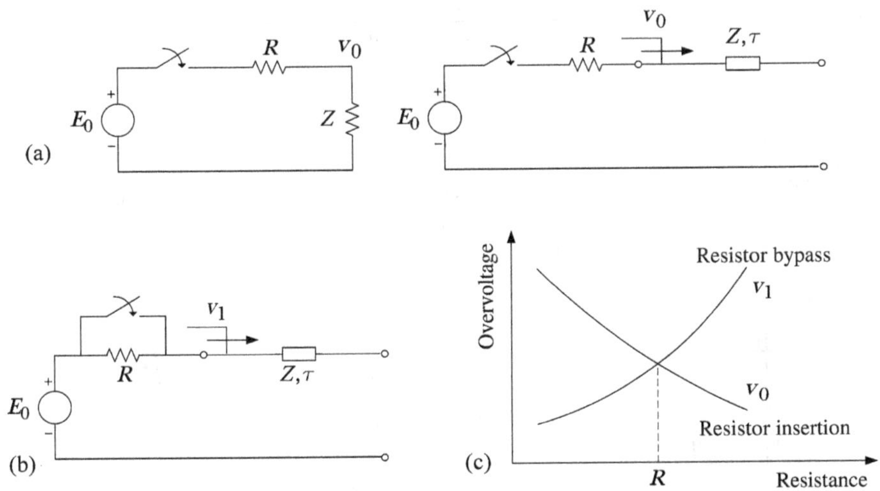

Figure 8.17 Insertion and by-pass of the resistor. (a) Insertion and applied voltage. (b) By-pass and applied coltage. (c) Overvoltages for insertion and by-pass of the resistor.

Voltage v_0 applied to the transmission line results from the voltage divider circuit, as follows:

$$v_0 = \frac{Z}{R+Z}E_0. \tag{8.18}$$

At first glance, one might think that by increasing the value of the resistor the overvoltages would be reduced even more. However, when the resistor is bypassed, a new switching surge, v_1, is introduced, which increases in proportion to the value of the resistor, Figure 8.17b.

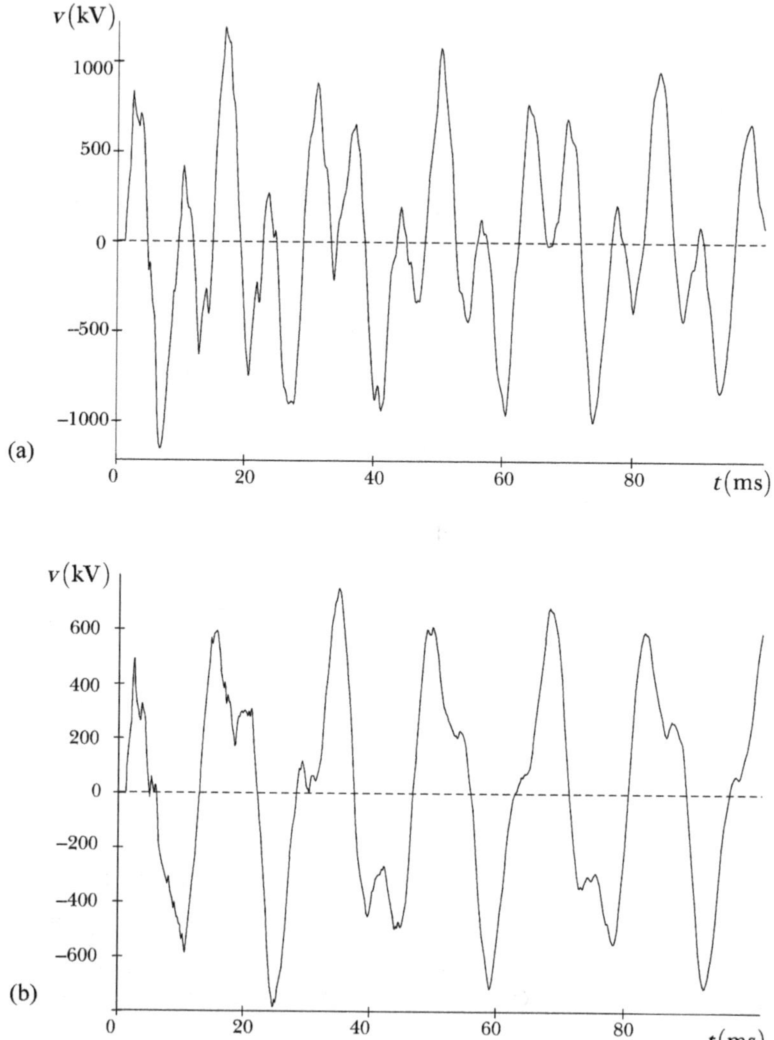

Figure 8.18 Energization overvoltage. (a) Without pre-insertion resistor. (b) With pre-insertion resistor.

The overvoltage levels during insertion and by-pass are presented in Figure 8.17c.

The energization overvoltage in phase a of a compensated 765-kV transmission line of 300 km is presented in Figure 8.18. The voltages at the receiving end of the line, without and with a preinsertion resistor are presented in Figure 8.18a and b, respectively.

In general, the optimal value of the resistor is on the order of magnitude of the surge impedance of the transmission line, assuming values ranging from 250 to 400 Ω.

A more detailed analysis of resistor insertion and by-pass is presented below, by the search for the appropriate value of the resistor and its insertion period.

Example 8.2 Obtain the overvoltages at the receiving end of a transmission line energized by circuit breakers with preinsertion resistors (Figure 8.19). Analyze the resistor values given below, as well as the effect of the resistor insertion time.

Data

S_1: inserts the resistor and energizes the line.
S_2: by-passes the resistor.
Z: surge impedance of the transmission line: 300 Ω.
τ: line time delay.
t_c: insertion time (t_c measured from the closing of S_1 – to the closing of S_2 that short-circuits R).
e_g: step voltage source

The cosine function of the generator voltage is approximated by a unitary step ($e_g = 1.0$).

Analyze the resistor by-pass for times greater or smaller than twice the time delay along the transmission line.

R is the preinsertion resistor:

a) 1000 Ω.

b) 300 Ω.

Solution

The overvoltages are analyzed with the aid of the lattice diagram, initially calculating the transmission and reflection coefficients at the sending and receiving ends of the transmission line.

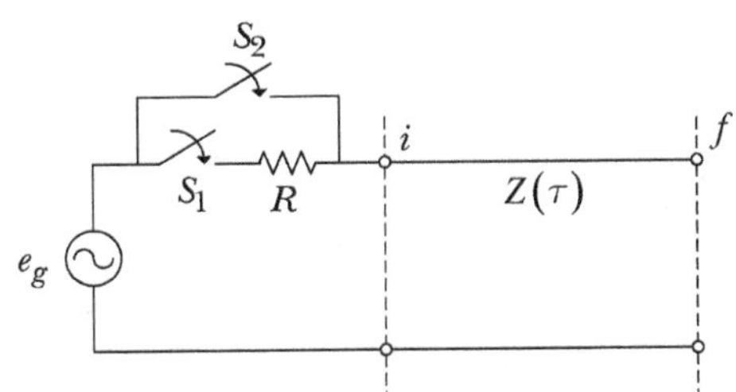

Figure 8.19 Transmission line and circuit breaker with preinsertion resistor.

- Reflection and transmission coefficients at i (sending-end) and f (receiving-end):

a) resistor connected to the network:

$$\Gamma_{r_i} = \frac{R-Z}{R+Z}, \quad \Gamma_{r_f} = 1,$$

$$\Gamma_{t_i} = \frac{2R}{R+Z}, \quad \Gamma_{t_f} = 2.$$

b) resistor by-passed:

$$\Gamma_{r_i} = -1, \quad \Gamma_{r_f} = 1,$$

$$\Gamma_{t_i} = 0, \quad \Gamma_{t_f} = 2.$$

Wave applied to the transmission line when the resistor is inserted, that is, at the closing of S_1:

$$v_0 = \frac{e_g Z}{R+Z}.$$

The voltages at the sending and receiving end of the transmission line are given by, Figure 8.20a:

$$v_i(t) = v_0 H(t) + \Gamma_{r_f}\Gamma_{t_i} v_0 H(t-2\tau) + \Gamma_{t_i}\Gamma_{r_f}^2\Gamma_{r_i} v_0 H(t-4\tau) + \ldots,$$

$$v_f(t) = \Gamma_{t_f} v_0 H(t-\tau) + \Gamma_{t_f}\Gamma_{r_i}\Gamma_{r_f} v_0 H(t-3\tau) + \ldots.$$

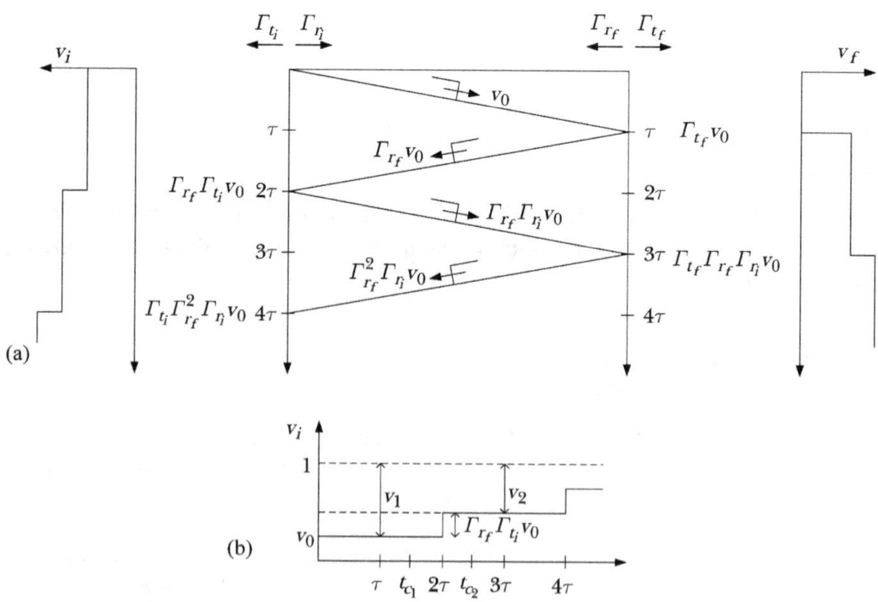

Figure 8.20 Resistor insertion. (a) Lattice diagram. (b) Receiving end time response.

The voltage waves applied at the sending-end are presented in Figure 8.20b for two conditions.

The resistor by-pass is analyzed first by closing the switch S_2 at $t_{c1} < 2\tau$. In this case, the voltage at the sending end varies from v_0 to the unitary value. Therefore, the voltage variation applied to the line is:

$$v_1 = 1 - v_0.$$

In the second case, the switch S_2 is closed at $t_{c2} > 2\tau$, and the voltage variation applied to the line is now given by:

$$v_2 = 1 - \left(v_0 + \Gamma_{r_f}\Gamma_{t_i}v_0\right).$$

Since $\Gamma_{r_f} = 1$, the equation is simplified to:

$$v_2 = 1 - v_0\left(1 + \Gamma_{t_i}\right).$$

The following lattice diagrams are obtained for t_{c_1} and t_{c_2}, in Figure 8.21a and b, respectively:

The expression for voltage v_{f_1}, at the receiving end of the line, with the resistor by-pass at t_{c_1} (see Figure 8.21a), is given by:

$$v_{f_1} = 2v_0 + 2v_1 = 2(v_0 + v_1).$$

Figure 8.21a shows that $v_0 + v_1 = 1.0$ or $v_{f_1} = 2.0$, resulting in the same value that would be obtained without the resistor. It can be concluded that by-pass of the resistor at $t_{c_1} < 2\tau$ has no effect on reducing the overvoltage, which remains doubled at the line's open terminal.

The following is a numerical analysis of the voltages at $t_{c_2} > 2\tau$, in this example, given that $2\tau < t_{c_2} < 4\tau$.

Figure 8.21b shows that:

$$v_{f_2} = 2v_0 + 2\Gamma_{r_i}v_0 + 2v_2,$$

Substituting for v_2:

$$v_{f_2} = 2\left[v_0\left(1 + \Gamma_{r_i}\right) + 1 - v_0\left(1 + \Gamma_{t_i}\right)\right],$$

$$v_{f_2} = 2\left[1 + v_0\left(\Gamma_{r_i} - \Gamma_{t_i}\right)\right].$$

Hence, as $\Gamma_{r_i} - \Gamma_{t_i} = -1$, (see Eq. (2.55) in Chapter 2):

$$v_{f_2} = 2(1 - v_0)$$

The following table summarizes the effects of resistor by-pass at t_{c_1} and t_{c_2}.

R(Ω)	Γ_{t_i}	Γ_{r_i}	v_0	v_1	v_2	v_{f_1}	v_{f_2}
(a) 1000	1.54	0.54	0.23	0.77	0.42	2.0	1.54
(b) 300	1.0	0	0.5	0.5	0	2.0	1.0

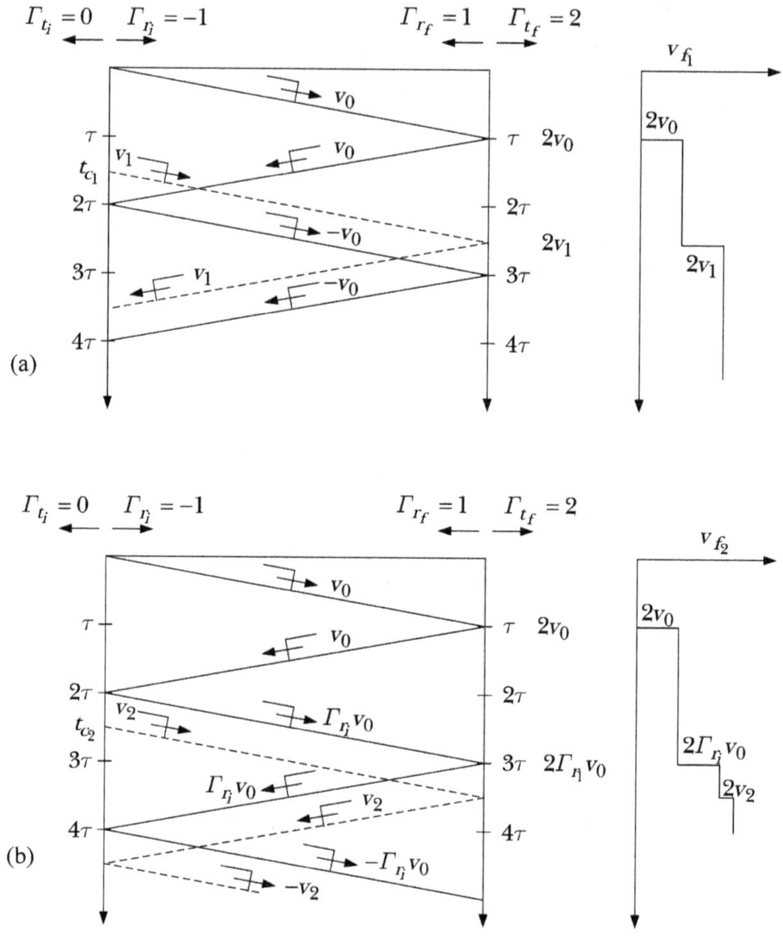

Figure 8.21 Resistor by-pass. (a) Resistor by-pass at t_{c1}. (b) Resistor by-pass t_{c_2}.

The results in this table can be compared with Figures 8.22 and 8.23. The value of v_{f_1} at $t_{c_1} < 2\tau$ is observed to always be 2.0, in accordance with previous considerations.

Regarding case (a), the v_{f_2} value is 1.54 pu and, in that case, it can be observed that although the applied voltage v_0 is small, the overvoltage occurs due to the resistor by-pass.

As for case (b), with the resistor equal to the line surge impedance $R = Z$ it is observed that no overvoltage occurs when the resistor is equal to the transmission line surge impedance.

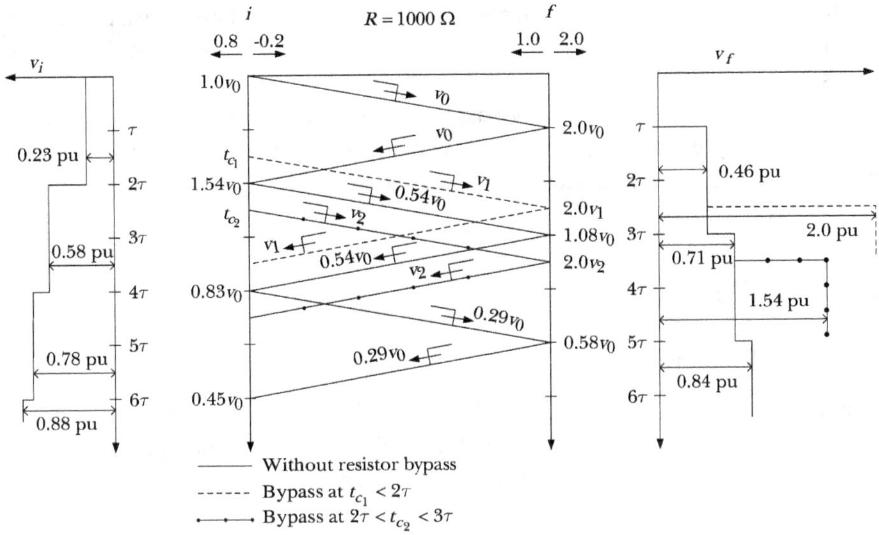

Figure 8.22 Voltages for $R = 1000\ \Omega$.

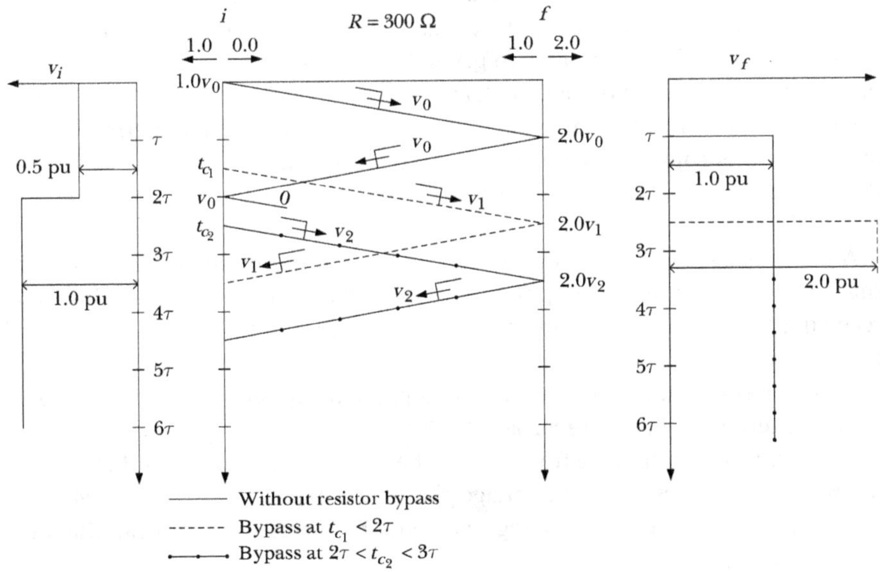

Figure 8.23 Voltages for $R = 300\ \Omega$.

Although the example is just an approximation to three-phase cases, the results explain why preinsertion resistors are used in real systems with values close to the line surge impedance.

Regarding the time of resistor by-pass, or the insertion time, the values specified for circuit breakers are in the range from 7 to 10 ms. Since long lines have time delays around 1 ms the resistors are by-passed at $t_{c_2} > 2\tau$.

8.4 Faults

8.4.1 AC Systems

The occurrence of a single-phase fault can lead to temporary overvoltages (as shown in Chapter 7), depending on the level of system grounding. With respect to the transient part composed of traveling waves, it is uncommon for very high overvoltages to occur [8].

If the fault occurs randomly at the line terminal, as in Figure 8.24a, at a given point on the voltage wave, as in Figure 8.24b.

The fault can be simulated by applying a voltage generator that cancels the pre-existing voltage without the fault, starting at time t_0, as in Figure 8.19c and d.

The applied voltage wave, with a sign opposite that of the former, propagates and must be superimposed with prefault voltages at each point in the system, canceling the voltage at the fault point and, in general, reducing the voltages on the faulted phases in other places. It is noted that, from the point of view of wave reflections in the network, the inclusion of an ideal voltage generator with an internal impedance of zero is consistent with the short circuit during the fault. The effect of the induced voltages in the healthy phases depends on the reflections at the line terminals.

A particular situation that may cause overvoltages, even in faulted phases, is that of very long systems subject to the phase delay between the sending and receiving terminals of the transmission system [9] due to the transmission of active power.

Take, for example, a radial circuit with a phase delay between the sending and receiving terminals in Figure 8.25a.

With active power flowing from A to B, the voltage phasor at point B is delayed by an angle α with respect to the voltage phasor at point A. A fault at $t = 0$, which is the time of a peak positive voltage at point B, represents a critical situation (Figure 8.25b).

The phasor diagram and rotating vectors are illustrated in Figure 8.25c.

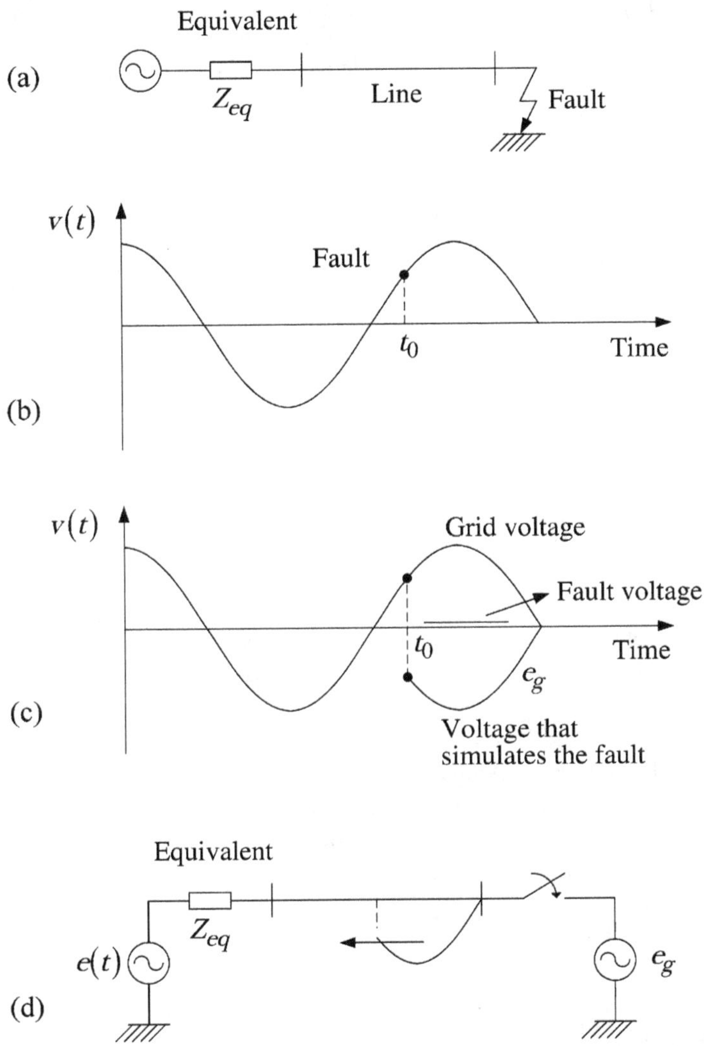

Figure 8.24 Occurrence of a fault. (a) Transmission line. (b) Fault application time. (c) Fault simulation. (d) Propagation of the voltage that simulates the fault.

The applied wave $e_g(t)$, with negative sign, begins to travel toward system point A and, during the time delay, voltage vector V_a rotates by an angle β. If the time delay corresponds to a certain angle β, constructive wave superposition could occur at the time of arrival of the traveling wave in A, because $e_g(t)$ and $v_a(t)$ would have the same sign.

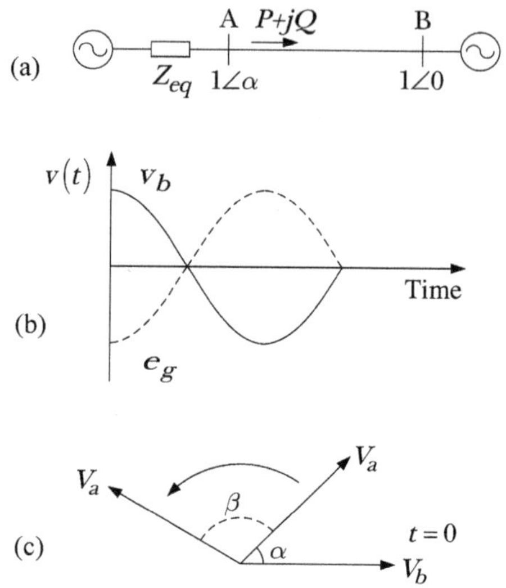

Figure 8.25 Fault application in a long radial system. (a) Steady-state. (b) Fault simulation. (c) Rotating vectors.

Example 8.3 Assuming that $\alpha = 70°$, analyze the superposition of waves when a fault occurs in a long radial system.

Assuming that the real part of V_a is -0.87 pu at the time of arrival of the traveling wave, then $\alpha + \beta = 150°$, that is, $\beta = 80°$. As 360° corresponds to 16.66 ms at a nominal frequency of 60 Hz, 80° corresponds to a time delay of 3.7 ms.

Assuming the speed of the traveling wave is 250 km/ms, which is slightly lower than the speed of light, the transmission line should have a length of 925 km. Smaller transmission systems should exhibit lower overvoltages, since modulus of the real part of V_a is lower.

8.4.2 DC Transmission Line

Energization of a DC transmission line can be smoothly performed by gradually triggering the thyristors on the rectifier bridge, and overvoltages do not occur.

The phase-to-ground short circuit is the event that causes the highest overvoltages in the DC link, which is important in the analysis of insulation coordination. Figure 8.26a illustrates a DC link bipole.

$E_a = +E$: voltage at the positive pole.
$E_b = -E$: voltage at the negative pole.
Z_t: impedance at the line terminals, which includes the effect of DC filters.

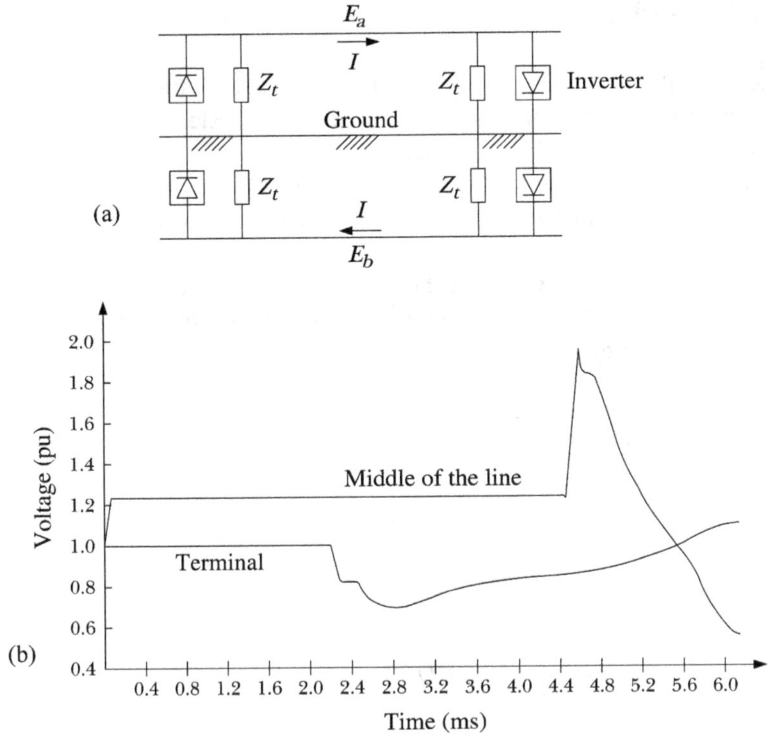

Figure 8.26 Phase-ground fault in a DC link. (a) DC link bipole. (b) Induced voltage at the healthy pole. *Source:* Adapted from Hingorani [10].

Since poles a and b are coupled, a fault in one of the poles induces a voltage in the other one that may cause overvoltages. The greatest overvoltages occur with faults in the middle of the DC line.

The induced voltage must be evaluated to minimize the possibility that the overvoltage causes a subsequent fault at the healthy pole, completely interrupting the transmission of energy. Modal analysis clarifies the origin of this overvoltage (Chapter 3, [10]). This analysis will not be done here, and only the basic idea will be accepted that a fault at one pole induces a voltage at the other, and that traveling waves result in overvoltages at the terminals and in the middle of the DC line.

In Figure 8.26b, a fault is applied in the middle of the line at the negative pole, with the impedances at the line terminals representing filters and smoothing reactors.

8.5 Fault Clearing

A fault is cleared with the opening of the circuit breakers at both transmission line terminals. The maneuver can introduce overvoltages on the system's healthy lines, primarily in the case of parallel lines or double circuits [8].

The phenomenon was partially discussed in Section 8.3.3, in relation to the opening sequence of the circuit breaker poles and current extinction at the zero crossing.

To understand the traveling waves in the network during a fault clearing, it is convenient to use a more elementary circuit, analyzing, for example, the opening of the switch connected to ground in a generic network, Figure 8.27a.

t_0: time of current interruption.
$i(t)$: current with the switch closed.
$i_g(t)$: injected current that cancels $i(t)$ and simulates the switch opening.

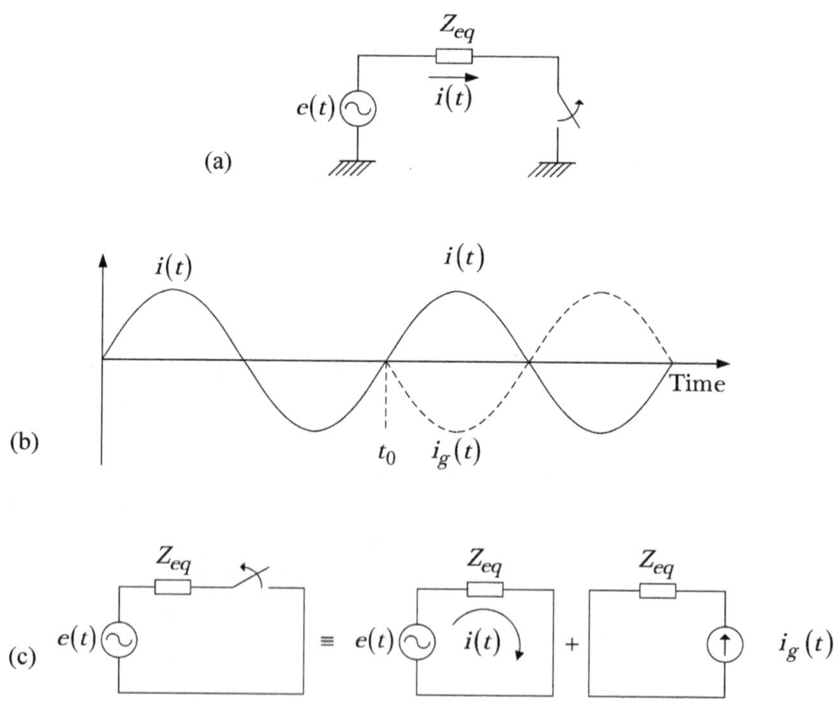

Figure 8.27 Fault clearing. (a) Switch opening. (b) Current injection with opposite sign. (c) Principle of superposition.

Current extinction at the switch, starting at time t_0, can be represented by a current generator that injects the same current waveform as the fault current, but opposite in sign (Figure 8.27b).

The method of current injection is based on the superposition principle and is examined in more detail in Chapter 12, which addresses the transient recovery voltages in circuit breakers, Figure 8.27c. Therefore, current interruption can be represented by the insertion of current generators, which may explain the traveling waves to the network.

In general, fault clearing does not represent the most critical overvoltage event in the system, and they are evaluated to have an overview of possible overvoltages.

8.6 Load Rejection

Much like the fault clearing analysis, the switching surges that take place due to load rejection can be understood by adopting a current generator that injects the same current waveform as the load current $i_l(t)$, but opposite in sign, at the time of circuit-breaker opening (Figure 8.28).

This transient should evidently depend on the impedance of the network and on the magnitude of the load current $i_l(t)$.

In strong systems, overvoltages will be low, and the same should occur if the load current is also low.

On the other hand, in weak systems, with a low short-circuit impedance, as for example, in the case of a radial circuit feeding a heavy load, high overvoltages can be expected. These overvoltages have a transient component and a steady-state component.

The steady-state component, which corresponds to a temporary overvoltage, was analyzed in Chapter 7 by considering the fundamental component at 60 Hz. The effect of possible harmonic distortion, due to transformer saturation, could increase the amplitude of the temporary parcel and of the entire transient.

In the case of high temporary overvoltages, the transformers saturate, and thus, the harmonic components can excite the natural oscillation frequencies of the network and further increase the overvoltages.

Temporary and switching overvoltages due to load rejection can cause high stress on arresters, which is an important topic to be considered in arrester specification. This issue is also discussed in Chapter 13.

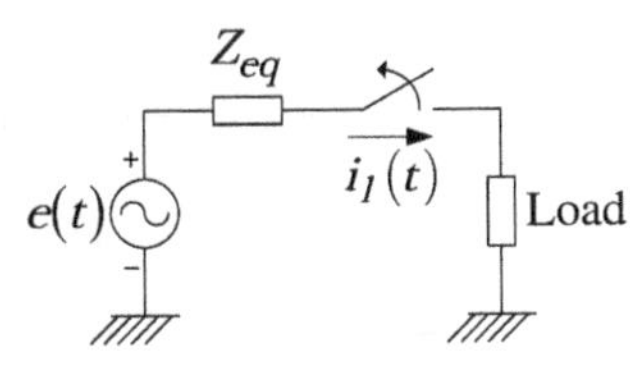

Figure 8.28 Load rejection.

8.7 Transformer Energization

The three-phase equation for the magnetization current during transformer energization is relatively complex in view of the nonlinearities involved.

The usual simplified treatment considers the single-phase case and extends the conclusions to the three-phase case [11, 12]. This simplified approach, despite a certain imprecision, allows an introduction to the basic concepts and will be adopted in the elementary analysis of this phenomenon.

Neglecting losses, the traditional model of the transformer consists of the short circuit reactance, the magnetization branch, and an ideal transformer, according to Figure 8.29.

This model is relatively accurate for two-winding transformers. In the analysis of three-phase transformers, depending on the types of winding and core used, application of more complex models, by means of the impedance matrix of the transformer, may be necessary [13]. For the energization of a single-phase transformer, as for example, through the primary winding, the secondary winding may be omitted, since there is no current in that winding, when i'_p and i_s are null, Figure 8.30a.

The magnetization branch of the transformer, which is determined by the $V \times I$ curve, also known as the magnetization curve or saturation curve, has two well-defined regions. The curve is obtained by means of measurements of rms voltages and currents, Figure 8.30b.

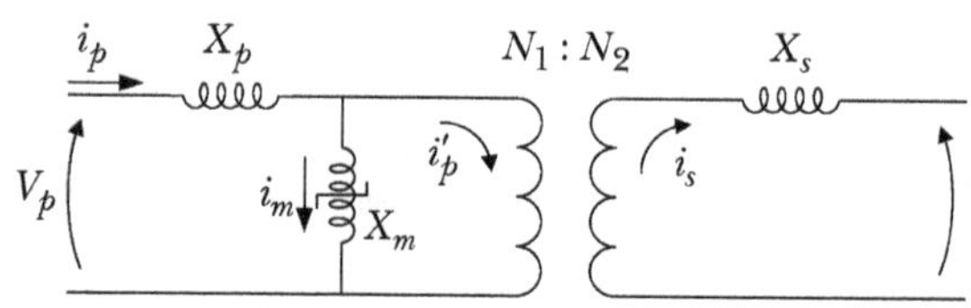

Figure 8.29 Single-phase transformer model.

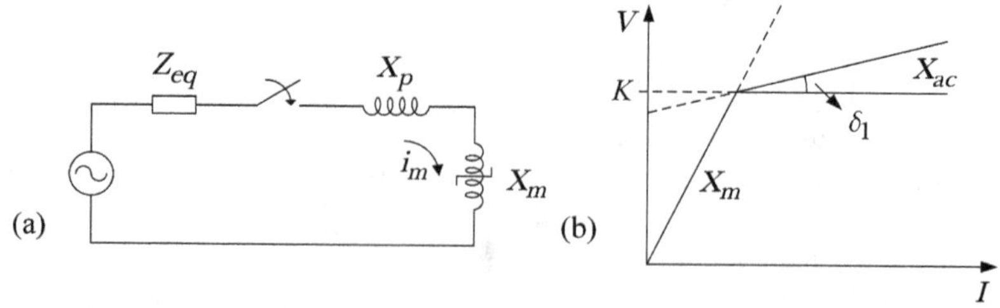

Figure 8.30 Transformer energization. (a) Simplified model. (b) Saturation curve.

Reactance X_m corresponds to the linear region of the magnetization branch under nonsaturated conditions. The transformer operates in that region with a magnetization current of usually less than 1% of the nominal current, in high-voltage systems.

Air core reactance X_{ac} defines a region in which the flux saturates the iron core and is linked with the windings through the air. In the saturated region, the current contains a large number of harmonics. As an estimate, X_{ac} is approximately twice the leakage reactance.

The intersection of the two straight lines, shown in Figure 8.30b, defines the knee point of the transformer, which, for cost reasons, is situated in the region from 1.1 to 1.2 pu of the nominal voltage of the transformer.

It is convenient to interpret the saturated currents, with the aid of an inductance model, based on the magnetization curve $\varphi \times i$ given by the peak values of flux and currents.

An ideal voltage source feeding a nonlinear coil is assumed. The flux linked to the coil is given by the integral of the applied voltage. Therefore, the flux φ is delayed with respect to the voltage $e(t)$.

The applied voltage is sinusoidal and, consequently, the flux has a sinusoidal behavior. Regions 1 and 3 of Figure 8.31 show, for a half cycle, the periods of nonsaturated flux and sinusoidal current. Region 2 shows the saturated region for the flux, in which currents can reach a high peak level.

Figure 8.31 Magnetization current in the saturated region.

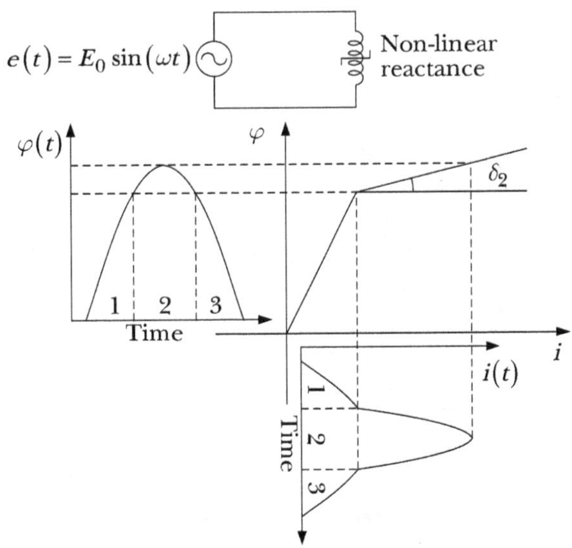

Note that slopes δ_1 and δ_2 in Figures 8.30b and 8.31, respectively, are not the same, because, in the first case in Figure 8.30b, voltage and current are given in rms values, and the current waveform consists of the fundamental and several harmonics. In the second case in Figure 8.31, the inductance relates the instantaneous values of flux and current.

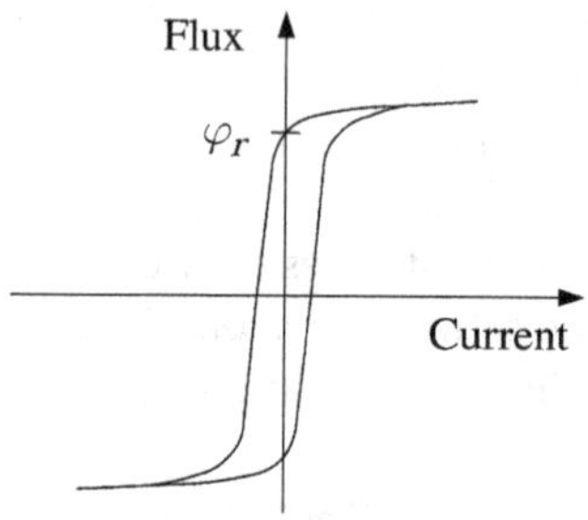

Figure 8.32 Hysteresis curve.

In [14], a method is presented for converting the saturation curve ($\varphi \times i$) from rms values to peak values. There are routines to make this conversion in the ATP program [15], as mentioned in Chapter 4.

In order to calculate the transients caused by transformer energization, the hysteresis curve has to be introduced into the model, in order to account for the remanent flux in the core, even when the transformer is unloaded. The $\varphi \times i$ curve, considering the hysteresis is exemplified in Figure 8.32.

For many switching surge calculations, hysteresis may be disregarded. However, in the analysis of transformer energization, representation of hysteresis is mandatory when the remanent flux is considered. For high switching-surge frequencies, the magnetization curve has a secondary role and approaches linearity.

The fundamentals of transformer energization may be understood by means of a very simple model using the energization of a linear inductance, Figure 8.33a.

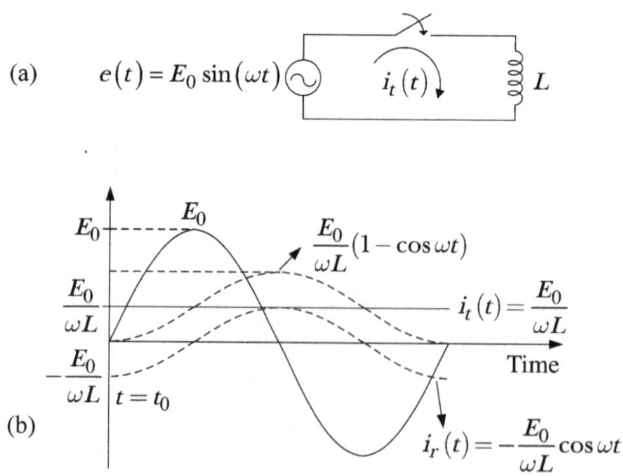

Figure 8.33 Energization of a linear inductance. (a) Elementary circuit. (b) Components of the transient current.

$$i(t) = \frac{1}{L}\int_0^t E_0 \sin(\omega t)dt, \tag{8.19}$$

$$i(t) = \frac{E_0}{\omega L}[1 - \cos(\omega t)]. \tag{8.20}$$

The circuit used to analyze energization of a reactance or to analyze a short circuit at the terminals of a generator is the same if the inductance is assumed linear. Thus, the energization of a transformer and a fault at the generator terminal are both subject to the same principle of noninstantaneous flux variation.

If the coil is energized at a time when the voltage is zero, the current under the steady-state regime would reach its maximum value. Figure 8.33b shows the voltage and current in the same verical axis. The sinusoidal current in the steady state has a negative peak at $t_0 = 0$:

$$i_r(t_0) = -\frac{E_0}{\omega L}. \tag{8.21}$$

Since the current and, consequently, the flux do not change instantaneously, a transient component arises identical in magnitude but opposite in sign to the steady state value that would exist. The transient component is also called the DC component. However, in this simplified case, damping is not observed since losses were neglected. Chapter 1 shows the equations for the RL circuit excited by a sinusoidal source, for instance, Eq. (1.51), and the analysis of transients under different conditions.

The presence of a transient component that depends on the time of the switch closing is observed. The time response of the current is like that of a fault at the terminals of synchronous generators, except for the transient variation in the reactance.

The transient behavior of the flux is analyzed below, without taking the remanent flux into consideration. In Figure 8.34, if the transformer is energized at time t_a, which is the time that the steady state flux reaches zero, the DC

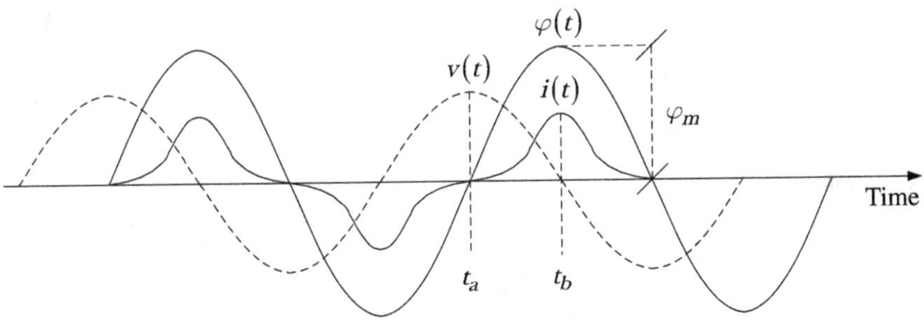

Figure 8.34 Voltage and flux.

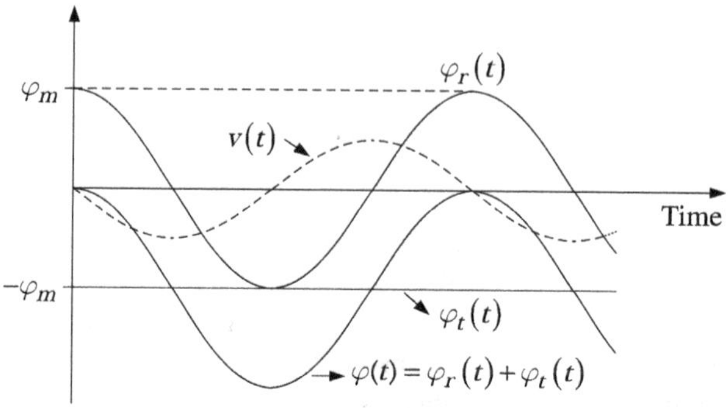

Figure 8.35 Influence of the initial time.

component of the flux is not generated, as there is no instantaneous flux variation at that time.

On the other hand, for the energization at timet_b, the point in time when the steady state flux is maximal, a transient DC component of the flux ($-\varphi_m$) arises that is identical in amplitude but opposite in sign in order to prevent instantaneous flux variation (Figure 8.35).

$\varphi_t(t)$: transient component of the flux (in this case, without damping).
$\varphi_r(t)$: flux component in the steady state.
$\varphi(t)$: total flux.

In this case, due to the adverse conditions at the time of energization, the total flux reaches $2\varphi_m$ in amplitude, thereby increasing the saturation level of the transformer.

The effect of the residual flux is like that of introducing an initial condition in the inductance, for instance, current i_0 in the linear inductor in Figure 8.33a.

When a remanent flux is assumed, the transient component can become even more critical. The flux behavior, assuming a remanent flux, as exemplified in Figure 8.36, is examined next.

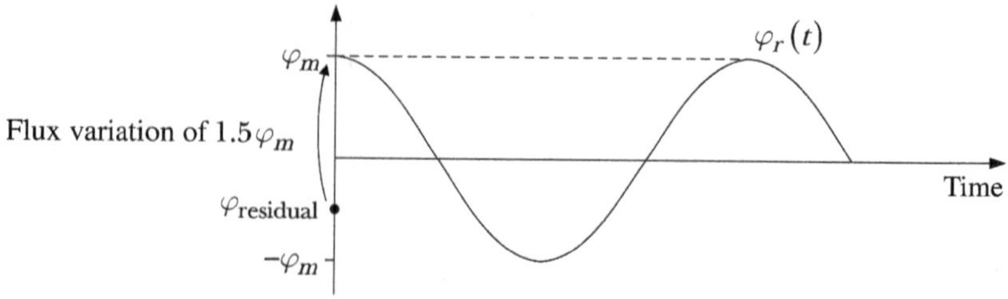

Figure 8.36 Influence of the residual flux.

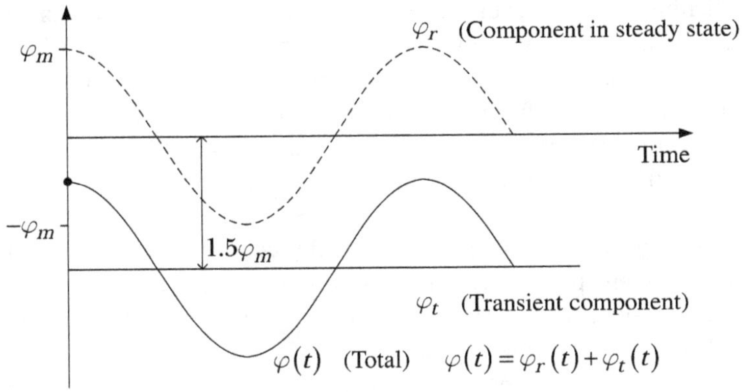

Figure 8.37 Flux waveform for a critical condition.

To prevent the flux from varying instantaneously from $-0.5\varphi_m$ to $1.0\varphi_m$, a $-1.5\varphi_m$ transient component arises, canceling the $1.5\varphi_m$ change that would otherwise occur (Figure 8.37).

In this case, the flux also does not show instantaneous variation, but the peak value reaches $2.5\varphi_m$, that is, with the remanent flux, even more saturated fluxes and distorted currents are obtained. Note that the transient component is, in fact, a DC component when losses are neglected, but an exponential decay would be expected in real cases.

So far, an ideal voltage generator was used as an excitation source, and it was observed that large currents could flow during energization of a transformer. However, cases are known in which the system's short-circuit power to energize the transformer is not high, which corresponds to a high value of the Thévenin equivalent impedance as well, for example, a high impedance Z_{eq} in Figure 8.30a.

Thus, lower currents are expected, but high overvoltages can be found at transformer terminals, especially if some of the harmonics generated during saturation interact with the natural frequencies of the network, creating resonant conditions.

8.8 Controlled Switching

Generally, switching surge analysis in transmission lines, reactors, and capacitors is focused on adverse conditions to investigate the most critical cases and their impact on the insulation level of the equipment.

Next, the controlled switching method will be analyzed, in which the opposite is done, that is, the method seeks to find the most favorable conditions for the circuit breaker switching while avoiding the deleterious effects of transients.

Controlled switching can replace some of the traditional equipment for transient mitigation, except for arresters that cannot be removed from the system for various reasons. However, synergistic effects between surge arresters and controlled switching can substantially benefit the electrical system.

The entire set, instrumentation, and circuit breaker must be reliable, reproduceable, and accurate both for the make and the break current, at specific points on the waveform.

Circuit breakers may or may not have individually controlled poles. Circuit breakers with independent poles have greater flexibility in performing optimal closings. In circuit breakers with a single operating mechanism, the poles can be staggered to close at certain times.

Random factors, such as environmental and operating variations, that can cause inaccuracy in real switching cases need to be identified and monitored, to compensate for them, and prestriking and arcing times must be considered as well.

An introduction is provided to controlled switching, and the details of the process can be found in references [16–18], as well as in the manufacturers' catalogs. Although the general rules must always be observed in the event of a real application, it is mandatory to document the details of each case with studies and simulations.

The strategies are summarized below with time points and time intervals represented by lowercase letters and capital letters, respectively.

8.8.1 Opening and Closing Switching

Common variables in the analysis of controlled opening and closing of circuit breakers are identified as:

t_{com}: time of random command signal.
T_{cont}: synchronization time interval to be calculated.
T_w: waiting time interval.
t_{start}: time of command to energize coil in opening or closing operations.
T_{total}: opening or closing time delay.
NT_{zero}: integer number of half cycles to achieve an appropriate T_{cont}.
ΔV: sinusoidal voltage across the circuit breaker.

To achieve favorable arcing or prestrike time intervals, T_{cont} is calculated with respect to relevant points on waves.

Therefore, Eqs. (8.22) and (8.23) can be written,

$$T_{tot} = T_w + T_{cont}, \tag{8.22}$$

$$t_{start} = T_{total} + t_{com}. \tag{8.23}$$

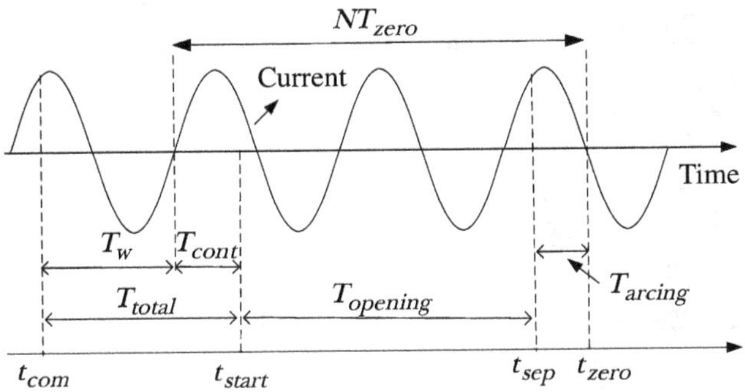

Figure 8.38 Opening sequence in the current waveform.

- Controlled Opening

  Although current is the most suitable signal for a controlled opening, eventually, the bus voltage can be used in the event of shunt reactor and shunt capacitor maneuvers, as the phase angle of the current can be calculated.

  The following variables are also defined (Figure 8.38):

  t_{sep}: time of contact separation.

  T_{arcing}: arcing time, between t_{sep} and the next current zero crossing.

  $T_{opening}$: time interval between t_{start} and t_{sep}.

  RRDS: rate of rise of dielectric strength when circuit breaker contacts are separating.

For arcing times smaller than a minimum value, the restrike probability is high; therefore, T_{cont} is calculated in Eq. (8.24), so that T_{arcing} is greater than this critical value.

$$T_{cont} = NT_{zero} - T_{arcing} - T_{opening} \tag{8.24}$$

The arcing time for each pole is adjusted by controlling t_{sep}.

- Controlled Closing

  For the controlled closing, the following variables are defined (Figure 8.39):

  t_{touch}: the instant that contacts touch, designated also as target.

  t_{make}: the instant that current starts to flow through an electric arc (making time).

  T_m: time interval between t_{make} and the next zero crossing of the current.

  $T_{prearcing}$: prearcing time, that is, the time interval between t_{make} and t_{touch}.

  T_{making}: time interval between t_{start} and t_{make}.

  $T_{closing}$: time interval between $t_{start}$and t_{touch}.

  V_{making}: voltage across the circuit breaker at t_{make}.

  RDDS: rate of decrease of dielectric strength when contacts are approximating.

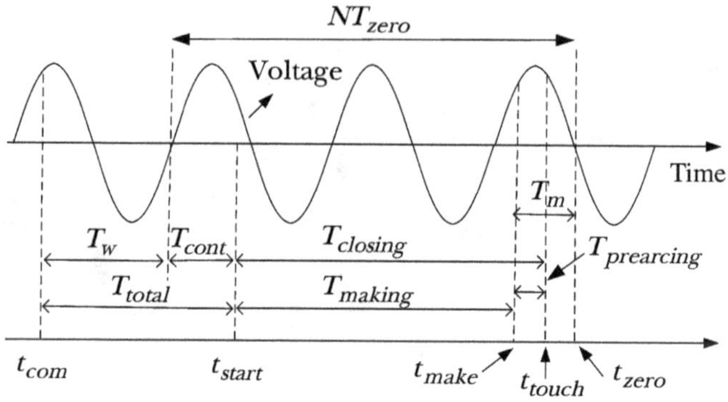

Figure 8.39 Closing sequence in the voltage waveform.

s: derivative of the system voltage across the circuit breaker.
s_0: is the maximum derivative of the system voltage, at the zero crossing of the voltage, $s_0 = \omega \Delta V_p$.
k: RDDS in per unit, takings_0 as a basis.
ΔV_p: peak voltage across the circuit breaker.
ω: system frequency in rd/s.

The controller introduces a time delay T_{cont} with respect to the relevant zero crossing that is calculated from (Figure 8.39):

$$T_{cont} = NT_{zero} - T_m - T_{making} \tag{8.25}$$

Prestrike control considers the applied voltage and the dielectric strength, the latter depending on the contact speed, which is not always the same.

Figure 8.40 illustrates two instances of making for the circuit breaker. The sine wave is the voltage across the circuit breaker, and the straight lines refer to the RDDS, defined by a rate $k \simeq 1$ in this figure.

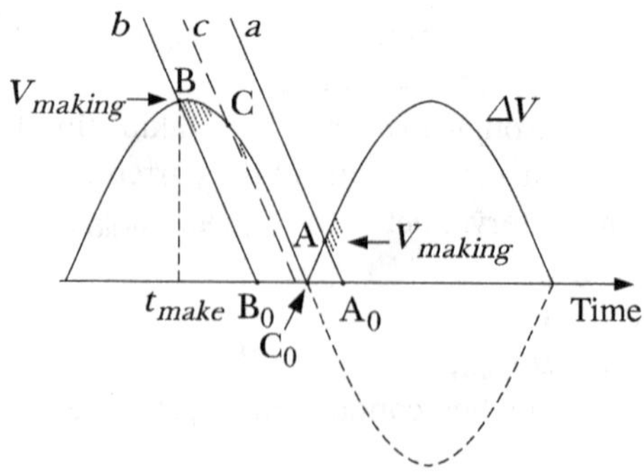

Figure 8.40 Prestrike at t_{make}. *Source:* CIGRE Working Group 13.07 [16].

As the polarity makes no difference, t_{make} and V_{making} are more evident in Figure 8.40 if ΔV is taken in absolute value.

The prestrike occurs when the sine wave across the breaker exceeds the dielectric strength given by straight lines *a* and *b*, at the beginning of the hatched areas, at points A and B, respectively.

The target points, A_0 and B_0, for two closing situations are defined by the zero crossing of lines *a* and *b*, respectively. The t_{make} occurs when the values on the line are lower than those of the absolute value of the sine wave, at points A and B.

The point C_0 defines the system's zero voltage across the contacts.

- Making at Voltage Zero
  Here is presented the case of line *a* with target at point A_0,

  Only the fundamental aspects of how mechanical variations can impact the making are presented below. A comprehensive analysis of the variables that influence the statistical behavior of circuit breaker closing is available in [17, 18].

  Mechanical scatter variations in the circuit breaker must be confined to a restricted time window. In addition, the tolerance range for RDDS slopes must also be considered.

  It can be seen in Figure 8.40 that if the target shifts a little to the left of C_0, as in line *c*, the variation in V_{making} is large, since the slopes of ΔV and RDDS are in agreement. On the other hand, moving the target to the right, the variation in V_{making} is much smoother, as the slopes are different now.

  Considering the asymmetry that exists at this point, to set the target window in order to reduce the error, it is necessary to locate this window to the right of point C_0 and that RDDS be as large as possible. It should be noted that *k* has variations, which can change the instant of prestrike.

  Successfully closing the circuit breaker at near-zero voltage is not trivial, requiring a circuit breaker with consistent closing times and a high RDDS.
- Making at the Voltage Peak
  Here is presented the case of line *b* with target at pointB_0. Point B illustrates the maximum making voltage value, in which T_m is a half cycle.

  Making at voltage peak is easier than at zero voltage, as the sine wave is smooth around the maximum, with voltages very close together within the target window.

8.8.2 Switching of Reactive Compensation and Transmission Lines

The switching of shunt capacitors and transmission lines have some similarities, as does the switching of shunt reactors and no-load transformers.

Tables that summarize the optimal controlled closing and opening times for different types of equipment and winding connections can be found in [16–18]. A summary of controlled switching is presented below.

- Shunt Capacitor Bank Switching
  - Optimal Opening
    The arc is easily extinguished because current and RRRV are low; however, since ΔV can reach 2 pu half a cycle later, it is necessary for the contacts to be sufficiently separated to withstand this overvoltage, thereby avoiding a restrike.
    A suitable strategy is to adjust the mechanical separation, setting the target just after the current zero crossing and letting T_{arcing} as big as possible in Figure 8.38. Thus, the arc will be extinguished almost half a cycle later, at the next zero current crossing. Since the maximum overvoltage occurs half a cycle after extinction, this gives the circuit breaker a period of almost one cycle to move the contacts away, minimizing the likelihood of restrike.
  - Optimal closing:
    As a rule, the best time is when $\Delta V = 0$.
    The shunt capacitor banks, grounded star connected, must be energized at $\Delta V = 0$ in each phase, delayed by 120°. In the case of delta and ungrounded-star connected capacitors, the making only occurs with the closing of two phases, to be carried out at the time of zero phase-to-phase voltage. The third phase is closed 90° later.

- Transmission Line Switching
  - Optimal Opening
    Except for some transmission line peculiarities, the transmission line and shunt capacitor targets are quite similar.
  - Optimal Closing:
    - Unloaded line switching
    - The strategy is like that used in shunt capacitors, with the target at the voltage zero crossing on the source side of the circuit breaker, by applying the smallest voltage step to the transmission line and minimizing waves traveling toward the opposite side of the line. Although the gain is significant, by controlling the closing of the first phase, the voltages induced in the other phases are not easily calculated, which makes an optimized solution for the closing of all phases difficult.
    - Line autoreclose

  Trapped charge for uncompensated or compensated lines, with DC and oscillatory patterns, respectively, were discussed in Section 8.3.3. Equipment remaining on the line can increase the damping of the trapped charge, as for example, inductive potential transformers. To determine the trapped charge, a sophisticated controller is required in compensated lines, to adjust the target in the valley periods of the oscillations.

- Shunt Reactor Switching
  - Optimal Opening
    The adverse effects of current chopping are well known. Current chopping causes high-frequency oscillations, and to avoid reignitions, it is necessary to increase the arcing time, which allows for a greater separation of contacts, but on the other hand, it increases overvoltage, which leads to an intermediate solution for the two effects.
  - Optimal closing:
    The making target is the maximum system voltage, as indicated in line *b* of Figure 8.40. Making at voltage peak mitigates the problems of DC offset current, but a steep wavefront is applied to the reactor insulation. Despite this drawback, this is the option often adopted.

    In the case of neutral-earthed reactors, each phase must be energized at voltage peak, delayed by 120°. In the case of unearthed reactors, the making of two phases must be performed at the moment of phase-to-phase system voltage peak. The third phase is closed 90° later. The case of reactors earthed by neutral reactors is also considered in [17].
- Unloaded Transformer Switching
  - Optimal Opening
    Although the opening of transformers and reactors are similar, in the case of transformers, the overvoltages and oscillation frequencies are not high and are relatively damped. Therefore, this maneuver frequently does not require a special control to reduce overvoltages. On the other hand, the control can make the residual flux more predictable, facilitating the setting of the target in a future energization.
  - Optimal closing:
    Certain factors such as magnetic coupling or winding connections, remanent flux, and relatively low saturation levels, can make the closing of transformers more complex than that of reactors.

    If the residual flux is ignored, the following strategies may be applied.

    For a transformer with its primary winding grounded star connected with uncoupled phases, the target is located at the maximum system voltage. If the primary winding is unearthed, the first closing should be performed at the maximum phase-to-phase system voltage with the third phase closed 90° later.

    In cases of phase coupling, energizing the first phase creates voltage and flux in the cores of the remaining phases. The first making is similar to the uncoupled transformer case but making of subsequent phases should occur at the moment that the fluxes induced in the cores of the open phases, as a result of the first making, coincide with the values that would exist in the steady state.

As for the residual flux, its calculation can be carried out by integrating the voltages measured in the transformer phases during the opening; however, it is necessary to know if the residual flux remained unchanged during the out-of-service period.

The controlled opening allows adjustment of the magnitude and polarity of the residual flux. Subsequent controlled closings can be performed based on this information, making the setting of contact targets more predictable.

8.9 VFTO – Very Fast Transient Overvoltages

8.9.1 Disconnector Operation in Gas-Insulated Substations

So far, the switching surges analyzed are relatively similar in their behavior with respect to slow wavefronts, and therefore, the modeling requirements for calculation are very much the same. Overvoltages are now presented that are also of internal origin but occur at very high frequencies, requiring models in a range of frequencies even higher than that of lightning surges. These overvoltages are treated as a special group of transients with very fast wavefronts [19] and are therefore designated as very fast transient overvoltages - VFTO.

When operating a disconnector switch in a substation, several arc restrikes occur, since the disconnector switches do not have extinction devices for electric arcs, such as circuit-breaker extinction chambers. Due to their construction characteristics, disconnector switches can only handle small capacitive currents related to the bus capacitance to ground and the stray capacitance of power apparatuses.

These overvoltages are analyzed not only because of their magnitude but also because of their repetitive behavior and high frequencies that can cause damage to the insulation of transformers.

To understand the electric arc restrikes during the operation of a disconnector switch, Figure 8.41 illustrates the disconnection of a bus segment.

$v_1(t)$: voltage at the source side.
$v_2(t)$: voltage at the bus side.

When the contacts are electrically disconnected, the voltage on side 2 remains constant in the capacitance, while the voltage on side 1 follows the sinusoidal voltage source. During the opening process, there are several current extinctions and

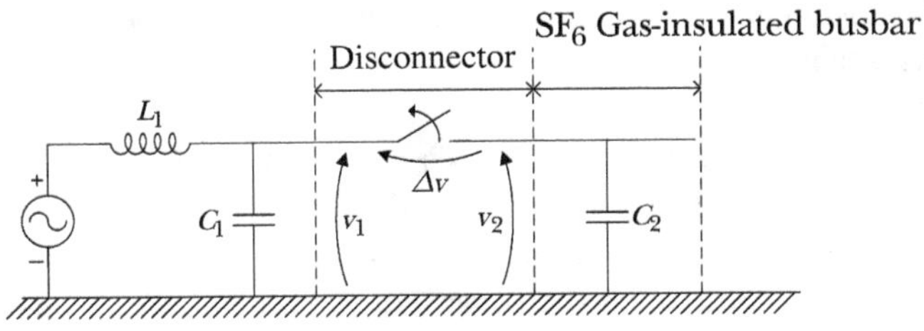

Figure 8.41 Disconnection of a bus segment.

restrikes. However, the overvoltages at each restrike are relatively short in duration, so the overvoltages do not overlap and can be studied separately.

During the opening of the disconnector switch, as the separation speed of the contacts is relatively slow, the dielectric recovery is also slow, allowing for several restrikes, whenever the difference $\Delta v = v_1(t) - v_2(t)$ is higher than the dielectric strength, v_w.

Figure 8.42 illustrates the conditions under which some of the arc restrikes occur; however, it must be considered that certain random conditions may affect the arc extinction.

The rate of rise of dielectric recovery voltage in Figure 8.42 is considered constant. The restrikes become less frequent over time and, at a given moment, the last restrike takes place, when the dielectric strength exceeds $2V_{max}$, which is twice the amplitude of the voltage source. The restrike transient, although with a much higher frequency, is also very similar to disconnection of capacitor banks, analyzed in detail in Chapter 10.

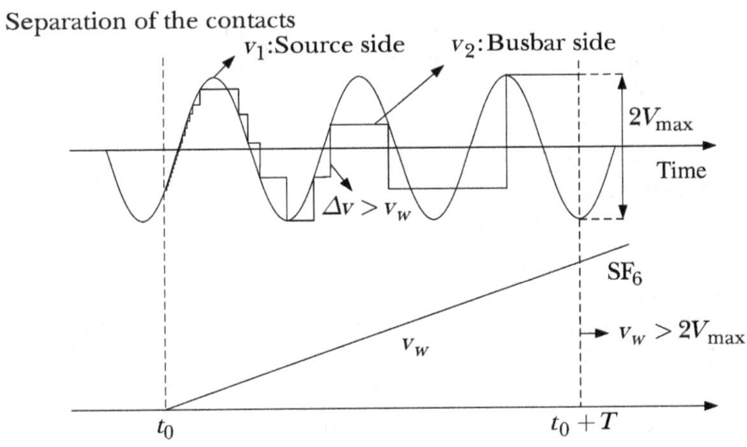

Figure 8.42 Arc restrikes upon disconnector opening. *Source:* Ragaller [20]/Springer Nature.

Although a capacitive model has been used on the bus side to explain the voltage stored during the restrike, a model based on the line surge impedance and its time delay is needed to analyze the wave reflections that generate high frequencies. Substation modeling is like that used in the analysis of insulation coordination for lightning surges, presented in Chapter 9 with a detailed representation of the buses and stray capacitance for all the equipment. However, detailed models may be needed to represent disconnectors and other GIS equipment at very high frequencies approaching MHz levels.

A factor worthy of careful consideration is the residual voltage $v_2(0) = v_0$ at the restrike condition because the residual value depends on the characteristics of the SF_6 busbars. The residual voltage of 1.0 pu is considered conservative and lower values may be adopted. Measurements of residual voltages can make the calculation more realistic. Studies show values ranging from 0.15 to 0.4 pu of the nominal voltage [21].

8.9.2 GIS Components Modeling

- *Transformer model*

  To calculate VFTO, the transformer model at very high frequencies is one of the critical points [22–24]. The simplified models at high frequencies consider a capacitance or a π-circuit capacitances, as an extension of the models for lightning analysis. Accurate models can be based on experimental tests to obtain the elements of the Y matrix of the transformer and their representations through rational functions, using the vector fitting method (see Chapter 5, Section 5.2.1). However, for very high frequencies, it is very likely the predominance of a capacitive behavior.

  These models are accurate when tuned in the high-frequency range, but they do not provide information about the internal parts of the transformer. To analyze internal overvoltages in the transformer, the model must be furnished by the manufacturer. The representation of the windings of a phase of a transformer may require a complex network of RLC elements, but models with distributed parameters may also be found. The representation of a transformer model in state variables can help in more detailed evaluations of interactions between the electrical network and the transformer, which are difficult to be carried out with tools in the time domain. The drawback of state variables is the modeling of transmission line segments, that must be fitted as π-circuits.
- *Models used for other GIS components.*

  The circuit breaker and disconnector models [25] are calibrated for high frequencies. In Figure 8.43, circuit breakers and disconnectors are represented by surge impedances and capacitors, whose values are defined for each specific equipment.

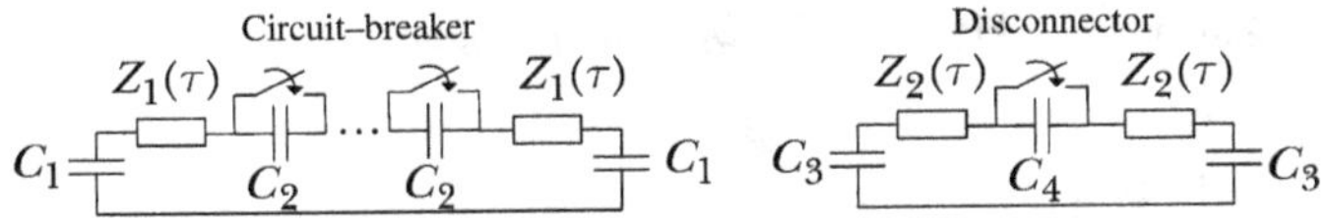

Figure 8.43 Circuit breaker and disconnector models for high frequencies.

The small segments of equipment and substation buses are modeled as transmission lines with lengths between $0.30 < \ell < 1.00$ m whose time delays (τ) are calculated with the propagation velocities. The typical surge impedance values for Z_1 and Z_2 are between 30 and 60 Ω and for capacitances C_1, C_3, and C_4 between 30 and 80 pF. For circuit breakers, grading capacitors are designated by C_2, and have values around 1 nF. A series resistor may also be considered in the disconnector, inserted between Z_2 and C_4.

Very short and long transmission lines can coexist in the GIS, which creates a size dimension conflict in EMT programs and long lines need to be represented by equivalents. It is not mandatory, but the principle of superposition can be convenient in calculating overvoltages using EMT programs, with long lines represented by equivalents.

Side 1 is energized, and the transmission lines are represented either as ideal lines of infinite length, or as Norton equivalents, as in Chapter 5. Side 2 is switched by the disconnector, but with residual voltage. Other GIS equipment may be included on side 1 of Figure 8.44, forming complex networks in case of large GIS.

The amplitude of the source is considered maximum when the electrical discharge occurs in the disconnector, $e(t) = E_0 \cos(\omega t)$. Furthermore, as the transient is very fast, the sinusoidal voltage is approximated by a DC generator, and thus, $e(t) \simeq E_0 H(t)$. The critical condition occurs when the residual voltage has an opposite polarity to the source voltage.

- $Z_1(s)$: network of energized side.
- $Z_2(s)$: GIS line segment switched by the disconnector.

From Figure 8.44, the voltage at the disconnector terminals can be calculated assuming residual voltage v_0 at side 2. If it is assumed that $v_0 = -0.5E_0$, the voltage across the disconnector is calculated as $\Delta v_0 = 1.5E_0$.

Figure 8.44 Predischarge equivalents.

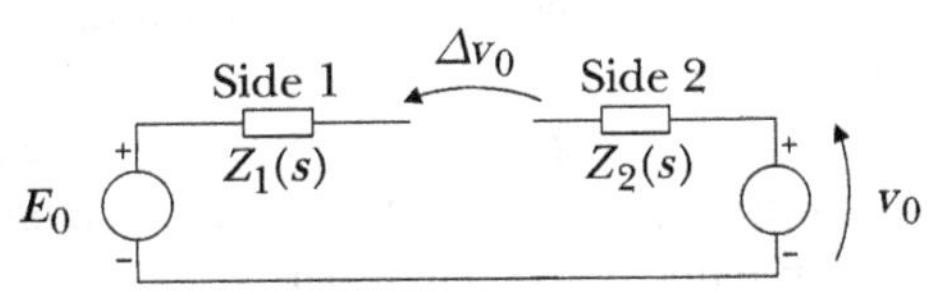

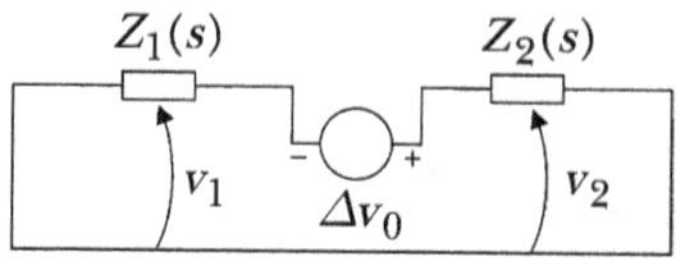

Figure 8.45 DC voltage application.

- *DC voltage application.*
  The closing of the switch represents the restrike, simulated with the application of a DC generator with opposite polarity to the preexisting voltage. If the EMT program does not allow voltage generators between nodes, voltage Δv_0 must be applied via an ideal transformer (Figure 8.45).
  To obtain the final values $v_{1t}(t)$ and $v_{2t}(t)$, on GIS sides 1 and 2, respectively, it is necessary to superimpose the calculated values with the preexisting voltages in the network, and thus:

$$v_{1t}(t) = v_1(t) + E_0 \tag{8.26}$$

$$v_{2t}(t) = v_2(t) + v_0 \tag{8.27}$$

References

1 CIGRE Working Group 13.05. (1979). The Calculation of Switching Surges III: transmission line representation for energization and re-energization studies with complex feeding networks. *Electra* (62), 45–78.

2 IEEE Committee Report. (1980). Bibliography of publications pertaining to switching surges in transmission systems. *IEEE Transactions on Power Apparatus and Systems*, PAS-99(1), 402–414.

3 CIGRE, Study Committee 33, & Working Group 02. (1990). *Guidelines for representation of network elements when calculating transients*. Paris. (Technical Brochure, 39).

4 CIGRE Working Group 13.05. (1971). The Calculation of Switching Surges I: a comparison of transient network analyser results. *Electra* (19), 67–78.

5 Sweeting, D. K. (1972). Overvoltages produced when energizing transmission lines. *Electra* (22), 63–107.

6 CIGRE Working Group 13.02. (1973). Switching overvoltages in EHV and UHV systems with special reference to closing and autoreclosure transmission lines. *Electra* (30), 70–122.

7 CIGRE Working Group 13.05. (1974). The Calculation of Switching Surges II: network representation for energization and re-energization studies on lines fed by an inductive source. *Electra* (32), 17–42.

8 Clerici, A., & Taschinj, A. (1970). Overvoltages due to line energization and reenergization versus overvoltages caused by faults and fault clearing in EHV systems. *IEEE Transactions on Power Apparatus and Systems*, PAS-89(5), 932–941.

9 Jardini, J. A., Kastrup Filho, O., Zanetta, L. C., Vian, A., & Frontin, S. O. (1978). Determinação das sobretensões de manobra do sistema de transmissão de Itaipu. In *4° Seminário Nacional de Produção e Transmissão de Energia Elétrica*, 1977, Rio de Janeiro. Rio de Janeiro: Comunicações técnicas.

10 Hingorani, N. G. (1970). Transient overvoltage on a bipolar HVDC overhead line caused by DC line faults. *IEEE Transactions on Power Apparatus and Systems*, PAS-89(4), 592–610.

11 Povh, D., & Schultz, W. (1978). Analysis of overvoltages caused by transformer magnetizing inrush current. *IEEE Transactions on Power Apparatus and Systems*, PAS-97(4), 1355–1365.

12 Westinghouse Electric Corporation. (1964). *Electrical transmission and distribution reference book* (4th ed.). East Pittsburgh: Westinghouse Electric Corporation.

13 Iravani, M. R., Chaudhary, A. K. S., Giesbrecht, W. J., Hassan, I. E., Keri, A. J. F., Lee, K. C., Martinez, J. A., Morched, A. S., Mork, B. A., Parniani, M., Sharshar, A., Shirmohammadi, D., Walling, R. A., & Woodford, D. A. (2000). Modeling and analysis guidelines for slow transients. III. The study of ferroresonance. *IEEE Transactions on Power Delivery*, 15(1), 255–265.

14 Dommel, H. W., Yan, A., & Wei, S. (1986). Harmonics from transformer saturation. *IEEE Transactions on Power Delivery*, 1(2), 209–215.

15 *ATP: Alternative Transients Program: Rule book*. (1987). Heverlee (BE): Leuven EMTP Center.

16 CIGRE Working Group 13.07. (1999). Controlled Switching of HVAC circuit breakers part-1. *Electra* (183), 43–73.

17 CIGRE Working Group 13.07. (1999). Controlled Switching of HVAC circuit breakers part-2. *Electra* (185), 37–57.

18 Goldsworthy, D., Roseburg, T., Tziouvaras, D., & Pope, J. (2008). Controlled Switching of HVAC circuit breakers: Application examples and benefits. In *61st Annual Conference for Protective Relay Engineers*, April 1–3, College Station, TX.

19 CIGRE, Study Committee 33, & Working Group 13.09. (1989). *Monograph on GIS very fast transients*. Paris. (Technical Brochure, 35).

20 Ragaller, K. (1980). Surges in high-voltage networks. In *Proceedings of the Brown Boveri Symposium on Surges in High-Voltage Networks, held at the Brown Boveri Research Center*, 1979 September 3–4, Baden, Switzerland. New York: Plenum Press.

21 Boggs, S. A., Chu, F. y., Fujimoto, N., Krenicky, A., Plessl, A., & Schlicht, D. (1982). Disconnect switch induced transients and trapped charge in gas-insulated substations. *IEEE Transactions on Power Apparatus and Systems*, PAS-101(10), 3593–3602.

22 Gustavsen, B. (2004). Wide band modeling of power transformers. *IEEE Transactions on Power Delivery*, 19(1), 414–422.

23 Gustavsen, B. (2004). Frequency-dependent modeling of power transformers with ungrounded windings. *IEEE Transactions on Power Delivery*, 19(3), 1328–1334.

24 Martinez, J. A., Walling, R., Mork, B. A., Martin-Arnedo, J., & Durbak, D. (2005). Parameter determination for modeling system transients-Part III: Transformers. *IEEE Transactions on Power Delivery*, 20(3), 2051–2062.

25 Lima Filho, J. F., Pazo Blanco, F. A. F., D'Ajuz, A., & Nascimento, L. F. A. (1995). Tucurui's 500 kV SF6 gas-insulated substation fast transients: Analysis, modeling and field test comparisons. In *Proceedings of the International Conference on Power Systems Transients (IPST),* 1995 September 3–7, Lisboa (pp. 65–69). IPST. https://www.ipstconf.org/papers/Proc_IPST1995/95IPST117-19.pdf

9

Lightning Surges

9.1 Introduction

Lightning surges have a considerable impact on the outage rates of transmission lines and equipment insulation. It is therefore important to analyze the performance of transmission lines and substations when subjected to lightning surges. Interruptions in the electric power supply impose damage and costs on utilities and costumers and reduce the system reliability [1–3].

These overvoltages can be classified as external in origin, as they result from an electrical interaction between a cloud and the power system and are also called overvoltages with fast wavefronts, with respect to the waveform.

The separation of charges in the cloud creates internal and external electric fields, which can result in the occurrence of flashes inside the cloud and between the cloud and earth. There is a vast body of literature addressing the lightning, as exemplified by [4–6].

In studies of lightning surges in electrical networks, the main concern is related to strokes with current flowing between the cloud and the ground since the discharges between clouds have no effect on the electrical system.

Two important phases of this phenomenon can be identified. The first corresponds to the formation of an ionized channel, called the leader, through the layer of air between the cloud and the earth. The second is related to an effective current flow through the ionized channel, called the return stroke current.

Different charge distributions are found in the cloud, and Figure 9.1 presents the case of a cloud with negative charges in its lower region. The discharge begins with the formation of a descending leader channel, which approaches the earth progressively, increasing the electric field and causing upward movement of charge of opposite polarity. The effective current flow between the cloud

Understanding Electromagnetic Transients in Power Systems, First Edition. Luiz Cera Zanetta Jr.

Published 2025 by John Wiley & Sons, Inc.

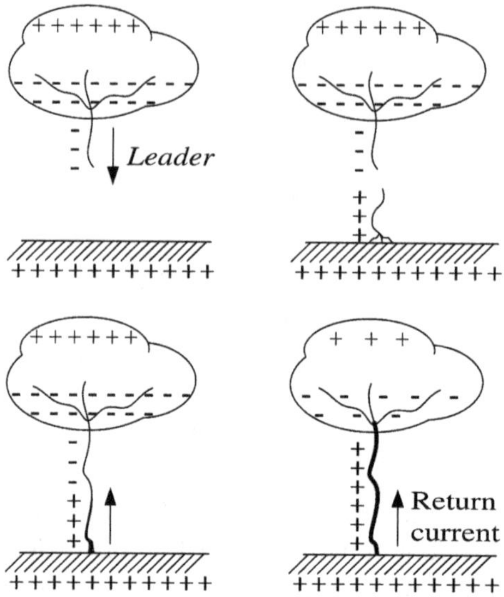

Figure 9.1 Lightning stroke.

and the earth occurs when the ascending and descending leaders contact each other.

An understanding of the physical nature of the phenomenon of lightning discharge is not essential for purposes of this text, and more details in this regard can be obtained in the suggested references. In the electrical system, surges can be reasonably calculated assuming a discharge modeled as the application of a current generator at the stroke point, Figure 9.2a. When necessary, better models of the leader channel [7–9] can be used or can even include an impedance in parallel with the current generator to represent the effect of the channel [10].

Among some possibilities, simplified triangular waveform can be used in many applications to represent the lightning current (Figure 9.2b).

Lightning surges have a short duration. They have relatively fast front times, in the range of a few microseconds, and decay times typically in the range of 30–300 μs, which provides an estimate of the time scale involved in lightning surge analysis.

For the insulation coordination of the equipment, an important reference value is the standard lightning impulse insulation level (IEC), known as the basic impulse level in (IEEE). Another important reference value for insulator strings is the CFO, critical flashover overvoltage. These reference voltages are obtained by testing the insulation using standard lightning impulse voltage waveforms.

Chapter 14 is especially dedicated to the study of insulation coordination.

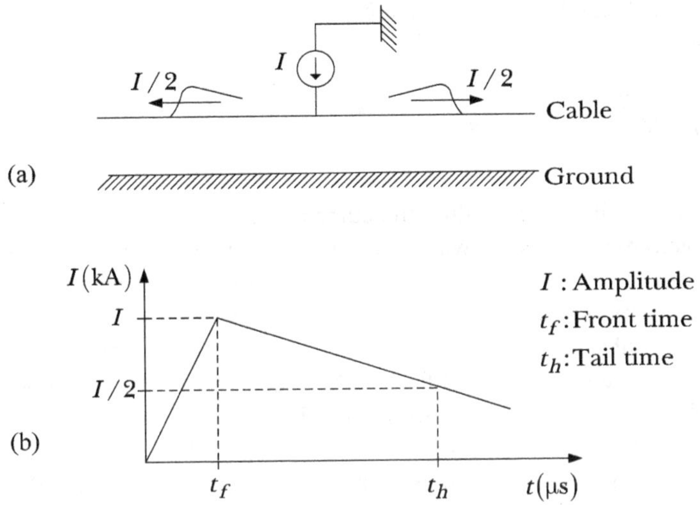

Figure 9.2 Lightning current. (a) Current generator model. (b) Simplified triangular waveform and main parameters.

9.2 Data to Calculate Lightning Surges

Substations and transmission lines are subject to a significant influence from the environment. For the design of lines and substations, in addition to electromechanical attributes, such as distances and cables, it is necessary to know how external factors, such as lightning, can interact with the electrical system, primarily for purposes of insulation coordination.

Information about lightning in this chapter is taken from refs. [10, 11] but new measurements can refine existing data, methodologies and models [3].

9.2.1 Lightning Current

For the calculation of lightning surges, the current stroke can be modeled as a current generator and the impedance of the channel can be inserted in parallel when more accuracy is necessary. Negative polarity is predominant in lightning strikes on transmission lines, that is, in most cases, the injection of negative charges is downward.

Several papers supply the probability curves for stroke current characteristics based on statistical analysis of measured values. Using the measurements in ref. [12], ref. [13] shows the probabilistic behavior of stroke current amplitudes of negative polarity, which can be approximated by the following equation for currents between 5 and 200 kA:

$$P_I = \frac{1}{1 + \left(\frac{I}{31}\right)^{2.6}}, \tag{9.1}$$

in which P_I is the probability of exceeding the current peak I.

From these measurements, it is known that strokes of negative polarity have a probability of occurrence of approximately 90%; however, strokes of positive polarity have larger amplitudes.

The first stroke may subsequently be followed by multiple strokes. Single strokes represent 45–55% of the total number of flashes. Multiple strokes rarely involve more than 10 strokes, and an average of 3 strokes can be expected per occurrence.

The magnitude of current associated with the first stroke usually is greater than that of subsequent strokes. Reference [10] uses a statistical approach to explain the distribution of current amplitudes. The logarithms of those amplitudes behave much like a normal distribution that can be fit by one or two straight lines by means of regression analysis, Figure 9.3.

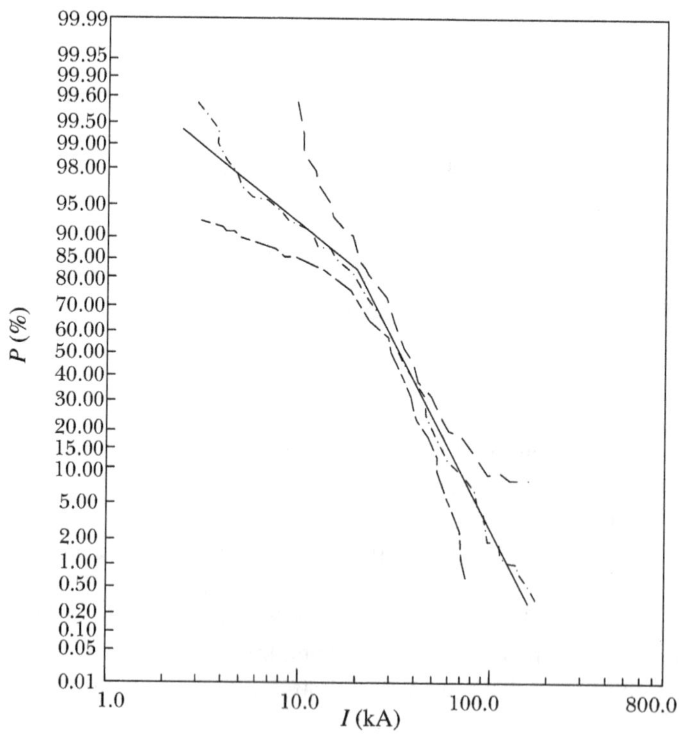

Figure 9.3 Distribution of the cumulative frequency of stroke current amplitudes. *Source:* CIGRE, Study Committee 33, & Working Group 01 [10] Figure 9.

Table 9.1 Distribution of current amplitudes for the first stroke of negative polarity.

Parameter	$I < 20$ kA	$I > 20$ kA
M	61	33.3
β	1.33	0.605

M: median value for ln(I).
β: standard deviation of ln(I).

That reference provides the parameters of the logarithmic distribution, M standing for the median and β for the standard deviation and defines the probability density function for the log-normal distribution. Reference [3] brings new information on this subject.

The following table summarizes parameters M and β as proposed for the distribution of current amplitudes for the first stroke of negative polarity (Table 9.1).

The two straight line segments are defined by median values and standard deviations, so that the integral of the two probability densities is unitary. In this reference, the same current distribution, approximated by just one straight line, can also be found, which avoids a discontinuity in the probability density.

The amplitude and wavefront time that define the rate of rise of the current are the two most important parameters needed to evaluate lightning surge behavior and the requirements of electrical equipment.

- Subsequent Strokes
  The subsequent strokes almost always have a smaller amplitude than that of the first stroke. On average, they reach values of approximately 40% of the first stroke, although subsequent strokes have previously been observed to show amplitudes higher than those of the first one. The logarithms of the current magnitude of the subsequent strokes fit a single straight line, and the following parameters are obtained:

$$M = 10.3, \qquad \beta = 0.53.$$

9.2.2 Wavefront and Tail Time

- WaveFront
  Although the lightning current has an upward concave waveshape, a linear wavefront is sufficiently accurate to perform the general calculation for lightning performance and more detailed waveforms for some specific calculation. The importance of accurately representing the maximum steepness just before the first peak is highlighted in [3], and this should be taken into account in the rise rate.

The statistical distribution of the rate of rise of the current is also available, as well as its correlation with the magnitude of the lightning current.

Reference [13] shows the probability function for the rate of rise of the stroke current of negative polarity, assuming a probabilistic behavior approximated by the simple expression:

$$P_{dI} = \frac{1}{1 + \left(\frac{dI}{dt}\frac{1}{24}\right)^4}, \tag{9.2}$$

in which P_{dI} is the probability of exceeding a given rate of rise, dI/dt.

In ref. [10], another statistical approach to the wavefront can be found, based on the front time t_f for the lightning current, Figure 9.4.

The current levels I_{10}, I_{30} and I_{90} correspond to 10%, 30%, and 90% of the peak current, respectively, according to Figure 9.4. The front times, t_{d10} and t_{d30}, are the time intervals from I_{10} to I_{90} and I_{30} to I_{90}, respectively. The front times are also used to calculate the wavefronts. Practical applications usually adopt the statistical behavior of the parameter t_{d30}. The distribution of this parameter on the first stroke is presented in Table 9.2:

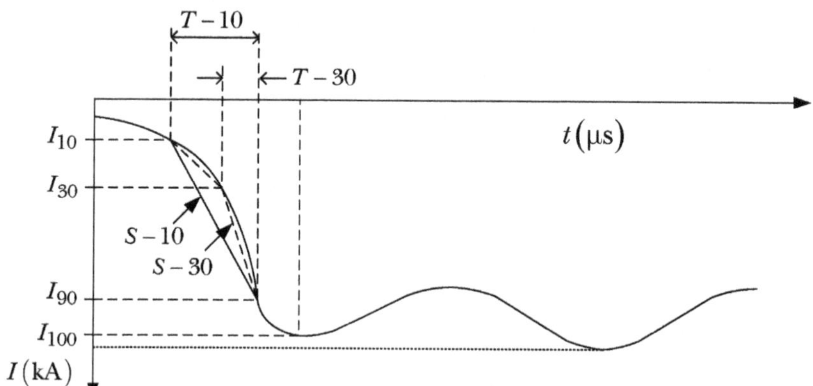

Figure 9.4 Current waveform.

Table 9.2 Front time t_{d30} for the first stroke.

Parameter	$3 < I < 20$ kA	$I > 20$ kA
M	$1.77 \times I^{0.188}$	$0.906 \times I^{0.411}$
β	0.494	0.494

The distribution of amplitudes and front times in the previous tables show statistical dependence, represented by the conditional probabilities. The statistical analysis can be performed by determining the joint probabilities of the normal distributions for the currents and front times, considering their correlation coefficients, according to Chapter 14, Eq. (14.33).

- Tail Time
  The tail time for the return stroke current t_c is defined as the time interval in which the current is reduced to half of its crest value. This time interval is important in calculations of energy absorbed by lightning arresters. According to [10], the median tail time is 77.5 μs on the first stroke and of 30.2 μs on the subsequent strokes.
  The probabilistic data of the stroke current is frequently updated as a function of new measurements. A summary of the most important parameters that aid in the calculation of lightning surges was presented. However, careful examination of these data and a broader knowledge of the published work are always necessary, and it should be kept in mind that the numbers presented here require constant updating, preferably with measurements in the region under study.

9.2.3 Ground Flash Density

An important piece of information for estimating the frequency of lightning strokes on electric systems is the ground flash density, which is the annual number of strokes to the ground per km^2 in the surrounding region of substations and transmission lines. Nowadays, these data are currently provided by modern lightning detection systems.

9.2.4 Topography and Soil Resistivity

The transmission line right of way can present a variety of topographical conditions that require different tower heights to provide the minimum distance from the conductors to the ground. According to the terrain profile, the conductors may assume different average heights, and a higher or lower incidence of flashes may occur.

The terrain is classified into three types:

a) Flat terrain: the conductor is assumed to be at an average height from the ground equal to its height at the tower minus two-thirds of the sag.
b) Uneven terrain: the conductor is considered to be at an average height equal to its height at the tower.
c) Mountainous terrain: the average height of the conductor is higher than its height at the tower. In this case, it is convenient to detail the terrain profile.

The soil resistivity usually varies over the right of way of a transmission line, and measurements are required to produce a suitable grounding design for the towers.

9.3 Models for Overvoltage Calculations

9.3.1 Lines and Cables

Lightning surges contain high frequencies in the range of hundreds of kHz and complex calculations can be performed only with electromagnetic transient programs. In the calculation of the parameters of the transmission line, it is necessary to take into account the coupling of conductors, the soil resistivity, and the frequency dependence of the parameters (see Chapter 5 and ref. [14]). In simplified cases, the assumption of lossless lines is an acceptable approach to the calculation of lightning surges, with the line represented as its surge impedance Z and its time delay τ.

A very simple example of a lightning stroke on a conductor cable, according to Figure 9.2, is presented next.

Example 9.1 An aerial cable is hit by a lightning stroke. The crest value of the lightning current is $I = 20$ kA and the surge impedance of the cable is $Z = 300\ \Omega$. Calculate the voltage at the conductor (see Chapter 2):

$$V = Z\frac{I}{2} = 3000\ \text{kV}.$$

The solution is very simple with the simplified model, which considers the surge impedance of the line and assumes that the current is divided into two halves traveling in each direction along the conductor. If the voltage exceeds the dielectric strength of the insulation, it is possible for a breakdown to occur.

Note that the voltage and current waveforms are the same, as are the peak values and this will be important in the analysis of direct strokes in cables, to be seen in Section 9.4.2. Alternatively, it is important to examine the lightning strokes that impact the tower or the shielding wires, as discussed in Section 9.4.3, focusing on backflashover phenomena. In such instances, both the towers and grounding systems play a role in the transient behavior, with some models presented below.

9.3.2 Towers

As detailed in [15], using numerical methods for surge calculations in transmission lines provides high accuracy but demands heavy computation. Therefore, EMT programs with simplified tower models can offer efficient and accurate solutions.

The towers are modeled as surge impedances and time delays. Certain parts of the tower can be modeled as inductances. Representation as an inductance makes it possible to increase the time step, which decreases the simulation time. Tower surge impedances can be calculated using empirical formulas and numerical methods, and the typical values obtained range from 70 to 250 Ω [11, 16–18]. In Figure 9.15:

h_t: tower height (m).
τ_t: time delay in the tower (μs).
r_1: radius of the tower top (m).
r_2: radius of the tower midsection (m).
r_3: radius of the tower base (m).
h_1: height from the base up to the midsection of the tower (m).
h_2: height from the midsection of the tower to the top (m).

Reference [11] suggests a surge propagation speed in towers at 80% of light speed, while [10] indicates it matches light speed. This topic is revisited in ref. [3], assuming:

$$\tau_t = \frac{h_t}{c}, \text{without crossarms,}$$

$$\tau_t = \frac{h_t}{0.85c} \text{ with crossarms.}$$

c: light speed 300 m/μs.

Calculation of the tower surge impedance depends on its geometry, as in Figure 9.5. In the same reference, the following expressions are presented to calculate the surge impedances in Ω:

- Conical tower [18].

$$Z_t = 60 \ln \cot\left(\frac{\theta}{2}\right), \tag{9.3}$$

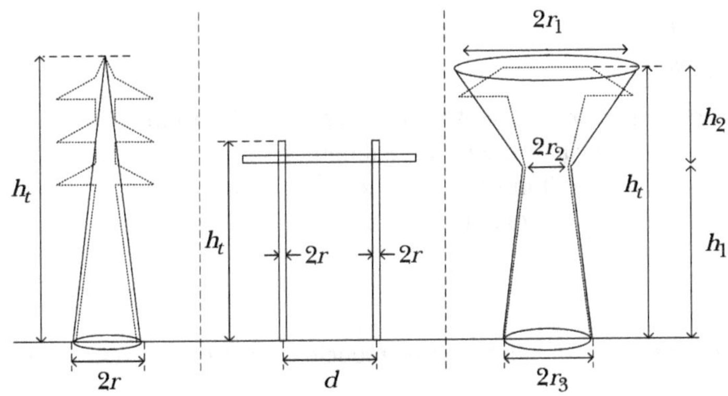

Figure 9.5 Typical tower shapes.

where,

$$\theta = \tan^{-1}\left(\frac{r}{h_t}\right).$$

For a cylindrical tower subtract 60Ω in (9.3)

- Portal [3].

$$Z_t = \frac{Z_1 \cdot Z_2}{Z_1 + Z_2}, \tag{9.4}$$

$$Z_1 = 60\left[\ln\left(\sqrt{2}\frac{2h_t}{r}\right) - 1\right],$$

$$Z_2 = \frac{d60\ln\left(\frac{2h_t}{r}\right) + h_t Z_1}{h_t + d}.$$

$$\tau_t = \frac{1}{cZ_t}\frac{h_t \cdot Z_1 \cdot (d + h_t) \cdot Z_2}{h_t \cdot Z_1 + (d + h_t) \cdot Z_2}$$

- Waist [10].

$$Z_t = 60\ln\cot\left(\frac{\theta}{2}\right), \tag{9.5}$$

where:

$$\theta = \tan^{-1}\left(\frac{r_{av}}{h_t}\right)$$

r_{av}: weighed mean of the tower radius.

$$r_m = \frac{r_1 h_2 + r_2 h_t + r_3 h_1}{h_t}, \tag{9.6}$$

All surge impedance equations presented above are valid for round shapes, as shown in Figure 9.5. For a structure with a rectangular shape with sides a and b, such as a typical lattice tower, it must be approximated with a cylindrical shape with the same perimeter, whose radius is given by:

$$r = \frac{a + b}{\pi}.$$

Towers having an irregular geometry can be modeled using the expressions in ref. [19].

9.3.3 Tower Grounding

The purpose of grounding is to facilitate the penetration of the lightning current into the soil and to reduce the voltage at the tower top.

Transmission line towers can be grounded with vertical rods or buried cables (also called counterpoise cables) distributed horizontally at some depth in the ground.

To obtain low values of the tower footing resistance, an increase in the number and in the length of cables may be necessary.

System grounding depends on soil resistivity, which can change along the right of way. In cases where the soil resistivity is low, more compact grounding systems can be adopted, composed of rods or short counterpoises, which can be designed in accord with the approaches presented in [3, 11, 20].

- Low frequency

  With respect to low-frequency currents, the resistance of a grounding rod and the resistance of one counterpoise wire were both described in ref. [21].

  Resistance of a grounding rod:

$$R_{lf} = \frac{\rho}{2\pi\ell} \ln\left(\frac{2\ell}{r}\right), \tag{9.7}$$

where:

r: radius of rod section.
ρ: soil resistivity in Ω m.
ℓ: rod length.

Resistance of one counterpoise wire:

$$R_{lf} = \frac{\rho}{\pi\ell}\left(\ln\frac{2\ell}{\sqrt{2rd}} - 1\right), \tag{9.8}$$

where:

r: radius of the cable cross section.
d: depth of the buried cable, parallel to the ground surface.

The grounding electrode model discussed in [22] includes both a geometric resistance and a contact resistance. This reference provides a comprehensive and detailed analysis of the contact resistance. The model considers the effects of burial depth and wire radius, providing accurate results compared to reference calculations. In addition to its generality, covering solid and wire-structured electrodes, the model can adapt as more grounding structures are added in parallel.

The formula for calculating the low-frequency resistance, R_{lf}, for typical transmission line earth electrodes, including vertical and horizontal buried wires, is presented below:

$$R_{lf} = \frac{\rho}{2\pi}\left[\frac{1}{g}\ln\left(\frac{11.8g^2}{A}\right) + \frac{1}{L_\omega}\ln\left(\frac{A}{F\cdot A_\omega}\right)\right], \tag{9.9}$$

where:

ρ: soil resistivity [Ω m].
g: electrode geometric radius equal to $g = \sqrt{r_x^2 + r_y^2 + r_z^2}$ [m].
A: surface area of electrode in contact with earth [m²].
L_ω: total length of the electrode wires [m].
A_ω: *wire* surface area. (For the same wire radius r_ω, $A_\omega = 2\pi r_\omega L_\omega$) [m²].
F: a constant, $F = 8/n^2$ for n radial wires F.

Equation (9.9) is valid for $A/(F\cdot A_\omega) > 1$.

- High frequency
  At higher frequencies, the behavior is no longer resistive, and the frequency response of the grounding system should be evaluated, mostly for long counterpoises.

  The long counterpoise cable is treated in a simplified manner in [1] with a traveling wave analysis in the counterpoise wire. At high frequency, increasing the length of the counterpoise too much is not very efficient due to the dissipation of the surge as it propagates along the wire.

  Working with a simplified model of the soil, ref. [23] developed a more elaborate set of equations, incorporating the frequency dependence of the resistivity and also the relative permittivity of the soil $\rho(\omega)$ and $\varepsilon_r(\omega)$, and adding elements neglected in analyses based only on traveling waves. From simplifications of more complex solutions, this reference provides expressions for the behavior of the grounding impedance as a function of time.

  A rigorous analysis should consider the solution of Maxwell's equations for buried conductors, to determine the distribution of voltages in the soil imposed by current excitation. The dependence of soil resistivity and relative permittivity on frequency should also be included. Therefore, it is possible to obtain the frequency response of the grounding impedance $Z(\omega)$ by means of numerical methods, assuming linear parameters. This approach improves the representation of the grounding impedance in specific studies.

  Reference [3] proposes a simplified method of approaching counterpoises, based both on the impulse impedance Z_p and on the impulse coefficient IC.

The impulse impedance is calculated by the ratio of the grounding potential rise (*GPR*), V_p, by the peak lightning current, I_p:

$$Z_p = \frac{V_p}{I_p}.$$

The impulse coefficient calculation, designated as the ratio

$$IC = \frac{Z_p}{R_{lf}},$$

is also described in this reference. Therefore, with R_{lf} and IC, the impulse impedance is calculated. Note that if the replacement of $Z(\omega)$ by Z_p is appropriate, it is reasonable to assume that when calculating the Fourier transform of voltage and current signals, the harmonic components are not very significant with respect to the fundamental components, all at high frequency, which should be verified.

- Soil ionization

  Another effect that can be added to these models is related to the soil ionization that reduces the grounding impedance. The model proposed in [10] can be applied in cases of grounding rods or short counterpoise cables, when the grounding system is small in length with a maximum effective diameter of 30 m.

$$R_i = R\left(1 + \frac{I}{I_g}\right)^{-1/2}, \; I_g = \frac{E_g \rho}{2\pi R^2}, \tag{9.10}$$

where:

R: grounding resistance (Ω).
R_i: tower footing resistance to current I, (Ω).
I: lightning current (kA).
I_g: ionization current limit (kA).
E_g: electric field corresponding to I_g (assumed equal to 400 kV/m in [10] and having a suggested value of 1000 kV/m in [11]).

Given the level of uncertainty concerning the soil's electrical behavior, care must be taken in order not to overestimate the ionization effect. The IEEE method assumes $R_i = R$ to be more conservative (Chapter 14).

For long counterpoise, the probability that the electric field reaches a given value of soil ionization is less, since the current is less concentrated due to its distribution on the wire.

9.3.4 Substation Equipment

The substations contain various pieces of equipment, such as transformers, reactors, potential transformers, current transformers, lightning arresters, and switches.

The models for some equipment operating at 60 Hz, such as transformers, for instance, are not appropriate at high frequencies because the inductive components take on very high values. In general, the substation equipment is represented by its equivalent stray capacitance.

For transformers and reactors, the equivalent stray capacitance represents the distributed capacitance of the windings. These capacitance values fall in the range from 1 to 4 nF. Another piece of equipment that can present a sizeable capacitance is the capacitive voltage transformer CVT. Stray capacitance values are lower for other types of equipment, ranging from 100 to 200 pF. A survey of the values of stray capacitance in several pieces of equipment can be found in refs. [2, 24].

9.3.5 Lightning Stroke Attachment

The lightning stroke attachment is addressed in refs. [3, 25]. In general, two approaches are adopted to evaluate the attachment process in transmission lines, the leader progression model and the electrogeometric model. The downward leader, which is predominant and originates from the cloud, and the upward leader, which starts from a grounded conductor, both evolve continuously with opposite charges.

The leader progression model [3, 26–29] is based on the attractive distance, which is the horizontal distance between the vertical axis of the descending leader and the conductor (or structure when the final jump enters its triggering zone). The model can be useful to explain specific cases of shielding failure, as for example, in UHV transmission lines.

The usual lightning analysis can be based on the electrogeometric model [30, 31]. Despite some specific shortcomings, the electrogeometric model exhibits very satisfactory statistical behavior, and it is adopted here as it is currently the simplest method for dealing with line shielding.

The electrogeometric model is calibrated using a significant number of records of lightning strokes on transmission lines. The model is based on the strike distance, defined as the distance from the tip of the downward leader to the target object (rather than the lateral distance). Within this region, lightning strikes the first object it encounters, Figure 9.6a.

The strike distance is a function of the amplitude of the prospective lightning stroke current. Different fittings can be found for the strike distance [3].

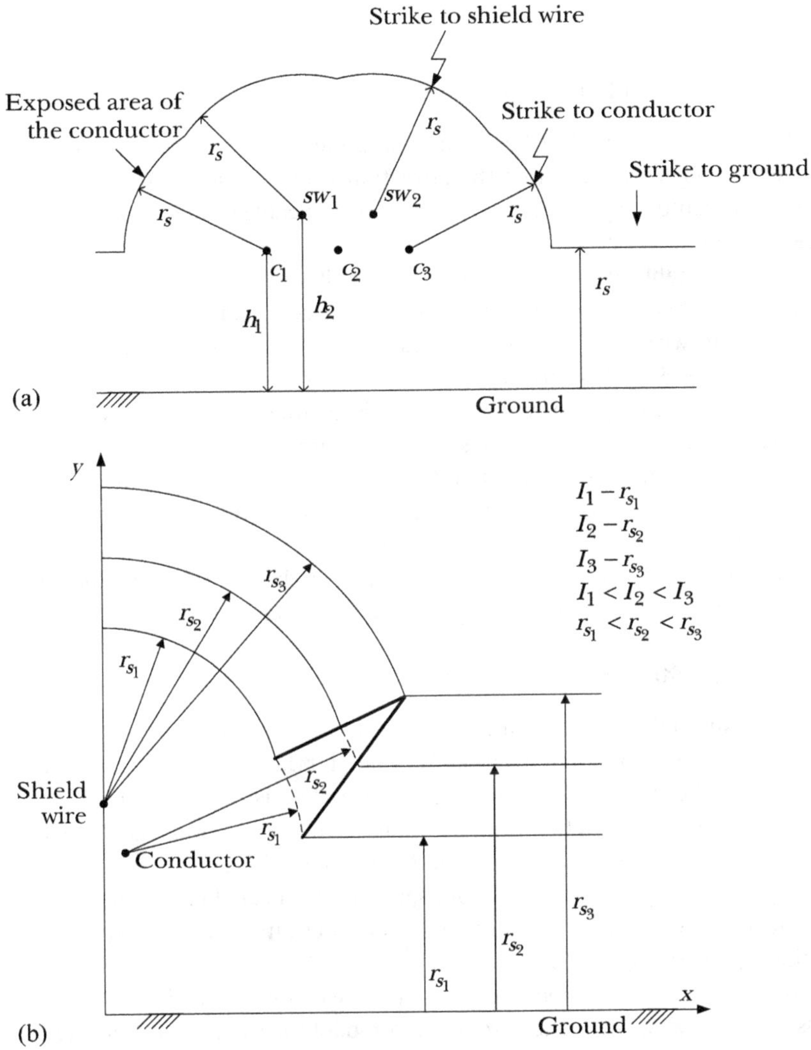

Figure 9.6 Electrogeometric model. (a) Strike distance to shield-wire, conductor and ground. (b) Shielding failure.

Reference [11] provides the following equation for the strike distance to conductor cables, shield wires and the ground:

$$r_s = 10I^{0.65}, \tag{9.11}$$

where:

r_S: strike distance (m).
I: lightning stroke amplitude - prospective (kA).

To determine the point of strike, to the line cable or to the ground, that is, to obtain the first object to penetrate the attraction zone, the solution is found by positioning the centers of the attraction sectors on the cables and defining a strike distance to ground.

Therefore, each cable of the transmission line has its circular sector, in which the lightning strikes this cable. The ground, which is also a conductor, has its own strike distance in which the lightning reaches the ground (Figure 9.6a).

The radius of the circular sector changes according to the current amplitude, and it can be shown that at high current levels, that is, by increasing the peak current, shield wire protection also increases. At a certain current level, the intersection between the shield wire sector and the horizontal line defined by the striking distance to the ground, offers complete protection to the phase conductors, directing strokes either to the shield wire or to the soil. At low current amplitudes, there are regions in which shield failure can occur, as shown in Figure 9.6b, with an increase in the exposed region of the phase conductors as the current is reduced.

9.3.6 Dielectric Strength of the Insulation

The dielectric strength of the insulation and its statistical behavior is discussed in Chapter 14. As the standard deviation is small for fast wavefront transients, assumed to be 3% of the average, this spread is usually disregarded, and it is sufficient to determine only whether the overvoltage is higher or lower than the CFO, which is the voltage at which the probability of flashover is 50%.

The dielectric strength of the insulators depends on the applied waveform. The results of laboratory tests are presented in the form of a curve depicting the amplitude of the applied voltage as a function of the time to breakdown of the insulation [10, 11, 13]. More elaborate models of insulator breakdown, applying integration methods or employing the leader progression model for the insulator string are presented in Chapter 14.

9.4 Transmission Line Analysis

Lightning performance analysis assesses the behavior of a transmission line subjected to lightning surges and furnishes its failure rate. Several evaluations are carried out to define a numerical index of the annual failure rate per unit length of the transmission line, usually focusing on a transmission line segment of 100 km [32]. Briefly, the calculations utilize the following variables: distance between conductors, coupling coefficients, the position of the shield wire and hence the shielding angle, insulation strength, tower model, tower footing resistances, soil resistivity, ground flash density, etc.

9.4.1 Lightning Strokes

Lightning strokes on transmission lines can happen at midspan or at the towers. The stroke can be a direct stroke to the phase conductor, or it can be a stroke to the shield wire. Distribution lines are also affected by lightning strikes reaching the soil close to the line structures. In this case, the lightning surges on the distribution lines are called induced overvoltages, which can also cause disruptive discharges.

9.4.2 Direct Stroke

The location of the shield wires in the tower should be such that shielding failures only occur at low current levels, which do not cause breakdown of the insulator. In the event of shielding failure, the maximum amplitude of the current that causes insulator breakdown is designated as the critical current; therefore, currents below this critical level should not cause the insulator breakdown.

Direct stroke surges, travel along the phase cable and last until they are damped due to conductor losses or corona. When the lightning stroke reaches a transmission line near a substation, it can affect the substation's equipment.

Example 9.2 The CFO of the insulator string in a transmission line is 900 kV and the surge impedance of the phase cable is 300 Ω. Determine the critical current for shielding failure:

$$I_{crit} = 2\frac{V}{Z} = 2 \times \frac{900}{300} = 6\text{ kA}.$$

This means that shielding failures are permissible at currents lower than 6 kA and the shield wires must be properly positioned for this purpose.

9.4.3 Back-Flashover

The back flashovers analysis assume that the lightning strikes the shield wire at the tower, causing a disruptive discharge in the insulator string. Based on the voltage at the tower top, V_t, which is coincident with the voltage at the shield wire, and also on the induced voltage at the phase conductor, kV_t, it is possible to calculate the voltage drop across the insulator string.

The coupling factor k, which depends on the spatial positions of conductors and shield wires, range from 0.15 to 0.30. Calculation of the coupling factor can be performed by considering the elements of the matrix of surge impedances of the transmission line. Assuming a current stroke I at shield wire i, the induced voltage at the phase conductor, j, is calculated using the matrix of surge impedances of the transmission line, according to the following equation:

$$\begin{bmatrix} V_1 \\ \vdots \\ V_i \\ \vdots \\ V_j \\ \vdots \\ V_n \end{bmatrix} = \begin{bmatrix} Z_{11} & \cdots & Z_{1i} & \cdots & Z_{1n} \\ \vdots & & & & \vdots \\ Z_{i1} & \cdots & Z_{ii} & \cdots & Z_{in} \\ \vdots & & & & \vdots \\ Z_{j1} & \cdots & Z_{ji} & \cdots & Z_{jn} \\ \vdots & & & & \vdots \\ Z_{n1} & \cdots & Z_{ni} & \cdots & Z_{nn} \end{bmatrix} \times \begin{bmatrix} 0 \\ \vdots \\ I \\ \vdots \\ 0 \\ \vdots \\ 0 \end{bmatrix} \tag{9.12}$$

where:

Z_{ji}: mutual surge impedance between shield wire i and phase conductor j.
Z_{ii}: surge impedance of the shield wire i.

$$V_i = Z_{ii} I, \tag{9.13}$$

thus,

$$I = \frac{V_i}{Z_{ii}}, \tag{9.14}$$

$$V_j = Z_{ji} I. \tag{9.15}$$

Substituting (9.14) in (9.15), the induced voltage at phase conductor j is calculated from:

$$V_j = \frac{Z_{ji}}{Z_{ii}} V_i. \tag{9.16}$$

The coupling factor between conductor j and shield wire i is given by:

$$k = \frac{Z_{ji}}{Z_{ii}}. \tag{9.17}$$

The voltage V_t can be calculated at the top of the tower or can be calculated at the crossarm, depending on the tower model used. In the case of lossless transmission lines, the equations employed to obtain (9.16) are presented in Chapter 3, Eqs. (3.60)–(3.67). Based on these equations, it can be shown that for a lossless line

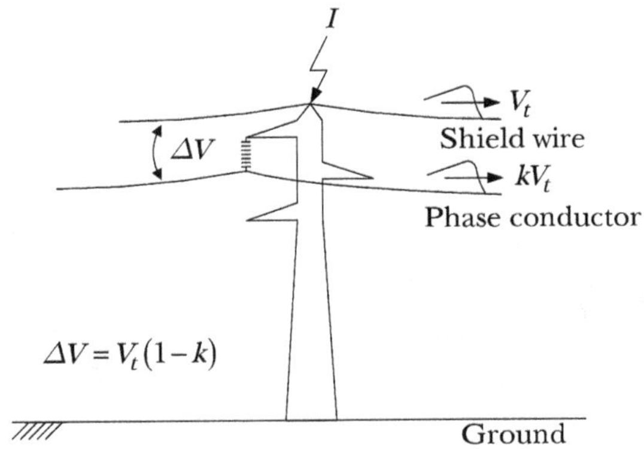

Figure 9.7 Voltage at the insulator string.

the same value for k in (9.17) is calculated by means of a capacitive voltage divider between shield wire i and phase conductor j, using only the capacitance matrix of the transmission line.

Using the tower top voltage V_t, and the induced voltage at the phase conductor, kV_t, the voltage drop on the insulator string may be written in (9.18), (see Figure 9.7). The case of a stroke in the meadle of the span is less significant due to corona effect, which increases k and reduces the voltage drop on the insulator. Moreover, the effect of the stroke in the middle of the span is incorporated into a factor applied to the tower stroke, simplifying the calculation.

$$\Delta V = (1 - k)V_t. \tag{9.18}$$

Since the shield wire is connected to the tower, calculation of V_t requires a traveling wave model of the tower, as previously described, but the above expressions remain valid.

In Figure 9.8a, the following variables are defined:

Z_{sw}: surge impedance of the shield wire or equivalent surge impedance if there are more than one shield wires.
Z_t: surge impedance of the tower.
τ: tower time delay.
R: tower footing resistance.

Simulation of some of the adjacent towers is necessary to improve the accuracy of the calculation because the waves reflected from these towers may reduce the voltage at the top of the tower, mainly for short spans.

In order to exemplify the calculation of the voltage at the top of the tower, the adjacent towers are not represented here, as it can be assumed that the time delay in the tower is much smaller than the time delay in the span between towers. For slow wavefronts, or even in the case of lines that are closer to adjacent towers, it may be necessary to represent the reflected waves at these towers in greater detail.

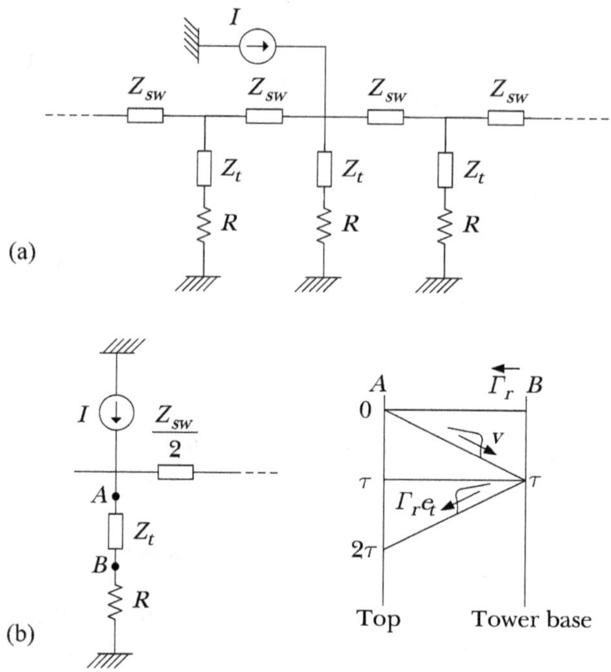

Figure 9.8 Model to calculate the voltage at the top of the tower. (a) Shield wire, conductor and tower footing resistance models. (b) Simplified circuit and lattice diagram.

The equivalent model in Figure 9.8b is adopted to calculate the traveling waves with the lattice diagram, where:

$Z_{sw}/2$: equivalent surge impedance assuming both sides of the transmission line are in parallel.
v: traveling wave that spreads to both sides of the shield wire toward the adjacent towers and in the tower.
$\Gamma_r e_t$: wave that returns to the top of the tower at twice the tower time delay.
τ: tower time delay.

The traveling wave in the tower is reflected from the tower footing resistance. The reflected wave has a sign opposite that of the incident wave e_t, if the tower footing resistance R is smaller than Z_t, since the reflection coefficient at the tower base is given by:

$$\Gamma_r = \frac{R - Z_t}{R + Z_t}. \tag{9.19}$$

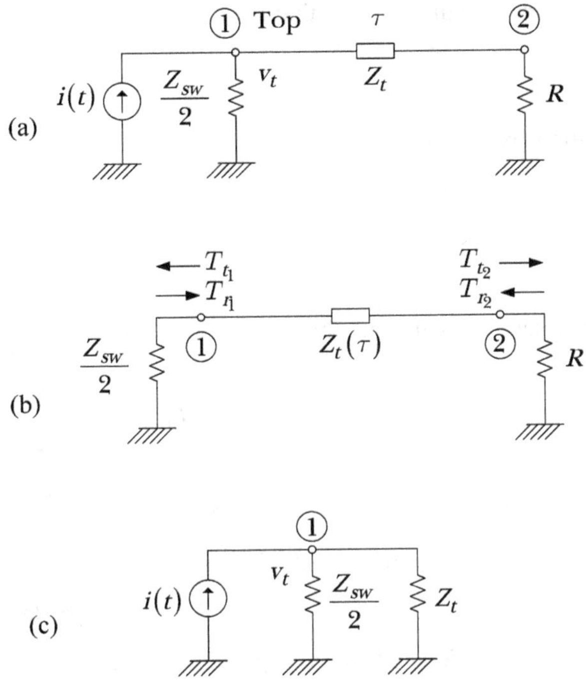

Figure 9.9 Voltage at the tower top. (a) Simplified circuit without adjacent towers. (b) Lattice diagram. (c) Simplified equivalent circuit.

Therefore, the tower footing resistance should be designed to be small, so that the tower top voltage is reduced, since towers with insufficient grounding could experience high tower-top voltages.

Figures 9.9a and b present the simplified circuit and the coefficients for calculating traveling waves, respectively. The model to associate the current and voltage applied to the tower is shown in Figure 9.9c.

The transmission and reflection coefficients are presented in Figure 9.9b and the applied voltage at the tower top can be calculated as in Figure 9.9c.

$$v(t) = i(t)Z_{eq}.$$

The surge impedance of the shield wire at the tower is given by $Z_{sw}/2$, due to the symmetrical conditions on both sides of the transmission line.

$$Z_{eq} = \frac{\frac{Z_{sw}}{2} Z_t}{\frac{Z_{sw}}{2} + Z_t} \Omega.$$

Assuming a linear wavefront for the current (note that k is the rate of rise of the current in the following equations):

$$i(t) = kt. \tag{9.20}$$

The voltage at the tower top is rewritten:

$$v(t) = \alpha t, \tag{9.21}$$

whose the rate of rise is given by:

$$\alpha = kZ_{eq}.$$

Equation (2.98) in Chapter 2 serves to calculate the tower top voltage, which is reproduced here:

$$v_t(t) = v(t) + \Gamma_{r_2}\Gamma_{t_1}v(t-2\tau) + \Gamma_{r_2}^2\Gamma_{r_1}\Gamma_{t_1}v(t-4\tau) + \Gamma_{r_2}^3\Gamma_{r_1}^2\Gamma_{t_1}v(t-6\tau) + \ldots \tag{9.22}$$

Thus, the tower top voltage can be written at times 2τ, 4τ, 6τ...

$$v_t(t) = \alpha\left[t + \Gamma_{r_2}\Gamma_{t_1}(t-2\tau) + \Gamma_{r_2}^2\Gamma_{r_1}\Gamma_{t_1}(t-4\tau) + \Gamma_{r_2}^3\Gamma_{r_1}^2\Gamma_{t_1}(t-6\tau) + \ldots\right] \tag{9.23}$$

After the wavefront time elapses, designated as t_f a negative rate of rise $-\alpha$ is applied to make the current wave flat. Note that the expression for the negative rate of rise is the same as that of the positive rate of rise, except for the time delay. The flat tail is not wrong, but just a little conservative since tail times are much greater than the front times. By increasing the negative rate of rise, the calculation of the tail effect can be improved.

$$v_t(t) = \alpha\begin{bmatrix} t + \Gamma_{r_2}\Gamma_{t_1}(t-2\tau) + \Gamma_{r_2}^2\Gamma_{r_1}\Gamma_{t_1}(t-4\tau) + \Gamma_{r_2}^3\Gamma_{r_1}^2\Gamma_{t_1}(t-6\tau) + \ldots \\ -(t-t_f) - \Gamma_{r_2}\Gamma_{t_1}(t-t_f-2\tau) - \Gamma_{r_2}^2\Gamma_{r_1}\Gamma_{t_1}(t-t_f-4\tau) - \ldots \end{bmatrix} \tag{9.24}$$

The voltage at the tower top allows calculation of the voltage drop in the insulator string and determines whether back-flashover occurs.

In (9.23) and (9.24) it may be observed that the peak voltage is proportional to α and consequently to the rate of rise of the current k.

Based on a simple model to qualitatively interpret the tower-top voltage, reference [3] models the tower as an inductance and the grounding electrode as a resistance, as shown in the following equation:

$$v_t(t) = L\frac{di(t)}{dt} + Ri(t).$$

Assuming the lightning current has a rise rate k, as in Eq. (9.21), the voltage drop along the tower body is a constant, given by kL, while the rise in ground potential is given by kRt. As in Eq. (9.23), the voltage at the top of the tower is proportional to the derivative of the current, emphasizing the need for an accurate representation of the rise rate.

It should be noted that the voltages at the crossarms are a little smaller than the tower-top voltage, which represents a reduction in V_t and thus lowers the value of ΔV. In ref. [13], interpolation between the voltages at the top and at the tower base is suggested, thus providing additional detail regarding the calculation. With improvement of the tower models, the crossarm voltage can also be calculated; see the crossarm modeling in [3]. Calculation of the probabilistic behavior of transmission lines subject to strokes employs either a probability density integration method or a Monte Carlo simulation method, as presented in Chapter 14. Calculations of the voltage at the top of the tower or the voltage drop at the insulator string can still contemplate a three-phase representation of all conductors and shield wires in the matrix of surge impedances. In the three-phase representation, it is possible to access the voltages at the insulators, and besides that, it is also possible to implement detailed models of the transmission line, towers, grounding system and the behavior of insulators subjected to non-standard waveforms.

So far, the discussion has been based on tower-top voltages that imply backflashovers; however, for each voltage V_t there is a corresponding applied current, which causes the same effect. The current levels that cause disruption of insulation are designated as critical currents. Thus, both approaches are equivalent, and the fundamentals of the probabilistic calculation are covered in Chapter 14, Section 14.6.1.

As mentioned before, the tower footing resistance plays an important role in the calculation of the critical current amplitude. An increase in this resistance implies lower levels of the critical current, increasing the probability of insulator breakdown.

Among several tools, the FLASH program, in [33], is a valuable resource for computing the lightning performance of transmission lines, a topic of focus in Chapter 14. Additionally, it offers significant support in analyzing the sensitivity of diverse parameters and variables influencing overvoltage behavior.

Electromagnetic transient programs allow a rather comprehensive representation of the electrical system [14, 34], permitting very complex modeling. With these programs, it is also possible to take into account the frequency dependence of the parameters as well as the corona effect. The availability of these programs has allowed three-phase representations, which have many advantages over single-phase simulations.

The three-phase representation is essential to the study of lightning surges when line arresters are installed in the transmission line. In this case, to determine the stress on the insulator string following a line-arrester intervention, it is necessary to represent the shield wires and the phase conductors, as in Section 9.4.4.

It is still necessary to consider the operating voltage of the phase conductor, at the nominal frequency, which influences the occurrence of flashover mainly in

EHV lines. The most distant conductor from the shield wire has the lowest coupling factor and is the most susceptible to back-flashover. However, when the operating voltage is significant, breakdown of the insulator string of a phase conductor can occur with a higher coupling factor, if the voltage drop at the insulator string is increased by that voltage.

In simulations using linear models, a proportionality factor relating voltages and currents can be applied. However, if nonlinear elements are included, the proportionality no longer exists.

The following example shows the numerical calculation of the tower-top voltage in a single-phase representation of the transmission line.

Example 9.3 Calculate the tower-top voltage waveform in a transmission line by means of the lattice diagram. Assume a lightning stroke to the tower (Figure 9.10).

Data

$I = 80$ kA: amplitude of the lightning stroke.
$t_f = 1.0$ μs: wavefront time of the current.
$Z_{sw} = 400\ \Omega$: surge impedance of the shield wire.
$Z_t = 70\ \Omega$: surge impedance of the tower.
$R = 20\ \Omega$: grounding resistance.
$d = 300$ m: distance between adjacent towers.
$h = 30$ m: tower height.

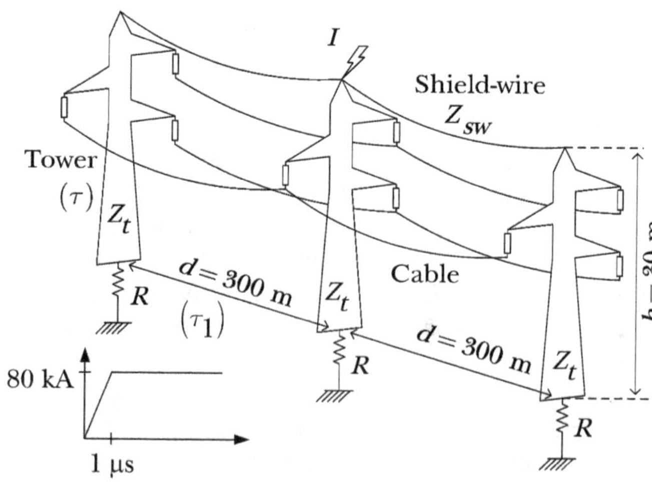

Figure 9.10 Stroke to tower in a transmission line.

Solution

The transmission line tower is simulated based on its surge impedance Z_t and time delay τ; only the voltage at the tower top is calculated, and the phase conductor is not represented.

For this transmission line, the distance between adjacent towers is 300 m and the tower height is 30 m; thus, the time delay between towers is 10 times greater than the time delay in one tower ($\tau_1 = 10\tau$).

Assuming the traveling waves propagate at the speed of light, the time delay between two towers (τ_1) is 1.0 μs and thus the adjacent towers only interfere with the calculation 2.0 μs after the start of the transient, corresponding to the round-trip time to the adjacent tower.

In this example, the peak current, and therefore the peak voltage, is reached within 1.0 μs, and hence, the shield wire may be represented as a transmission line of infinite length, as in Figure 9.9a.

Voltage applied at the tower top:

$$v(t) = i(t)Z_{eq}.$$

The voltage wave travels to the tower base and the reflected wave returns to the top of the tower with a time delay $2\tau = 0.2$ μs.

$$Z_{eq} = \frac{200 \times 70}{270} = 51.85\ \Omega,$$

$$\Gamma_{t_1} = \frac{2 \times 200}{270} = 1.48, \qquad \Gamma_{t_2} = \frac{2 \times 20}{90} = 0.44,$$

$$\Gamma_{r_1} = 0.48, \qquad \Gamma_{r_2} = -0.56.$$

Substituting the numerical values in (9.24):

$$v_t(t) = v(t) - 0.83v(t - 2\tau) + 0.22v(t - 4\tau) - 0.06v(t - 6\tau).$$

A current pulse with amplitude i, traveling in the tower, has a corresponding voltage pulse,

$$v = iZ_{eq}.$$

For each voltage pulse, the corresponding pulses at the tower top are shown in Figure 9.11. Any current waveform can be analyzed by subdividing it into current pulses with superposition of the responses.

In the example, the solution can be simplified with (9.21) for:

$$\alpha = 4148\ \text{kV}/\mu s.$$

The tower top voltage can be calculated using (9.24).

For $1 \le n \le 5$, ($t \le 10\tau$).

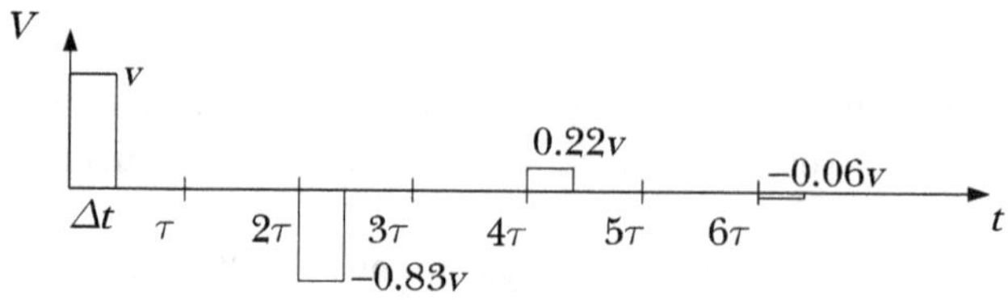

Figure 9.11 Voltage pulses at the tower top.

$$v_t(t) = \alpha \sum_{i=1}^{n} \left[t + \Gamma_{r_2}^{i} \Gamma_{r_1}^{i-1} \Gamma_{t_1} (t - 2i \times \tau) \right].$$

After the front time elapses, ($t_f = 5 \times 2\tau$), the second ramp must be applied, for $n \geq 6$, ($t > 10\tau$).

$$v_t(t) = \alpha \sum_{i=1}^{n} \left[t + \Gamma_{r_2}^{i} \Gamma_{r_1}^{i-1} \Gamma_{t_1} (t - 2i \times \tau) \right]$$
$$- \alpha \sum_{i=6}^{n} \left[(t - t_f) + \Gamma_{r_2}^{i-5} \Gamma_{r_1}^{i-6} \Gamma_{t_1} (t - 2i \times \tau - t_f) \right].$$

Substituting numerical values, the following table may be assembled, with results plotted in Figure 9.12:

Time (μs)		Voltage (kV)
0	t_0	0
0.2	$t_0 + 2\tau$	829.30
0.4	$t_0 + 4\tau$	976.44
0.6	$t_0 + 6\tau$	1305.90
0.8	$t_0 + 8\tau$	1586.44
1.0	$t_0 + 10\tau$	1880.16
1.2	$t_0 + 12\tau$	1340.70
1.4	$t_0 + 14\tau$	1485.00
1.6	$t_0 + 16\tau$	1385.89

As the current wave was assumed flat after 1.0 μs, the results for the tower top voltage are conservative since there would be a reduction in the current according to the tail time. The maximum tower-top voltage occurs before the arrival of the reflections from the neighboring towers in this example.

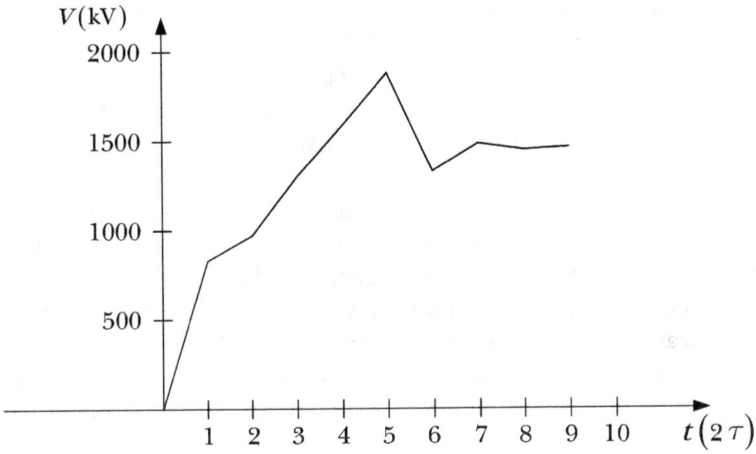

Figure 9.12 Voltage at the top of the tower.

9.4.4 Line-Arrester Application

Transmission lines with high failure rates are usually located in regions with high ground flash densities. The direct strokes are caused by a lack of shielding on phase cables, and the back-flashovers are associated with inadequate tower grounding. The grounding problem does not allow a satisfactory solution in many cases, especially when the soil resistivity and groung flash density are high.

Installation of line arresters in parallel with the insulator string is an attempt to improve the lightning performance of transmission lines with high outage rates [35] (Figure 9.13). This approach has proven to be more efficient than the classical

Figure 9.13 Line arrester in the tower.

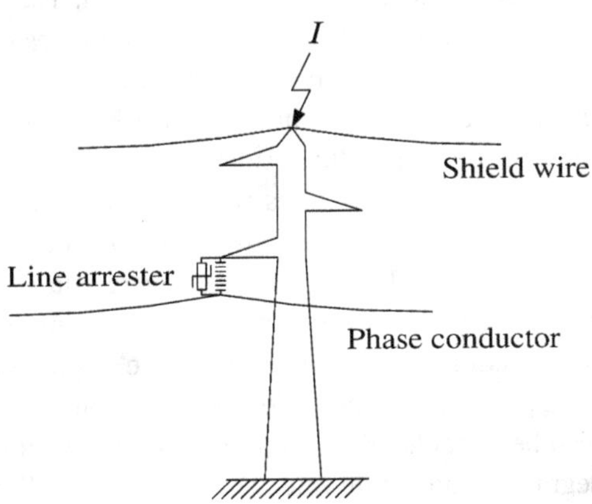

methods for correcting the shielding angle or improving the grounding system when high soil resistivity and ground flash density are high.

That idea is not new because the literature mentions attempt to apply expulsion-type lightning arresters designed for this purpose in the 1930s [36, 37]. These pioneering applications were not successful due to inadequate performance of the lightning arresters.

Current use of line arresters is linked to two basic factors arising from technological developments: manufacture of line arresters using metal oxides that give a more robust performance in the face of overvoltages and housing made of polymeric materials that are lighter, such that the weight of the line arresters does not have a significant impact on the tower structure.

Studies on the application of line arresters should thus determine a cost-effective number of units to be installed along the route of a transmission line, and their locations in the tower structures, thereby protecting the phases that are most susceptible to flashovers.

It is therefore necessary to identify the critical segments of the transmission line, in relation to their failure rate, and to analyze the best alternatives to locate the line-arresters.

To define alternative line-arrester locations at specific towers and phases on the towers, the possibility of back-flashover due to lightning strokes to the tower or flashovers due to direct strokes to phase conductors should be investigated, in view of the randomness of crest currents and wavefronts. An analysis of the energy and maximum current encountered by line arresters should be performed to establish the requirements for arrester specification.

The procedure to be adopted is quite like back-flashover analysis without line arresters, but extrapolation of current amplitudes is not possible because the arrester intervention causes the system to become nonlinear. The purpose of these studies is to obtain, for a given front time, the corresponding current amplitude that could cause insulator breakdown. Having determined the critical current level, lightning performance with the arresters may be calculated, following the approach described in Chapter 14. In fact, the critical currents can also be used in the case of lines without line arresters, as mentioned before, as an alternative to the tower-top voltage calculation.

This analysis addresses the widespread application of line arresters along the length of a transmission line or at specific critical line segments, as for example, at towers with high tower footing resistances and subject to frequent lightning strokes. Application of line arresters is quite common in all three phases of the transmission line within certain line segments, which may be showing poor lightning performance behavior, but as few as two (or even one) arresters per tower can also be considered. This procedure allows for suitable protection in that specific segment if adequate line arresters are installed. An additional evaluation of insulator behavior in the towers close to the protected segment is also recommended.

In the case of widespread application of line arresters in the three phases along the entire transmission line, it may be assumed that the number of faults will be greatly reduced, making lightning performance studies unnecessary. However, evaluation of the energy and current levels in the line arresters and possible strokes to the middle of the span should also be investigated.

The following example shows the overvoltages calculated by an electromagnetic transient program [34, 38], for a 138-kV transmission line. A comparison of the results for the transmission line with or without line arresters is shown.

Example 9.4 Figure 9.14 shows the tower geometry adopted for the simulations.

For a stroke to a tower without any line arresters, Figure 9.15a gives the voltages at the crossarm of the tower, at the lowest phase *c*, and at the insulator string. The results are calculated, assuming the occurrence of a lightning stroke with a crest value of 20 kA, a front time of 1.0 μs, and a tower footing resistance of 70 Ω.

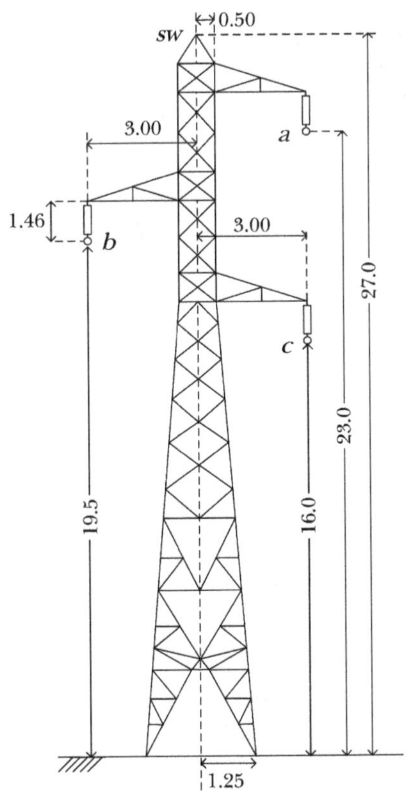

9 insulators
Conductor: LINNET
Shield wire: 3/8″ EHS
Conductors sag: 8.0 m
Shield wire sag: 5.5 m
Span: 350 m

Figure 9.14 Tower with a 138-kV line, single circuit arrester.

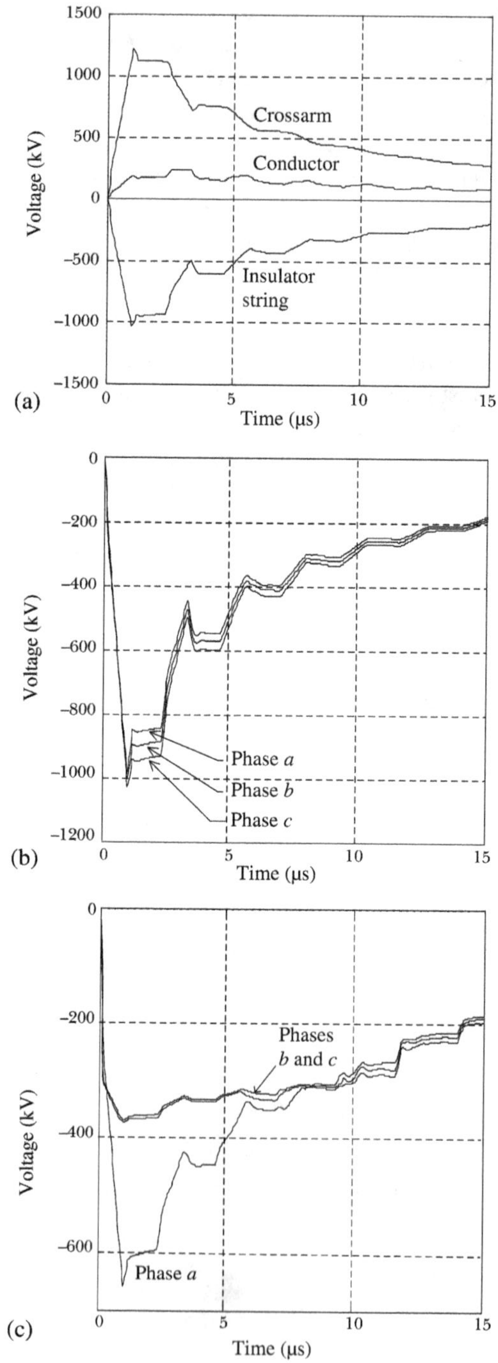

Figure 9.15 Stroke to tower. (a) Overvoltages at the crossarm, conductor, and insulator string without line arrester. (b) Overvoltages at insulator strings, without line arrester. (c) Overvoltage at phase *a* insulator string with line arrester on phases *b* and *c*.

Under the same conditions, Figure 9.15a and b presents the overvoltages on the insulator strings for the three phases, without line arresters installed on the transmission line.

Figure 9.15c presents the overvoltages at the insulator strings for the three phases, under the previous conditions, but this time with the line arresters installed on the two lower phases.

Comparison of Figure 9.15b and c shows that the application of two line-arresters to phases *b* and *c* reduces the maximum voltage on the insulator strings from approximately 1000 to 650 kV, for the same level of stroke current (i.e., 20 kA).

The critical currents for flashovers are shown in Figure 9.16, as a function of the current front times, for a tower footing resistance of 20 Ω. The line arresters are installed on every tower at the lower phase or at the two lower phases. Both alternatives are compared with the case without line arresters.

When the number of arresters is increased, the amplitude of the critical currents also increases, reducing the probability of flashover in the transmission line.

However, installation of line arresters increases the level of the critical currents. Note that the tower footing resistance has a significant influence on these current values. The risk of insulator breakdown evidently depends on the position of the critical current curve in relation to the joint probability density function of the current amplitudes and front times.

Another important topic is the energy requirements for the application of line arresters on transmission lines. These should be investigated in relation to strokes

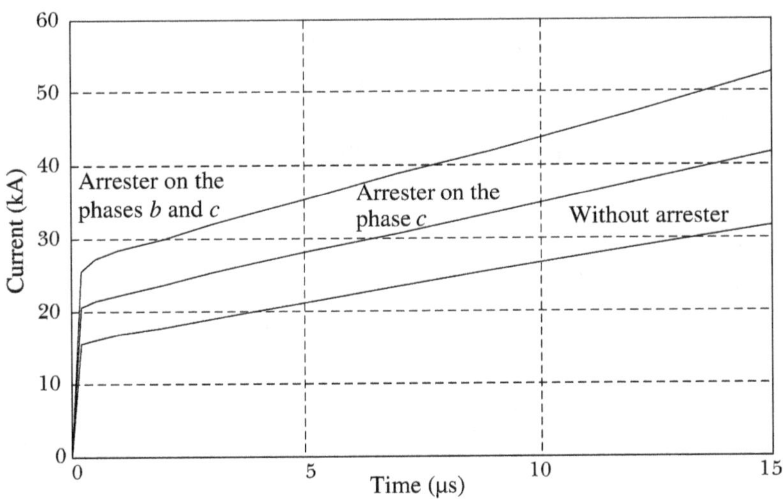

Figure 9.16 Critical currents for a tower footing resistance of 20 Ω.

to phase conductors or the tower, to establish equipment specifications. The line arresters will be stressed the most in lines without shield wires, since the current through the arresters is higher in this case. Besides their shielding effect, the presence of shield wires also offers the possibility of current distribution throughout the tower and the grounding system. The number of line arresters in each tower is also important, because they divide the dissipated energy between them.

A simplified interpretation of the energy absorbed by metal oxide line arresters is presented next. For that purpose, an elementary circuit is analyzed, disregarding the reflections from adjacent towers and assuming the residual voltage at the line arrester to be constant. The voltage drop at the tower is significant initially, during the lightning current wavefront when the voltage at the top of the tower is high, but after some initial reflections, this voltage drop is practically null. To calculate arrester energy during the current tail period, the tower voltage drop is neglected for simplicity.

The circuit in Figure 9.17 shows the pathways by which the lightning could reach both the conductor, with switch S_1 closed, and the shield wire, with switch S_2 closed. The line is protected by shield wires and line arresters. The conductor cable and the shield wire are represented by their surge impedances.

Z_c, Z_{sw}, Z_m: surge impedances of the conductor, shield wire, and mutual impedance, respectively.
R: tower footing resistance.

An ideal line arrester is assumed, which has an approximately constant residual voltage v_p, and the following variables are defined:

I_d: current due to the direct stroke applied to the phase conductor.
I_s: current due to the stroke to shield wire and tower.
i_r: current in the line arrester, without the effect of the residual voltage, v_p.
i_a: current reduction in the line arrester due to the residual voltage, v_p.

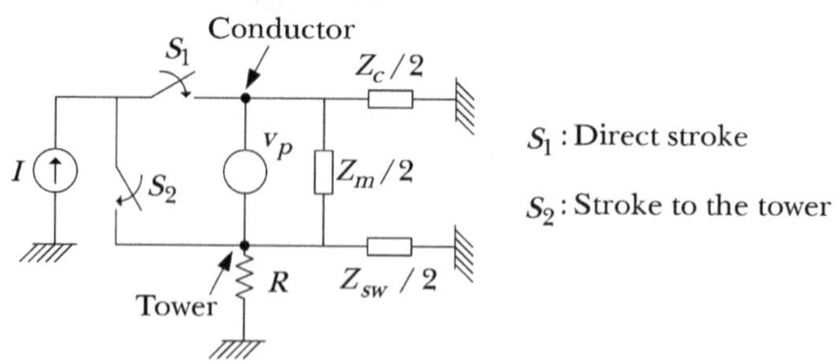

Figure 9.17 Stroke to the conductor or to the tower.

Applying the superposition principle to Figure 9.17, the following analysis may be presented:

a) Line with shield wire.
For a direct stroke on the phase conductor, with S_1 closed, the currents are not expected to be very high due to the shielding. R_e is the equivalent resistance of R in parallel with one-half of Z_{sw}, as shown in Figure 9.18a:

$$R_e = \frac{Z_{sw}\, R}{Z_{sw} + 2\,R}, \tag{9.25}$$

Current i_a is obtained with a direct stroke to the shielded line and is equal to:

$$i_a = 2v_p\left(\frac{1}{Z_c + 2R_e} + \frac{1}{Z_m}\right), \tag{9.26}$$

Current i_a is due to the arrester voltage acting as a counterelectromotive force.

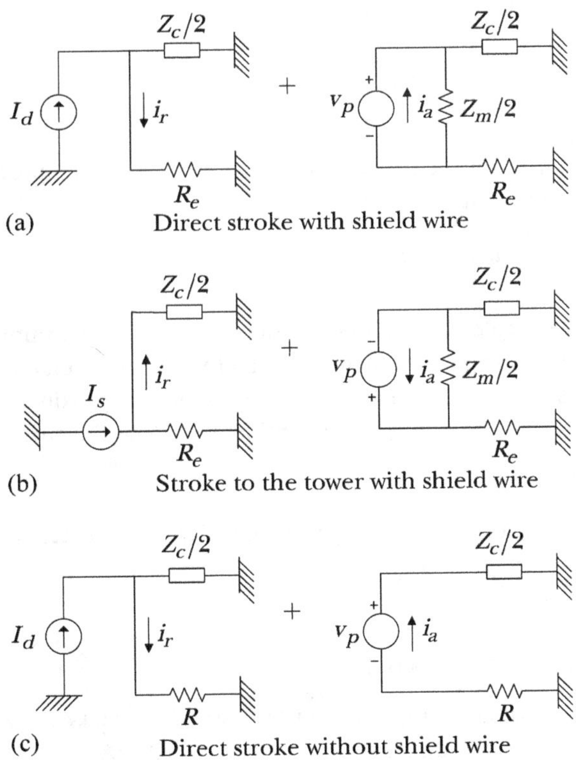

Figure 9.18 Current in the line arrester. (a) Diret stroke with shield wire. (b) Stroke to tower with shield wire. (c) Direct stroke without shield wire.

Current i_r can be calculated based on the current divider between R_e and one-half of Z_c.

$$i_r = \frac{Z_c}{Z_c + 2R_e} I_d, \tag{9.27}$$

For a stroke to the shield wire connected to the tower, with S_2 closed, Figure 9.18b:

$$i_r = \frac{2R_e}{Z_c + 2R_e} I_s. \tag{9.28}$$

Moreover, usually $i_a \ll i_r$, in cases of line arresters with low ratings, as calculated in (9.26).

b) Line without shield wire.
   In this case, Figure 9.18c, for a direct stroke to the phase conductor (bearing in mind that the current can attain high values):

$$R_e = R, \tag{9.29}$$

$$i_a = \frac{2v_p}{Z_c + 2R}, \tag{9.30}$$

$$i_r = \frac{2R}{Z_c + 2R} I_d. \tag{9.31}$$

The total current in the line arrester, in all previous cases, is calculated by superposition:

$$i_p = i_r - i_a. \tag{9.32}$$

Example 9.5 A 138-kV transmission line is examined in this example, with a line arrester installed in parallel with the insulator string. The residual voltage is assumed to be 400 kV and to remain approximately constant during the time of the line arrester intervention. Calculate the current in the line arrester:

Data

Tower footing resistance values: 20 and 100 Ω;

$Z_c = 480\,\Omega$

$Z_{sw} = 500\,\Omega$

$Z_m = 2000\,\Omega$

Assumed stroke current to tower: $I_b = 200$ kA.
Assumed direct stroke current: $I_d = 20$ kA.

Solution

The following results are based on Eqs. (9.25)–(9.32).

a) Line with shield wire.
Using a tower footing resistance of 20 Ω,

$$R_e = 18.64\,\Omega.$$

$$i_a = 1.95\,\text{kA}.$$

Current in the line arrester assuming direct stroke: $i_{ad} = 16.61$ kA.
Current in the line arrester assuming back-flashover: $i_{ab} = 10.5$ kA.
Using a tower footing resistance of 100 Ω.

$$R_e = 73.33\,\Omega.$$

$$i_a = 1.68\,\text{kA}.$$

Direct stroke: $i_{ad} = 13.64$ kA.
Back-flashover: $i_{ab} = 45.13$ kA.

b) Line without shield wire

$$I_d = 200\,\text{kA}.$$

Tower footing resistance: 20 Ω.

$$i_a = 1.54\,\text{kA}.$$

Direct stroke: $i_{ad} = 183.08$ kA.
Tower footing resistance: 100 Ω.

$$i_a = 1.18\,\text{kA}.$$

Back-flashover: $i_{ab} = 140.00$ kA.

Since the residual voltage is assumed to be practically constant, the energy absorbed by the arresters is the integral of the current, and it can be shown that in transmission lines without shield wires, direct strokes are more severe than back-flashover. For direct strokes, it may be observed that the energy rises with the reduction in the tower footing resistance, and for back-flashover, the energy rises with the increase in this resistance.

The simplified qualitative analysis presented above gives an overview of the different stress conditions on line arresters due to lightning strokes. An accurate energy evaluation should be performed, since certain limitations of the simplified model can affect the numerical results, such as the influence of adjacent towers on the wave reflections, the number of arresters per tower, disruption of insulator strings, the residual voltage on the arrester, and other factors.

9.4.5 Induced Overvoltages in Transmission Lines

These overvoltages occur during lightning strokes near the transmission lines, without direct contact with shield wires or phase conductors (Figure 9.19).

The magnitude of the induced overvoltages is less than 300 kV, which has a larger impact on distribution systems with a nominal voltage of up to 69 kV. Therefore, their effect is not considered in the transmission lines above 138 kV.

On the other hand, the study of induced overvoltages is of greater importance in the analysis of distribution lines, in which the effect is more significant because although the amplitude of the voltages is not high, flashovers can occur. The complexity of the phenomenon instigates researchers to look for models that can reproduce waveforms of field measurements.

Calculation of induced voltages in transmission lines is relatively complex, involving the electromagnetic fields produced by the return stroke current. Different models exist attempting to explain the induced voltages [7, 39–51]. Calculation of the induced voltages is performed in three stages: (a) modeling of the lightning channel to obtain the temporal and spatial current waveforms; (b) calculation of the electromagnetic field created by the lightning current and its propagation in time and space; (c) modeling of the coupling between the electromagnetic fields and the transmission line.

- Model for the Lightning Channel
  In the calculation of the induced voltages, the model of the current in the lightning channel requires greater accuracy than the simple current generator approach adopted so far. Some physical models attempt to explain the behavior of the current, according to the review presented in ref. [7]. Within the time frame of interest, the transmission line representation of the lightning channel (TL – transmission line model), yields reasonable results and simplifies the

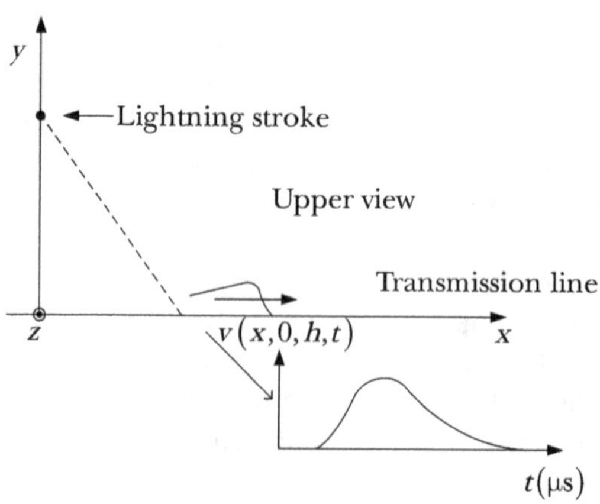

Figure 9.19 Voltages induced in transmission lines.

calculation. The lightning channel is assumed perpendicular to the ground and the current propagates without distortion.

$$i(x,t) = i\left(0, t - \frac{x}{\nu}\right). \tag{9.33}$$

ν: propagation velocity of the current.

- Electromagnetic Field
  The electromagnetic field propagation using Maxwell's equations was rigorously formulated in [50]. Complexity increases when considering the soil resistivity in the components of the electric field.
  The calculation of the electromagnetic field is presented in several works and the discussion on the charge distribution along the channel can be found in ref. [42].
  Several methods are proposed to compute the electromagnetic field, especially the horizontal component of the electric field.
  To overcome the complex formulations in the frequency domain in [50], an efficient proposal in [40, 42, 48] introduces the concept of the surface impedance of the ground, which can be used to calculate the horizontal field accurately for ground conductivity above 0.001 S/m [43].
- Electromagnetic Coupling Model
  The electromagnetic fields in Cartesian coordinates are presented in [41].
  It is assumed that the total electric field, $\vec{E}$, and magnetic density field, $\vec{B}$, are obtained by the sum of the incident fields, $\vec{E}^i$ and $\vec{B}^i$, that would exist without the transmission line and the scattered fields, $\vec{E}^s$ and $\vec{B}^s$, respectively, created by induction effects on the line.

$$\vec{E} = \vec{E}^i + \vec{E}^s. \tag{9.34}$$

$$\vec{B} = \vec{B}^i + \vec{B}^s. \tag{9.35}$$

Writing a generic field:

$$\vec{F} = \vec{F}_x\vec{x} + \vec{F}_y\vec{y} + \vec{F}_z\vec{z}, \tag{9.36}$$

$\vec{x}$, $\vec{y}$, $\vec{z}$: unit vectors in the x, y and z direction, respectively.

All components of fields are functions of (x, y, z, t) and may be written in the same way of Eq. (9.36).

The electric and magnetic fields generated by the lightning flash at the power line can be calculated from the scalar potential, ϕ^i, and vector potential, $\vec{A}^i$, of the electromagnetic field. The incident electric and magnetic density fields, $\vec{E}^i$ and $\vec{B}^i$, are given by:

$$\vec{E}^i = -\text{grad}\phi^i - \frac{\partial \vec{A}^i}{\partial t}, \tag{9.37}$$

$$\vec{B}^i = \nabla \times \vec{A}^i. \tag{9.38}$$

The vector grad ϕ^i is written:

$$\text{grad } \phi^i = \frac{\partial \phi^i(x,y,z,t)}{\partial x}\vec{x} + \frac{\partial \phi^i(x,y,z,t)}{\partial y}\vec{y} + \frac{\partial \phi^i(x,y,z,t)}{\partial z}\vec{z}. \tag{9.39}$$

The power cable is located on the x axis and the horizontal component E_x^i of the incident electric field can be exemplified:

$$E_x^i(x,y,z,t) = -\frac{\partial \phi^i(x,y,z,t)}{\partial x} - \frac{\partial A_x^i(x,y,z,t)}{\partial t}. \tag{9.40}$$

Next, the coupling model of the electromagnetic fields with the electrical system will be obtained through transmission line equations. The following coupling model was proposed in [39], whose equations are based on scattered variables in the transmission line equations, but excited by incident electromagnetic fields. Although improvements may be proposed, this is the most usual model in describing the induced voltages in transmission lines.

Using Maxwell's law for total electromagnetic field,

$$\nabla \times \vec{E} = -\frac{\partial \vec{B}}{\partial t}. \tag{9.41}$$

According to Figure 9.20, integrating (9.41),

$$\int_C \vec{E}\cdot\vec{d\ell} = -\frac{\partial}{\partial t}\int_S B_y\vec{y}\cdot d\vec{S}. \tag{9.42}$$

$\vec{S}$: rectangular surface in the $x-z$ plane ($dS = dxdz$).

C: counterclockwise contour encircling $\vec{S}$.

Due to the cable location in x axis, only the main coordinates in plane x-z are used (coordinate y and variable t may be omitted).

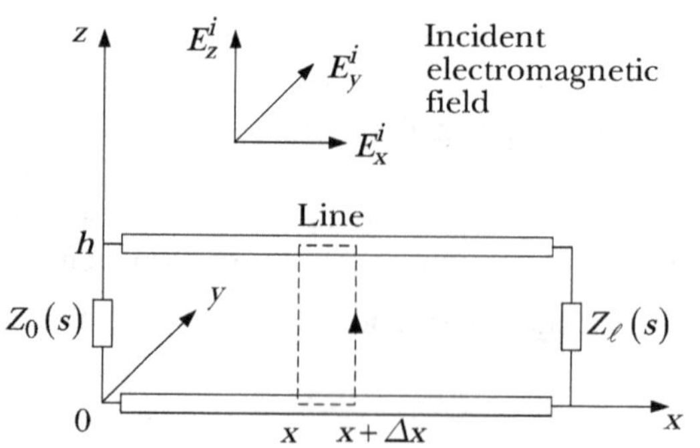

Figure 9.20 Electromagnetic field and circuit for calculation of the induced voltages.

$$\int_0^h E_z(x+\Delta x,z)dz - \int_0^h E_z(x,z)dz - E_x(x,h)\Delta x + E_x(x,0)\Delta x = -\frac{\partial}{\partial t}\Delta x\int_0^h B_y(x,z)dz. \tag{9.43}$$

Dividing (9.43) by Δx and taking the limit as $\Delta x \to 0$:

$$\frac{\partial}{\partial x}\int_0^h E_z(x,z)dz - \{E_x(x,h) - E_x(x,0)\} = -\frac{\partial}{\partial t}\int_0^h B_y(x,z)dz. \tag{9.44}$$

The aerial conductor and the reference conductor are separated by a vertical distance h and the scattered line voltage V^s between them is given by:

$$V^s = -\int_0^h E_z^s(x,z)dz. \tag{9.45}$$

Defining the following variables in (9.44),

$$\Theta = \frac{\partial}{\partial x}\int_0^h E_z(x,z)dz, \tag{9.46}$$

$$\Xi = E_x(x,h) - E_x(x,0), \tag{9.47}$$

$$\Psi = \frac{\partial}{\partial t}\int_0^h B_y(x,z)dz, \tag{9.48}$$

substituting (9.46), (9.47), and (9.48) in (9.44), Eq. (9.49) is obtained:

$$\Theta - \Xi = -\Psi. \tag{9.49}$$

Note that (9.44) and (9.49) are defined for total fields but can be applied to scattered and incident fields individually, as in (9.34) and (9.35). Thus, the total variables Θ, Ξ, and Ψ, written as $\Theta = \Theta^s + \Theta^i$, $\Xi = \Xi^s + \Xi^i$ and $\Psi = \Psi^s + \Psi^i$ may be substituted and in (9.49) by the sum of incident and scattered fields and placing their scattered variables on the left-hand side and incident variables on the right-hand side, as in [39]. However, it is easier to get the same result in [39] writing (9.49) only for scattered variables:

$$-\Theta^s - \Psi^s + \Xi^s = 0. \tag{9.50}$$

Substituting (9.45),

$$\Theta^s = -\frac{\partial}{\partial x}V^s, \tag{9.51}$$

in (9.50),

$$\frac{\partial}{\partial x}V^s - \Psi^s + \Xi^s = 0. \tag{9.52}$$

Adding the incident field E^i both on right and left-hand sides of (9.52), the final equation proposed in [39] is obtained:

$$\frac{\partial}{\partial x}V^s - \Psi^s + \Xi = \Xi^i. \tag{9.53}$$

Equation (9.53), written for one conductor, can be generalized for multi-conductors, associating $-\Psi^s$ to inductive voltage drops, $-\underset{\sim}{\Psi}^s = L'\frac{\partial}{\partial t}\underset{\sim}{I}(x)$, and the total electric field, Ξ to resistive voltage drops, $\underset{\sim}{\Xi} = R'\underset{\sim}{I}(x)$, by means of the per-unit-length inductance and resistance matrices L' and R', respectively. In case of multiple conductors, the equation used in [39] is presented below:

$$\frac{\partial}{\partial x}\underset{\sim}{V}^s + L'\frac{\partial}{\partial t}\underset{\sim}{I}(x) + R'\underset{\sim}{I}(x) = \underset{\sim x}{E}^i(x,h) - \underset{\sim x_0}{E}^i(x,0). \tag{9.54}$$

Note that the forcing function that excites the line is the horizontal electric field that appears only on the right-hand side, which is a significant contribution to the numerical solution of the illuminated transmission line.

Next step is to substitute the return conductor with the ground and consider the dependence on frequency of the series parameters of the transmission line [47].

The second transmission-line equation is derived from Maxwell's equation.

$$\nabla \times \vec{H} = \vec{J} + \frac{\partial \vec{D}}{\partial t}. \tag{9.55}$$

Integrating (9.55) over an external cylindrical surface of the conductor with radius r and length Δx:

$$\oint \vec{J}.d\vec{S} + \oint \frac{\partial \vec{D}}{\partial t} d\vec{S} = I(x+\Delta x) - I(x) + \Delta x\frac{\partial Q}{\partial t}. \tag{9.56}$$

On the right-hand side, the first two parcels refer to the integral of the current density on the two flat surfaces of the cylinder.

The third parcel is the displacement term, where Q is the total net charge per unit length on the conductor.

Dividing (9.56) by Δx and taking the limit as $\Delta x \rightarrow 0$,

$$\frac{\partial}{\partial x} I(x) + \frac{\partial}{\partial t} Q = 0. \tag{9.57}$$

Equation (9.57) can be generalized for its matrix form in (9.58).

$$\frac{\partial}{\partial x} \underset{\sim}{I}(x) + C' \frac{\partial}{\partial t} \underset{\sim}{V^s}(x) = 0, \tag{9.58}$$

where C' is the capacitance matrix of the conductors in per unit length of the line.

In (9.58) the voltage on the line is the scattered voltage and the total voltage V on a generic conductor is given by (Figure 9.21):

$$V(x) = V^s(x) + V^i(x) = V^s(x) - \int_0^h E_z^i(x, z) dz. \tag{9.59}$$

The incident vertical field is assumed approximately constant in the interval $0 - h$, simplifying the integral in (9.59).

Only the first steps in calculating induced voltages were derived here, and the solution to Eqs. (9.54) and (9.58) by the finite difference method as well as more details of the model formulation are found in [39].

There are different approaches for solving the model equations coupled to an electrical system and a well-tested code is presented in [45, 46]. Coupling codes interfaced to EMT programs have been developed in [43, 52], the main advantage of which is the coupling of the induced voltage model to the available models in the programs, thus allowing analysis of complex networks.

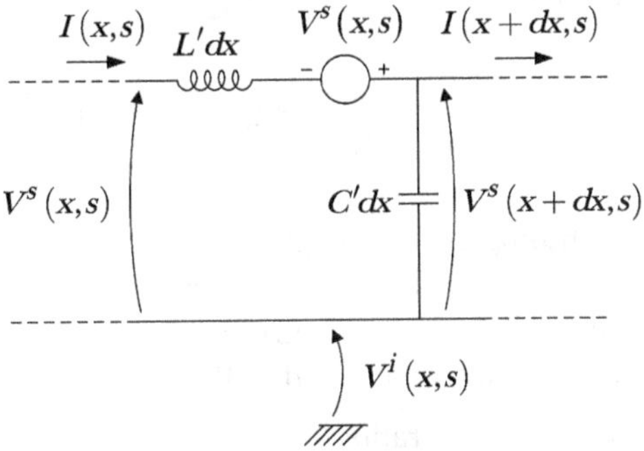

Figure 9.21 Calculation of induced voltages.

The case of a lossless line above a ground plane assumed to be a perfect conductor is frequently considered to address the main effects of induced voltages and is presented below. Assuming this simplification, the resulting transmission line equations are given by:

$$\frac{\partial V^s(x,t)}{\partial x} + L'\frac{\partial I(x,t)}{\partial t} = -\frac{\partial \phi_i(x,z,t)}{\partial x} - \frac{\partial A_x^i(x,z,t)}{\partial t}, \tag{9.60}$$

$$\frac{\partial I(x,t)}{\partial x} + C'\frac{\partial V^s(x,t)}{\partial t} = 0. \tag{9.61}$$

The electrical field on the right-hand side in (9.60) is written in its equivalent form given in (9.40) and Eq. (9.58) is repeated in (9.61). Equations (9.60) and (9.61) can be extended to the multiconductor case, assuming L' and C' to be the per-unit-length inductance and capacitance matrices, respectively, of the aerial conductors.

The first model to calculate induced voltages in transmission lines was presented in ref. [49] and it is still useful to perform calculations under the simplified assumptions of lossless system and lightning channel perpendicular to the soil.

Reference [41] demonstrates the equivalence of the results in [39] and [49], if the effect of the incident potential vector A_x^i in (9.60) is neglected for lossless transmission lines.

In [49], the induced voltage on a transmission line, caused by a lightning current step I, is obtained from the following expression:

$$V = V_1 + V_2, \tag{9.62}$$

where:

$$V_1 = Z_0 Ihb\frac{ct-x}{y^2 + b^2(ct-x)^2}\left(1 + \frac{x + b^2(ct-x)}{\sqrt{b^2c^2t^2 + (1-b^2)(x^2+y^2)}}\right), \tag{9.63}$$

V: induced voltage,

$$Z_0 = \frac{1}{4\pi}\sqrt{\frac{\mu_0}{\varepsilon_0}} = 30\,\Omega. \tag{9.64}$$

I: lightning current step.
h: line height above ground (m).
v: speed of the return current.
$c = 300 \times 10^6$ m/s, speed of light.
$b = \frac{v}{c}$: velocity ratio.
t: time after the return stroke current (S).

y: distance between the lightning contact point and the transmission line (m).
x: distance from the origin (the origin is located at the transmission line obtained from a perpendicular straight line passing through the stroke point).

V_2 is obtained from Eq. (9.63) by replacing x with$-x$.

The maximum voltage at the origin, $x = 0$, is calculated using the following expression:

$$V_{0\,max} = \frac{Z_0 I h}{y}\left(1 + \frac{1}{\sqrt{2}}\frac{b}{\sqrt{1-\frac{b^2}{2}}}\right). \tag{9.65}$$

Equation (9.65) furnishes the maximum induced voltage in a long aerial cable, calculated using a current step. To consider the effect of the front time of the return stroke, the wavefront can be calculated with the aid of the integral of Duhamel by superposition of responses to a sequence of steps [2].

Applying a unitary current step $H(t)$, the induced voltage $v(t)$ is calculated at a certain point on the transmission line, Figure 9.22a.

The time response to a generic current excitation is obtained by the superposition of the following incremental steps with their corresponding time delays.

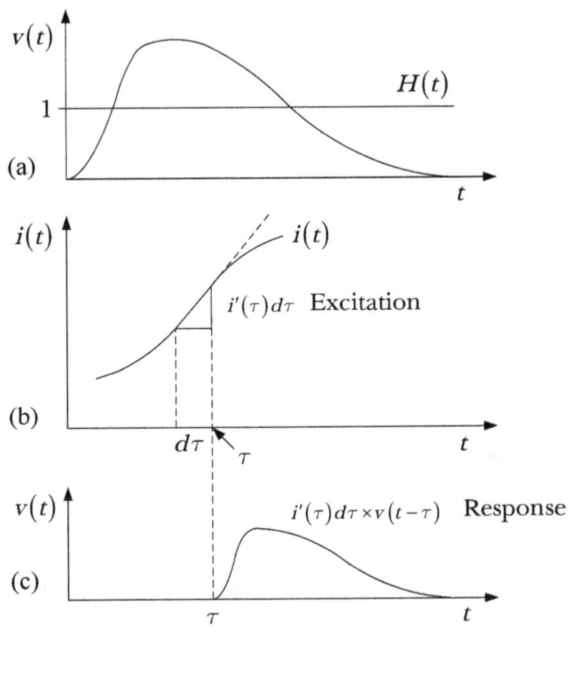

Figure 9.22 Superposition of steps. (a) Unitary step response. (b) Incremental step for a generic current excitation. (c) Time response for the incremental step excitation.

If $i'(\tau)$ is the derivative of the current at time τ, the amplitude of the incremental parcel for a step of interval $d\tau$ is given by $i'(\tau)d\tau$, Figure 9.22b. The corresponding time response to this incremental excitation, at $t \geq \tau$, may be written as $i'(\tau)d\tau \times v(t-\tau)$, Figure 9.22c. The induced voltage is obtained using the convolution integral:

$$v_t(t) = \int_0^t i'(\tau)v(t-\tau)d\tau, \tag{9.66}$$

which can be rewritten as:

$$v_t(t) = \int_0^t i'(t-\tau)v(\tau)d\tau. \tag{9.67}$$

If the current wavefront is assumed linear, with a constant derivative k, Eq. (9.67) may be rewritten as:

$$v_t(t) = k\int_0^t v(\tau)d\tau. \tag{9.68}$$

The integral in (9.68) can also be easily understood in the frequency domain, as the voltage in [49] was calculated for the step response, $(1/s)$. With the ramp, the same system should respond to (k/s^2), which is the integral of the step response multiplied by k.

Since this is the voltage at point x on the transmission line, a time delay needs to be considered; thus, $v_x(t) = v_t(t - t_d)$.

$$t_d = \frac{\sqrt{y^2 + x^2}}{c}.$$

t_d: time delay.

The effect of the return-stroke current tail can be obtained by superposition of another injected current but of negative slope, with the same approach in Example 9.3.

9.4.6 Underground Cables

Single-phase underground cables have models quite similar to those of aerial transmission lines, except that shielding and grounding also have to be represented. The surge impedances of the cables are smaller than those of transmission lines, and initial attenuation of the waves transmitted from aerial transmission lines to underground cables should be expected, but overvoltages can occur later due to traveling waves. The surge impedances of the cables are frequently lower than 60 Ω, and the propagation speeds are about one-third the speed of light due to the high dielectric constant used in the insulation.

Cable models are complex because they take into account the stranded core conductor, the insulation layers, the bedding, and metallic and plastic sheaths. However, accurate cable models are available in electromagnetic transient programs.

9.4.7 Corona

Corona occurs in shield wires and conductors when they are subjected to voltage surges of high amplitude. Several references address the corona effect on transmission lines, with the phenomenon reviewed in refs. [10, 13, 24], but only the case of overvoltages with fast wavefronts is addressed here.

The ionization process begins with the electric field accelerating free electrons in the air. The accelerated electrons, colliding with nitrogen and oxygen atoms in the air, may release new free electrons and positive ions. The new electrons hit other atoms triggering an avalanche mechanism known as ionization by electron impact, explained by the Townsend avalanche process.

The collisions shift electrons to a higher energy level, and later, the return of the electrons to their prior states liberate the excess energy acquired in the form of light emission, which is the visible corona, and electromagnetic waves, such as audible noise, radio interference, and other energy-dissipating effects.

The factors that can affect the corona, such as air pressure, humidity, voltage polarity, conductor roughness, etc., are discussed in [13]. Irregularities in the conductor's surface may intensify the voltage gradient, which become sources of corona.

Initially, corona was studied in relation to the formation of space charges around the conductor and energy loss. With respect to voltage surges, there was also interest in the delay and attenuation of the wavefront. The delay occurs at voltages above the threshold level for the corona together with an increase in the equivalent capacitance of the conductor, represented as an extra capacitance added to the geometric capacitance of the line. Thus, a nonlinear relationship between the charge and the voltage in the conductor is observed.

The nonlinear model of the corona in electromagnetic transient programs can be delimited in three stages, according to Figure 9.23.

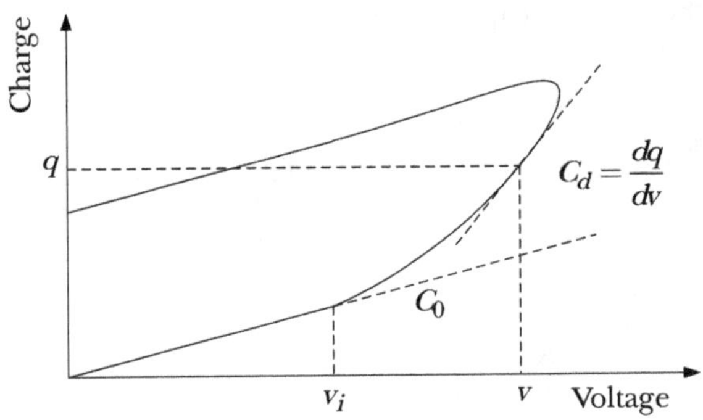

Figure 9.23 Charge in function of the voltage $q = f(v)$.

Equation (9.69) gives the threshold voltage for corona inception in a cable, [24], indicating that above this voltage level, the electric field surpasses the critical field:

$$v_i = 23\left(1 + \frac{1.22}{r^{0.37}}\right) r \ln\left(\frac{2h}{r}\right) \text{ kV}, \tag{9.69}$$

h: conductor height.
r: conductor radius.

Within the first region, before voltage v_i is reached, the curve $q = f(v)$ is a straight line defined by the geometric capacitance of the line. This capacitance is already included in the transmission line models.

In the second stage, with $v \geq v_i$, the nonlinear relationship between charge and voltage is given by (9.70), [10]:

$$q = [C_0 + C_i + k(v - v_i)]v. \tag{9.70}$$

C_0: geometric capacitance of the line.
C_i: initial capacitance jump.
v: voltage at the conductor.
k: corona constant.

The capacitance C_i takes on values lower than 1 pF/m; however, values ranging from 1 to 3 pF/m can be found [10]. This capacitance is difficult to determine because it requires an accurate evaluation of the voltage v_i according to the line geometry.

The constant k depends on the conductor diameter, the number of sub-conductors, and the polarity of the applied wave. The values for negative polarity are lower than those for positive polarity.

The nonlinear relationship between charge and voltage can be modeled as two straight lines. Above voltage v_i, there is an increase in the slope of curve $q = f(v)$, which implies an additional capacitance that depends on the voltage, but a linear approximation for this second stage shows sufficient accuracy. The equation for the dynamic capacitance is:

$$C_d = \frac{dq}{dv} = C_0 + C_i - kv_i + 2kv. \tag{9.71}$$

The capacitance in addition to the geometric one, is given by:

$$C_\Delta = C_i - kv_i + 2kv. \tag{9.72}$$

And the capacitive current to ground can be formulated as:

$$i = \frac{dq}{dt} = \frac{\partial q}{\partial v}\frac{dv}{dt}. \tag{9.73}$$

For the decrease in voltage after the voltage peak, the third stage is defined by a straight line with the same slope as the geometric capacitance. The inductances associated with the conductors remain unchanged during the corona.

The corona effect damps the traveling waves and increases the coupling between conductors. It reduces the voltage amplitude by about 5–20%, and a decrease in the wavefronts is also observed. An approximate procedure to take the corona effect into account is to assume an increase only in the equivalent radius of the conductors and shield wires, as a function of the electric field.

Corona models of three-phase lines [53, 54] can be implemented in electromagnetic transient programs; however, accurate models are not trivial, mainly due to the lack of measurements in multiconductor transmission lines [24].

Taking into account the complexity of the corona models, it is acceptable to perform the calculation without representing the corona, which leads to conservative results [24], or even assume simplified modeling for the radius of conductors. However, the corona effect should be considered in some cases to explain the reduction in the voltage levels, for instance, when analyzing lightning strokes to shield wires in the midspan of the transmission line.

9.5 Substations Studies

Lightning surges only reach the substation equipment when they occur in transmission lines near the substation due to the damping of high-frequency overvoltages. In direct strokes, a half of the current travels along the line and the other half travels toward the substation. The voltage waveform may be obtained from the current waveform, as mentioned in Section 9.2, and is defined by the statistical distributions of front times, amplitudes, and tail times of the current. The corona effect may also be considered in the wavefront attenuation.

The voltage waveforms toward the substation in the case of back-flashovers are the result of a complex interaction between phase conductors, shield wires, tower, and tower footing resistance (Figure 9.24).

v_{sw}: voltage at the shield wire (voltage at tower top), or at the tower crossarm
v_c: voltage at the phase conductor.
Δv: voltage drop in the insulator string.

The voltage waveform applied to the conductor, soon after the back-flashover, behaves initially almost as a step voltage. The steep front propagates toward the substation, but it changes due to damping and possible insulator breakdown in adjacent towers, and the severity of the event depends on how far the stroke is from

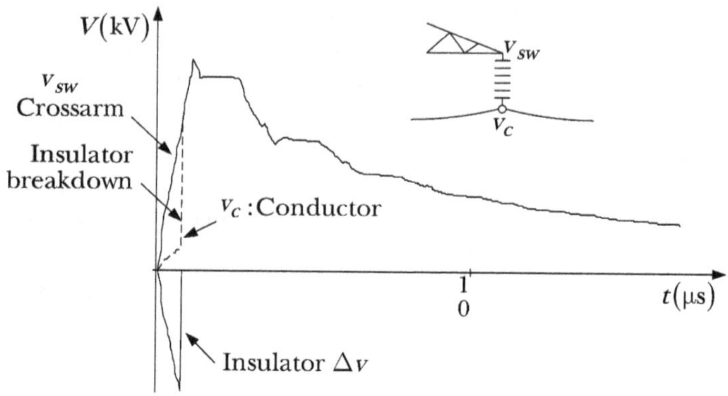

Figure 9.24 Voltages at the tower crossarm, phase conductor, and breakdown of the insulator string.

the substation. A step voltage would represent a severe condition for the insulator strings and power apparatus since standard laboratory tests are limited to maximum wavefronts of 1200–2000 kV/μs. However, step voltages are not expected to reach the substation, since the backflashover is unlikely to occur at the line entrance of the substation due to its low ground resistance.

A reduction in the wavefront moving in a cable toward the substation should be considered, for instance, to have a time delay of about 1 μs/km according to field measurements. In Chapter 3, the analysis of modal propagation in multiphase lines and differences in the speed of the aerial and ground modes partially explain wavefront attenuation. Other phenomena such as high-frequency damping and the corona effect should also be considered.

The amplitude is also subject to damping due to line losses, and the maximum overvoltage that reaches the substation is limited by the dielectric strength of the transmission line insulator string.

In substation analysis, detailed simulations can be performed by means of three-phase representation of transmission lines, including the capacitive and inductive coupling of phase conductors and shield wires as well as tower modeling. A lightning stroke to the tower or shield wire can be simulated, and the waveform that reaches the substation can be calculated as close as possible to reality.

Three-phase models for air-insulated substations are more difficult to obtain than for transmission lines due to geometric irregularities, different conductor heights from the ground, and the short length of busbars and leads. On the other hand, the surge impedance for aerial cables and bus do not change substantially, thus allowing an approximate modeling of the substation.

Another point to be considered is the lack of suitable high-frequency data and models for three-phase equipment. Therefore, it is acceptable in some cases to use a single-phase representation of the substation, but three-phase representation should be more accurate. Another variable that should be introduced into EHV substation analysis as well as transmission line studies is the system operating voltage.

9.5.1 Air Insulated Substations

The traveling waves that reach the substation spread through the busbars, cables, and leads, with multiple reflections that result in different overvoltage levels at various internal points within the substation. The single-line diagram of a very simple substation arrangement in 138 kV is presented in Figure 9.25a. It provides insight into the complexity of representing a large substation with varying layouts

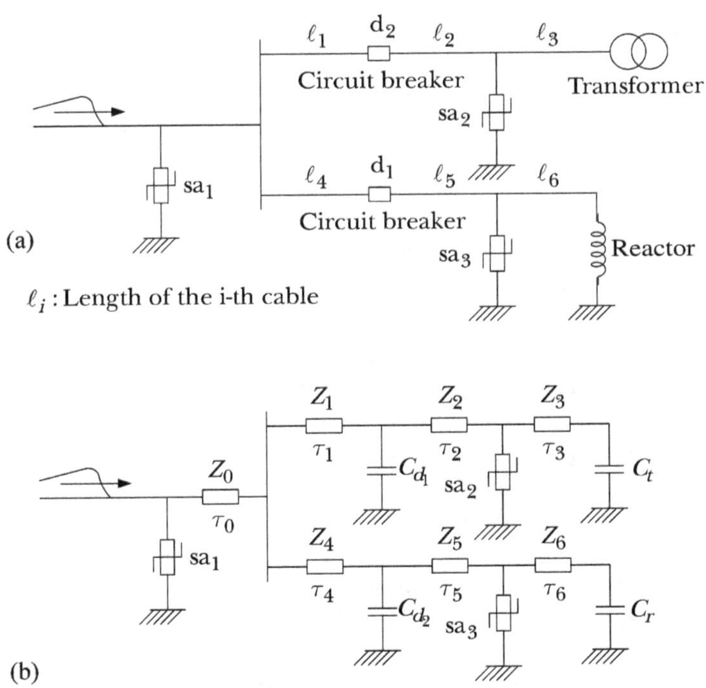

Figure 9.25 Overvoltages inside a substation. (a) Single line diagram. (b) Substation modeling.

and voltage levels, for instance, substation yards with nominal voltages 500/230/138 kV, featuring different busbar arrangements.

Assuming lossless transmission lines, the surge impedance of substation cables, Z_i, can be calculated using the following equation (see Chapter 3, Eqs. (3.60)–(3.67)):

$$Z_i = 60\ln\left(\frac{2h_i}{r_i}\right), \tag{9.74}$$

where:

r_i: equivalent radius of the conductor bundle.
h_i: conductor height from the ground.

The time delays, τ_i, can be obtained, assuming the speed of light.

These calculations provide an analysis of the effective equipment protection by surge arresters as well as of their best locations.

Cables and equipment modeling is exemplified in Figure 9.25b, where:

sa: surge arrester.
Z_i: surge impedance in segment i.
τ_i: time delay.
C_{d_i}: switch stray capacitance.
C_t: transformer stray capacitance.
C_r: reactor stray capacitance.

In Chapter 14, the behavior of overvoltages at certain internal points of a substation with a one-and-a-half breaker layout is exemplified.

The calculation of overvoltages resulting from lightning strokes reaching the substation is accomplished by representing the conductors as their surge impedances and time delays corresponding to the cable segments.

Due to the limited length of cables, the transmission line model may not be suitable for achieving high precision. In such scenarios, it becomes essential to employ specialized software programs that rigorously simulate electromagnetic fields. This necessity also applies to grounding modeling for substations. However, it is important to note that the surge impedances of overhead lines, even with variations in height and configuration, remain relatively consistent. As a result, assuming an average surge impedance for multiple cable segments is a practical approach that aligns reasonably well with the objectives of insulation coordination, offering acceptable accuracy.

Such an analysis can be achieved using electromagnetic transient programs. The simulation of the lightning surges on the power apparatus is conducted to determine its basic impulse level while also considering high-frequency models, as outlined in Section 9.3.4.

Example 9.6 In this example, a transformer bay in a 138-kV substation is assumed, according to Figure 9.26a. Simulated equipment and cables may be seen in Figure 9.26b. A simplified single-phase model is analyzed with the transformer disconnected from the substation and with switch S in its open state.

The power apparatus as regards the transformer and the capacitive voltage transformer, are represented by their stray capacitances.

Capacitive voltage transformer, $C_1 = 5$ nF.
Surge impedance of substation cables, $Z = 370\ \Omega$.
Capacitance of the transformer $C_2 = 3$ nF (not used in this example).
CFO of the insulator string to lightning impulse, $V_{50\%} = 908$ kV.
The surge arrester has a spark-gap in series with the SiC nonlinear resistor.
Maximum sparkover voltage of the arrester to lightning impulse, $V_{\max} = 320$ kV.
Residual voltage: 3 kA-216 kV; 10 kA-266 kV.
Time delay τ.

Assuming waves traveling at the velocity of the light, $c = 300$ m/μs, over a distance of 15 m, the time delay is 0.05 μs, and the time delay over a length of 45 m is 0.15 μs.

- Magnitude of the overvoltages.
  In this example, the lightning surge in the phase conductor, reaching the substation, is limited by the maximum withstand voltage of the insulator string.

$$V_{\max} = V_{50\%}(1 + 3\sigma),$$

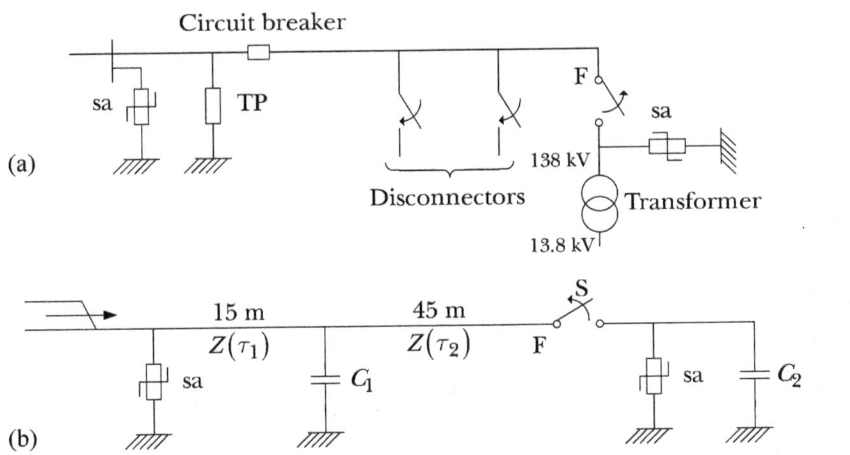

Figure 9.26 138 kV substation. (a) Single line diagram. (b) Transformer bay modeling.

The standard deviation is assumed to be $\sigma = 3\%$.

$$V_{\max} = 908 \times 1.09 = 990\,\text{kV}.$$

The wavefront of the lightning surge entering the substation is assumed to have a rate of rise of 500 kV/μs which means a time front of 1.98 μs .

The steepness of the wave due to back flashover is most critical and can be considered infinite at the stroke point on the line, as in Figure 9.24, but the wavefront that arrives at the substation depends on the distance traveled. The slope of the wave is reduced by the corona as the wave travels toward the substation, and this attenuation can be evaluated using empirical data.

With the capacitive voltage transformer connected, the arrester sparkover effect may be seen in Figure 9.27. It may be observed that the voltage at the open terminal is higher than the voltage near the arrester, and the protection zone of the arrester with respect to the equipment should be considered. The fact that the voltage at the open terminal is not the same as that at the arrester is very important for the insulation coordination of the substation equipment under conditions of lightning impulses. The presence of capacitance as well as high-frequency models introduces another level of complexity to the traveling wave calculation, which should be performed with an EMT program. In this case, if the transformer were connected, its capacitance would change the value of the calculated overvoltage.

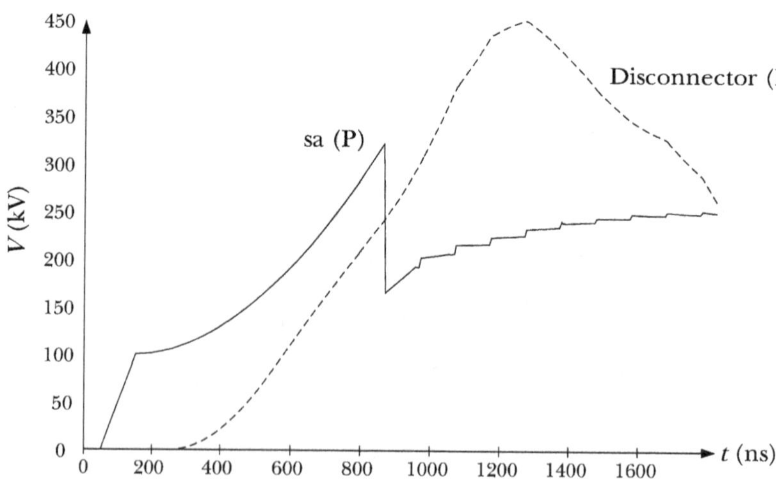

Figure 9.27 Overvoltage results.

Without the capacitive voltage transformer and arrester sparkover, the wave doubles at the open disconnector, reaching 1980 kV. In this case, considering the arrester sparkover, the voltage is reduced to 650 kV at this terminal. The analysis of overvoltages near surge arresters is also addressed in Chapter 13.

A simplified analysis was presented in Example 9.6, showing the basic calculations and modeling to be used in an electromagnetic transient program.

Much more complex arrangements can be analyzed with electromagnetic transient programs. To perform calculations under more realistic operating conditions, the lightning return stroke current should be applied to the transmission line so that the surge reaching the substation is the result of the interaction of the surge waves with the discontinuities in the transmission line. The breakdown of the insulator strings should also be simulated using models based on the disruptive effect or the leader progression approach to the insulator string.

9.5.2 Gas Insulated Substations-GIS

Gas-insulated substations are addressed next. The SF_6 gas is pressurized inside enclosures, increasing their dielectric constant, and therefore allowing distance reduction in the busbars and switchgear, Figure 9.28a.

Initially, a decrease in the surge wave that gets into the substation can be observed, since the surge impedances of the bus and equipment are smaller than

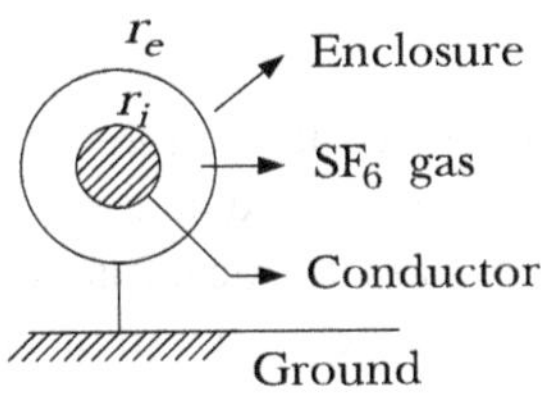

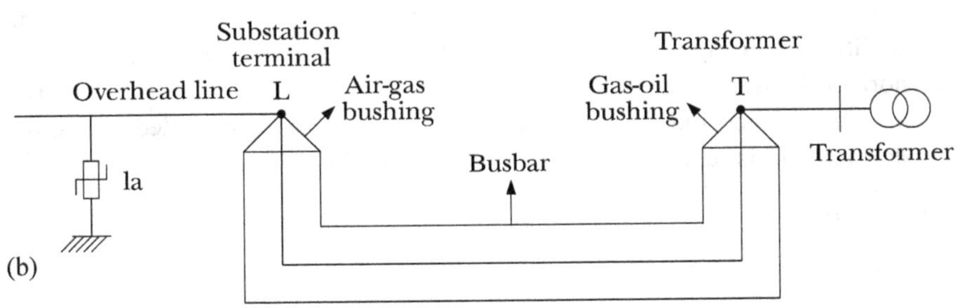

Figure 9.28 GIS connection. (a) Cross-section of the GIS busbar. (b) GIS layout.

that of the line. However, as the substation is very compact, after a relatively short period of time, the voltage increases due to internal reflections. Therefore, overvoltages may be expected, although a reduction in the wavefront could occur initially. Surge arresters generally provide more effective protection to GIS equipment than to air substation equipment because the distances are shorter; however, lightning surge analyses are required when investigating the minimum number of arresters to use, especially in large SF6 shielded installations with many line bays and transformers.

Since the distances inside the GIS are very short, the travel times are small, which may lead to wave reflections at high frequencies, causing overvoltages with steep wavefronts.

A very simplified arrangement is presented in Figure 9.28b, which could represent a section of a GIS connected to an overhead line. The arrangement is similar to that of an underground cable connection, as cable surge impedances and GIS surge impedances are very similar, but the distances along cables can exceed the distances in a GIS.

L: Line terminal.
T: transformer.

Unlike overhead substations, GIS buses have a regular geometry and can be calculated using (9.75):

$$Z = 60 \ln\left(\frac{r_e}{r_i}\right). \tag{9.75}$$

The time delays are calculated assuming the velocity of the light.

The following example assumes an elementary configuration of a GIS busbar connected to an overhead line, showing the initial reduction in the waves that penetrate the busbar and their subsequent growth after a time delay.

Example 9.7 This example shows the traveling waves from an aerial line getting into a GIS using a highly simplified arrangement, that is, a line connected to a transformer, Figure 9.29a. The surge is assumed to be a step, and the stray capacitance of the transformer and other equipment units are disregarded. Thus, the transformer lead is represented as an open terminal.

Overhead line:

$$h = 10\,\text{m}, \quad r = 2\,\text{cm}.$$

$$Z_1 = 60 \ln\left(\frac{2 \times 10}{2 \times 10^{-2}}\right) = 415\,\Omega.$$

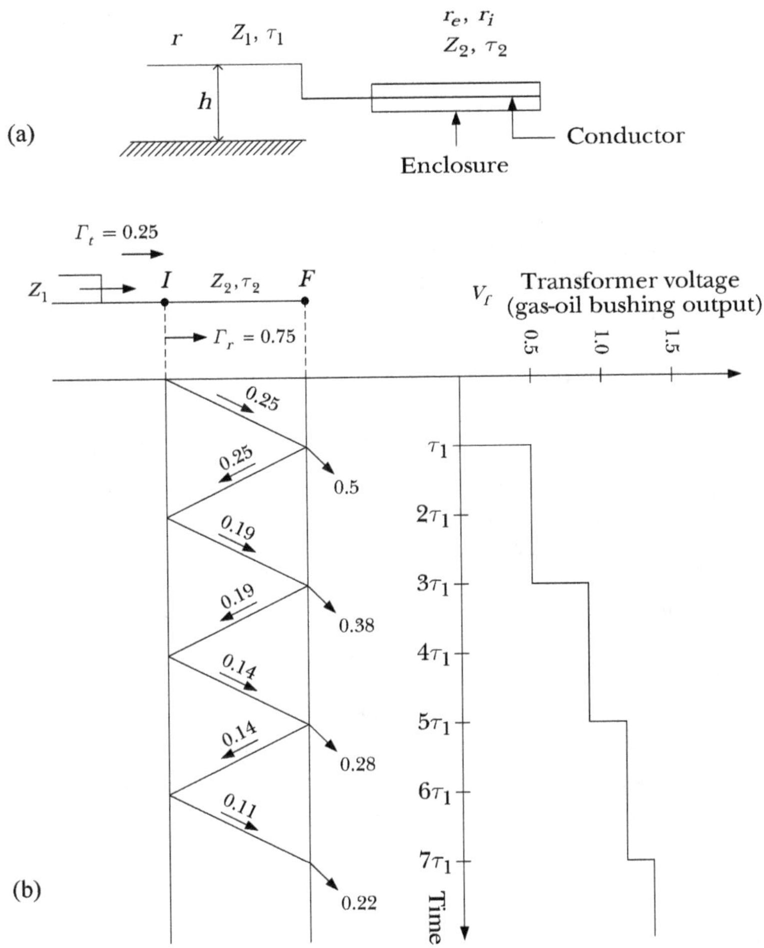

Figure 9.29 Voltage rising in the substation. (a) GIS layout. (b) Lattice diagram.

The line is assumed to have an infinite length, allowing reflections at the sending end terminal to be disregarded.

SF_6 busbar :

$$\frac{r_e}{r_i} = e = 2.71828$$

$$Z_2 = 60\ln(e) = 60\,\Omega.$$

The time delay τ_2 is calculated based on the length of the SF_6 busbar.

The transmission coefficient from aerial line to SF_6 cable may be approximated by:

$$\Gamma_t = 0.25,$$

and the reflection coefficient from SF_6 cable to the aerial line by:

$$\Gamma_r = 0.75.$$

Initially, a decrease in the surge wave that gets into the substation can be observed, since the surge impedance of the cable is smaller than that of the line. However, as the substation is very compact, after a relatively short period of time, the voltage increases due to internal reflections, but overvoltages are expected, although with reductions in the wavefronts.

The voltage at the connection point of the line with the substation rises as if there were a capacitor at this point (Figure 9.29b). The GIS bus is a line segment with a surge impedance and time delay well below the respective values of the overhead line. The small line segment acts as a capacitor and in old traveling wave programs, the simulation of capacitors was done in this way, known as the stub line model.

The analysis of lightning surges reaching a GIS is the same as that of air-insulated substations. The source of the overvoltage is the same, located in the aerial transmission lines, but the compact design achieved in GIS reduces the distances considerably. The methodology also requires circuit representation by means of surge impedances and time delays. To obtain a detailed model, it may be necessary to incorporate the stray capacitances of the substation equipment and switchgear, spacers, arresters, transformers, and bushings.

References

1 Bewley, L. V. (1963). *Travelling waves on transmission systems* (2nd ed.). New York: Dover.

2 Greenwood, A. (1991). *Electrical transients in power systems* (2nd ed.). New York: John Wiley & Sons.

3 CIGRE Working Group C4.23. (2021). *Procedures for estimating the lightning performance of transmission lines: new aspects*. Paris. (Technical Brochure, no. 839).

4 Uman, M. A. (1987). *The lightning discharge*. Orlando: Academic Press. (International Geophysics Series, 39).

5 Prinz, H. (1977). Lightning in history. In R. H. Golde (Ed.), *Lightning* (Vol. 1, pp. 1–21). London: Academic Press.

6 Moore, C. B., & Vonnegut, B. (1977). The thundercloud. In R. H. Golde (Ed.), *Lightning* (Vol. 1, pp. 51–98). London: Academic Press.

7 Nucci, C. A., Diendorfer, G., Uman, M. A., Rachidi, F., Ianoz, M., & Mazzetti, C. (1990). Lightning return stroke current models with specified channel-base current: A review and comparison. *Journal of Geophysical Research: Atmospheres*, 95(D12), 20395–20408.

8 Nucci, C. A., Rachidi, F., Ianoz, M. V., & Mazzetti, C. (1993). Lightning-induced voltages on overhead lines. *IEEE Transactions on Electromagnetic Compatibility*, 35(1), 75–86.

9 Rakov, V. A., & Uman, M. A. (1998). Review and evaluation of lightning return stroke models including some aspects of their application. *IEEE Transactions on Electromagnetic Compatibility*, 40(4), 403–426.

10 CIGRE, Study Committee 33 & Working Group 01. (1991). *Guide to procedures for estimating the lightning performance of transmission lines*. Paris: CIGRE.

11 Whitehead, J. T., & Chisholm, W. A. (1993). Estimating lightning performance of transmission lines. II. Updates to analytical models. *IEEE Transactions on Power Delivery*, 8(3), 1254–1267.

12 Anderson, R. B., & Eriksson, A. J. (1980). Lightning parameters for engineering application. *Electra*(69), 65–102.

13 Electric Power Research Institute. (1982). *Transmission line reference book: 345 kV and above* (2nd ed.). Palo Alto: Electric Power Research Institute.

14 Dommel, H. W. (1986). *Electromagnetic transients program reference manual: (EMTP) theory book*. Portland: Bonneville Power Administration.

15 CIGRE Working Group C4.37. (2019). *Electromagnetic computation methods for lightning surge studies with emphasis on the FDTD method*. (Technical Brochure, no. 785).

16 Wagner, C. F., & Hileman, A. R. (1960). A new approach to the calculation or the lightning performance or transmission lines III – a simplified method: stroke to tower. *Transactions of the American Institute of Electrical Engineers. Part III: Power Apparatus and Systems*, 79(3), 589–603.

17 Sargent, M. A., & Darveniza, M. (1969). Tower surge impedance. *IEEE Transactions on Power Apparatus and Systems*, PAS-88(5), 680–687.

18 Chisholm, W. A., Chow, Y. L., & Srivastava, K. D. (1983). Lightning surge response of transmission towers. *IEEE Transactions on Power Apparatus and Systems*, PAS-102(9), 3232–3242.

19 Chisholm, W. A., Chow, Y. L., & Srivastava, K. D. (1985). Travel time of transmission towers. *IEEE Transactions on Power Apparatus and Systems*, PAS-104 (10), 2922–2928.

20 Chisholm, W. A., & Janischewskyj, W. (1989). Lightning surge response of ground electrodes. *IEEE Transactions on Power Delivery*, 4(2), 1329–1337.

21 Sunde, E. D. (1968). *Earth conduction effects in transmission systems.* New York: Dover.

22 Chisholm, W. A. (2015). Evaluation of simple models for the resistance of solid and wire-frame electrodes. *IEEE Transactions on Industry Applications*, 51(6), 5123–5129.

23 Devgan, S. S., & Whitehead, E. R. (1973). Analytical models for distributed grounding systems. *IEEE Transactions on Power Apparatus and Systems*, PAS-92(5), 1763–1770.

24 Imece, A. F., Durbak, D. W., Elahi, H., Kolluri, S., Lux, A., Mader, D., McDemott, T. E., Morched, A., Mousa, A. M., Natarajan, R., Rugeles, L., & Tarasiewicz, E. (1996). Modeling guidelines for fast front transients. *IEEE Transactions on Power Delivery*, 11(1), 493–506.

25 CIGRE Working Group C4.26. (2017). *Evaluation of lightning shielding analysis methods for EHV and UHV DC and AC transmission lines.* (Technical Brochure, no. 704).

26 Dellera, L., & Garbagnati, E. (1990). Lightning stroke simulation by means of the leader progression model. I. Description of the model and evaluation of exposure of free-standing structures. *IEEE Transactions on Power Delivery*, 5(4), 2009–2022.

27 Eriksson, A. J. (1987). An improved electrogeometric model for transmission line shielding analysis. *IEEE Transactions on Power Delivery*, 2(3), 871–886.

28 Eriksson, A. J. (1987). The incidence of lightning strikes to power lines. *IEEE Transactions on Power Delivery*, 2(3), 859–870.

29 Rizk, F. A. M. (1990). Modeling of transmission line exposure to direct lightning strokes. *IEEE Transactions on Power Delivery*, 5(4), 1983–1997.

30 Armstrong, H. R., & Whitehead, E. R. (1968). Field and analytical studies of transmission line shielding. *IEEE Transactions on Power Apparatus and Systems*, PAS-87(1), 270–281.

31 Brown, G. W., & Whitehead, E. R. (1969). Field and analytical studies of transmission line shielding: part II. *IEEE Transactions on Power Apparatus and Systems*, PAS-88(5), 617–626.

32 Darveniza, M., Sargent, M. A., Limbourn, G. J., Choy, L. A., Caldwell, R. O., Currie, J. R., Holcombe, B. C., Stillman, R. H., & Frowd, R. (1979). Modelling for lightning performance calculations. *IEEE Transactions on Power Apparatus and Systems*, PAS-98(6), 1900–1908.

33 IEEE PES Lightning Performance of Overhead Lines Working Group: 15.09.08. (2013). *IEEE Flash v2.05.* San Diego: SourceForge Headquarters. Retrieved from https://sourceforge.net/projects/ieeeflash

34 Leuven EMTP Center *ATP: Alternative Transients Program: Rule book.* (1987). Heverlee (BE): Leuven EMTP Center.

35 Hileman, A. R., Stenstrom, L., Gaunt, C. T., Volker, O., Kawamura, T., & Comber, M. G. (1992). Application of arresters to transmission lines In *Progress Report Cigre 33 – WG11* (Vol. 92). Paris.

36 Bewley, L. V. (1935). Effect which different distributions of expulsion gaps have on direct-stroke protection of transmission lines. *General Electric Review*, 38(11), 505–510.

37 Sporn, P., & Gross, I. W. (1935). Expulsion Protective Gaps on 132 kV Lines. *Transactions of the American Institute of Electrical Engineers*, 54(1), 66–73.

38 Zanetta Junior, L. C., & Moreira, F. A. (1997). Improvement of the lightning performance evaluation of lines with arresters. In *International Conference on Power Systems Transients. Seattle, 1997* (pp. 307–312). Seattle: University of Washington.

39 Agrawal, A. K., Price, H. J., & Gurbaxani, S. H. (1980). Transient response of multiconductor transmission lines excited by a nonuniform electromagnetic field. *IEEE Transactions on Electromagnetic Compatibility*, EMC-22(2), 119–129.

40 Cooray, V. (1992). Horizontal fields generated by return strokes. *Radio Science*, 27(4), 529–537.

41 Cooray, V. (1994). Calculating lightning-induced overvoltages in power lines. A comparison of two coupling models. *IEEE Transactions on Electromagnetic Compatibility*, 36(3), 179–182.

42 Cooray, V. (1998). Some considerations on the 'Cooray–Rubinstein' approximation used in deriving the horizontal electric field over finitely conducting ground. In *Proceedings of International Conference on Lightning Protection*, Birmingham, UK, September 14–18 (pp. 282–286).

43 Hoidalen, H. K. (2003). Analytical formulation of lightning-induced voltages on multiconductor overhead lines above lossy ground. *IEEE Transactions on Electromagnetic Compatibility*, 45(1), 92–100.

44 Master, M. J., & Uman, M. A. (1983). Transient electric and magnetic fields associated with establishing a finite electrostatic dipole. *American Journal of Physics*, 51(2), 118–126.

45 Nucci, C. A., & Rachidi, F. (2014). Interaction of electromagnetic fields with electrical networks generated by lightning. In V. Cooray (Ed.), *The lightning flash* (2nd ed., pp. 559–610). London: IET.

46 Paolone, M., Nucci, C. A., & Rachidi, F. (2001). A new finite difference time domain scheme for the evaluation of lightning induced overvoltage on multiconductor overhead lines. In *International Conference on Power System Transients IPST'01, Rio de Janeiro, June 2001*.

47 Rachidi, F., Nucci, C. A., & Ianoz, M. (1999). Transient analysis of multiconductor lines above a lossy ground. *IEEE Transactions on Power Delivery*, 14(1), 294–302.

48 Rubinstein, M. (1996). An approximate formula for the calculation of the horizontal electric field from lightning at close, intermediate, and long range. *IEEE Transactions on Electromagnetic Compatibility*, 38(3), 531–535.

49 Rusck, S. (1958). Induced lightning over-voltages on power transmission lines with special reference to the over-voltage protection of low-voltage networks. In *Transactions of the Royal Institute of Technology* (Vol. 120). Stockholm.

50 Sommerfeld, A. (1909). Über die Ausbreitung der Wellen in der drahtlosen Telegraphie. *Annalen der Physik*, 333(4), 665–736.

51 Uman, M. A., & McLain, D. K. (1969). Magnetic field of lightning return stroke. *Journal of Geophysical Research (1896–1977)*, 74(28), 6899–6910.

52 Brignone, M., Delfino, F., Procopio, R., Rossi, M., & Rachidi, F. (2017). Evaluation of power system lightning performance, part I: Model and numerical solution using the PSCAD-EMTDC platform. *IEEE Transactions on Electromagnetic Compatibility*, 59(1), 137–145.

53 Guillier, J. F., Poloujadoff, M., & Rioual, M. (1995). Damping model of travelling waves by corona effect along extra high voltage three phase lines. *IEEE Transactions on Power Delivery*, 10(4), 1851–1861.

54 Jesus, C., & Barros, M. T. C. (1994). Modelling of corona dynamics for surge propagation studies. *IEEE Transactions on Power Delivery*, 9(3), 1564–1569.

10

Transients in Systems with Shunt Capacitors

10.1 Introduction

Shunt-capacitor banks introduce high-frequency transients that may stress the substation equipment [1–6]. During the switching of shunt-capacitor banks, both overvoltages and overcurrents can occur. A general analysis of switching surges is the subject of Chapter 8, but the specific characteristics of high frequencies caused by shunt capacitors are covered in this chapter.

The shunt-capacitor switching analysis in this chapter is primarily aimed at high-frequency transients. The major types of electromagnetic transients that can occur in relation to the presence of shunt capacitors are approached here using single-phase elementary circuits.

The single-phase equations can be extended to three-phase networks when the positive sequence prevails in the system. More complex cases, including the three-phase analysis and coupling related to the feeder system, will most likely require electromagnetic transient programs to perform the calculation.

However, for comparison with the single-phase circuits, a simplified analysis involving three-phase unearthed shunt-capacitor banks is also described.

Several pieces of equipment, such as arresters, circuit breakers, and current transformers, are subject to high-frequency stresses due to the presence of shunt capacitors. Therefore, a brief introduction to the main considerations in equipment specification is also provided in this chapter.

10.2 High-Frequency Current and Voltage Transients

In a substation with multiple shunt-capacitor banks, switching surges can occur during energization or circuit breaker restrike involving a single bank. Therefore, the following analysis focuses on the switching of one shunt-capacitor bank, assuming the others are not in operation.

Understanding Electromagnetic Transients in Power Systems, First Edition. Luiz Cera Zanetta Jr.

Published 2025 by John Wiley & Sons, Inc.

10.2.1 Energization of Shunt-Capacitor Banks

In a substation, switching surges can occur during energization of the first bank of capacitors. The energization event causes an oscillation between the shunt capacitors and the feeder system. A simplified representation of the network, by means of an equivalent inductance, is employed, so that the single-phase circuit can simulate the oscillations in voltages and currents. Since the system inductance is relatively high compared to the inductance of the cables and buses, the current is mainly limited by this inductance; nevertheless, switching surges can occur in the substation.

The system is represented by means of its equivalent inductance, L_s. This is a quite reasonable approach when the shunt capacitors are connected to the transmission lines through transformers. If the shunt capacitors are directly connected to transmission lines, the model should reproduce the traveling waves generated during their interaction with the transmission system.

In Figure 10.1, L represents the small internal inductance of the capacitors, within the range from 5 to 10 μH. The inductance L could also represent current-limiting reactors, which are installed in series with the shunt capacitors; however, for purposes of this initial analysis, the inductance L can be included in the system inductance, L_s, or it may be neglected.

Applying Kirchhoff's second law to the circuit in Figure 10.1:

$$e(t) = L_s \frac{di(t)}{dt} + \frac{1}{C}\int i(t)dt. \tag{10.1}$$

Applying the Laplace transform, and assuming an initial current $i(0) = 0$ and an initial voltage across the capacitor $v_c(0) = v_0$:

$$E(s) = \left(sL_s + \frac{1}{sC}\right)I(s) + \frac{v_0}{s}, \tag{10.2}$$

It is known that:

$$Z(s) = sL_s + \frac{1}{sC}, \tag{10.3}$$

$$\omega_0{}^2 = \frac{1}{L_sC},$$

$$\omega_0 = \frac{1}{\sqrt{L_sC}}, \tag{10.4}$$

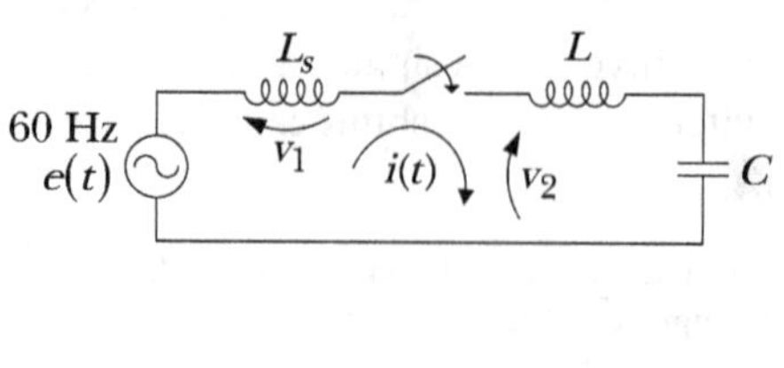

Figure 10.1 Shunt-capacitor energization.

and

$$Z_0 = \sqrt{\frac{L_s}{C}} \tag{10.5}$$

where ω_0 and Z_0 represent the natural frequency of oscillation and the surge impedance of the circuit, respectively.

Rewriting (10.3), using (10.4) and (10.5):

$$Z(s) = \frac{Z_0(s^2 + \omega_0{}^2)}{s\omega_0}, \tag{10.6}$$

To analyze a critical condition, the capacitor energization is considered when a cosine excitation source reaches its peak value, with Laplace transform $E(s) = \frac{E_0 s}{s^2 + \omega^2}$.

Substituting the excitation source and Eq. (10.6) into (10.2):

$$\frac{E_0 s}{s^2 + \omega^2} = \frac{Z_0(s^2 + \omega_0{}^2)}{s\omega_0} I(s) + \frac{v_0}{s}, \tag{10.7}$$

$$I(s) = \frac{E_0 \omega_0 s^2}{Z_0(s^2 + \omega^2)(s^2 + \omega_0{}^2)} - \frac{v_0 \omega_0}{Z_0(s^2 + \omega_0{}^2)}, \tag{10.8}$$

As can be seen, the natural oscillation frequency ω_0 superimposes on the nominal frequency ω of the system at 60 Hz.

Considering the high frequencies and the fact that most of the transient occurs just after closing the switch, an additional simplification may be introduced, if during this initial period, the voltage $e(t)$ is assumed constant.

A constant voltage source is then applied, and the circuit in Chapter 1, Section 1.5.2, can be used.

With the circuit in Figure 10.2, where v_0 is the residual voltage corresponding to the charge stored in the capacitor, the current may be analyzed with the aid of:

$$\frac{E_0}{s} = \frac{Z_0(s^2 + \omega_0{}^2)}{s\omega_0} I(s) + \frac{v_0}{s}, \tag{10.9}$$

$$I(s) = \frac{E_0 - v_0}{s} \frac{s\omega_0}{Z_0(s^2 + \omega_0{}^2)}. \tag{10.10}$$

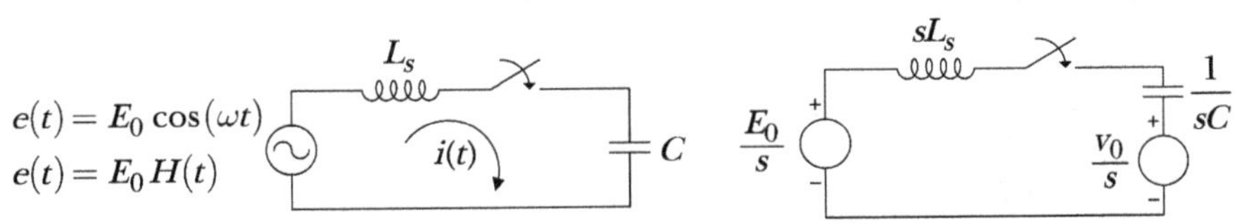

Figure 10.2 Energization using a constant voltage excitation source.

Taking the inverse transform of (10.10):

$$i(t) = \frac{E_0 - v_0}{Z_0}\ \sin(\omega_0 t). \tag{10.11}$$

The trapped charge can be incorporated into bank energization or not. Large resistors are used to discharge the shunt capacitor in preparation for the next energization, but it takes a few minutes for complete discharge. Reclosing the capacitor for a shorter time will probably result in a high residual voltage level.

Example 10.1 A 230 kV, 60 Hz substation with a short-circuit current of 20 kA has a 50 MVAr shunt-capacitor bank to be energized. Calculate the natural oscillation frequency of the circuit and the amplitude of the current, assuming the capacitor to have previously been discharged, at $v_0 = 0$.

Solution

The system inductance L_s can be obtained from:

$$L_s = \frac{230}{\sqrt{3}\,20 \times 377} = 17.6\ \text{mH}.$$

Capacitance of the bank:

$$C = \frac{Q}{\omega V^2},$$

$$C = \frac{50}{230^2 \times 377} = 2.51\ \mu\text{F}.$$

The natural oscillation frequency can be calculated from the inductance and capacitance:

$$\omega_0 = \frac{1}{\sqrt{17.6 \times 10^{-3} \times 2.51 \times 10^{-6}}} \text{ and } f_0 = \frac{1}{2\pi\sqrt{17,6 \times 2.51 \times 10^{-9}}} = 750\ \text{Hz}.$$

Thus, the natural oscillation frequency can be shown to be higher than the nominal frequency of 60 Hz, by a factor of more than 12.

$$E_0 = \frac{230\sqrt{2}}{\sqrt{3}}\text{kV},$$

$$E_0 = 188\ \text{kV},$$

$$Z_0 = \sqrt{\frac{17.6 \times 10^{-3}}{2.51 \times 10^{-6}}}\ \Omega,$$

$$Z_0 = 83.74\ \ \Omega.$$

Thus, the transient inrush current may be calculated:

$$I_0 = \frac{188}{83.74},$$

$$I_0 = 2.25 \text{ kA}.$$

This simplified representation gives an estimate of the transient inrush current [7, 8]. Currents and frequencies observed during energization are usually lower than the withstand levels of the substation equipment. The next section covers the residual voltage trapped in the shunt-capacitor bank in order to analyze currents and overvoltages.

10.2.2 Restrike and Trapped Charge

- Trapped Charge in the Capacitor

  Assuming the presence of trapped charge, the single-phase circuit used to analyze energization with trapped charge in the capacitor is the same as the circuit used to analyze circuit breaker restrike after disconnecting the capacitor.

  In restrike analysis, the trapped charge may be higher than during energization with residual voltage, since the worst restrike scenario occurs a half cycle after current extinction, leaving no time for the capacitor to discharge.

  The electric circuit is the same as that in Figure 10.2, except for the closing of switch S, where the voltage stored in the capacitor is due to the prior opening of the circuit breaker.

  Disregarding losses, the circuit contains only inductances and capacitances. In this circuit, at the current zero crossing in t_1, the opening of the switch occurs, and simultaneously, the voltage across the capacitor assumes its negative peak (Figure 10.3a and b), as the phase angle of the current is 90° ahead of the voltage.

  In fact, the voltage amplitude at the capacitor is a little higher than the voltage amplitude at the source due to the voltage divider that contains inductive and capacitive elements (see equation (7.16) in Chapter 7 and Example 10.2 in this chapter). However, in Figure 10.3c, residual voltage and source voltage amplitudes are assumed approximately equal, but opposite in sign, that is, $v_0 = -E_0$, when the voltage across the inductance is neglected.
- Transient Inrush Current

  By assuming that restrike occurs one half cycle after the opening time, t_2, it can be shown that the current reaches its maximum value when v_0 and E_0 have

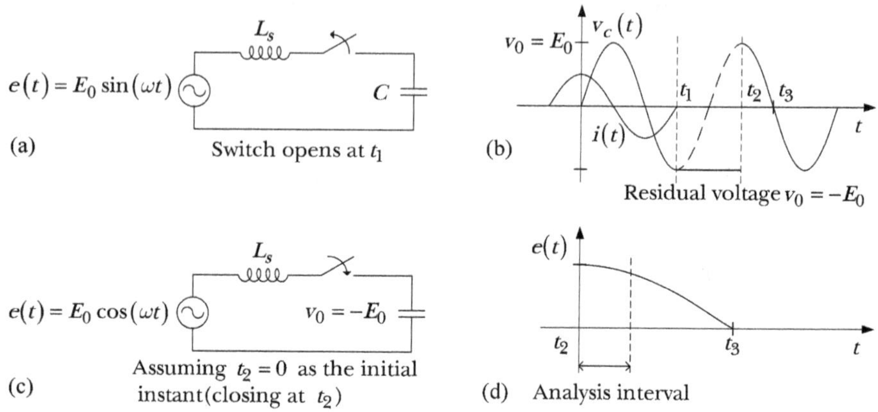

Figure 10.3 Disconnection of capacitor and residual voltage. (a) Equivalent circuit. (b) Current and voltage waveforms at switch opening and closing times. (c) Equivalent circuit with trapped charge. (d) Source voltage at switch closing time.

opposite polarities. Note that similar conditions as in Figure 10.3c may be assumed for an energization with trapped charge.

Since the natural frequency is much higher than the excitation frequency, assuming excitation to be a cosine function and taking $t_2 = 0$ as the origin, the excitation voltage can again be supposed to be approximately constant.

$$i(t) = \frac{2E_0}{\sqrt{\frac{L_s}{C}}} \sin\,(\omega_0 t). \tag{10.12}$$

In this case, at the closing time analyzed, the current can double with respect to an energization without trapped charge. Although the currents are even greater, the values are not so high as to damage the equipment. However, the current increase can cause overvoltages at the substation and affect the arresters, which will be examined more specifically in the next section.

The occurrence of restrike depends on the dielectric strength of the circuit breaker when the contacts are separated. In cases in which the dielectric strength is not sufficient, restrike may occur. The maximum current occurs, in the case of restrikes, one-half cycle after arc extinction. The breakers adopted for capacitor switching are designed to avoid this phenomenon and restrike-free breakers should be specified. Although a rare situation, there is no guarantee that it will never occur.

It is noted that restrike and reignition are terms used to denote reestablishment of the current within a period greater or less than 1/4 of a cycle after arc interruption, respectively. This chapter deals with breaker restrike.

10.2.3 Overvoltages and Arresters

In studies of a single shunt-capacitor bank to be energized, simulations by means of electromagnetic transient programs are essential to accurately determine overcoltages and intervention of surge arresters.

The pre-insertion resistors, and more recently, the circuit breakers with voltage synchronization devices can be used to control the energization switching surges, minimizing the currents and transient overvoltages.

The details of network representaion for analysing energizations also apply to restrikes; however, neither of the traditional methods of overvoltage control through pre-insertion resistors or synchronization methods are effective to restrikes.

The following equations address restrike or even energization with residual voltage and the corresponding stress on arresters. The following single-phase analysis, while simplified, helps to illustrate the stress on arresters during the opening or energization of a shunt-capacitor bank [3] (Figure 10.4).

The voltage at the capacitor before the operation of the surge arrester is given by:

$$v_c(t) = \frac{1}{C}\int_0^t i(t)dt + v_0. \tag{10.13}$$

Substituting (10.11) into (10.13):

$$v_c(t) = \frac{1}{C}\frac{(E_0 - v_0)}{\sqrt{\frac{L_s}{C}}}\left[-\frac{\cos\,(\omega_0 t)}{\omega_0}\right]_0^t + v_0, \tag{10.14}$$

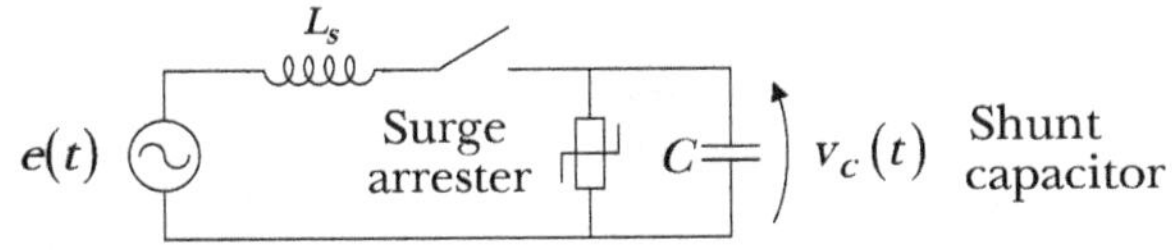

Figure 10.4 Opening of shunt capacitor with arrester.

$$v_c(t) = \frac{1}{C}\frac{(E_0 - v_0)}{\sqrt{\frac{L_s}{C}}}\sqrt{L_sC}\,[1 - \cos\,(\omega_0 t)] + v_0, \tag{10.15}$$

$$v_c(t) = (E_0 - v_0)[1 - \cos(\omega_0 t)] + v_0. \tag{10.16}$$

The voltage is identical to that which would occur in a substation without an arrester.

In the case of energization (Section 10.2.1), when the voltage across the capacitor is $v_0 = 0$, the maximum amplitude of the voltage in (10.16) is $2E_0$. This is the usual case, since the banks have parallel resistors, allowing the capacitors to discharge during operation timeouts.

In the case of energization with trapped charge, or even if a restrike (Section 10.2.2) occurs, assuming $v_0 = -E_0$, the maximum amplitude of the voltage is $3E_0$ in (10.16).

The switching surges and the energy absorbed by the metal oxide surge arrester (MOSA) are now analyzed. A more detailed description of the surge arrester is presented in Chapter 13.

This analysis addresses metal oxide resistors due to their extensive application in power systems.

The MOSA has a very flat $V \times I$ curve, and it may be represented as a voltage source for one half-cycle of the transient, without a change in polarity.

Using the simplified model, the equations for the high-frequency transient in the LC circuit can be written, adopting the DC source $e(t) = E_0$. As for the surge arrester, a constant voltage v_p is also assumed during this intervention. The voltage source v_p is switched on and off at voltage levels higher or lower than v_p, respectively.

During the interval without surge-arrester intervention, current and voltage rise can be observed. When the intervention level of the surge arrester is reached, the voltage across the capacitor remains constant, and current commutation between the capacitor and the surge arrester occurs.

Time t_1 when the current, $i(t)$, through the capacitor is zero can be calculated.

$$i(t) = C\frac{dv(t)}{dt} = 0,$$

in this case:

$$v_c(t_1) = v_p. \tag{10.17}$$

From Eq. (10.16):

$$(v_p - v_0) = (E_0 - v_0)[1 - \cos(\omega_0 t_1)], \tag{10.18}$$

or:

$$\cos(\omega_0 t_1) = \frac{E_0 - v_p}{E_0 - v_0}, \tag{10.19}$$

with:

$$\sin(\omega_0 t_1) = \sqrt{1 - \left(\frac{E_0 - v_p}{E_0 - v_0}\right)^2}, \tag{10.20}$$

substituting (10.20) into (10.11), the current at this time is given by:

$$I_m = \frac{E_0 - v_0}{\sqrt{\frac{L_s}{C}}}\sqrt{1 - \left(\frac{E_0 - v_p}{E_0 - v_0}\right)^2} = \frac{\sqrt{(E_0 - v_0)^2 - (E_0 - v_p)^2}}{\sqrt{\frac{L_s}{C}}}. \tag{10.21}$$

Before the arrester intervention, the waveforms for the current and voltage are presented from time zero to time t_1 in Figure 10.5b.

When $i(t_1) = I_m$ the voltage at the capacitor is $v_c(t_1) = v_p$, and at this instant, the current is commutated to the surge arrester.

After time t_1, using the circuit in Figure 10.5a, the equation for the current is derived:

$$E_0 - v_p = L_s \frac{di(t)}{dt}, \tag{10.22}$$

$$i(t) = \int_{t_1}^{t} \frac{E_0 - v_p}{L_s}\, dt + i_0, \tag{10.23}$$

where $i_0 = I_m$.

$$i(t) = \frac{E_0 - v_p}{L_s}(t - t_1) + I_m, \tag{10.24}$$

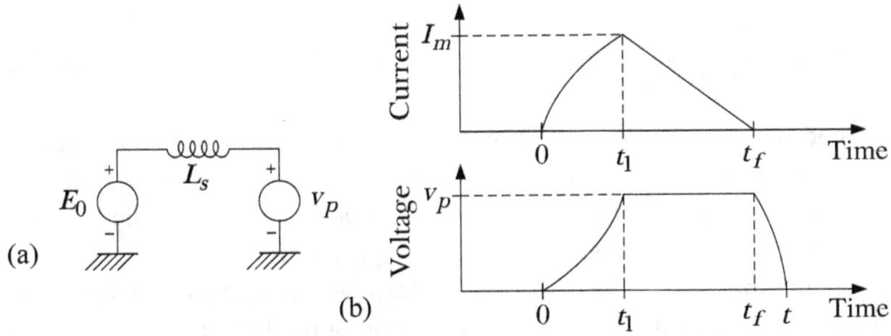

Figure 10.5 Surge arrester intervention. (a) Equivalent circuit with arrester inserted. (b) Voltage and current waveforms before and during arrester intervention.

For $t > t_1$, the interval Δt is defined by:

$$\Delta t = t - t_1. \tag{10.25}$$

This equation represents a straight line with a negative derivative since v_p is greater than E_0, as shown in Figure 10.5b. The surge arrester behaves as an opposing voltage to E_0 in the time interval t_1 to t_f, in which the current declines to zero.

$$i(t_f) = \frac{E_0 - v_p}{L_s}(t_f - t_1) + I_m. \tag{10.26}$$

When $i(t_f) = 0$ the interval $\Delta t_1 = t_f - t_1$, is determined by:

$$\Delta t_1 = \frac{L_s I_m}{v_p - E_0}, \qquad v_p > E_0. \tag{10.27}$$

Thus, Δt_1 represents the conduction time for the surge arrester to be used in calculations of the absorbed energy. The following expression for the energy can be written as:

$$E = \int_{t_1}^{t_f} v_p i(t)dt = v_p \int_{t_1}^{t_f} i(t)dt, \tag{10.28}$$

$$E = \frac{1}{2} I_m \Delta t_1 v_p. \tag{10.29}$$

Substituting (10.21) and (10.27) into (10.29):

$$E = \frac{1}{2}\frac{L_s}{v_p - E_0}\frac{(E_0 - v_0)^2 - (v_p - v_0)^2}{\frac{L_s}{C}} v_p, \tag{10.30}$$

$$E = \frac{1}{2}C\frac{(E_0 - v_0)^2 - (v_p - v_0)^2}{v_p - E_0} v_p. \tag{10.31}$$

In the worst case, $v_0 = -E_0$, and the maximum energy may be written as:

$$E_{\max} = \frac{1}{2}C\frac{4E_0^2 - (v_p - E_0)^2}{v_p - E_0} v_p. \tag{10.32}$$

As mentioned before, since preinsertion resistors cannot be applied to control overvoltages caused by restrikes, it is convenient to apply restrike-free breakers that have a low probability of restrike. Nevertheless, surge arresters able to withstand the energy of a possible restrike should be specified.

Accurate calculations should be performed, using electromagnetic transient programs to make a detailed three-phase representation of the feeder system, as well as of the equipment and transmission lines connected to the substation. Delta and star connections to shunt-capacitor banks should be analyzed. Ungrounded capacitor banks can increase the switching surges and even require more expensive switching devices. However, ungrounded shunt capacitors do not contribute to high currents in single-phase faults, thus relieving the switches and other equipment from these stresses.

10.2.4 Voltage Amplification

Capacitor-bank switching at one point in the network can increase the transient voltages in banks at other network locations [4, 6].

A simplified circuit that explains this phenomenon is presented in Figure 10.6. The basic conditions under which voltage amplification may occur are:

a) $C_1 \gg C_2$.
b) $L_1 \ll L_2$, i.e., the short circuit power in loop 1 is greater than that in loop 2.
c) $f_1 \approx f_2$.

$$f_1 = \frac{1}{2\pi\sqrt{L_1C_1}}, \; f_2 = \frac{1}{2\pi\sqrt{L_2C_2}}$$

The phenomenon develops if loop 1, with an oscillatory natural frequency, f_1, behaves as an excitation source for loop 2, with a natural frequency f_2, approaching a resonance condition (see Chapter 1, Section 1.6). In this case, the largest overvoltages may not occur at the switched capacitor, but instead, they may appear amplified at another location in the network.

10.2.5 Lightning Surges

The following account gives a brief description of the overvoltages originating from lightning surges in shunt-capacitor banks with a star connection that has a neutral point grounded. The ungrounded capacitor, with delta or star connections, is not addressed here but can be analyzed using the same approach; however, the incoming surge in one phase will probably be transferred to the others through the capacitance of the bank. The general treatment of lightning surges is presented in Chapter 9, so only a few comments are made because the

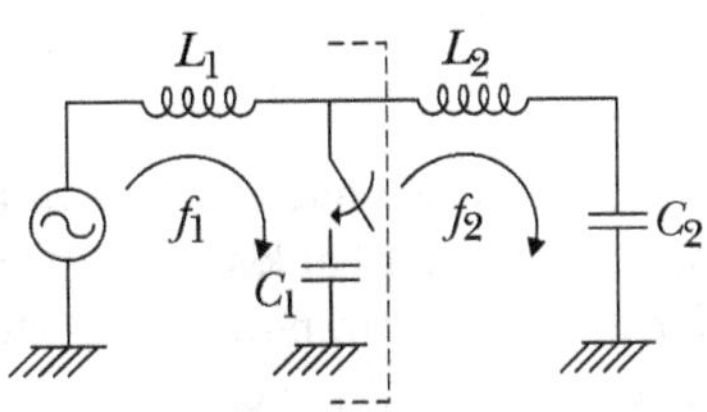

Figure 10.6 Voltage amplification.

capacitance of the banks is substantially higher than the stray capacitance of the other equipment.

Shunt capacitors without limiting reactors are initially considered. The equation that relates the stored charge and the voltage at the capacitor terminals is given by:

$$Q = Cv. \tag{10.33}$$

When a current impulse strikes the shunt capacitor, the charge is given by:

$$Q_t = Q_0 + Q_i \tag{10.34}$$

Q_0: charge due to the operating voltage, where $Q_0 = C\,v_0$.
Q_i: charge due to the lightning surge.

Assuming that V_L is the maximum line operating voltage, in order to take the worst-case scenario at the time of the stroke into account, the initial voltage across the capacitor is calculated from:

$$v_0 = \frac{V_L\sqrt{2}}{\sqrt{3}}. \tag{10.35}$$

A reference value for the charge contained in the current stroke, Q, should be assumed, and the following voltage would be developed across the capacitor:

$$v_c = v_0 + \frac{Q}{C}. \tag{10.36}$$

The voltage decreases when the capacitance of the bank increases. Assuming a maximum voltage level at the bank terminals of $BIL/1.2$, where BIL is the basic impulse level of the insulation with respect to lightning surges, and factor 1.2 refers to the safety margin, the capacitance of the shunt-capacitor bank can be calculated from:

$$C = \frac{Q}{\frac{NBI}{1.2} - v_0}. \tag{10.37}$$

Once the capacitance C is known, the reactive power of the shunt capacitor can be determined.

$$S_r = \frac{\omega Q {V_L}^2}{\frac{NBI}{1.2} - v_0} \tag{10.38}$$

This expression furnishes a first indication of the nominal reactive power of the shunt-capacitor bank so that the voltage does not surpass its basic insulation level, for a given lightning charge, which has a probabilistic behavior. This approach assumes that all the charge is stored in the capacitor, disregarding any traveling waves inside the substation. On the other hand, the shunt-capacitor banks can absorb part of the charge, reducing the wavefront and amplitude of the lightning surge inside the substation.

Given the nominal line-to-line voltage $V_L = 230$ kV, $NBI = 900$ kV, and a lightning-current charge, Q, the required reactive power such that the shunt capacitor is self-protected from lightning surges is calculated as $S_r = 36Q$ MVAr, using Eq. (10.38).

It is recommended that the substation be analyzed in detail with electromagnetic transient programs as per Chapters 9 and 14. This may also be convenient when the capacitor features an inrush current-limiting reactor connected in series. In this case, it is necessary to assess whether a parcel of the surge is transferred to the capacitor through the stray capacitance of the reactor, or whether the surge occurs mainly at the reactor terminal, whichever is the most likely.

10.3 Back-to-Back Shunt Capacitor

10.3.1 Transient Inrush Currents

The following analysis assumes the existence of more than one shunt-capacitor bank operating in the substation. The same considerations concerning single-phase and three-phase circuits as in Section 10.2 remain valid.

It was noted above that during energization of the first shunt-capacitor bank, even though transient currents can surpass the rated value at 60 Hz, the relatively high reactance of the system limits the current to levels that the equipment can bear.

However, higher transient inrush currents may occur during energization of shunt-capacitor banks, when more than one of the banks are involved in the energization. This bank configuration is known as back-to-back. High-frequency currents can also arise in the substation if a restrike occurs while opening shunt capacitors operating in a back-to-back configuration. The most critical condition occurs when the first shunt-capacitor bank is deenergized, while the other banks are operating. In that case, the high current values can resemble those of faults, when the reactive power of the operating shunt capacitors is high.

The equipment that may be subjected to high currents during the switching or fault involving shunt capacitors are the shunt capacitors themselves, circuit breakers, current transformers, and possibly power line carriers (PLC). The PLC can be stressed during faults at the line terminals, but it is not stressed by faults inside the substation. At high frequencies, the coil behaves predominantly as an inductance and the voltage at the PLC is proportional to the high-frequency current.

The approach to be applied to the substation equipment should consider the standard limitations, which may depend on how often the switching or fault takes place.

10.3.2 Back-to-Back Energization

Energization of the last shunt-capacitor bank to be inserted into the system is considered. The inductance L_i represents the bus inductance, including the lead that connect the shunt capacitor (Figure 10.7a).

The equations may or may not consider the inductance of the feeder system, according to Section 10.2.1. However, it is assumed that the frequencies are high and that $L_s \gg L_i$, because the bus inductance is much lower than the inductance of the equivalent system. The current path occurs preferentially through the bus inductance, with the system behaving practically like an open circuit at high frequencies. The main transient occurs much like that of a free oscillatory LC circuit, composed of only the shunt capacitor to be energized and the shunt capacitors already operating. The stored charge in the operating capacitors depends on the switching time, and the capacitor to be energized or reclosed may or may not have a stored charge. An analysis of the free *LC* circuit was presented in Chapter 1, Section 1.5.2.

In the following set of equations, some approximations are made, but they should be evaluated on a case-by-case basis. The residual charge trapped in the case of capacitor reclosing will be considered in the next section. The approach to the single-phase case is to assume that the shunt-capacitor banks are identical:

$$L_1 = L_2 \cdots L_n = L. \tag{10.39}$$

$$C_1 = C_2 \cdots C_n = C. \tag{10.40}$$

$$L_{n-1} = \frac{L}{n-1}, \tag{10.41}$$

$$C_{n-1} = (n-1)C. \tag{10.42}$$

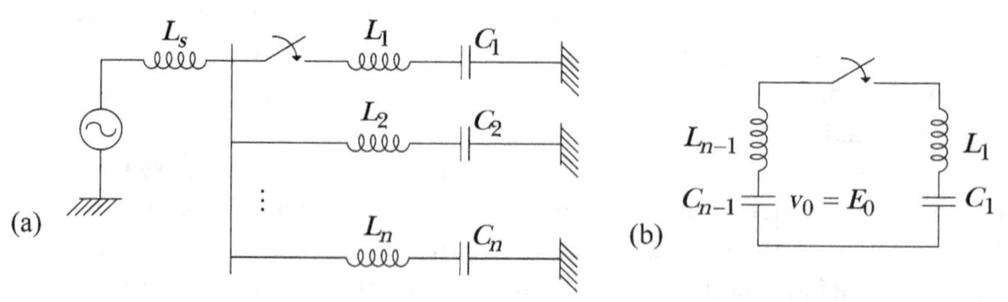

Figure 10.7 Back-to-back energization. (a) Substation with operating shunt capacitors. (b) Approximated equivalent circuit.

The equivalent circuit for the energized banks is obtained in Figure 10.7b:

Energization can occur at several moments in time. An event considered critical is when the voltage at the bus reaches its maximum amplitude, E_0. It is also assumed that the initial voltage, v_0, on the capacitor to be energized is approximately equal to E_0, and the following equivalent circuit is proposed:

$$L_{eq} = L + L_{n-1}, \tag{10.43}$$

$$L_{eq} = L + \frac{L}{n-1} = \frac{nL}{n-1}, \tag{10.44}$$

$$C_{eq} = \frac{C\,C_{n-1}}{C + C_{n-1}}, \tag{10.45}$$

$$C_{eq} = \frac{n-1}{n}C. \tag{10.46}$$

The maximum current is calculated from the following expression:

$$I_m = \frac{E_0}{\sqrt{\frac{n}{n-1}\frac{L}{\frac{n-1}{n}C}}} = \frac{n-1}{n}\frac{E_0}{\sqrt{\frac{L}{C}}}, \tag{10.47}$$

The natural oscillatory frequency of the equivalent circuit is equal to the natural frequency of just one shunt-capacitor bank.

$$f_0 = \frac{1}{2\pi\sqrt{LC}}. \tag{10.48}$$

Using the same assumptions made previously, it can be shown that energization of a capacitor involves energy exchange primarily between capacitors, as the grid's contribution is minimal. However, if the system's contribution is significant, the substation and system should be modeled in full detail.

So far, the losses in inductances and capacitances were neglected. Regarding the damping effect, it may be necessary to include these losses to avoid overestimating the requirements that the system equipment must meet. With the losses included, the time constant of the damped RLC circuit is given by $\tau = 2L/R$.

Small reactors in series with the capacitors are quite common, increasing the inductance of the circuit and limiting inrush currents. The reduction in the current peak is proportional to the square root of the total inductance, given by the sum of the previous inductance in the circuit and the reactor inductance. On the other hand, an increase in the time constant and a reduction in damping occur.

Another point of interest is that in back-to-back switching, high overvoltage levels are not observed in the substation. Most of the energy exchange occurs between the capacitors, and high currents can be expected; however, the inductances of buses and leads are low, as are the overvoltages at the substation.

Just a small part of the current flows through the system inductance, causing only moderate switching surges at the substation.

10.3.3 Restrike

The similarity of the circuits used to analyze energization of shunt capacitors with trapped charge and restrike in the circuit breakers were discussed previously. Hence, only the critical case of restrike is analyzed for shunt capacitors operating in a back-to-back configuration .

It is assumed that at the moment of opening the voltage in the capacitor is $v_0 = -E_0$, which is approximately equal to the system voltage. It is also assumed that, in the case of Figure 10.3, the restrike occurs one half-cycle after current extinction, when the voltage across the operating capacitors reaches an amplitude of E_0. At this moment, there is a voltage difference of $2E_0$ across the loop formed in the equivalent circuit, thereby doubling the energization current.

From this analysis, it can be shown that the worst-case scenario for the occurrence of a restrike is realized on opening the first shunt-capacitor bank, when all banks in the substation are operating (Figure 10.8).

10.3.4 Faults

High currents are also expected due to faults near or inside the substation, known as outrush currents. Although faults can lead to higher current levels with respect to those observed with energization in the back-to-back configuration, the probability of a fault is much lower.

In the case of faults inside the substation, the total current consists of the contribution of the system at 60 Hz and the transient inrush currents due to capacitor discharge.

In the circuit in Figure 10.9, the current contribution of the RL circuit of the system is practically the same, independently of fault location inside the substation. The solution of the circuit may be obtained by the approach in Chapter 1, Section 1.4.1, assuming a sinusoidal excitation voltage:

$$e(t) = E_0 \sin(\omega t + \theta), \tag{10.49}$$

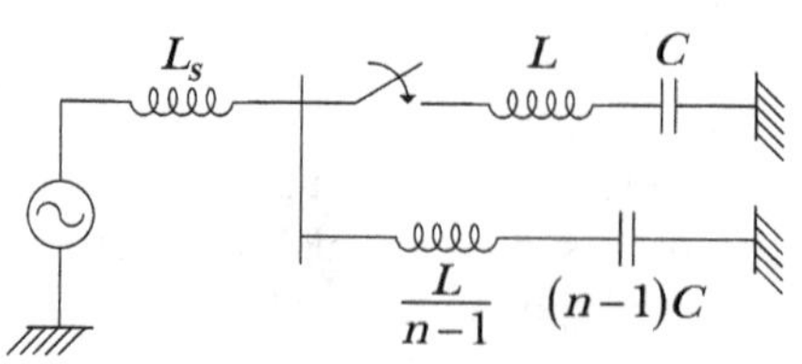

Figure 10.8 Restrike in the circuit breaker.

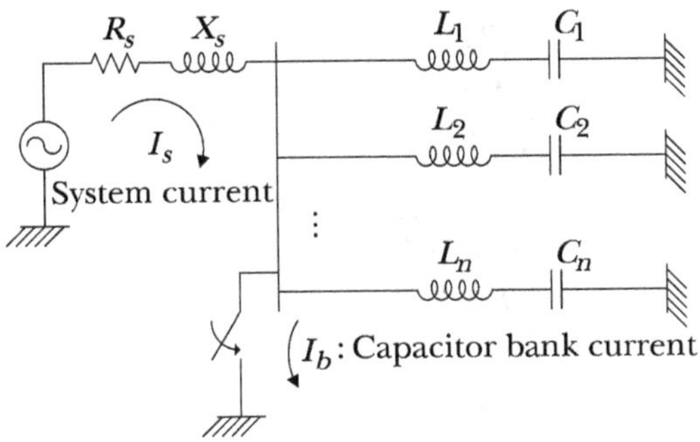

Figure 10.9 Short-circuit.

where:

$X_s = \omega L_s$ and $\omega = 377$ rad/s.

$$i(t) = \frac{E_0}{|Z|}\left(\sin(\omega t + \theta - \phi) - e^{-\frac{t}{\tau}} \sin(\theta - \phi)\right), \tag{10.50}$$

$$|Z| = \sqrt{R^2 + \omega L_s^2}, \tag{10.51}$$

$$\phi = \text{atan}\left(\frac{\omega L_s}{R}\right), \tag{10.52}$$

$$\tau = \frac{L_s}{R_s}. \tag{10.53}$$

The system contribution includes the alternating current component and also the dc-offset component:

To simplify the analysis, $\phi \simeq 90°$ is assumed.

For a fault at maximum voltage amplitude, a minimal contribution of the DC offset may be expected. However, if the fault occurs at a voltage of zero, the maximum DC offset happens.

The high-frequency oscillation originates with the discharge of the capacitors. The same simplifications will be assumed as before, that is, the capacitors are considered to have the same parameters ($L = L_i$, $C = C_i$, $1 \leq i \leq n$). In addition, the voltage across the capacitor is assumed with the same excitation voltage level, $E_0 = v_0$, disregarding the voltage drop at 60 Hz in the inductance.

The following expressions may be written:

$$L_{eq} = \frac{L}{n}, \tag{10.54}$$

$$C_{eq} = nC, \tag{10.55}$$

$$I = \frac{E_0}{\sqrt{\frac{L_{eq}}{C_{eq}}}}, \tag{10.56}$$

$$f_0 = \frac{1}{2\pi\sqrt{L_{eq}C_{eq}}}. \tag{10.57}$$

$$I = n\frac{E_0}{\sqrt{\frac{L}{C}}}, \tag{10.58}$$

$$f_0 = \frac{1}{2\pi\sqrt{LC}}. \tag{10.59}$$

The natural frequency of oscillation of the equivalent circuit for all capacitors is the same as that of only one capacitor, and the total current is the sum of the contributions of all capacitors.

The maximum amplitude of the high-frequency current is obtained when the fault occurs at the time of voltage peak, when the energy stored in the shunt capacitors $CE_0^2/2$ is maximal. It is noted that, on the other hand, this case implies a minimal contribution of the DC-current component, owing to the feeder system.

In any analysis performed using electromagnetic transient programs, it is convenient to determine the maximum current, by considering the proportions of the high-frequency current and the 60 Hz component, when a sinusoidal excitation is applied.

As the high-frequency component due to capacitors is often greater than that of the system, the above model is satisfactory, and the worst-case scenario occurs if there is a fault when $e(t) = E_0$. However, specific cases should be analyzed in more detail.

The equations for the capacitors should consider losses, including the resistance in the circuit. Accurate analysis of the fault current should also include the inductance of the bus and leads. However, when the capacitors have current-limiting reactors, the simplifications above can be applied because the inductance of the reactors is frequently much higher than that of the bus and leads.

When equivalent inductances or capacitances cannot be approached using Eqs. (10.54) or (10.55), the complete system of equations should be established in each specific case. Example 10.5 shows one case of detailed circuit representation.

Regarding faults, some additional comments are made below. The type of bank connection (star versus delta) and its grounding should also be considered when analyzing unbalanced faults.

The single-phase short circuit is the most common, and in the case of an ungrounded system, the fault current is zero, but a large voltage imbalance is expected in the substation.

If a fault occurs within the substation, it is necessary to identify the fault-clearing circuit breakers and whether the circuit breaker at the capacitor bank will be tripped. For a fault internal to the capacitor bank, the circuit breaker associated with the bank will clear the fault, although if there are inrush reactors, the RRRV can be very high.

Substations related to ungrounded systems may have a grounding transformer and, in this case, a small phase-to-earth fault is expected. In the case of fault-clearing, the opening sequence for the phases and the oscillations occurring between the shunt capacitor and the grounding transformer should be investigated, since overvoltages can occur.

Inrush reactors are used to limit currents during energization and restrikes and can also be used for the purpose of limiting outrush fault currents. The reactors should limit the amplitude of the current, but on the other hand, they increase the transient duration. Closing resistors limit the inrush currents and energization overvoltages but are inactive in restrikes and faults. High-frequency currents can damage circuit-breaker contacts, mainly in cases of frequent energizations. Another issue for inrush reactors is the lightning surges discussed in Section 10.2.5.

Example 10.2 Consider a bank of capacitors with a nominal reactive power of 4 MVAr and a nominal line-to-line voltage of 11.8 kV, connected to a system having a short-circuit power of 478 MVA. Analyze the transient currents for a three-phase fault at the circuit breaker terminals.

L_S: system inductance.
L: internal inductance of the capacitor.
C: capacitance of the shunt capacitor.

Data

$$Q = 4\,\text{MVAr}, V_n = 13.8\ \text{kV}.$$

$$L = 10\,\mu\text{H}.$$

$$e(t) = E_0 \sin(\omega t + \theta).$$

Solution

The shunt capacitors can be connected in star or in delta, and at this voltage level it is quite common to find ungrounded capacitors. If a three-phase fault occurs, only the positive sequence exists and a star-connected capacitor bank can be assumed.

$i_f = i_1 + i_2$: current through the switch
i_1: system component according to Eq. (10.50):

$$i_1(t) = \frac{E_0}{|Z|}\left[\sin(\omega t + \theta - \varphi) - e^{-\frac{t}{\tau}}\sin(\theta - \varphi)\right],$$

Assuming the system impedance to be purely inductive, that is $\varphi = \pi/2$, and that the voltage at the capacitor is maximal at $t = 0$, $\theta = \pi/2$.

It can be assumed that $L \ll L_s$, and therefore, the 60 Hz voltage at the capacitor is practically the same as the voltage at the source, $v_0 = E_0$.

Note that the high-frequency component i_2 has opposite signs in Figure 10.10a and b. The sign of i_2 in Figure 10.10b follows the convention for current entering the LC branch.

$v_i(t)$: voltage at the inductor.
$v_c(t)$: voltage at the capacitor.

$$0 = v_i(t) + v_c(t),$$

$$0 = L\frac{di_2(t)}{dt} + \frac{1}{C}\int i_2(t)dt,$$

or:

$$0 = sLI_2(s) - i_0 + \frac{I_2(s)}{sC} + \frac{v_0}{s}.$$

Assuming that $i_0 = 0$, the inverse transform is obtained:

$$i_2(t) = -\frac{v_0}{\sqrt{\frac{L}{C}}}\sin(\omega_0 t).$$

It can be observed that the former expression is identical to Eq. (10.8), after excitation is removed, making $E_0 = 0$. However, the negative sign of the current, $i_2(t)$, means that it is feeding the short-circuit, using the convention in Figure 10.10a.

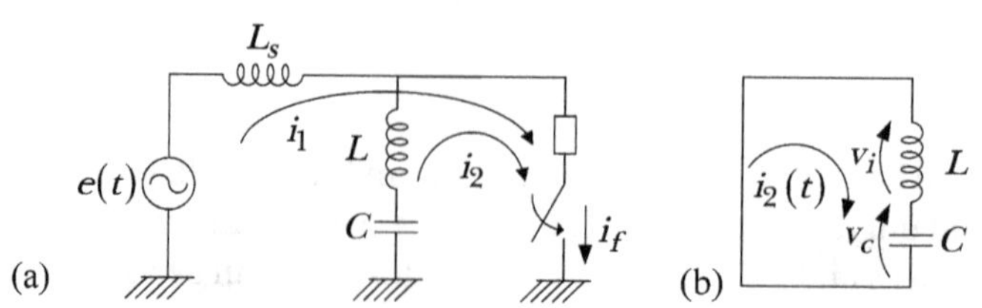

Figure 10.10 Fault near the circuit-breaker. (a) System and shunt capacitor currents. (b) Shunt capacitor loop.

Using the previous assumptions, $v_0 = E_0$, $\theta = \pi/2$ and $\varphi = \pi/2$, current i_1 may be expressed as:

$$i_1(t) = \frac{E_0}{\omega L_s} \sin(\omega t) \ (Z = j\omega L_s).$$

Numerical values are calculated.

$$\omega\, L_s = X_s = \frac{13.8^2}{478}.$$

$$X_s = 0.398 \ \ \Omega.$$

$$i_1(t) = \frac{E_0}{0.398} \sin(377t).$$

$$i_1(t) = 28.3 \sin(377t).$$

$$C = \frac{4}{\omega \times 13.8^2} = 55.7\ \mu\text{F}.$$

$$E_0 = \frac{13,8}{\sqrt{3}}\sqrt{2} = 11.27\ \text{kV}.$$

$$\sqrt{\frac{L}{C}} = \sqrt{\frac{10 \times 10^{-6}}{55.7 \times 10^{-6}}} = 0.424 \ \ \Omega.$$

$$f_0 = \frac{10^6}{2\pi\sqrt{10 \times 55.7}} = 6743\ \text{Hz}.$$

$$i_2(t) = \frac{11.27}{0.424} \sin(42371t).$$

$$\omega_0 = 42371\ \text{rad/s}.$$

$$\omega = 377\ \text{rad/s}.$$

The relation between ω_0 and ω is:

$$\frac{\omega_0}{\omega} \simeq 110.$$

The high-frequency component is:

$$i_2(t) = 26.6\ \sin(42371t).$$

In this example, with $\theta = \pi/2$, the DC component is null, and the high-frequency and 60 Hz contributions are comparable.

Regarding the voltage at the capacitor, $v_0 = E_0$, it may be observed that the voltage at the bus is actually a little different from the voltage at the source, since there is a voltage divider in the steady-state regime:

$$V_c = \frac{-\dfrac{j}{\omega C}}{j\left(\omega L_s - \dfrac{1}{\omega C}\right)} E,$$

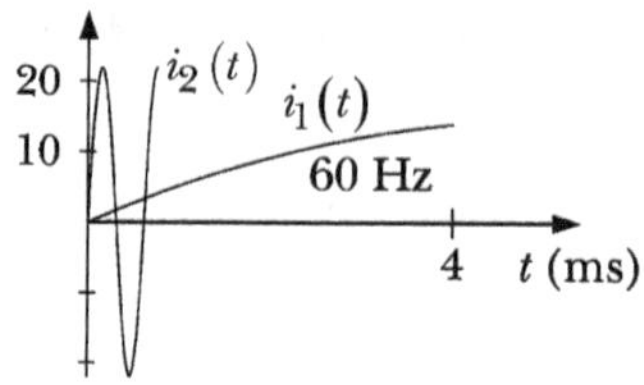

Figure 10.11 Short circuit current.

$$V_c = \frac{-47.61}{1.9 - 47.61} = 1.04E,$$

where E is the phasor corresponding to the excitation, $e(t)$; in other words, $v_0 = 1.04E_0$, resulting in the high-frequency component (Figure 10.11):

$$i_2(t) = 1.04 \times 26.6 \sin(42371t) = 27.66 \sin(42371t).$$

As can be seen, the two components will superimpose their maximum values at 4 ms, at which time the total current can then reach:

$$I_m = 27.66 + 28.3 = 56\,\text{kA}.$$

This occurs only in undamped circuits.

A detailed model of the capacitor would require the representation of losses in the dielectric, and the RLC circuit should be adequate.

Example 10.3 Analyze a fault applied at the terminal of a shunt-capacitor bank by means of an RLC circuit, adopting the same data used in the previous example and assuming that $R = 0.2\ \Omega$.

Solution

The basic equation for the RLC circuit excited by a voltage source was presented in Chapter 1. Here the case of a fault is examined, assuming null excitation.

$$0 = L\frac{di}{dt} + Ri + \frac{1}{C}\int_{t_0}^{t} i\,dt.$$

R: resistance of the capacitor bank.
L: inductance of the capacitor bank.

Applying the Laplace transform:

$$0 = sLI(s) - Li_0 + RI(s) + \frac{1}{sC}I(s) + \frac{v_0}{s},$$

and since damping can be assumed very low, with $i_0 \simeq 0$ at the time of maximum voltage:

$$0 = sLI(s) + RI(s) + \frac{1}{sC}I(s) + \frac{v_0}{s},$$

Considering the complex roots for the underdamped circuit:

$$s_{1,2} = -\frac{R}{2L} \pm \sqrt{\left(\frac{R}{2L}\right)^2 - \frac{1}{LC}},$$

$$\sigma = -\frac{R}{2L}, \qquad \omega_d = \sqrt{\frac{1}{LC} - \left(\frac{R}{2L}\right)^2},$$

$$s_{1,2} = \sigma \pm j\omega_d, \qquad \omega_0 = \frac{1}{\sqrt{LC}}.$$

The residues are calculated, as follows:

$$I_2(s) = \frac{k_1}{[s-(\sigma + j\omega_d)]} + \frac{k_1^*}{[s-(\sigma - j\omega_d)]},$$

$$k_1 = \left.\frac{1}{[s-(\sigma - j\omega_d)]}\right|_{s=\sigma+j\omega_d} = \frac{1}{2j\omega_d},$$

$$k_1 = -j\frac{1}{2\omega_d}, \qquad k_1^* = j\frac{1}{2\omega_d}.$$

The time response is given by:

$$2|k_1|e^{\sigma t}\cos\left(\omega_d t - \frac{\pi}{2}\right).$$

The inverse transform of $I_2(s)$ is given by:

$$i_2(t) = -\frac{v_0}{L}2\frac{1}{2\omega_d}e^{\sigma t}\sin(\omega_d t),$$

$$i_2(t) = -\frac{v_0}{\sqrt{\frac{L}{C} - \frac{R^2}{4}}}e^{-\frac{R}{2L}t}\sin\left(\sqrt{\frac{1}{LC} - \left(\frac{R}{2L}\right)^2}t\right).$$

The time constant and the maximum current in the circuit may be written as:

$$\tau = \frac{2L}{R}, \qquad I_{2m} = \frac{v_0}{\sqrt{\frac{L}{C} - \frac{R^2}{4}}}.$$

For the free oscillatory circuit, as in Example 10.2:

$$\sqrt{\frac{L}{C}} = 0.424, \qquad \omega_0 = \frac{1}{\sqrt{LC}} = 42371\ \text{rad/s}.$$

The damped oscillation frequency shows a small reduction with respect to the natural oscillation frequency.

$$\omega_d = \sqrt{\frac{1}{10\times10^{-6}\times55.7\times10^{-6}} - \left(\frac{0.2}{20\times10^{-6}}\right)^2} = 41174\ \text{rad/s}.$$

$$\sqrt{\frac{L}{C} - \frac{R^2}{4}} = \sqrt{\frac{10\times10^{-6}}{55.7\times10^{-6}} - \frac{0.2^2}{4}} = 0.412\ \ \Omega,$$

The amplitude of $i_2(t)$ is calculated, assuming the same correction factor, 1.04, and a slightly higher amplitude, compared to the case in Example 10.2, is obtained:

$$I_{2m} = \frac{1.04\times11.27}{0.412} = 28.45\ \text{kA}.$$

The time constant is:

$$\tau = \frac{2L}{R} = \frac{20\times10^{-6}}{0.2} = 100\ \mu\text{s}.$$

However, after three time constants, that is, 300 μs, the high-frequency transient almost vanishes, and only the 60 Hz component remains.

Example 10.4 In the circuit in Example 2, calculate the transient current assuming a limiting reactor of 100 μH.

Solution

In this case, disregarding the losses, the maximum current is given by:

$$L = 100 + 10 = 110\ \mu\text{H},$$

$$I_{2m} = \frac{E_0}{\sqrt{\frac{L}{C}}},$$

$$Z_0 = \sqrt{\frac{L}{C}} = \sqrt{\frac{110\times10^{-6}}{55.7\times10^{-6}}} = 1.405\ \Omega,$$

$$I_{2m} = \frac{11.27\times1.04}{1.405} = 8.34\ \text{kA},$$

$$\omega_0 = \frac{1}{\sqrt{LC}} = 12\,775.4\ \text{rad/s}\quad (f_0 = 2\ \text{kHz}),$$

$$\tau = \frac{2\times110\times10^{-6}}{0.2} = 1100\ \ \mu\text{s}.$$

The time constant is proportional to the increase in inductance, and for larger reactors, superposition on the 60 Hz component can be expected.

In the previous examples, based on Figure 10.10, it was possible to calculate the low-frequency fault current, as the contribution of the system, and the high-frequency current as the contribution of the capacitor bank in simple terms because the topology allows calculation of the current in two independent loops.

This simplification depends on the case analyzed. Inrush reactors, whose reactance prevails over bus reactance, allow for interesting simplifications in topology.

For the same reason, the fault current contribution due to the system can be considered as independent of the fault location in the substation because generally the feeder system reactance is much higher than any other reactance.

A more complex topology for calculating the high-frequency fault component is presented in the next example [2].

Example 10.5 Analyze the fault in the system in Figure 10.12, which has two shunt-capacitor banks.

Solution

To maximize the transient component, the fault occurs when the voltage is at its maximum. In this case, the DC offset of the 60 Hz component is null. To analyze the influence of the DC offset, the phase angle of the voltage must be changed.

a) Approximate current component at 60 Hz.

$$i_1(t) = \frac{E_0}{\omega L_s} \sin(\omega t).$$

b) High-frequency transient component, Figure 10.13a.

Inductances L_x and L_y incorporate the internal inductance of the capacitors, the inductance of the inrush reactors, and the inductance of the lead cables. The capacitance C_x may also be different from C_y.

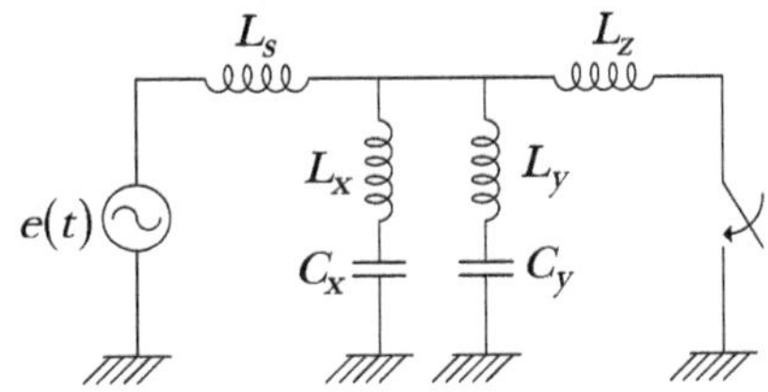

Figure 10.12 Two shunt-capacitor banks.

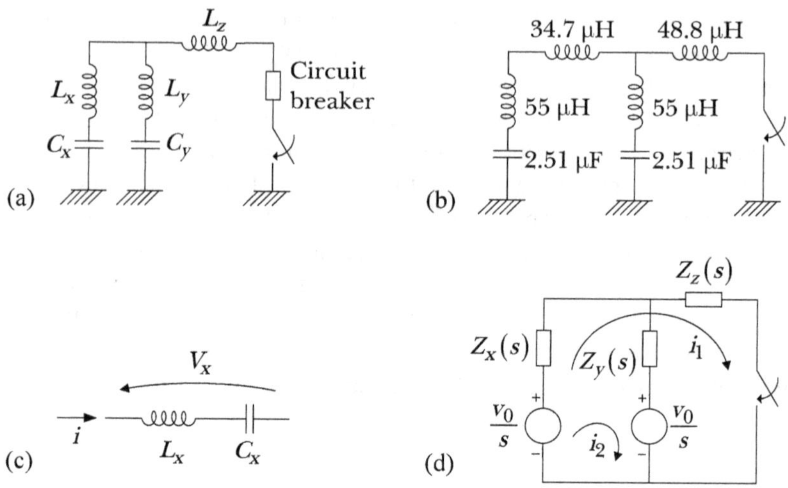

Figure 10.13 Circuit for high-frequency analysis. (a) Fault near the shunt capacitor banks. (b) Data. (c) Capacitor-inductor element. (d) Loop currents.

Equation for the LC branch, Figure 10.13c:

$$v_x(t) = L_x \frac{di(t)}{dt} + \frac{1}{C_x}\int i(t)dt,$$

$$V_x(s) = sL_xI(s) - L_xi(0) + \frac{1}{C_xs}I(s) + \frac{v_{0x}}{s}.$$

A situation is assumed, in which the voltage at the bank is maximal, and the current is null at the instant of fault, that is, $i(0) = 0$.

$$V_x(s) = \left(sL_x + \frac{1}{C_xs}\right)I(s) + \frac{v_{0x}}{s} = Z_x(s)I(s) + \frac{v_{0x}}{s}.$$

Given that at the nominal frequency, the values for L_x and L_y are negligible with respect to the reactance of the feeder system, it can be assumed that $v_{0x} = v_{0y} = v_0$.

The equivalent circuit in the frequency domain is presented in Figure 10.13d.

Using the convention for current entering at branches, the loop equations can be set up with simplified notation:

$$Z_x(I_1 + I_2) + Z_zI_1 = \frac{v_0}{s}, \tag{e.1}$$

$$Z_x(I_1 + I_2) + Z_yI_2 = 0. \tag{e.2}$$

From (e.2):

$$I_2 = -\frac{I_1Z_x}{Z_x + Z_y}.$$

Substituting I_2 into (e.1):

$$I_1\left(Z_x + Z_z - \frac{Z_x^2}{Z_x + Z_y}\right) = \frac{v_0}{s},$$

$$I_1 = \frac{v_0}{s}\frac{Z_x + Z_y}{Z_x Z_y + Z_z(Z_x + Z_y)},$$

where:

$$Z_x = sL_x + \frac{1}{sC_x},\ Z_y = sL_y + \frac{1}{sC_y},\ Z_z = sL_z,$$

$$I_1 = \frac{v_0}{s}\frac{Z_x + Z_y}{\left(\frac{1}{sC_x} + sL_x\right)\left(\frac{1}{sC_y} + sL_y\right) + sL_z\left(\frac{1}{sC_x} + sL_x + \frac{1}{sC_y} + sL_y\right)}.$$

Designating the denominator as $D(s)$ and the numerator as $N(s)$, according to the nomenclature in reference [2], the denominator $D(s)$ is analyzed as follows:

$$D(s) = \frac{1 + s^2(C_xL_x + C_yL_y) + s^4C_xL_xC_yL_y}{s^2C_xC_y} + \frac{C_xL_z + C_yL_z + s^2(L_x + L_y)L_zC_xC_y}{C_xC_y}.$$

By developing the denominator:

$$D(s) = \frac{L_xL_y + L_xL_z + L_yL_z}{s^2}\left(s^4 + Vs^2 + W\right),$$

$$V = \frac{C_x(L_x + L_z) + C_y(L_y + L_z)}{C_xC_y(L_xL_y + L_xL_z + L_yL_z)},$$

$$W = \frac{1}{C_xC_y(L_xL_y + L_xL_z + L_yL_z)}.$$

The numerator, $N(s) = Z_x + Z_y$, is now examined.

$$N(s) = \frac{1}{sC_x} + sL_x + \frac{1}{sC_y} + sL_y = \frac{C_y + C_x + s^2(L_x + L_y)C_xC_y}{sC_xC_y}.$$

Thus:

$$I_1(s) = \frac{N(s)}{D(s)},$$

$$I_1(s) = \frac{v_0(L_x + L_y)}{L_xL_y + L_xL_z + L_yL_z}\frac{s^2 + \frac{C_x + C_y}{C_xC_y(L_x + L_y)}}{s^4 + Vs^2 + W},$$

$$I_1(s) = Kv_0\frac{s^2 + U}{s^4 + Vs^2 + W}. \tag{e.3}$$

Setting:

$$K = \frac{L_x + L_y}{L_x L_y + L_x L_z + L_y L_z} \text{ and } U = \frac{C_x + C_y}{C_x C_y (L_x + L_y)}.$$

To obtain the time response, it is necessary to find the inverse transform of (e.3).

If the circuit is considered lossless, the response is oscillatory and has two natural frequencies. Even if losses are included in the calculation, in the case of shunt capacitors, the response remains oscillatory.

Then, it follows that:

$$s^4 + Vs^2 + W = (s^2 + a^2)(s^2 + b^2),$$

The residues can be calculated as follows:

$$I_1(s) = Kv_0 \frac{s^2 + U}{s^4 + Vs^2 + W} = Kv_0 \frac{s^2 + U}{(s^2 + a^2)(s^2 + b^2)},$$

$$I_1(s) = Kv_0 \left(\frac{k_1}{s + ja} + \frac{k_1^*}{s - ja} + \frac{k_2}{s + jb} + \frac{k_2^*}{s - jb} \right),$$

$$k_1 = \left. \frac{U - a^2}{(s - ja)(b^2 - a^2)} \right|_{s = -ja} = \frac{U - a^2}{-2ja(b^2 - a^2)}.$$

Similarly, k_2 can be obtained. The inverse transform of (e.3) may be written as:

$$i_1(t) = Kv_0 2|k_1| \cos\,(at + \angle k_1) + Kv_0 2|k_2| \cos(bt + \angle k_2).$$

Data

A substation, with a nominal line-to-line voltage of 230 kV, is analyzed with two shunt capacitors separated by lead cables and bus. The fault occurs at some distance from the capacitors, and numerical values if inductances and capacitances are given in Figure 10.13b.

The inductance values of 34.7 and 48.8 μH are related to aerial cables. The typical inductance of the cable is 1 μH/m, in other words, the separation distance between the capacitor banks is approximately 35 m, and the fault happens about 50 m from one of the two banks.

The shunt capacitors are assumed to have the same nominal reactive power, that is, $C_x = C_y = 2.51$ μF:

$$Q = 377 \times 2.51 \times 10^{-6}(230)^2 = 50 \text{ MVAr}.$$

The two shunt capacitors are placed about 45 m from the bus. Assuming an internal inductance of 10 μH for the capacitors, an inductance of approximately 55 μH in series with each capacitor is obtained.

Numerical Results

Substituting the numerical values:

$$L_x = 55 + 34.7 = 89.7\ \mu\text{H},$$

$$L_y = 55\ \mu\text{H},$$

$$C_x = C_y = 2.51\ \mu\text{F},$$

$$L_z = 48.8\ \mu\text{H},$$

$$v_0 = \frac{230\sqrt{2}}{\sqrt{3}} = 187.79\ \text{kV}.$$

It is assumed that $v_0 = E_0$.

- Calculation of the constants U, V, W, and K:

$$K = \frac{(55 + 89.7) \times 10^{-6}}{(55 \times 89.7 + 55 \times 48.8 + 89.7 \times 48.8) \times 10^{-12}} = 12063.50.$$

$$U = \frac{2 \times 2.51 \times 10^{-6}}{2.51 \times 2.51 \times 10^{12}(55 + 89.7) \times 10^{6}} = 5506.65 \times 10^{6}.$$

$$L_xL_y + L_xL_z + L_yL_z = 0.0119949 \times 10^{-6}.$$

$$V = \frac{2.51 \times 10^{-6}(55 + 2 \times 48.8 + 89.7) \times 10^{-6}}{(2.51 \times 10^{-6})^2 \times 0.0119949 \times 10^{-6}} = 8047.936 \times 10^{6}.$$

$$W = \frac{1}{(2.51 \times 10^{-6})^2 \times 0.0119949 \times 10^{-6}} = 13232.971 \times 10^{15}.$$

Roots of $s^4 + Vs^2 + W$:

$$a^2 = -2.3036979 \times 10^{9}, \qquad b^2 = -5.7442112 \times 10^{9}.$$

$$a = \pm j47997, \qquad b = \pm j75791.$$

$$k_1 = j\,0.969798 \times 10^{-5}, \qquad k_2 = j\,0.0455568 \times 10^{-5}.$$

Using the convention of currents leaving the bipole, that is, supplying the fault, the high-frequency current is written as:

$$i_1(t) = I_a \sin(at) + I_b \sin(bt):$$

$$I_a = 12063.43 \times 187.79 \times 2 \times |k_1|.$$

$$I_a = 43.94\ \text{kA}.$$

$$I_b = 12063.43 \times 187.79 \times 2 \times |k_2|.$$

$$I_b = 2.06\ \text{kA},$$

$$i_1(t) = 43.94 \sin(47997t) + 2.06 \sin(75791t).$$

10.4 Three-Phase Circuits

High-frequency transients involving inrush currents during reclosing and restrike are examined here. Regarding three-phase circuits, it is of interest to analyze in greater detail the voltages due to trapped charges during three-phase openings, thereby complementing the single-phase approach employed in items 10.3.2 and 10.3.3. It is therefore important to clarify the correspondence between single-phase circuits and three-phase circuits.

10.4.1 Stored Charges in Ungrounded Shunt Capacitors

Ungrounded shunt capacitors are quite common in substations up to 69 kV.

- Star connection

To write equations for charges and voltages at the moment of three-phase capacitor bank opening, it may be assumed that the voltage imposed by the network on the substation does not change in amplitude. For ungrounded capacitors, there is no zero-sequence path and thus the feeder network is represented only by its positive sequence. Figure 10.14a presents the phasors for the three-phase source voltages ($E_0 = \sqrt{2}|E_a|$).

$$e_a(t) = E_0 \cos(\omega t).$$

$$e_b(t) = E_0 \cos(\omega t - 120°).$$

$$e_c(t) = E_0 \cos(\omega t + 120°).$$

Equations for the capacitor in star connection, Figure 10.14b and c, can be written, simplifying the notation for phase-to-earth voltages $v_a = v_a(t), \;\; v_b = v_b(t)$, etc.

$$v_a = v_n + \frac{1}{C}\int i_a dt + v_{a0}, \tag{10.60}$$

$$v_b = v_n + \frac{1}{C}\int i_b dt + v_{b0}, \tag{10.61}$$

$$v_c = v_n + \frac{1}{C}\int i_c dt + v_{c0}. \tag{10.62}$$

Given that:

v_{a0}, v_{b0}, v_{c0}: phase-to-neutral initial voltages across the capacitors
C: capacitance of each phase of the bank.
C_p: stray capacitance

The reference to ground after the three-phase opening may somehow depend on stray capacitances [1], but they will not be used.

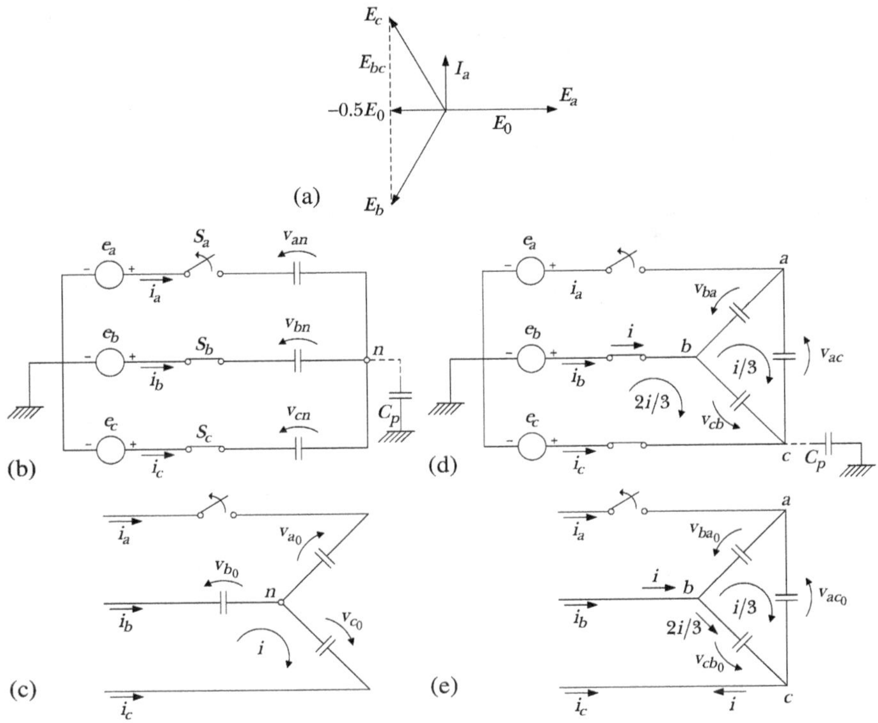

Figure 10.14 De-energization of the three-phase shunt-capacitor bank. (a) Current and voltage phasors. (b) First pole to open, star-connection. (c) Current path for star-connection. (d) First pole to open, delta-connection. (e) Current path for delta-connction.

At the opening time, t_0, for switch S_a, the following voltages across the capacitors are obtained:

$$v_{a0} = E_0, v_{b0} = -\frac{E_0}{2}, v_{c0} = -\frac{E_0}{2}. \tag{10.63}$$

It is assumed that the voltage at phase a is at its maximum value, according to the phasors in Figure 10.14a.

Applying the voltage between phases, $e_{bc} = e_b(t) - e_c(t)$, to phases b and c, while S_b and S_c remain closed, the following expressions are developed:

$$v_a = v_n + E_0, \tag{10.64}$$

$$v_b = v_n + \frac{1}{C}\int i_b dt - \frac{E_0}{2}, \tag{10.65}$$

$$v_c = v_n + \frac{1}{C}\int i_c dt - \frac{E_0}{2}, \tag{10.66}$$

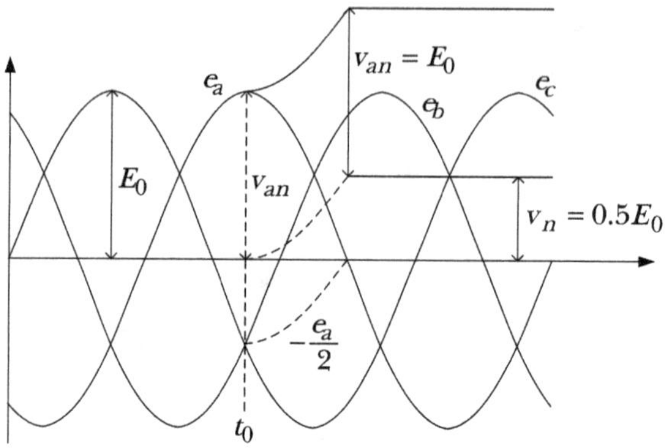

Figure 10.15 Opening of phases, *a*, *b*, *c*.

$$v_n = \frac{v_b + v_c}{2} + \frac{E_0}{2}, \tag{10.67}$$

with $i = i_b = -i_c$.

Adding (10.65) and (10.66):

As $e_b = v_b$, $e_c = v_c$, $e_a = -(e_b + e_c)$, the voltage at the neutral point is calculated (Figure 10.15).

$$v_n = \frac{E_0}{2} - \frac{E_0}{2}\cos(\omega t) = \frac{E_0}{2}[1 - \cos(\omega t)]. \tag{10.68}$$

After one-half cycle elapses, the voltage between phases e_{bc} reaches its maximum value and currents $i_b = -i_c$ decay to zero. Assuming that opening of switches S_b and S_c occurs at that instant, then:

$$e_b = v_b = \frac{\sqrt{3}}{2}E_0, \tag{10.69}$$

$$e_c = v_c = -\frac{\sqrt{3}}{2}E_0, \tag{10.70}$$

$$v_n = \frac{E_0}{2}. \tag{10.71}$$

Resulting in the following voltages across the capacitors with all switches open:

$$v_a = 1.5E_0, \qquad v_{a0} = E_0, \tag{10.72}$$

$$v_b = \frac{\sqrt{3}}{2}E_0, \qquad v_{b0} = \left(\frac{\sqrt{3}-1}{2}\right)E_0, \tag{10.73}$$

$$v_c = -\frac{\sqrt{3}}{2}E_0, \qquad v_{c0} = -\left(\frac{\sqrt{3}+1}{2}\right)E_0. \tag{10.74}$$

And the voltage between phases are given by :

$$v_{ab0} = \frac{3-\sqrt{3}}{2}E_0; v_{ac0} = \frac{3+\sqrt{3}}{2}E_0; v_{bc0} = \sqrt{3}E_0. \tag{10.75}$$

- Delta connection

Regarding the voltages during de-energization of the capacitors, the delta connection gives the same results. Considering the same power for the capacitor bank, the nodal voltages will be the same for the system in both connection cases, delta, and star. The nodal admittance matrix with star connection will have only the neutral as an additional node that can be eliminated by Kron reduction, leading to the same nodal admittance matrix in delta connection. In the phases of the banks, the current can be different, but the current interaction with the system is the same, in both cases of connection.

As an example, the voltage on phase a can be evaluated considering the opening switch S_a at the same time, t_0. The voltages across the capacitors connected in delta, Figure 10.14d can be written as:

$$v_{ba0} = -1.5E_0, v_{ac0} = 1.5E_0, v_{cb0} = 0. \tag{10.76}$$

In Figure 10.14e, the current at phase b is designated as i_b ($i_b = 3i$ and i has no relation to previous analysis). Thus, the current on the upper side of the delta, flowing by the two capacitors in series, is given by i and therefore Eqs. (10.77) and (10.78) can be written as follows:

$$v_a - v_c = 1.5E_0 + \frac{1}{C}\int \frac{i}{3}\,dt, \tag{10.77}$$

$$v_b - v_a = -1.5E_0 + \frac{1}{C}\int \frac{i}{3}\,dt. \tag{10.78}$$

Subtracting (10.78) from (10.77):

$$v_a = \frac{v_b + v_c}{2} + \frac{3}{2}E_0, \tag{10.79}$$

but since

$$\frac{e_a}{2} = -\frac{v_b + v_c}{2},$$

the nodal voltage at phase a can be written in the time domain until the opening of phases b and c:

$$v_a = \frac{E_0}{2}[3 - \cos(\omega t)]. \tag{10.80}$$

After a half cycle elapses, $v_a = 1.5E_0$ is obtained, which is the same calculated value for the opening with the star connection.

10.4.2 Trapped Charges in Grounded Shunt Capacitors

In the grounded star connection, after opening the first phase, the capacitor bank behaves as a grounded two-phase load and then, with the opening of the second phase, as a single-phase load. During this period, small voltage fluctuations may arise at the shunt-capacitor terminals, and only in the exceptional case of a weak system will the voltages differ substantially from the nominal values. In the case of single-phase faults, possible increases in voltages and trapped charges must be investigated.

10.4.3 Reclosing and Restrike in Three-phase Circuits

The sinusoidal source voltages e_a, e_b, e_c make up a three-phase, symmetrical and balanced system, which can be approximated by DC voltages during the transient period. The polarities indicated in Figures 10.16–10.18 help in the formulation of the loop equations.

- Ungrounded Capacitors

  The same comments on losses in the previous sections can be made. Thus, a lossless approach can be assumed to analyze the three-phase equations and only the star connection is presented, as the treatment for the delta connection is similar.

  The basic equations are developed for transient conditions with two or three switches closed. The analysis for one closed switch is not performed since there is no current. In fact, the small current flow through the stray capacitances can be neglected. The Laplace transform is applied to the loop equations.
- Two closed switches

  The models presented in Chapter 1 for the inductance and capacitance are adopted here in the frequency domain.

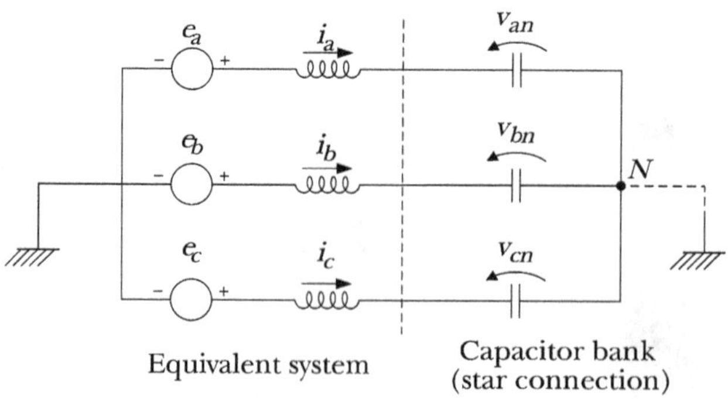

Figure 10.16 Ungrounded capacitor in star connection.

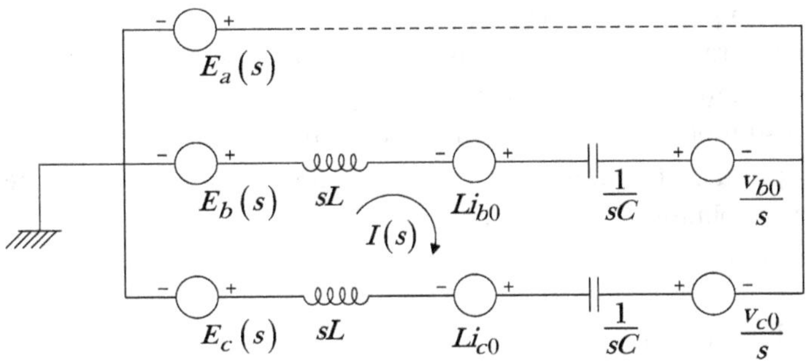

Figure 10.17 Two closed switches.

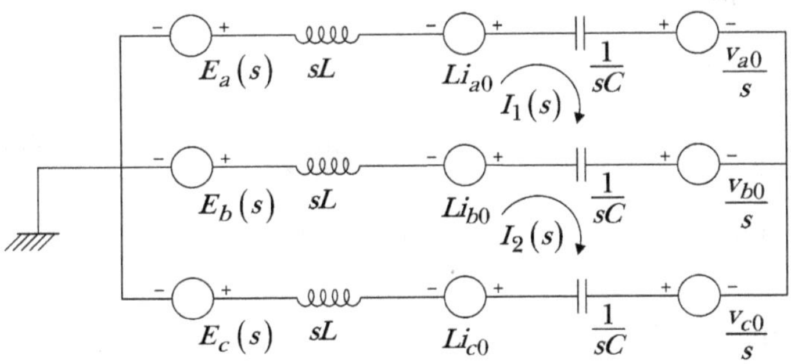

Figure 10.18 Three closed switches.

Initially, the possibility of only two closed switches is examined (phases b and c): The system inductance is designated as L, and the loop equation is given in (10.81):

$$E_b(s) - E_c(s) + \frac{v_{c0} - v_{b0}}{s} + L(i_{b0} - i_{c0}) = 2\left(sL + \frac{1}{sC}\right)I(s), \tag{10.81}$$

noting that $i(t) = i_b(t) = -i_c(t)$ and that initial phase voltages to neutral at the capacitors v_{bn0} and v_{cn0} are written simply as v_{b0} and v_{c0}, respectively. Recalling (10.2) and (10.3):

$$Z(s) = sL + \frac{1}{sC},$$

thus:

$$I(s) = \frac{1}{2Z(s)}\left[E_b(s) - E_c(s) + \frac{v_{c0} - v_{b0}}{s} + L(i_{b0} - i_{c0})\right]. \tag{10.82}$$

For the closing of two phases, the surge impedance of the loop is given by $2Z_0$, and the initial currents are zero since there is no current flow when the first switch closes. Assuming the same hypothesis of polarity opposition as in (10.11) and (10.12)—that is, the sum of the amplitudes of the applied phase-to-phase voltage $\sqrt{3}E_0$, with the amplitude of residual voltage at the capacitors in (10.75)—the maximum currents are obtained for the three cases:

Closing for phases a and b.

$$i(t) = \frac{\sqrt{3}+3}{4Z_0}E_0\sin(\omega_0 t), \tag{10.83}$$

Closing for phases b and c.

$$i(t) = 2\sqrt{3}E_0\sin(\omega_0 t). \tag{10.84}$$

Closing for phases a and c.

$$i(t) = \frac{3\sqrt{3}+3}{4Z_0}E_0\sin(\omega_0 t), \tag{10.85}$$

showing a higher amplitude for (10.85) and slightly greater than the result in (10.12).

- three closed switches

  Applying the Laplace transform to the loop equations.

$$Z(s)\begin{bmatrix} 2 & -1 \\ -1 & 2 \end{bmatrix} \times \begin{bmatrix} I_1(s) \\ I_2(s) \end{bmatrix} = \begin{bmatrix} E_a(s) - E_b(s) + \dfrac{v_{b0} - v_{a0}}{s} - L(i_{b0} - i_{a0}) \\ E_b(s) - E_c(s) + \dfrac{v_{c0} - v_{b0}}{s} + L(i_{b0} - i_{c0}) \end{bmatrix}. \tag{10.86}$$

$$\begin{bmatrix} I_1(s) \\ I_2(s) \end{bmatrix} = \frac{1}{3Z(s)}\begin{bmatrix} 2 & 1 \\ 1 & 2 \end{bmatrix} \times \begin{bmatrix} E_a(s) - E_b(s) + \dfrac{v_{b0} - v_{a0}}{s} - L(i_{b0} - i_{a0}) \\ E_b(s) - E_c(s) + \dfrac{v_{c0} - v_{b0}}{s} + L(i_{b0} - i_{c0}) \end{bmatrix}, \tag{10.87}$$

The loop currents $I_1(s)$ and $I_2(s)$ are related to the phase currents by $i_a = i_1$, $i_b = i_2 - i_1$ and $i_c = -i_2$:

$$I_1(s) = \frac{1}{3Z(s)}\begin{bmatrix} 2E_a(s) - E_b(s) - E_c(s) + \dfrac{-2v_{a0} + v_{b0} + v_{c0}}{s} + \\ L(2i_{a0} - i_{b0} - i_{c0}) \end{bmatrix}. \tag{10.88}$$

$$I_2(s) = \frac{1}{3Z(s)} \begin{bmatrix} E_a(s) + E_b(s) - 2E_c(s) + \dfrac{-v_{a0} - v_{b0} + 2v_{c0}}{s} + \\ L(i_{a0} + i_{b0} - 2i_{c0}) \end{bmatrix}. \quad (10.89)$$

It is known that:

$$E_a(s) + E_b(s) + E_c(s) = 0, \quad (10.90)$$

The loop currents can be written:

$$I_1(s) = \frac{1}{3Z(s)} \left[3E_a(s) + \frac{-2v_{a0} + v_{b0} + v_{c0}}{s} + L(2i_{a0} - i_{b0} - i_{c0}) \right]. \quad (10.91)$$

$$I_2(s) = \frac{1}{3Z(s)} \left[-3E_c(s) + \frac{-v_{a0} - v_{b0} + 2v_{c0}}{s} + L(i_{a0} + i_{b0} - 2i_{c0}) \right]. \quad (10.92)$$

The following cases are considered:

- Simultaneous closing of the three phases.
  Energization with the initial conditions given by (10.72), (10.73), and (10.74) is analyzed. The initial currents are null, and the voltage sources and the trapped charge are imposed by the positive sequence, allowing the following analysis.

  Assuming a constant voltage source during the high-frequency transient, $e_a = E_0$, which means that $3E_a(s) = \frac{3E_0}{s}$ in (10.91).

  For the residual voltages in the capacitors, since $v_{a0} + v_{b0} + v_{c0} = 0$, it may be concluded that $-2v_{a0} + v_{b0} + v_{c0} = -3E_0$.

  However, if a displacement of 180° is assumed for the opening of the capacitor phases, the signs of all voltages change and therefore $-2v_{a0} + v_{b0} + v_{c0} = 3E_0$. Assuming the same hypothesis of polarity opposition for the amplitudes, the maximum loop current in Eq. (10.93), which is equal to the current in phase a, is identical to the single-phase current analyzed in (10.12).

$$I_1(s) = \frac{2E_0}{Z(s)}. \quad (10.93)$$

  For simultaneous closing of the switches with no residual voltages in the capacitors, $v_{a0} = v_{b0} = v_{c0} = 0$ the current is:

$$I_1(s) = \frac{E_0}{Z(s)}. \quad (10.94)$$

- Nonsimultaneous closing of the three phases
  Analyses using electromagnetic transient programs show that the highest currents occur with the closing sequence *acb,* using the previous closing ac, which shows the largest phase-to-phase current in (10.85). Closure of the third

phase b should occur when the source voltage e_b is in opposition to the capacitor voltage. Since the source voltage is sinusoidal at nominal frequency, and the capacitor-side voltage oscillates at a high frequency, there will be times when this opposition occurs during a time interval in which the amplitude of the source voltage can be assumed nearly constant. The worst-case scenario in phase b closure occurs when the current in phase b reinforces the current in phase a, at the moment when the current in phase a is zero. However, this critical current in the analyzed cases of nonsimultaneous closure is only slightly higher than that of simultaneous closure.

- back-to-back

In the case of shunt capacitors operating back-to-back, the equations are very similar to the previous cases, as indicated in Figure 10.19.

where:

v_{a0}, v_{b0}, v_{c0}: voltages across energized shunt capacitors at t_0.
L: inductance of bus and limiting reactor.
C_{n-1}: equivalent capacitance of $n-1$ energized banks.

The equivalent capacitance of the circuit is given by:

$$C_e = \frac{C_{n-1}C}{C_{n-1}+C}. \tag{10.95}$$

In three-phase analysis, it is important to have a detailed analysis of the system, representing the transmission lines, transformers, etc.

Switching of capacitor banks with ungrounded star or delta connections is only influenced by the positive sequence, and the transient is like the single-phase case.

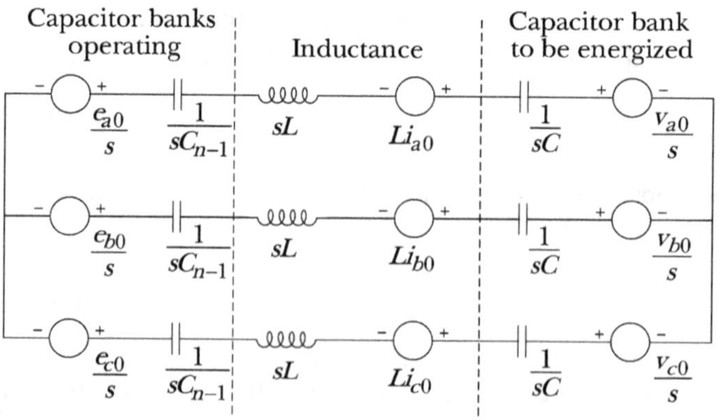

Figure 10.19 Shunt-capacitor switching in back-to-back.

- Grounded Shunt Capacitors
  When there is only one shunt-capacitor bank operating in the substation, after closing of the first phase, the capacitor behaves initially as a single-phase load, and after closing of the second phase, as a two-phase grounded load. The transient contains frequencies higher than 60 Hz, and the currents are higher than the steady-state values, leading to overvoltages in the substation. If the coupling between the phases in the feeding system is low or if the substation is fed by a delta-connected transformer, only the positive sequence takes place, and the transient is like the single-phase case. However, if the coupling is significant, the phases can interact, and the transient requires a more complex evaluation that should be performed with the statistical tools available in electromagnetic transient programs.

Example 10.6 This example presents a single-phase and three-phase simulation of an ungrounded capacitor connected to a 69 kV substation.

- Equivalent inductance of the system: 26.5 mH.
- Reactive power of the capacitors: 4.28 MVAr.

Figure 10.20a shows the voltage upon reclosing the shunt capacitors, in a single-phase simulation. Reclosing occurs one-half cycle after de-energization with trapped charge in the capacitor, which is equivalent to a circuit-breaker restrike.

Figure 10.20b shows the time response of the current in a single-phase and three-phase simulation with simultaneous pole closing, presenting identical results to the current in phase a. It is worth mentioning that the sinusoidal excitation was approximated by a constant voltage in previous equations, and thus the steady-state current component is omitted, as it is typically much smaller than the transient current. However, in this example, since the source voltage is sinusoidal, the steady-state component at 60 Hz is observable in this figure.

Figure 10.21 shows the voltages during the opening of the shunt capacitor. The results for capacitor banks connected in delta or in star are similar.

10.5 High-Frequency Requirements for Substation Equipment

In this section, some of the criteria and equipment characteristics are summarized, without intending to reproduce the standard requirements in full detail. The main features exemplified are in reference to capacitor switching, assuming a 245 kV substation containing equipment at this nominal voltage.

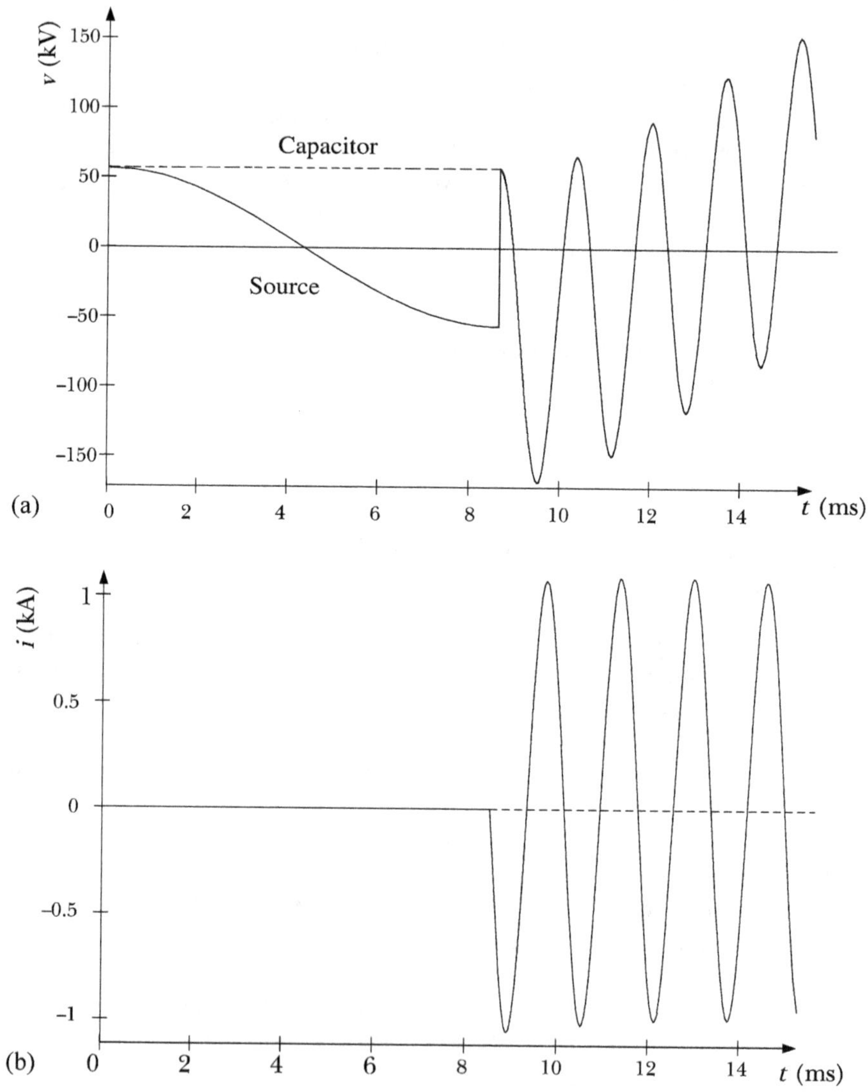

Figure 10.20 Reclosing with trapped residual voltage. (a) Single-phase simulation. (b) Current for single-phase and three-phase simulation.

10.5.1 Circuit Breakers

Circuit breakers should be able to withstand the opening of shunt capacitors without restrikes. At the time of capacitor opening, there is a voltage drop between the breaker contacts estimated at $2E_0$, according to Section 10.2.2. Under certain

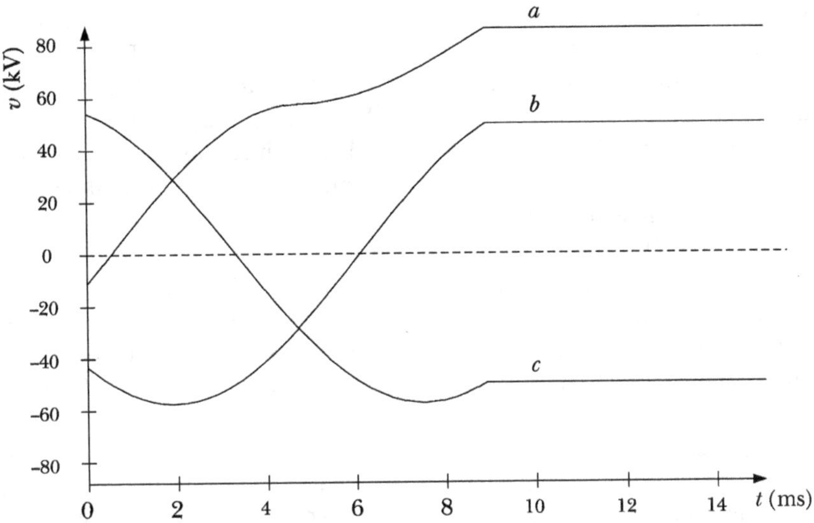

Figure 10.21 Phase-to-ground voltages during the three-phase opening of the capacitors.

abnormal conditions, the system voltage can be higher, raising the amplitude of E_0, as in the case of load rejection, and consequently the system imposes greater stress.

References [7, 9] make a distinction between general purpose and definite-purpose breakers.

The bank is considered isolated if the maximum derivative of the current does not exceed:

$$\left(\frac{di}{dt}\right)_{\max} = \sqrt{2}\omega I_{sc} \text{ A/s}, \tag{10.96}$$

where:

I_{sc}: rated short-circuit current in amperes (rms value).
ω: nominal system frequency in rad/s.

For a breaker having a rated short-circuit current of 40 kA at 60 Hz, the maximum derivative of the current should be:

$$\left(\frac{di}{dt}\right)_{\max} = 21.3 \text{ A/}\mu\text{S}.$$

If the current derivative exceeds this value, the application is assumed in back-to-back, and a definite-purpose circuit breaker should be used.

Currently, under this standard, restrike is rated probabilistically during type tests, abandoning the previously used term restrike-free [9].

Class C0 general-purpose circuit breakers are not suitable for back-to-back switching, and therefore, the restrike probabilities are not reported. These circuit breakers are used in isolated capacitor banks, as defined in Eq. (10.96).

Circuit breakers of definite purpose, identified as class C1 or C2, are indicated for back-to-back switching. Class C1 is recommended for medium voltage systems with infrequent switching. Class C2 is recommended for capacitor bank circuit breakers and frequent switching.

When energizing the capacitor bank, the allowable values of the inrush current and the maximum frequency can be taken from the preferred ratings in Table 14 in [9] or Table 21 in [8], which in the case of the nominal voltage of 245 kV are 20 kA and 4.3 kHz. However, alternative values such as 60 kA peak and 8.5 kHz are mentioned in the same table. Other values can be agreed on between the manufacturer and the user [9].

The peak current during a fault, for Class C0 circuit breakers subjected to transient currents from nearby capacitor banks, should be no less than 1.41 times the short-circuit current or 50 kA peak. For class C1 and C2 circuit breakers subjected to the same stress, the peak current must not exceed the closing and latching capability. As this event can be considered infrequent, it is assumed that the circuit breaker can withstand the stress on two occasions during its lifetime, without the need for contact maintenance.

It is noted that superposition of the high-frequency components on the fundamental current at 60 Hz may occur if inrush reactors are used in conjunction with the shunt capacitor.

a) IEEE Stds [8, 9] give the Closing and Latching Capability, obtained using a factor of 2.6 times the rated short-circuit current.
b) IEC Standard [10] gives the making current, obtained using a factor of 2.5 times the rated short-circuit current. For instance, a 40 kA circuit breaker has a making current of 100 kA.

10.5.2 Current Transformers

High-frequency currents through the primary winding of a current transformer induce high voltages in the secondary (Figure 10.22).

At high frequencies, if the load in the secondary winding is predominantly inductive, with $Z_c \cong jX_L$, the voltage at the secondary, according to the IEEE [7] Standard is obtained from:

$$V_2 = \frac{f_1}{f_n} X_L \left(\frac{N_1}{N_2}\right) I_1, \tag{10.97}$$

Figure 10.22 Currents and voltages in a CT.

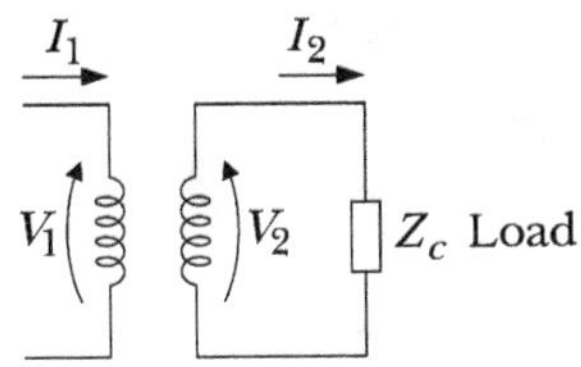

where:

V_2: voltage at the secondary winding.
I_1: current in the primary.
f_1: frequency of the transient current in the primary.
f_n: nominal frequency of the system.
X_L: load reactance at the nominal frequency.
$\frac{N_1}{N_2}$: transformation ratio for the CT.

Example 10.7 Calculate the voltage at the secondary winding of a CT feeding an electro-mechanical relay.

Data

$$X_L = 1\,\Omega,\ f_n = 60\,\text{Hz},\ \frac{N_1}{N_2} = \frac{1}{200},\ I_1 = 10\,\text{kA},\ f_1 = 5\,\text{kHz}.$$

Solution

Applying (10.97):

$$V_2 = \frac{5000}{60} \times 1 \times \frac{1}{200} \times 10 = 4.17\,\text{kV}.$$

The current transformers should be protected from high-frequency currents, in such a way that the voltages induced in the secondary have permissible values. The first reference value that can be used for secondary voltages, although obtained from tests at 60 Hz, is the standardized 3.0 kV test, which corresponds to a dielectric withstand of 7 kV (peak value).

Formula (10.97) introduces simplifications into the CT secondary circuit, furnishing values that are indicative of the abovementioned requirements, calculated from the product of the high-frequency current times the reactance. A detailed representation of the secondary circuit, mainly for long secondary cables, may show different voltage levels from (10.97).

Note that digital relays have low burden values, but some situations involving CT secondary overvoltages can present a very difficult problem to solve satisfactorily. The use of varistors offers the possibility of controlling these overvoltages, but the generalized use of varistors may be a cumbersome solution.

10.5.3 Shunt Capacitors

For frequent energizations, IEC Std [11] has the following limit:

$$I_{peak} < 100 \times I_n, \tag{10.98}$$

where I_n is the rated capacitor current (rms).

IEEE Stds [12, 13] permit higher values, depending on the frequency of each event. For regularly occurring transients, transient current capability and transient peak overvoltage capability of capacitor units can be found in tables 5 and 6 in [13].

10.5.4 Surge Arrester

Overvoltages in the substation are controlled by surge arresters, and their integrity must be guaranteed with respect to the dissipated energy and maximum current.

In cases of capacitor-bank switching, the current level is higher, but the duration is shorter than typical switching surge currents.

The values for the energy capacity of a MOSA, in relation to switching surges, can be found in arresters datasheet (see Chapter 13).

References

1 Greenwood, A. (1991). *Electrical transients in power systems* (2nd ed.). New York: John Wiley & Sons.

2 McCauley, T. M., Pelfrey, D. L., Roettger, W. C., & Wood, C. E. (1980). The impact of shunt capacitor installations on power circuit breaker application. *IEEE Transactions on Power Apparatus and Systems*, PAS-99(6), 2210–2222.

3 McGranaghan, M. F., Reid, W. E., Law, S. W., & Gresham, D. W. (1984). Overvoltage protection of shunt-capacitor banks using MOV arresters. *IEEE Transactions on Power Apparatus and Systems*, PAS-103(8), 2326–2336.

4 Mikhail, S. S., & McGranaghan, M. F. (1986). Evaluation of switching concerns associated with 345 KV shunt capacitor applications. *IEEE Transactions on Power Delivery*, 1(2), 221–230.

5 D'ajuz, A., & Zanetta Junior, L. C. (1992). *Instalação de bancos de capacitores em derivação em AT e EAT: Discussão dos critérios, premissas e futuros procedimentos.* In *III Simpósio de Especialistas em Planejamento da Operação e Expansão Elétricas*, Belo Horizonte, 1992. (III SEPOPE, Belo Horizonte, CEMIG, 1992, SP-04.1-4.).

6 Shultz, A. J., Johnson, J. B., & Shultz, N. R. (1979). Magnification of switching surge. *IEEE Transactions on Power Apparatus and Systems*, 98(2), 1418–1426.

7 IEEE. (2005). *C37.012-2005: IEEE application guide for capacitance current switching for AC high-voltage circuit breakers.* New York: IEEE.

8 IEEE. (2009). *C37.06-2009: IEEE Standard for AC high-voltage circuit breakers rated on a symmetrical current basis: Preferred ratings and related required capabilities for voltages above 1000 V.* New York: IEEE.

9 IEEE. (2018). *C37.04-2018: IEEE Standard rating structure for AC high-voltage circuit breakers.* New York: IEEE.

10 IEC. (2008). *IEC-62271-108: High-voltage switchgear and control gear part 100 edition 2: Alternating current circuit breaker.* Geneve: IEC.

11 IEC. (2014). *IEC-60871-1: Shunt capacitors for A. C. power systems having a rated voltage above 1000 V.* Geneva: IEC.

12 IEEE. (2002). *IEEE 18-2002: IEEE Standard for shunt power capacitors.* New York: IEEE.

13 IEEE. (2010). *IEEE Std 1036-2010: IEEE guide for application of shunt power capacitors.* New York: IEEE.

11

Transients in Systems with Series Capacitors

11.1 Introduction

The application of series capacitor banks (SCBs) in transmission systems helps to improve the efficiency of the transmission system [1–4]. This analysis includes steady-state, electromechanical, and electromagnetic transient investigations. On the basis of an overall analysis of the results, one can define the ratings for the equipment and protection system.

While SCB may provide remarkable benefits, certain situations and potential problems should be carefully analyzed.

One of the known benefits of the SCB is to reduce the reactance between two systems, improving power flow and stability conditions.

Studies of the specifications of SCBs consist of the following topics:

- Steady-state studies define reactive support under normal and emergency conditions, allowing improvement of the voltage profile, division of flow in the system, generation scheduling, etc. The steady-state analysis should also confirm the increase in fault currents due to the presence of SCBs.
- Electromechanical transient studies address the stability conditions of the system based on operating criteria and hypotheses for the potential fault conditions in the system. At this stage, some characteristics established under steady-state conditions can be redefined.
- The electromagnetic transient calculations allow analysis of the protective scheme for the capacitor bank and its performance against overvoltages.
- Finally, it is necessary to verify the adequacy of transmission-line protection by series capacitors.

Series capacitors can include thyristors to switch capacitor modules and also to insert parallel reactors to control the total capacitance of the bank. The trend in

Understanding Electromagnetic Transients in Power Systems, First Edition. Luiz Cera Zanetta Jr.

Published 2025 by John Wiley & Sons, Inc.

power electronics applications in electrical systems, known by the acronym FACTS (flexible alternating current transmission systems) [5], is to incorporate other similar equipment, such as static synchronous series compensators. FACTS equipment is not covered here in view of the breadth of the subject; however, they are briefly addressed in Chapter 6.

This chapter deals with some basic aspects of electromagnetic transient studies involving SCBs. A significant number of literature studies describe the theoretical and practical aspects of the application of SCBs to power systems, and some of them are suggested as supplementary reading.

11.2 Protection Schemes for Series Capacitor Banks

SCBs are important devices for the stability of interconnected power systems during transient phenomena, reducing the reactance between the systems and the amplitude of angular oscillations.

As the compensation banks are connected in series with the transmission line, they are subject to the same load currents and system fault currents, but the fault currents become even greater when SCBs are present.

The bank nominal voltage is defined for a given nominal current in the transmission line, under operating conditions in the steady-state regime.

$$V_n = X_c I_n, \tag{11.1}$$

where X_c is the capacitive reactance of the SCB.

The fault current can be much greater than the nominal current, and the voltage across the capacitor bank can surpass permissible values with respect to the insulation of the capacitor cells.

$$V_{fault} = X_c I_{fault}, \tag{11.2}$$

where V_{fault} is much greater than V_n.

Protection devices may be applied to limit overvoltages, thus retaining acceptable insulation levels. The protection schemes, as well as the details for several bank components, can be found in references [6–8].

The purpose of the protection devices is to limit the voltages at high current levels. When a fault occurs in the system, critical conditions can be created, causing instability and electromechanical oscillations. The currents can rise due to a fault or even due to angular oscillations. On the other hand, the protection should not by-pass the operating bank in a situation in which its presence in the transmission system would be critical for maintaining stability.

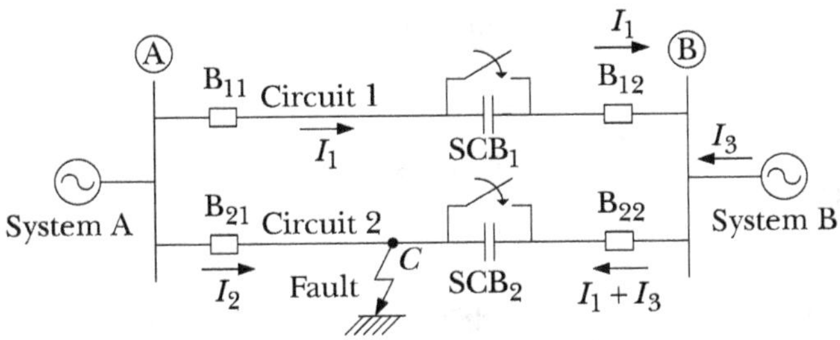

Figure 11.1 Fault currents in a double circuit.

Figure 11.1 shows systems A and B interconnected by transmission lines compensated by SCBs.

Assuming that for a fault at point C in circuit 2, the SCBs SCB_1 in circuit 1 and SCB_2 in circuit 2 may be by-passed by their protection schemes.

The fault in circuit 2 can be removed by disconnecting circuit 2; therefore, on opening breakers B_{21} and B_{22}, systems *A* and *B* will remain interconnected only through circuit 1 during the electromechanical transient. Electromechanical transient analysis should indicate the necessity for SCB_1 to remain operational and, if by-passed, its reinsertion during the transient period to ensure the stability of the transmission system.

When the series capacitor is by-passed, through spark gaps or breakers, very large currents may occur due to capacitor discharge, and therefore, damping circuits and current limiters are provided in the protection devices.

11.2.1 Protection by Spark Gaps

Figures 11.2a and b show protection devices with spark gaps designed for fast reinsertion [9–11]. Spark gap triggering is controlled by very accurate equipment units, called pilot spark gaps allowing an accurate triggering set point.

In Figure 11.2a, the SCB is initially by-passed by the spark gap.

Dielectric recovery and arc extinction in this spark gap are implemented by closing the by-pass circuit breaker. Fast reinsertion depends on the breaker closing speed and the elapsed time during which the spark gap remains by-passed and recovers its dielectric properties.

When the by-pass provided by the circuit breaker is very fast, 60–80% of the dielectric withstand level of the spark gap recovers within a short period of time, and the series capacitor can be reinserted within a time interval in the range of 170–200 ms after the fault. With ordinary circuit breakers, only slow reinsertions are attained, at times greater than 300 ms.

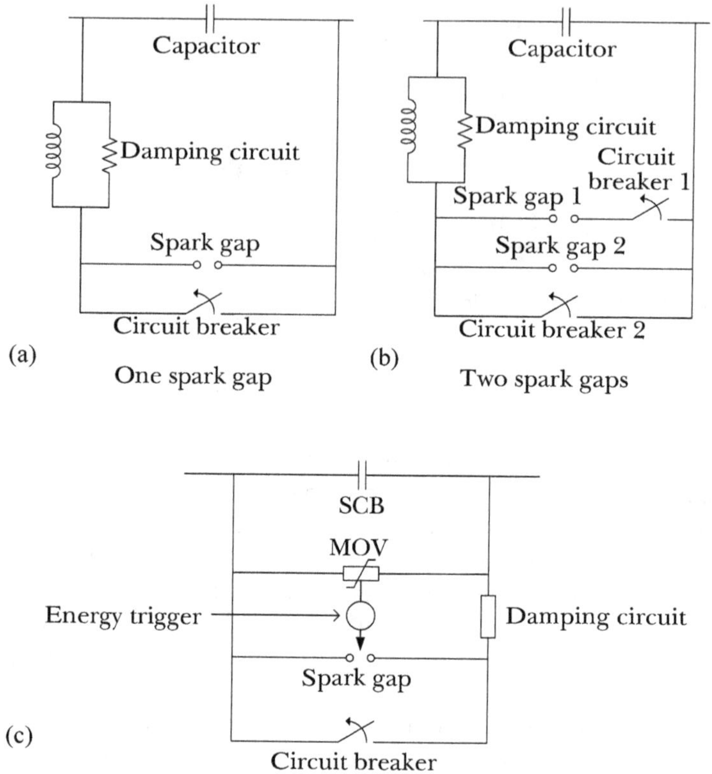

Figure 11.2 Protection schemes. (a) One spark gap. (b) Two spark gaps. (c) Protection with MOV.

The protection scheme in Figure 11.2b has two spark gaps. Spark gap 1 has its triggering level adjusted to a lower level than that of spark gap 2. During the fault, spark gap 1 is triggered first. Arc extinction is accomplished by opening circuit breaker 1, and the bank is protected temporarily by spark gap 2, until breaker 1 closes again. As spark gap 2 still has not been actuated, the bank can quickly be reinserted, since there is no reduction in the dielectric strength of spark gap 2.

Initial protection schemes based on spark gaps, despite their ingenuity, were replaced by protection schemes using varistors, to be described below.

11.2.2 Protection by Metal Oxide Varistor

In Chapter 13, the fundamental properties of metal oxide surge arresters (MOSA) are described for power-system applications. Protection by varistors, designated in this chapter by MOV (metal oxide varistor), has been most widely applied in series compensation banks. The protection layout can be seen in Figure 11.2c [12].

The voltage limits at the series capacitor terminals are determined by the MOV intervention when the fault current, and therefore, the voltage at the bank terminals rise. With the MOV intervention, its energy absorption should be controlled to within permissible levels.

In order for the varistor design to be economically viable, it is necessary to limit the energy dissipated by the MOV by means of the intervention of an additional spark gap controlled by an electronic circuit, whose function is to by-pass the MOV, and consequently, the SCB, thereby removing the MOV from the circuit.

Analyses are carried out to determine the acceptable level of energy absorbed by the MOV and spark gap intervention. The next Section deals with the strategies used to set the triggering levels of protection devices.

11.3 Protection Schemes Performance

11.3.1 Triggering Levels for Spark Gaps

Protection systems with gaps have been discontinued by manufacturers, although they are still used in some important transmission systems and may be updated soon. Since they share many methodological aspects with varistors, gap-based systems will be briefly presented, providing useful insights for varistor adjustment methods.

For the analyzed transmission line segment, the faults in the transmission system are classified according to their internal or external location, as shown in Figure 11.3.

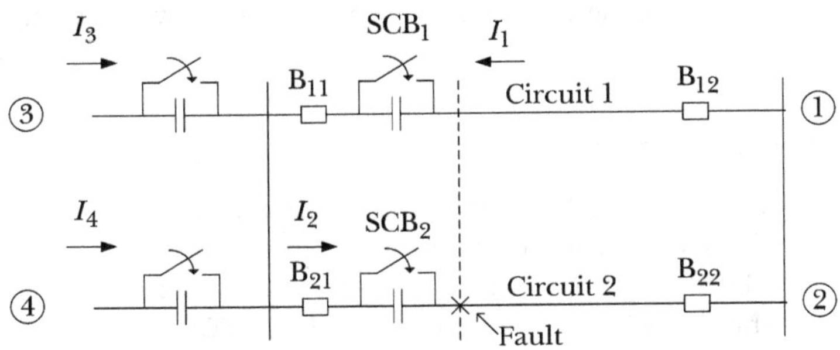

Figure 11.3 Internal and external faults.

The fault is internal with respect to circuit 2, and this line segment is disconnected by means of breakers B_{21} and B_{22} responsible for fault clearing. As for circuit 1, the fault is external, and this circuit remains connected to the network after fault clearing.

Faults critical to system stability are reproduced in electromagnetic transient programs that take into account fault type, location, and application time, with respect to the sine wave voltage.

To adjust the trigger level of the spark gaps, it is necessary to determine the overvoltages across the series capacitor located in the lines that will remain on the network after fault clearing. The question whether the series capacitors can be by-passed in external faults requires a considerable effort to analyze, since it is closely linked to other network constraints based, for example, on stability studies. Lowering the set points reduces the required insulation levels; however, it may degrade the transient performance of the system due to frequent interventions, or even exclude the capacitor when its presence would be essential to maintain stability.

Internal faults due to high current levels near the series capacitor usually cause the intervention of protection devices. Figure 11.3 shows that the current related to the internal fault is higher due to the contribution of other transmission lines, that is, $I_2 = I_1 + I_3 + I_4$. Thus, the internal fault may cause the intervention of the protection device related to SCB_2, in transmission line 2. Note that this intervention does not impact the transmission system because the faulted transmission line is naturally removed by the transmission line protection. However, careful analysis of the protection schemes related to external faults is required. For example, from the perspective of SCB_1, this is an external fault, and its protection device should not be activated.

Overvoltages due to external faults in an SCB are outlined in reference [2]. The system, with a double-circuit configuration akin to Figure 11.3, is used for assessing these overvoltages. This evaluation encompasses various adjustment levels for triggering spark gaps related to the external fault circuit associated with the analyzed SCB.

11.3.2 Reinsertion Overvoltages

In the event of an external fault, SCBs can be by-passed and must be reinserted as soon as possible according to system stability requirements.

The following simplified single-phase equations aim to illustrate the fundamental principles of transients that arise during reinsertion of SCBs.

The transmission line is modeled as an equivalent inductance in Figure 11.4a. It is also assumed that the inductance of the equivalent circuits in systems A and B are included in the total inductance of the network.

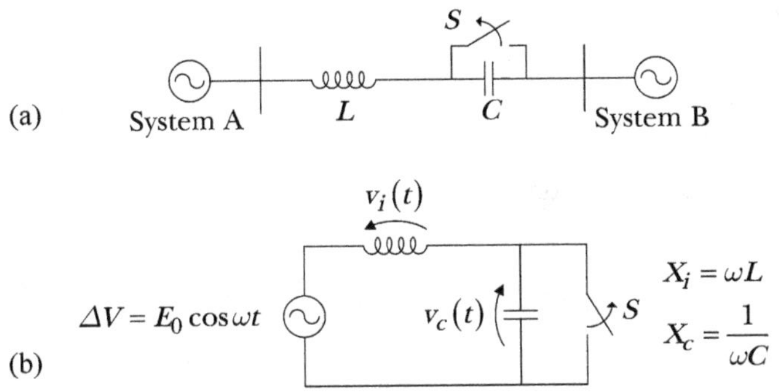

Figure 11.4 Reinsertion of the series capacitor bank. (a) Transmission line and SCB. (b) Equivalent circuit.

Thus, the model in Figure 11.4b is used, where ΔV is the difference between the voltages in the Thévenin equivalent circuits, held constant during the opening of the switch. This circuit is similar to the ones studied in Chapter 1, sections 1.5.1 and 1.5.2.

$$\frac{d^2 v_c(t)}{dt^2} + \frac{v_c(t)}{LC} = \frac{E_0 \cos(\omega t)}{LC}. \tag{11.3}$$

Applying the Laplace transform, and assuming quiescent initial conditions, with $i(0) = 0$, the following can be written:

$$\omega_0^2 = \frac{1}{LC}, \tag{11.4}$$

$$s^2 V_c(s) + \omega_0^2 V_c(s) = \omega_0^2 E_0 \frac{s}{s^2 + \omega^2}, \tag{11.5}$$

thus, Eq. (11.6) is obtained, which is the same as (1.66) in Chapter 1 if $\Theta = \pi/2$ and $R = 0$:

$$V_c(s) = \frac{\omega_0^2 E_0 s}{(s^2 + \omega^2)(s^2 + \omega_0^2)}, \tag{11.6}$$

or:

$$V_c(s) = \frac{\omega_0^2 E_o}{(\omega_0^2 - \omega^2)} \left[\frac{s}{(s^2 + \omega^2)} - \frac{s}{(s^2 + \omega_0^2)} \right]. \tag{11.7}$$

Taking the inverse transform:

$$v_c(t) = \frac{\omega_0^2 E_0}{(\omega_0^2 - \omega^2)}[\cos(\omega t) - \cos(\omega_0 t)]. \tag{11.8}$$

Rearranging the equation slightly, one can write:

$$\omega_0^2 = \frac{1}{LC} = \frac{\omega^2}{\omega L \omega C}, \tag{11.9}$$

$$\omega_0^2 = \omega^2 \frac{X_c}{X_i}, \tag{11.10}$$

where X_i is the sum of the line plus system reactance and X_c is the capacitive reactance.

Setting:

$$k = \frac{X_c}{X_i}, \tag{11.11}$$

here, k refers to the compensation level for the entire system. However, the degree of transmission-line series compensation would be used if the reactance of the equivalent systems is neglected; thus:

$$\omega_0 = \sqrt{k}\omega, \tag{11.12}$$

and,

$$v_c(t) = \frac{kE_0}{(k-1)}\left[\cos(\omega t) - \cos\left(\sqrt{k}\omega t\right)\right]. \tag{11.13}$$

In this simplified equation, note that the reinsertion overvoltage increase with the degree of series compensation, and a subsynchronous frequency is also observed since $k < 1$.

Example 11.1 Analyze the reinsertion overvoltage for the SCB with the same circuit as in Figure 11.4a, with the following data:

Data

The impedances of the Thévenin equivalents are neglected.
Transmission line length, $\ell = 400$ km,
Reactance of transmission line, $X' = 0.3\ \Omega/\text{km}$,
Degree of series compensation:

a) $k = 0.3$ (30%).
b) $k = 0.6$ (60%).

Solution

$$L = \frac{0.3 \times 400}{377} = 0.318 \text{ H}$$

Applying Eq. (11.13):

$$v_{ca}(t) = -0.43E_0[\cos(377t) - \cos(206.5t)],$$
$$v_{cb}(t) = -1.5E_0[\cos(377t) - \cos(292.0t)].$$

In this example, increasing the degree of compensation from 30% to 60% causes the overvoltage during reinsertion to rise 3.5-fold. Note that the simplifying assumption was made that the Thévenin equivalent voltages are constant.

Reinsertion overvoltages depend on the current levels in the transmission line before reinsertion.

In Figure 11.5a, a fault in transmission line 2 and a capacitor-bank by-pass in line 1 are assumed. All power flows through transmission line 1 after the fault in line 2 is cleared, thus increasing the current and consequently the reinsertion transient in the SCB serving transmission line 1. The increase in the reinsertion overvoltage can be explained by the current-injection method, whose details are presented in Chapter 12. Reinsertion can be simulated by placing a current generator between points 1 and 2 with the same waveform and current amplitude that exists during capacitor by-pass; however, note that the applied current is opposite in sign to the preexisting current. The superposition of the prior current on that of the current generator results in zero current, thus simulating reinsertion of the bank.

Figure 11.5b shows the current generator applied to the equivalent impedance of the circuit, as seen from points 1 and 2, thus showing that in this elementary case, the voltage is proportional to the prior current in transmission line 1.

When more than one SCB is reinserted into the same transmission line, or when one SCB is subdivided into sub-banks and reinsertion is done sequentially, as in Figure 11.6a, a DC offset voltage can be observed.

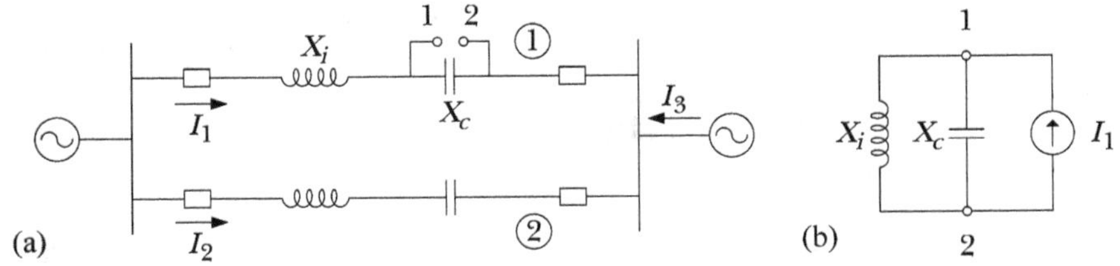

Figure 11.5 Capacitor bank reinsertion. (a) Double circuit line. (b) Equivalent circuit.

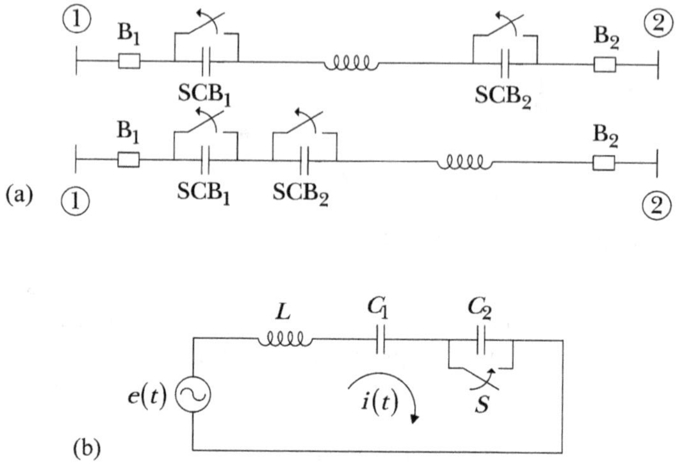

Figure 11.6 Capacitor bank reinsertion. (a) Two SCBs or two sub-banks in the same transmission line. (b) Equivalent circuit.

An interpretation of the DC offset can be made using state variables, Figure 11.6b. With SCB_2 by-passed and taking the current through the inductor and the voltage across the capacitors as state variables, the following equation can be written:

$$\frac{d}{dt}\underset{\sim}{x}_1 = A_1\underset{\sim}{x}_1 + \underset{\sim}{b}_1 e(t), \tag{11.14}$$

where:

$$\underset{\sim}{x}_1 = \begin{bmatrix} i \\ v_{c1} \end{bmatrix}, \quad A_1 = \begin{bmatrix} 0 & -\dfrac{1}{L} \\ \dfrac{1}{C_1} & 0 \end{bmatrix}, \quad \underset{\sim}{b}_1 = \begin{bmatrix} \dfrac{1}{L} \\ 0 \end{bmatrix}. \tag{11.15}$$

The eigenvalues are complex conjugates:

$$\det(sI - A_1) = s^2 + \frac{1}{LC_1} = 0, \tag{11.16}$$

$$s_{1,2} = \pm j\frac{1}{\sqrt{LC_1}}. \tag{11.17}$$

After opening switch S, the second SCB is reinserted, and the new system of state variables may be written:

$$\frac{d}{dt}\underset{\sim}{x}_2 = A_2\underset{\sim}{x}_2 + \underset{\sim}{b}_2 e(t), \tag{11.18}$$

where:

$$\underset{\sim}{x}_2 = \begin{bmatrix} v_{c2} \\ i \\ v_{c1} \end{bmatrix}, \quad A_2 = \begin{bmatrix} 0 & \frac{1}{C_2} & 0 \\ -\frac{1}{L} & 0 & -\frac{1}{L} \\ 0 & \frac{1}{C_1} & 0 \end{bmatrix}, \quad \underset{\sim}{b}_2 = \begin{bmatrix} 0 \\ \frac{1}{L} \\ 0 \end{bmatrix}. \tag{11.19}$$

The linear dependence of state matrix lines 1 and 3 is responsible for a null eigenvalue:

$$\det(sI - A_2) = s\left[s^2 + \frac{1}{L}\left(\frac{1}{C_1} + \frac{1}{C_2}\right)\right] = 0. \tag{11.20}$$

In the expansion into partial fractions, this null eigenvalue is responsible for the parcel k/s, which introduces a unidirectional component in the circuit.

It can also be seen that reinsertion of the second SCB, or second sub-bank, is more critical than the first, since the current increases after reinsertion of the first bank, according to Figure 11.6.

11.3.3 Protection Schemes with MOV

One of the advantages of protecting SCBs with MOV, compared to the use of spark gaps, is the reduction of switching transients in SCBs.

As the voltage across the capacitor terminals increases, the current is commutated to the MOV when its intervention level is reached, thus limiting the voltage rise. With a reduction in the voltage level, the SCB is naturally reinserted into the circuit, without a reinsertion transient, characteristic of circuit-breaker opening. Another advantage is that reinsertion is instantaneous, since the delays in the schemes employing gaps are no longer necessary.

Hence, some benefit can be achieved in system stability, which should be evaluated during the process of estimating capacitor bank size and cost.

With respect to system analysis, from the point of view of protection scheme performance, the studies are similar to those in the spark-gap setting for internal and external faults.

These studies report the energy dissipated by the MOV protecting the bank from internal and external faults (Figure 11.7a). The MOV is sized with an energy dissipation capacity such that the protection scheme does not need to operate in the event of external faults.

For the energy levels in internal faults, the operation of the protection scheme is allowed, establishing a limit on the maximum energy level of the MOV. The

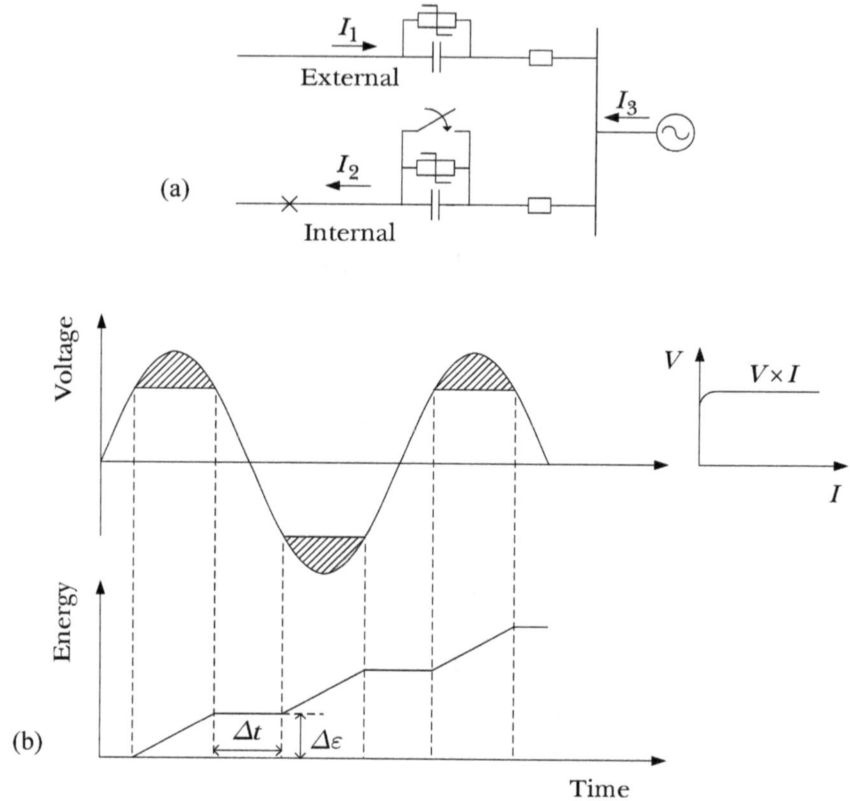

Figure 11.7 MOV protection. (a) Double circuit internal and external faults. (b) Energy rise.

protection device currently monitors the fault current and energy level, characterizing its severity and by-passing the MOV in cases of critical internal faults.

The analysis consists of simulation of single-phase, two-phase, and three-phase faults, for the purpose of evaluating the energy and current demands on the MOV.

To reduce the insulation levels of the capacitors, it is necessary to lower the MOV residual voltage curve. However, very low residual voltages can increase the frequency of operation of the MOV in external faults.

The studies are accomplished by estimating the residual curve for the MOV, considering the intervention level, which is usually within the range of 2.0–2.5 pu of the SCB-rated voltage. The number of parallel columns depends on the MOV energy capacity.

For internal faults, the SCB is disconnected from the system by the line-protection and circuit-breaker operation. The rise in the energy to be dissipated is proportional to the duration of exposure to the fault. When the energy capacity is reached, the protection spark gap of the MOV is triggered.

The following factors should be considered in the setting of protection devices, taking into account variables such as voltage, current, and energy, under conditions of internal and external faults:

- Fault clearing time, delimited by the line protection.
- Protection scheme operation during internal and external faults.
- Residual voltage of the MOV and energy dissipation capacity.
- System assumptions such as type of faults, line reclosing after faults, currents in the steady-state regime, overloads, and power oscillations.

Figure 11.7b illustrates the rise in energy and operation of the MOV during internal and external faults. The delay in the operation of the protection scheme should be considered, since it has a significant impact on the amount of energy transferred to the MOV. When the rate of rise of the energy is high, it is important that the performance of the protection scheme be as fast as possible.

Figure 11.8 shows the results of studies of series capacitors in the system analyzed in [2]. In Figure 11.8a, the voltage across the terminals of the SCBs is limited by the MOV intervention. Figure 11.8b shows the current through the MOV during its operation, and Figure 11.8c shows the energy rise during the intervention period.

Since the voltage at the terminals of the SCB is practically constant during MOV operation, the current during this time interval is commutated from the SCB to the MOV.

During internal faults, the current peaks, and the rate of rise of the energy dissipated by the MOV is highly dependent on the short-circuit power of the substation adjacent to the SCB. As mentioned above, when the short-circuit power is high, to keep the energy dissipated by the MOV at permissible levels, the MOV protection gap must spark over as quickly as possible to limit the transferred energy.

The following is an example of the theoretical calculation of voltages, currents, and energy in a single-phase circuit, which may represent a three-phase fault involving only the positive sequence. This example adopts some simplifications, and the results of the calculations are compared with calculations that use an electromagnetic transient program and realistic system parameters.

This example clarifies the commutation of current between the capacitor and MOV.

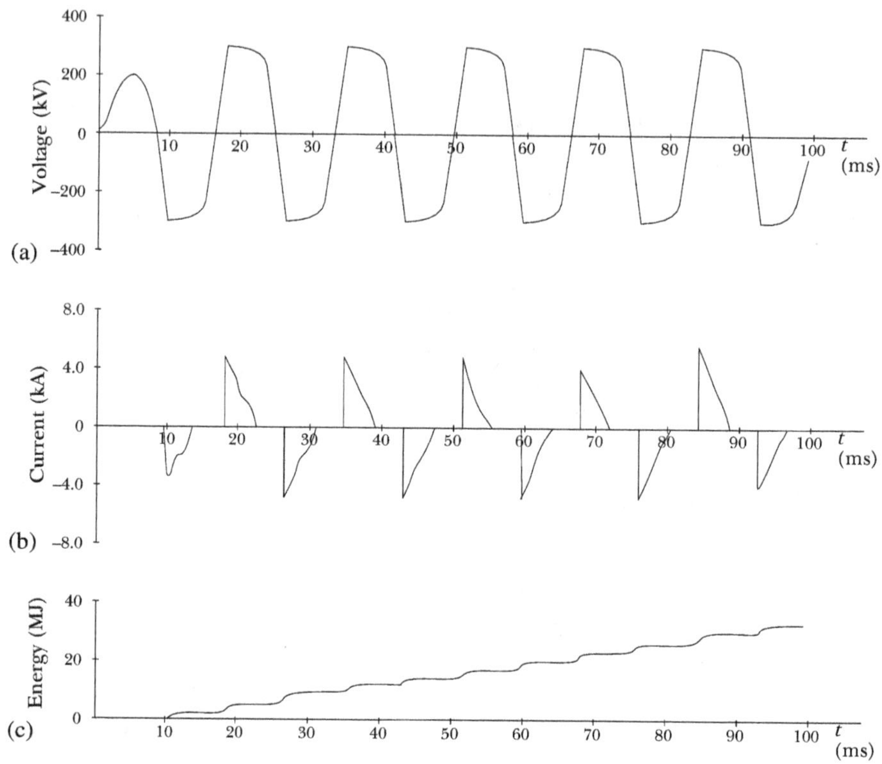

Figure 11.8 MOV intervention during an internal fault. (a) Voltage. (b) Current. (c) Energy.

Example 11.2 Calculate the transient fault application in a transmission line compensated by series capacitors protected by MOV, as in Figure 11.9a.

Data

Rated line-to-line voltage, 525 kV.

The fault is applied at one-third of the line's length: $L = 0.066$ H, (25 Ω).

Series capacitor: $C = 70.85$ μF, (37.44 Ω).

MOV: assume a simplified model with a constant residual voltage of 200 kV.

Solution

In this example, the location of the fault leads to a near-resonance condition between the reactance of the transmission line and the reactance of the series capacitor.

Following the procedures in Chapters 10 and 13, the residual voltage of the MOV is assumed flat and can be approximated by a DC voltage (Figure 11.9b).

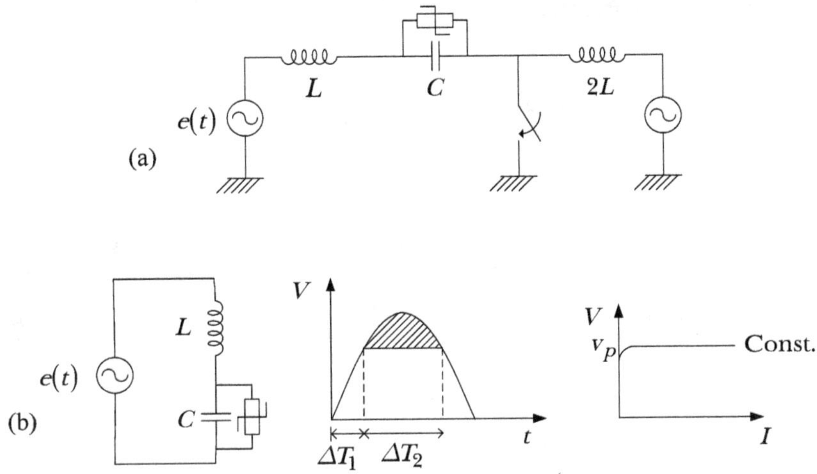

Figure 11.9 Series capacitors and MOV intervention. (a) Fault application. (b) Equivalent circuit and MOV intervention.

The transient is decomposed into two intervals, the first, without the MOV intervention, and the second, with the MOV intervention.

a) Interval ΔT_1, without MOV intervention, Figure 11.10a:

$$e(t) = E_0 \sin(\omega t),$$

$$E_0 = \frac{525\sqrt{2}}{\sqrt{3}} = 428 \text{ kV},$$

$$Y(s) = \frac{sC}{s^2LC + 1},$$

$$E(s) = \frac{E_0\omega}{s^2 + \omega^2},$$

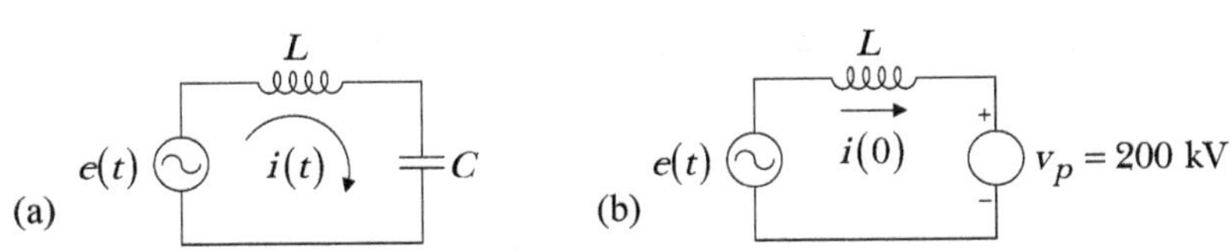

Figure 11.10 Equivalent circuits. (a) Without MOV. (b) With MOV active.

Resulting in the current:

$$I(s) = \frac{E_0\omega}{s^2 + \omega^2} \frac{sC}{LC(s^2 + \omega_0^2)},$$

$$\omega_0^2 = \frac{1}{LC},$$

$$I(s) = \frac{E_0}{L} \frac{s\omega}{(s^2 + \omega^2)(s^2 + \omega_0^2)}.$$

Taking the inverse transform:

$$i(t) = \frac{E_0\omega}{L(\omega_0^2 - \omega^2)}[\cos(\omega t) - \cos(\omega_0 t)],$$

$$i(t) = 34.31[\cos(377t) - \cos(461.4t)].$$

As for the voltage at the capacitor:

$$v_c(t) = \frac{1}{C}\int_0^t i(t)dt,$$

$$v_c(t) = \frac{E_0\omega}{LC(\omega_0^2 - \omega^2)}\left[\frac{\sin(\omega t)}{\omega} - \frac{\sin(\omega_0 t)}{\omega_0}\right].$$

Substituting the numerical values:

$$v_c(t) = 484262.53\left[\frac{\sin(377t)}{377} - \frac{\sin(461.4t)}{461.4}\right].$$

Note that, since the fault occurs at 1/3 the length of the line and because the equivalent reactance of the system is neglected, with this simplification $\omega_0 = 461.4$ rad/s, which is higher than the synchronous frequency; however, this is not a typical subsynchronous situation.

The expression for $v_c(t)$ is valid up to a voltage level of 200 kV with the MOV intervention. Numerical values of voltage and current at the capacitor in the time interval $\Delta T_1 = 3.55$ ms are given below:

Time (ms)	Voltage at capacitor (kV)	Current (kA)
2	42.50	4.30
3	131.00	8.24
3.5	195.52	10.04
3.55	202.66	10.20

b) Interval ΔT_2, with MOV intervention, Figure 11.10b.
With the MOV intervention, represented by a DC voltage source, the circuit in Figure 11.10b is proposed. The current commutation from the capacitor to the MOV is clear since the current in the capacitor is zero for a constant voltage across its terminals. Consequently, the capacitor can be removed from the circuit.

$$i(t) = C\frac{dv_c(t)}{dt} = 0. \tag{11.21}$$

The initial condition for the current through the inductance is 10.2 kA in the new circuit.

$$i(t) = \frac{1}{L}\int_{t_0}^{t}(E_0\sin(\omega t) - v_p)dt,$$

$$i(t) = \frac{E_0}{\omega L}[\cos(\omega t_0) - \cos(\omega t)] - \frac{v_p}{L}(t - t_0) + i_0.$$

For $t = t_0$, $i_0 = 10.2$ kA,

$$i(t) = 17.12[0.23 - \cos(377t)] - 3016(t - t_0) + 10.2.$$

The following current values in the MOV are obtained:

Time (ms)	Current (kA)
4	11.70
5	15.06
6	17.67
7	18.74
8	17.71
9	12.30
10	8.54

Figure 11.11 gives the calculated values using the equations above. Figure 11.12 shows the time response obtained using the ATP program for comparison with the actual residual curve for the MOV.

The validity of MOV representation by a DC source is restricted to a time interval in which the sign of the current does not change.

In detailed simulations, the current switching is not instantaneous, according to Figure 11.8, because the voltage across the model of the MOV is not perfectly constant.

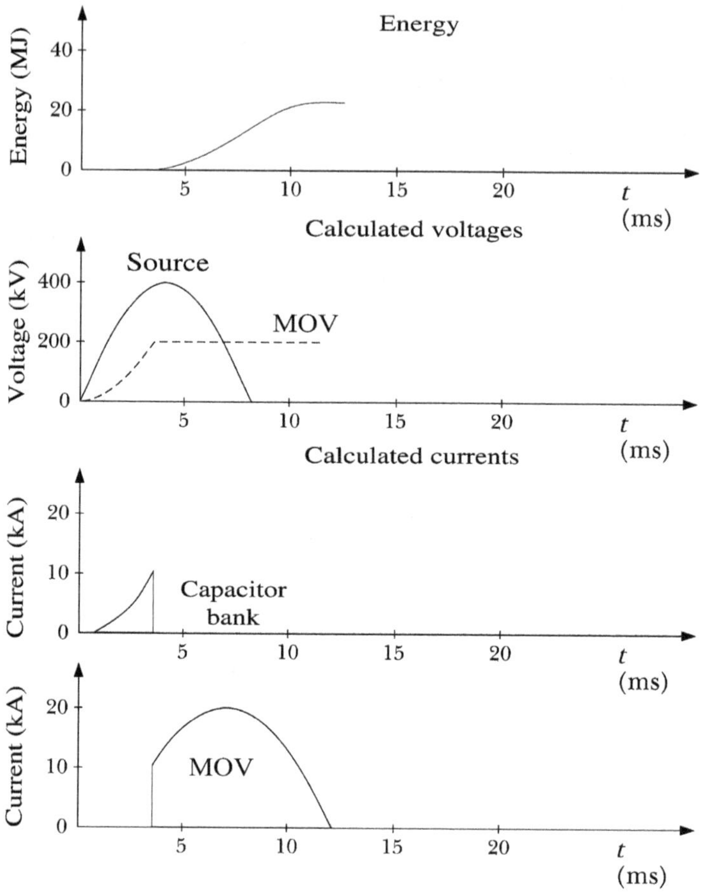

Figure 11.11 Energy, voltage, and current in the approximated circuit.

11.4 Complementary Studies

In the previous sections, the main transient studies of transmission lines equipped with series capacitors and their protection schemes were presented. The following details additional analyses to be carried out with respect to SCBs.

- Transmission Line Energization and Autoreclosure
  The SCBs usually are of little influence as regards energization and autoreclosure of transmission lines, because the currents involved are very small. However, fast reclosing with permanent faults can give rise to a severe condition,

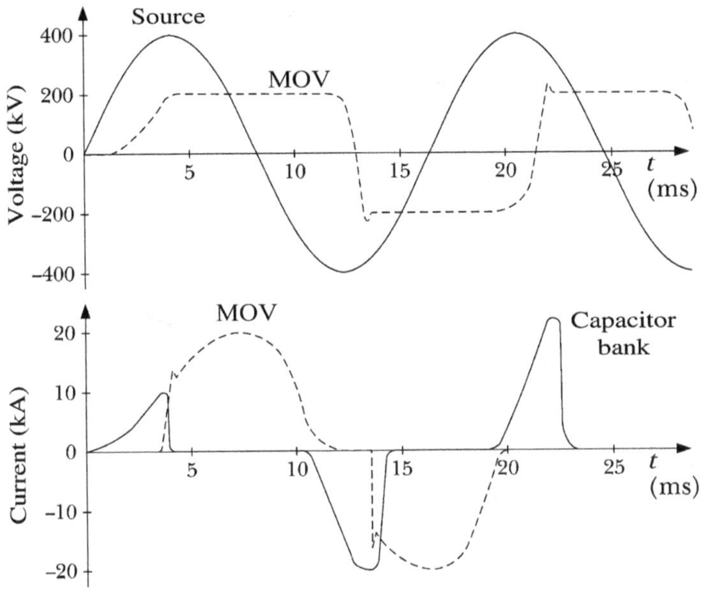

Figure 11.12 Voltages and currents obtained from ATP simulations.

since the energy absorbed in the fault added to the energy absorbed during reclosing practically doubles the energy stress on the model.

- Transient Recovery Voltage

  The SCBs can change the transient recovery voltages in the circuit breakers due to faults over the transmission line, and this issue is approached in Chapter 12. For terminal faults near the substation, capacitor by-pass can frequently be assumed and therefore, the system behaves as if there were no capacitors in the transmission line.

- Secondary Arc Current

  The fault current after fault clearing is designated secondary arc current, and it is supplied by stored charges or induction from other phases of the transmission line.

  This item shows the influence of SCBs on the secondary arc current (Figure 11.13).

I_s: Secondary arc current after the opening of breakersB$_1$ and B$_2$.
C: line capacitance.
L: inductance of the transmission line up to the fault point.
C_m: mutual capacitance with other phases of the transmission line or other transmission lines.

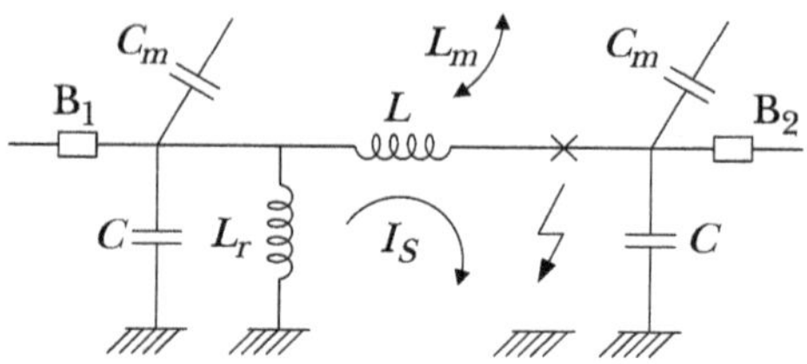

Figure 11.13 Coupling to a secondary arc current.

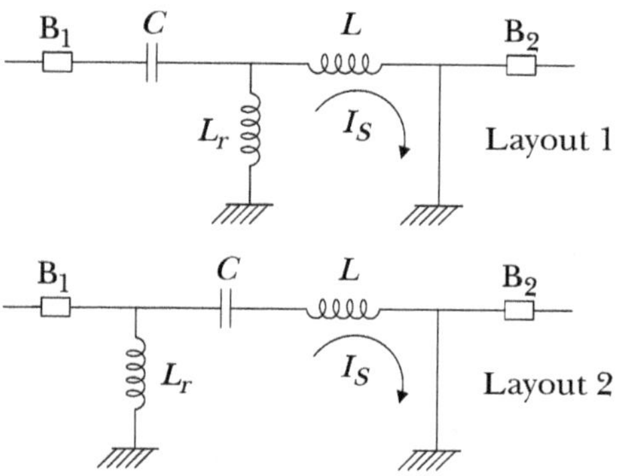

Figure 11.14 Participation of the series capacitor in the secondary arc current.

L_m: mutual inductance with other phases of the transmission line or other transmission lines.

L_r: inductance of the transmission line reactor.

If the capacitance of the SCB is included, it predominates in the circuit and preferentially oscillates with the inductance of the transmission line reactor (Figure 11.14).

In layout 1, the series capacitor does not affect the secondary arc current, but it is not always possible to adopt one, especially when the use of capacitors was not foreseen in the planning stage. Layout 2 depicts an oscillatory circuit composed of a series capacitor, a reactor, and the transmission line inductance. The charge stored in the series capacitor introduces an additional source of secondary arc current, postponing the extinction of the secondary arc.

In attempting to damp the additional subsynchronous current that may be introduced by series capacitors, several methods can be proposed. Resistors connected to the neutral point on the reactor L_r, saturable reactors with resistors in parallel with the series capacitors, and grounding switches can be proposed. Switches or spark gaps, whose operation is controlled so as to by-pass the series capacitor may be an effective solution, thus withdrawing the bank from the circuit.

However, the by-pass should occur at an optimal time when the voltage across the capacitor is at its maximum value, thereby withdrawing its stored energy from the circuit [2].

The secondary arc extinction is a specific topic that will not be discussed here.

References

1 Iliceto, F., Cinieri, E., Cazzani, M., & Santagostino, G. (1976). Transient voltages and currents in series-compensated e.h.v. lines. *Proceedings of the Institution of Electrical Engineers*, 123(8), 811–817.

2 Pinheiro, R. P., Fernandes, J. H., Tannuri, J. G., Patrão, J., Zanetta, L. C., Ferreira, C., & Ross, R. P. D. (1986). Sistema de 500 kV da Eletronorte com compensação série: Estudos de sobretensoes nos bancos de capacitores série e análise das sobretensões no sistema. In *VIII Seminário Nacional de Produção e Transmissão de Energia Elétrica, São Paulo*. São Paulo: CESP.

3 Breuer, G. D., Rustebakke, H. M., Gibley, R. A., & Simmons, H. O. (1964). The use of series capacitors to obtain maximum EHV transmission capability. *IEEE Transactions on Power Apparatus and Systems*, 83(11), 1090–1102.

4 Iliceto, F., & Cinieri, E. (1977). Comparative analysis of series and shunt compensation schemes for AC transmission systems. *IEEE Transactions on Power Apparatus and Systems*, 96(6), 1819–1830.

5 Hingorani, N. G., & Gyugyi, L. (1999). *Understanding FACTS*. Piscataway: IEEE Press.

6 IEC. (2015). *IEC-60143-1: Series capacitors for power systems: Part 1 "General"*. Geneve.

7 IEC. (2012). *IEC-60143-2: Series capacitors for power systems: Protective equipment for series capacitors banks*. Geneve.

8 IEEE. (2005). *IEEE-824-2005: Standard for series capacitors in power systems*. Washington.

9 Fahlén, N. T. (1981). Reactive compensation. EHV series capacitor equipment protection and control. *IEE Proceedings C (Generation, Transmission and Distribution)*, 128(6), 394–401.

10 Jancke, G., Fahlen, N., & Nerf, O. (1975). Series capacitors in power systems. *IEEE Transactions on Power Apparatus and Systems*, 94(3), 915–925.

11 Batho, J. L., Hardy, J. E., & Tolmunen, N. (1977). Series capacitor installations in the B. C. hydro 500 kV system. *IEEE Transactions on Power Apparatus and Systems*, 96(6), 1767–1776.

12 Hamann, J. R., Miske, S. A., Johnson, I. B., & Courts, A. L. (1981). A zinc oxide varistor protective system for series capacitors. *IEEE Transactions on Power Apparatus and Systems*, PAS-100(3), 929–937.

12

Transient Recovery Voltage

12.1 Introduction

Faults in power systems can occur for several reasons, and proper operation of interconnected power systems is only possible with reliable circuit breakers.

Successful elimination of faults depends on the current level and on the voltage developed between the contacts of the circuit breaker, after extinction of the electric arc. After fault clearing, the system may exhibit transient voltages, depending on its configuration and operating equipment [1–6].

The extinction of the electric arc results from a relatively complex physical phenomenon. In the face of the difficulties in modeling the arc, the standards do not explicitly contemplate this issue. However, solid experience in the form of standardized test parameters and capability envelopes are provided, delimitating successful regions of current interruption and application of circuit breakers on a symmetrical current basis.

Calculation of the TRV focuses on the prospective stresses inherent to the system that would eventually result without the modifications introduced by a circuit breaker, such as its capacitance and the arc voltage, assuming an ideal circuit breaker.

The transient behavior of the voltage can be obtained via electromagnetic transient programs. Once the transient voltage is calculated, one can select the appropriate circuit breaker according to the recommendations of IEEE or IEC standards [7–11].

This chapter presents the fundamental concepts governing the analysis of the transient recovery voltage (TRV) and provides the first steps in the specification of the circuit breakers.

12.1.1 Fault Currents

The steady-state fault current is designated as the symmetrical fault current, neglecting the asymmetry introduced by the DC component.

Understanding Electromagnetic Transients in Power Systems, First Edition. Luiz Cera Zanetta Jr.

Published 2025 by John Wiley & Sons, Inc.

The fault currents in power systems are predominantly inductive. The inductive behavior is exemplified in the circuit as in Figure 12.1a, by means of a fault applied to the equivalent circuit of the power system or even a transmission line represented by its π- equivalent circuit.

The maximum asymmetrical fault current and the DC component, as well as its damping, can be calculated using the equivalent RL circuit. The steady-state component is called the symmetrical fault current and the total current is the asymmetrical fault current. An example of the calculation can be found in Section 1.4.1 in Chapter 1.

A simplified analysis of faults at the rated frequency may neglect losses by assuming that $R \ll X_l \ll X_c$, which results in the equivalent circuit for the fault described in Figure 12.1b.

12.1.2 Extinction of the Fault Current

Regarding interruption of the fault current, a distinction is made between the mechanical separation of the circuit breaker contacts and the effective extinction of the arc that persists in the ionized path until the current passes through zero (Figure 12.2a).

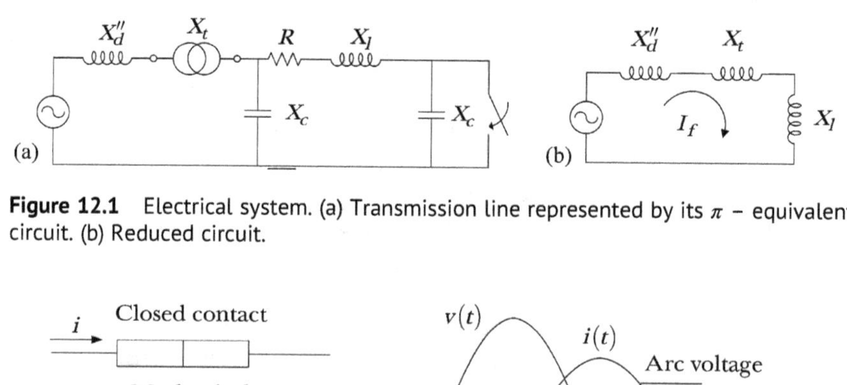

Figure 12.1 Electrical system. (a) Transmission line represented by its π - equivalent circuit. (b) Reduced circuit.

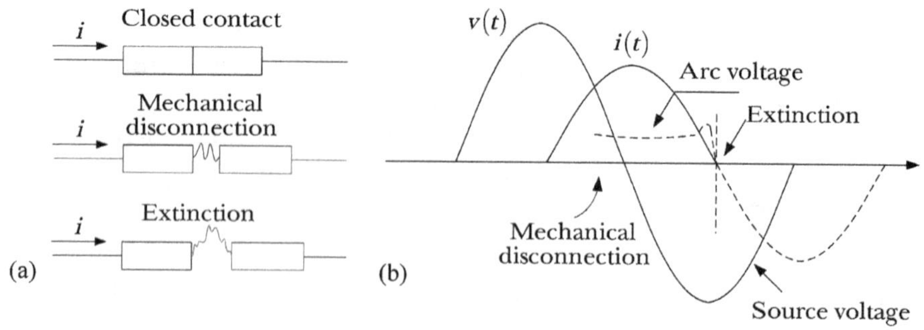

Figure 12.2 Mechanical separation and current interruption. (a) Electric arc. (b) Arc extinction.

The current cannot change instantaneously in an inductive circuit, and for successful interruption of the current, after mechanical separation of the contacts, the most propitious point in time for extinction of the electric arc is at the zero crossing of the current, also known as the thermal phase of interruption.

Reignitions can occur, and circuit breaker extinction chambers are designed so that successful interruption of the current occurs in the first few cycles after contact separation.

Regarding faults in a purely inductive circuit, the current lags the voltage by 90°. At the current's zero crossing, the applied voltage is at its maximum amplitude and immediately after the current interruption. The equation for this elementary circuit can be seen in item 12.4.1.

After arc extinction, the circuit breaker chambers attempt to recover the dielectric characteristics of the insulating medium, while at the same time, the voltage between the contacts acts to reestablish the current. The design of the extinction mechanism of the circuit breaker seeks to deal with this conflict.

The electric arc conductance subjected to the fault current develops an arc voltage, as shown in Figure 12.2b.

The current reestablishment immediately after arc extinction is designated as reignition and if it happens after 1/4 of a cycle, its designation is restrike.

12.2 Transient Recovery Voltage

The TRV that arises between the contacts of the circuit breaker after interruption of the fault current, is restricted to a time interval in the range of 1–2 ms. The voltage between the contacts of the circuit breaker can be obtained from the difference in the phase to earth voltages at each circuit breaker terminal.

$$\Delta v = v_1(t) - v_2(t). \tag{12.1}$$

The solution to the differential equation that determines the TRV has a steady-state component and a transient component.

12.2.1 Steady-State Component and Transient Component

The steady state component can be calculated using traditional methods of analysis of three-phase networks with unbalanced faults, through assembly of the nodal admittance matrix.

In order to analyze unbalanced conditions, composed simultaneously of open phases and faults in the network, three-phase nodal analysis is more convenient than application of symmetrical components.

Voltages v_1 and v_2 arise at the circuit breaker terminals as does the voltage drop, Δv, with a steady state component that depends on the fault type, location, and system grounding.

The transient component is related to the energy stored in capacitors and inductors during the process of accommodation from one topology to another.

$$\Delta v = \Delta v_s + \Delta v_t = (v_1 - v_2)_s + (v_1 - v_2)_t, \tag{12.2}$$

where:

s: steady-state component,
t: transient component.

The voltage across the switch terminals shows transient behavior similar to that of Figure 12.3a and b.

In the first stage, the separation of the contacts is very small, and the temperature is very high in the ionized channel. At this stage, it is necessary to examine the behavior of the rate of rise of the voltage, which is designated as rate of rise of recovery voltage (RRRV).

In the second stage, with the contacts further apart and the chamber more deionized, the focus is to make sure that the voltage amplitude does not exceed the dielectric strength of the still fragile insulating medium.

The transient voltage depends on network characteristics, such as the operating voltage, fault current, transmission lines and equipment in operation, stray capacitance, and type and location of the fault, etc.

12.2.2 Opening Sequence for the Circuit Breaker Poles

The sequence for the extinction of the currents in phases *a*, *b* and *c* is not known in advance. For balanced faults, the voltages only show a phase rotation, according to the opening sequence of the circuit breaker poles. In the case of unbalanced faults, the opening sequence can change the voltage values at the circuit breaker poles.

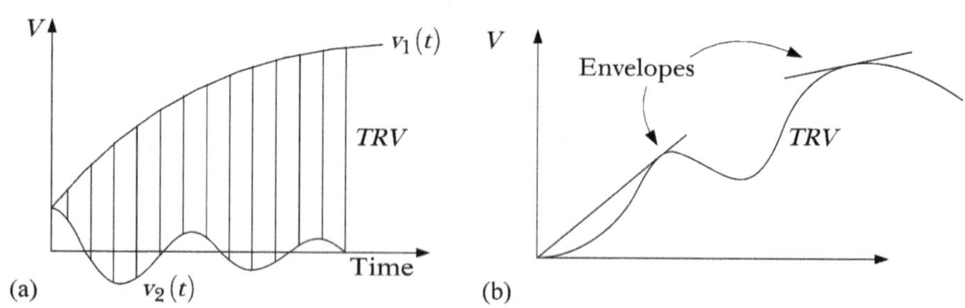

Figure 12.3 TRV waveform. (a) Waveforms on sides 1 and 2. (b) TRV and standard envelopes.

12.3 Calculation of the Transient Recovery Voltage

Calculation of the TRV can be accomplished by means of electromagnetic transient programs, which simulate the network in detail, including the effect of traveling waves in transmission lines, thus making it possible to obtain accurate rate of rise and peak voltage values.

Simplified models are useful in solving the basic differential equations, thus providing the elements for a qualitative understanding of the transients. In spite of the simplifications made, the basic equations are essential for analysis of complex networks.

12.3.1 Current Injection Method and Principle of Superposition

The method of current injection is very useful in the analysis of voltages that arise when a switch is opened. Initially, a generic network is investigated with current $i(t)$ flowing from terminals 1–2 of a closed switch, S. The current could be, for instance, the fault current through a circuit breaker (Figure 12.4a).

Opening of the switch between terminals 1 and 2 occurs when the alternating current $i(t)$ crosses zero, at time t_0, as shown in Figure 12.4b.

The same interrupting effect is achieved if current $i(t)$ is cancelled from that moment on, by an equal and opposite sign injected current between terminals 1 and 2, according to Figure 12.4b. The cancellation can be applied to any function $i(t)$ not just to sinusoidal functions from Figure 12.4b.

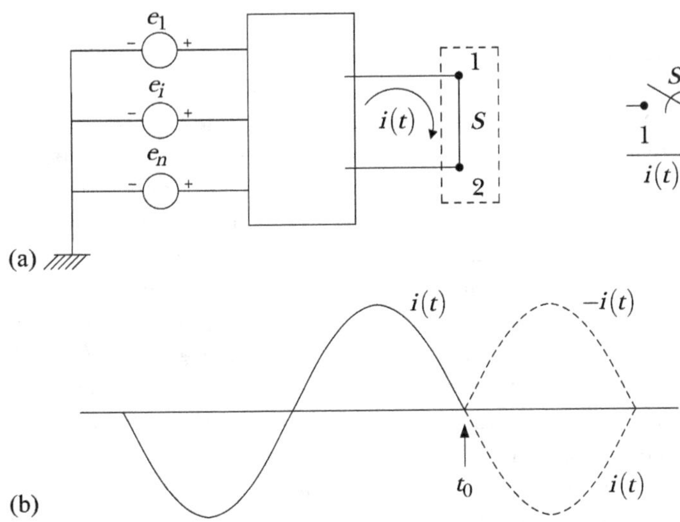

Figure 12.4 Current between points 1 and 2. (a) Switch S closed. (b) Current zero crossing.

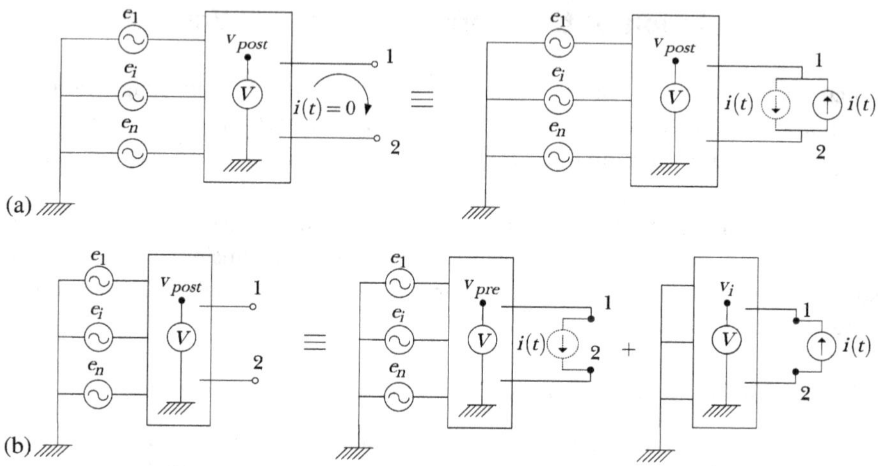

Figure 12.5 Opening of switch S. (a) Open switch S and equivalent circuit with two opposing current sources. (b) Superposition principle.

Separating nodes 1 and 2 might seem to create a conflict, as it results in two distinct circuits: one before the opening of switch S (Figure 12.4a) and another after the opening (Figure 12.5a). However, the superposition principle is applied only to the post-opening circuit, ensuring linearity. The network remains linear on both sides of the switch, maintaining the consistency of the analysis.

To simulate the open switch S from time t_0, two opposing current generators with the same function $i(t)$ are used as in Figure 12.5a. These generators ensure zero current between nodes 1 and 2, in accordance with Kirchhoff's first law.

In the first situation, with only the voltage generators and the auxiliary current generator in the dotted line turned on, as shown in Figure 12.5a, the same voltages and currents as in Figure 12.4a are reproduced (as if nodes 1 and 2 were still short-circuited). Although this current generator could be eliminated (short-circuited), it is kept in place to facilitate the application of the superposition principle. In the second situation, by connecting only the current generator that effectively interrupts the current in a continuous line (simulating the interruption of current), with the other generators turned off, the voltages in the network on both sides of this second case are calculated and designated as v_i. The voltages at nodes 1 and 2 are calculated in the post-opening network using the superposition principle, which sums the effects of the two generator conditions shown in Figure 12.5b. The voltages at nodes 1 and 2, before and after switch S opening, are designated as v_{pre} and v_{post}, respectively. Next the voltages at nodes 1 and 2 (indicated between parentheses) are calculated.

$$v_{post}^{(1)} = v_{pre}^{(1)} + v_i^{(1)},$$

$$v_{post}^{(2)} = v_{pre}^{(2)} + v_i^{(2)}.$$

Designating the voltage drop across the switch:

$$\Delta v_{post} = v_{post}^{(1)} - v_{post}^{(2)}, \tag{12.3}$$

but, since the nodes 1 and 2 have the same voltage when the switch is closed (or when current generator with dotted line in Figure 12.5b is active):

$$v_{pre}^{(1)} = v_{pre}^{(2)},$$

it can be concluded that the voltage across the switch, after disconnection, is obtained simply by applying the current generator that simulates the current interruption, which greatly simplifies the solution:

$$\Delta v_{post} = v_i^{(1)} - v_i^{(2)},$$

$$\Delta v_{post} = \Delta v_i. \tag{12.4}$$

Simulation of a current generator between terminals 1 and 2 can also be achieved using two grounded units, according to Figure 12.6.

The current injection method is very useful in the calculation of RRRV, allowing simplified treatment of the network.

12.3.2 Calculation with Electromagnetic Transient Programs

As mentioned previously, the TRV calculation is performed using electromagnetic transient programs, thus providing suitable changes in the admittance matrix when integrating the differential equations. Circuit breakers are simulated as switches. The solution of the network with closed switch is used as the initial condition for the solution of the differential equations governing the new network with open switch.

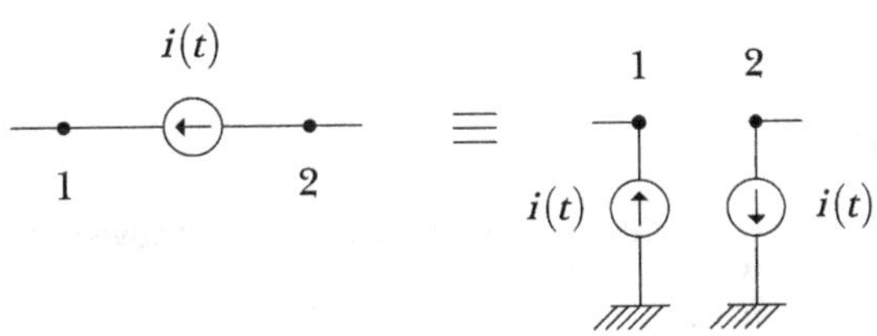

Figure 12.6 Series current generator and two grounded current generators.

12.4 TRV in Single Phase Inductive Circuits

In this item, single-phase basic circuits are analyzed. Subsequently the same circuits will be extended to three phase cases.

12.4.1 Current Interruption in Inductances

- Current Interruption

  First, interruption of the sinusoidal current in a purely inductive circuit is considered, modeled as an inductance that represents, for example, a generator connected to a step-up transformer, with a fault on the high-voltage side of the transformer, Figure 12.7a.

  Assuming a cosine-function for the voltage source, the following equations can be written:

$$e(t) = E_0 \cos(\omega t), \tag{12.5}$$

$$i(t) = \frac{E_0}{\omega L} \sin(\omega t). \tag{12.6}$$

  Applying the current injection method in the frequency domain:

$$V(s) = I(s)Z(s). \tag{12.7}$$

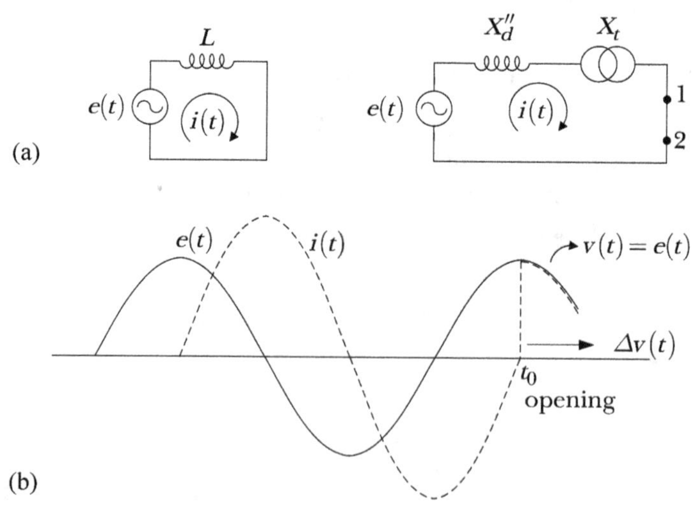

Figure 12.7 Current interruption in the inductive circuit. (a) Equivalent circuit. (b) Voltage at the switch terminals.

$$V(s) = sL\frac{E_0\omega}{\omega L(s^2 + \omega^2)}. \tag{12.8}$$

$$V(s) = E_0\frac{s}{s^2 + \omega^2}. \tag{12.9}$$

Taking the inverse transform of (12.9):

$$v(t) = E_0\cos(\omega t). \tag{12.10}$$

Which is equal to the source voltage in (12.5).

The same result is also obtained by the switch opening method, according to Figure 12.7b, with switch opening taking place at the zero crossing of the current, which is the point in time that the voltage is maximum.

In an inductive circuit, it is known that the voltage at the circuit breaker terminals at time t_0^+, instantaneously rises from zero to E_0, thus, the RRRV is infinite in this case.

The RRRV imposes severe limitations on the successful extinction of an electric arc, which makes it difficult to interrupt the current in predominantly inductive circuits with high rates of rise of the voltage at the circuit breaker.

- Simplification of the Injected Current
  An additional simplification for calculating the TRV (mainly the RRRV) is to approximate the sine current as current ramp, as shown in Figure 12.8a.

$$i(t) = I_0\sin(\omega t), \tag{12.11}$$

$$\left.\frac{di(t)}{dt}\right|_{t=0} = \omega I_0\cos(\omega t)|_{t=0} = \omega I_0, \tag{12.12}$$

$$i'(t) = \omega I_0 t. \tag{12.13}$$

Setting:

$$k = \omega I_0, \tag{12.14}$$

Equation (12.15) can be written:

$$i'(t) = kt, \tag{12.15}$$

where $i'(t)$ approximates the current ramp in Figure 12.8a.

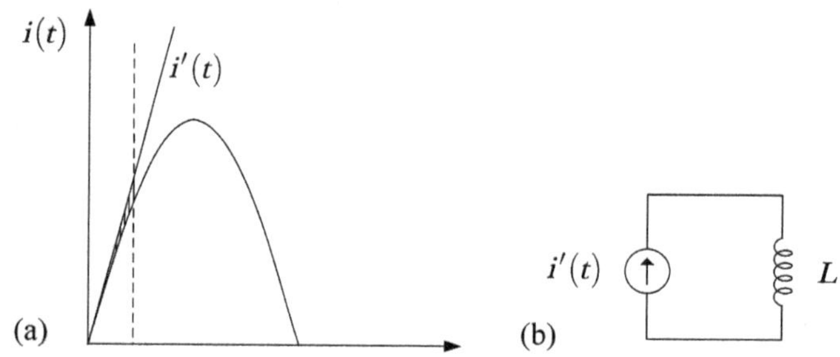

Figure 12.8 Simplified fault current. (a) Current approximation. (b) Inductive circuit.

Regarding current interruption in the same inductive circuit, Figure 12.8b, the simplified current is applied to calculate the TRV. In this case, from (12.6) and (12.11):

$$I_0 = \frac{E_0}{\omega L}.$$

Given that,

$$k = \frac{E_0}{L},$$

the voltage across the inductance, which is equal to L times the derivative of the current ramp $i'(t)$, may be calculated as follows:

$$v(t) = E_0 H(t). \tag{12.16}$$

The constant voltage in (12.16) is a fair and sligthly conservative approximation of the cosine-function voltage in (12.10) during the initial period relevant to the TRV (1-2ms), showing a close match between the results of the sine wave and the current ramp.

For the calculation of the RRRV, the current ramp approximation is even more accurate.

12.4.2 Inductance and Capacitance

Power system equipment, such as transformers, reactors, generators, etc., may be modeled by a reactance at moderate frequencies. The transmission line can also be assumed predominantly inductive at nominal frequency; however, a more convenient approach to travelling waves has to be taken with respect to higher frequencies.

With the higher frequencies in the transient, representing only the inductance is not accurate, making mandatory to include in the model a capacitive parcel, designated as stray capacitance, which always exist in the equipment.

A discussion of the capacitances in substation equipment and its typical values can be found in references [2, 7].

In the transformer windings, there are capacitances between turns and also between the windings and the tank, so that relatively complex oscillations inside the equipment may occur. Both the oscillations mentioned and the models that account for internal capacitances were analyzed in [2]. The most accurate procedure would involve using a rational fitted model if the frequency response of the transformer is available. However, this is a special test, not included in the transformer datasheet. Therefore, the most practical solution is to combine the nominal frequency model with internal capacitances obtained from impulse tests or calculated by the transformer manufacturer.

This subject is covered in a specific standard [11], the most significant example of which is the transformer limited fault (TLF), which deserves special attention. Since the stray capacitance of the transformers is relatively small, some applications may present RRRV that exceed the standard values for terminal faults, and therefore specific rise times for the voltage at the circuit breakers should be specified.

A terminal fault limited by a transformer is depicted in Figure 12.9a. Different accuracy levels in the representation of the electrical system can be assumed and this is not always an easy issue to address. Two typical situations can be considered. If the equivalent impedance of the electrical system is much lower than that of the transformer, the Thévenin equivalent of the system would be a very reasonable representation. Otherwise, it would be appropriate to represent the system in detail or even adopt a frequency-dependent equivalent.

The following simplified analysis assumes that the predominant impedance is that of the transformer, including losses and stray capacitance, which results in the circuit of Figure 12.9b.

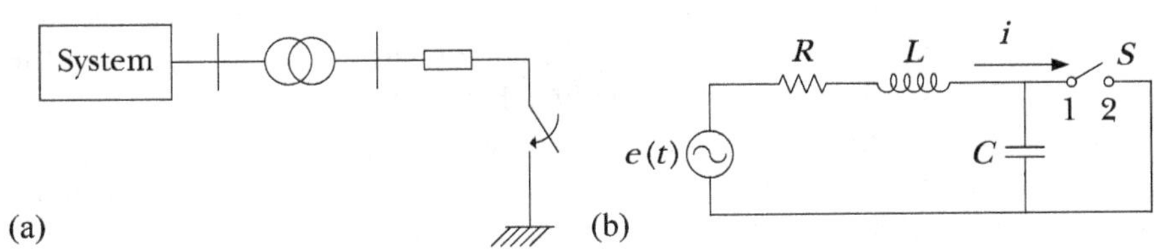

Figure 12.9 Fault near a transformer. (a) Equivalent circuit for fault current calculation. (b) Stray capacitance representation.

This RLC circuit has already been studied in Chapter 1, Example 1.4, representing the π circuit for a transmission line with the elimination of the line's sending end capacitance.

Equation (1.148) is reproduced with small changes in (12.18), where the voltage across the capacitance provides the TRV between the circuit breaker contacts. Assuming that the impedance of the transformer is practically a reactance, the time response can be analyzed with the LC circuit.

To make sure that the initial instant is coincident with the zero crossing of the current, an excitation with cosine waveform is adopted in (1.143), where $\theta = \frac{\pi}{2}$.

$$e(t) = E_0 \cos(\omega t). \tag{12.17}$$

The TRV in the LC circuit may be written as:

$$V_c(s) = \frac{E_0 s}{LC(s^2 + \omega^2)(s^2 + \omega_0^2)}. \tag{12.18}$$

$$\omega_0 = \frac{1}{\sqrt{LC}}. \tag{12.19}$$

Taking the inverse transform, the oscillatory response is given by:

$$v_c(t) = E_0[\cos(\omega t) - \cos(\omega_0 t)]. \tag{12.20}$$

Equation (12.20) assumes that the steady-state amplitude of the voltage across the capacitor is equal to the amplitude of the source, E_0. Since the stray capacitance is very small, virtually all source voltage is applied to the capacitor. In fact, this amplitude should be slightly above E_0, because of the voltage divider between the inductance and the capacitance, as discussed in Chapter 7. Since the stray capacitance from the transformer is very small, it can be assumed that $\omega_0 >> \omega$, and a further simplification can be made by assuming that, for the small time interval corresponding to the first peak of the TRV, the source amplitude is practically constant, with $\cos(\omega t) \cong 1$, resulting in Eq. (12.21):

$$v_c(t) = E_0[1 - \cos(\omega_0 t)]. \tag{12.21}$$

Equations (12.20) and (12.21) were obtained by solving the differential equations for two circuits, in which the solution of the circuit with the switch open uses the initial conditions of the circuit with the switch closed.

With the same simplifying assumptions that were adopted for the approximated LC circuit, a comparison is made with the current injection method, in which the current with the switch closed may be expressed, as in (12.6):

$$i(t) = \frac{E_0}{\omega L} \sin(\omega t). \tag{12.22}$$

The sinusoidal current generator is simulated by a ramp current generator, and the voltage across the capacitor in the LC parallel circuit is given by:

$$V_c(s) = I'(s)Z(s). \tag{12.23}$$

Applying Eq. (12.15):

$$I'(s) = \frac{k}{s^2}, \tag{12.24}$$

$$Z(s) = \frac{sL\dfrac{1}{sC}}{sL + \dfrac{1}{sC}} = \frac{1}{C}\frac{s}{\left(s^2 + \dfrac{1}{LC}\right)}, \tag{12.25}$$

thus:

$$V_c(s) = \frac{k}{sC}\frac{1}{(s^2 + \omega_0^2)}. \tag{12.26}$$

Expanding the equation into partial fractions:

$$V_c(s) = \frac{k_1}{s} + \frac{k_2}{s - j\omega_0} + \frac{k_2^*}{s + j\omega_0}. \tag{12.27}$$

$$k_1 = \frac{k}{C}\frac{1}{s^2 + \omega_0^2}\bigg|_{s=0} = kL. \tag{12.28}$$

$$k_2 = \frac{k}{Cs}\frac{1}{s + j\omega_0}\bigg|_{s=j\omega_0} = -\frac{k}{2}L. \tag{12.29}$$

Taking the inverse transform, the voltage at the switch terminal is expressed in the time domain:

$$v_c(t) = kL - kL\cos(\omega_0 t) = kL[1 - \cos(\omega_0 t)], \tag{12.30}$$

recalling that for sinusoidal excitation, $k = \dfrac{E_0}{L}$:

$$v_c(t) = E_0[1 - \cos(\omega_0 t)], \tag{12.31}$$

which is the same expression as (12.21).

With the inclusion of the stray capacitance, the derivative of the voltage at time t_0^+ is no longer infinite, and the voltage derivative does not change abruptly at time t_0^+.

$$v_c'(t) = E_0\omega_0\sin(\omega_0 t), \tag{12.32}$$

$$v_c'(0) = 0. \tag{12.33}$$

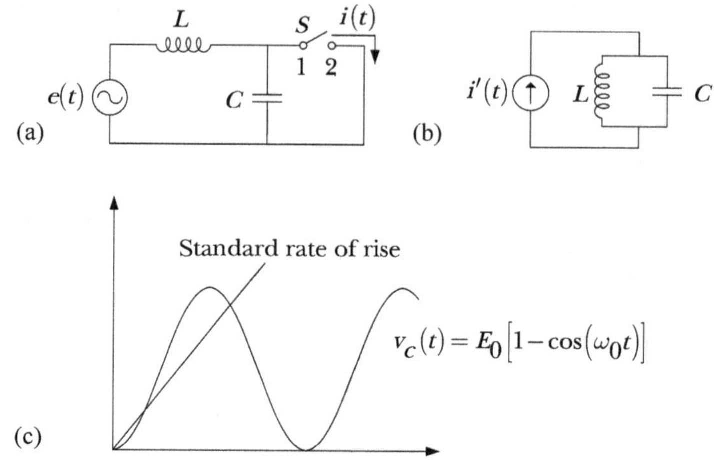

Figure 12.10 Current interruption in the LC circuit. (a) Switch opening. (b) Equivalent circuit with current injection. (c) Oscillatory TRV.

However, if the oscillation frequency is high, the voltage rise is also high, and the result should be compared to the standard envelopes, Figure 12.10c.

In the circuit in Figure 12.10a, only the contribution of the TRV on side 1 of the switch was considered since side 2 was short circuited. Thus, with this simplification, it is possible to observe the oscillatory component of the TRV, known as (1 – cos) curve (one minus cosine) in circuits analogous to those in Figure 12.10.

In the circuit in Figure 12.11a, side 2 of the switch is no longer short-circuited and now its contribution to the TRV should be considered. Side 2 of the switch could represent another piece of equipment, such as a transformer for instance, or even a transmission line connected to the transformer with a fault a few kilometers from the substation. Although the π circuit model of the transmission line may be questioned, the circuit is intended to exemplify the possibility of TRV inputs on both sides of the circuit breaker. The current injection method may be applied to both sides of the switch in Figure 12.11b.

The voltage across the switch terminals can be calculated from:

$$\Delta v = v_1 - v_2. \tag{12.34}$$

Two natural frequencies are present, one on each side of the switch, showing that there may be different voltage patterns on either side. The total voltage drop at the switch terminals is composed of two functions. Neglecting the resistance, these functions are presented in (12.35) and (12.36):

$$v_1(t) = kL_1[1 - \cos(\omega_{01}t)], \tag{12.35}$$

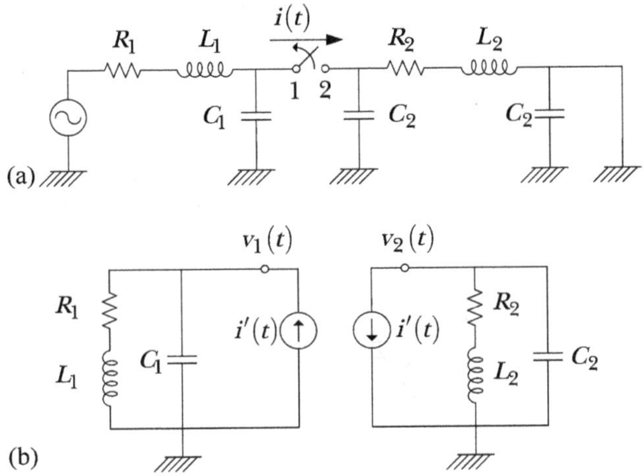

Figure 12.11 Short circuit at some distance from the switch. (a) Simplified equivalent circuit. (b) Different natural frequencies on both sides of the circuit breaker.

$$v_2(t) = -kL_2[1 - \cos(\omega_{02}t)]. \tag{12.36}$$

With an increase in the number of loops on both sides of the switch, more natural oscillation frequencies are expected. As the circuit could become more complex, calculation using electromagnetic transient programs is recommended for a more detailed analysis of the TRV.

12.4.3 Transient Recovery Voltage with Transmission Lines

The TRV is analyzed for the very common topology in Figure 12.12a, assuming that the fault is fed by the transformer and by the transmission line.

The difference with respect to Figure 12.9a is the transmission line connecting the two equivalent systems. It is assumed that the line is relatively long so that reflections back to the switch from the system-2 equivalent does not interfere during the restricted period of the TRV analysis.

An equivalent circuit that takes into account the stray capacitance from equipment on side 1 of the circuit breaker with side 2 grounded is proposed. In a simplified approach that neglects losses, the system equivalent in series with the transformer is simulated by an inductance. It is assumed that surge waves introduced on opening the circuit breaker travel through the long transmission line. The lossless transmission line is represented by its surge impedance, which is a more accurate representation at high frequencies than the π circuit.

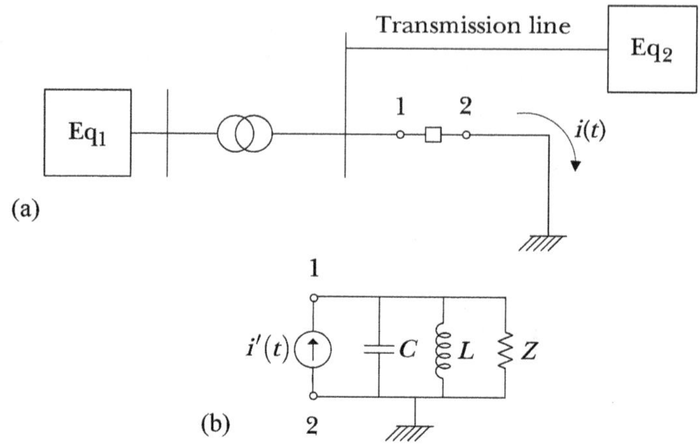

Figure 12.12 Electrical system with transmission lines. (a) Substation with transmission lines. (b) Equivalent circuit for the TRV calculation.

The equivalent circuit is depicted in Figure 12.12b.
where:

C: equipment stray capacitance.
L: equivalent inductance for the transformer and system 1.
Z: surge impedance of the transmission line.

Equivalent impedance in the frequency domain:

$$Z_e(s) = \frac{1}{sC + \frac{1}{Z} + \frac{1}{sL}}, \tag{12.37}$$

$$Z_e(s) = \frac{s}{C\left(s^2 + \frac{s}{ZC} + \frac{1}{LC}\right)}. \tag{12.38}$$

Using the current injection method, by multiplying the equivalent impedance of the parallel RLC circuit by the current in (12.24), the voltage at the switch terminals is obtained:

$$V(s) = \frac{k}{Cs\left(s^2 + \frac{s}{ZC} + \frac{1}{LC}\right)}. \tag{12.39}$$

To obtain the inverse transform, the poles of the rational function are calculated:

$s_1 = 0,$

$$s_{2,3} = -\frac{1}{2ZC} \pm \sqrt{\left(\frac{1}{2\,ZC}\right)^2 - \frac{1}{LC}}.$$

In the case of an oscillatory response:

$$\left(\frac{1}{2ZC}\right)^2 < \frac{1}{LC}.$$

Setting:

$$\sigma = -\frac{1}{2ZC} \text{ and } \omega_d = \sqrt{\frac{1}{LC} - \left(\frac{1}{2ZC}\right)^2},$$

thus:

$$v(t) = kL\left[1 - e^{\sigma t}\left(\cos(\omega_d t) - \frac{\sigma}{\omega_d}\sin(\omega_d t)\right)\right]. \tag{12.40}$$

A damped response is usually found when transmission lines are operating in the substation:

$$\left(\frac{1}{2ZC}\right)^2 > \frac{1}{LC}.$$

Setting:

$$a = -\frac{1}{2ZC} \text{ and } b = \sqrt{\left(\frac{1}{2ZC}\right)^2 - \frac{1}{LC}},$$

the voltage at the switch terminals in the damped system may be written as:

$$v(t) = kL\left[1 - e^{at}\left(\cosh(bt) - \frac{a}{b}\sinh(bt)\right)\right]. \tag{12.41}$$

Neglecting the effect of the stray capacitance, a simplified circuit is obtained. This conservative simplification makes calculation easier, allowing only the critical results to be detailed representing the capacitance.

$$Z_e(s) = \frac{ZsL}{Z + sL}, \tag{12.42}$$

$$V(s) = \frac{k}{s^2}\,\frac{Zs}{\left(s + \frac{Z}{L}\right)} = \frac{kZ}{s\left(s + \frac{Z}{L}\right)}. \tag{12.43}$$

Taking the inverse transform:

$$v(t) = kL\left(1 - e^{-\frac{Z}{L}t}\right). \tag{12.44}$$

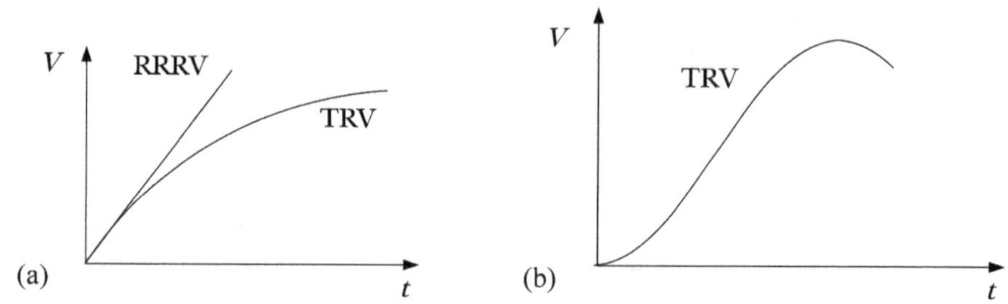

Figure 12.13 TRV without and with capacitance. (a) Without capacitance. (b) With capacitance.

From expression (12.44) the maximum value for the voltage is calculated:

$$V_{\text{max}} = kL. \tag{12.45}$$

It may be noted that the presence of the capacitance in the circuit does not change the maximum value of the voltage at the switch.

The derivative of the voltage at zero time is then calculated using (12.41), including the capacitance effect:

$$\left.\frac{dv(t)}{dt}\right|_{t=0} = 0, \tag{12.46}$$

from Eq. (12.44):

$$\left.\frac{dv(t)}{dt}\right|_{t=0} = kZ. \tag{12.47}$$

Thus, for the simplified circuit without capacitance, the inductance at zero time behaves like an open circuit, and all current is injected into the surge impedance of the transmission line. The behavior of the voltage at time t_0^+ is coincident with the RRRV, Figure 12.13a.

When the capacitance is included in the circuit, the derivative at time t_0^+ is zero, and the RRRV is calculated from the time response according to Figure 12.13b. The inclusion of the capacitance prevents instantaneous variation of the voltage, which thus contributes to the RRRV reduction. References [7, 11] propose capability envelopes (limit curves) to address oscillatory time responses in circuit breakers.

So far, only single-phase circuits have been analyzed, and three-phase circuits are addressed below.

12.5 Calculation of the TRV in Three-Phase Circuits

The opening of circuit breaker poles is examined for a three-phase ungrounded fault in a substation bus, near the circuit breaker terminals.

In item 12.2.2, it was mentioned that for unbalanced faults the opening sequence for the phases should be investigated; however, this is a task that should be performed with EMTP-type programs.

The intent here is to perform an analysis of the first pole to clear an ungrounded three-phase fault using the current injection method. The method is suitable in this case because the current fault is assumed to be sinusoidal with a frequency of 50 or 60 Hz and can be calculated using traditional methods of network analysis. In the first few cycles of the fault, the asymmetrical current is composed of a steady state sinusoidal current, and a damped DC offset.

As a critical TRV event associated with fault clearing, standards suggest a three-phase ungrounded fault on the substation bus, opening of the first pole of the circuit breaker and symmetrical current.

Note that, although this is a critical fault there is no guarantee that worse cases can happen as in cases where the single-phase fault current is higher than the three-phase fault current. The DC-offset is an additional piece of information to be provided in the circuit breaker specifications. After the first pole, the subsequent poles to clear are subject to transient currents, making it difficult to apply the method to open those poles. For the subsequent poles, it is advisable to perform the calculation with simulation tools, unless they are also examined on a symmetrical basis.

Thus, TRV can be calculated using relatively simple equations, which have validity up to the return of the first waves reflected from the shortest transmission line connected to the substation. If the reflections are positive, the traveling waves should be taken into account in the calculation; however, if the reflections are negative, the calculation even without traveling waves, already leads to worse cases.

In the context of the circuit shown in Figure 12.12, the primary characteristics of the TRV can be explored. However, to delve into the complexities of networks that include transmission lines with different lengths and substation equipment, computational methods are essential.

12.5.1 Three-phase Ungrounded Fault in the Transmission Line

A three-phase ungrounded fault is assumed on the line-side terminal of circuit breaker B, and its first pole to clear is analyzed (Figure 12.14).

$Z = Z_1 \,||\, Z_2 \,||\, ... \,||\, Z_n$: equivalent surge impedance for n operating transmission lines without faults.

L: equivalent inductance of the power source.

Z_ℓ: surge impedance of the transmission line with a fault.

C: stray capacitance of the equipment and substation bus.

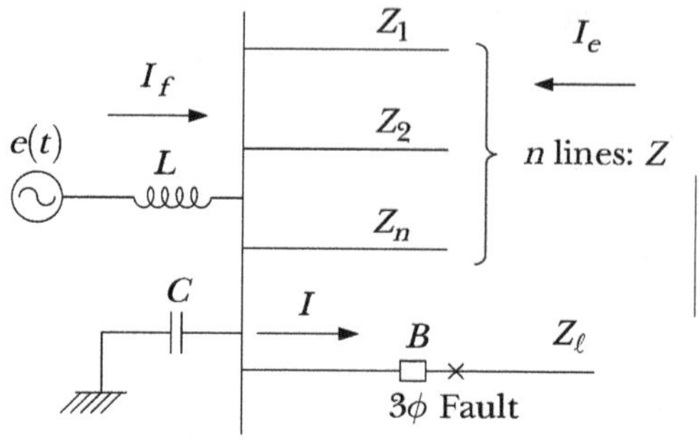

Figure 12.14 Three-phase electrical network.

The fault current is given by:

$$I = I_f + I_e, \tag{12.48}$$

where:

I_f: current through the equivalent inductive source.
I_e: Sum of the currents in the transmission lines.

The three-phase description of the circuit can be examined in Figure 12.15, during the opening of the first pole. Before opening the first pole, the following currents in each pole of the circuit breaker can be assumed:

$$i_a(t) = \sqrt{2}I\sin(\omega t), i_b(t) = \sqrt{2}I\sin(\omega t - 120°) \text{ and } i_c(t) = \sqrt{2}I\sin(\omega t + 120°).$$

The method of current injection is applied using a current generator, $i(t) = \sqrt{2}I\sin(\omega t)$, at phase a, at the time of opening of the first pole.

The current generator is added to the circuit for fault-current cancellation.

Current injection may be split in the two generators, as in Figure 12.6.

The equivalent circuits on sides 1 and 2 are shown in Figure 12.16a and b, respectively.

The TRV, between circuit breaker terminals 1 and 2, is shown below and refers to the circuit breaker pole of phase a:

$$\Delta v = v_1 - v_2, \tag{12.49}$$

where:

$$V_1(s) = I(s)Z_1(s), \tag{12.50}$$

$$V_2(s) = -I(s)Z_2(s), \tag{12.51}$$

thus:

$$\Delta V(s) = I(s)[Z_1(s) + Z_2(s)]. \tag{12.52}$$

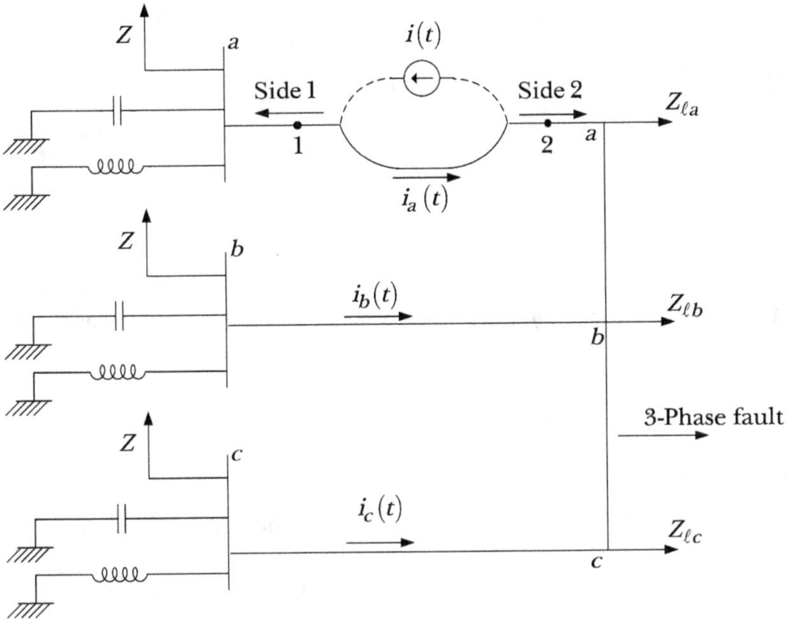

Figure 12.15 First pole to clear.

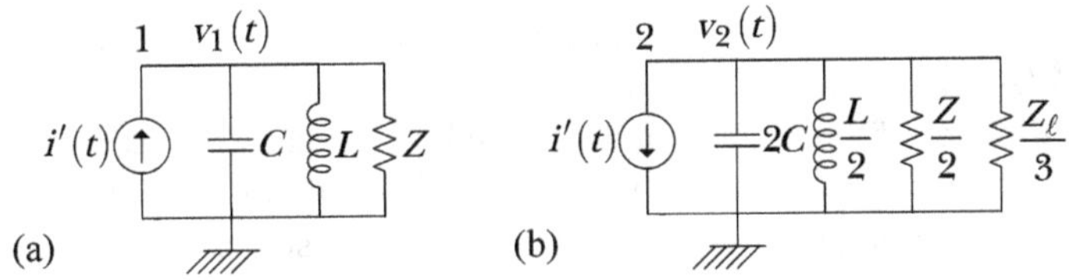

Figure 12.16 Circuit on sides 1 and 2. (a) Side 1. (b) Side 2.

Equation (12.52) can also be obtained by changing the sign of the current generator, according to Figure 12.17a, and summing the voltages on both sides of the circuit breaker.

In Figure 12.17a:

$$Z_e = \frac{Z}{2} \parallel \frac{Z_\ell}{3}. \tag{12.53}$$

In the case of a fault at the circuit breaker, line-side terminal, the solution of two distinct circuits is needed, that is, on system sides 1 and 2. To simplify the analysis, the capacitance can be neglected, as in Figure 12.17b:

$$v_1(t) = kL\left(1 - e^{-\frac{Z}{L}t}\right) = kL\left(1 - e^{-\frac{t}{\tau_1}}\right), \tag{12.54}$$

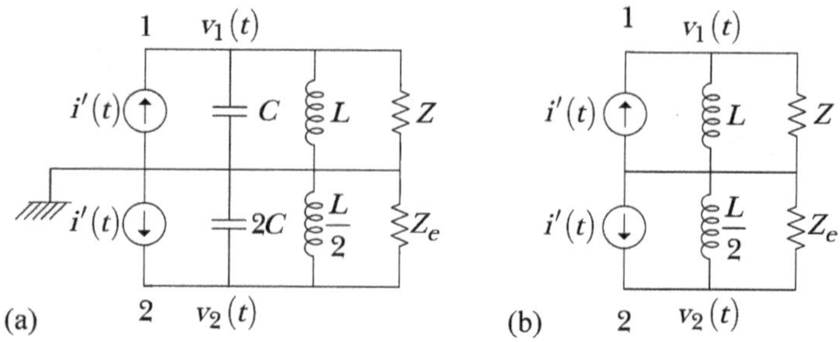

Figure 12.17 Equivalent circuit. (a) With stray capacitances. (b) Without stray capacitances.

$$v_2(t) = \frac{kL}{2}\left(1 - e^{-\frac{2Z_e t}{L}}\right) = \frac{kL}{2}\left(1 - e^{-\frac{t}{\tau_2}}\right), \tag{12.55}$$

$$\Delta v = v_1 + v_2. \tag{12.56}$$

The summation in (12.56) implies that the sign of $v_2(t)$ has already been switched in (12.55).

As $k = I_0\omega$ and $I_0 = \sqrt{2}I$, where I_0 is the amplitude of the phasor of the fault current, thus:

$$k = \sqrt{2}I\omega. \tag{12.57}$$

In the case of several operating lines in the substation, it can be assumed in Figure 12.17, that $Z/2 \ll Z_\ell/3$, thus $Z_e \cong Z/2$.

In this case, the influence of the faulted transmission line is greatly reduced, and the condition is like a fault at the substation bus near the terminals of the circuit breaker. The case of a three-phase ungrounded fault at a substation bus is analyzed next.

Note that oscillatory systems can also be analyzed using the circuit in Figure 12.17 by simply omitting the transmission line surge impedance.

12.5.2 Three-Phase Ungrounded Fault in the Substation Bus

Removing the transmission line in Figure 12.14, Z_ℓ, a three-phase ungrounded fault at the substation bus is analyzed in Figure 12.18a.

The equivalent circuit is obtained in a manner like the previous one, simply by eliminating $Z_\ell/3$, Figure 12.18b.

Disregarding the stray capacitance C, the following simplified expressions for the voltages are obtained:

$$v_1(t) = kL\left(1 - e^{-\frac{Z t}{L}}\right), \tag{12.58}$$

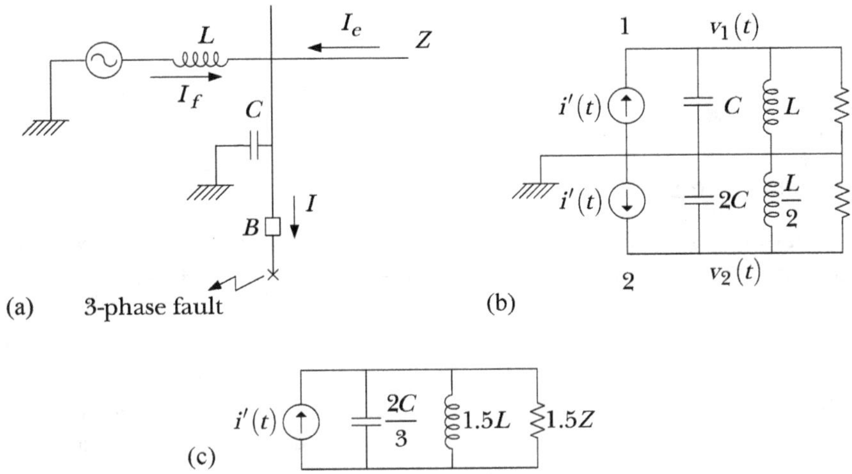

Figure 12.18 Three-phase fault. (a) Ungrounded three-phase fault at the substation bus. (b) Equivalent circuit for TRV calculation. (c) Simplified circuit.

$$v_2(t) = \frac{kL}{2}\left(1 - e^{-\frac{Z}{L}t}\right), \tag{12.59}$$

Now, sides 1 and 2 of the circuit breaker have the same time constant, and the amplitude of the voltage on side 2 is half of the amplitude of the voltage on side 1, thus:

$$\Delta v(t) = 1.5\,kL\left(1 - e^{-\frac{Z}{L}t}\right). \tag{12.60}$$

The circuit can also be simplified as shown in Figure 12.18c.

It is convenient to comment that the surge impedance, Z, was taken to be a phase impedance, without any reference to the coupling between phases, but it will be clarified in item 12.5.4 that it actually represents the positive sequence impedance.

12.5.3 Rate of Rise of the Recovery Voltage – RRRV

In three-phase circuits, particularly in the case of three-phase ungrounded faults, the RRRV can be calculated as the sum of the rates of rise on sides 1 and 2 of the circuit breaker.

When the capacitance is neglected, the following expressions are obtained:

- Three Phase Ungrounded Fault at Substation Bus

  Taking the derivative of (12.60):

$$RRRV = 1.5\,kZ, \tag{12.61}$$

or,

$$RRRV = 1.5\sqrt{2}I\omega Z. \tag{12.62}$$

- Three Phase Ungrounded Fault at Circuit Breaker Line Side Terminal
  Taking the derivatives of (12.54) and (12.55):

$$RRRV = \sqrt{2}I\omega(Z + Z_e). \tag{12.63}$$

If necessary, the model can be detailed, including the stray capacitances, and calculating the RRRV from the time response given by Eq. (12.41).

Example 12.1 Analyze the TRV in the circuit of Figure 12.19a, which is analogous to the case presented in a past edition of the IEEE standard [7].

A quite similar equivalent circuit is assumed here, which is obtained with the three transmission lines in parallel, $Z_1 \parallel Z_2 \parallel Z_3$, considering the stray capacitance of the equipment and transformer windings.

$$Z_1 = Z_2 = Z_3 = 450\ \Omega$$

$$Z = Z_1 \parallel Z_2 \parallel Z_3 = 150\ \Omega.$$

$$\mathrm{L} = 0.01035\ \mathrm{H}.$$

$$\mathrm{C} = 12.2\ \mathrm{nF}.$$

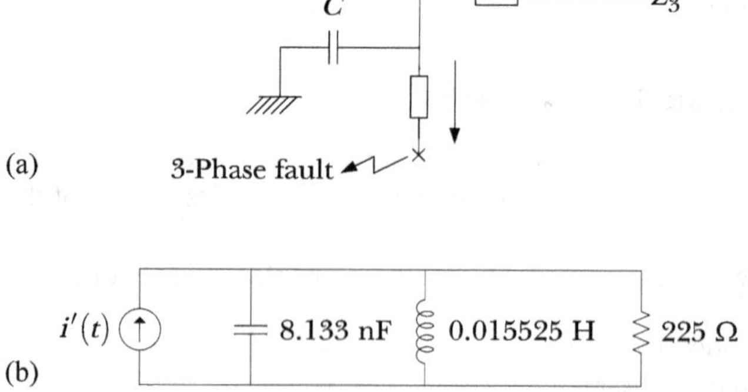

Figure 12.19 Analyzed network in example 12.1. (a) Three-phase ungrounded fault at the substation bus. (b) Equivalent circuit.

For the first pole to clear, the equivalent circuits in Figure 12.18b and c may be used.

Figure 12.19b shows the circuit used for the calculation of TRV, substituting the numerical values.

$$i'(t) = kt, \text{where } k = \sqrt{2} \times 377 \times 17.$$

According to Eq. (12.41):

$$\Delta v(t) = kL\left[1 - e^{-at}\left(\cosh(bt) - \frac{a}{b}\sinh(bt)\right)\right].$$

The circuit is not oscillatory, because:

$$\left(\frac{1}{2\,ZC}\right)^2 \cong 7.46 \times 10^{10} > \frac{1}{LC} = 7.92 \times 10^9,$$

Constants a and b are calculated:

$$a = -\frac{1}{2ZC} = -2.7323 \times 10^5 \text{s}^{-1},$$

$$b = \sqrt{\left(\frac{1}{2\,ZC}\right)^2 - \frac{1}{LC}} = \sqrt{6.6738 \times 10^{10}} = 258336.28 \text{ rad/s},$$

$$kL = \sqrt{2} \times 377 \times 17 \times 0.015525 = 140.714 \text{ kV}.$$

Resulting in the following numerical values for $\Delta v(t)$:

$$t = 50\ \mu\text{s}, \qquad \Delta v = 71.97 \text{ kV}.$$

$$t = 100\ \mu\text{s}, \qquad \Delta v = 108.07 \text{ kV}.$$

$$t = 150\ \mu\text{s}, \qquad \Delta v = 125.21 \text{ kV}.$$

Assuming a circuit without any capacitance, Eq. (12.60) is applied, taking into consideration the fact that a factor of 1.5 is already included in the model in Figure 12.19b, based on the circuit in Figure 12.18c. It should be noted that the same result can be obtained by utilizing side 1 of the circuit, and then applying the factor of 1.5.

$$\Delta v(t) = kL\left(1 - e^{-\frac{Z}{L}t}\right),$$

$$\Delta v(t) = 140.714\left(1 - e^{-14492.7t}\right).$$

Using the former expression, the maximum value for the transient voltage is calculated, without considering the reflections from the transmission line terminals:

$$\Delta v_{\text{max}} = 140.714 \text{ kV},$$

The value for the RRRV is:

$$RRRV = kZ,$$
$$RRRV = \sqrt{2} \times 377 \times 17 \times 225 = 2.04 \times 10^6\ \text{kV/s},$$

or,

$$RRRV = 2.04\ \text{kV}/\mu\text{s}.$$

Several considerations were shown to enter into the calculation of the TRV and determination of the parameters primarily for the equivalent winding capacitance of the transformers in use. The example in this chapter is not intended to provide a complete calculation in every detail, but only to illustrate the application of the equations presented previously.

12.5.4 Analysis with Symmetrical Components

- Three-Phase Ungrounded Fault at the Substation Bus

  The analysis of the first pole to clear with symmetrical components first examines the case of three-phase ungrounded faults.

  Considering the simultaneous fault composed of a three-phase short circuit and the opening of one pole of the circuit breaker, in this particular case, since the short-circuit is balanced, only the opening of the first pole of the circuit breaker can be interpreted as a fault. Figure 12.20a shows the opening of phase a in a generic three-phase network with surge impedances in both sides.

  In phase domain transmission line surge impedances are designated as, Z_a, Z_b,Z_c in sides 1 and 2. In symmetrical components, with respect to surge

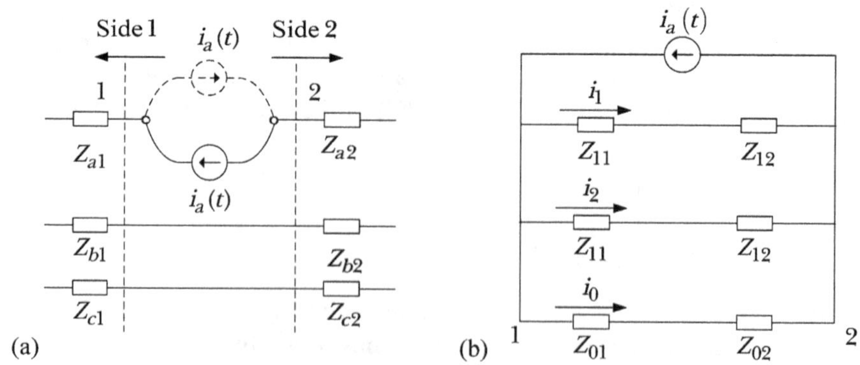

Figure 12.20 Analysis with symmetrical components. (a) Opening of the first pole. (b) Equivalent circuit in symmetrical components.

impedance Z_{ik}, the index i represents the sequence, and k represents the side of the circuit breaker. The same index designation applies to voltages v_{ik}.

The upper current generator, in dashed line, represents the current in phase a with the switch closed and the lower generator is the one that cancels the current simulating the switch opening, and therefore the TRV can be obtained from the circuit in Figure 12.20b, as discussed in Section 12.3.1. In this circuit, it is known that the positive sequence impedance is equal to the negative sequence impedance, $Z_{11} = Z_{21}$ and $Z_{12} = Z_{22}$, and it is also known that:

$$I_0 + I_1 + I_2 = I_a. \tag{12.64}$$

The circuits representing the sequences are independent, according to references [5, 12, 13]:

$$\Delta v = \Delta v_0 = \Delta v_1 = \Delta v_2, \tag{12.65}$$

$$\Delta v_0 = v_{01} - v_{02}, \tag{12.66}$$

$$\Delta v_1 = v_{11} - v_{12}, \tag{12.67}$$

$$\Delta v_2 = v_{21} - v_{22}. \tag{12.68}$$

The TRV at the first pole to clear is obtained from the following expression:

$$\Delta v_a = \Delta v_0 + \Delta v_1 + \Delta v_2 = 3\Delta v. \tag{12.69}$$

The network impedances can be seen on both sides of the circuit in Figure 12.21a, in the case of a three-phase ungrounded fault at the substation bus, on side 2.

The impedances of the sequence networks, on side 1 and 2, are presented in Figure 12.21b and c, respectively.

If the positive, negative and zero-sequence networks are connected as illustrated in Figure 12.20b, it may be observed that there is no zero-sequence path, since Z_{02} is infinite. Thus, it is necessary to consider just the positive and negative sequences, both with the same impedance values, simplifying the calculation.

$$\Delta V(s) = I_a(s)\frac{1}{2}Z_{11}(s), \tag{12.70}$$

$$\Delta V_a(s) = \frac{3}{2}I_a(s)Z_{11}(s). \tag{12.71}$$

According to the convention previously adopted, Z_{11} is the positive sequence impedance on side 1 of the circuit breaker. It is concluded that the surge impedance used previously in equation (12.60) is in fact the positive sequence impedance.

Disregarding the stray capacitance and assuming $Z_{11}(s)$ to be the complex impedance of the RL circuit in Eq. (12.42), the following equation can be written:

$$\Delta v_a(t) = 1.5kL\left(1 - e^{-\frac{Z_t}{L}t}\right). \tag{12.72}$$

This equation is equivalent to that obtained in (12.60).

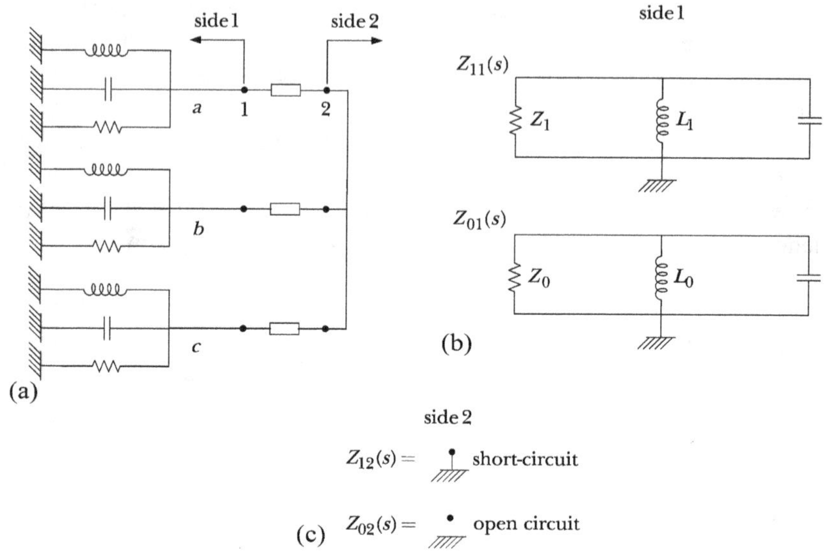

Figure 12.21 Analysis with symmetrical components. (a) Three-phase ungrounded fault. (b) Sequence diagrams for side 1. (c) Sequence diagrams for side 2.

- Grounded and Ungrounded Three-Phase Faults

  The fault current is the same in both cases. However, in the case of grounded three-phase faults, the zero-sequence impedance is zero, lowering the equivalent impedance in Figure 12.20b and consequently the transient voltage also becomes smaller.

  For a three-phase grounded or ungrounded fault, the steady-state voltage across the open pole of phase a can be analyzed as follows.

  In Figure 12.22a, on opening the first pole in the case of a grounded three-phase fault with the switch at point M closed, the voltage between points 1 and 2 is equal to the phase voltage, since point 2 is grounded, $\Delta v = v_{a1} = v_a$.

  For an ungrounded three-phase fault, the voltage at point 2 is given by $v_{a2} = -0.5v_a$, as in Figure 12.22b, thus increasing the voltage between the circuit breaker contacts to $\Delta v = v_{a1} - v_{a2} = 1.5v_a$.

Example 12.2 Calculate the TRV for the generator connected to a step-up transformer. The three-phase fault is located at the substation bus, on the high voltage side (500 kV) of the transformer.

- Sub transient reactance of the generator: $X_d'' = 20\%$.
- Short circuit reactance of the transformer: $X_t = 10\%$.
- Stray capacitance of the transformer on the high voltage side: $C = 5$ nF.

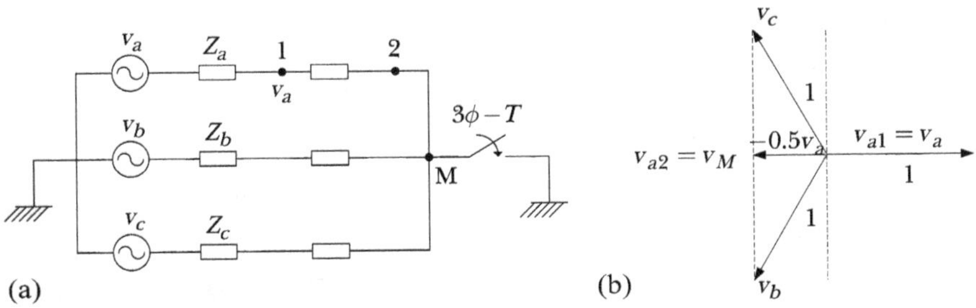

Figure 12.22 Grounded fault. (a) Grounded and ungrounded three-phase fault. (b) Voltage on side 2 for ungrounded fault.

- Rated power of the transformer and generator: 100 MVA.
- Neglect the capacitance of the substation bus.

Solution

The oscillatory three-phase circuit in Figure 12.23 is analyzed using symmetrical components.

Reactance calculated on the 500 kV side.

$$X = \left(X_d'' + X_t\right)\frac{500^2}{100} = 0.3 \times \frac{500^2}{100} = 750\ \Omega,$$

$$\text{Fault current}: I = \frac{500}{\sqrt{3} \times 750} = 0.385\ \text{kA}.$$

Sequence diagrams (Figure 12.24):

$$L_1 = \frac{750}{377} = 1.99\ \text{H}, C_1 = 5\ \text{nF}.$$

According to Eq. (12.71):

$$\Delta V_a(s) = 1.5 I_a(s) Z_{11}(s)$$

The inverse Laplace transform for the product $I_a(s)Z_{11}(s)$ is obtained from Eq. (12.31), by including the factor of 1.5.

Figure 12.23 Ungrounded three-phase fault at the high voltage side of the transformer.

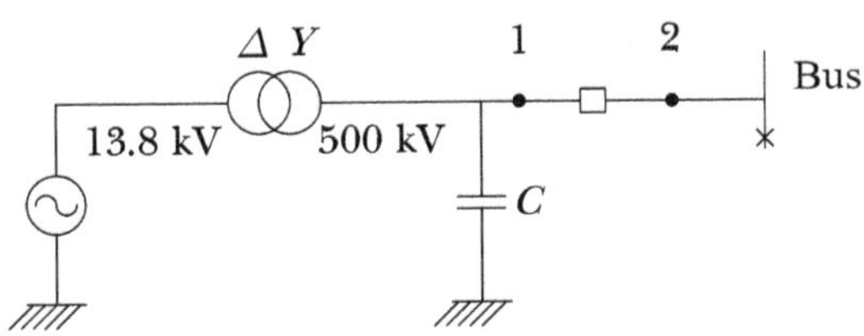

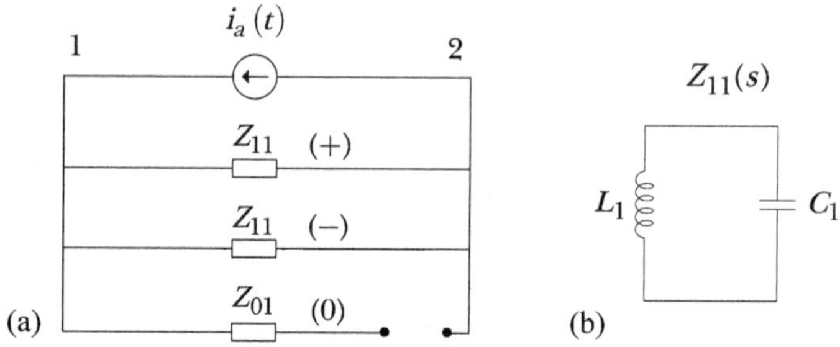

Figure 12.24 Sequence diagrams. (a) Sequence impedances connected in parallel. (b) Positive sequence circuit.

$$\Delta v_a(t) = 1.5E_0(1 - \cos(\omega_0 t)),$$

$$E_0 = kL = \sqrt{2} \times 377 \times 0.385 \times 1.99 = 408.48\ \text{kV},$$

$$\Delta v_a(t) = 612.72(1 - \cos(\omega_0 t)),$$

$$\omega_0 = \frac{1}{\sqrt{1.99 \times 5 \times 10^{-9}}} = 10025\ \text{rad/s},$$

$$f_0 = 1595\ \text{Hz}.$$

The maximum value of the voltage at the switch terminals is:

$$\Delta V_{a\ \max} = 1225.44\ \text{kV}.$$

- Three-Phase Ungrounded Fault at the Line Side of the Circuit Breaker
  As previously examined, the opening of the first pole causes the following association between the diagrams of sequence impedances (Figure 12.25a). Note that the impedances in each diagram are the sum of the individual impedances on both sides of the circuit breaker.
  Neglecting the capacitances in each diagram, the simplified circuit is obtained in Figure 12.25b, without capacitance.

$$\frac{Z_1(s)}{2} = \frac{1}{2}\left(\frac{sL_1 Z_1}{Z_1 + sL_1}\right), \tag{12.73}$$

$$Z_0(s) = \frac{sL_0 Z_0}{Z_0 + sL_0}, \tag{12.74}$$

$$\Delta V_a(s) = \frac{\frac{1}{2}\frac{sL_1 Z_1}{Z_1 + sL_1}\left(\frac{sL_0 Z_0}{Z_0 + sL_0} + Z_{\ell 0}\right)}{\frac{1}{2}\frac{sL_1 Z_1}{Z_1 + sL_1} + \frac{sL_0 Z_0}{Z_0 + sL_0} + Z_{\ell 0}} \frac{3I_a}{s^2}. \tag{12.75}$$

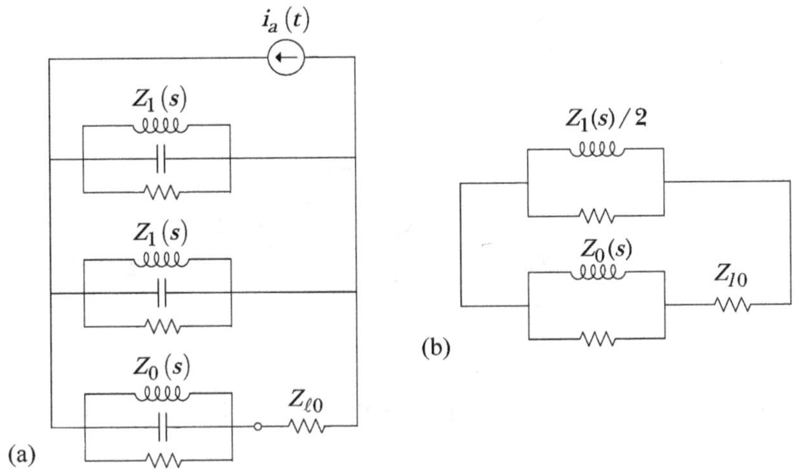

Figure 12.25 Three-phase ungrounded fault at the line side of the circuit breaker. (a) Sequence impedances connected in parallel. (b) Equivalent circuit.

Using Eq. (12.75), the TRV can be calculated, taking into account the impedances at the circuit breaker terminals for the positive sequence, Z_1, and zero sequence, Z_0, and employing the same approach to the current in phase a as in (12.15).

12.5.5 Traveling Waves

So far, the transmission line has been modeled only by its surge impedance, however, the transmission line model should take into account the traveling waves developed at the terminals of the circuit breaker (Figure 12.26a).

Z: surge impedance.
τ: time delay.

The time response of the traveling waves depends on the simulation details, as for example, how the transmission lines, stray capacitance and substation equipment were represented. The reflections at the remote terminals of the transmission lines, after the corresponding time delays, return to the circuit breaker terminals, thus changing the shape of the TRV.

In the case of an ungrounded three-phase fault, Figure 12.26b illustrates the equivalent circuit for the substation connected to several transmission lines. The stray capacitance is neglected.

Designating $v_1(t)$ as the wave that travels through every transmission line, which is the same in all lines, the first reflection to come back is from the shortest line, after twice its time delay elapses. Only the circuit breaker side 1 is analyzed (Figure 12.27).

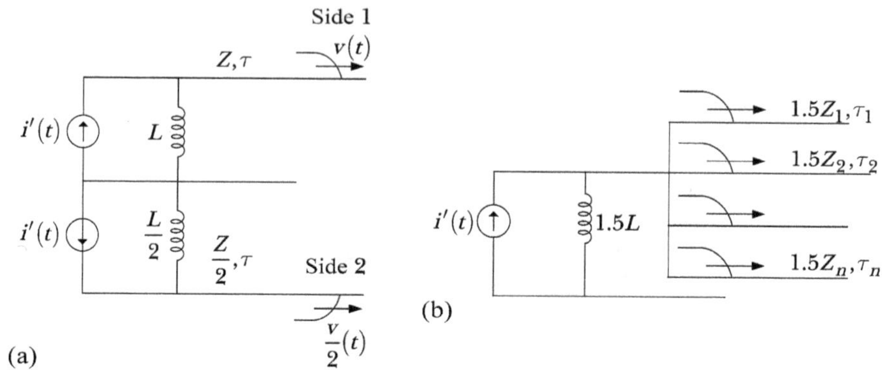

Figure 12.26 Traveling waves. (a) Traveling waves on sides 1 and 2. (b) Equivalent circuit.

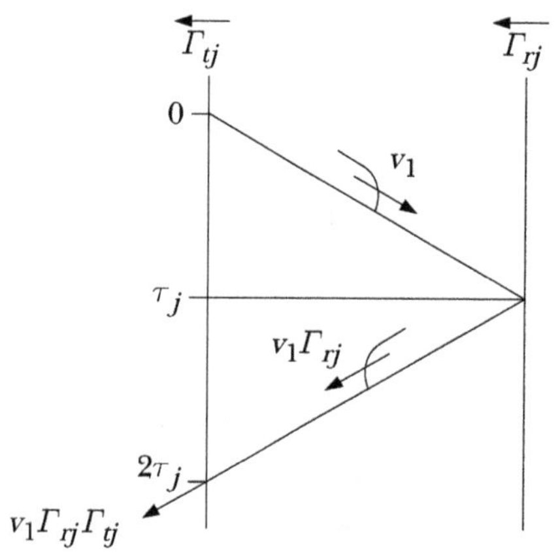

Figure 12.27 Reflection at the remote terminal of line *j*.

Traveling waves in multiphase transmission lines should be calculated by means of matrices of the transmission line's reflection coefficients and transmission coefficients, according to Chapter 3. A simplified approach is taken here for one phase of the transmission line, neglecting any mutual coupling. However, for the ungrounded three-phase fault at the substation bus, only positive sequence can be undertaken.

The reflection coefficient at the remote terminal of the transmission line j may be written:

$$\Gamma_{rj} = \frac{Z_{rj} - Z_j}{Z_{rj} + Z_j}, \tag{12.76}$$

where:

Z_j: surge impedance of transmission line j.
Z_{rj}: impedance on the remote side of transmission line j.

The reflected wave is multiplied by the transmission coefficient Γ_{tj} on the substation side of the transmission line, furnishing the traveling-wave component of the first reflection back to the circuit breaker terminal for phase a, that is:

$$v_{1t}(t) = \Gamma_{tj}\Gamma_{rj}v_1(t - 2\tau_j). \tag{12.77}$$

The voltage at the circuit breaker terminal is given by superposition of $v_1(t) + v_{1t}(t)$ (Figure 12.28b), thus:

$$v_{b1}(t) = v_1(t) + \Gamma_{tj}\Gamma_{rj}v_1(t - 2\tau_j). \tag{12.78}$$

Taking a conservative approach, the shortest line is considered to be an open wire transmission line, or perhaps a line connected to a transformer, assuming that the transformer can initially respond as an open circuit, in which case, $\Gamma_{rj} = 1.0$. Calculation of the transmission coefficient Γ_{tj}, on the substation side of the transmission line, can also be simplified by disregarding the effect of the inductance, which is treated as an open circuit.

$$\Gamma_{tj} = \frac{2Z_{eq}}{Z_{eq} + Z_j}, \tag{12.79}$$

where Z_{eq} is the equivalent of the surge impedance of the other transmission lines connected to the substation (Figure 12.28a).

The effect of any equivalent inductance at the substation on the transmission coefficient can easily be elaborated in the frequency domain (see Example 12.3 and Chapter 2).

The voltage on side 2 is obtained from $v_2(t) = 0.5v_1(t)$ and therefore the total TRV at the circuit breaker terminals is $\Delta V = 1.5v_1(t)$.

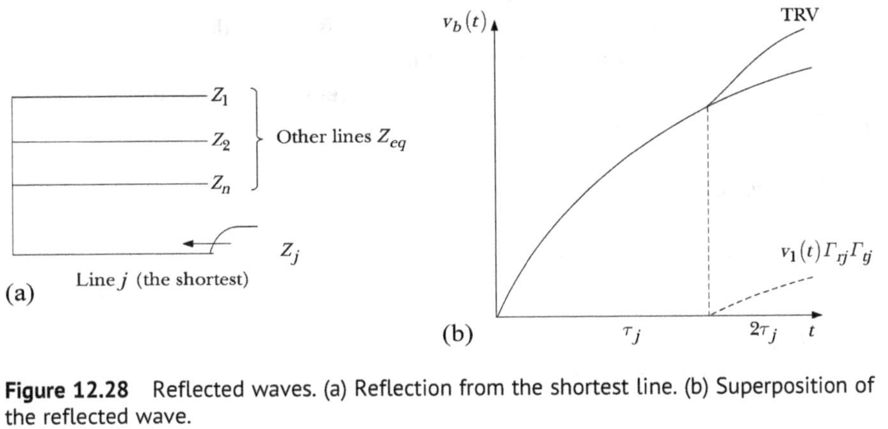

Figure 12.28 Reflected waves. (a) Reflection from the shortest line. (b) Superposition of the reflected wave.

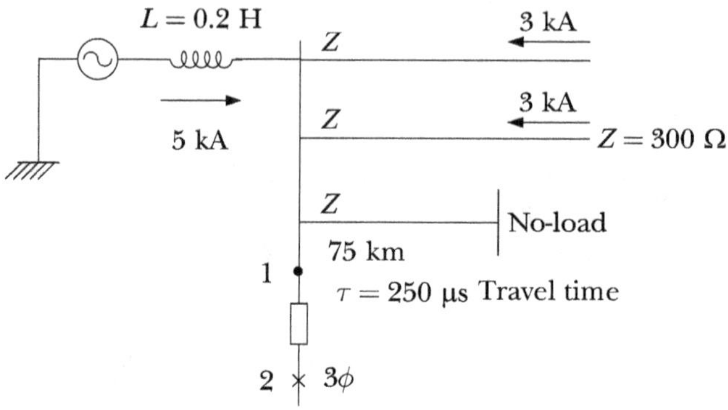

Figure 12.29 Circuit for Example 12.3.

Example 12.3 Calculate the TRV for an ungrounded three-phase fault at the substation bus, assuming the first reflection to be from a 75-km transmission line operating without a load at the remote terminal (Figure 12.29).

Solution

The voltages on sides 1 and 2 of the circuit breaker are calculated.

Equation for phase a, side 1, takes the equivalent of the three lines and the source inductance:

$$V_1(s) = \frac{k\dfrac{Z}{3}}{s\left(s + \dfrac{Z}{3L}\right)} \text{ or } V_1(s) = \frac{100k}{s(s+500)},$$

$$v_1(t) = 0.2k\left(1 - e^{-500t}\right).$$

The voltage is considered to travel at the speed of light in the 75 km transmission line and returns to the substation after $2\tau = 500$ μs. The reflection coefficient is assumed to be unity at the remote terminal.

The voltage due to the first reflection from the remote terminal encounters the parallel of the complex impedances at the substation:

$$\frac{Z}{2} \text{ and } sL.$$

The transmission coefficient at point 1 is given by:

$$\Gamma_t(s) = \frac{2 \times \dfrac{150 \times 0.2s}{150 + 0.2s}}{300 + \dfrac{150 \times 0.2s}{150 + 0.2s}},$$

or,

$$\Gamma_t(s) = \frac{2}{3}\frac{s}{s+500}.$$

The voltage component reflected from the remote terminal and multiplied by the transmission coefficient at the substation is expressed as:

$$V_{1t}(s) = V_1(s)\Gamma_t(s) = \frac{66.6k}{(s+500)^2}.$$

Obtaining the inverse transform:

$$v_{1t}(t) = 66.6kt\, e^{-500t}.$$

Taking into account the 2τ delay:

$$v_{1t}(t-2\tau) = 66.6k(t-2\tau)e^{-500(t-2\tau)}H(t-2\tau),$$

thus, the voltage on side 1 of the circuit breaker may be rewritten as:

$$v_{b1}(t) = 0.2k\left(1-e^{-500t}\right) + 66.6k(t-2\tau)\, e^{-500(t-2\tau)}H(t-2\tau).$$

From Eq. (12.57), the constant k is calculated: $k = 5864.6$.

Figure 12.30 shows the superposition of the exponential component on the reflected wave.

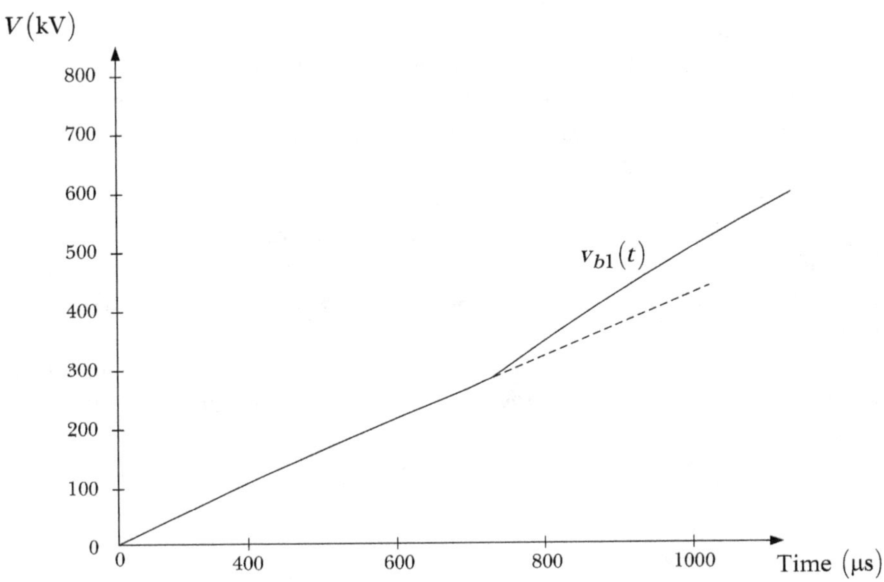

Figure 12.30 Transient recovery voltage with the reflected component.

The voltage on side 2 of the circuit breaker is obtained from $v_{b2}(t) = 0.5v_{b1}(t)$. The total TRV at the circuit breaker terminals is $\Delta V_b = 1.5v_{b1}(t)$.

12.5.6 TRV Analysis in the Frequency Domain

Accurate calculation of the TRV can be performed using electromagnetic transient programs. The analysis performed using the symmetrical components in item 13.5.4 was intended to clarify the behavior of some basic TRV circuits, but the connections of the sequence diagrams are relatively complex in certain cases of simultaneous faults.

The aim of this item is to show how easily calculation of the TRV with simultaneous faults can be evaluated in phase components (also desigmated as phase domain), complementing the analyses using symmetrical components. The short circuit represents one fault and the first pole to clear represents the other.

The equations describing the TRV in the first pole to clear are presented in the case of single-phase, double phase, double-phase to earth, three-phase grounded and three-phase ungrounded faults, Figure 12.31.

In the equations that follow, $[Z]$ is the three-phase nodal impedance matrix for the Thévenin equivalent of all the system equipment connected to the circuit breaker that eliminates the fault. Although the nomenclature is similar, the matrix $[Z]$ contains elements that are more general than line-surge impedances, which have also been represented so far by Z (see Section 12.4.3). These elements can be simplified to surge impedances only in the very specific cases of simplified RRRV equations.

The matrix includes the network elements connected to the bus, such as inductances, stray capacitances and transmission lines. The network representation should be as accurate as possible, including the dependence of the parameters on the frequency. In this way, it is possible to calculate the TRV, including travelling waves, in all transmission lines in the system. Since the approach is implemented in the frequency domain, the solution in the time domain is obtained by inversion of the Fourier transform, Reference [14] applies the method to a large system with several transmission lines, and the results are quite accurate compared to the results of EMT program.

The current injection method is applied using Eq. (12.11), providing more general expressions for the TRV. However, the approximation in (12.15) can also be used.

$$i(t) = I_0 \sin(\omega t), \tag{12.80}$$

$$I(\omega) = \frac{I_0 \omega}{s^2 + \omega^2}, \tag{12.81}$$

where I_0 is the fault current amplitude at the first pole to clear.

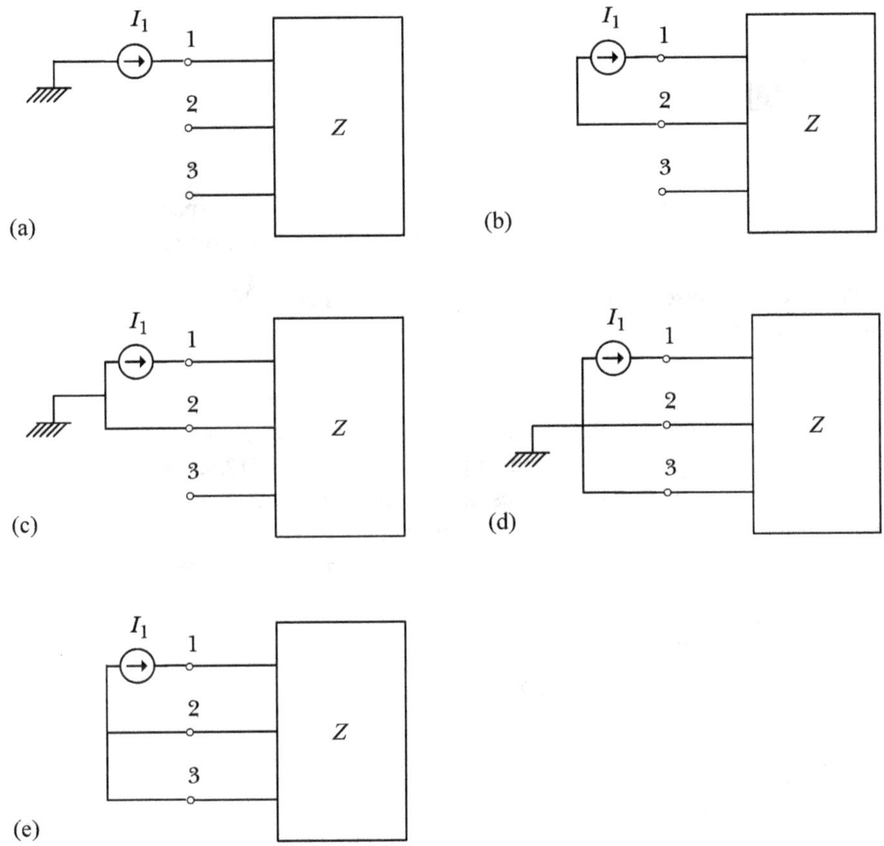

Figure 12.31 System faults. (a) Single-phase fault. (b) Double-phase fault. (c) Double-phase to ground fault. (d) Three-phase to ground fault. (e) Ungrounded three-phase fault.

Phases a, b and c are related to nodes 1, 2 and 3, respectively, and all elements that are a function of ω, as for example, $z_{ij}(\omega)$, are written simply as z_{ij}.

The first pole to clear is assumed to be in phase a. According to the nomenclature discussed above, the TRV is calculated in the frequency domain from $\Delta V = V_1 - V_2$, and when pole 2 of the circuit breaker is grounded, $\Delta V = V_1$.

Figure 12.31 shows the fault clearing with the correspondent injected currents for different system faults.

- Single-Phase Fault

  The following relationships between voltages and nodal currents can be written with the aid of the nodal impedance matrix, and Figure 12.31a.

$$\begin{bmatrix} V_1 \\ V_2 \\ V_3 \end{bmatrix} = \begin{bmatrix} z_{11} & z_{12} & z_{13} \\ z_{21} & z_{22} & z_{23} \\ z_{31} & z_{32} & z_{33} \end{bmatrix} \times \begin{bmatrix} I_1 \\ 0 \\ 0 \end{bmatrix}, \tag{12.82}$$

where:

z_{11}, z_{22}, z_{33}: diagonal elements related to phases *a*, *b*, and *c*, respectively.
z_{12}, z_{13}, z_{23}: mutual elements between phases *ab*, *ac*, and *bc*, respectively.

The TRV in phase a, in the frequency domain, is given by (12.83), and in this case $\Delta V = V_1$:

$$\Delta V = z_{11} I_1. \tag{12.83}$$

In accord with the previous comments on RRRV, a simplified expression may also be obtained by replacing the impedance values in (12.82) with the surge impedance of the transmission lines, neglecting the inductance and capacitance and approximating the sinusoidal current using (12.15).

In the case of transposed lines and balanced matrices, expression (12.82) is rewritten as:

$$\begin{bmatrix} V_1 \\ V_2 \\ V_3 \end{bmatrix} = \begin{bmatrix} z_p & z_m & z_m \\ z_m & z_p & z_m \\ z_m & z_m & z_p \end{bmatrix} \times \begin{bmatrix} I_1 \\ 0 \\ 0 \end{bmatrix}. \tag{12.84}$$

The TRV in phase *a* is:

$$\Delta V = z_p I_1. \tag{12.85}$$

- Double-Phase Fault

  Double-phase fault can be seen in Figure 12.31b.

$$\begin{bmatrix} V_1 \\ V_2 \\ V_3 \end{bmatrix} = \begin{bmatrix} z_{11} & z_{12} & z_{13} \\ z_{21} & z_{22} & z_{23} \\ z_{31} & z_{32} & z_{33} \end{bmatrix} \times \begin{bmatrix} I_1 \\ -I_1 \\ 0 \end{bmatrix}. \tag{12.86}$$

From Figure 12.31b, the TRV, $\Delta V = V_1 - V_2$, is calculated by subtracting row 2 from row 1.

$$\Delta V = (z_{11} - 2z_{12} + z_{22}) I_1. \tag{12.87}$$

For balanced matrices, $\Delta V = 2(z_p - z_m) I_1$, and since $z_1 = z_p - z_m$:

$$\Delta V = 2z_1 I_1, \tag{12.88}$$

where z_1 is the positive sequence impedance.

- Double-Phase to Ground Fault
  Figure 12.31c shows the first pole to clear in phase a for a double-phase to earth fault.
  Assuming again the current injected in phase a:

$$\begin{bmatrix} V_1 \\ 0 \\ V_3 \end{bmatrix} = \begin{bmatrix} z_{11} & z_{12} & z_{13} \\ z_{21} & z_{22} & z_{23} \\ z_{31} & z_{32} & z_{33} \end{bmatrix} \times \begin{bmatrix} I_1 \\ I_2 \\ 0 \end{bmatrix}. \tag{12.89}$$

  Expanding the first two rows of (12.89):

$$V_1 = z_{11}I_1 + z_{12}I_2, \tag{12.90}$$

$$0 = z_{21}I_1 + z_{22}I_2. \tag{12.91}$$

  By eliminating I_2 in Eqs. (12.90) and (12.91) and considering that $\Delta V = V_1$:

$$\Delta V = \left(z_{11} - \frac{z_{12}^2}{z_{22}}\right) I_1. \tag{12.92}$$

  In the case of balanced matrices:

$$\Delta V = \left(z_p - \frac{z_m^2}{z_p}\right) I_1. \tag{12.93}$$

- Three-Phase to Ground Fault
  In Figure 12.31d a three-phase to earth fault is presented.

$$\begin{bmatrix} V_1 \\ 0 \\ 0 \end{bmatrix} = \begin{bmatrix} z_{11} & z_{12} & z_{13} \\ z_{21} & z_{22} & z_{23} \\ z_{31} & z_{32} & z_{33} \end{bmatrix} \times \begin{bmatrix} I_1 \\ I_2 \\ I_3 \end{bmatrix}. \tag{12.94}$$

  Eliminating rows 2 and 3 through the Kron reduction:

$$V_1 = \left\{ z_{11} - \begin{bmatrix} z_{12} & z_{13} \end{bmatrix} \times \begin{bmatrix} z_{22} & z_{23} \\ z_{32} & z_{33} \end{bmatrix}^{-1} \times \begin{bmatrix} z_{21} \\ z_{31} \end{bmatrix} \right\} I_1. \tag{12.95}$$

  Rearranging (12.95):

$$\Delta V = \left[z_{11} - \frac{z_{33}z_{12}^2 - 2z_{12}z_{13}z_{23} + z_{22}z_{13}^2}{z_{22}z_{33} - z_{23}^2} \right] I_1. \tag{12.96}$$

  For balanced matrices:

$$\Delta V = \left(z_p - \frac{2z_m^2}{z_p + z_m}\right) I_1. \tag{12.97}$$

- Three-Phase Ungrounded Fault
  The three-phase ungrounded fault is analyzed using Figure 12.31e.

$$\begin{bmatrix} V_1 \\ V_2 \\ V_2 \end{bmatrix} = \begin{bmatrix} z_{11} & z_{12} & z_{13} \\ z_{21} & z_{22} & z_{23} \\ z_{31} & z_{32} & z_{33} \end{bmatrix} \times \begin{bmatrix} I_1 \\ -I_1 - I_3 \\ I_3 \end{bmatrix}. \tag{12.98}$$

Since $I_2 = -I_1 - I_3$ in (12.98) and subtracting row 3 from row 2, current I_3 can be obtained as a function of I_1.

$$I_3 = -\frac{z_{22} + z_{31} - z_{21} - z_{32}}{z_{22} + z_{33} - 2z_{23}} I_1. \tag{12.99}$$

From rows 1 and 2 of (12.98), voltages V_1 and V_2 are obtained.

$$V_1 = (z_{11} - z_{12})I_1 + (z_{13} - z_{12})I_3. \tag{12.100}$$

$$V_2 = (z_{21} - z_{22})I_1 + (z_{23} - z_{22})I_3. \tag{12.101}$$

Subtracting (12.101) from (12.100) and substituting (12.99), the voltage difference $\Delta V = V_1 - V_2$ is obtained in (12.102):

$$\Delta V = \begin{bmatrix} (z_{11} - 2z_{12} + z_{22}) - \\ (z_{13} - z_{12} - z_{23} + z_{22}) \dfrac{z_{22} + z_{31} - z_{21} - z_{32}}{z_{22} + z_{33} - 2z_{23}} \end{bmatrix} I_1, \tag{12.102}$$

For balanced matrices:

$$\Delta V = 1.5\left(z_p - z_m\right)I_1, \tag{12.103}$$

where $(z_p - z_m)$ is equal to the positive sequence impedance of the nodal impedance matrix and therefore Eqs. (12.71) and (12.103) are equivalent.

12.6 Short Line Fault

In previous discussions of the TRV, terminal faults in the substation bus were adopted. Faults occurring in the transmission line are analyzed next. These faults are conventionally known as short line faults, since they are considered to be in the vicinity of the substation, thus retaining the high current levels of terminal faults.

12.6.1 Time Domain Analysis

Short line faults are assumed to occur in the transmission lines near a substation and therefore the current levels are equivalent to those at the substation bus. For more distant faults, for instance at points B and C in Figure 12.32, as the distance

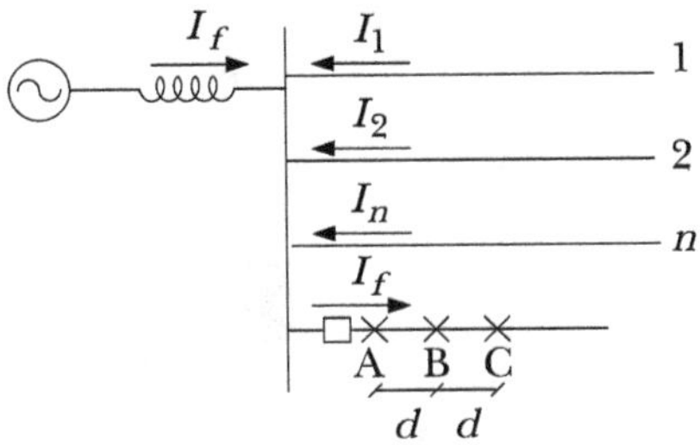

Figure 12.32 Short-line fault.

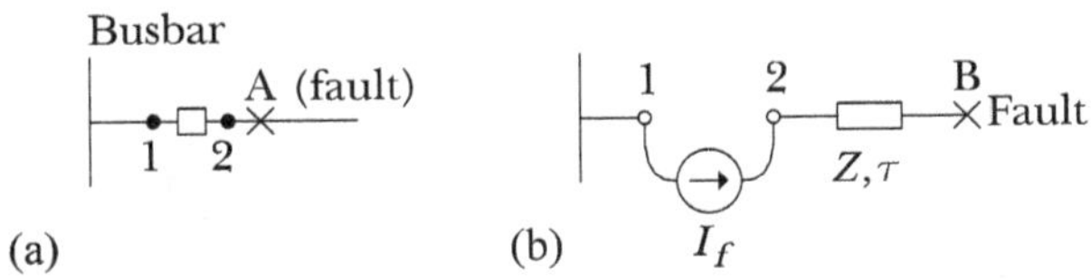

Figure 12.33 Fault on the line side of the circuit breaker. (a) At the circuit-breaker terminal (point A). (b) Fault on the transmission line near the circuit breaker (point B).

between the fault and the substation is small, the transmission line impedance introduces a slight reduction in the fault current, which thus remains high.

For phase to ground faults on the line side of the circuit breaker, as in point A in Figure 12.33a, when the circuit breaker is opened, the voltage rises only on side 1.

For a fault occurring at point B, located very near to the substation, the current amplitude is slightly reduced; however, a rise in the voltage on side 2 appears due to the surge impedance of the transmission line, Figure 12.33b.

An increase in the distance from the fault point slightly reduces the current and also the TRV on side 1, while the natural oscillation frequencies remain unchanged.

Prior to the opening of the pole, terminals 1 and 2 have the same phase to earth voltage, as there is a small impedance to the fault point. This voltage is relatively low and can be superimposed on both sides of the open pole, according to the current injection method, but it cancels out in the TRV calculation.

Considering the high frequencies that can occur in short line faults, a suitable model for side 2 should take into account the traveling waves along the transmission line segment. The same effect of current reduction with an increase in the fault distance is also observed, which leads to a slight reduction in the voltage. However, the negative reflections from the fault take more time to reach point 2 and consequently the voltage tends to rise.

An interpretation of the lattice diagram in Figure 12.34 can be found in Figure 12.35, showing the TRV components on side 2, for faults at distances d

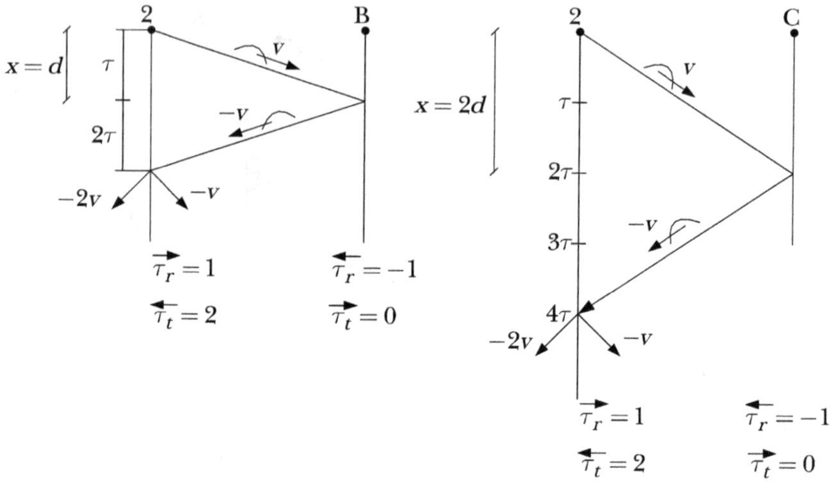

Figure 12.34 Short line fault at points B and C.

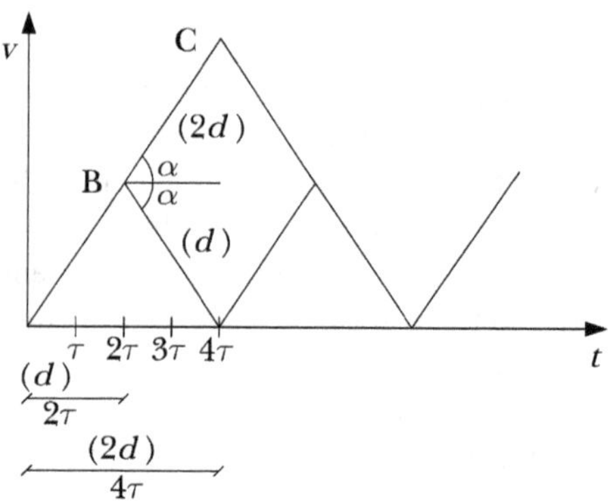

Figure 12.35 TRV component on side 2.

and 2*d* from the circuit breaker, at points B and C in Figure 12.32, respectively. A minor inaccuracy is introduced by assuming that the current level remains the same at the two distances, whereas at distance 2*d* the slope should be slightly less due to the current reduction.

Applying the current injection method, for a fault at point B, the slope of the rate of rise of the voltage on side 2 is designated as α:

$$i'(t) = kt, \tag{12.104}$$

$$v_2(t) = Zi'(t) = kZt, \tag{12.105}$$

thus:

$$\alpha = kZ. \tag{12.106}$$

At $t = 2\tau$, the wave reflected back from the short circuit has a slope of $-\alpha$ which, when it reaches point 2, increases to -2α. For the slope $-\alpha$, reflected at point 2, the cycle repeats but with a negative sign and so on, explaining the triangular shape of the waveform on side 2 of the circuit breaker. For a fault at point C, the time delay doubles, and the voltage may be higher on side 2. When distance d increases significantly, as for example, when a fault occurs in the middle of the line, the effect of the current reduction predominates relative to the delay due to the transit time.

The reflection coefficient $\Gamma_r = -1$ is related to grounded faults. For ungrounded faults, this coefficient needs to be calculated, taking into account the number of phases connected by the fault. However, the reflected wave should also be negative, and the waveforms would be similar.

Thus, the behavior of the TRV component on side 1 is similar to that previously studied in Sections 12.4 and 12.5. The shape of the TRV component on side 2 reflects the triangular waveform at higher frequencies than those of side 1. The superposition of the components on both sides is presented in Figure 12.36a, showing the triangular behavior of the TRV for short line faults, as described in [7].

In evaluating the damping on side 2, the waveform in Figure 12.36b is expected.

A relevant aspect of short line fault calculations is the rate of rise of the TRV. The RRRV can be calculated with the aid of the sequence networks in Figure 12.37 for the first pole to clear. The fault may be of any type. Since the fault is located at a distance d from the circuit breaker, until arrival of the first reflection, modeling of the transmission lines using their surge impedances is sufficient for an accurate calculation.

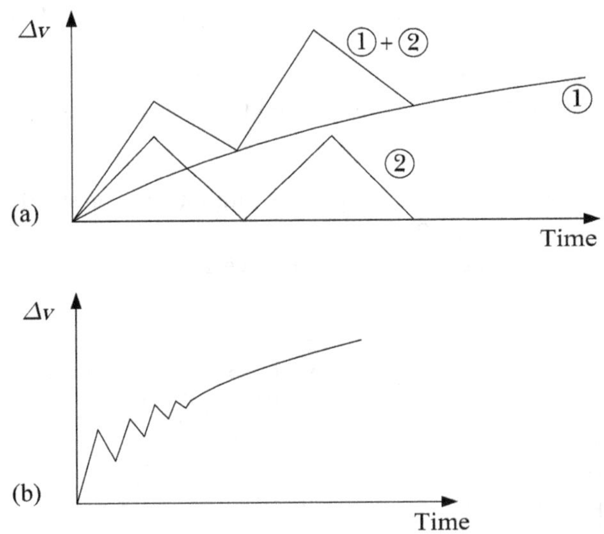

Figure 12.36 Short line fault. (a) Superposition of the components on sides 1 and 2. (b) Typical damped TRV waveform for short line fault.

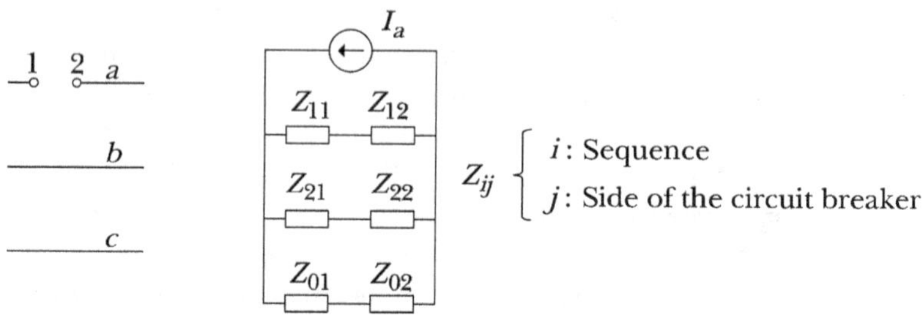

Figure 12.37 Sequence impedance for one phase open.

Figure 12.37 shows the surge impedances of the equivalent circuit representing the transmission lines on side 1 of the system and the surge impedance of the faulted line on side 2, neglecting the inductance and capacitance. The positive-sequence and zero-sequence impedance are presented.

The approach to simultaneous faults using symmetrical components is not simple [12, 13] but, even for unbalanced faults, the circuit in Figure 12.37 remains valid, because the time delay decouples the simultaneous faults corresponding to the opening of one phase of the circuit breaker and the unbalanced fault at some point along the transmission line.

Single-phase faults near the substation can cause the highest current level, when the zero-sequence impedance of the Thévenin equivalent is lower than the positive sequence impedance. Thus, the circuit in Figure 12.37 can be used to calculate the RRRV until the arrival of the reflected waves from the fault, although inclusion of the reflections needs to be carefully examined, particularly when the fault is unbalanced.

The simplified circuit used to calculate the RRRV for a short line fault is shown in Figure 12.38, taking into consideration the surge impedance of the transmission lines.

Z_{eqi}: equivalent surge impedance of the transmission lines without fault, on side 1, for positive and zero sequences ($0 \leq i \leq 2$).

$Z_{\ell i}$: surge impedances of the faulted line, on side 2, for positive and zero sequences ($0 \leq i \leq 2$).

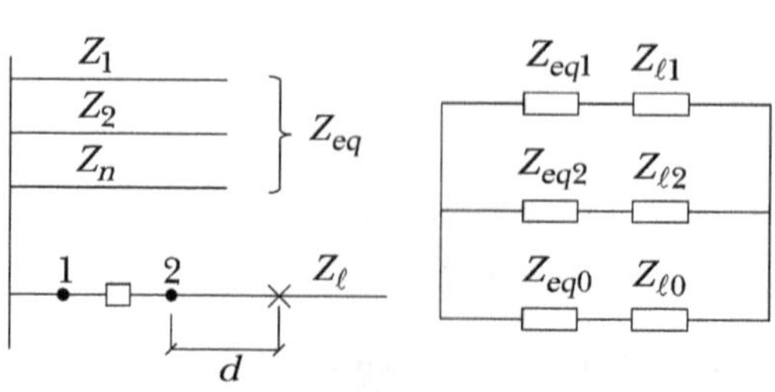

Figure 12.38 Simplified model with surge impedances.

Example 12.4 Calculate the RRRV for a three-phase short line fault of 20 kA. The transmission line constants are the same and the surge impedance of each line is $Z_1 = 250\ \Omega$ for positive sequence and $Z_0 = 642\ \Omega$ for zero sequence (Figure 12.39).

$$I = I_1 + I_2 + I_3,$$

$$I = 20\ \text{kA rms},$$

$$i_a(t) = i'(t) = 20 \times \sqrt{2} \times 377t.$$

The equivalent circuit shown in Figure 12.40 takes into account only the surge impedance of the transmission lines, neglecting the source inductance.

$$\Delta v_a = \Delta v_0 + \Delta v_1 + \Delta v_2 = 3\ \Delta v,$$

$$\Delta v_a = 3 \times \left(20 \times \sqrt{2} \times 377 \times 156,94\right)t, \qquad Z = 156.94\ \Omega,$$

$$RRRV = 3 \times \left(20 \times \sqrt{2} \times 377 \times 156.94\right),$$

$$RRRV = 5.02\ \text{kV}/\mu\text{s}.$$

The RRRV at the circuit breaker terminal calculated in this example considered the representation of both sides of the circuit breaker (sides 1 and 2), illustrating the use of a model that incorporates positive and zero-sequence components.

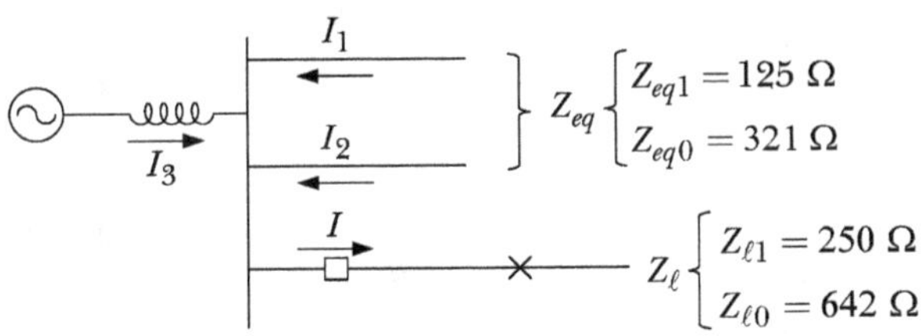

Figure 12.39 Short line fault.

Figure 12.40 Equivalent circuit and numerical values.

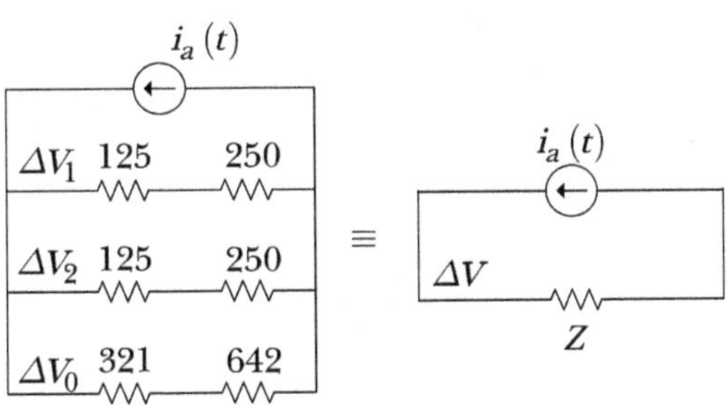

However, it is important to note that, standard [8] suggests a differentiated approach for side 1 (source side) and side 2 (line side), in this specific case of a short-line fault. The TRV on side 1 (source) is compared to the standard normalized TRV envelope (for terminal faults), while the RRRV on side 2 (line) is compared to a special limit applicable to short-line faults.

12.6.2 Analysis with Two-Port Network

According to reference [2], an elegant interpretation of the short line fault can be made in the frequency domain, using a single-phase circuit.

The voltages and currents at both transmission line terminals are related by the two-port network (Figure 12.41):

$$\begin{bmatrix} V_S \\ I_S \end{bmatrix} = \begin{bmatrix} A & B \\ C & D \end{bmatrix} \times \begin{bmatrix} V_r \\ I_r \end{bmatrix}, \tag{12.107}$$

where:

$$A = \cosh(\gamma\ell), B = Z\sinh(\gamma\ell), C = \frac{\sinh(\gamma\ell)}{Z}, D = \cosh(\gamma\ell).$$

Z: characteristic impedance (or surge impedance for a lossless transmission line).
γ: propagation constant.
ℓ: transmission line length.

For the short line fault, at the fault point, $V_r = 0$:

$$V_S = BI_r, \tag{12.108}$$

$$I_S = DI_r. \tag{12.109}$$

From (12.108) and (12.109):

$$V_S = I_S \frac{B}{D}, \tag{12.110}$$

and defining:

$$Z_S(s) = \frac{B}{D} = Z\tanh(\gamma\ell). \tag{12.111}$$

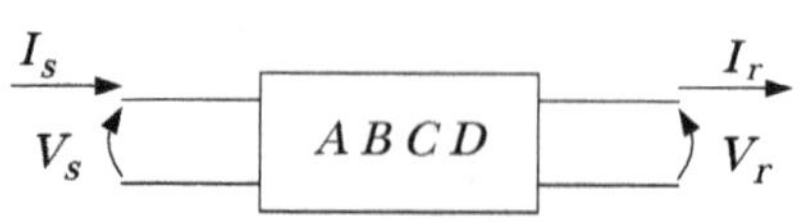

Figure 12.41 Voltages and currents in a two-port network.

Assuming a lossless transmission line, the surge impedance and the propagation function may be calculated:

$$Z = \sqrt{\frac{L}{C}} \text{ and } \gamma\ell = s\tau,$$

where τ is the time delay (Chapter 2),

$$V_s(s) = I_s(s)Z_s(s). \tag{12.112}$$

Substituting (12.111) into (12.112):

$$V_s(s) = I_s(s)Z\tanh(\tau s), \tag{12.113}$$

$$V_s(s) = I_s(s)Z\frac{e^{\tau s} - e^{-\tau s}}{e^{\tau s} + e^{-\tau s}}, \tag{12.114}$$

or:

$$V_s(s) = I_s(s)Z\frac{1 - e^{-2\tau s}}{1 + e^{-2\tau s}}. \tag{12.115}$$

Expanding (12.115) into a series:

$$V_s(s) = I_s(s)Z\left(1 - 2e^{-2\tau s} + 2e^{-4\tau s} - 2e^{-6\tau s} + \ldots\right). \tag{12.116}$$

Using the approximation of the current in (12.15), $I_s(s) = I'(s) = \frac{k}{s^2}$,

$$V_s(s) = kZ\left(\frac{1}{s^2} - 2\frac{e^{-2\tau s}}{s^2} + 2\frac{e^{-4\tau s}}{s^2} - 2\frac{e^{-6\tau s}}{s^2}\right). \tag{12.117}$$

Obtaining the inverse transform of (12.117):

$$v_s(t) = kZ[t - 2(t - 2\tau)H(t - 2\tau) + 2(t - 4\tau)H(t - 4\tau) - 2(t - 6\tau)H(t - 6\tau) + \ldots], \tag{12.118}$$

which is the equation for a triangular wave, according to the analysis of the lattice diagram shown in Section 12.6.1.

12.7 TRV in Systems with Series Capacitors

The application of series capacitors in power systems is analyzed in Chapter 11.

In TRV analysis, which consists of the application of faults within the substation and short line faults, the series capacitors do not introduce noticeable modifications into the procedures previously described. The protection schemes provide capacitor by-pass at high levels of fault current, such that when the series

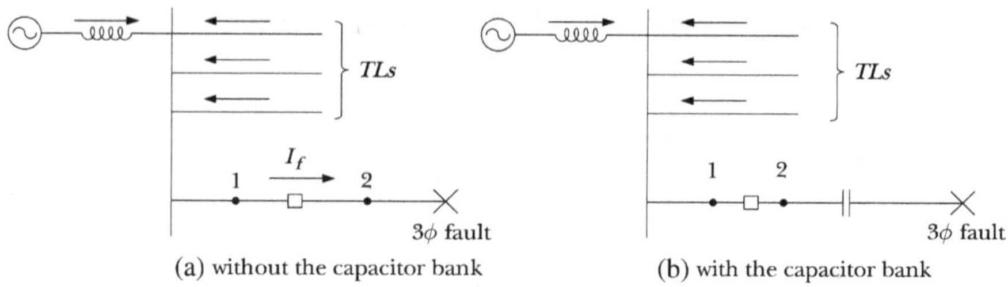

Figure 12.42 TRV in series compensated transmission lines. (a) Series capacitors bypassed. (b) Series-capacitors inserted.

capacitors are located close to the fault, the intervention of the protection scheme is very likely to occur, thus removing the capacitors from the circuit. Hence, for TRV, the capacitors would be excluded from the circuit (Figure 12.42a). As the fault moves along the transmission line with decreasing current levels, the capacitors are likely not bypassed, remaining in series with the surge impedance of the transmission lines (Figure 12.42b).

The proportionality between fault currents and TRV is evident by the method of current injection.

The source voltage on side 1 of the circuit breaker does not depend on the series capacitors, but the amplitude of the fault current does and may be higher in the case of the capacitors. The voltage on side 2, however, may depend on the series capacitors, as this voltage will rise until the arrival of the negative reflections from the fault.

In transmission lines without series capacitors, the TRV and fault current are drastically reduced by the line impedance; therefore, critical cases of faults occur close to the substation. As in the case of short line faults, two opposing effects can be observed with faults along the transmission line. With increasing distance, the voltage rises due to the delay in the return of negative reflections from the fault point, but on the other hand there is a reduction in the fault current level, which is limited by the increased impedance of the transmission line and leads to a reduction in the voltage on side 2.

In transmission lines with series capacitors, however, the reduction in fault current does not occur, because the negative reactance of the series capacitors cancels the positive reactance of the transmission line, keeping the short-circuit level close to that of the substation, and thus the critical TRV can be obtained from faults along the transmission line.

Figure 12.43 shows the TRV results for a three-phase short circuit along a 500 kV transmission line [15] with and without triggering of the series capacitor by-pass. In TRV analysis, it is convenient to determine the instances of fault application that do not trigger the series capacitor by-pass.

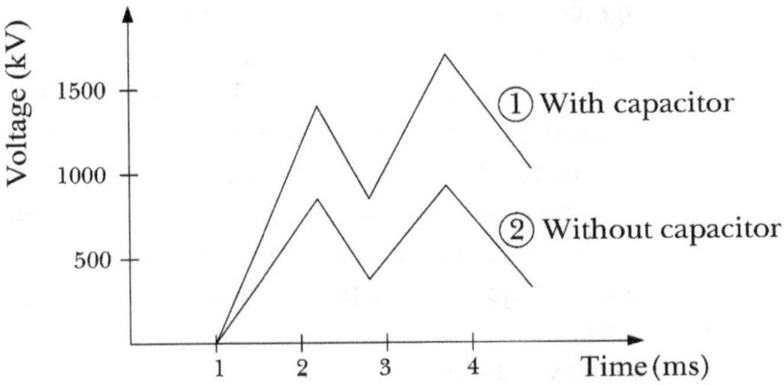

Figure 12.43 TRV with and without series capacitors.

12.8 Electric Arc

Simulation of an electric arc can be considered a special topic in the study of electromagnetic transients. Physical modelling of the arc and specific information for each circuit breaker design is required.

In gas circuit breakers, after mechanical separation of the contacts, the electric current flows in an ionized channel, with an approximately cylindrical shape and subjected to an electric field in the axial direction.

During the period of current flow, the ionized path is sustained by the electric power supplied by the source, which exchanges heat with the surrounding cylindrical surface or through cross-sections of the channel in the axial direction. After the current has passed through zero, there is a residual current through the heated channel, designated the post-arc current. Extinction of the arc, without reignition, depends on the mechanisms of deionization of the circuit breaker, which during the extinction period, must extract heat at a higher rate than it is introduced by the electric power. Electric current interruption is successful when the conductance of the arc is null after the zero crossing of the current, thereby avoiding arc reignition.

Since the phenomenon is very complex, the proposed models of the electric arc in circuit breakers can be approached under well-defined conditions, but generic models are not available in electromagnetic transient programs.

Reference [16] classifies the existing models into three families:

- Physical Models of the arc detailing the plasma behavior according to the laws of local thermal equilibrium, mass conservation, momentum, and energy, as well as heat transfer by convection and radiation.
- Black Box Models that use functions relating the voltage and current of the electric arc through mathematical formulations.
- Diagrams and formulas that relate parameters obtained from experiments.

Physical models, despite the high computation times, are very useful in circuit breaker design. The main physical principles can be found in references as in [1, 17], which provide descriptions starting from the initial black box models all the way to physical models. With the recent research in computational fluid dynamics (CFD), electric field intensity and flow field variables can be calculated, which allows to envisage the use of physical arc models to analyze the interrupting performance of gas circuit breakers. However, physical models are not yet available to efficiently couple the arc equations with the network models in electromagnetic transient programs.

The mechanisms of heat exchange by radiation, convection and conduction are also considered to be dependent on current amplitude.

A fundamental aspect of arc extinction has to do with thermal phenomena, near the zero crossing of the current. The thermal inertia of the plasma is very small and thus the arc near the current's zero crossing is not significantly affected by the period during which the sinusoidal current is high. On the other hand, higher current intensities imply higher derivatives at the zero crossing of the current, making extinction more critical.

After extinction, the electric power supplied to the arc is reduced during the period of post-arc current and tends to zero at later times. The possibility of dielectric reignition is linked to the high temperatures remaining in the arc channel, which reduce the dielectric strength.

As for arc-current extinction, the focus of the standards is to avoid the electric arc modelling, offering as an alternative the capability envelopes, which, through robust experimentation, offer limits to circuit breaker application.

Black-box models are derived from the pioneering work of Cassie and Mayr [18, 19]. Initially used in the development of circuit breakers, these models were gradually replaced by physical models developed in the 1970s.

The black-box models describe the behavior of the arc from the perspective of the voltage and current variables, without detailing the physical process involved and disregarding the radial variation in the properties of the arc. Despite their limitations, since they are described by relatively simple differential equations, these arc models remain an effective approach for the analysis of the dynamic interaction between the electric arc and the network in electromagnetic transient programs.

The electric arc shows a nonlinear dynamic behavior and is subject to random factors. The equations describing this nonlinear behavior have time constants that depend on the operating point of the arc but are assumed constant in some ranges of voltage and current variation.

The conductance is analyzed assuming a cylindrical shape for the arc channel and a given length (in per unit length). The following variables are defined:

- i: arc current.
- u: arc voltage.

- g: conductance.
- p: power supplied to the arc.
- p_r: arc power released by cooling.
- q: energy stored in the plasma.
- A: cross-sectional area of the arc.

With the previous definitions and in a very concise manner, the set of equations in (12.119) is presented, describing the energy content q and the arc conductance g (expressing g as a function of q), as well as their differential forms.

$$\begin{aligned} q &= \int (p - p_r)dt, \\ g &= f(q), \\ \frac{\partial g}{\partial q} &= f'(q) \cdot \text{and} \cdot \frac{dg}{dt} = \frac{\partial g}{\partial q} \cdot \frac{dq}{dt}, \end{aligned} \tag{12.119}$$

From this set of equations, the basic formulation for obtaining the Cassie and Mayr models was proposed initially in [20],

$$\frac{1}{g}\frac{dg}{dt} = \frac{f'(q)}{f(q)} = (p - p_r). \tag{12.120}$$

12.8.1 Cassie's Model

The premises adopted in Cassie's model are [18]:

- constant temperature in the radial direction
- arc cooling by convection in the axial direction.
- variable cross-sectional area of the arc, $A(t)$.

The following variables are defined at the equilibrium point:

g_0: conductance.
p_0: dissipated power.
q_0: energy content.

In a very concise way, with the definitions and assumptions above the function between g and q is derived as $g = (g_0/q_0)q$. Using the arc voltage as variable, obtained from the relation $p = gu^2$, and designating θ as the time constant in Cassie's model, which is given by $\theta = q_0/p_0$, the model equation is given in (12.121).

$$\frac{1}{g}\frac{dg}{dt} = \frac{1}{\theta}\left(\frac{u^2}{u_0^2} - 1\right). \tag{12.121}$$

For sinusoidal current, this model best fits the high amplitude period at temperatures above 8000 K.

12.8.2 Mayr's Model

The following hypotheses are adopted [19]:

- constant arc cross-sectional area, $A(t) = A$ (constant radius).
- heat loss by conduction in the radial direction.
- constant power removal from the arc, $p_r = p_0$.

The model proposed by Mayr assumes that heat loss occurs by conduction in the arc radial direction, with the arc conductance varying exponentially with the stored energy.

$$g = f(q) = k\,e^{q/q_0}. \tag{12.122}$$

Taking the derivative of (12.122):

$$f'(q) = \frac{1}{q_0} f(q). \tag{12.123}$$

Dividing (12.123) by (12.122) and applying (12.120):

$$\frac{1}{g}\frac{dg}{dt} = \frac{1}{q_0}(ui - p_0), \tag{12.124}$$

or,

$$\frac{1}{g}\frac{dg}{dt} = \frac{p_0}{q_0}\left(\frac{ui}{p_0} - 1\right). \tag{12.125}$$

The arc thermal constant is defined as $\theta = \frac{q_0}{p_0}$.

Replacing the thermal constant in (12.125) the final equation is:

$$\frac{1}{g}\frac{dg}{dt} = \frac{1}{\theta}\left(\frac{ui}{p_0} - 1\right). \tag{12.126}$$

The time constants in the models of Cassie and Mayr correspond to the thermal inertia of the arc, but the type of heat exchange with the external medium is different.

With the arc stabilized, it can be shown that $ui = p_0$, assuming the derivative of the conductance to be null in Eq. (12.126). The static model has a hyperbolic characteristic, suitable for the low-current range.

For small currents, the conductance decreases with reductions in current. If the product ui is smaller than p_0, the derivative of the conductance remains negative and favors extinction of the current. For the arc to reignite, a voltage increase is required such that the product ui is greater than p_0.

12.8.3 Stability of the Electric Arc for Small Currents

The linearized model of Mayr is usually employed in the analysis of the instability of electric arcs in the low-current range.

Reactor disconnection causes current chopping in circuit breakers, the effect of which is to cause an overvoltage at the reactor terminal (Chapter 1, Eq. (1.82)). Since current chopping occurs at low current levels, Mayr's model is used to analyze this transient in power systems. The interruption of small inductive currents was analyzed in depth in references [21, 22]. Regarding the current chopping, an alternative way to evaluate the electric arc stability using the Nyquist criterion is presented in [23].

12.9 Comments on Asymmetrical Faults and ITRV

To calculate the TRV a symmetrical fault current is assumed, according to the standards in [7, 8]

12.9.1 Asymmetrical Current

A concern that may raise doubts about an approach based on symmetrical current is that faults usually have a DC component making the sinusoidal current function asymmetrical (Figures 12.44 and 12.45).

The time constant for the damping of the DC component depends on the X/R ratio at the fault point, which is calculated by dividing the reactance of the Thévenin equivalent circuit for the system by the resistance. At low X/R ratios, the delays in the protection system and circuit breaker can significantly damp the DC component, leading to the interruption of an almost symmetrical current. However, at high X/R ratios, large asymmetries in the fault currents can be found. Therefore, although TRV calculations may be proposed on a symmetrical basis, asymmetries must be considered by manufacturers on the basis of the X/R ratio.

Figure 12.44 Asymmetrical fault current.

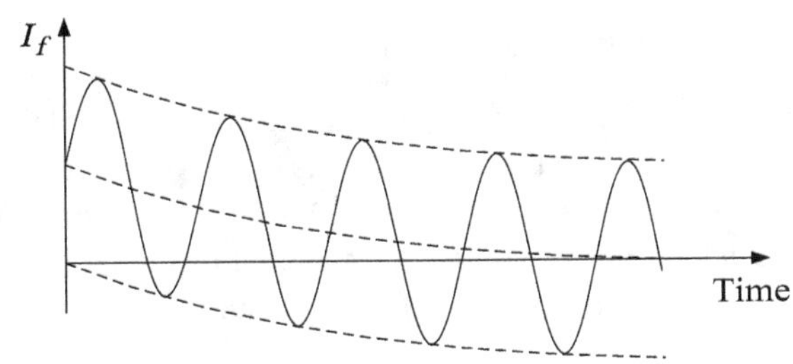

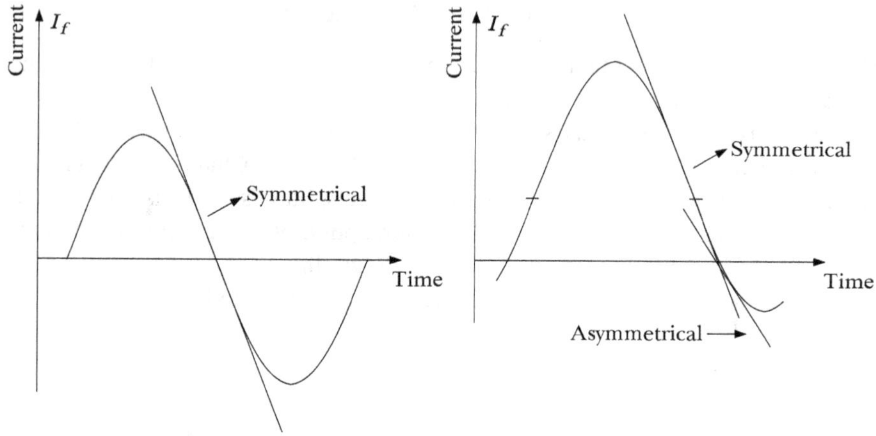

Figure 12.45 Derivative of symmetrical and asymmetrical fault currents.

When this ratio is not known, the standard [8] suggests that $X/R = 17$, which corresponds to a time constant of 45 ms, at 60 Hz.

A critical condition for the TRV is attained with symmetrical current, since the derivative of the sinusoidal function is at a maximum, at the zero crossing of the current, which can be observed by the application of the current injection method. For asymmetrical faults the current is shifted, and the zero-crossing point has lower derivative values. However, other factors are also considered in the standards to take the asymmetrical currents into account.

As a final remark, for high X/R ratios, as for example, when the fault is at the terminal of the step-up transformer of the generator and the circuit is oscillatory, very large asymmetries in the fault currents can be observed. There are exceptional cases in which the current does not pass through zero during some cycles, which represents a very severe condition for the circuit breaker [24]. Zero-delay analysis is not considered here, and it can only be accurately calculated using Park's model of the generator.

12.9.2 Initial Transient Recovery Voltage

In the first few microseconds after the extinction of the arc, the voltage waves propagate inside the substation. In this initial period, the voltage is known as the initial transient recovery voltage (ITRV). The frequencies that are present during this period are very high due to the small bus lengths in the substation. Therefore, a more detailed analysis of the substation equipment and substation layout is required.

12.10 Standards for Transient Recovery Voltage

The following comments refer to topics that are relevant to TRV analysis but are not intended to replace a thorough study of the standards that are essential for circuit breaker specification.

Technical standards address the application of circuit breakers to power systems, and some of them are mentioned here [7–11].

The required asymmetrical current breaking capability is obtained based on the nominal value of the symmetrical current and the DC component of the current.

To establishing reference parameters for TRV limits, capability envelopes are presented in [7, 8, 10, 11], providing guidelines for the thermal and dielectric strength of the circuit breakers.

For circuit breaker applications, the TRV levels should not surpass the capacity limits defined by the standards. In cases in which the TRV exceeds capability, the most common solution is to use a circuit breaker with a higher capability or try to make changes in the parameters of the network. When the RRRV is too high, a careful analysis of the stray capacitance in the substation is required. However, if the RRRV limits are exceeded, a circuit breaker with higher capability or installation of additional capacitors may be required, but new equipment could introduce major complications in the layout of a high-voltage substation.

Faults limited by the transformer are dealt with in [11], which provides the preferred TRV parameters for fast transient recovery voltage rise times.

IEC [8] proposes standardized parameters defined by straight line segments in order to establish limits, determined by 2 or 4 parameters, according to Figure 12.46. The standards seek to cover situations in which the TRV waveform reaches its maximum value with or without an intermediate peak.

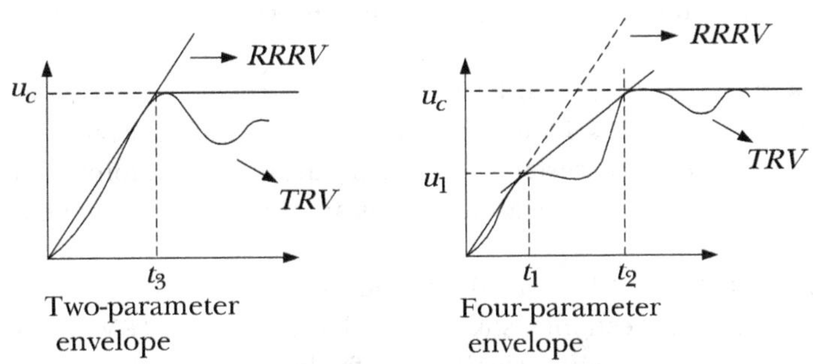

Figure 12.46 TRV parameters of IEC for terminal faults.

TRV limits with two parameters are defined basically by u_c, t_3 and the limits with four parameters are defined by u_1, t_1, u_c, t_2.

There are other auxiliary parameters for test purposes, such as u', t', t_d. The parameter for the time delay t_d is particularly important in order to characterize delays in the rise of the voltage.

The parameters refer to current levels corresponding to 10%, 30%, 60%, and 100% of the rated short-circuit current of the circuit breaker. The four-parameter TRV is associated with the first pole factor, which depends on the system grounding conditions and is defined by the steady-state voltage at the terminals of the first pole to clear. For example, in Figure 12.22b, the first pole factor is 1.5 with respect to the rated voltage of the system. This standard applies to first pole factor assumed to be 1.3 or 1.5. The factor of 1.3 is applied to grounded systems.

The following table exemplifies the values on a TRV curve with four parameters, assuming the first pole factor to be 1.5.

Fault current	u_1 (kV)	u_c (kV)	t_1 (μs)	t_2 (μs)	RRRV (kV/μs)
30%	$1.5U\sqrt{2/3}$	$1.5u_1$	$u_1/RRRV$	$7.5t_1$	5
60%	$1.5U\sqrt{2/3}$	$1.5u_1$	$u_1/RRRV$	$4.5t_1$	3
100%	$1.5U\sqrt{2/3}$	$1.4u_1$	$u_1/RRRV$	$3t_1$	2

Source: From IEC (cited as [8]).

where:

U: rated voltage of the circuit breaker in kV.
RRRV: rate of rise of the recovery voltage in kV/μs.

Envelopes related to different fault current levels from those specified in this table are also presented in the standards [7, 10].

For short-line faults, complementary parameters should be defined in the circuit breaker tests, making a distinction between the source and the line sides [8].

Previous editions of the IEEE standard assumed a TRV envelope defined as 1 − cos (one minus cosine) for oscillatory systems and circuit breakers with rated voltages up to 72.5 kV. For systems with some degree of damping and rated voltages above 121 kV, the TRV was defined as an exp - cosine, (exponential minus cosine) envelope.

For a long time, the IEEE and IEC standards had different formats for TRV envelopes. Recently, there has been intensive work to harmonize the TRV envelopes provided by these standards, which significantly facilitates the application of circuit breakers in power systems.

References

1 Browne, T. E. (1984). *Circuit interruption: Theory and techniques*. New York: Marcel Dekker.

2 Greenwood, A. (1991). *Electrical transients in power systems* (2nd ed.). New York: John Wiley & Sons.

3 Naef, O., Zimmerman, C. P., & Beehler, J. E. (1965). Proposed transient recovery voltage ratings for power circuit breakers. *IEEE Transactions on Power Apparatus and Systems*, 84(7), 580–608.

4 Hedman, D. E., & Lambert, S. R. (1976). Power circuit breaker transient recovery voltages. *IEEE Transactions on Power Apparatus and Systems*, 95(1), 197–207.

5 Wagner, C. L., & Smith, H. M. (1984). Analysis of transient recovery voltage (TRV) rating concepts. *IEEE Transactions on Power Apparatus and Systems*, PAS-103(11), 3353–3363.

6 Catenacci, G. (1979). Transient recovery voltage in extra-high voltage networks (362 kV and above). *Electra*(63), 37.

7 IEEE. (2019). *C37.011-2019: IEEE guide for the application of transient recovery voltage for AC high-voltage circuit breakers with rate maximum voltage above 1000 kV*. New York: IEEE.

8 IEC. (2008). *IEC-62271-100: High-voltage switchgear and control gear part 100 edition 2: Alternating current circuit breaker*. Geneva: IEC.

9 IEEE. (2018). *C37.04-2018: IEEE Standard for ratings and requirements for AC high-voltage circuit breaker with rated maximum voltage above 1000 kV*. New York: IEEE.

10 IEEE. (2009). *C37.06-2009: IEEE standard for AC high-voltage circuit breakers rated on a symmetrical current basis: Preferred ratings and related required capabilities for voltages above 1000 V*. New York: IEEE.

11 IEEE. (2017). *C37.06.1-2017: IEEE recommended practice for standard for preferred ratings for high-voltage (>1000 V) AC circuit breakers designated definite purpose for fast transient recovery voltages above rise times*. New York: IEEE.

12 Westinghouse Electric Corporation. (1964). *Electrical transmission and distribution reference book* (4th ed.). East Pittsburgh: Westinghouse Electric Corporation.

13 Anderson, P. M. (1973). *Analysis of faulted power systems*. Ames: Iowa State University Press.

14 Haleplian, J. P., & Zanetta, L. C. (2018). Cálculo da tensão de restabelecimento transitória no domínio da frequência. In M. S. G. Tsuzuki & F. Junqueira (Eds.), *Proceedings of the 13th IEEE International Conference on Industry Applications INDUSCON 2018, São Paulo* (pp. 1784–1789). Piscataway: IEEE.

15 Pinheiro, R. P., Fernandes, J. H., Tannuri, J. G., Patrão, J., Zanetta, L. C., Ferreira, C., & Ross, R. P. D. (1986). Sistema de 500 kV da Eletronorte com compensação série: Estudos de sobretensoes nos bancos de capacitores série e análise das sobretensões

no sistema. In *VIII Seminário Nacional de Produção e Transmissão de Energia Elétrica, São Paulo*. São Paulo: CESP.

16 CIGRE Working Group 13.01. (1993). Applications of black box modelling to circuit breakers. *Electra*(149), 41–71.

17 Ragaller, K. (1977). *Current interruption on high voltage network*. New York: Plenum Press.

18 Cassie, A. M. (1939). *A new theory of rupture and circuit severity*. Paris: Cigre. (Cigre Rp. 102).

19 Mayr, O. (1943). Beitrage zur theorie des statischemd denamschen lichtbogens. *Archiv fur Electrotechnik*, 37, 589–609.

20 Simon, H. T. (1905). Uber die dynamik der lichtbogenvorgänge und uber lichtbogenhystresis. *Physikalische Zeitschrift*, 6(10), 297.

21 CIGRE Working Group 13.02. (1985). Interruption of small inductive currents. *Electra* (101), 13–39.

22 Rizk, F. A. M. (1963). *Interruption of small inductive currents with air-blast circuit breakers*. Gothenburg. Transactions Chalmers University of Technology, n. 37.

23 Zanetta, L. C., & Bergerman, M. (1995). Analysis of the interruption of small inductive currents using the nyquist criterion. In *9th International Symposium on High Voltage Engineering, 9, Graz, 1995* (pp. 6323). Graz: Institute of High Voltage Engineering, Graz University of Technology.

24 Graber, W., & Gysel, T. (1980). The interruption of fault currents with delayed current zeros by HV circuit-breakers. *Brown Boveri* Review(4), 237–243.

13

Surge Arrester

13.1 Introduction

Electrical networks are subject to temporary overvoltages, switching surges, and lightning surges. The surge arrester has the primary function of limiting switching and lightning surges. Surge arresters are not designed to limit temporary overvoltages, although they can be considered for such a purpose in exceptional cases.

The insulation coordination of transmission lines and substations, as well as the economical design of the equipment are closely related to the application of surge arresters.

The simplest form of surge protection employs only spark gaps. When the voltage exceeds the insulation level of the gap, sparkover occurs, with current flowing to ground through an electric arc.

A protection device of this kind is very simple and economical, but it has the following drawbacks:

- With short-circuit current, the electric arc in the gap can only be extinguished after opening the circuit breakers, thereby interrupting the electrical power supply.
- Setting the trigger level of the spark gap is not simple because the spark voltage depends on atmospheric conditions and voltage polarity.

Although under certain well-defined conditions, spark gaps can be proposed because their cost is low compared to other alternatives, they are no longer used, and the most efficacious solution to overvoltage control employs surge arresters, without the need to open circuit breakers.

Initially, arresters consisted of spark gaps in series with nonlinear resistors. The triggering voltage became more accurate with sealed gaps, thus reducing the sensitivity to atmospheric conditions. In the most advanced designs, the arresters consisted of nonlinear SiC resistors in series with current limiting gaps, capable of

Understanding Electromagnetic Transients in Power Systems, First Edition. Luiz Cera Zanetta Jr.

Published 2025 by John Wiley & Sons, Inc.

developing an arc voltage that contributed to energy dissipation and facilitated current extinction.

In the late 1970s, a new generation of surge arresters was introduced, known as MOSA, in which a nonlinear resistor was composed of metal oxides. In view of the relatively high proportion of zinc oxide in the MOSA composition, the arresters are also referred to as ZnO arresters.

The exceptional nonlinear characteristic of these surge arresters allowed the suppression of the spark gap, making the equipment more robust and reliable while reducing investment and maintenance costs.

This chapter deals with the application of surge arresters in overvoltage control in power systems, emphasizing the basic equations without concern for the technological description or design. The examples presented are intended to show the analytical treatment of the surge arrester intervention, using simplified models, to exhibit the fundamental principles of the interaction between this equipment and the transmission line, since accurate calculations must be made with the available electromagnetic transient tools.

The standards for the application of surge arresters in power systems can be found in references [1–3].

13.2 Overvoltage Control – Basic Concepts

The methodology of including the nonlinearities in electrical circuits was introduced in Chapters 2 and 4. This chapter extends that approach for application in power systems and makes a useful simplification to the MOSA model.

13.2.1 Analysis Using the Thévenin Equivalent Circuit

Suppose that a surge arrester is connected at point P, in Figure 13.1a. The voltage that would exist at this point, with the surge arrester disconnected from the

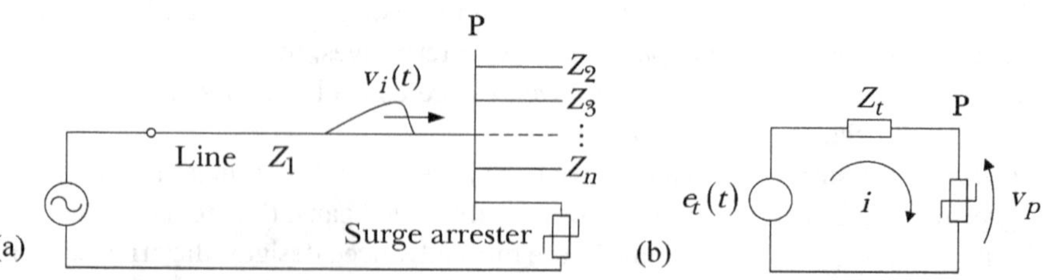

Figure 13.1 Analysis with the Thévenin equivalent. (a) Original system. (b) Thévenin equivalent.

network, is the Thévenin equivalent voltage. This voltage is also known as the prospective voltage.

The voltage, $e_t(t)$, with the surge arrester disconnected, is calculated from the product of the incident voltage, $v_i(t)$, times the transmission coefficient in P:

$$e_t(t) = \Gamma_t v_i(t). \tag{13.1}$$

The Thévenin equivalent impedance, as seen by the surge arrester, is:

$$Z_t = Z_1 \parallel Z_2 \parallel \dots \parallel Z_n. \tag{13.2}$$

After insertion of the surge arrester into the network, the current through the arrester is calculated from Figure 13.1b:

$$i(t) = \frac{e_t(t) - v_p}{Z_t}, \tag{13.3}$$

where the residual voltage at the surge arrester is given by a nonlinear function of the current:

$$v_p = f(i). \tag{13.4}$$

The residual voltage, also refered to as the impulse discharge voltage, is presented in more detail in Section 13.3.2. The graphical solution at a certain point in time is presented in Figure 13.2a and b for different residual voltages. The slope of the line with respect to the vertical axis is inversely proportional to Z_t which is the Thévenin equivalent impedance. That is, if Z_t increases, the slope will be smaller and the current through the nonlinear element will be limited to a greater extent. By contrast, if the impedance, Z_t, decreases, the slope will be higher and the amplitude of the current through the nonlinear element will also be higher:

It is also observed that the voltage developed across the nonlinear resistor behaves as a kind of counter-electromotive force to limit the current that flows through the arrester to earth.

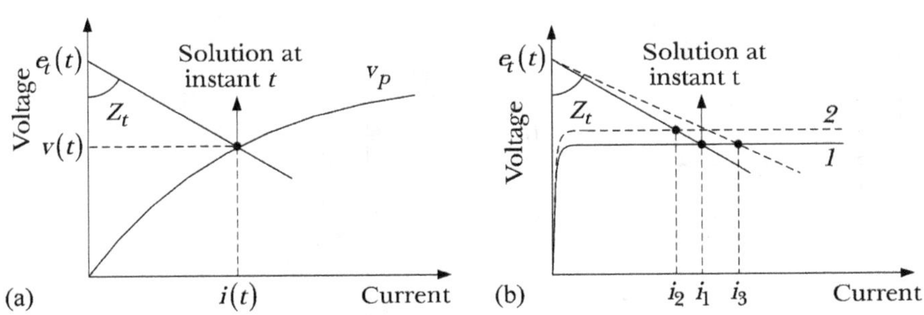

Figure 13.2 Graphical solution. (a) Thévenin equivalent and residual curve. (b) Analysis of different Thévenin equivalents and residual curves.

If the arrester has a higher residual voltage, there is a reduction in the value of the current (i_2). By contrast, at a lower residual voltage, the current is higher (i_1). With an increase in the number of lines at the substation, a reduction in the impedance of the Thévenin equivalent occurs accompanied by a rise in the current through the surge arrester (i_3), and at a higher impedance, the current is lower (i_2).

The numerical solution for the nonlinear system can be obtained iteratively by Newton Raphson's method, or by applying the Thévenin equivalent (Chapters 2 and 4).

In a simplified hand calculation, if there are several transmission lines connected to point P, the prospective voltage, or the Thévenin equivalent voltage in (13.1), can be used for a period corresponding to twice the time delay over the shortest transmission line without any correction (Figure 13.3). After this period, it is necessary to correct the prospective voltage because reflections of the waves created by insertion of the surge arrester into the system return to point P. In electromagnetic transient programs, these changes are implicitly considered in the method employed to find the solution, and the programs have different solution methods to deal with nonlinear elements, such as arresters.

The following variables are defined:

$e_t(t)$: voltage obtained without the surge arrester.
τ: travel time.
t_0: time of connection of the surge arrester to point P.
$t_0 < t < t_0 + 2\tau$: valid period of prospective voltage without corrections.

For $t > t_0 + 2\tau$, the prospective voltage needs to consider the reflections from the transmission-line terminals.

The effect of the arrester on the system can be interpreted as equivalent to a voltage opposite in polarity to that of the incident wave applied to the system.

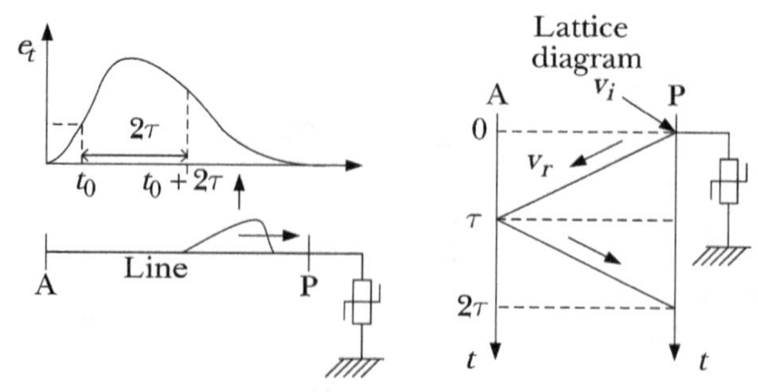

Figure 13.3 Wave reflected to point P.

Assuming an incident wave, $v_i(t)$, at point P of the system, which is higher than the voltage, $v_p(i)$, developed at the surge arrester, Eq. (13.5) may be written as:

$$v_i(t) + v_r(t) = v_p(i) \qquad (13.5)$$

Thus, the reflected wave is given by:

$$v_r(t) = v_p(i) - v_i(t). \qquad (13.6)$$

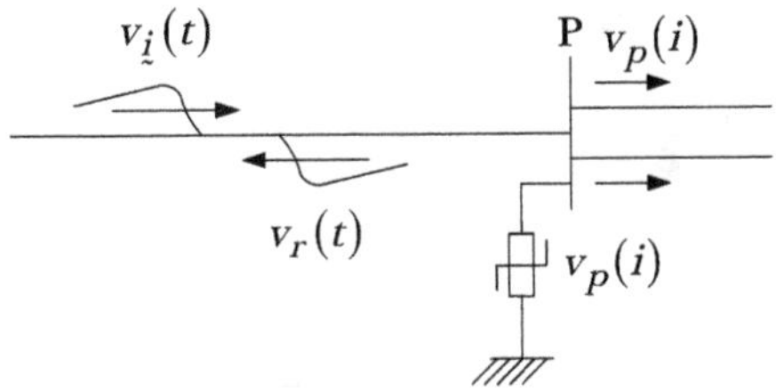

Figure 13.4 Transmitted and reflected waves after surge arrester intervention.

The transmitted wave is reduced, since it is equal to the voltage at the surge arrester $v_p(i)$ and the reflected wave is also opposite in polarity to that of the incident wave; therefore, it behaves as if a wave of opposite polarity were applied to the system after the arrester intervention (Figure 13.4) (see Example 13.6).

13.2.2 Three-Phase Transmission Line

In three-phase transmission lines, coupling between the phases should be considered in the solution of the nonlinear equations, according to Chapter 3 and Figure 13.5, but if the arrester is inserted into the network in one phase only, the solution is quite simple.

Consider the Thévenin equivalent of a balanced three-phase transmission line at the time of arrester insertion into phase a. Using the phase and mutual surge impedances, Eq. (13.7) is written in terms of phase components:

$$\begin{bmatrix} e_{ta} \\ e_{tb} \\ e_{tc} \end{bmatrix} - \begin{bmatrix} Z & Z_m & Z_m \\ Z_m & Z & Z_m \\ Z_m & Z_m & Z \end{bmatrix} \times \begin{bmatrix} i_a \\ i_b \\ i_c \end{bmatrix} = \begin{bmatrix} v_a = v_p \\ v_b \\ v_c \end{bmatrix}, \qquad (13.7)$$

where:

e_{ta}, e_{tb}, e_{tc}: Thévenin equivalent voltages with surge arresters disconnected.
v_a, v_b, v_c: voltages with surge arresters connected to the circuit.
i_a, i_b, i_c: currents in phases a, b, and c.

With a single-phase surge-arrester intervention in phase a, currents i_b and i_c are assumed null, thus:

$$e_{ta} - Zi_a = v_p, \qquad (13.8)$$

$$v_p = f(i_a). \qquad (13.9)$$

The voltage, v_p, and the current, i_a, can be calculated by a numerical method.

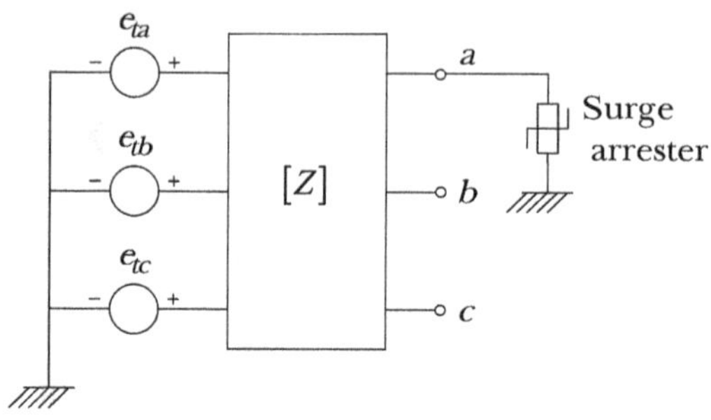

Figure 13.5 Three-phase Thévenin equivalent.

For balanced three-phase systems, Eqs. (13.10) and (13.11), may be used:

$$Z_0 = Z + 2Z_m, \tag{13.10}$$

$$Z_1 = Z - Z_m, \tag{13.11}$$

Resulting in:

$$Z_m = \frac{Z_0 - Z_1}{3}, \tag{13.12}$$

$$Z = \frac{Z_0 + 2Z_1}{3}, \tag{13.13}$$

and the Thévenin equivalent impedance for phase a can be obtained from the $[Z]$ matrix or calculated using the positive and zero-sequence surge impedances, as in (13.13).

13.3 Types and Characteristics of Surge Arresters

13.3.1 Silicon–Carbide Surge Arrester

This arrester has fallen out of common use in modern times. Its inclusion here serves the purpose of illustrating the evolution of technology and the function of the spark gap within it. The arrester has a spark gap connected in series with a nonlinear resistor made of silicon carbide, as depicted in Figure 13.6a.

The sealed design of the spark gap serves to minimize the impact of atmospheric conditions, resulting in a narrower range of sparkover levels. Nonetheless, it is crucial to consider both the maximum and minimum triggering levels, with a particular emphasis on the maximum level, especially in the context of insulation coordination.

Before triggering, the spark gap essentially behaves like an open circuit. After triggering, it behaves like a short circuit, leading to resistor insertion. However, there are spark gaps that develop a controlled arc voltage. This controlled arc

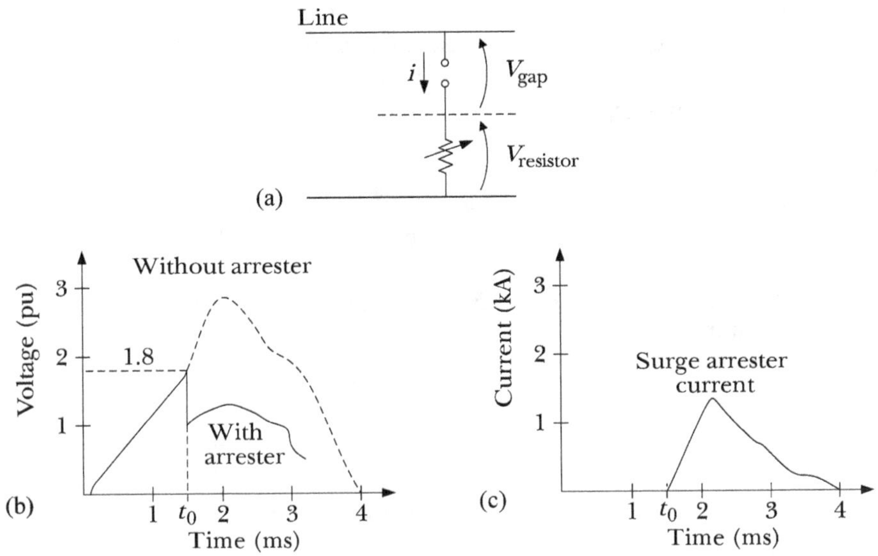

Figure 13.6 Intervention of SiC arrester in switching surges. (a) Surge arrester model. (b) Voltage with and without surge arrester. (c) Current waveform.

voltage aids in limiting the current flowing through the arrester and dissipating a portion of the total energy. Consequently, it assists in reducing the energy dissipated by the nonlinear resistor. Such spark gaps are referred to as active spark gaps.

In situations where the transmission system experiences temporary overvoltage, these spark gaps prove useful in preventing unnecessary operation of the surge arrester.

The typical performance of a silicon carbide surge arrester can be seen in Figure 13.6b and c in the case of a switching overvoltage.

The discontinuity in the voltage at the beginning of arrester insertion is due to disruption of the spark gap, which instantaneously inserts a nonlinear resistor into the system.

Surge arresters made of silicon carbide were replaced by metal oxide surge arresters and for this reason their analysis is not detailed in this text.

13.3.2 Metal Oxide Surge Arrester (MOSA)

Since the end of the 1970s, a remarkable growth in the utilization of MOSA in high-voltage systems has occurred, replacing the use of silicon carbide surge arresters. Although metal oxide surge arresters can employ spark gaps in special applications, only the more common cases of surge arrester use without a spark gap are addressed below. In this case, a smoother insertion of the surge arrester into the grid, avoiding the discontinuity observed in Figure 13.6b.

Some concepts and definitions are presented next.

- Residual Voltage

  The residual voltage is the voltage at the terminals of an arrester due to a discharge current. This voltage depends on both the magnitude and the waveform of the current.

  The MOSA shows residual voltage behavior as a function of the current, according to Figure 13.7a.

  Figure 13.7b, on a logarithmic scale, shows a pronounced difference in the voltages developed by metal oxide and silicon carbide disks.

  The disks are piled one on top of the other, forming a column that makes operation at higher voltages possible.

  Impulse discharge voltages [3], or residual voltages [2], are furnished by manufacturers for lightning current levels from 1500 to 40 000 A. Switching current levels are given as 500 and 2000 A. The residual voltage depends on the current waveform applied during the test, both for MOSA and SiC. For lightning impulse and switching impulse, the current front-time is 8 and 45 μs, respectively.

  The voltage developed by a MOSA describes an almost flat curve when compared to the highly current-dependent voltage of the SiC arrester.

  Reducing the current front-time, the residual voltage across the surge arrester increases, and thus lightning impulses produce residual voltages higher than the switching impulses.
- Arrester Basic Characterization

  Distribution class arresters have the primary function of limiting the lightning surges and should preferably be used in systems with a rated voltage below 52 kV.

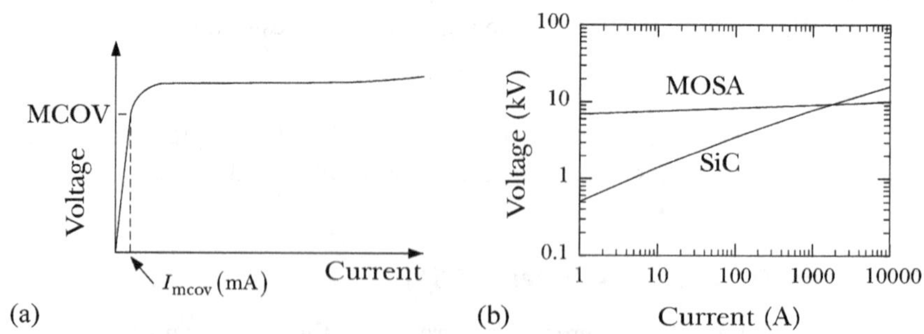

Figure 13.7 Residual voltage. (a) MOSA residual voltage and MCOV. (b) MOSA and SIC residual voltages.

Station class arresters are typically, but not exclusively, intended to be applied in systems with rated voltages above 72.5 kV for the purpose of limiting lighting and switching surges.

This class is defined by a specific set of tests, meant to be applied to the arrester [2, 3], which will not be detailed here.

The nominal discharge current I_n is the amplitude of the lightning current impulse with a wave shape of 8 × 20 μs. This current is applied during the Operating Duty Test and is subdivided into levels of 2.5–20 kA for station and distribution arresters [2]. Similarly, tables 3 and 4 in [3] assign the impulse classifying current a range of 5–20 kA.

In addition to the standard discharge current, arrester performance needs to consider electrically transferred charge and dissipated energy, as defined next.

- The thermal charge transfer rating Q_{th} and the thermal energy rating W_{th}, are used to specify distribution arresters and station arresters, respectively. The tests define the charge transfer capability and the energy dissipation capability, followed by the thermal recovery of the arrester.
- The repetitive charge transfer rating Q_{rs}, is used both for distribution and station arresters. This is the charge per run that the arrester must withstand in a series of interspersed tests with sample cooling time periods.

- Maximum System Voltage

The maximum system voltage, U_s, is the maximum phase-to-phase voltage during normal service.

- Arrester Rated Voltage

The arrester-rated voltage, U_r, is a reference for tests and permissible operative stresses. Additionally, the arrester must withstand the rated voltage U_r for 10 seconds in [2]. In [3], the rating has gradually been replaced by the MCOV.

The rating selection is closely related to MCOV and TOV, as described below.

- Operating Voltage

For a gapless arrester, very small currents flow through the arrester in steady state, in the range of mA. This current is negligible in a power system but may present heating and thermal stability problems for the MOSA.

To prevent this current from causing arrester aging, a maximum continuous operating voltage is defined, also known by the acronym MCOV, indicated in Figure 13.7a. In [2], this voltage is designated as the continuous operating voltage, U_c. As a first guess, the U_c level is about 80% of the rated voltage of the arrester.

Thus, if the maximum operating voltage of the system is known, the minimum rating for the arrester can be defined.

The influence of harmonics needs to be considered following the standard procedures. In systems with a slow fault clearing time, a temporary overvoltage may be assumed to be equal to the MCOV, mainly in ungrounded systems.

- Temporary Overvoltages – TOV
  Temporary overvoltages exceed the MCOV level of the surge arrester, increasing the current and the temperature. Phase-to-ground faults and load rejection can be mentioned as causes of temporary overvoltages. If, initially, the temporary overvoltage is preceded by an energy stress due to a switching surge, the effect can be further aggravated, raising the internal temperature even more. This initial energy stress is also designated as prior duty.

  The surge arresters have permissible temporary overvoltage levels associated with the corresponding elapsed times, which may or may not take the energy of a previous switching surge into account. If these values are exceeded, the thermal runaway of the surge arrester can occur. Some typical permissible values of temporary overvoltages and their corresponding durations are presented per unit of the arrester voltage rating U_r.

$1.33U_r$	1 s
$1.26U_r$	10 s
$1.21U_r$	15 s
$1.11U_r$	10 000 s

The actual values can be extracted from arrester datasheets. Note that ref. [3] changed the voltage basis for TOV from rated voltage (designated as V_r) to MCOV.

- Energy Capability
  Evaluation of the energy in station arresters or electrical charge in distribution arresters is very similar and thus, this section covers the dissipated energy in station arresters.

  Each disk has a cylindrical shape whose volume, given by its height and diameter, defines its capacity to dissipate energy. Increasing the volume increases the disk's energy rating. The energy that the surge arrester can dissipate increases with the total volume of the blocks. The rating increases with the number of blocks in series and thereby the energy capability of the arrester. With the same number of blocks in series, the energy capability rises if disks with larger diameters or more disk columns in parallel are used.

  For similar block dimensions, the MOSA has a higher energy rating than the SiC.

  The thermal energy ratings W_{th}, related to station arresters, may be found in Table 5 in [3]. The energy class is defined from A-N, corresponding to a range from 3 to 30 kJ/kV-MCOV minimum thermal energy withstand rating. Note that the base voltage is the MCOV [3]. Similar values are proposed in [2], but the voltage basis is U_r.

In specific cases, an increase in the energy capability can be achieved through the installation of columns in parallel, but the division of current among the columns is not uniform since the columns are not identical. If more columns are placed in parallel, the most stressed one reaches a limit while the others not, but the parallel group of columns is limited by the most stressed column.

The dissipated energy is given by the expression:

$$E_n = \int_{t_i}^{t_f} v(t)i(t)\, dt, \tag{13.14}$$

where t_i and t_f are the start and end times of MOSA discharge.

The energy dissipated during lightning and switching surges must be investigated. Lightning surges have a short duration but high amplitude and fast-front waves. Switching surges have lower amplitudes and slow-front waves, but their durations are longer.

Evaluation of the energy requirements of surge arresters is important mainly in certain events such as energization and autoreclosing of long transmission lines, restrikes upon disconnection of shunt capacitors or cables, lightning strokes near the surge arresters, etc.

13.4 Surge Arrester Application

The choice of arrester involves an organized series of steps, taking into account the characteristics of the system and overvoltages.

The following voltage levels should be considered:

- Maximum system voltage.
- Temporary overvoltages.
- Switching surges.
- Lightning surges.
- Magnitudes and probabilities of switching surges and lightning surges.

The general procedure for surge arrester selection involves the following topics:

- MCOV (U_c).
- Arrester rating based on the MCOV and TOV.
- Rated discharge current.
- Charge and energy capability.
- Switching and lightning surge protection levels.
- Arrester location.

The following are some basic concepts guiding the selection of surge arresters and their implications in insulation coordination, complementing the discussion in Chapter 14.

13.4.1 Rating Selection

The choice of the rating, U_r, considers the maximum continuous operating voltage and temporary overvoltages, which may lead to thermal stress and accelarate the aging of the equipment.

For ungrounded systems, during phase-to-ground faults, the voltage of the healthy phases reaches the line-to-line voltage. Without earth fault protection, this voltage can persist for a long period and therefore the minimum continuous operating voltage of the arrester should be equal to or greater than the line-to-line voltage of the system. High impedance grounded systems may exhibit similar temporary overvoltages in ungrounded systems, but if earth fault protection is available, the exposure time of the arrester to the overvoltage is significantly lower; therefore, the TOV-time curve should be considered to define the arrester rating.

EHV transmission systems can be considered grounded with lower overvoltage levels, but single-phase faults at the end of long transmission lines can increase the overvoltage. Although faster protection trips are expected, backup protection times should be assumed for overvoltage duration. The rating is selected using the same approach, based on maximum continuous operating voltage with a complementary check on TOV. The following example addresses the choice of arrester rating.

Example 13.1 Select the arrester rating for an ungrounded 69 kV system, which operates in the range of ±5% of the rated system voltage, given that the temporary phase-to-ground overvoltage reaches 1.73 pu with an elapsed time of 10 seconds.

Solution

The maximum operating voltage is 72.5 kV and the MCOV is assumed to be 80% of the arrester rating. Thus, the following relation may be written:

$$0.8V_n > \frac{72.5}{\sqrt{3}},$$

thus, $U_r > 52.3$ kV rms.

For a temporary overvoltage, the phase-to-ground voltage is given by:

$$1.73 \times \frac{72.5}{\sqrt{3}} = 72.5 \text{ kV}.$$

The permissible overvoltage level for 10 seconds, exemplified in Section 13.3.2, is assumed to be $1.26U_r$, thus:

$$U_r > \frac{72.5}{1.26}\ \text{kV, or } U_r > 57.54\ \text{kV}.$$

Note that this information must be extracted from the arrester catalog and that more recently, the basis for per unit values may be the MCOV, as in [3].

In this example, the rating is determined by the temporary overvoltages and it should be higher than 58 kV.

For an isolated system, if the temporary overvoltage lasts for a long time, a higher rating must be adopted and the MCOV voltage becomes 72.5 kV, with the rating rising to:

$$U_r > \frac{72.5}{0.8}\ \text{kV, or } U_r > 90.63\ \text{kV}.$$

On the other hand, if the system is grounded and the temporary overvoltages are less than 1.4 pu, the rating is likely to be governed by the MCOV.

13.4.2 Protection Levels and Insulation Coordination

In the case of surge arresters with gaps, the protection level is defined by the voltage that causes gap sparkover with the insertion of the nonlinear element into the system and changes in the current through the arrester from zero to effective amplitudes. The surge arrester intervention is well defined by a discontinuity and a sudden reduction in overvoltage at the point where the surge arrester is connected to the system.

In the case of gapless arresters, insertion into the grid is smoother, and it can be assumed that the intervention effectively occurs when the current increases substantially, from approximately zero to more significant values, in the range of hundreds of Amperes or more. The protection level of the gapless arrester is defined by the residual voltages developed during the current flow in the arrester.

Surge arresters, with or without gaps, have different protection levels for overvoltages with slow or fast-front waves, that is, for switching surges or lightning surges. Note that the gap sparkover voltage, for arrester with gap, and the residual voltage, for gapless arrester, increase as the surge wavefront increases. These voltages are also designated as discharge voltages in [3].

- Temporary Overvoltages
  In grounded systems, gapless metal oxide arresters perform well against temporary overvoltages preceded by switching surges. However, in ungrounded systems with high temporary overvoltages, arresters may need higher ratings.

- Lightning Surge Protection Level
  Substations, despite being protected by rods and shield wires, can be affected by lightning surges originating in transmission lines. The effect of lightning surges will be more pronounced in the case of lines without shield wires, implying greater analysis in the application of surge arresters. The surge arrester can offer protection for various pieces of equipment; however, inside the substation, the main focus is on the transformer. Circuit-breaker and disconnector operations that change the configuration of the substation must be considered in positioning surge arresters, so that the equipment remains protected under all conditions.

  The protection level of a surge arrester is designated U_{pl} in [2, 4] and is obtained from the manufacturer's datasheet for arresters with or without gaps. The lightning surge protection level for a gapless surge arrester is the residual voltage at high currents and fast-front waves. Suggested reference values for station arresters are the residual voltages at current amplitudes of 10 and 20 kA, 8×20 μs, whereas 5 and 10 kA may be assumed for distribution arresters.

  Surge arrester application for transformer protection against lightning surges implicitly limits the maximum level of a representative overvoltage for the transformer. The coordination voltage, considering the distance between the surge arrester and the transformer can also be obtained, as mentioned in Chapter 14.

  Lightning surges are associated with traveling waves with high rates of rise that may cause overvoltages in the protected equipment. These overvoltages can be higher than the residual voltage in the surge arrester depending on the distance between arrester and equipment. The overvoltages in the substation equipment are calculated using electromagnetic transient programs. In the absence of these calculations, the formulas presented in [1, 2, 5] can be adopted to obtain the coordination withstand voltage against lightning surges.
- Switching Surge Protection Level
  The protection level of the surge arrester (with or without gap) used for slow-front waves is designated as U_{ps} in [4] and can also be obtained from manufacturers' catalogs.

  For slow wavefronts, the effect of traveling waves is not significant within the substation. Thus, it can be assumed that the equipment and surge arresters are at the same voltage level. The effect of traveling waves on switching surges is only observable over longer distances as in the case of transmission lines. Therefore, the representative voltage for slow-front waves is quite well established for equipment in substations protected by surge arresters (Chapter 14).

  Due to the remarkable nonlinearity of the MOSA, the residual voltage of the surge arrester is practically constant within the expected range of currents from 0.5 to 2 kA, as in the case of switching surges. The protection level, U_{ps}, can be assumed to be equal to the residual voltage at 2 kA for gapless arresters, and it can be estimated to be slightly more than twice the arrester rating.

Independently of the grounding type, if surge arresters are installed phase-to-earth, the phase-to-phase overvoltages will be limited to twice the arrester's phase-to-earth protection level.

13.5 Performance of Surge Arresters

As previously mentioned, an accurate analysis of the interaction between the electrical network and surge arresters can be performed using electromagnetic transient programs. The application of nonlinear elements to the reduction of overvoltages may be found in several chapters in this book. Additional examples are presented in this section to complement some of the fundamental principles governing the performance of arresters in electrical networks.

13.5.1 Simplified Model of the Surge Arrester

An appropriate model of the surge arrester is important for each phenomenon analyzed, but in this item, only MOSA modeling is considered. In switching surge studies, the residual voltage described in Section 13.3.2 is sufficient to properly model the surge arrester. In analyzing overvoltages with fast-front waves, such as those due to lightning surges, it is necessary to consider the voltage drop across the arrester leads, since a few meters of connection lead could have a significant influence on the voltage at the connection point (see Example 13.5).

The grounding resistance of the surge arresters should also be considered and would cause the voltage at the connection point to rise even higher.

A substantial contribution of the lead to the arrester connection may exist in distribution systems. In transmission systems, the influence of the lead is relatively smaller but cannot be neglected, much like the resistance of the grounding.

At very high frequencies, the surge arrester models become more sophisticated, as in the model in reference [6].

For moderate frequencies, the almost flat residual voltage of the MOSA, despite pronounced variations in the current, helps to simplify the model. It is assumed that the residual curve may be replaced by a DC voltage source, as long as the current does not change sign. Curiously, in early TNAs, arresters were modeled roughly as DC sources. The residual voltage variations can be modeled by means of a small resistance in series with the voltage source.

13.5.2 Arrester Energy Dissipation

- Energy Transfer in an Elementary Circuit

  In Chapter 10, the performance of the arrester that protects a shunt capacitor bank was analyzed assuming a high-frequency overvoltage caused by the restrike of the capacitor bank's breaker. The purpose of the following analysis is a little different and aims to analyze the transfer of energy from the electrical system to the arrester. Initially, the very simple network will be assumed, composed of an RLC circuit, in order to clarify the behavior of the arrester in more complex networks. The resistance can represent losses in the network and grounding or even simulate the slope of the residual voltage of the arrester as a function of the current, also designated as apparent resistance. The inductance is associated with transformers and inductance and capacitor may represent transmission lines that can store energy to be partially transferred to the arrester during an overvoltage.

  Consider the circuit in Figure 13.8a, from the moment the arrester is effectively inserted into the network.

  The following equation may be written, with v_0 representing the initial voltage across the capacitor and v_p denoting the residual voltage across the arrester. The simplified model of the arrester is depicted in Figure 13.8b, where the residual voltage is represented by a DC source and an apparent resistance, R_a, is introduced to account for the slight slope of the residual voltage curve. This resistance

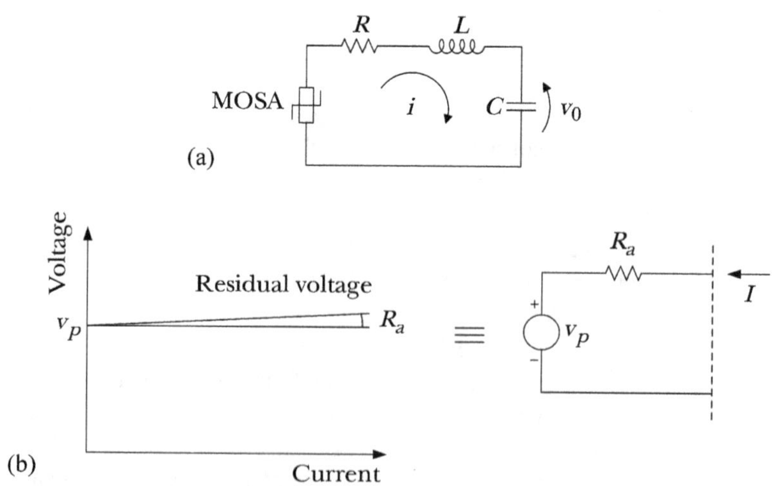

Figure 13.8 Transfer of energy form the capacitor to the arrester. (a) Simplified circuit. (b) MOSA simplified model.

can be assumed to be zero if the residual voltage curve is considered flat. Alternatively, it may be included in the resistance of the RLC circuit.

$$v_p = Ri + L\frac{di}{dt} + \frac{1}{C}\int i\,dt. \tag{13.15}$$

Applying the Laplace transform and assuming $i_0 = 0$:

$$\frac{v_p - v_0}{s} = \left(R + sL + \frac{1}{sC}\right)I(s), \tag{13.16}$$

$$I(s) = \frac{v_p - v_0}{L}\frac{1}{s^2 + \frac{R}{L}s + \frac{1}{LC}}. \tag{13.17}$$

The inverse transform of (13.17), assuming an oscillatory circuit, can be written as (Chapter 1):

$$i(t) = \frac{v_p - v_0}{L}2Ke^{-\frac{R}{2L}t}\cos(\omega_d t + \phi), \tag{13.18}$$

where:

$$\omega_0^2 = \frac{1}{LC}, \tag{13.19}$$

$$\omega_d = \sqrt{\frac{1}{LC} - \left(\frac{R}{2L}\right)^2}. \tag{13.20}$$

$K\angle\phi$: residue of the partial fraction expansion.

Setting:

$$f(t) = \frac{2K}{L}e^{-\frac{R}{2L}t}\cos(\omega_d t + \phi). \tag{13.21}$$

The energy dissipated by the surge arrester is:

$$E_n = \int_0^T -v_p\left(v_p - v_0\right)f(t)dt. \tag{13.22}$$

In Eq. (13.22), the sign of the current is flipped in accordance with the current entering in the surge arrester, resulting in:

$$E_n = v_p\left(v_0 - v_p\right)\int_0^T f(t)dt. \tag{13.23}$$

Setting:

$$\eta(T) = \int_0^T f(t)dt, \tag{13.24}$$

where T is the period of analysis, with a duration lasting from the beginning of the first discharge to the first zero crossing of the current, and $\eta(T)$ is a factor that just depends on network parameters, such as inductance, capacitance, resistance, etc. With this simplified model, only the energy of the first discharge is calculated, and this is considered the most significant. If there are subsequent discharges, the calculation remains similar.

$$E_n = v_p\left(v_0 - v_p\right)\eta(T). \tag{13.25}$$

The integral in (13.24) can easily be obtained by integrating by parts, but that is not the main purpose here. Assuming an overvoltage with $v_0 > v_p$, a positive energy is obtained from this expression. It can also be shown that energy as a function of arrester voltage is shaped like a concave down parabola (Figure.13.9).

When $v_p = 0$, the surge arrester does not dissipate any energy. It also happens when $v_p > v_0$, since the shunt capacitors cannot discharge into the surge arrester, which explains the two points of zero energy.

- Arrester and Transmission Lines

  The equation is now extended to the case of a transmission line. To facilitate analysis, the transmission line is divided into sections modeled as π – circuits, analogous to simulations of transient network analyzers. In addition to this approximation, the line is assumed to have a constant voltage over its length. Despite these simplifications, the equation obtained is very useful for understanding the energy transferred to surge arresters by transmission lines. Two simplified cases are highlighted next:

  a) Short circuit in DC converters, discharging the stored energy of the transmission line into the surge arrester of the converter (Figure 13.10),
  b) Transmission line discharge test (AC),

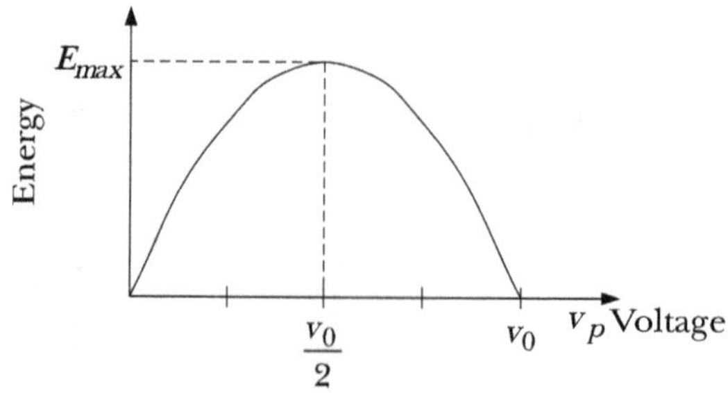

Figure 13.9 Energy dissipated in the surge arrester.

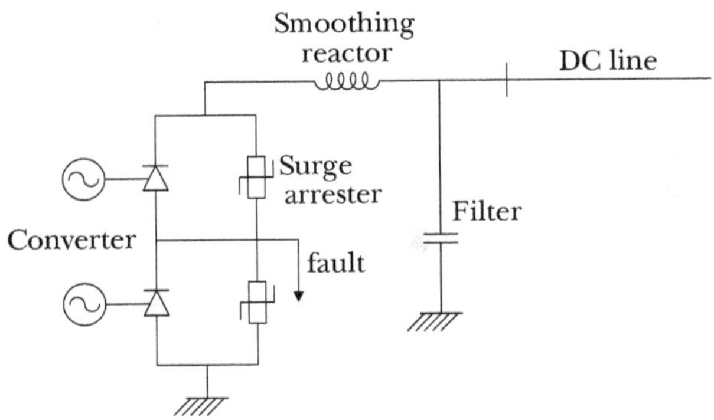

Figure 13.10 DC converter and transmission line.

where:

R_1: Losses in the inductor or apparent resistance of the surge arrester.
L_1: Smoothing reactor in the case of DC lines.
C_1: capacitance at terminals, which may include the DC line filter capacitance, $(C/2 + C_f)$.
R, L, C: resistance, inductance, and capacitance of the π circuits, respectively.

This was an important standardized test in the analysis of energy discharges into surge arresters, Figure 13.11a. Although it has been discontinued [2], analysis of the test is still of interest for a better understanding of the energy transfer from transmission line to arrester. In this case, the smoothing reactor in series with the transmission line may be removed from the model.

In the loop equations, Figure 13.11b, the initial values of the capacitors cancel out, resulting in:

$$\frac{v_p - v_0}{s} = \left(R_1 + sL_1 + \frac{1}{sC_1}\right) I_1(s) - \frac{1}{sC_1} I_2(s), \tag{13.26}$$

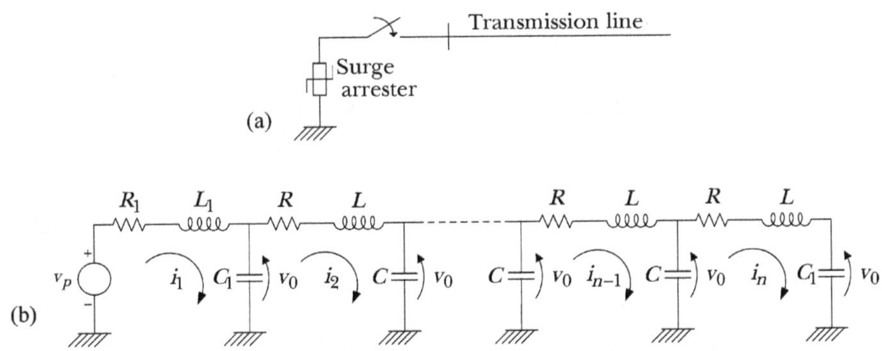

Figure 13.11 Transmission line discharge test. (a) Transmission line and surge arrester. (b) Transmission line model with distributed π-circuits.

$$0=\left(R+sL+\frac{C_1+C}{sC_1C}\right)I_2(s)-\frac{1}{sC}I_3(s)-\frac{1}{sC_1}I_1(s),\tag{13.27}$$

$$0=\left(R+sL+\frac{2}{sC}\right)I_{n-1}(s)-\frac{1}{sC}I_n(s)-\frac{1}{sC}I_{n-2}(s),\tag{13.28}$$

$$0=\left(R+sL+\frac{C_1+C}{sC_1C}\right)I_n(s)-\frac{1}{sC}I_{n-1}(s).\tag{13.29}$$

From (13.29), the following is obtained:

$$I_n(s)=\frac{1}{sC}\frac{1}{\left(R+sL+\dfrac{C_1+C}{sC_1C}\right)}I_{n-1}(s),\tag{13.30}$$

which substituted into (13.28), gives:

$$I_{n-1}(s)=F_1(s)I_{n-2}(s).\tag{13.31}$$

Repeating the process, with the elimination of the currents from I_2 to I_n, results in:

$$I_1(s)=(v_p-v_0)F(s).\tag{13.32}$$

This expression is like the discharge of shunt capacitors; however, the function, $F(s)$, and its inverse transform, $f(t)$, are now much more complex.

Equation (13.32) shows that the current remains proportional to the difference between the initial transmission line voltage, v_0, and the residual voltage of the surge arrester, v_p.

To obtain the energy function, analogous to Eq. (13.23), (13.32) is integrated with the opposite current sign:

$$E_n=v_p(v_0-v_p)\int_0^T f(t)dt.\tag{13.33}$$

Therefore, if one case of energy absorption is known, i.e., given E_n, v_0, v_p, the constant $\eta(T)$ can be obtained from (13.33) without integration. The value of this integral is approximately constant for a given transmission line, if the discharge period T does not change substantially:

$$\eta(T)=\int_0^T f(t)dt.\tag{13.34}$$

$$E_n=v_p(v_0-v_p)\eta(T)\tag{13.35}$$

Thus, the energy at different values of v_0, v_p can be estimated from the known result. Again, the curve of energy as a function of v_0 has a concave down parabolic shape. From this curve, it may be appreciated that, if $v_p > v_0/2$, the energy increases with a reduction in v_p, which is the case for surge arresters in AC transmission lines. Therefore, a common practice in the conservative calculation of the energy dissipated by arresters in AC transmission lines is to consider the minimum residual voltages. However, if $v_p < v_0/2$, the energy increases with an increase in v_p, which is the case for MOSA interventions in protection schemes using series-capacitor or surge-arrester interventions in the converter bridges of DC links.

- Simple Formula for Energy Estimate

The following formula for the approximate calculation of energy dissipation in arresters may be found in [2]:

$$W = \frac{U_{ps}\left(U_{rp} - U_{ps}\right)}{Z_s} 2\tau \tag{13.36}$$

Which can be translated into the same nomenclature adopted so far:

$$E_n = v_p \frac{\left(v_0 - v_p\right)}{Z} 2\tau \tag{13.37}$$

Assuming a model for a lossless transmission line and given that $Z = \sqrt{\frac{L'}{C'}}$; $\nu = \frac{1}{\sqrt{L'C'}}$ and $\tau = \ell/\nu$, where:

ℓ: line length.

L', C': inductance and capacitance of the transmission line, per unit length, respectively.

Therefore, Eq. (13.37) may be rewritten as:

$$E_n = v_p\left(v_0 - v_p\right)2C \tag{13.38}$$

C: total capacitance of the transmission line.

Under extreme conditions, it may be assumed that the energy dissipated in the arrester would be the difference in the energy stored in the transmission line capacitance between the initial and final state. Equation (13.39) presents even more conservative results and is rarely used.

$$E_n = \frac{C}{2}\left(v_0^2 - v_p^2\right), \tag{13.39}$$

The simplified models, examined above in (13.37) and (13.39), neglected the line losses and assumed a constant voltage in the surge arrester, which understates the damping and energy that would be dissipated in the transmission line, resulting in a more conservative calculation of the energy. Other model inaccuracies can be mentioned, but despite some limitations, the equations may be useful in a first analysis of energy transfer from the electrical system to the arrester.

Therefore, for accurate results, it is always more appropriate to calculate charge transfer and energy dissipation using EMT programs, as previously mentioned.

Example 13.2 The DC transmission system in Figure 13.12 with a rated voltage of 600 kV is simulated in an electromagnetic transient program. A fault-to-ground applied at point 1 gives a maximum current of 4.8 kA and an energy value of 13.7 MJ, with the line discharging into the surge arrester that protects the upper converter bridge. This surge arrester during the discharge has an approximate residual voltage of 300 kV, with the same polarity. The voltage along the line is 1200 kV in the load rejection condition.

Obtain the current through and energy dissipated by the arresters for faults at points 2 and 3, given that the residual voltages developed by the surge arresters are known to be 600 and 900 kV, respectively, with other arrester units placed in series, thereby increasing the voltage.

Solution

The period of arrester intervention, T, is assumed to be approximately constant in all three cases.

The currents are estimated in proportion to the difference between v_0 and v_p, for different values of v_p:

300 kV,	4.8 kA;
600 kV,	3.2 kA;
900 kV,	1.6 kA.

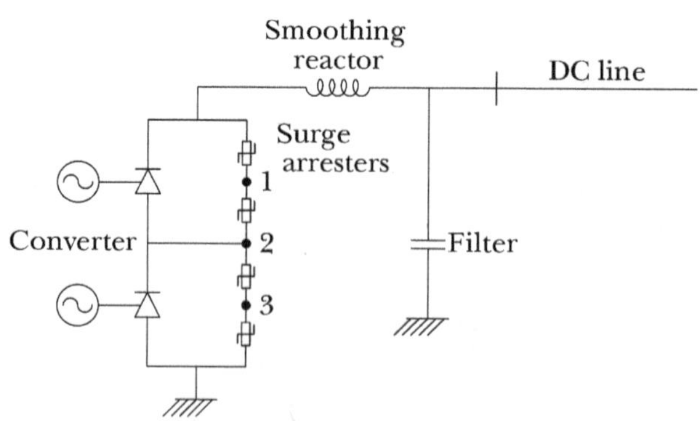

Figure 13.12 Fault in the DC converter.

The constant, η, is calculated:

$$\eta = \frac{13.7}{300 \times 900} = 5.07 \times 10^{-5} \frac{\text{MJ}}{\text{kV}^2}.$$

Applying (13.33):

300 kV,	13.7 MJ;
600 kV,	18.3 MJ;
900 kV,	13.7 MJ.

The results are shown in Figure 13.13. Note that the model is valid as long as the voltage polarity does not change.

Although the high overvoltage level is favorable for clarifying the behavior of Equation (13.35), it may be questionable in terms of system planning. For example, if load rejection overvoltage level of 1.4 pu and residual voltage of 300 kV are assumed, based on the same η, the expected energy absorption by the arrester is 8.2 MJ.

- Transmission Line Model with Surge Impedances
  So far, the arrester interaction with the system used a very approximate representation for the transmission line, by means of lumped parameters and π circuits.

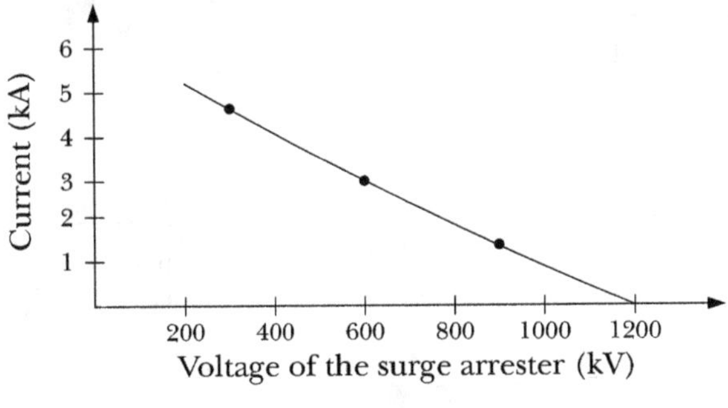

Figure 13.13 Current and energy in the arresters.

The appropriate representation of the transmission line, according to Chapters 2 and 3, assumes distributed parameters, with the line modeled as a surge impedance and time delay. This analysis, using the Thévenin equivalent, is very interesting to clarify the interaction between the arrester and the transmission line, as pointed out previously.

The following example shows the discharge of a DC transmission line into the converter arrester, while representing the transmission line as its surge impedance, according to Section 13.2.2.

Example 13.3 For a fault to ground in a converter bridge (like in Example 13.2), calculate the current through the surge arrester (Figure 13.14a).

Data

v_p: Constant residual voltage across the surge arrester – 150 kV.
v_0: DC transmission line voltage – 600 kV.
L: Smoothing reactor – 0.3 H.
C: Capacitance of the DC filter – 2 μF.
Z: Surge impedance of the transmission line – 300 Ω.

Assume a long transmission line (900 km) without interference from reflections.

Solution

The Thévenin equivalent of the transmission line and filter connected to the surge arrester model is presented in Figure 13.14b.

$$Z_{eq}(s) = sL + \frac{Z\frac{1}{sC}}{Z + \frac{1}{sC}} = \frac{sL + s^2ZCL + Z}{1 + ZCs},$$

$$I(s) = \frac{v_0 - v_p}{s}\frac{1}{ZLC}\frac{1 + ZCs}{s^2 + \frac{s}{ZC} + \frac{1}{LC}},$$

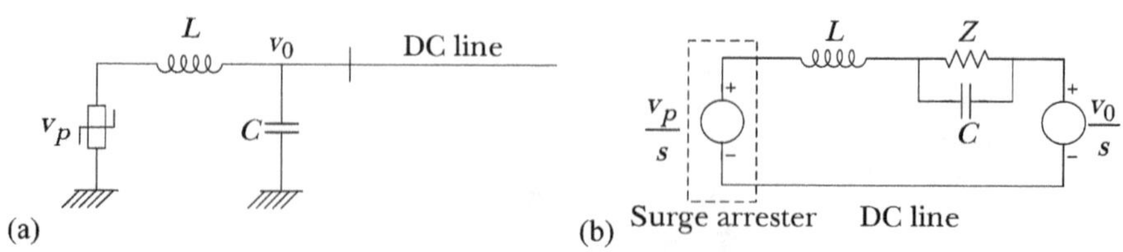

Figure 13.14 Fault in the DC converter. (a) DC transmission line and surge arrester. (b) Equivalent circuit.

$$I(s) = \frac{v_0 - v_p}{L} \frac{s + \dfrac{1}{ZC}}{s\left(s^2 + \dfrac{s}{ZC} + \dfrac{1}{LC}\right)}.$$

Substituting the numerical values:

$$I(s) = 1500 \frac{s + 1666.7}{s(s^2 + 1666.7s + 1.6667 \times 10^6)}$$

Expanding into partial fractions:

$$I(s) = 1500\left(\frac{10^{-3}}{s} + \frac{k_1}{(s - p_1)} + \frac{k_1{}^*}{(s - p_1{}^*)}\right),$$

$$k_1 = 5.07909 \times 10^{-4} \angle -170.4059°,$$

$$p_1 = 833.33 + j986.01.$$

Figure 13.15 shows the time response of the current.

Since the Thévenin equivalent voltage is constant, after the initial period, the current peak decreases to a steady value around 1.5 kA, because the inductance behaves as a short circuit and the capacitance as an open circuit:

$$i = \frac{600 - 150}{300} = 1.5\,\text{kA}.$$

This value can be observed in Figure 13.15 from 4 to 6 ms. Note that 6 ms corresponds to 2τ, twice the time delay of the line, and the Thévenin equivalent

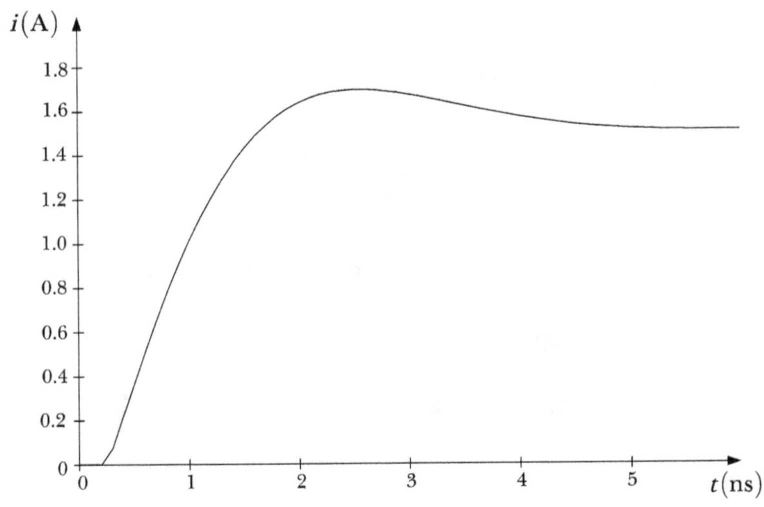

Figure 13.15 Current through the surge arrester.

voltage is known during this time interval, without reflections from the opposite terminal. After this period, reflections due to the arrester intervention begin to arrive. The maximum current can be calculated, but there are limitations of the energy calculation with this simplified method, due to the absence of losses. In general, the discharge of long transmission lines into converter arresters is associated with high intervention times and marked dissipation of energy by the arresters; in addition, more than one discharge may also occur in critical cases.

13.5.3 Arrester and Switching Surges

The SiC surge arrester's intervention in a typical switching surge is illustrated in Figure 13.6a. When using a MOSA, the intervention is smoother, without a sudden voltage variation. However, in qualitative terms, both surge arresters exhibit similar behaviors and durations.

Example 13.4 refers to the surge arrester's intervention during switching surges.

Example 13.4 Calculate the currents through the surge arrester located at the end of a three-phase AC transmission line of 500 kV, during the line energization.

Data

The prospective overvoltages at the open terminal and their corresponding times were calculated without arresters and are furnished for phase *a*.

Voltage (kV)	Time (ms)
650	0.5
940	1.0
1100	1.2
770	1.5

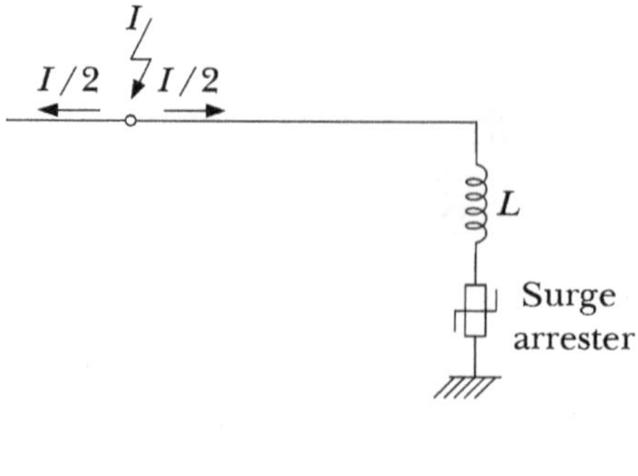

Figure 13.16 Surge arrester with lead.

The three-phase matrix of surge impedances of the transmission line is given as follows:

$$Z = \begin{bmatrix} 398.7 & 141.3 & 141.3 \\ 141.3 & 398.7 & 141.3 \\ 141.3 & 141.3 & 398.7 \end{bmatrix}.$$

Transmission line length, $\ell = 300$ km.
Surge-arrester residual voltage, $v_p = 800$ kV.

Solution

The model is presented in Figure 13.5. The prospective overvoltage furnished without arresters is used as Thévenin equivalent voltage. The arrester intervention occurs only at time 1.0 and 1.2 ms, when the voltage of the Thévenin equivalent exceeds 800 kV. Considering a round-trip time delay of 2 ms for reflections after the surge arrester's intervention, the prospective voltages do not require correction within the 1 to 3 ms interval.

The equation for the Thévenin equivalent is:

$$\begin{bmatrix} e_{ta} \\ e_{tb} \\ e_{tc} \end{bmatrix} - \begin{bmatrix} 398.7 & 141.3 & 141.3 \\ 141.3 & 398.7 & 141.3 \\ 141.3 & 141.3 & 398.7 \end{bmatrix} \times \begin{bmatrix} i_a \\ i_b \\ i_c \end{bmatrix} = \begin{bmatrix} v_p \\ v_b \\ v_c \end{bmatrix}.$$

Variables are the same as in Section 13.2.2 and the equations used are (13.7) and 8. Taking the first line of the system of equations and assuming that $i_b = i_c = 0$:

$$e_{ta} - Zi_a = v_p$$

$$e_{ta} - 398.7i_a = 800.$$

Next, the currents are calculated.

At $t = 1.0$ ms:

$$940 - 398.7i_a = 800, \qquad i_a = 0.351 \text{ kA}.$$

At $t = 1.2$ ms:

$$1100 - 398.7i_a = 800, \qquad i_a = 0.752 \text{ kA}.$$

The currents during the intervention of a surge arrester on an AC transmission line were evaluated in this example, using the precalculated prospective overvoltages. The complete simultaneous calculation of voltages and currents requires improvement in the models, but this involves more complicated analysis, which needs electromagnetic transient programs.

Compared to DC transmission lines, AC lines have shorter lengths and therefore both the arrester energy and conduction time are also lower. The discharge period in AC transmission lines is about 1–3 ms.

13.5.4 Surge Arrester and Fast-Front Overvoltages

Next, Examples 13.5–13.7 are related to fast-front waves using simplified models for the purpose of presenting the analytical calculation of surge arrester intervention in transmission lines and substations.

Example 13.5 addresses the voltage drop across the lead that connects the surge arrester to the substation cable or busbar.

Example 13.5 Figure 13.16 presents a lightning stroke to a transmission line with an arrester at the line terminal. Calculate the voltage at the line terminal and the voltage drop at the arrester lead.

Data

k_i: Rate of rise of lightning current, 10 kA/μs.
Z: Surge impedance of the transmission line, 400 Ω.
L: Surge arrester connection cable inductance, 2 μH.
v_p: Residual voltage at the surge arrester, 50 and 1000 kV.

Solution

The rate of rise of the current, after it divides at the stroke point is 0.5 k_i (5 kA/μs), which implies a voltage rate of rise of $k_v = 400 \times 5 = 2000$ kV/μs.

The rate of rise of the voltage without the surge arrester, which is the rate of rise of the Thévenin equivalent, doubles at the line terminal, thus rising to $k_v = 4000$ kV/μs.

For the purpose of simplifying the solution, consider the initial time, t_0, at which the voltage reaches the value of the arrester residual voltage in Figure 13.17a,

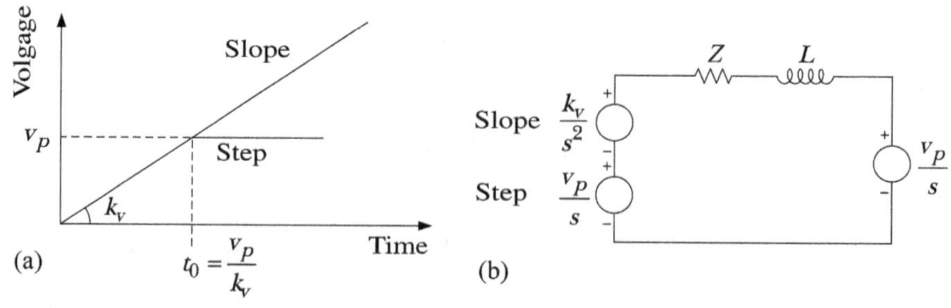

Figure 13.17 Insertion of the surge arrester. (a) Time for arrester insertion in the circuit. (b) Equivalent circuit.

resulting in the post- t_0 Thévenin equivalent voltage, which has a DC component and a ramp component with a rate of rise of k_v.

The Thévenin equivalent impedance is given by the surge impedance of the transmission line in series with the inductance of the arrester lead. The equivalent circuit is shown in Figure 13.17b, with the arrester lead.

Thévenin equivalent voltage post – t_0:

$$e_t = \frac{v_p}{s} + \frac{k_v}{s^2}$$

The current is calculated using only the ramp component (with rate of rise, k_v), since the step voltages cancel.

$$I(s) = \frac{k_v}{s^2(Z + sL)} = \frac{k_v}{s^2 L\left(s + \frac{Z}{L}\right)}.$$

The inverse transform of $\frac{1}{s^2(s + a)}$,

is:

$$\frac{1}{a^2}(e^{-at} + at - 1)$$

resulting in:

$$i(t) = \frac{k_v L}{Z^2} e^{-\frac{Z}{L}t} + \frac{k_v}{Z} t - \frac{k_v L}{Z^2},$$

$$i(t) = \frac{k_i L}{Z}\left(e^{-\frac{Z}{L}t} - 1\right) + k_i t.$$

The time constant is calculated from:

$$\frac{L}{Z} = \frac{2 \times 10^{-6}}{400}\ (5\ \text{ns}),$$

$$i(t) \cong k_i t.$$

This time constant is very small and is due to simplification of the models; thus, only the slope of the $k_i\, t$ parcel remains, which coincides with the original lightning current. This current is divided by 2 when it reaches the transmission line. When the surge arrester is effectively inserted, its effect is equivalent to connecting a DC generator, and in this case, the transmission coefficient for the current wave is 2, which is identical to that of a short circuit, thus doubling the current.

As the inductance of the lead is very small, it may be assumed that it does not interfere with the rate of rise of the current, independently of the arrester rating.

The voltage drop through the inductance is calculated from:

$$\Delta v(s) = sLI(s) = \frac{k_v}{s\left(s + \frac{Z}{L}\right)}$$

Taking the inverse transform:

$$\Delta v(t) = k_v \frac{L}{Z}\left(1 - e^{-\frac{Z}{L}t}\right) = k_i L\left(1 - e^{-\frac{Z}{L}t}\right).$$

Due to the small time constant, the voltage drop across the lead is given by:

$$\Delta v(t) = k_i L,$$

which is an approximately constant voltage, corresponding to $L\,\frac{di}{dt}$.

Substituting the numerical values:

$$\Delta v = 2 \times 10 = 20\ \text{kV},$$

which corresponds to an increase of 40% of the residual voltage of 50 kV for a distribution-line arrester. For a transmission line of 500 kV, the voltage drop through the lead would correspond to 2.5% of the residual voltage of 800 kV. In fact, in EHV transmission lines, the leads are longer, and the voltage drops are in the range of 5–10% of the arrester residual voltage.

These illustrative results were calculated for the lightning current and transmission-line parameters specific to this example and should be evaluated on a case-by-case basis. However, despite the use of a simplified model, the main effect of the lead can be seen in this example.

Example 13.6 demonstrates that the arrester not only provides a protection zone for equipment situated behind it but also for equipment located ahead, in the path of the lightning surge.

Example 13.6 Assuming a surge arrester located at point P, and a lightning stroke at point A, calculate the voltage at point M, which is located upstream on the transmission line, as in Figure 13.18a.

Designating:

v_a, v_m, v_{pr}: voltages at points A, M, P.
t: Time from the instant that the lightning surge reaches point M.
t_f: Front-time of the lightning current.
k_i: Rate of rise of the lightning current.

At the time of current division, the rate of rise is $k_i/2$ on each side of the line. The current waveform is shown in Figure 13.18b, with a slope and constant value

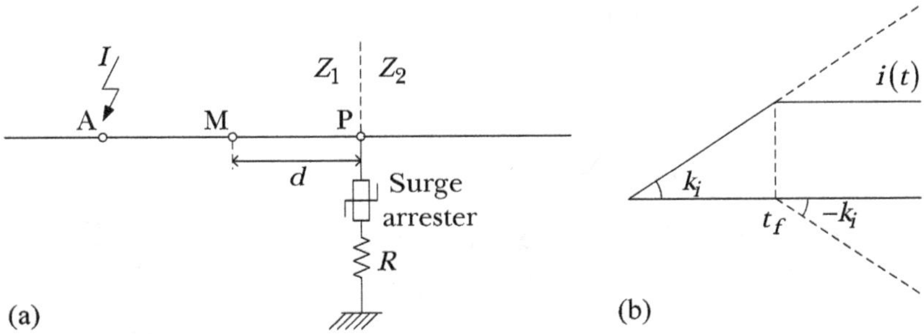

Figure 13.18 Protection upstream of the arrester on the transmission line. (a) Stroke at point A. (b) Lightning current waveform.

post-t_f. The constant value is obtained by applying a rate of rise with opposite sign, $-k_i$, at t_f.

Z_1 and Z_2: surge impedances of the transmission lines connected to point P.
R: Grounding resistance of the surge arrester.
k_v: Rate of rise of the voltage at point A moving in the direction of point M ($k_v = k_i Z/2$)
k_{v2}: Rate of rise of the voltage at point P, without the surge arrester (Thévenin equivalent voltage, $k_{v2} = \Gamma_t k_v$).
Γ_t: Voltage transmission coefficient, without the surge arrester, $\Gamma_t = \Gamma_r + 1$.
Γ_r: Voltage reflection coefficient, without the surge arrester.
d: Distance between points M and P.
τ: Time delay between points M and P ($\tau = d/c$, where c is the speed of light).
v_p: Surge-arrester residual voltage.
τ_1: Elapsed time between surge arrival and the arrester intervention.
v_{pr}: Voltage at point P.
i_p: Current through the surge arrester.

The voltage at point M is analyzed at four different intervals (Figure 13.20).

$$\text{Interval } 1 : 0 \leq t \leq 2\tau.$$

In this interval, the rate of rise of the voltage at point M is $k_{v1} = k_v$:

$$v_m\,(t) = k_v\,t.$$

The voltage reflections from point P in the direction of point M begin after time τ, changing the rate of rise of the voltage at M. If $Z_1 = Z_2$ no reflection occurs.

$$\text{Interval } 2 : 2\tau \leq t \leq 2\tau + \tau_1.$$

Rate of rise of the voltage at point P:

$$k_{v2} = k_v + \Gamma_r k_v = \Gamma_t k_v.$$

Voltage at point M:

$$v_m(t) = k_v t + \Gamma_r k_v(t - 2\tau).$$

The Thévenin equivalent impedance at P is given by:

$$Z_e = \frac{Z_1 Z_2}{Z_1 + Z_2}.$$

The current analysis in the surge arrester after $\tau + \tau_1$, is done using the Thévenin equivalent (Figure 13.19).

After $\tau + \tau_1$, the voltage has a slope and a constant value, v_p.

By solving the Thévenin equivalent circuit, the current through the arrester and the grounding resistance is obtained at a time greater than $\tau + \tau_1$ (the equation is valid in intervals 2, 3 and 4):

$$i_p(t) = \frac{k_{v2}}{Z_e + R}(t - \tau - \tau_1).$$

The voltage at point P, designated v_{pr}, after the surge arrester intervention, is obtained from the sum of v_p and the voltage drop across the grounding resistance, R, given by $Ri_p(t)$:

$$v_{pr}(t) = v_p + \frac{Rk_{v2}}{Z_e + R}(t - \tau - \tau_1).$$

The voltage that reaches point P after $\tau + \tau_1$ is calculated from the following equation:

$$v_i(t) = k_v(t - \tau - \tau_1) + v_p,$$

where, $v_p = k_{v2}\tau_1$, and it is possible to calculate $\tau_1 = \dfrac{v_p}{k_{v2}}$.

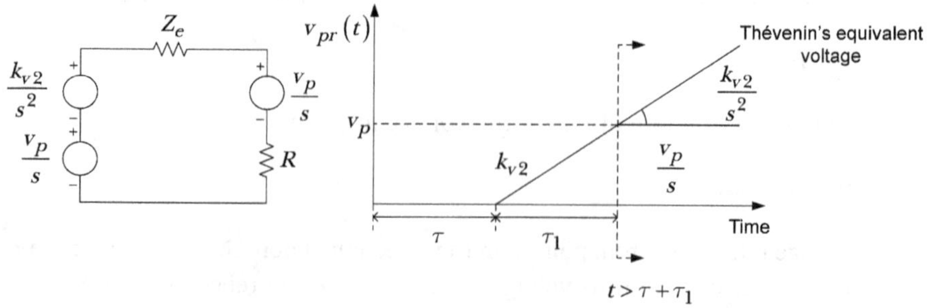

Figure 13.19 Thévenin equivalent after $\tau + \tau_1$.

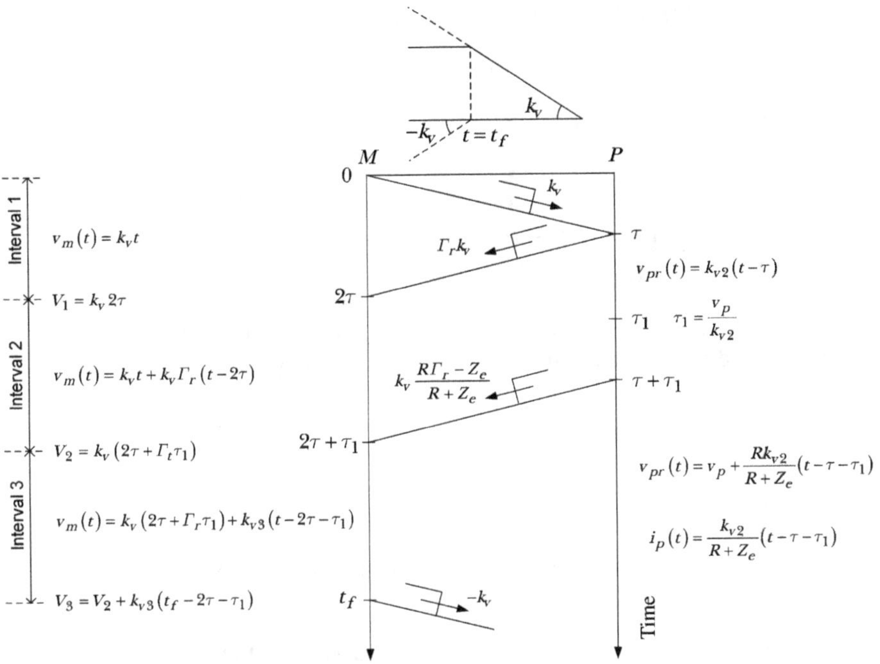

Figure 13.20 Lattice diagram, $t_f > 2\tau + \tau_1$.

The reflected wave at point P, after the surge arrester intervention, is given by:

$$v_i(t) + v_r(t) = v_{pr}(t).$$

Substituting the previous equations:

$$v_r(t) = k_v \frac{R(\Gamma_t - 1) - Z_e}{R + Z_e}(t - \tau - \tau_1) = k_v \frac{R\Gamma_r - Z_e}{R + Z_e}(t - \tau - \tau_1).$$

This reflected wave changes the rate of rise of the voltage at M. At the end of interval 2, the voltage at M is:

$$v_m(t)\big|_{t = 2\tau + \tau_1} = [k_v t + \Gamma_r k_v (t - 2\tau)]\big|_{t = 2\tau + \tau_1}$$

$$v_m(t)\big|_{t = 2\tau + \tau_1} = k_v(2\tau + \Gamma_t \tau_1)$$

Setting:

$$V_2 = k_v(2\tau + \Gamma_t \tau_1)$$

Interval 3 : $2\tau + \tau_1 \leq t \leq t_f$.

$$k_{v3} = k_v + k_v \frac{R\Gamma_r - Z_e}{R + Z_e} = k_v \frac{R\Gamma_t}{R + Z_e},$$

$$v_m(t) = k_v(2\tau + \Gamma_t\tau_1) + k_{v3}(t - 2\tau - \tau_1).$$

At the end of interval 3, the voltage at M is:

$$v_m(t)\big|_{t=t_f} = V_2 + k_{v3}(t_f - 2\tau - \tau_1).$$

Setting:

$$V_3 = V_2 + k_{v3}(t_f - 2\tau - \tau_1).$$

Interval 4 : $t \geq t_f$.

The waveform of the current was shown in Figure 13.18b, with a front time, $t_f > 2\tau + \tau_1$, and at that moment, the rate of rise of the voltage at M becomes:

$$k_{v4} = k_{v3} - k_v = k_v \frac{R\Gamma_r - Z_e}{R + Z_e},$$

$$v_m(t_f) = k_v\left(2\tau + \frac{v_p}{k_v}\right) + k_{v4}(t_f - 2\tau - \tau_1).$$

Figures 13.21 and 13.22 present the voltages at point M for the common parameters:

$v_p = 250$ kV; $Z_1 = 400\ \Omega$; $\tau = 0.2$ μs; $t_f = 3$ μs; $k_v = 400$ kV/μs; $R = 50\ \Omega$.

The following calculations refer to Figure 13.21.

$Z_e = 133.33\ \Omega$; $\Gamma_t = 0.667$; $\Gamma_r = -0.333$.

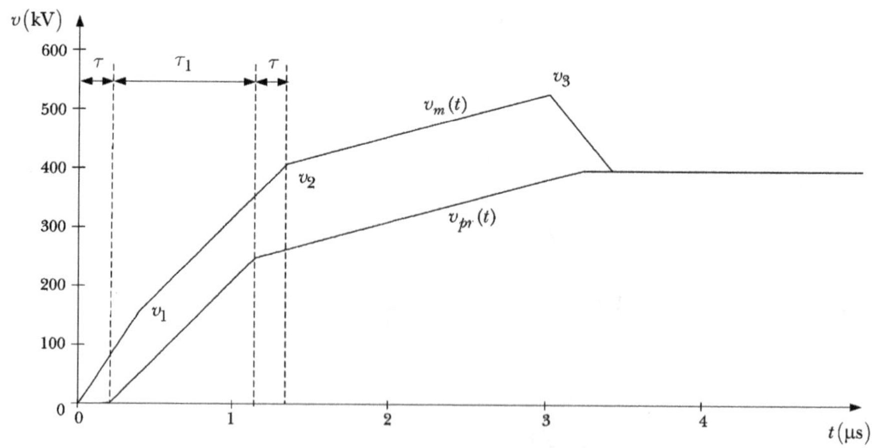

Figure 13.21 Surge impedance downstream the surge arrester Z_2 = 200 Ω.

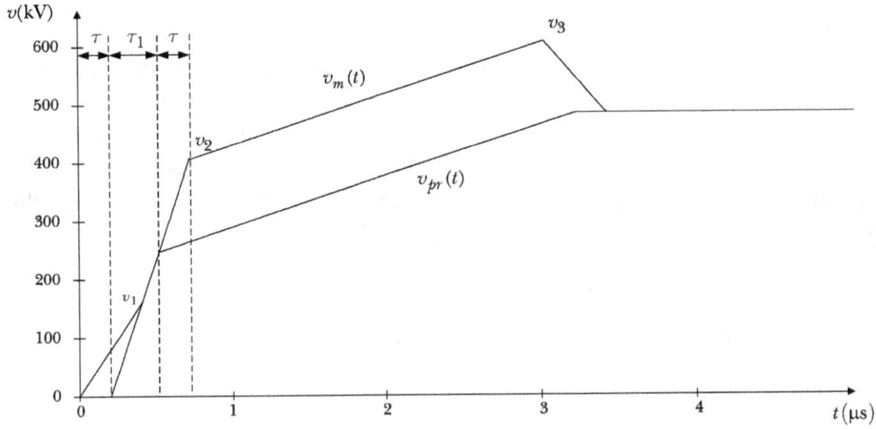

Figure 13.22 Open line terminal, $Z_2 = \infty$.

$$\tau_1 = \frac{v_P}{\Gamma_t k_v} = \frac{250}{0.667 \times 400} = 0.937\ \mu\text{s},$$

$$V_1 = 160\ \text{kV}; V_2 = 410\ \text{kV},$$

$$k_{v3} = 72.76\ \text{kV}/\mu\text{s},$$

$$V_3 = 410 + 72.76 \times 1.663 = 531\ \text{kV}.$$

The following calculations refer to Figure 13.22.

$$Z_e = 400\ \Omega; \Gamma_t = 2; \Gamma_r = 1.$$

$$\tau_1 = \frac{v_P}{\Gamma_t k_v} = 0.3125\ \mu\text{s},$$

$$V_1 = 160\ \text{kV}; V_2 = 410\ \text{kV},$$

$$k_{v3} = 88.90\ \text{kV}/\mu\text{s},$$

$$V_3 = 410 + 88.90 \times 2.2875 = 613\ \text{kV}.$$

The arrester offers protection upstream in the transmission line, preventing the voltage from rising too much at point M. In this case of infinite impedance on side 2 the voltage would reach 2400 kV at M.

Example 13.7 complements Example 13.6 but focuses on the analysis of the equipment located behind the arrester in the surge trajectory. A simple approach to dealing with lightning surge wavefront is employed.

Example 13.7 For Figure 13.23, obtain the voltage at the transmission line terminal F, protected by a surge arrester located at point P at a certain distance from F. The voltage levels supplied are in pu of the nominal system voltage.

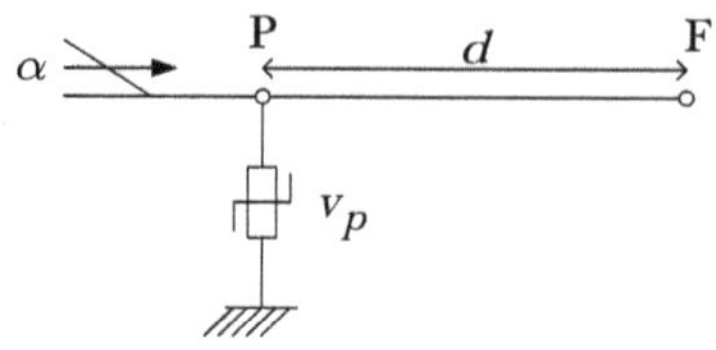

Figure 13.23 Transmission line with surge arrester.

Data

- ideal model of surge arrester maintaining a constant voltage $v_p = 2.0$ pu;
- incident voltage wave with rate of rise $\alpha = 4$ pu/μs, traveling at the speed of light.
- distance from the surge arrester to the line's end:

a) $d = 150$ m ($\tau = 0.5$ μs).

b) $d = 37.5$ m ($\tau = 0.125$ μs).

Solution

a) Separation distance $d = 150$ m.
Figure 13.24 shows the voltage increase at point P with or without the surge arrester. At $t = 0.5$ μs, the surge arrester intervention starts, maintaining a constant voltage at point P, $v_p = 2.0$ pu, which travels to point F. This behavior can be simulated by injecting a voltage wave with rate of rise $-\alpha$ at that moment, superimposed on the previously existing voltages.

Figure 13.25 shows the voltage at F without the surge arrester. The rate of rise at F is 2α. After $t = \tau$, the rate of rise at P also becomes 2α, with the superposition of incident and reflected waves.

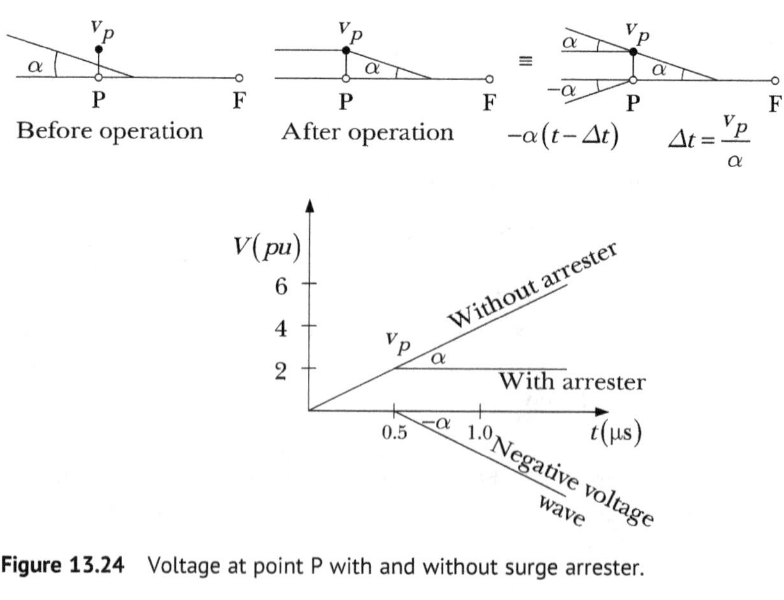

Figure 13.24 Voltage at point P with and without surge arrester.

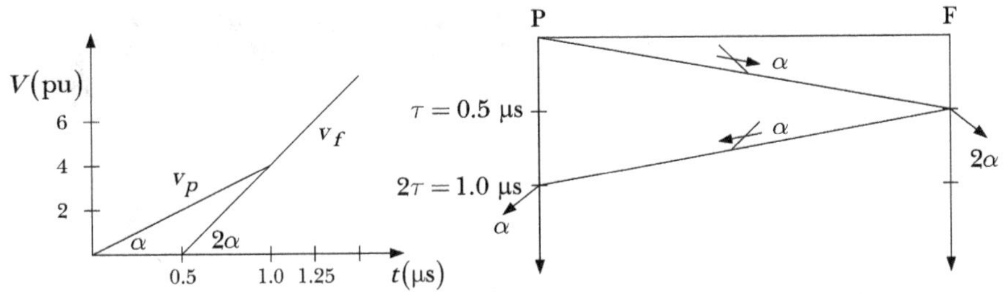

Figure 13.25 Voltage at points P and F, without surge arrester.

On the assumption that the surge arrester intervention occurs with time delay $\Delta t = v_p/\alpha$ (in this particular case $\Delta t = \tau$), the voltage transmitted from point P toward the line terminal is given by $\alpha t - \alpha(t - \Delta t)\,H(t - \Delta t)$.

Figure 13.26 shows the voltages at points P and F. The surge arrester intervention starts at time $t = 0.5$ μs, and it all happens as if the arrester had sent a wave with a negative rate of rise $-\alpha$, which becomes -2α at F, causing a constant voltage at F from that instant on (which can also be interpreted as if the 2.0 pu step, constrained by the arrester, doubled ant the end of the line reaching 4.0 pu).

At $t = 2\tau$, the wave reflected from point F arrives at point P. When it reaches P with a rate of rise, α, since the voltage at the surge arrester is constant, everything happens as if the wave reached an ideal voltage source, and the wave, reflected to F, at this time, has a rate of rise of $-\alpha$.

It is noted that, at 150 m, the voltage at the line terminal is twice the voltage at the surge arrester.

b) Separation distance $d = 37.5$ m.

Figure 13.27 presents the lattice diagram of the line segment between points P and F.

Within the interval $0 - \tau$, the rate of rise at P is α.

$$v_P(t) = \alpha t.$$

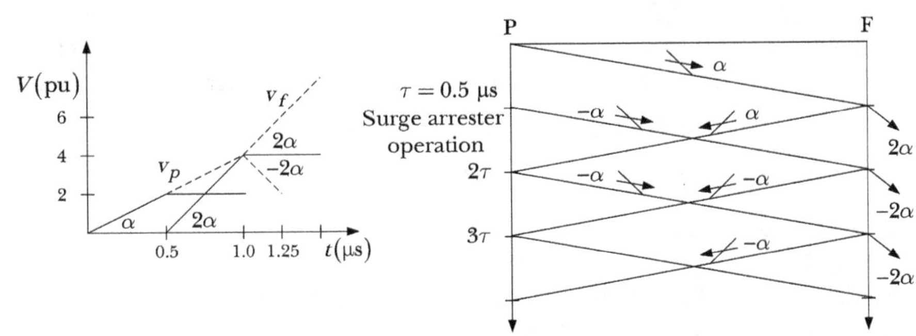

Figure 13.26 Surge arrester intervention at $d = 150$ m.

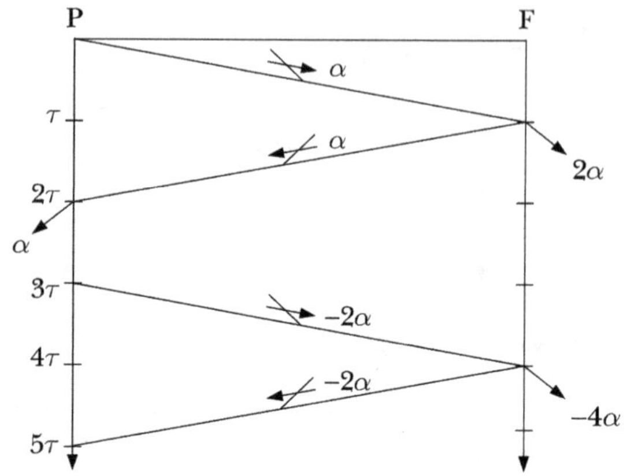

Figure 13.27 Operation of the surge arrester at $d = 37.5$ m.

Within the $\tau - 2\tau$ interval, the rates of rise are α at P and 2α at F.

$$v_p(t) = \alpha t,$$
$$v_f(t) = 2\alpha(t-\tau)H(t-\tau).$$

In the interval, $2\tau < t < 3\tau$, the rates of rise are 2α at P and 2α at F. The reflected voltage, returning from point F with a slope of α, and the incident wave are superimposed at P.

$$v_p(t) = \alpha t + \alpha(t-2\tau)H(t-2\tau), v_f(t) = 2\alpha(t-\tau)H(t-\tau).$$

At time $t = 3\tau$, the surge arrester intervention occurs when the voltage reaches 2.0 pu, which remains constant. With the same analysis as item (a), a wave with rate of rise of -2α propagates toward F.

$$v_p(t) = \alpha t + \alpha(t-2\tau)H(t-2\tau) - 2\alpha(t-3\tau)H(t-3\tau),$$
$$v_f(t) = 2\alpha(t-\tau)H(t-\tau).$$

At point F and time $t = 4\tau$, the rate of rise becomes -4α, which when superimposed on the pre-existing 2α, results in a net rate of rise of -2α.

$$v_p(t) = \alpha t + \alpha(t-2\tau)H(t-2\tau) - 2\alpha(t-3\tau)H(t-3\tau),$$
$$v_f(t) = 2\alpha(t-\tau)H(t-\tau) - 4\alpha(t-4\tau)H(t-4\tau).$$

In relation to the previous case, a reduction in the voltage at F is observed, now reaching 3.0 pu.

Figures 13.28 and 13.29 present the results obtained with the ATP program at distances of 150 and 37.5 m, respectively.

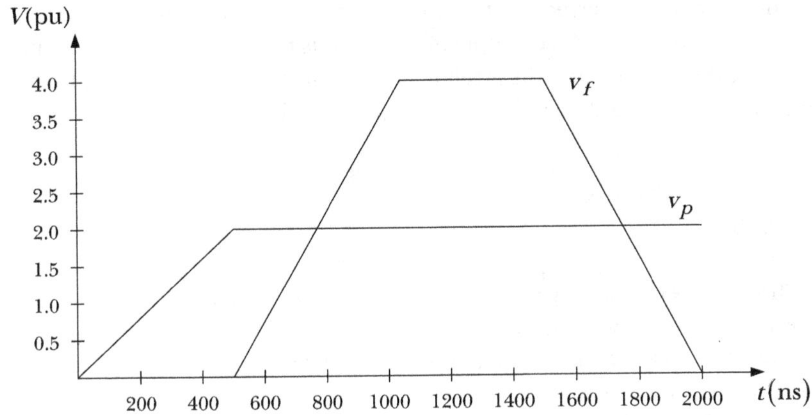

Figure 13.28 Voltages with separation distance $d = 150$ m.

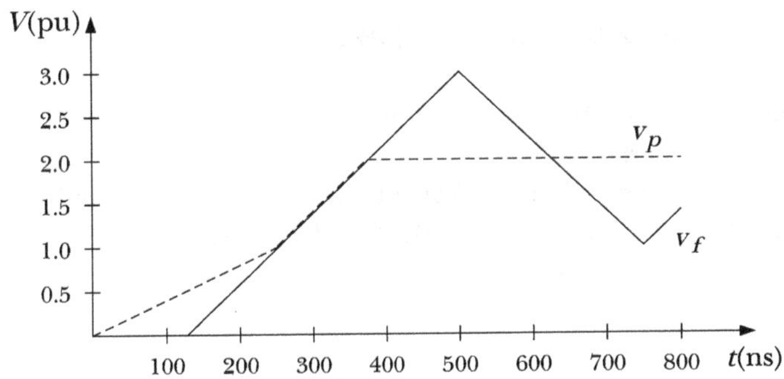

Figure 13.29 Voltages with separation distance $d = 37.5$ m.

This example illustrates the behavior of lightning surges in substations with an equipment protected by a surge arrester at a certain distance d. The overvoltage decreases with the reduction of the distance d. This effect has implications for the insulation coordination of the equipment, which are discussed in Chapters 9 and 14.

The triangle waveform is repetitive and nonattenuating in this ideal case, but in real cases, such as a transformer at point F, the triangle waveform is smoothed by the transformer capacitance (see Figure 14.15b in Chapter 14), and it is attenuated by the high-frequency losses.

To conclude this chapter, it is also worth commenting on the aspect of energy transfer to surge arrester by lightning surges. Although the energy transferred to the arrester by switching surges is more decisive for its sizing, it may be necessary to know the energy transferred by lightning surges, calculated with appropriate models in EMTP programs. An example of this need is related to the sizing of line arresters, since the number of units is relatively large in a transmission line. In this case a probabilistic methodology may be necessary, as in [7].

References

1 IEC. (2018). *IEC 60099-5: Surge arresters. Part 5: Selection and application recommendations.* Geneva: IEC.

2 IEC. (2014). *IEC-60099-4: Surge arresters. Part 4: Metal oxide surge arresters without gaps for A.C. systems.* Geneva IEC.

3 IEEE. (2005). *C62.11-2005: IEEE Standard for metal-oxide surge arresters for AC power circuits.* New York IEEE.

4 IEC. (2006). *IEC-60071-1: Insulation co-ordination. Part 1: Definitions, principles and rules.* Geneva IEC.

5 IEC. (2018). *IEC-60071-2: Insulation co-ordination. Part 2: Application guide.* Geneva IEC.

6 IEEE Working Group 3.4.11. (1992). Modeling of metal oxide surge arresters. *IEEE Transactions on Power Delivery*, 7(1), 302–309.

7 Zanetta, L. C. (2003). Evaluation of line surge arrester failure rate for multipulse lightning stresses. *IEEE Transactions on Power Delivery*, 18(3), 796–801.

14

Insulation Coordination of Transmission Lines and Substations

14.1 Introduction

Chapters 7–9 dealt with temporary overvoltages, switching surges, and lightning surges, respectively. The need to know and quantify these overvoltages is aimed at the specification of the various levels of equipment insulation organized under the insulation coordination guidelines.

This chapter does not intend to provide a complete description of the insulation coordination of transmission lines and substations presented in [1–7] because of the considerable breadth of the subject. The purpose here is to introduce the topic and understand the fundamental concepts needed to address the effect of overvoltages on the insulation levels of power system equipment.

The known stresses on power equipment are:

- Operating voltages at the nominal frequency.
- Temporary overvoltages.
- Switching surges with slow-front overvoltages.
- Lightning surges with fast-front overvoltages.

Analysis of the interaction between overvoltage stresses and insulation withstand levels is performed, considering the technical-economic aspects in the adequate design of electrical systems. The intent behind insulation coordination is to achieve the best fit between the lower withstand levels of the insulation and the most representative overvoltages in the power system [1], to meet security and economic requirements. The objective is to establish the procedure for the selection of the phase-to-ground and phase-to-phase insulation levels of equipment but in this introduction to the subject, the text focuses only on phase-to-earth insulation.

In this chapter, overvoltage stresses are designated as U and withstand voltages are designated as V, to make it easier to understand the statistical approach employed. However, standard nomenclature is always used when necessary.

Understanding Electromagnetic Transients in Power Systems, First Edition. Luiz Cera Zanetta Jr.

Published 2025 by John Wiley & Sons, Inc.

14.2 Basic Probabilistic Concepts

14.2.1 Elementary Concepts

Overvoltages and insulation withstand levels exhibit random behavior, and a statistical approach is needed to analyze these variables. Thus, some elementary concepts regarding probabilistic events and the theory of probability that will be useful in this chapter are introduced hereinafter.

Assuming A and B are two generic events with probabilities $P(A)$ and $P(B)$, respectively, in the specific case of insulation coordination, both events could be associated with the breakdown of two identical insulator strings, A and B, subjected to the stress of the same overvoltage level, and in this case $P(A) = P(B)$ (Figure 14.1a).

The probability of occurrence of a disruptive discharge in at least one insulator string, that is, occurrence of event A or B is calculated from:

$$P(A + B) = P(A) + P(B) - P(A \cap B), \tag{14.1}$$

where $P(A \cap B)$ is the probability of occurrence of both events simultaneously, causing the breakdown of both insulator strings, Figure 14.1b.

Assuming independent events, the following expression can be written:

$$P(A + B) = P(A) + P(B) - P(A)P(B), \tag{14.2}$$

because

$$P(A \cap B) = P(A)P(B).$$

Another way to obtain the same result is by calculating the probabilities of the complementary events $\overline{A}$ and $\overline{B}$, in which there is no breakdown.

$$P(\overline{A}) = 1 - P(A), \tag{14.3}$$

$$P(\overline{B}) = 1 - P(B). \tag{14.4}$$

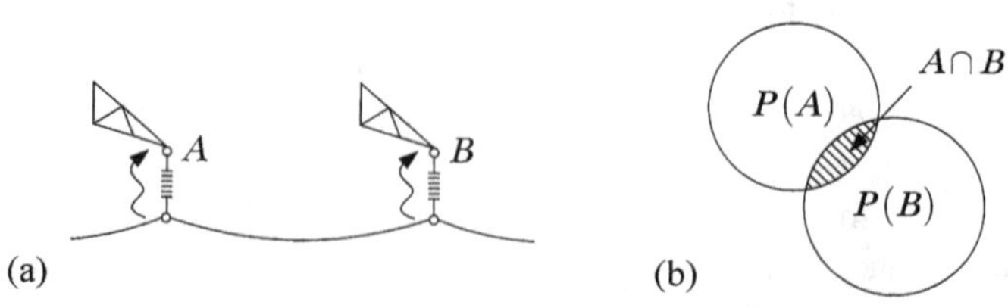

Figure 14.1 Breakdown of the insulator string. (a) Insulator strings A and B. (b) Probability of simultaneous events.

The withstand probability of both insulators is given by the simultaneous occurrence of $\overline{A}$ and $\overline{B}$. Thus, assuming independent events, the probability that both insulator strings simultaneously withstand the overvoltage can be evaluated using the product $P(\overline{A})P(\overline{B})$:

$$P(\overline{A} \cap \overline{B}) = [1 - P(A)][1 - P(B)], \tag{14.5}$$

$$P(\overline{A} \cap \overline{B}) = 1 - P(A) - P(B) + P(A)P(B). \tag{14.6}$$

However, the breakdown probability of at least one insulator is the complementary event of all insulators withstanding the same overvoltage level, thus:

$$P\left(\overline{\overline{A} \cap \overline{B}}\right) = 1 - P(\overline{A} \cap \overline{B}), \tag{14.7}$$

and,

$$P\left(\overline{\overline{A} \cap \overline{B}}\right) = P(A) + P(B) - P(A)P(B). \tag{14.8}$$

Note that (14.7) and (14.2) give the same result, since:

$$P\left(\overline{\overline{A} \cap \overline{B}}\right) = P(A + B) \tag{14.9}$$

To calculate the breakdown probability in transmission lines, when many identical insulators at the same overvoltage level may be assumed, the possible combinations of insulator breakdown can be calculated, as in (14.2), according to the binomial distribution, which can be a cumbersome task. However, considering the simultaneous event of no insulator breakdown and calculating the probability of its complementary event, that is, at least one failure, leads to the same result in a much simpler way, as shown in (14.7).

14.2.2 Probability Density Function and Distribution Function

Insulation coordination frequently utilizes Gaussian functions, but certain specific events can assume other distributions, such as a Weibull function, for instance. In this introductory approach to probabilistic calculation, Gaussian distributions are chosen for their widespread use and treatment analogous to other distributions.

Let $f(x)$ be the normal probability density function of a certain event, defined by its mean and standard deviation, as for example, the withstand voltage distribution of an insulator string subjected to an overvoltage.

In Figure 14.2a, the following notation is used:

$\overline{x}$: mean,

σ: standard deviation.

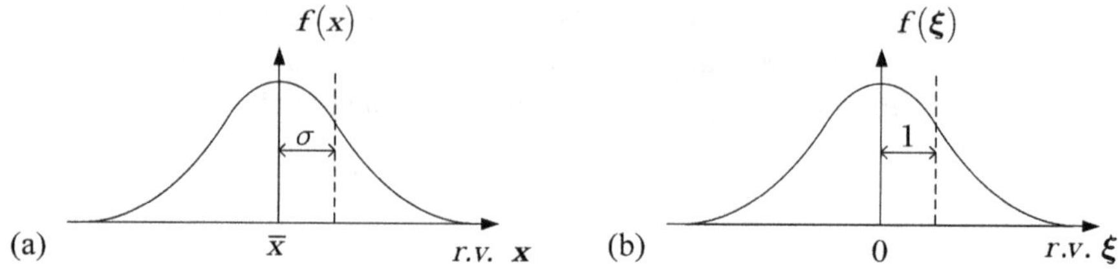

Figure 14.2 Normal probability density function. (a) Mean and standard deviation. (b) Standard normal distribution

In statistical distributions, the median is the 50% probability value and in Gaussian distributions, the mean and the median have the same value. Standard normal distribution may be obtained by converting the random variable into a standard normal variable, as in Eq. (14.10) and Figure 14.2b:

$$\xi = \frac{\boldsymbol{x} - \overline{x}}{\sigma}, \tag{14.10}$$

with a mean of zero and a standard deviation of 1.

With reference to the normal density function, the distribution function furnishes the probability that any random variable, $\boldsymbol{x}$, is less than or equal to a given value $x = X$. Thus, the probability $P(\boldsymbol{x} \leq X)$, is obtained from:

$$P(\boldsymbol{x} \leq X) = \int_{-\infty}^{X} f(x)dx. \tag{14.11}$$

This value is obtained from the distribution function, Figure 14.3a:

$$F(x) = \int_{-\infty}^{x} f(u)du. \tag{14.12}$$

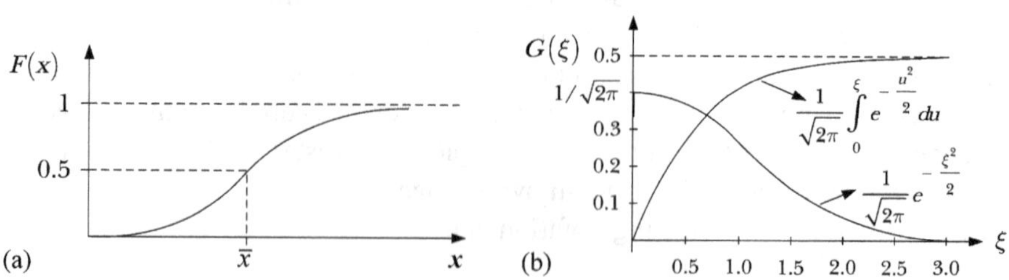

Figure 14.3 Integral of the probability density function. (a) Distribution function. (b) Integral of the normalized density function.

The normal probability density function is defined as:

$$g(x) = \frac{1}{\sigma\sqrt{2\pi}} e^{-\frac{1}{2}\left(\frac{x-\overline{x}}{\sigma}\right)^2} \tag{14.13}$$

Working with normalized variables, Figure 14.3b shows the function g(ξ) and its integral, $G(\xi)$, for $\xi \geq 0$:

$$g(\xi) = \frac{1}{\sqrt{2\pi}} e^{-\frac{\xi^2}{2}}, \tag{14.14}$$

$$G(\xi) = \frac{1}{\sqrt{2\pi}} \int_0^{\xi} e^{-\frac{u^2}{2}} du. \tag{14.15}$$

It is also known that:

$$G(\infty) = \frac{1}{\sqrt{2\pi}} \int_0^{\infty} e^{-\frac{u^2}{2}} du = 0.5. \tag{14.16}$$

Therefore, calculation of $P(\boldsymbol{x} \leq X)$, according to Figure 14.4a, can be extracted from the distribution function, by setting $P(\boldsymbol{x} \leq X) = F(\xi)$, which is equal to the numerical complement of $G(\xi)$:

$$F(\xi) = 0.5 - G(\xi). \tag{14.17}$$

Given the symmetry of the normal function, the calculation of $P(\boldsymbol{x} \leq X)$, according to Figure 14.4b, can be obtained from:

$$P(\boldsymbol{x} \leq X) = 0.5 + G(\xi). \tag{14.18}$$

These calculations are only illustrative for calculating the distribution function, as they can be performed more efficiently and directly using mathematical software that computes the cumulative distribution function (CDF) through the error function (erf).

The classical problem of calculating probability presented below will be useful in discussing the probabilistic aspects of insulation coordination.

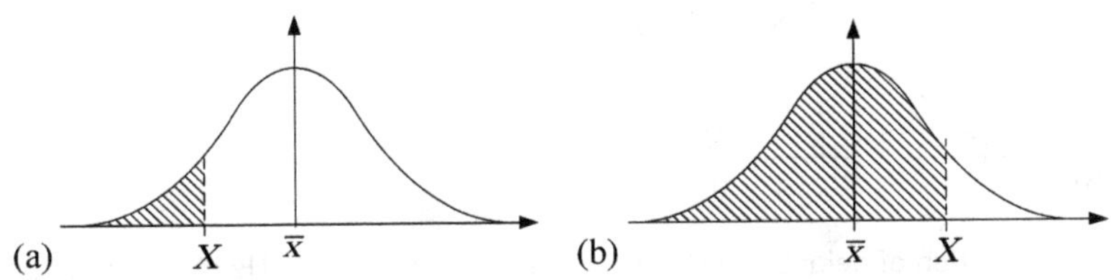

Figure 14.4 Calculation of the probabilities. (a) Probability of $x \leq X$ when $X < \overline{x}$. (b) Probability of $x \leq X$ when $X > \overline{x}$.

The two probability density functions of heights of men and women are known and are defined by their respective means and standard deviations, as in Figure 14.5a:

men: $f(h), \overline{h}, \sigma_h$.
women: $f(m), \overline{m}, \sigma_m$.

It is desired to know the probability that a woman is taller than a man.

The problem is solved by building a new probability density function given by the random variable, $\boldsymbol{x}$, as in Figure 14.5b, which is obtained from the difference between the random variables, $\boldsymbol{m}$–$\boldsymbol{h}$, which is still a Gaussian variable defined by:

$$\mathbf{x} = \boldsymbol{m} - \boldsymbol{h}, \tag{14.19}$$

with the mean value,

$$\overline{x} = \overline{m} - \overline{h}, \tag{14.20}$$

and if the number of samples is high enough, its variance can be approximated by:

$$\sigma_x^2 = \sigma_h^2 + \sigma_m^2. \tag{14.21}$$

To calculate the probability that a woman is taller than a man, it is sufficient to find, based on the difference function, the probability that the random variable $\boldsymbol{x}$ is positive by calculating the probability that $P(\boldsymbol{x} > 0)$. Due to the symmetry of the normal function, the probability that $P(\boldsymbol{x} > 0)$ coincides with the calculation of the distribution function given by (14.17).

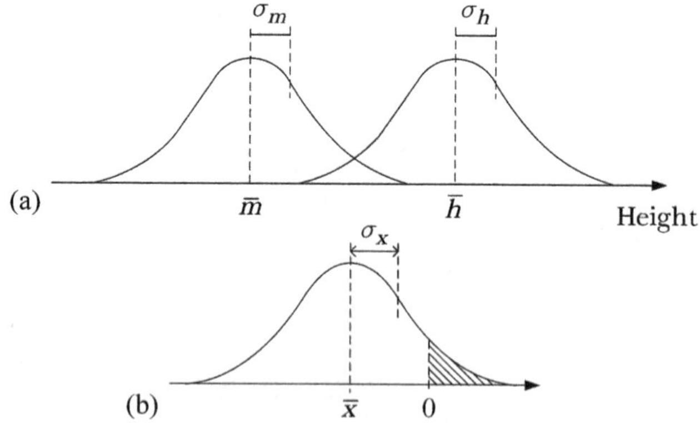

Figure 14.5 Distribution of heights. (a) Heights of men and women. (b) Heights difference function.

Example 14.1 Calculate the probability that a woman is taller than a man using the following normal distributions:

Data

men: $\overline{h} = 1.8646\ \text{m}, \sigma_h = 0.1054\ \text{m}$.
women: $\overline{m} = 1.600\ \text{m}, \sigma_m = 0.0800\ \text{m}$.

Solution

Writing the difference function $\boldsymbol{x} = \boldsymbol{m} - \boldsymbol{h}$:

$$\overline{x} = \overline{m} - \overline{h} = -0.2646\ \text{m},$$

$$\sigma_x^2 = \sigma_h^2 + \sigma_m^2,$$

$$\sigma_x = 0.1323\ \text{m},$$

$$\xi_x = \frac{0 - (-0.2646)}{0.1323} = 2.$$

According to Figure 14.5b, the calculated probability is given by $P(\boldsymbol{x} > 0) = 2.28\%$.

In insulation coordination, a similar problem arises when the probability that an overvoltage surpasses the insulation withstand level needs to be calculated, which will be addressed in Sections 14.4.2 and 14.6.2.

- WEIBULL DISTRIBUTION

The Weibull distribution finds several applications in insulation coordination, and an approximation to this distribution, which approaches the shape of the normal distribution [2], is presented in (14.22). This distribution is truncated at the minimum value of four standard deviations from the mean value with an assumed probability of zero below the minimum.

$$\xi = \frac{\boldsymbol{x} - \overline{x}}{\sigma},$$

$$W(\xi) = 1 - 0.5^{\left(1 + \frac{\xi}{4}\right)^5}. \tag{14.22}$$

With this distribution, the probabilistic approach to transmission lines and equipment can be simplified. Equation (14.22) can be used to define insulation withstand distributions, given that the Weibull function can be truncated between a minimum and a maximum value, which would better approach what happens physically in specific cases. For further details, see Annex B of [2].

14.2.3 Function of Random Variable

In insulation coordination of lightning surges, a statistical treatment can be found that uses the logarithmic values of random variables. The logarithmic functions behave like a normal distribution, even though the distributions of the original random variables are not normal.

For a function of a random variable, for example, $\boldsymbol{y} = g(\boldsymbol{x})$ which increases monotonically much like the logarithmic function, $g(\boldsymbol{x}) = \ln(\boldsymbol{x})$, it is known that (Figure 14.6):

$$P(-\infty < \boldsymbol{x} < X) = P(-\infty < \boldsymbol{y} < Y) = P(-\infty < g(\boldsymbol{x}) < g(X)). \tag{14.23}$$

This means that the probability that $\boldsymbol{x}$ is less than X is equal to the probability that $g(\boldsymbol{x})$ is less than $g(X)$:

$$P(\boldsymbol{x} \le X) = P(\boldsymbol{y} \le Y) \tag{14.24}$$

The median of the random variable, $\boldsymbol{x}$ is designated as x_m, thus:

$$F(x_m) = P(x \le x_m) = 0.5. \tag{14.25}$$

From the previous equations, it can be shown that $g(x_m)$ is the median of $\boldsymbol{y} = g(\boldsymbol{x})$.

Assuming that $\boldsymbol{y} = \ln(\boldsymbol{x})$ has a normal probability density function, the mean and median are the same; thus $\bar{y} = \ln(x_m)$. Therefore, since x_m is the median of $f_x(x)$, $\ln(x_m)$ is the median of the function, $\ln(x)$.

It is also known that the relationship between $f_x(x)$ (the probability density of the random variable $\boldsymbol{x}$), and $f_y(y)$ (the probability density of the random variable $\boldsymbol{y}$), is given by [8]:

$$f_x(x) = f_y(y)\left|\frac{dy}{dx}\right|. \tag{14.26}$$

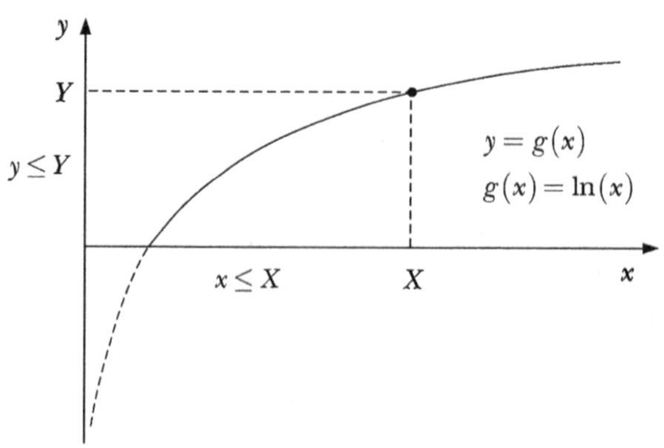

Figure 14.6 Function of random variable.

In this analyzed case, for $\boldsymbol{y} = \ln(\boldsymbol{x})$,

$$\left|\frac{dy}{dx}\right| = \frac{1}{x} \text{ for } x > 0, \tag{14.27}$$

thus,

$$f_y(\ln(x)) = \frac{1}{\sigma_{\ln(x)}\sqrt{2\pi}} e^{-\frac{1}{2}\left(\frac{\ln(x)-\overline{\ln(x)}}{\sigma_{\ln(x)}}\right)^2}, \tag{14.28}$$

$$f_x(x) = \frac{1}{x\sigma_{\ln(x)}\sqrt{2\pi}} e^{-\frac{1}{2}\left(\frac{\ln(x)-\ln(x_m)}{\sigma_{\ln(x)}}\right)^2}. \tag{14.29}$$

14.2.4 Joint Probability Density Function and Distribution with Two Random Variables

For a probability density function with two random variables $\boldsymbol{x}$ and $\boldsymbol{y}$, the distribution function is given by [8]:

$$F_{xy}(x,y) = \int_{-\infty}^{x}\int_{-\infty}^{y} f_{xy}(x,y)dxdy. \tag{14.30}$$

If the function, $f_{xy}(x, y)$, is known, $f_x(x)$ can be found from:

$$f_x(x) = \int_{-\infty}^{+\infty} f_{xy}(x,y)dy; \tag{14.31}$$

the procedure is the same for the random variable $\boldsymbol{y}$.

For conditional probabilities, the following equation may be written [8]:

$$f_y(y/\boldsymbol{x} = x) = \frac{f_{xy}(x,y)}{f_x(x)}, \tag{14.32}$$

which is the probability density function of the random variable $\boldsymbol{y}$ for a given value of variable $\boldsymbol{x}$.

The expression for the normal probability density function of two random variables, $\boldsymbol{x}$ and $\boldsymbol{y}$, is given in (14.33):

$$f_{xy}(x,y) = \frac{1}{2\pi\sigma_x\sigma_y\sqrt{1-r^2}} e^{-\frac{1}{2(1-r^2)}\left(\frac{(x-\bar{x})^2}{\sigma_x^2} - \frac{2r(x-\bar{x})(y-\bar{y})}{\sigma_x\sigma_y} + \frac{(y-\bar{y})^2}{\sigma_y^2}\right)}, \tag{14.33}$$

where r is the correlation coefficient.

Integrating (14.33) as in (14.31) and then dividing (14.33) by the result of this integration, $f_y(y/\boldsymbol{x} = x)$ is obtained in (14.34). This function will be useful in Section 14.6.1 to account for the statistical dependence between the front time and the amplitude of lightning currents.

$$f_y(y/\boldsymbol{x} = x) = \frac{1}{\sigma_y\sqrt{2\pi(1-r^2)}} e^{-\frac{1}{2\sigma_y^2(1-r^2)}\left(y-\bar{y}-\frac{r\sigma_y(x-\bar{x})}{\sigma_x}\right)^2}, \tag{14.34}$$

with median:

$$\bar{y} + \frac{r\sigma_y(x-\bar{x})}{\sigma_x}, \tag{14.35}$$

and standard deviation:

$$\sigma_y\sqrt{1-r^2}. \tag{14.36}$$

14.3 Insulation Strength

Insulation breakdown depends on several factors, such as magnitude, waveform, duration, polarity, electric field distribution, type, and aging of the insulating medium.

Positive polarity usually has a lower insulation breakdown level than negative polarity, so that its use is more critical for calculations.

The overvoltage waveform is a crucial parameter, both for the analysis of insulation breakdown and for the final selection of the insulation levels to be adopted. The dependence on waveform would imply a very complex analysis of insulation breakdown, entailing inspection of every overvoltage over a wide variety of situations in each system. The calculation tools are quite advanced today, and this would no longer be a constraint. However, there is still a neeed for better consolidation of breakdown models and endorsement by standards.

To address this issue, the standards classify overvoltages according to their wavefront, that is, into fast and slow front-time overvoltages, which allows a simplification for the purpose of insulation coordination.

A significant simplification in the analysis of slow front-time surges comes from the possibility of considering only the amplitude of the overvoltage.

The analysis of fast front-time overvoltages stressing self-restoring insulation, for instance, an insulator string, can be performed with breakdown models, as discussed in Section 14.3.3. The same analysis for non-self-restoring insulation, for instance, the internal insulation of a transformer, can only be performed in a simplified way, comparing the calculated waveforms with full-wave and chopped-wave tests.

Insulation performance at standard test levels is the source of information used to verify the adequacy of insulation against power system overvoltages, details of which can be found in the high voltage techniques [9].

A brief description of full wave standardized tests and considerations entering into insulation classification are presented next.

14.3.1 Impulse Tests for Lightning and Switching Surges

Impulse tests on system components are carried out in high-voltage laboratories using standard waveforms to assess the behavior of insulation subjected to overvoltages.

The lightning impulse test, for fast-front overvoltages, has the standard waveform in Figure 14.7a with a front time of 1.2 μs, and the magnitude declines to 50% after 50 μs.

The switching surge impulse test, for slow-front overvoltages, has a far greater front time. The front time in the standard switching impulse test is 250 μs, with the peak value declining to 50% in 2500 μs, Figure 14.7b.

14.3.2 Self-Restoring and Non-Self-Restoring Insulation

The insulation used by electrical systems can be classified according to its ability to restore its dielectric capacity.

Self-restoring insulation can restore its dielectric properties after a disruptive discharge. These properties are restored by renewal of the dielectric medium, such as air insulation, for example, in the case of insulator strings in transmission lines.

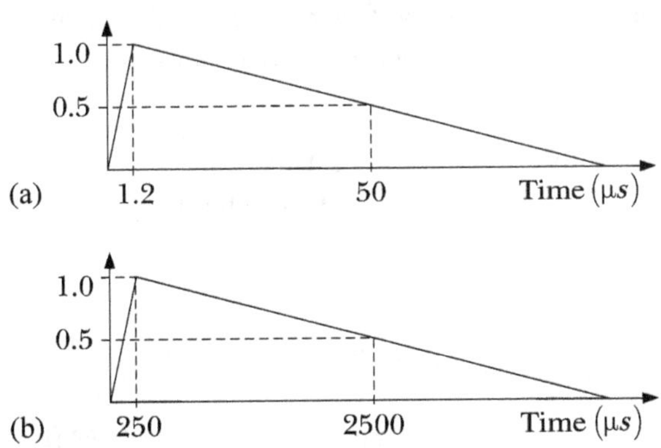

Figure 14.7 Standard impulse tests. (a) Lightning impulse test. (b) Switching surge impulse test.

The dielectric properties of non-self-restoring insulation deteriorate after a flashover and therefore cannot be restored, as in the case of equipment internally insulated with oil-impregnated paper or even plastic materials.

There are insulation levels related to switching surges and lightning surges. The switching surge level, related to slow-front overvoltages, is known as the standard switching impulse withstand voltage. The corresponding term in American standards is basic switching level (BSL). The withstand value related to lightning surges with fast-front waveforms is known as the standard lightning impulse withstand level, whose corresponding term in American standards is the basic impulse level – BIL.

In the case of non-self-restoring insulation, as in the interior of transformers, the withstand values, BIL and BSL, are treated by a deterministic approach.

With respect to power equipment, successive tests may damage it, and the statistical performance of the insulation is unknown; although a distribution function can be supposed to exist, the BIL and BSL should be far from breakdown levels, Figure 14.8a.

Thus, a negligible probability of insulation breakdown is assumed for the equipment, so that below the withstand level, BIL and BSL, the disruptive discharge probability is assumed 0, and above it, the probability is unknown but may be assumed 1.

The non-self-restoring insulation has a predominantly deterministic design, although a probabilistic approach based on a safe reliability criterion may be assumed for the equipment. The selected reliability criterion determines the characteristics of the incoming surges as well as the overvoltages inside the substation. The calculation method is described in Clause 5 in [4] and is based on the mean time between failures.

In the case of self-restoring insulation, as for example, in overhead transmission lines, the approach is different, and the insulation may assume an acceptable probability level for the disruptive discharge. For self-restoring insulation, it is important to know the voltage at which the probability of electrical breakdown, termed CFO (critical-flashover overvoltage), is 50%, which in a Gaussian probability density function coincides with the mean value.

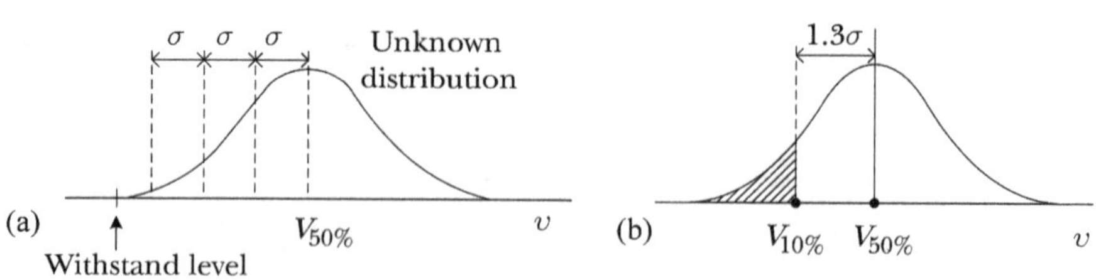

Figure 14.8 Standard impulse withstand levels. (a) Non-self restoring insulation. (b) Self-restoring insulation.

For self-restoring insulation, especially regarding slow-front overvoltages, an important reference value is the voltage at which the probability of breakdown is 10%, equivalent to a 90% withstand probability, as shown in Figure 14.8b. In this case, the BSL assumes a 10% probability of a disruptive discharge.

To obtain the withstand levels, tests are carried out, but high voltage techniques are not explored here.

The insulation levels defined by $V_{10\%}$, or BSL, and $V_{50\%}$, or CFO, can be related by the standard deviation σ of the normal distribution.

$$V_{10\%} = V_{50\%}(1 - 1.3\sigma). \tag{14.37}$$

Here are some values for the probability of insulation breakdown with respect to position on the normal distribution. Some values of the probability of breakdown and their position on the normal distribution are illustrated in Figure 14.9.

$$V_{0.13\%} = V_{50\%}(1 - 3\sigma), \tag{14.38}$$

$$V_{2.3\%} = V_{50\%}(1 - 2\sigma), \tag{14.39}$$

$$V_{16\%} = V_{50\%}(1 - \sigma), \tag{14.40}$$

$$V_{84\%} = V_{50\%}(1 + \sigma), \tag{14.41}$$

$$V_{97.7\%} = V_{50\%}(1 + 2\sigma), \tag{14.42}$$

$$V_{99.87\%} = V_{50\%}(1 + 3\sigma). \tag{14.43}$$

The standard deviation is calculated in pu relative to the mean value $V_{50\%}$, and the case of voltage $V_{16\%}$ is exemplified. When the mean value $V_{50\%}$ and the standard deviation, σ', are originally given in kV, for instance, Eq. (14.44) can be written as:

$$V_{16\%} = V_{50\%} - \sigma'. \tag{14.44}$$

To obtain the standard deviation, σ , in pu relative to the mean, Eq. (14.45) is applied:

$$\sigma = \frac{\sigma'}{V_{50\%}}. \tag{14.45}$$

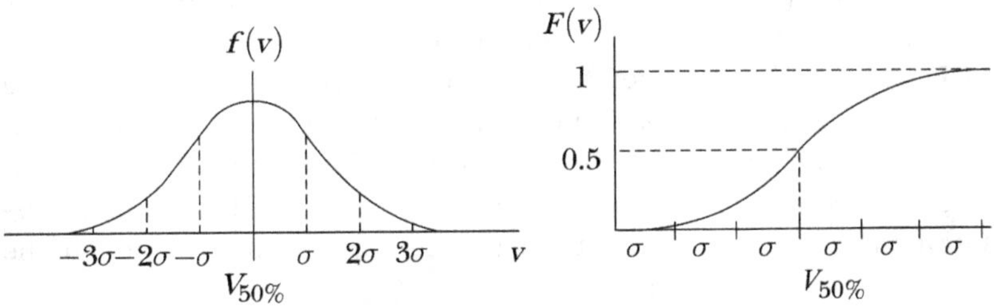

Figure 14.9 Some breakdown values.

Thus, working with σ in pu relative to the mean $V_{50\%}$, $V_{16\%}$ is obtained from:

$$V_{16\%} = V_{50\%}(1-\sigma). \tag{14.46}$$

14.3.3 Withstand Levels for Self-Restoring Insulation

- Withstand to Slow-Front Overvoltages
Numerous laboratory tests have made it possible to obtain curves and approximated formulas for the insulation strength of air, considering the shape of the electrodes, length of the gap, and polarity of the applied wave. Relevant contributions on withstand levels were discussed in [10, 11], taking into account the polarity and wet and dry conditions. The influence of electrode shape on the behavior of air gaps, both with and without insulator strings between the electrodes, is also determined.

A useful approximation for the average withstand level, $V_{50\%}$, is given by:

$$V_{50\%} = K500d^{0.6}, \tag{14.47}$$

where:

d: length of the gap in m.
K: *gap* factor that takes into account the shape of the electrodes.

For instance, $K = 1$ for rod-plane gaps and $K = 1.45$ for conductor-crossarm gaps.

The standard deviation as a percentage of $V_{50\%}$ is assumed to be σ = 6% (5% according to [4]). This formula applies to positive polarity and dry conditions, which gives lower withstand levels than negative polarity, and is more representative of air gaps 2–7 m long. Insulators can decrease the breakdown level, but the effect is small for dry cap and pin insulators.

The strength of the insulation also depends on the front time; thus, the worst-case scenario for the withstand level should be used. In determining the withstand voltage using the standard voltage waveform of 250/2500 μs, this effect is somehow included, since, for EHV distances, the critical front times range from 100 to 250 μs.

Positive polarity pulses together with the critical front times, result in a minimum CFO value for air-porcelain insulation.

The CFO can also be estimated from Eqs. (14.11)–(14.15) in [4] for V-string and I-string insulators. I-strings can consider the swing angle. For wet conditions, which are the most usual, the suggested reduction in the CFO is about 4%. The strike distance to a tower-window is the smallest of the following distances from the cable: to the upper truss, to the tower side, or the length of the insulator string divided by 1.05.

Some gaps are called compound because they have more than one probability of breakdown. For a cable that the breakdown can occur to the upper structure, with probability p_1, or to the lateral structure, with probability p_2; the breakdown probability for the compound *gap* is given by:

$$p_c = 1 - (1 - p_1)(1 - p_2).$$

The gap with the greater p value usually prevails over the others.

- Withstand to Fast-Front Overvoltages
Insulation level in air can be approximated by a linear function of gap length and it is lower for the positive polarity [10, 11]. The CFO is estimated to be 489 kV/m in [12], 530 kV/m in [2], and 560 kV/m in [4] for positive polarity. References [13, 14] estimate the CFO 525 kV/m for a rod-plane gap and positive polarity. To calculate the CFO considering the gap factor K, (the same K used in switching surges), the following expression can be used:

$$CFO = (0.75 + 0.25K) \times 525\ \text{kV/m}.$$

Gaps with insulator strings depend on the type of insulator used and a conservative value to be assumed is 560 kV/m.

More information on negative polarity and insulation levels can be found in [13, 14]. The typical standard deviation for lightning surges is $\sigma = 3\%$.

Insulation performance depends on the duration of the wavefronts and applied voltage. For lightning surges, an increase in impulse duration causes breakdowns of lower magnitude.

This fact is evident by the voltage × time to breakdown ($v \times t$) curve, which is obtained using test applications with standard waveforms. Initially, at low applied voltage levels, as for example, amplitude V_2 in Figure 14.10a, breakdown occurs at the tail of the waveform and, as the peak value increases, as for example, in the case of amplitude V_1, breakdowns shift to the wavefront.

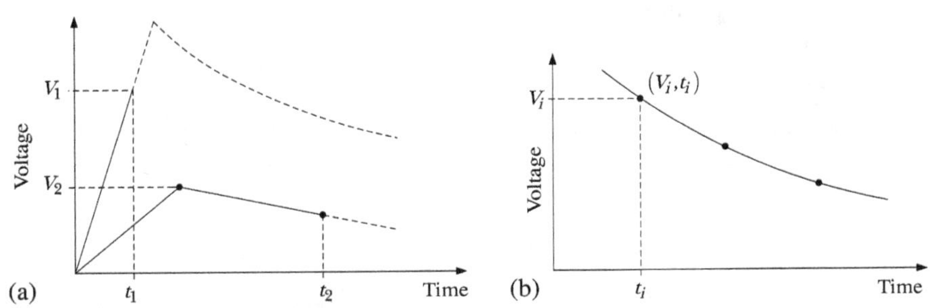

Figure 14.10 Voltage-time to breakdown. (a) Registered voltage-time pairs (V_i, t_i). (b) Voltage-time curve ($v \times t$).

The peak value of the voltage and the time to breakdown are recorded (V_i, t_i) and plotted, as in Figure 14.10b.

References [5, 15, 16] show the voltage x time curve, developed in [17], for various lengths of insulator strings.

$$\frac{V}{d} = 400 + \frac{710}{t^{0.75}}, \tag{14.48}$$

where:

V: peak value of applied voltage (kV).
d: length of the insulator (m).
t: time to breakdown (μs).

However, it is inaccurate to assume flashover, when the voltage waveform intersects the volt-time characteristic, since it only relates the peak voltage of the standard impulse to the time to breakdown.

The insulation of transmission lines and substations is subject to a great deal of waveform variability, and therefore, a more precise breakdown criterion than that provided by the voltage-time curve is necessary.

One of the approaches to evaluation of the insulation withstand, depending on the applied waveform, is through methods of integration and calculation of the disruptive effect (*de*), according to Refs. [13, 15, 18, 19].

The purpose of this calculation is to predict the insulation behavior in response to nonstandard waveforms based on flashover data from standard impulse tests and this is done using the following integral:

$$de = \int_{t_0}^{t} (u(t) - u_0)^n dt, \tag{14.49}$$

where:

$u(t)$: applied voltage waveform.
u_0: reference voltage (onset voltage) to be fitted.
n: coefficient to be fitted.

Integration is performed only at intervals in which $u(t) > u_0$.

The constants can be obtained by taking at least two points (1, 2) from the $v \times t$ curve to find the solution to the equation:

$$de = \int_{t_0}^{t_1} (u_1(t) - u_0)^n dt = \int_{t_0}^{t_2} (u_2(t) - u_0)^n dt, \tag{14.50}$$

using $u_1(t)$ and $u_2(t)$, according to Figure 14.10.

For a good fit, it is of interest for points 1 and 2 to be widely spaced along this curve, and according to [19], for an insulator string with eight insulators, the parameters of best fit were $u_0 \approx 0.9V_{50\%}$ (639 kV) and $n = 1$. Insulator string tests provide the best-fitting parameters, but this is not always feasible.

The leader progression model is also used to analyze insulator breakdown [15, 20]. This model considers physical hypotheses for dielectric breakdown, fitted according to empirical formulas and analytical formulations. The advantage of this method is its flexibility to analyze polarity and waveform of the surge, as well as the presence of insulators filling the air gap.

The time to breakdown is given by the sum:

$$t = t_i + t_s + t_\ell,$$

where:

$t_i \approx 0$: corona inception time.
t_s: streamer propagation time.
t_ℓ: leader propagation time.

The empirical formulation for t_s, in μs, is given by:

$$\frac{1}{t_s} = 1.25\frac{E}{E_{50}} - 0.95,$$

where:

E: electric field across the gap before breakdown (kV/m).
E_0: electric reference field for insulator string, 520 kV/m at positive polarity and 600 kV/m at negative polarity. The similarity of E_0 to the estimates of CFO/m at positive polarity (see above) should be noted.

$$E_{50} = \frac{CFO}{d}.$$

As the propagation time of the streamer depends on the electric field, its correct evaluation can only be done a posteriori. However, it is acceptable to assume that the streamer phase is complete when the electric field surpasses E_0. Additionally, the model's consistency with the voltage-time curve should be verified.

In the leader stage, a reasonable expression for the speed of propagation is given by:

$$\frac{d\ell(t)}{dt} = ku(t)\left(\frac{u(t)}{d - \ell(t)} - E_0\right). \tag{14.51}$$

where:

d: insulator string length.
ℓ: channel length.

For cap and pin insulators, the suggested values for the constant, k, is 1.2 at positive polarity and 1.3 at negative polarity, in units of $\frac{\mathrm{m}^2}{\mathrm{kV}^2\mathrm{s}}$.

The length of the channel only increases at positive speeds in (14.51), remaining stationary at negative speeds. Breakdown occurs when $\ell(t)$ reaches d.

The parameters of best fit to the model are discussed in [20]. A review of some different implementations of the model is also presented in Ref. [13].

- Atmospheric Correction Factor

  The ionization and rupture of air insulation depends on atmospheric conditions, such as temperature, humidity, relative air density, pressure, etc.

  The behavior of these parameters randomly changes the levels of air insulation, introducing another factor of probabilistic nature.

  Data collected in meteorological stations, by means of numerous records, are organized in the form of histograms for insulation coordination purposes. There is a standard methodology for treatment of atmospheric data, resulting in a factor that contains probabilistic variables.

  This insulation correction factor, also known as the atmospheric correction factor (*acf*), is applied to the standardized voltage levels obtained in laboratory tests.

$$V_{corr} = acf \times V_{test}.$$

  This factor is obtained by analysis of variables, such as air density and humidity, according to specific standards [9]. However, as suggested in [2], for purposes of insulation coordination, the effects of humidity and temperature cancel each other out. Thus, only the altitude, along with its correction factor, K_a, which is related to the pressure, needs to be considered in the calculation:

$$K_a = e^{m\left(\frac{h_a}{8150}\right)},$$

  where:

  m: coefficient obtained from Figure 9 of Ref. [2] for switching surges and is equal to 1 for lightning surges.
  h_a: altitude above sea level.

  For further details, see Annex H of [2].

14.4 Insulation Coordination Methods

Insulation coordination at maximum operating voltages and temporary overvoltages are not presented in this book and the application methodology can be found in the technical standards. Therefore, only basic aspects of insulation coordination focusing on slow-front and fast-front overvoltages are presented here. The procedure described in [2] serves as a foundation and its nomenclature is summarized below, with some definitions. In this chapter, a distinction was made between overvoltage stress, *U*, and withstand voltage, *V*, but in [2] both voltages are designated only by the letter *U*. Some of the definitions in [2] are mentioned below:

U_m: highest voltage for equipment.
U_{cw}: co-ordination withstand voltage of equipment.
U_ω: standard withstand voltage.
U_{e2}: phase-to-earth overvoltage with a 2% probability of exceedance.
U_{et}: truncation value of the distribution function of phase-to-earth overvoltages.
U_{pl}: lightning impulse protection level of an arrester.
U_{ps}: switching impulse protection level of an arrester.
U_{rp}: amplitude of the representative overvoltage.
$U_{r\omega}$: required withstand voltage.
K_c: coordination factor.
K_{cd}: deterministic coordination factor.
K_{cs}: statistical coordination factor.
K_s: safety factor.

- Representative Voltages and Overvoltages (U_{rp}, U_{rs})
  The representative continuous power-frequency voltage may be considered equal to the highest system voltage.
  Calculations and analysis of the system provide the values of temporary overvoltages, switching surges, U_{rs}, and lightning surges, U_{rp}, with different waveforms. Representative overvoltage, regardless of the evaluated waveform, assumes its peak value with a standard waveform.
  It should be noted that the best characterization of overvoltages is obtained with appropriate models and calculation tools based on detailed studies of the electrical system. When evaluating overvoltages, it is essential to take into account the limiting devices such as surge arresters.
  The subject is quite extensive, and standards [2, 4] provide a comprehensive discussion of system overvoltage stresses and how to obtain representative voltages of switching and lightning surges. In this book, the subject is addressed in Chapters 7–9.
- Co-Ordination Withstand Voltage ($U_{c\omega}$)
  Insulation performance depends on the dielectric properties of the materials used, on the amplitude and on the waveform of the applied overvoltage. Breakdown in air also depends on atmospheric conditions.
  The coordination voltage is associated with a withstand level that meets a certain performance criterion for the equipment or insulator string and for self-restoring and non-self-restoring insulation, as described in Section 14.3.2. This voltage also considers the probability of a disruptive discharge of the insulation.

For non-self-recovering insulation, the probability of breakdown is considered negligible, if the stress is lower than the withstand level, and 100%, if it is higher.

The performance of self-restoring insulation has a statistical interpretation. Although the Weibull distribution can be used [2], breakdowns are considered here using the Gaussian distribution.

- Required Withstand Voltage ($U_{r\omega}$)
The overvoltages during the lifetime of the equipment are different from the standard test application. The required withstand voltage must consider this difference and also consider the circumstances that can reduce the level of equipment insulation under actual operating conditions, such as aging and environment, ensuring that the coordination withstand voltage is met in the field. Therefore, correction factors for the environment and safety factors should be employed.
The safety factors proposed are 1.15 for internal insulation of the equipment and 1.05 for external insulation, both for fast-front and slow-front overvoltages.
- Standard Withstand Voltage U_{ω}
The standard withstand voltages may be found in the tables in Ref. [1], according to the highest voltages for the equipment, U_m. There is more than one insulation level for a certain system nominal voltage, and the higher levels may be intended for equipment lacking the protection of surge arresters. A similar procedure is adopted to determine BSL and BIL in [4], but considering small differences in safety margins.
Two basic methods of insulation coordination, the deterministic and the statistical method, are presented next.

14.4.1 Deterministic Method

To obtain the co-ordination voltage, U_{cw}, this method applies a deterministic co-ordination factor, K_{cd}, to a representative voltage. The deterministic co-ordination factor considers uncertainties in the withstand and representative voltages.

$$U_{c\omega} = K_{cd} U_{rp} \text{ or } U_{c\omega} = K_{cd} U_{rs}. \tag{14.52}$$

This method is applied to the non-self-restoring insulation of the internal parts of the equipment, which may require a greater safety margin if the equipment is not protected by surge arresters. No reference to the probability of equipment failure is possible when this method is applied.

14.4.2 Statistical Method

The statistical method is applied to self-restoring insulation, in the case of insulator strings or even the external parts of equipment. In this case, the probabilistic approach is very convenient when defining tower distances and insulator string length, which facilitates the analysis and comparison of alternatives.

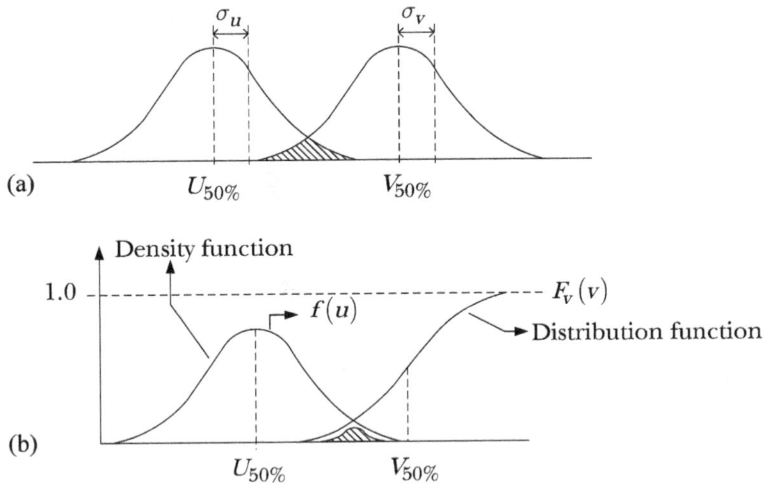

Figure 14.11 Calculation of the failure risk. (a) Overvoltage and withstand voltage probability densities. (b) Failure risk.

In the treatment of self-restoring insulation, a small risk of insulation breakdown is permissible, which is delimitated by the performance criteria.

If the statistical distribution of slow-front overvoltages ($U_{50\%}$, σ_u) and insulation withstand ($V_{50\%}$, σ_v) are known, a suitable analytical approach may be proposed. Gaussian probability densities are assumed here, but the same concepts are applicable to other probabilistic distributions, such as the distribution of Weibull.

The statistical approach relies on the relative position of the probability density functions of the overvoltages and withstand voltages, such that the risk of insulation breakdown is acceptable, Figure 14.11a.

The calculation of the probability of insulation breakdown is shown below. For the withstand voltages, it is convenient to work with the distribution function, $F_v(v = V)$, that gives the probability of breakdown at overvoltages lower than V, as in Figure 14.11b

To work with the probability density curve, incremental mass probabilities are assumed (Figure 14.12).

The probability of insulation breakdown, at a certain voltage level, u, is the probability of occurrence of two simultaneous events, one of which is the voltage, u, with an incremental probability of $f(u)d(u)$, and the other is the insulation breakdown, given by the distribution function, $F_v(u)$ in Eq. (14.12).

Since these two events may be assumed independent, the incremental probability of insulation breakdown is calculated from the product of the two probabilities.

$$dR = F_v(u)f(u)du. \tag{14.53}$$

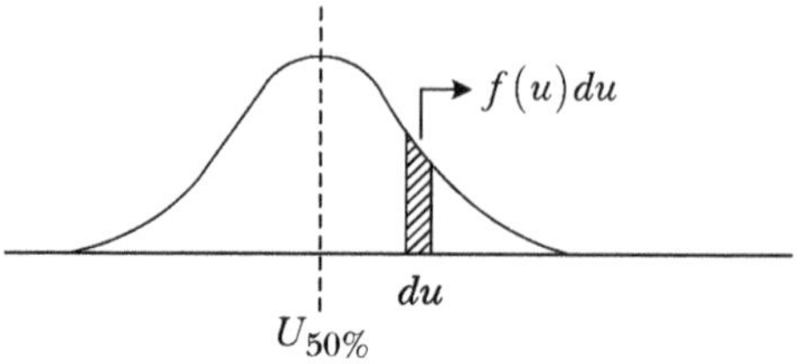

Figure 14.12 Incremental mass probability.

The breakdown probability can be calculated for any distribution function by numerical methods and is obtained from:

$$R = \int_{-\infty}^{+\infty} f(u)F_v(u)du. \tag{14.54}$$

In the case of a Gaussian distribution, it is possible to solve the problem by introducing a significant simplification into the calculation. This problem is identical to the problem of height in men vs. women in Example 14.1, replacing female height with overvoltage and male height with the withstand voltage; the probability that an overvoltage value exceeds the withstand voltage value needs to be obtained.

To calculate the probability that the overvoltage exceeds the withstand voltage, the difference function is obtained:

$$\overline{x} = U_{50\%} - V_{50\%}, \tag{14.55}$$

$$\sigma_x^2 = \sigma_u^2 + \sigma_v^2, \tag{14.56}$$

calculating

$$P(\boldsymbol{x} > 0), \tag{14.57}$$

or alternatively

$$\overline{x} = V_{50\%} - U_{50\%}, \tag{14.58}$$

$$P(\boldsymbol{x} \leq 0), \tag{14.59}$$

which is the probability of finding a negative difference between withstand voltage and an overvoltage.

In general, when Gaussian densities cannot be assumed, equation (14.54) should be used.

The Monte Carlo simulation method can be applied [18, 21] by combining overvoltage and breakdown probabilities, using a shot-by-shot approach, with risk of failure analysis of slow-front and fast-front waves. In Section 14.6, the Monte Carlo method is applied to the study of fast-front overvoltages. So far, the probability of

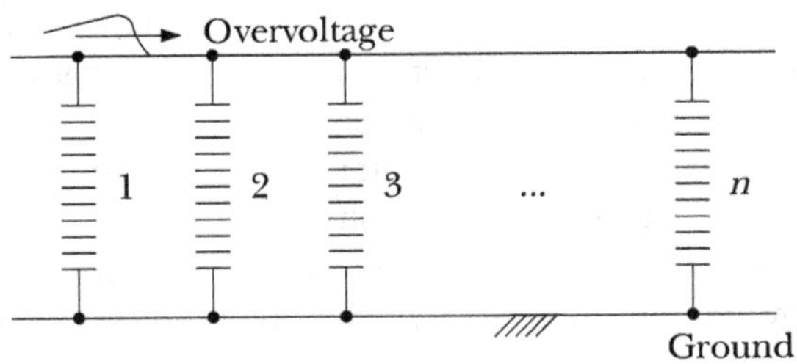

Figure 14.13 Several gaps in parallel.

breakdown of one gap was examined. The case of several gaps in parallel in a substation or in a transmission line subjected to the same overvoltage stress will be analyzed next (Figure 14.13).

The probability of breakdown within the set of parallel gaps means the breakdown in at least one gap or insulator string (Section 14.2.1).

Given p_1, the probability that breakdown occurs in one gap, the complementary probability of withstanding the stress is calculated as $(1 - p_1)$.

Assuming n identical gaps subjected to the same overvoltage, the probability that all of them withstand the stress, simultaneously, is given by $(1 - p_1)^n$.

The probability of breakdown of at least one insulator string is obtained from the complementary probability that all insulator strings withstand the stress, given by the expression:

$$p_t = 1 - (1 - p_1)^n. \tag{14.60}$$

This calculation makes no distinction between positive and negative polarity, each corresponding to 50% of the number of shots. As the insulation strength at negative polarity is greater than that at positive polarity, it contributes little to the calculation and can be neglected [4]; therefore, a 50% reduction in the final rate can be applied.

In the case of unequal gaps, the treatment is analogous, that is, by taking the product of the probabilities and assuming independent events.

The case of gaps in parallel is found in transmission lines and substations, and this procedure is useful for slow-front overvoltages. In this case, the whole substation is assumed to have the same surge level, considering the small distances and the relatively slow wavefronts, as in the case of nominal frequency. Therefore, since the frequencies are not very high, the small bus lengths do not show appreciable voltage drops, which is easily understood with the π model for short cables (10–100 m) that show the same voltage at both extremities over this moderate range of frequencies.

In a substation, to calculate the risk of failure, all similar gaps can be grouped and subjected to the same overvoltage.

In a transmission line, a rigorous calculation should take into account the overvoltages in the insulator strings over the length of the transmission line, which would be a cumbersome task.

Assuming all the insulator strings are equal, an alternative approach may be used to calculate the overvoltage distribution over the line and group the parallel gaps subjected to the same overvoltage levels in the respective line sections. The overall transmission line failure rate can then be more easily obtained.

14.4.3 Simplified Statistical Method

It is possible to use the simplified method that provides a preprocessing relating the risk of failure with the statistical coordination factor, K_{cs}, [2]. This factor, in turn, establishes the relationship between U_{e2} and $U_{c\omega}$ (see previous definitions), allowing the appropriate selection of insulation level for a certain failure risk.

$$K_{cs} = \frac{U_{c\omega}}{U_{e2}}. \tag{14.61}$$

Based on the U_{e2} value and the assumed risk of failure of the insulation, the factor, K_{cs}, is extracted from Figure 14.17 and Figure 8 in [2]. A similar approach can be found in [4].

14.4.4 Further Comments on Slow-Front and Fast-Front Overvoltages

Only slow-front and fast-front transients are analyzed below. For details on the proper treatment of the nominal system operating voltage and temporary overvoltages, see Ref. [2].

- Slow-Front Overvoltages
  a) Deterministic method

As previously mentioned, the deterministic method is more suitable for non-self-restoring insulation. When surge arresters are used, the maximum representative voltage is given by the protection level for the slow-front waves of the arrester, that is, $U_{rs} = U_{ps}$, and the deterministic coordination factor in [2] is adopted as a function of the ratio, $r = U_{ps}/U_{e2}$:

$$U_{c\omega} = K_{cd} U_{ps} \; K_{cd} = f(r), \tag{14.62}$$

where:

$$r \le 0.7 \; K_{cd} = 1.1.$$

$$0.7 < r \le 1.2 \; K_{cd} = 1.24 - 0.2r.$$

$$1.2 < r \; K_{cd} = 1.0.$$

The maximum protection level of the arrester determines U_{et} as well as the truncation of the overvoltage probability distribution, assuming $K_{cd} = 1$.

b) Statistical method

The maximum amplitude of the overvoltage is used in the elaboration of the probability density function. The insulation of the transmission line is defined by matching the failure rate with the performance criterion adopted. The flashover rate for switching surges is designated by SSFOR in [4] given by the ratio of expected number of flashovers per number of switching operations. It is also possible to use the simplified statistical method.

- Fast-Front Ovarvoltages
  a) Deterministic method.
  When non-self-restoring equipment is protected by arresters, the deterministic coordination factor, K_{cd}, has a unitary value, and thus the co-ordination voltage is equal to the representative voltage. A formula to calculate, $U_{c\omega}$, is presented in [2], which takes into account several factors, as well as the arrester's lightning impulse protection level, U_{pl}, (see Example 14.2). However, accurate results should be used if detailed simulations are available.
  b) Statistical method.
  The lightning surges and the insulation depend on random factors. Since the standard deviation of the insulation is small, it is acceptable in very simplified approaches to analyze the likelihood of breakdown by checking only if the maximum overvoltage exceeds the insulation CFO. However, a more accurate failure analysis can be done with the calculation of the lightning surge waveform along with the leader progression model.
  Examples of statistical applications for self-restoring insulation are described in Sections 14.5 and 14.6.

14.5 Insulation Coordination of Substations

For systems with a rated voltage of less than 245 kV, group I in [2], the maximum operating voltage and fast-front overvoltages are in general sufficient to prevail over the requirements for slow-front overvoltages. For systems with a rated voltage above 245 kV, group II in [2], the overvoltage requirements for slow-front overvoltages in general supersede the requirements for maximum operating voltage and fast-front overvoltages.

The insulation coordination for substations considers the maximum operating voltage, temporary, slow-front, and fast-front overvoltages. Ref. [2] presents all the details and procedures needed to perform the insulation coordination of substations in groups I and II, and gives illustrative examples, which will not be repeated here, since only a few of the most relevant topics are addressed in this section.

Shield wires and masts are used to effectively protect the equipment. The substation shield design is described in [22].

14.5.1 Power-Frequency Voltage

The insulation behavior at the rated frequency is dependent on environmental conditions such as humidity, degree of pollution of insulators, etc. The technical standards define the characteristics of the insulators at different pollution levels [2].

The behavior of the insulation under standard test conditions for different gap sizes is shown in Figure 14.14 [5].

14.5.2 Fast-Front Overvoltages

The substations are protected from lightning surges by shield wires and lightning rods. Analysis of the shielding is like the case of transmission lines. The lightning surges at internal points in the substation, created by lightning strokes to transmission lines, are calculated with the aid of electromagnetic transient programs, according to the procedure mentioned in Chapter 9.

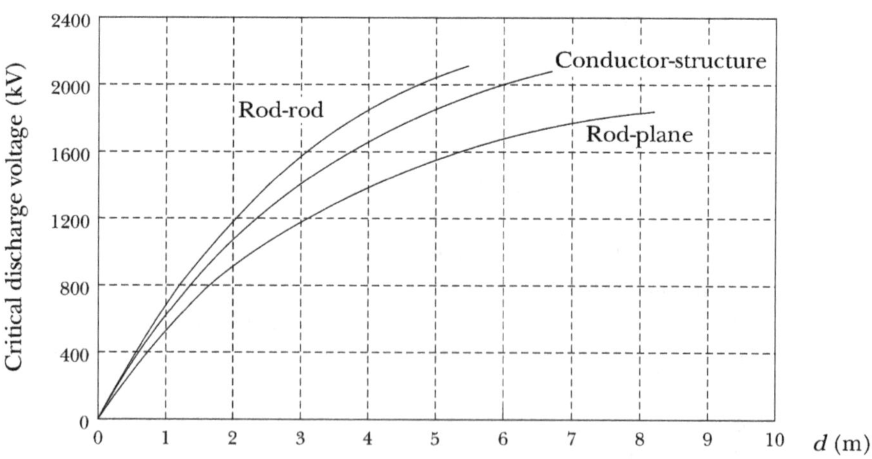

Figure 14.14 Insulation strength at power frequency. *Source:* [5]/Electric Power Research Institute.

In the analysis of traveling waves, cables are represented by surge impedances and the time delays are determined by the length of buses and wave speeds.

The equipment is generally represented by the internal capacitances, since the inductive elements, at high frequencies, can almost be treated as open circuits.

The front-time of the lightning surges that reach the substation is the result of several factors, such as the waveform of the return stroke, the stroke point on the line, breakdown of insulators, etc. Inside the substation, the waveform is altered by discontinuities in buses and cables, in addition to interactions with equipment.

An accurate approach is to calculate the waveforms that reach the substation using EMT-type programs. A three-phase representation of the transmission line and buses can be performed, incorporating the frequency dependence of the line parameters, but models of the corona must be provided by the user. Single-phase studies on a conservative basis are useful to provide the main transient effects. The selection of the incoming surge can take into account both shield failures and backflashes. As backflash rates frequently prevail, effectively only the backflash is considered. For a backflash, the steepness of the surge at the flashover location may be assumed infinite and the step becomes a ramp with a slope that decreases due to the corona as the distance traveled increases. A discussion on the use of wavefronts associated with the mean time between failures and backflashover rates can be found in [4]. This approach is implicitly assumed in Example 14.2 below [2].

Steeper wavefronts increase overvoltages, but the insulation strength also increases in this case. However, the worst cases are usually found with the sharpest wavefronts.

The results in Figure 14.15 exemplify the calculation for a 500-kV substation with a breaker-and-a-half layout, which is subjected to an overvoltage originating from a stroke to the transmission line. The arresters are located at the line entrance at the substation and at the transformer, thus limiting the overvoltage at the transformer and at the substation's internal points.

For lightning surges in a substation, there is a need to analyze the protection distance of lightning arresters in relation to the existing equipment (see Chapter 13).

The representative voltage has a standard waveshape, which is different from the lightning waveform that stress the equipment. Thus, it is necessary to find a standardized wave that presents a stress equivalent to the lightning wave analyzed, mainly in the case of GIS and transformers.

In specifying the BIL, this decision is very difficult with approximate procedures. Annex I, recently introduced in [2], is based on experiments focused on this topic and provides a remarkable contribution on this matter.

This annex describes five typical waveforms, from A to E, to guide de BIL selection. As an example, the nonstandard waveform in Figure 14.15 is likely to be

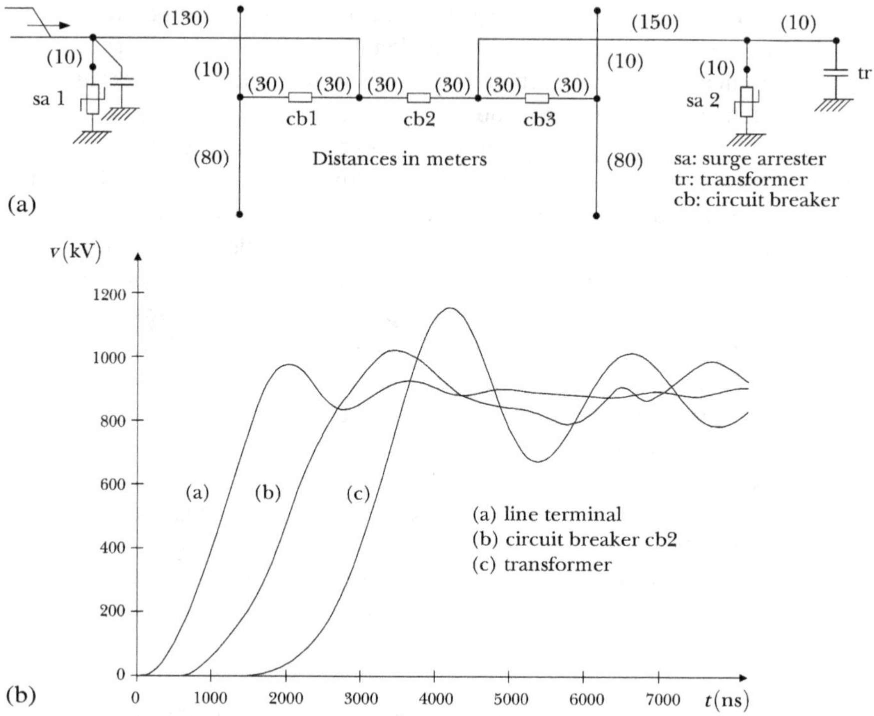

Figure 14.15 Lightning surges in substations. (a) One-and-half circuit breaker layout. (b) Voltage waveforms.

found in transformers, with an initial peak followed by oscillations of smaller amplitude, as in Example 13.7 in Chapter 13. This waveform can be represented by an equivalent standardized wave, following the steps in this annex.

14.5.3 Slow-Front Overvoltages

As shown previously, switching surges have slow wavefronts and it is possible to assume that the entire substation is subject to the same overvoltage level, generally limited by the level of protection offered by the surge arresters.

Insulator strings with identical gap factors are grouped together for purposes of calculation of the probability of breakdowns inside the substation. Reference [2] gives detailed examples of insulation coordination for substations in groups I and II. The following is a partial example of the method proposed in [2] for a substation in group II, addressing mainly slow and fast-front surges, thereby complementing the discussion in Chapters 8 and 9.

Example 14.2 This example shows the insulation coordination of a transformer in a 525-kV substation in group II, using the simplified procedure [2, 23], covering Sections 14.5.2 and 14.5.3 above. The substation is fed by two transmission lines, and the transformer is protected by a surge arrester with a rating of 420 kV (Figure 14.16).

Data

U_{ps} = 860 kV: switching impulse protection level of the surge arrester.
U_{pl} = 950 kV: lightning impulse protection level of the surge arrester.
U_{e2} = 832 kV: overvoltage phase-to-earth with a 2% exceedance probability.
$N = 2$: number of lines connected to the substation.
L_t = 30 m: total length, $d + d_1 + d_2 + d_A$.
L_{sp} = 400 m: length of the span between towers.

R_a = 0.25% failures/year is the acceptable failure rate for equipment and suggested values are between 0.1% and 0.4% failures/year.

R_{km} = 0.5 failures/100 km/year. The overhead line outage rate is discussed in Section 14.6.1 and here it should be referred to the first kilometer adjacent to the substation.

A = 11 000 kV is a factor given in table E.2 [2].
Altitude: 800 m above sea level.
d: length of the lead connecting the surge arrester to the protected equipment.
d_1: length of the lead connecting the surge arrester to the transmission line.
d_2: length of the lead connecting the surge arrester to earth.
d_A: length of the active part of the surge arrester.
Z_g: grounding impedance.
T: protected equipment.
U: impinging overvoltage surge.

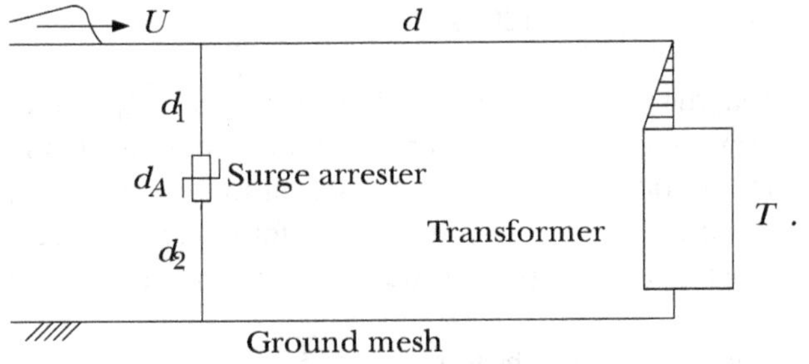

Figure 14.16 Transformer bay.

I) Representative overvoltages are obtained through analysis of the system and electromagnetic transient studies of surges with slow and fast wavefronts, as well as arrester data. This example assumes that calculations were previously performed only with respect to slow wavefronts and that for fast wavefronts, the simplified formulas in [2] will be used in the coordination of the phase-to-earth insulation of the transformer.

II) Co-ordination voltages

II.1 Internal insulation

a) Slow-front overvoltages.

$$r = U_{ps}/U_{e2} = 1.031,$$

$$K_{cd} = 1.034 \text{ see equations in Section 14.4.4,}$$

$$U_{cw} = 1.034 \times 860 = 889 \text{ kV}.$$

b) Fast-front overvoltages.

Under the simplified procedure in Annex E.4, Eq. (E.19) in [2] is adopted below, which considers the separation distance between the arrester and the protected equipment and gives the coordination voltage, U_{cw}.

$$U_{cw} = U_{pl} + \frac{A}{N}\frac{L_t}{L_{sp} + L_f}.$$

$L_f = R_a/R_{km}$ is the overhead line section with an outage rate equal to the acceptable failure rate of the equipment.

Since $N = 2$, the outage rates must be added together.

$$R_{km} = 0.005 \times 2 = 0.01 \text{ failures/1 km/year},$$

$$L_f = \frac{0.0025}{0.01} = 0.250 \text{ km (250 m)},$$

thus,

$$U_{cw} = 950 + \frac{11000}{2} \times \frac{30}{400 + 250} = 1204 \text{ kV}.$$

It should be noted that the parcel added to U_{pl} is proportional to L_t. Therefore, by reducing L_t, lower coordination voltage levels are achieved. On the other hand, higher failure rate and lower span adjacent to the substation, would lead to higher coordination voltage levels. The formula used is also based on a critical substation arrangement, which results in conservative coordination voltages.

A similar probabilistic approach can be found in [4], Section 6.3.3, in a simplified example of substation insulation coordination. The L_f, calculation is associated with the distance between the backflashover and the substation and, consequently, to the incoming wavefront. The distance is increased so that the backflashover occurs at the next tower of the transmission line.

Note that the approach is suggested when representative overvoltages, obtained from analyzes with EMTP-type programs, are not available. However, it is not an easy task to arrive at a comprehensive solution, given all the random factors involved.

II.2 External insulation

a) Slow-front overvoltages.
According to the statistical procedure, the statistical co-ordination factor, K_{cs}, from Figure 8 in Ref. [2] (see Figure 14.17), corresponds to a risk of 10^{-4} (assuming a risk of 10^{-3}, the factor would be very close to $K_{cs} = 1.15$).

$$K_{cs} = 1.18,$$

$$U_{cw} = K_{cs} U_{e2},$$

$$U_{cw} = 1.18 \times 832 = 982 \text{ kV}.$$

b) Fast-front overvoltages
Determination of U_{cw} is not necessary in the case of fast-front overvoltages, since, in group II, the insulation protecting against slow-front overvoltages confers sufficient withstand against fast-front overvoltages.

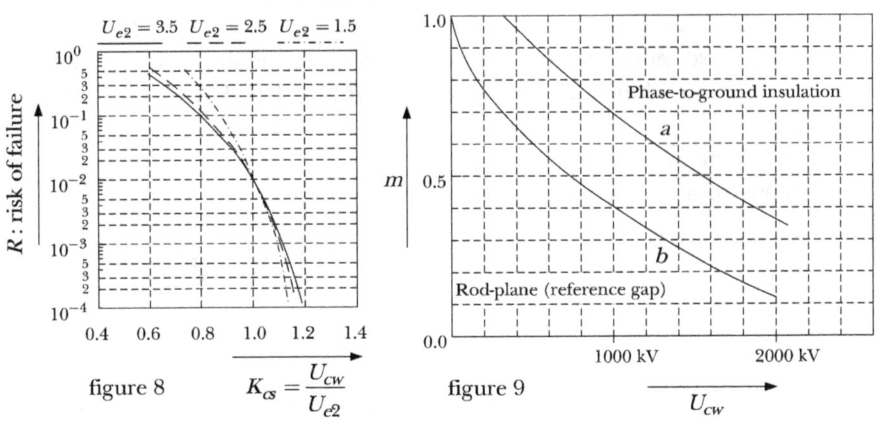

Figure 14.17 Figures 8 and 9. *Source:* [2]/The International Electrotechnical Commission (IEC).

III) Required voltagesThe safety factors, K_s, are given by:
$K_s = 1.15$ for internal insulation,
$K_s = 1.05$ for external insulation,

$$U_{rw} = K_s U_{cw}.$$

III.1 Internal insulation

a) Slow-front waves.

$$U_{rw} = 1.15 \times 889 = 1022\,\text{kV}.$$

b) Fast-front waves.

$$U_{rw} = 1.15 \times 1204 = 1384\,\text{kV}.$$

III.2 External insulation

a) Slow-front overvoltages.
The atmospheric correction due to altitude is made with the factor, K_a. At an altitude of 800 m, the factor $m = 0.7$ is obtained from Figure 9 in Ref. [2] (see Figure 14.17):

$$K_a = e^{0.7\left(\frac{800}{8150}\right)} = e^{0.069} = 1.07,$$

giving:

$$U_{rw} = 1.05 \times 1.07 \times 982 = 1103\,\text{kV}.$$

b) Fast-front overvoltages.
Not necessary, as in Item II, part (b), above.

IV) Standard withstand voltage The standard voltage levels, U_w, are obtained from the required voltage values and are presented in tables 2 and 3 in Ref. [1]. The standard values take the highest voltage for equipment U_m. The test procedures to adopt the requirements of groups I and II can be found in Ref. [2].

IV.1 Internal insulation

a) Slow-front waves (level above 1022 kV).

$$U_w = 1050\,\text{kV}.$$

b) Fast-front waves (level above 1384 kV).

$$U_w = 1425\,\text{kV}.$$

Note that with less-conservative hypotheses, it is very plausible to achieve 1300 kV as the insulation level for the transformer in a substation with a breaker-and-half arrangement.

IV.2 External insulation.

a) Slow-front overvoltages (level above 1103 kV).
   $U_w = 1175\,\text{kV}$.

b) Fast-front overvoltages.
   The usual procedure is to go a step higher than the internal insulation level, which would be 1550 kV.

The overvoltage distribution in Figure 8, in [2], takes the maximum values in phases *a*, *b*, and *c*, at each switching event; this is referred to as the phase-peak method. There is also the case-peak method, which selects only the maximum value in the three phases, at each switching event.

14.6 Insulation Coordination of Transmission Lines

The standard insulation coordination approach allows one to define the geometry of the transmission-line towers and to establish the phase-to-ground distances from the conductors to the towers and the phase-to-phase distances between conductors (Figure 14.18).

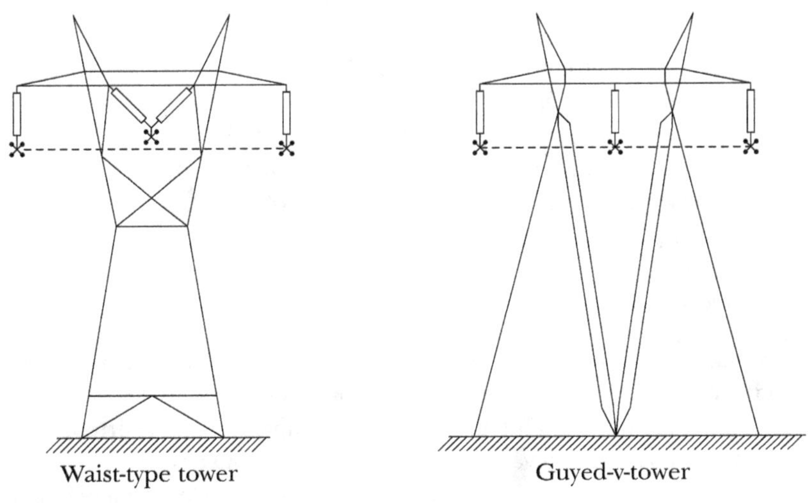

Figure 14.18 Examples of tower structures.

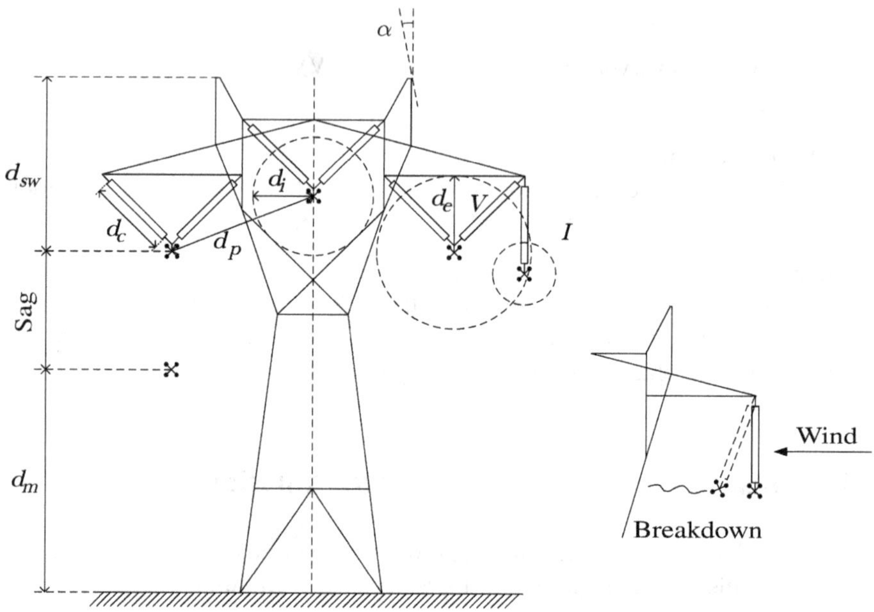

Figure 14.19 Distances and geometry to define.

There are several distances that define the towers of a transmission line, and the following structures are exemplified. Figure 14.19 shows the main distances to be defined:

- minimum clearance to ground (d_m).
- conductor and shield wire sags.
- length of insulator string (d_c), types (*I* or *V*), number of insulators (*N*).
- conductor-structure distance for the external phases (d_e).
- conductor-structure distance for the internal phases (d_i).
- distance between phases (d_p).
- bundle of subcondutors.
- shield wire location (d_{sw}).

Some distances are defined by taking into account the mechanical characteristics of conductors subjected to environmental conditions such as temperature, wind, etc. The sags are not constant and depend on temperature and current with implication in clearances and tower heigths. The free-swinging design for external phases is cost-effective, but raises concerns about insulator string movement under wind pressure and the risk of insulation breakdown.

Another consideration, as mentioned above, is the change in withstand levels in relation to altitude and atmospheric factors.

Various aspects can be considered in the insulation coordination of transmission lines, however this section will focus on the impact of switching surges and lightning surges on the dimensions of transmission line towers.

14.6.1 Insulation Coordination for Lightning Surges

The performance of transmission lines at operating voltages below 245 kV is strongly dependent on lightning surges. Fast-front overvoltage analysis allows one to evaluate distances in transmission-line tower design. The performance index is based on an acceptable failure rate, which depends on the impact of the failure on the electrical system.

The objective is to obtain a representative index of the lightning performance of the transmission line, considering the highly random nature of lightning strokes. This calculation, discussed in Chapter 9, depends on the shape of the lightning current, mainly on the amplitude and front time, which are assumed to be random variables (as seen in Eq. (14.33)). As an increase in pulse duration can cause breakdowns of lower magnitude, the tail time can be approached conservatively.

The performance criterion is specified as the number of flashovers per 100 km-years, which can vary from 0.5 to 1 in EHV systems and even more in HV systems.

One of the important factors in the analysis of the lightning performance of the overhead transmission line is the design of the shield-wires to lower the failure rates to acceptable levels.

Next, the calculation of the lightning performance of transmission lines will be presented, covering direct strokes, backflashovers, and induced voltages.

- Direct Strokes

  Direct strokes, as seen in Chapter 9, are analyzed assuming the incidence of flashes to the phase conductors of a transmission line.

  Direct strokes to transmission lines can occur with or without shield wires. In transmission lines with shield wires, it is necessary to evaluate if a shielding failure occurs and whether the current is sufficient to cause insulation breakdown.

  The shielding should allow only direct strokes with surges below the insulation level, thus minimizing line failures caused by such strokes. Hence, direct strokes with overvoltages above the withstand level of the insulator string should be intercepted by the shield wires. The calculation consists of the following steps:

  a) Critical current calculation

     The critical current defines a threshold level, below which flashovers are not likely to occur, calculated based on the CFO, here referred to as $V_{50\%}$, assuming negative polarity, and the surge impedance of the phase conductor Z. The critical current is obtained from (14.63), [15]:

$$I_c = \frac{2V_{50\%}}{Z}, \tag{14.63}$$

Equal division of the stroke current on the two sides of the transmission line is assumed in (14.63).

$$Z = 60 \ln\left(\frac{2h}{R_{eq}}\right), \tag{14.64}$$

$$R_{eq} = \sqrt[n]{rnR^{n-1}}, \tag{14.65}$$

where:

$V_{50\%}$: CFO of the insulator string.
Z: surge impedance of the bundle of conductors.
R_{eq}: equivalent radius of the bundle.
r: radius of a single subcondutor.
n: number of subconductors in the bundle, assuming a symmetrical arrangement.
R: radius of the circle circumscribing the bundle.

More detailed calculations can be performed with surge impedances extracted from line constant routines. Using electromagnetic transient programs, representations of corona and withstand characteristics, according to the leader progression model, may also be considered.

b) Electrogeometric model
The strike distance related to the critical current is calculated as (see Chapter 9 and [15]):

$$r_s = 10I^{0.65}. \tag{14.66}$$

c) Shielding angle and shielding failure rate
The height of the conductors relative to the ground plays an important role in their exposure to strokes. The average height above the ground must be evaluated according to the type of terrain, which may be considered flat, wavy, or mountainous.

The shield wires are placed, according to the electrogeometric model, to protect the most exposed phases from the strokes, and a check of the best-protected phases is made. In the examples illustrated in the figures below, the most exposed phases are the external phases.

Note that the upper sectors of the circles in the cross section of the transmission line represent the exposed surface along the length of the line.

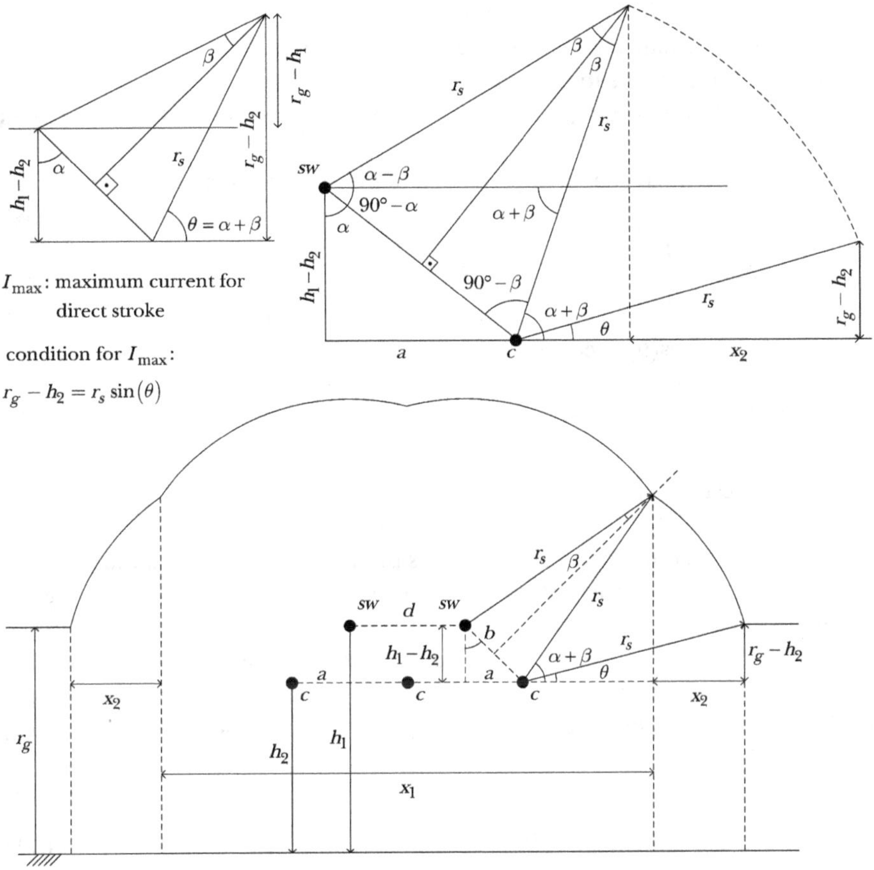

Figure 14.20 Shielding angle and horizontal projection of the exposed area for vertical strokes.

Simple geometric solutions may be found if an inclination angle is considered for the leader, however vertical strokes are normally assumed. Stroke to shield wire is calculated using the ground flash density related to the projection of its exposed surface into the ground. A rectangular projection is defined by one side given by $x_1 = x_1(I)$, in the cross-section orthogonal to the line axis in Figure 14.20, and by the other side with length L and parallel to the line axis. The same approach is used for direct stroke in the phase conductor, using $x_2 = x_2(I)$.

From Figure 14.20, the following variables can be defined:

α: shielding angle.
h_1: average height of shield wires.

h_2: average height of the phase conductors.
$r_s = r_s(I)$: strike distance to shield wires and conductors.
$r_g = r_g(I)$: strike distance to ground.
$x_1 = x_1(I)$: side of the rectangular projection of the exposed area of the shield wires to the stroke.
$x_2 = x_2(I)$: side of the rectangular projection of the exposed area of each conductor.

$$b = \frac{\sqrt{(h_1 - h_2)^2 + a^2}}{2}. \tag{14.67}$$

$$x_2(I) = r_s[\cos(\theta) - \cos(\alpha + \beta)]. \tag{14.68}$$

$$\theta = \arcsin\left[\frac{(r_g - h_2)}{r_s}\right]. \tag{14.69}$$

$$\beta = \arcsin\left(\frac{b}{r_s}\right). \tag{14.70}$$

Although a difference between r_s and r_g is proposed [15, 16], it is quite common to assume that $r_s = r_g$. The distance r_g can be considered slightly smaller than r_s [15, 16]. A review of the relationship between these distances can be found in [3].

At point P, in Figure 14.21, the maximum direct stroke current is calculated, beyond which the stroke extends to the shield wire, and it is known that the smallest stroke current observed is 3 kA [15]. If vertical strokes are assumed, the shielding failure rate can be calculated from:

$$s_{fr} = 2N_gL \int_3^{I_{max}} x_2(I) f_I(I) dI. \tag{14.71}$$

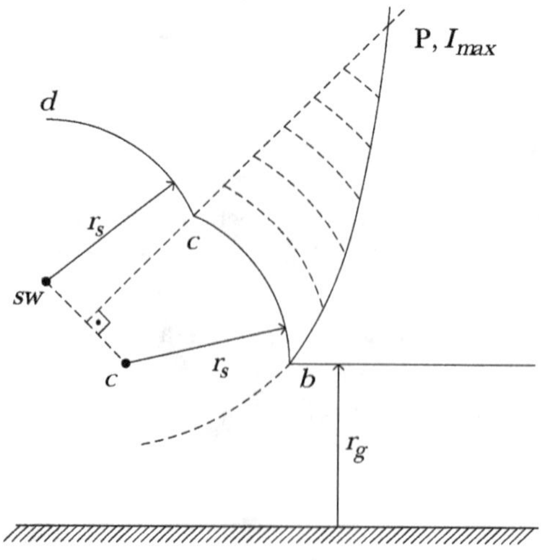

Figure 14.21 Exposure surfaces for different return stroke currents.

where:

I_{max}: maximum current for direct stroke.
N_g: annual ground flash density (ground flashes/km^2/year).
$f_I(I)$: probability density of lightning strokes with crest value I.
L: transmission line length (assumed 100 km).

In (14.71), the area of the projection of exposed surface for a current amplitude of I is defined as $Lx_2(I)$.

Shielding is considered effective when no flashover occurs due to direct strokes. In cases without effective shielding, it is possible to calculate the shielding failure flashover rate in response to direct strokes, for currents that cause flashovers in both external insulator strings, that is, $I_c < I < I_{max}$, where I_c is the critical current defined in (14.63).

$$s_{ffor} = 2N_g L \int_{I_c}^{I_{max}} x_2(I) f_I(I) dI. \tag{14.72}$$

As can be seen in the upper right inset in Figure 14.20, for $\theta < \alpha + \beta$ direct strokes can occur because $x_2(I) \geq 0$.

This failure rate due to direct strokes takes into account only the first stroke. The impact of subsequent strokes is not considered in [15], and this reference mentions that only 12% of the subsequent strokes have crest values higher than the initial stroke, but the rate of rise is higher. Reference [16], which presents an analysis of the failure rate considering subsequent strokes, shows that it is difficult to achieve perfect shielding and thereby completely avoid failures.

In [15, 16], there is a more comprehensive treatment of direct strokes and their relation to the electrogeometric model.

The maximum current is associated with a strike distance r_s which can be calculated using the same inset in Figure 14.20. In this case $\theta = \alpha + \beta$, thus:

$$\sin(\alpha + \beta) = \frac{r_s - h_2}{r_s}, \tag{14.73}$$

Substituting (14.70) into (14.73) r_s can be found by numerical methods, since α is known. A fair approximation to the calculation of r_s is proposed in [15], assuming $b \ll r_s$ and $\beta \approx 0$, again using the inset in Figure 14.20:

$$r_s = \left(\frac{h_1 + h_2}{2}\right) \times \frac{1}{(1 - \sin(\alpha))} \tag{14.74}$$

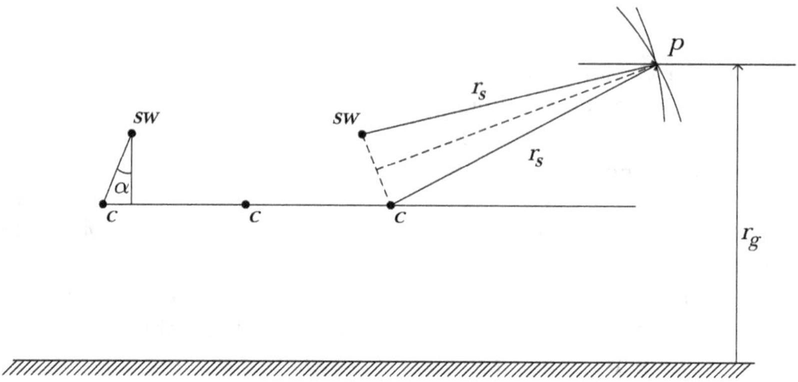

Figure 14.22 Effective shielding for the critical current.

As can be seen in the same inset, the effective shielding angle, at which a direct stroke is unlikely to occur, can be obtained using (14.75). In this equation, r_s is calculated using the critical current in (14.63) (Figure 14.22):

$$\alpha_e = 0.5\left[\sin^{-1}\left(\frac{r_g - h_1}{r_s}\right) + \sin^{-1}\left(\frac{r_g - h_2}{r_s}\right)\right], \tag{14.75}$$

where:

α_e: effective shielding angle.
r_s, r_g: strike distances for the critical current (I_c).

The first parcel in (14.75) corresponds to $\alpha_e - \beta$ and the second parcel corresponds to $\alpha_e + \beta$. Therefore, by averaging these angles, the equation can be verified.

A statistical analysis of backflashovers is performed in the next section.

- Backflashover

The location of shield wires at the towers may not ensure complete protection against backflashovers, as shown in Chapter 9. A stroke to the shield wires induces a voltage in the conductors and the voltage drop across the insulator string may cause insulation breakdown. Since the most probable lightning strokes have a negative polarity and the induced voltage at the conductor is less than the voltage at the tower, the voltage drop across the insulator can be assumed to have a positive polarity.

To reduce the failure rate due to backflashovers, certain procedures are adopted, such as the selection of an adequate number of insulators and the reduction of the tower footing resistance.

Initially, the following transmisssion line data is collected.

- Phase conductors and shield wires.
- Tower type and main distances.
- Insulation strength to fast-front overvoltages.
- Tower footing resistances.
- Ground flash density along the length of the transmission line.

An evaluation of the lightning performance of overhead transmission lines should allow a consistent analysis of tower distances in the face of a phenomenon subject to parameters that vary randomly.

The random variables present in the lightning surges are:

- Crest value of the stroke current.
- Front-time.
- Location of the stroke.
- Leader angle.
- Tower footing resistance.
- Operating voltage.
- Insulation withstand.

Probabilistic studies assume a simplified treatment of the tower footing resistances, which are considered in groups exhibiting similar behavior. The instantaneous operating voltage can increase the voltage drop across the insulator strings and should be taken into account when significant [15]. The insulation withstand has a small standard deviation and the simplified approach can use only the CFO, although the more accurate analyses should make use of the voltage waveform, according to Section 14.3.3.

For the calculation of lightning performance rates, the integration method or the Monte Carlo simulation method can be used.

a) Flashover in insulator strings

In simplified single-phase studies, tables are set up, using the voltages at the top of the tower, which are the same as the voltages at the shield wires, as a function of the crest and front-time of the stroke current. The voltages in the conductors are obtained from coupling coefficients, and subsequently, the voltage drops across insulator strings are calculated. This calculation can be further improved by evaluation of the voltages at the crossarms, interpolated between the voltages at the top and base of the tower. An equivalent procedure consists of the calculation of the critical currents that imply insulation breakdown as a function of the front-time of the stroke current [24, 25].

In three-phase studies, the procedure is the same, except that the voltages across the insulators are automatically furnished by the tower and transmission line models. If the system is linear, for a given front-time, t_f, a current with an intensity of 1 kA is applied and the voltage drop at the insulator string is obtained. The current leading to the voltage drop across the insulator string that causes the breakdown is obtained proportionally, and thus, the critical current curve, as a function of the front-time, is obtained. In the same way, the table of voltages at the tower top, as a function of the front-time, can also be obtained.

When the system is nonlinear, a search of the intensity versus front-time of the current that implies the breakdown of the insulation must be performed. Some examples of nonlinearities in the system are the breakdown of insulator strings, as calculated by the leader progression model, the corona effect, or even the inclusion of line arresters.

b) Line arresters
If the transmission line is protected by line arresters, a significant nonlinearity is present, and the procedure is somewhat different.

The installation of metal oxide arresters in parallel with insulator strings is intended to reduce the line failure rate due to direct strokes or backflashovers.
The following are the steps in the installation of line arresters:

- Identification of the critical zones in relation to the number of failures, establishing the best locations for line arrester application.
- Analysis of system requirements for the specification of line arresters and selection of alternative locations in relation to the phases and towers of the transmission line. Calculation of energy and current stresses on the arresters.

c) Evaluation of the outage rate with integration method
In Ref. [15], an equation like (14.76) for the calculation of outage rates due to backflashovers can be found, assuming vertical strokes:

$$r_{bf} = 0.6N_gL \int_0^{\infty} \int_{I_c}^{\infty} x_1(I) f_{I,t_f}(I/t_f) f_t(t_f) dI dt_f. \tag{14.76}$$

where:

$x_1(I)$: as in Figure 14.20.
$f_{I,t_f}(I/t_f)$: conditional probability density of current as a function of front-time.
$f_{t_f}(t_f)$: probability density of the front-time.
0.6 N_gL: considers the annual ground-flash density and the effect of strokes on the tower and midspan since strokes at midspan have a lower impact on the outage rate.

I_C: critical current for backflashover, that is, current amplitude above this level causes insulation breakdown, which is not related to the critical current for direct stroke.

These current levels can take into account the waveform of the lightning current. The voltages developed across the insulator string can be obtained using the leader progression model or the CFO.

d) Evaluation of the line outage rate using the Monte Carlo simulation method
A certain number of tests, n, is initially proposed, large enough for an accurate statistical representation of the phenomenon. For each test, generation of the following random variables is performed using the flow chart:

- Point on the ground to be struck by lightning.
- Current amplitude.
- Angle of incidence of the lightning.
- Check the stroke point, using the electrogeometric model. If it hits the ground, generate a new stroke point again. If it hits the conductor, check the insulator breakdown for direct stroke account. If it hits the shield wire, go to the next step.
- Generate of random variable for the front-time, whose probability depends on the amplitude of the current.
- Generate a random variable for the instantaneous operating voltage.
- Using the amplitude and the front-time, check the insulator breakdown by referring to the table in item (a) and, in the event of failure, the variable, s, is incremented.
- After the n tests, the probability of backflash is obtained from the quotient $p = s/n$.

The results are weighted according to the histogram of probabilities for the different tower footing resistances found along the transmission line.

The final evaluation of lightning performance also considers the annual ground-flash density along the route of the transmission line.

Two simple examples of integration and simulation methods are presented next, which are intended to introduce the basic steps in the calculation of lightning performance. For greater accuracy, the calculations must use specialized tools.

Example 14.3 This example presents a simplified calculation of the lightning performance of a transmission line to backflashes.

In order to clarify the fundamental steps of the calculation method, the example assumes very elementary probability functions, allowing a very simple calculation.

Assumptions

- Only shield wires are considered in the electrogeometric model and direct strokes are not calculated.

- Vertical strokes are assumed.
- All towers have the same tower footing resistance.
- Probability distributions are considered discrete. Currents, wavefronts, or front-times have discrete probabilities. Probability mass functions are calculated in the interval $(x, x + dx)$, as $f(x)dx$ [8]. In this example, currents and wavefronts are discretized in intervals.
- The distributions of current amplitudes and wavefronts are assumed independent.
- Strokes to the tower are assumed with application of the factor, 0.6, in [15].
- Insulation breakdown is calculated at $V_{50\%}$, neglecting the standard deviation.

Data

- Crest value of the current.
  A discrete distribution of lightning currents is assumed for illustrative purposes, consisting of three groups of amplitude and probability of occurrence:

I (kA)	*P* (I)
100	5%
65	30%
20	65%

- Wavefront of current.
  The wavefronts are also divided into three groups, using average values:

w_f (kA/μs)	$P(w_f)$
70	5%
40	25%
18	70%

In this example, the wavefront probabilities are used, but the conversion of wavefronts to front-times is trivial.

- $N_g = 4.8$ annual ground flash density (flashes/km^2 year) along the route of the transmission line.
- The ratio of the tower top voltage to stroke amplitude, in kV/kA, is given as a function of the wavefront (or front-time; see Example 9.3 in Chapter 9).

Using the same current amplitude of 1 kA for each front time t_f, the tower top voltages V_t are calculated per kA. Since the system is linear, for any amplitude of the stroke and for a given front-time, the corresponding tower-top voltage is obtained proportionally from the previously calculated data. These previously calculated values depend only on the characteristics of the transmission line, such as surge impedance, distances, tower footing resistance, etc.

The tower top voltage, V_t, in this example is given by an equation fitted to the hypothetical case:

$$\frac{V_t}{\text{kA}} = \left(20 + \frac{30}{t_f}\right).$$

- CFO: $V_{50\%} = 2000\,\text{kV}$.
- Coupling coefficient of shield wire and phase conductor, $k = 0.24$.
- $L = 100\,\text{km}$ (line length).

Solution

- The strokes that cause backflashes are such that:

$$V_{50\%} < (1-k)V_t, \text{thus } V_t > \frac{V_{50\%}}{0.76}.$$

For $V_{50\%} = 2000\,\text{kV}$, insulation breakdown occurs when the tower top voltage exceeds 2600 kV ($V_t > 2600\,\text{kV}$).

The possible conditions are presented in the following table.

I(kA)	w_f(kA/μs)	t_f(μs)	V_t(kV)	Breakdown	$P(I)P(w_f)$
	70	1.43	4100	Yes	$0.05 \times 0.05 = 0.0025$
100	40	2.5	3200	Yes	$0.05 \times 0.25 = 0.0125$
	18	5.56	2540	No	—
	70	0.93	3400	Yes	$0.30 \times 0.05 = 0.015$
65	40	1.62	2500	No	—
	18	3.61	1840	No	—
	70	0.29	2500	No	—
20	40	0.50	1600	No	—
	18	1.11	940	No	—

The probability of a stroke to the shield wire is calculated. Initially, the strike distance, r_s, is obtained from the expression:

$$r_s = 9I^{0.65}.$$

Note that the expression recommended in [15] is somewhat different, that is, $r_s = 10I^{0.65}$, but it does not interfere with this elementary example.

To simplify it is also assumed that $r_g = r_s$.

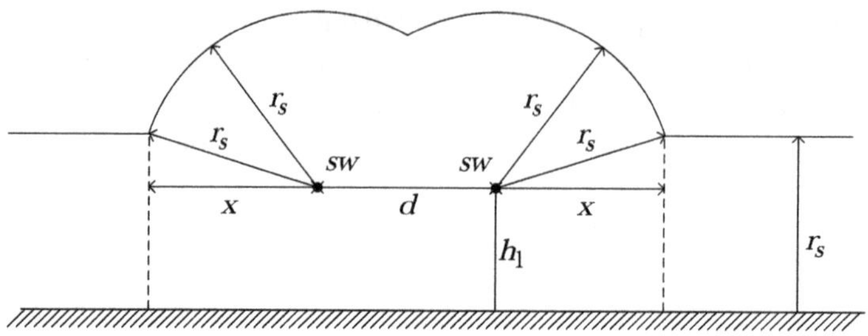

Figure 14.23 Horizontal projection of the exposed arc given by the shield wire to stroke current, *I*.

$$x = \sqrt{r_s^2 - (r_s - h_1)^2}, \quad r_s = r_s(I).$$

- Horizontal projection of the exposed sector of the circle defined by the strike distance to the shield wire with current, *I*, according to Figures 14.20 and 14.23.

$$x_1(I) = d + 2\sqrt{r_s^2 - (r_s - h_1)^2}.$$

- Assuming $d = 16.5$ m, $h_1 = 35$ m, $x_1(I)$ is calculated for 100, 65, and 20 kA (Figure 14.24).

$$x_1(100) = 229 \text{ m}, \quad (0.229 \text{ km}),$$
$$x_1(65) = 198 \text{ m}, \quad (0.198 \text{ km}),$$
$$x_1(20) = 129 \text{ m}, \quad (0.129 \text{ km}).$$

- Expected annual number of flashes of amplitude *I* reaching $x_1(I)$, for a line length of 100 km.

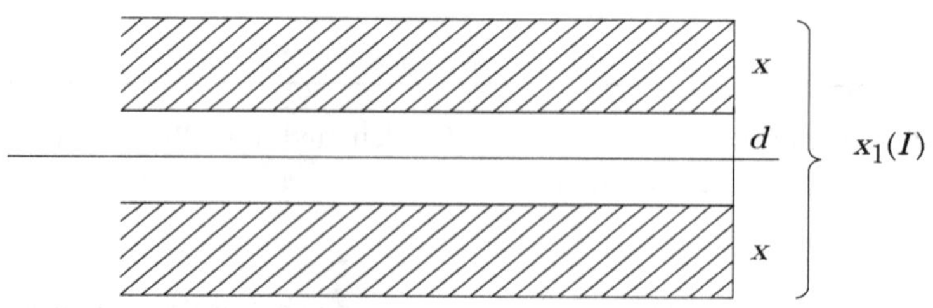

Figure 14.24 $x_1(I)$ for return strokes of amplitude *I*.

$$N_1(I) = N_g x_1(I) L.$$

Considering currents and wavefronts as independent events, the failure rate of backflashes at a line length of 100 km is calculated from the sum:

$$r_{bf} = \sum N_1(I) P(I) P(w_f),$$

where r_{bf} is the backflash rate.

Thus:

$$r_{bf} = N_g L[0.229 \times (0.0025 + 0.0125) + 0.198 \times 0.015],$$

$$r_{bf} = N_g 0.6405 \cong 3.07 \text{ backflashes}/100\text{ km/year}.$$

Applying the factor of 0.6, outage rate is estimated to be:

$$r_{bf} = 1.84 \text{ backflashes}/100\text{ km/year}.$$

Example 14.4 Solve Example 14.3 under the same hypotheses but use the Monte Carlo simulation method.

Solution

In this solution, using the simplified assumptions in Example 14.3, the application of the Monte Carlo method is shown.

For the random variable generated, $0 \leq \boldsymbol{a} \leq 1$, the corresponding current, I, is obtained:

$$0 \leq \mathbf{a} < 0.05 \Rightarrow I = 100 \text{ kA},$$

$$0.5 \leq \mathbf{a} < 0.35 \Rightarrow I = 65 \text{ kA},$$

$$0.35 \leq \mathbf{a} \leq 1.00 \Rightarrow I = 20 \text{ kA}.$$

Similarly, for the generated random variable, $\boldsymbol{b}$, the corresponding wavefront is obtained.

$$0 \leq \boldsymbol{b} < 0.05 \Rightarrow w_f = 70 \text{ kA}/\mu\text{s},$$

$$0.05 \leq \boldsymbol{b} \leq 0.30 \Rightarrow w_f = 40 \text{ kA}/\mu\text{s},$$

$$0.30 < \boldsymbol{b} \leq 1 \Rightarrow w_f = 18 \text{ kA}/\mu\text{s}.$$

The location of the stroke, inside $x_1(I)$, at width, ℓ, is verified using the random variable **c**. The width, ℓ, is assumed to be 200 m in this example.

Using the assumptions for the simulation method of Monte Carlo, the following flowchart is proposed.

For a certain number of experiments, n, the number of backflashes, s, is obtained. For this, variable, s has an initial value of zero, in step 1.

1) $s = 0$.
2) Do $1 \leq j \leq n$
3) Generate $\boldsymbol{a}$ and obtain I.
   Calculate r_s and $x_1(I)$.
4) Generate $\mathbf{c}$ and verify if the stroke reaches the shield wire.
   If yes, go to step 5.
   otherwise, increase j in step 2.
5) Generate $\boldsymbol{b}$ and obtain w_f.
   Calculate $t_f = \dfrac{I}{w_f}$.
   Calculate $V_t = I\left(20 + \dfrac{30}{t_f}\right)$.
   Verify the backflash.
   $V_t > 2600$.

   If yes, do $s = s + 1$ and increment j in step 2.
   otherwise, increment j in step 2.
   The probability of backflashes is given by:

$$p = \frac{s}{n}.$$

The backflash rate is:

$$r_{bf} = N_g L 2\ell p\,.$$

For 10 000 experiments, the probability p found is:

$$p = 0.0165,$$

$$r_{bf} = 4.8 \times 0.4 \times 100 \times 0.0165 = 3.17\,\text{backflashes}/100\,\text{km}/\text{year}.$$

Applying the factor of 0.6:

$$r_{bf} = 1.90\,\text{backflashes}/100\,\text{km}/\text{year},$$

that is very close to the result in Example 14.3, considering that by increasing the number of tests the results could be even closer.

Comments

Although the example adopted simplifying assumptions, the calculation algorithm does not differ substantially from actual cases.

The aim is to introduce the probabilistic calculation, and some points to be improved are:

- Probability distributions according to updated data from Refs. [5, 13, 15].

- Dependence of the probabilities of currents and wavefronts can easily be introduced into a computer program. Another possibility is to integrate the probability density function with two random variables, according to Eq. (14.33), Section 14.2.4.
- In calculating the horizontal projection x, it is necessary to include the effect of the phase conductors, using a simple trigonometric procedure.
- To calculate the flashovers, the adopted procedure used the tower top voltage, but use of the critical currents, that cause flashovers, in function of front times may be more efficient.
- Analysis of different tower footing resistances.
- The corona effect, the dependence of the parameters on frequency, and the operating voltage should also be considered.
- Since a large number of tests may be necessary, a well-optimized program is required, running as few arithmetic operations as possible.

- Induced Voltages

  The induced voltages are significant to transmission lines with nominal voltages up to 69 kV and distribution lines, with minimum impact on the insulation coordination of EHV transmission lines. While it requires complex models not detailed here, it is assumed that useful calculations can be performed using the equations provided in Chapter 9.

  A simplified analysis of the flashover rate caused by induced overvoltages in distribution lines is presented, assuming:

- Vertical strokes.
- Distribution line without discontinuities and without shield wires.
- The point of maximum voltage induced in the line is as close as possible to the stroke location.

It is assumed that strokes very near the distribution line result in direct strokes. In Figure 14.25a, a direct stroke is assumed at a current amplitude of I and a distance of $y < y_d(I)$ from the external conductor.

$$y_d(I) = \sqrt{r_s(I)^2 - [r_s(I) - h]^2}, \tag{14.77}$$

where $r_s(I) = 10I^{0.65}$ and to simplify $r_g = r_s$.

The voltage induced by a stroke at a distance, y, from the line and a return stroke current of amplitude, I, is designated as $U(y, I)$.

The breakdown due to induced overvoltages occurs when $U(y, I) > V_{50\%}$, within a range of distances from the lateral phase conductor, as shown in Figure 14.25b. For $y < y_d(I)$, direct discharge occurs. At a distance designated as $y_{max}(V_{50\%}, I)$, the induced voltage becomes lower than $V_{50\%}$, preventing breakdown. Therefore, the relevant range is within these limits, that is, $y_d(I) < y < y_{max}(V_{50\%}, I)$.

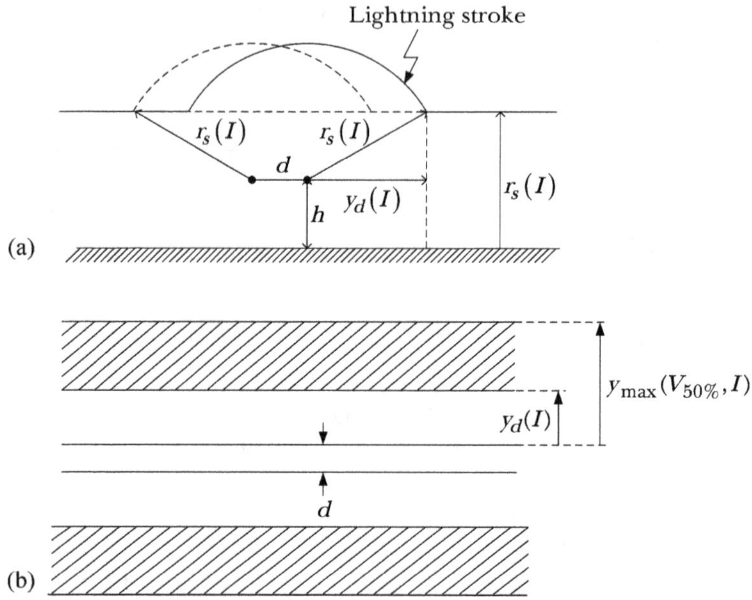

Figure 14.25 Induced overvoltages. (a) Region for direct stroke. (b) Region for induced voltages surpassing $V_{50\%}$.

The area defined for lightning strikes of amplitude, I, exceeding $V_{50\%}$, is calculated assuming a transmission line of length L.

$$A(V_{50\%}, I) = 2L(y_{max}(V_{50\%}, I) - y_d(I)). \tag{14.78}$$

For strokes of amplitude I, the incremental value for the rate of induced voltages higher than $V_{50\%}$, is given by:

$$dr_{ind} = \underbrace{N_g A(V_{50\%}, I)}_{\substack{\text{Probability of voltages}\\ \text{higher than } V_{50\%}\text{, for strokes}\\ \text{of amplitude } I\text{, reaching}\\ \text{area } A}} \times \underbrace{f_I(I)dI.}_{\substack{\text{Probability of current}\\ \text{strokes in the interval}\\ I < I < I + dI}} \tag{14.79}$$

N_g: ground flash density.

The expected outage rate, over the entire range of induced voltages exceeding $V_{50\%}$ is given by the integral:

$$r_{ind}(V_{50\%}) = 2N_g L \int_{I_{min}}^{\infty} [y_{max}(V_{50\%}, I) - y_d(I)] f_I(I) dI. \tag{14.80}$$

I_{min} defines the lower limit for the range in which $y_{max}(V_{50\%}, I) > y_d(I)$.

The front time, t_f, can be included, when the equations for the induced voltage take this variable into account, and only a small change in the equation is necessary, this time having to do with the distribution of joint probabilities:

$$\underbrace{f_{I,t_f}(I, t_f)dIdt_f}_{\substack{\text{Probability of current strokes within} \\ I < I < I + dI \text{ and } t_f < t_f < t_f + dt_f}}.$$

Calculation of the area considers the distance, y, within the range $y_d(I) < y < y_{max}(V_{50\%}, I, t_f)$, in which the pair (I, t_f) produces a voltage higher than $V_{50\%}$.

Thus, the modified equation is:

$$r_{ind}(V_{50\%}) = 2N_g L \int_0^{\infty} \int_{I_{min}}^{\infty} [y_{max}(V_{50\%}, I, t_f) - y_d(I)] f_{I,t_f}(I, t_f) dI dt_f. \tag{14.81}$$

For each voltage level, $V_{50\%}$, the numerical calculation can be done, according to the following steps:

- The current, I, and front time, t_f, are incremented.
- Calculation of $y_{max}(V_{50\%}, I, t_f)$ and $y_d(I)$.
- Calculation of the area, $2L[y_{max}(V_{50\%}, I, t_f) - y_d(I)]$ and the number of overvoltages with respect to the currents and front time increments.

  Calculation of the outage rate for overvoltages higher than $V_{50\%}$. In addition to calculating the outage rate to induced overvoltages, it is necessary to calculate the outage rate for direct strokes, which is obtained from the area:

$$A' = L(2y_d(I) + d), \tag{14.82}$$

considering only the current to be a random variable.

$$r_{ds} = N_g L \int_{I_{min}}^{\infty} (2y_d(I) + d) f_I(I) dI. \tag{14.83}$$

Example 14.5 Using the simplified overvoltage calculation, as in Ref. [26], calculate the lightning performance of a distribution line at a height of 10 m from the ground, as a function of the insulation level [13, 27], with the aid of the statistical data proposed in Refs. [15, 28].

Data

Assuming:

$$N_g = 1 \text{ ground flash/km}^2\text{/year}, h = 10 \text{ m and } d = 2 \text{ m}.$$

- Flash.
  a) Reference [15]:
  For $I < 20$ kA: median $I_m = 61$ kA and standard deviation $\sigma_{\ln I} = 1.33$.
  For $I > 20$ kA: median $I_m = 33$ kA and standard deviation $\sigma_{\ln I} = 0.605$.
  b) Reference [28]:
  Median $t_{fm} = 3.8$ μs and standard deviation $\sigma_{\ln t_f} = 0.553$.
  Correlation coefficient between I and t_f: $r = 0.47$.

Solution

The failure rate due to direct strokes is calculated using Eq. (14.83), resulting in $r_{ds} = 8.8$ failures/100 km/year. It is assumed that any stroke will result in an insulation breakdown, regardless of the $V_{50\%}$ level.

Equation (14.81) is used to calculate the failure rate for induced overvoltages as a function of the voltage level, $V_{50\%}$, [27].

The joint density function for a Gaussian distribution of logarithmic values of current and front time is provided in [28] and can also be obtained from (14.33) (see [8]).

$$f_{I,t_f}(I,t_f) = \frac{1}{2\pi I t_f \sigma_{\ln I}\sigma_{\ln t_f}\sqrt{1-r^2}} e^{-\frac{1}{2(1-r^2)}\left[\left(\frac{\ln I - \ln I_m}{\sigma_{\ln I}}\right)^2 - \frac{2r(\ln I - \ln I_m)(\ln t_f - \ln t_{fm})}{\sigma_{\ln I}\sigma_{\ln t_f}} + \left(\frac{\ln t_f - \ln t_{fm}}{\sigma_{\ln t_f}}\right)^2\right]},$$

where I_m and t_{fm} are median values.

The equations in Chapter 9 are used to calculate the induced overvoltages. For a line without discontinuities and assuming the ground to be a lossless conductor, the approximate formulas proposed in [26] facilitate the calculation and give satisfactory results for the purpose of this example. The maximum induced voltage values were calculated at point $x = 0$, setting $u = u_1 + u_2$, but since $u_1 = u_2$, $u = 2u_1$.

The following formulas refer to the maximum voltage induced at point $x = 0$, the point closest to the stroke, that is, on the straight line that passes through the stroke point and is perpendicular to the transmission line.

$$b = \frac{1}{\sqrt{1 + \frac{500}{I}}}, \; A = \frac{Z_o I\, h}{y}, \; Z_0 \approx 30\ \Omega.$$

$$B = 1 + \frac{b}{\sqrt{2 - b^2}}\ (\text{for}\ x = 0), \; C = \frac{b\, c\, t_f}{y}\ (t_f\ \text{front time}),$$

$$D = \sqrt{1 + \left(\frac{C}{2}\right)^2} + \frac{C}{2}, \; E = \sqrt{1 + \left(\frac{C}{2}\right)^2} - \frac{C}{2},$$

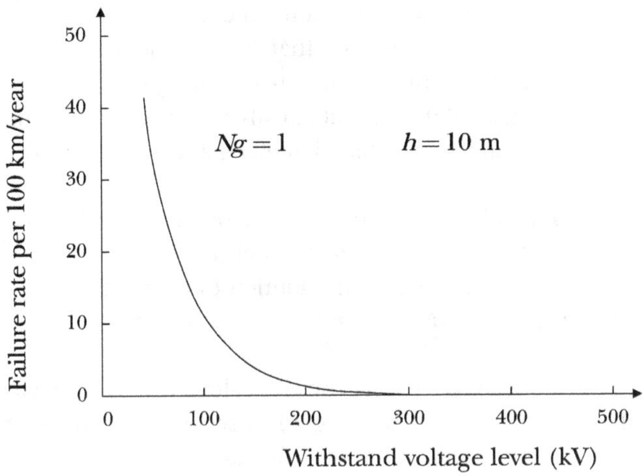

Figure 14.26 Failure rate for induced voltage, per 100 km/year.

$$u_1 = \frac{AB}{2C} \ln \frac{1 + D^2}{1 + E^2}.$$

The expression for u_1 is a good approximation to the more exact formulas discussed in Chapter 9 .

$$u = 2u_1.$$

The total number of failures can be obtained from the sum of direct strokes and induced overvoltages (Figure 14.26). In this example, only the first stroke was considered, but the results can be improved by including subsequent strokes.

14.6.2 Insulation Coordination for Switching Surges

Analysis of the insulation coordination for switching surges determines the geometry of the towers of the transmission lines and the insulation distances of substations, with respect to phase-to-ground and phase-to-phase insulation, mainly at voltages above 245 kV.

The insulation must consider the atmospheric corrections that have a probabilistic behavior. As previously mentioned, Ref. [2] endorses an approach using only corrections for altitude.

Another probabilistic factor to consider is the wind. Economical designs often adopt type *I* insulator strings (free-swinging) in the external phases (see Figure 14.19), which can swing due to the pressure exerted by the wind. The statistical treatment of conductor swing is more significant in relation to the insulation level at nominal frequencies, but it is not common to consider this effect in switching surges.

The swing angle of the conductors depends on the component of the wind speed perpendicular to the axis of the transmission line, that is, the calculated force applied to the conductor cables. Due to the random nature of wind-gust frequency, intensity, and duration, meteorological data are statistically analyzed. The swing angle is calculated taking into account the spans, diameters, and weights of the conductors.

In the simplified approach, to calculate the failure rate of a transmission line, an important factor is the number of insulator strings in parallel, and hence subjected to the same voltage. Therefore, in the statistical distribution of overvoltages, the profile of overvoltages along the line is obtained using sections that have the same overvoltage level.

Increasing the number of sections makes the failure risk calculation more accurate, but acceptable results can be obtained by using a few sections, or even just one, and adopting conservative assumptions for overvoltages.

The following example illustrates the statistical method for phase-to-earth switching surges.

Example 14.6 Analyze the failure rate for a transmission line due to phase-to-earth switching surges. Neglect *acf* and altitude effect.

Data

Transmission line operating voltage: 550 kV.
Transmission line length: 300 km.
Span between towers: 400 m.
Number of insulators per string: 20.
Cap and pin insulators: 5 1/4 in.
Gap factor: 1.45, Ref. [2].

Assume a simplified overvoltage profile, as in Figure 14.27, but with two transmission line sections with 150 km each:

$$u^1_{50\%} = 1.6\,\text{pu}, \sigma_u = 10\% \text{ of the mean value;}$$

$$u^2_{50\%} = 1.8\,\text{pu}, \sigma_u = 10\% \text{ of the mean value.}$$

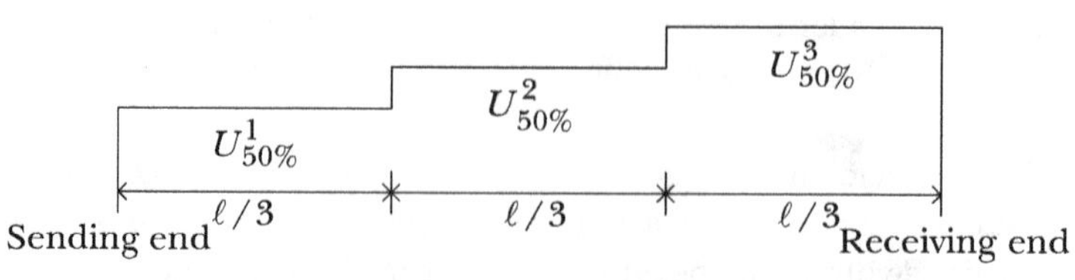

Figure 14.27 Estimated overvoltage distribution over the transmission line.

An approximate statistical method is used in the calculation of failure risk. The distance between the conductor and crossarm is given by the number of insulators and by the size of each insulator.

$$d = 20 \times 5.25 \times 2.54\,\text{cm}.$$

$$d = 2.67\,\text{m}.$$

The $V_{50\%}$ value in kV is calculated as:

$$V_{50\%} = 1.45 \times 500(2.67)^{0.6} = 1307\,\text{kV},$$

which in pu of the operating voltage peak value is given by:

$$\frac{550\sqrt{2}}{\sqrt{3}} = 449\,\text{kV},$$

$$v_{50\%} = \frac{1307}{449} = 2.91\,\text{pu}.$$

$$\sigma_v = 0.06 \times v_{50\%}, \sigma_u = 0.10 \times u_{50\%}.$$

The standard deviation of the difference function, that is, overvoltage stress minus the withstand voltage is:

$$\sigma_X = \sqrt{\sigma_u^2 + \sigma_v^2}.$$

- Calculation of the failure probability of an insulator string in a line section with the same overvoltage level:

$$\overline{x} = u_{50\%} - v_{50\%},$$

$$\xi_x = \frac{0 - \overline{x}}{\sigma_X}.$$

The probability of finding a positive number in the difference function, as described in the male versus female height problem (see Figure 14.11), coincides with the calculation of the distribution function given by (14.17). This probability is called p_1 and refers to the flashover of one insulator string.

The probability of flashover in one phase of a 150-km line section, with 375 gaps in parallel (150/0.4), is given by:

$$p_{ph} = 1 - (1 - p_1)^{375}.$$

The probability of flashover in one of the three phases is given by:

$$p_t = 1 - (1 - p_1)^{3 \times 375}$$

The calculations are conservative since they do not take into account the polarity of the overvoltages. As 50% of the overvoltages are likely to be of negative polarity with a higher withstand level compared to positive polarity, a reduction in the probability p_1 of 50% should be applied [4], and had this been done, the p_t values would be even smaller.

The gap factor window-conductor for the central phase is slightly different from the conductor-crossarm gap factor, but all phases are assumed to have the same gap factor.

The calculation of the probability for a 300-km line length is given by:

p_{t_1}: probability of flashover in the first section.
p_{t_2}: probability of flashover in the second section.

The probability of transmission-line failure is:

$$p_t = 1 - \left(1 - p_{t_1}\right)\left(1 - p_{t_2}\right).$$

The following results are obtained, with a predominant second interval.
Section 1:

$$\xi_x = 5.5316; p_1 = 1.5870 \times 10^{-8}; p_{t_1} = 1.7854 \times 10^{-5}.$$

Section 2:

$$\xi_x = 4.4264; p_1 = 4.7913 \times 10^{-6}; p_{t_2} = 0.0054.$$

And the probability of failure for the association of the two sections is calculated as:

$$p_t = 1 - \left(1 - 1.7854 \times 10^{-5}\right) \times (1 - 0.0054) = 0.0054.$$

That would correspond to one flashover every 185 switchings, and assuming 30 switchings per year, a flashover every 6.17 years is likely to occur.

Example 14.7 Calculate the number of insulators in an insulator string such that the risk of failure of one phase of a line section with 100 strings in parallel is less than 1%.

Data

- operating line voltage: 550 kV;
- height of the insulator disc: 14 cm;
- gap factor: 1.45;
- overvoltage distribution: $u_{50\%} =$ 1.65 pu, σ_u = 10% of mean value.

Solution

Setting $v = v_{50\%}$ and assuming a standard deviation: $\sigma_v = 6\%$;

The standard deviation of the difference function of overvoltage (stress minus the withstand voltage) is:

$$\sigma_x = \sqrt{(0.1 \times 1.65)^2 + (0.06 \times v)^2}.$$

- Calculation of the risk of flashover in one insulator string:

$$\overline{x} = 1.65 - v,$$

$$\xi_x = \frac{0 - \overline{x}}{\sigma_x}.$$

or,

$$\xi_x = \frac{v - 1.65}{\sqrt{(0.1 \times 1.65)^2 + (0.06 \times v)^2}}. \tag{e.1}$$

The ξ_x value for one insulator string can be obtained from the imposed condition:

$$0.01 = 1 - (1 - p_1)^{100},$$

resulting in the probability of flashover in one insulator string, $p_1 = 1.004983 \times 10^{-4}$, which corresponds to $\xi_x = 3.7178$, calculated through the inversion of the distribution function.

With this ξ_x substituted into (e.1), a second-order equation is obtained:

$$0.9502v^2 - 3.3v + 1.8758 = 0,$$

whose roots are:

$$v = 2.7567 \text{ and } v = 0.7161.$$

A viable solution is $v = 2.7567$ pu and therefore:

$$V_{50\%} = 2.7567 \times \frac{550\sqrt{2}}{\sqrt{3}} = 1238.0\,\text{kV}.$$

Recalling (14.42):

$$1238 = 1.45 \times 500 \times (d)^{0.6},$$

$$d = 2.44\,\text{m},$$

$$n = \frac{2.50}{0.14} = 17.42.$$

Giving 18 insulator units for each insulator string.

The same comment on the 50% reduction factor in Example 14.6 applies here.

Examples 14.6 and 14.7 employed a statistical approach to assess the number of insulators in a string of insulators in a transmission line. They are intended to illustrate the calculation but should not be regarded as a design rule. As mentioned earlier, other methods can be adopted, such as the method proposed in [5].

The number of decimal places used in the results is not necessary and serves solely the purpose of enabling result reproduction. In reality, the calculation is merely an indicator of the transmission line's performance concerning switching overvoltage and aids in selecting the best alternative from the perspective of insulation coordination.

Coordination of the phase-to-phase insulation between conductors takes into account a factor that considers the percentage of overvoltages of negative polarity in relation to the total phase-to-phase overvoltage. Statistical calculations of phase-to-phase overvoltages are carried out to determine the behavior of this new variable α, which is used to calculate the failure rate. Further details of this procedure are given in Ref. [2].

References

1 IEC. (2006). *IEC-60071-1: Insulation co-ordination. Part 1: Definitions, principles and rules*. Geneva: International Electrotechnical Commission (IEC).

2 IEC. (2018). *IEC-60071-2: Insulation co-ordination. Part 2: Application guide*. Geneva: International Electrotechnical Commission (IEC).

3 IEEE. (2010). *C62.82.1-2010: IEEE Standard for insulation coordination: Definitions, principles, and rules*. New York: Institute of Electrical and Electronics Engineers (IEEE).

4 IEEE. (1999). *IEEE std 1313.2™-1999 (R2005): IEEE guide for the application of insulation coordination*. New York: Institute of Electrical and Electronics Engineers (IEEE).

5 Electric Power Research Institute. (1982). *Transmission line reference book: 345 kV and above* (2nd ed.). Palo Alto: Electric Power Research Institute.

6 Hileman, A. R. (1999). *Insulation coordination of power systems*. Boca Raton: Taylor & Francis.

7 Diesendorf, W. (1974). *Insulation coordination in high voltage electric power systems*. London: Butterworth.

8 Papoulis, A. (1984). *Probability, random variables and stochastic processes* (2nd ed.). Auckland: McGraw-Hill.

9 IEC. (1989). *IEC 60. High-voltage test techniques. Part 1.* Geneva: International Electrotechnical Commission (IEC).

10 Paris, L., & Cortina, R. (1968). Switching and lightning impulse discharge characteristics of large air gaps and long insulator strings. *IEEE Transactions on Power Apparatus and Systems*, PAS-87(4), 947–957.

11 Thione, L. (1984). Evaluation of switching impulse strength of external insulation. *Electra*, 77–95.

12 IEEE Working Group on Estimating Lightning Performance of Transmission Lines. (1985). A simplified method for estimating lightning performance of transmission lines. *IEEE Transactions on Power Apparatus and Systems*, PAS-104(4), 918–932.

13 CIGRE Working Group C4.23. (2021). *Procedures for estimating the lightning performance of transmission lines: new aspects.* Paris: CIGRE (International Council on Large Electric Systems). (Technical Brochure, no. 839).

14 CIGRE, Study Committee 33, Working Group 33.07. (1992). *Guidelines for the evaluation of the dielectric strength of external insulation.* Paris: CIGRE (International Council on Large Electric Systems). (Technical Brochure, no. 72).

15 CIGRE, Study Committee 33, Working Group 01. (1991). *Guide to procedures for estimating the lightning performance of transmission lines.* Paris: CIGRE (International Council on Large Electric Systems).

16 Whitehead, J. T., & Chisholm, W. A. (1993). Estimating lightning performance of transmission lines. II. Updates to analytical models. *IEEE Transactions on Power Delivery*, 8(3), 1254–1267.

17 Darveniza, M., Popolanski, F., & Whitehead, E. R. (1975). Lightning protection of UHV transmission lines. *Electra*(41), 39–69.

18 Darveniza, M., Sargent, M. A., Limbourn, G. J., Choy, L. A., Caldwell, R. O., Currie, J. R., Holcombe, B. C., Stillman, R. H., & Frowd, R. (1979). Modelling for lightning performance calculations. *IEEE Transactions on Power Apparatus and Systems*, PAS-98(6), 1900–1908.

19 Darveniza, M. (1988). The generalized integration method for predicting impulse volt-time characteristics for non-standard wave shapes-a theoretical basis. *IEEE Transactions on Electrical Insulation*, 23(3), 373–381.

20 Imece, A. F., Durbak, D. W., Elahi, H., Kolluri, S., Lux, A., Mader, D., McDemott, T. E., Morched, A., Mousa, A. M., Natarajan, R., Rugeles, L., & Tarasiewicz, E. (1996). Modeling guidelines for fast front transients. *IEEE Transactions on Power Delivery*, 11(1), 493–506.

21 Anderson, J. G. (1961). Monte Carlo computer calculation of transmission-line lightning performance. *Transactions of the American Institute of Electrical Engineers. Part III: Power Apparatus and Systems*, 80(3), 414–419.

22 IEEE. (1996). *IEEE Std 998-1996: IEEE Guide for direct lightning stroke shielding of substations*. New York: Institute of Electrical and Electronics Engineers (IEEE).

23 Eriksson, A. J., & Weck, K. H. (1988). Simplified procedures for determining representative substation impinging lightning overvoltages, Paris, 1988. In *International Conference on High Voltage Engineering* (pp. 33–16). (Proceedings, Paris, CIGRÉ, 1988).

24 Kastrup, O., Zanetta Junior, L. C., Nigri, A., Brasil, D. O. C., Maldonado, P., & Moreira, F. A. (1996). Lightning performance assesment with line arresters. In *IEEE/PES Transmision and Distribution Conference, Los Angeles, 1996* (pp. 288–293). New York: IEEE.

25 Zanetta Junior, L. C., & Moreira, F. A. (1997). Improvement of the lightning performance evaluation of lines with arresters. In *International Conference on Power Systems Transients. Seattle, 1997* (pp. 307–312). Seattle: University of Washington.

26 Rusck, S. (1958). Induced lightning over-voltages on power transmission lines with special reference to the over-voltage protection of low-voltage networks. In *Transactions of the Royal Institute of Technology* (Vol. 120). Stockholm.

27 Matsuo, N. M., & Zanetta Junior, L. C. (1997). Frequency of occurrence of lightning overvoltages on distribution lines In *14th International Conference & Exhibition on Electricity Distribution, Birmighan, 1997* (pp. 2.18). London: Institution of Electrical Engineers.

28 Chowdhuri, P. (1989). Analysis of lightning-induced voltages on overhead lines. *IEEE Transactions on Power Delivery*, 4(1), 479–492.

Index

Understanding Electromagnetic Transients in Power Systems, First Edition. Luiz Cera Zanetta Jr.

Published 2025 by John Wiley & Sons, Inc.

d

m

IEEE Press Series on Power and Energy Systems

Series Editor: Ganesh Kumar Venayagamoorthy, Clemson University, Clemson, South Carolina, USA.

The mission of the IEEE Press Series on Power and Energy Systems is to publish leading-edge books that cover a broad spectrum of current and forward-looking technologies in the fast-moving area of power and energy systems including smart grid, renewable energy systems, electric vehicles and related areas. Our target audi- ence includes power and energy systems professionals from academia, industry and government who are interested in enhancing their knowledge and perspectives in their areas of interest.

1. *Electric Power Systems: Design and Analysis, Revised Printing*
   Mohamed E. El-Hawary
2. *Power System Stability*
   Edward W. Kimbark
3. *Analysis of Faulted Power Systems*
   Paul M. Anderson
4. *Inspection of Large Synchronous Machines: Checklists, Failure Identification, and Troubleshooting*
   Isidor Kerszenbaum
5. *Electric Power Applications of Fuzzy Systems*
   Mohamed E. El-Hawary
6. *Power System Protection*
   Paul M. Anderson
7. *Subsynchronous Resonance in Power Systems*
   Paul M. Anderson, B.L. Agrawal, J.E. Van Ness
8. *Understanding Power Quality Problems: Voltage Sags and Interruptions*
   Math H. Bollen
9. *Analysis of Electric Machinery*
   Paul C. Krause, Oleg Wasynczuk, and S.D. Sudhoff
10. *Power System Control and Stability, Revised Printing*
    Paul M. Anderson, A.A. Fouad
11. *Principles of Electric Machines with Power Electronic Applications, Second Edition*
    Mohamed E. El-Hawary
12. *Pulse Width Modulation for Power Converters: Principles and Practice*
    D. Grahame Holmes and Thomas Lipo
13. *Analysis of Electric Machinery and Drive Systems, Second Edition*
    Paul C. Krause, Oleg Wasynczuk, and S.D. Sudhoff
14. *Risk Assessment for Power Systems: Models, Methods, and Applications*
    Wenyuan Li

15. *Optimization Principles: Practical Applications to the Operations of Markets of the Electric Power Industry*
    Narayan S. Rau

16. *Electric Economics: Regulation and Deregulation*
    Geoffrey Rothwell and Tomas Gomez

17. *Electric Power Systems: Analysis and Control*
    Fabio Saccomanno

18. *Electrical Insulation for Rotating Machines: Design, Evaluation, Aging, Testing, and Repair*
    Greg C. Stone, Edward A. Boulter, Ian Culbert, and Hussein Dhirani

19. *Signal Processing of Power Quality Disturbances*
    Math H. J. Bollen and Irene Y. H. Gu

20. *Instantaneous Power Theory and Applications to Power Conditioning*
    Hirofumi Akagi, Edson H. Watanabe and Mauricio Aredes

21. *Maintaining Mission Critical Systems in a 24/7 Environment*
    Peter M. Curtis

22. *Elements of Tidal-Electric Engineering*
    Robert H. Clark

23. *Handbook of Large Turbo-Generator Operation and Maintenance, Second Edition*
    Geoff Klempner and Isidor Kerszenbaum

24. *Introduction to Electrical Power Systems*
    Mohamed E. El-Hawary

25. *Modeling and Control of Fuel Cells: Distributed Generation Applications*
    M. Hashem Nehrir and Caisheng Wang

26. *Power Distribution System Reliability: Practical Methods and Applications*
    Ali A. Chowdhury and Don O. Koval

27. *Introduction to FACTS Controllers: Theory, Modeling, and Applications*
    Kalyan K. Sen, Mey Ling Sen

28. *Economic Market Design and Planning for Electric Power Systems*
    James Momoh and Lamine Mili

29. *Operation and Control of Electric Energy Processing Systems*
    James Momoh and Lamine Mili

30. *Restructured Electric Power Systems: Analysis of Electricity Markets with Equilibrium Models*
    Xiao-Ping Zhang

31. *An Introduction to Wavelet Modulated Inverters*
    S.A. Saleh and M.A. Rahman

32. *Control of Electric Machine Drive Systems*
    Seung-Ki Sul

33. *Probabilistic Transmission System Planning*
    Wenyuan Li

34. *Electricity Power Generation: The Changing Dimensions*
    Digambar M. Tagare

35. *Electric Distribution Systems*
    Abdelhay A. Sallam and Om P. Malik

36. *Practical Lighting Design with LEDs*
Ron Lenk, Carol Lenk

37. *High Voltage and Electrical Insulation Engineering*
Ravindra Arora and Wolfgang Mosch

38. *Maintaining Mission Critical Systems in a 24/7 Environment, Second Edition*
Peter Curtis

39. *Power Conversion and Control of Wind Energy Systems*
Bin Wu, Yongqiang Lang, Navid Zargari, Samir Kouro

40. *Integration of Distributed Generation in the Power System*
Math H. Bollen, Fainan Hassan

41. *Doubly Fed Induction Machine: Modeling and Control for Wind Energy Generation Applications*
Gonzalo Abad, Jesús López, Miguel Rodrigues, Luis Marroyo, and Grzegorz Iwanski

42. *High Voltage Protection for Telecommunications*
Steven W. Blume

43. *Smart Grid: Fundamentals of Design and Analysis*
James Momoh

44. *Electromechanical Motion Devices, Second Edition*
Paul Krause Oleg, Wasynczuk, Steven Pekarek

45. *Electrical Energy Conversion and Transport: An Interactive Computer-Based Approach, Second Edition*
George G. Karady, Keith E. Holbert

46. *ARC Flash Hazard and Analysis and Mitigation*
J.C. Das

47. *Handbook of Electrical Power System Dynamics: Modeling, Stability, and Control*
Mircea Eremia, Mohammad Shahidehpour

48. *Analysis of Electric Machinery and Drive Systems, Third Edition*
Paul C. Krause, Oleg Wasynczuk, S.D. Sudhoff, Steven D. Pekarek

49. *Extruded Cables for High-Voltage Direct-Current Transmission: Advances in Research and Development*
Giovanni Mazzanti, Massimo Marzinotto

50. *Power Magnetic Devices: A Multi-Objective Design Approach*
S.D. Sudhoff

51. *Risk Assessment of Power Systems: Models, Methods, and Applications, Second Edition*
Wenyuan Li

52. *Practical Power System Operation*
Ebrahim Vaahedi

53. *The Selection Process of Biomass Materials for the Production of Bio-Fuels and Co-Firing*
Najib Altawell

54. *Electrical Insulation for Rotating Machines: Design, Evaluation, Aging, Testing, and Repair, Second Edition*
Greg C. Stone, Ian Culbert, Edward A. Boulter, and Hussein Dhirani

55. *Principles of Electrical Safety*
Peter E. Sutherland

56. *Advanced Power Electronics Converters: PWM Converters Processing AC Voltages*
Euzeli Cipriano dos Santos Jr., Edison Roberto Cabral da Silva

57. *Optimization of Power System Operation, Second Edition*
Jizhong Zhu

58. *Power System Harmonics and Passive Filter Designs*
J.C. Das

59. *Digital Control of High-Frequency Switched-Mode Power Converters*
Luca Corradini, Dragan Maksimoviæ, Paolo Mattavelli, Regan Zane

60. *Industrial Power Distribution, Second Edition*
Ralph E. Fehr, III

61. *HVDC Grids: For Offshore and Supergrid of the Future*
Dirk Van Hertem, Oriol Gomis-Bellmunt, Jun Liang

62. *Advanced Solutions in Power Systems: HVDC, FACTS, and Artificial Intelligence*
Mircea Eremia, Chen-Ching Liu, Abdel-Aty Edris

63. *Operation and Maintenance of Large Turbo-Generators*
Geoff Klempner, Isidor Kerszenbaum

64. *Electrical Energy Conversion and Transport: An Interactive Computer-Based Approach*
George G. Karady, Keith E. Holbert

65. *Modeling and High-Performance Control of Electric Machines*
John Chiasson

66. *Rating of Electric Power Cables in Unfavorable Thermal Environment*
George J. Anders

67. *Electric Power System Basics for the Nonelectrical Professional*
Steven W. Blume

68. *Modern Heuristic Optimization Techniques: Theory and Applications to Power Systems*
Kwang Y. Lee, Mohamed A. El-Sharkawi

69. *Real-Time Stability Assessment in Modern Power System Control Centers*
Savu C. Savulescu

70. *Optimization of Power System Operation*
Jizhong Zhu

71. *Insulators for Icing and Polluted Environments*
Masoud Farzaneh, William A. Chisholm

72. *PID and Predictive Control of Electric Devices and Power Converters Using MATLAB®/Simulink®*
Liuping Wang, Shan Chai, Dae Yoo, Lu Gan, Ki Ng

73. *Power Grid Operation in a Market Environment: Economic Efficiency and Risk Mitigation*
Hong Chen

74. *Electric Power System Basics for Nonelectrical Professional, Second Edition*
Steven W. Blume

75. *Energy Production Systems Engineering*
Thomas Howard Blair

76. *Model Predictive Control of Wind Energy Conversion Systems*
Venkata Yaramasu, Bin Wu

77. *Understanding Symmetrical Components for Power System Modeling*
J.C. Das

78. *High-Power Converters and AC Drives, Second Edition*
Bin Wu, Mehdi Narimani

79. *Current Signature Analysis for Condition Monitoring of Cage Induction Motors: Industrial Application and Case Histories*
William T. Thomson, Ian Culbert

80. *Introduction to Electric Power and Drive Systems*
Paul Krause, Oleg Wasynczuk, Timothy O'Connell, Maher Hasan

81. *Instantaneous Power Theory and Applications to Power Conditioning, Second Edition*
Hirofumi, Edson Hirokazu Watanabe, Mauricio Aredes

82. *Practical Lighting Design with LEDs, Second Edition*
Ron Lenk, Carol Lenk

83. *Introduction to AC Machine Design*
Thomas A. Lipo

84. *Advances in Electric Power and Energy Systems: Load and Price Forecasting*
Mohamed E. El-Hawary

85. *Electricity Markets: Theories and Applications*
Jeremy Lin, Jernando H. Magnago

86. *Multiphysics Simulation by Design for Electrical Machines, Power Electronics and Drives*
Marius Rosu, Ping Zhou, Dingsheng Lin, Dan M. Ionel, Mircea Popescu, Frede Blaabjerg, Vandana Rallabandi, David Staton

87. *Modular Multilevel Converters: Analysis, Control, and Applications*
Sixing Du, Apparao Dekka, Bin Wu, Navid Zargari

88. *Electrical Railway Transportation Systems*
Morris Brenna, Federica Foiadelli, Dario Zaninelli

89. *Energy Processing and Smart Grid*
James A. Momoh

90. *Handbook of Large Turbo-Generator Operation and Maintenance, Third Edition*
Geoff Klempner, Isidor Kerszenbaum

91. *Advanced Control of Doubly Fed Induction Generator for Wind Power Systems*
Dehong Xu, Dr. Frede Blaabjerg, Wenjie Chen, Nan Zhu

92. *Electric Distribution Systems, 2nd Edition*
Abdelhay A. Sallam, Om P. Malik

93. *Power Electronics in Renewable Energy Systems and Smart Grid: Technology and Applications*
Bimal K. Bose

94. *Distributed Fiber Optic Sensing and Dynamic Rating of Power Cables*
Sudhakar Cherukupalli, and George J. Anders

95. *Power System and Control and Stability, Third Edition*
Vijay Vittal, James D. McCalley, Paul M. Anderson, and A. A. Fouad

96. *Electromechanical Motion Devices: Rotating Magnetic Field-Based Analysis and Online Animations, Third Edition*
Paul Krause, Oleg Wasynczuk, Steven D. Pekarek, and Timothy O'Connell

97. *Applications of Modern Heuristic Optimization Methods in Power and Energy Systems*
Kwang Y. Lee and Zita A. Vale

98. *Handbook of Large Hydro Generators: Operation and Maintenance*
Glenn Mottershead, Stefano Bomben, Isidor Kerszenbaum, and Geoff Klempner

99. *Advances in Electric Power and Energy: Static State Estimation*
Mohamed E. El-Hawary

100. *Arc Flash Hazard Analysis and Mitigation, Second Edition*
J.C. Das

101. *Maintaining Mission Critical Systems in a 24/7 Environment, Third Edition*
Peter M. Curtis

102. *Real-Time Electromagnetic Transient Simulation of AC-DC Networks*
Venkata Dinavahi and Ning Lin

103. *Probabilistic Power System Expansion Planning with Renewable Energy Resources and Energy Storage Systems*
Jaeseok Choi and Kwang Y. Lee

104. *Power Magnetic Devices: A Multi-Objective Design Approach, Second Edition.*
Scott D. Sudhoff

105. *Optimal Coordination of Power Protective Devices with Illustrative Examples*
Ali R. Al-Roomi

106. *Resilient Control Architectures and Power Systems*
Craig Rieger, Ronald Boring, Brian Johnson, and Timothy McJunkin

107. *Alternative Liquid Dielectrics for High Voltage Transformer Insulation Systems: Performance Analysis and Applications*
Edited by U. Mohan Rao, I. Fofana, and R. Sarathi

108. *Introduction to the Analysis of Electromechanical Systems*
Paul C. Krause, Oleg Wasynczuk, and Timothy O'Connell

109. *Power Flow Control Solutions for a Modern Grid using SMART Power Flow Controllers*
Kalyan K. Sen and Mey Ling Sen

110. *Power System Protection: Fundamentals and Applications*
John Ciufo and Aaron Cooperberg

111. *Soft-Switching Technology for Three-phase Power Electronics Converters*
Dehong Xu, Rui Li, Ning He, Jinyi Deng, and Yuying Wu

112. *Power System Protection, Second Edition*
Paul M. Anderson, Charles Henville, Rasheek Rifaat, Brian Johnson, and Sakis Meliopoulos

113. *High Voltage and Electrical Insulation Engineering, Second Edition*
Ravindra Arora and Wolfgang Mos

114. *Modeling and Control of Modern Electrical Energy Systems*
Masoud Karimi-Ghartemani

115. *Control of Power Electronic Converters with Microgrid Applications*
Armdam Ghosh and Firuz Zare

116. *Coordinated Operation and Planning of Modern Heat and Electricity Incorporated Networks*
Mohammadreza Daneshvar, Behnam Mohammadi-Ivatloo, and Kazem Zare

117. *Smart Energy for Transportation and Health in a Smart City*
Chun Sing Lai, Loi Lei Lai, and Qi Hong Lai

118. *Wireless Power Transfer: Principles and Applications*
Zhen Zhang and Hongliang Pang

119. *Intelligent Data Mining and Analysis in Power and Energy Systems: Models and Applications for Smarter Efficient Power Systems*
Zita Vale, Tiago Pinto, Michael Negnevitsky, and Ganesh Kumar Venayagamoorthy

120. *Introduction to Modern Analysis of Electric Machines and Drives*
Paul C. Krause and Thomas C. Krause

121. *Electromagnetic Analysis and Condition Monitoring of Synchronous Generators*
Hossein Ehya and Jawad Faiz

122. *Transportation Electrification: Breakthroughs in Electrified Vehicles, Aircraft, Rolling Stock, and Watercraft*
Ahmed A. Mohamed, Ahmad Arshan Khan, Ahmed T. Elsayed, Mohamed A. Elshaer

123. *Modular Multilevel Converters: Control, Fault Detection, and Protection*
Fuji Deng, Chengkai Liu, and Zhe Chen

124. *Stability-Constrained Optimization for Modern Power System Operation and Planning*
Yan Xu, Yuan Chi, and Heling Yuan

125. *Interval Methods for Uncertain Power System Analysis*
Alfredo Vaccaro

126. *Practical Partial Discharge Measurement on Electrical Equipment*
Greg Stone, Andrea Cavallini, Glenn Behrmann, and Claudio Angelo Serafino

127. *Graph Database and Graph Computing for Power System Analysis*
Renchang Dai and Guangyi Liu

128. *The Power of Artificial Intelligence for the Next-Generation Oil and Gas Industry: Envisaging AI-inspired Intelligent Energy Systems and Environments*
Pethuru Raj Chelliah, Venkatraman Jayasankar, Mats Agerstam, B. Sundaravadivazhagan and Robin Cyriac

129. *Microgrids: Theory and Practice*
Peng Zhang

130. *Smart Cyber-Physical Power Systems, Volume 1: Fundamental Concepts, Challenges, and Solutions*
Ali Parizad, Hamid Reza Baghaee and Saifur Rahman

131. *Smart Cyber-Physical Power Systems, Volume 2: Solutions from Emerging Technologies*
Ali Parizad, Hamid Reza Baghaee and Saifur Rahman

132. *Understanding Electromagnetic Transients in Power System*
Luiz Cera Zanetta Jr.

www.ingramcontent.com/pod-product-compliance
Lightning Source LLC
Chambersburg PA
CBHW051609300825
31485CB00008B/20
* 9 7 8 1 3 9 4 2 4 0 5 5 5 *